T0221233

Toxicology
of the Pancreas

TARGET ORGAN TOXICOLOGY SERIES

Series Editors
A. Wallace Hayes, John A. Thomas, and Donald E. Gardner

(Continued)

Target Organ Toxicology Series

Toxicology of the Pancreas

Edited by

Parviz M. Pour

informa

healthcare

New York   London

First published in 2006 by CRC Press.

This edition published in 2010 by Informa Healthcare, Telephone House, 69-77 Paul Street, London EC2A 4LQ, UK.

Simultaneously published in the USA by Informa Healthcare, 52 Vanderbilt Avenue, 7th Floor, New York, NY 10017, USA.

Informa Healthcare is a trading division of Informa UK Ltd. Registered Office: 37–41 Mortimer Street, London W1T 3JH, UK. Registered in England and Wales number 1072954.

A CIP record for this book is available from the British Library.

Library of Congress Cataloging-in-Publication Data available on application

ISBN-13: 9780415320719

Orders may be sent to: Informa Healthcare, Sheepen Place, Colchester, Essex CO3 3LP, UK
Telephone: +44 (0)20 7017 5540
Email: CSDhealthcarebooks@informa.com
Website: http://informahealthcarebooks.com/

For corporate sales please contact: CorporateBooksIHC@informa.com
For foreign rights please contact: RightsIHC@informa.com
For reprint permissions please contact: PermissionsIHC@informa.com

Printed and bound by CPI Group (UK) Ltd, Croydon, CR0 4YY

Transferred to Digital Print 2012

FOREWORD

Although all pancreatic diseases have serious consequences and the pancreas appears to be sensitive to a variety of chemicals, this tissue has not been a subject of toxicology. The reason for this could be the "unreachable" location of the pancreas deep in the abdomen and the lack of experimental models. The development of animal models for acute pancreatitis was initiated during the last decade and focused primarily on the normal and abnormal function of acinar cells. The data gained from animals, however, could not be translated to the human disease because spontaneous acute pancreatitis in the laboratory animals was almost nonexistent and the methods of acute pancreatitis induction was hardly applicable to humans. It remained entirely unclear why alcohol, the etiological factor in acute and chronic pancreatitis, was ineffective in inducing pancreatic injury in animals. Studies on chronic pancreatitis were limited to human material as there was no adequate animal model for this illness. The diseases of the endocrine pancreas were entirely out of the focus of acute pancreatitis research and remained in the interest of endocrinologists, particularly in the European countries.

The increased incidence of pancreatic cancer in the Western world and its grave prognosis resulted in an urgency program of research in this area by governmental health agencies, especially in the United States and Japan. The establishment of several animal models intensified research into the pancreas. Based on the experimental results that provided a controversial concept on the origin of pancreatic cancer from acinar cells, ductal cells, or islet cells, all cell components of the pancreas became the subject of interest. Toxicological studies, which were initially and primarily restricted to the liver, were initiated in the pancreas just recently.

This book will highlight some fundamental research performed in the last 10 years, on the toxicology of the pancreas. Needless to say, understanding the basic structure, embryology, and physiology of every tissue is paramount in toxicological research. Although the anatomy and fine structure of the pancreas have been described in sufficient detail in the past, it seemed pertinent to highlight some of the components of the tissue that appear to be of significant importance in understanding toxicological events. This particularly applies to the little understood vascular structure of the pancreas, which is vital in regulating the normal function of the pancreas and the normal nutritional and biochemical requirements for the intact function of the pancreas. Also, the overseen function of some pancreatic cell elements involved in the regulation of the normal enzyme secretion and that play a role in injuries had to be presented. Similarly, the imbalances in certain nutritional and molecular events that can irreversibly damage this tissue have been mentioned. Chapter 1 through Chapter 13 deal with these issues.

Toxicological studies in the past were primarily concentrated on the identification and distribution of drug-metabolizing enzymes, which play a significant role in the protection of cells from harmful agents, in the liver. The pancreas was not included in these studies, as it was believed that pancreatic injury occurs by the drugs that were activated in the liver by the drug-metabolizing enzymes. Recent studies, however, clearly showed that the pancreas has its own specific drug-metabolizing enzyme system. As presented in Chapter 7 and Chapter 8, contrary to the previous belief, the pancreas contains many cytochrome p450 and GST metabolizing enzymes, which remarkably vary in type and quantity between the individual pancreatic cells, both in humans and animals. Chapter 14 through Chapter 18 discuss the role of these enzymes in pancreatic injuries.

It has become increasingly clear that environmental factors, and dietary practices can adversely affect the pancreas and eventually lead to serious consequences. A few examples of these have been presented in Chapter 19 through Chapter 23.

Not all pancreatic injuries are induced. Examples of immunological and genetic defects are given in Chapter 23 through Chapter 26. Any break in the normal intrauterine development of the pancreas can cause immediate damage or inherent susceptibility to certain injuries in adult life. Diseases of certain nonpancreatic tissues can also severely affect the pancreas. Recent intense studies on molecular biological levels identify that many pancreatic diseases, both of exocrine and endocrine tissues, are due to abnormalities of one gene or another. Examples of this include hereditary chronic pancreatitis, pancreatic cancer, endocrine abnormalities, and mucovicidosis that are discussed in Chapter 25 and Chapter 26.

Despite intensive research, the cause and treatment of diabetes mellitus have remained elusive. Experimental data suggest that the sensitivity of the endocrine islets to toxic agents leading to diabetes is limited, restricted, and remarkably species dependent (Chapter 27). Chapter 28 is an update of the current concept of the human endocrine pancreas. Chapter 29 discloses the long neglected fact that the endocrine pancreas is a fundamental part of the exocrine pancreas both in normal and disease conditions. This chapter also discusses the controversial view on the target cells of pancreatic carcinogens.

The subject on pancreatic cancer was limited because the rapid progress in this area concerned with chemical and metabolic patterns of pancreatic carcinogens, the development of a variety of models, and the bulk of molecular biological findings requires a separate book. Nevertheless, this book should present an incentive for further research into the toxicology of the pancreas by a combined effort of researchers from various fields of biology.

Parviz M. Pour
Editor

CONTRIBUTORS

Reid Aikin
Department of Surgery
McGill University
Montreal, Canada

Mahefatiana Andrianifahanana
Department of Biochemistry and Molecular
  Biology
Eppley Institute for Research in Cancer and
  Allied Diseases
University of Nebraska Medical Center
Omaha, Nebraska

Surinder K. Batra
Department of Biochemistry and Molecular
  Biology
Eppley Institute for Research in Cancer and
  Allied Diseases
University of Nebraska Medical Center
Omaha, Nebraska

Dale E. Bockman
Department of Cellular Biology and
  Anatomy
Medical College of Georgia
Augusta, Georgia

Paige M. Bracci
Department of Epidemiology and
  Biostatistics
University of California at San Francisco
San Francisco, California

Randall Brand
Feinberg School of Medicine
Norhwestern University
Chicago, Illinois

Amy E. Brix
Experimental Pathology Laboratories
Research Triangle Park, North Carolina

Markus W. Büchler
Department of General Surgery
University of Heidelberg
Heidelberg, Germany

Ta-min Chang
Rochester Institute for Digestive Diseases
  and Sciences
Rochester, New York

William Y. Chey
Rochester Institute for Digestive Diseases
  and Sciences
Rochester, New York

Carolyn A. Deters
Hereditary Cancer Institute
Creighton University School of Medicine
Omaha, Nebraska

Eric J. Duell
Norris Cotton Cancer Center
Department of Community and Family
  Medicine
Dartmouth Medical School
Lebanon, New Hampshire

Karam El-Bayoumy
Institute for Cancer Prevention
Valhalla, New York

Gordon Flake
Laboratory of Experimental Pathology
National Institute of Environmental Health
  Sciences
Research Triangle Park, North Carolina

Helmut Friess
Department of General Surgery
University of Heidelberg
Heidelberg, Germany

Vay Liang W. Go
David Geffen School of Medicine at UCLA
University of California at Los Angeles
Los Angeles, California

Ahmed Guweidhi
Department of General Surgery
University of Heidelberg
Heidelberg, Germany

Stephan L. Haas
Department of Medicine II
  (Gastroenterology, Hepatology, Infectious
  Diseases)
University Hospital of Heidelberg at
  Mannheim
Mannheim, Germany

Stephen Hanley
Department of Surgery
McGill University
Montreal, Canada

Joseph K. Haseman
Biostatistics Branch
National Institute of Environmental Health
  Sciences
Research Triangle Park, North Carolina

Masahiko Hirota
Departments of Gastroenterological Surgery
Graduate School of Medical Sciences
Kumamoto University
Honjo, Kumamoto, Japan

Dietrich Hoffmann
Institute for Cancer Prevention
Valhalla, New York

Elizabeth A. Holly
Department of Epidemiology and
  Biostatistics
University of California at San Francisco
San Francisco, California

Tomoko Inagaki
The First Department of Pathology
Showa University of Medicine
Tokyo, Japan

Michael P. Jokinen
Pathology Associates — A Charles River
  Company
Durham, North Carolina

Masahiko Kawamoto
Department of Surgery and Oncology
Graduate School of Medical Sciences
Kyushu University
Fukuoka, Japan

Jörg Kleeff
Department of General Surgery
University of Heidelberg
Heidelberg, Germany

Günter Klöppel
Department of Pathology
University of Kiel
Kiel, Germany

Steven D. Leach
Departments of Surgery and Oncology
Johns Hopkins Medical Institutions
Baltimore, Maryland

Markus M. Lerch
Division of Gastroenterology,
  Endocrinology, and Nutrition
Ernst-Moritz-Arndt-Universität Greifswald
Greifswald, Germany

John W. Lin
Department of Surgery
Johns Hopkins Medical Institutions
Baltimore, Maryland

Mark Lipsett
Department of Surgery
McGill University
Montreal, Canada

Daniel S. Longnecker
Department of Pathology
Dartmouth Medical School
Lebanon, New Hampshire

Albert B. Lowenfels
Departments of Surgery and Community
  and Preventive Medicine
New York Medical College
Valhalla, New York
Division of Epidemiology and Biostatistics
European Institute of Oncology
Milan, Italy

Henry T. Lynch
Department of Preventitive Medicine
Creighton University
Omaha, Nebraska

Jane F. Lynch
Creighton University School of Medicine
Hereditary Cancer Institute
Omaha, Nebraska

Patrick Maisonneuve
The Department of Community and
  Preventive Medicine
New York Medical College
Valhalla, New York
Division of Epidemiology and Biostatistics
European Institute of Oncology
Milan, Italy

Seiki Matsuno
Division of Gastroenterological Surgery
Graduate School of Medicine
Tohoku University
Sendai, Japan

Yukio Mikami
Division of Gastroenterological Surgery
Graduate School of Medicine
Tohoku University
Sendai, Japan

Nicolas Moniaux
Department of Biochemistry and Molecular
  Biology
Eppley Institute for Research in Cancer and
  Allied Diseases
University of Nebraska Medical Center
Omaha, Nebraska

Toshio Morohoshi
The First Department of Pathology
Showa University of Medicine
Tokyo, Japan

Takuro Murakami
Department of Human Morphology,
  Functional Physiology, Biophysiological
  Science
Graduate School of Medicine and Dentistry
Okayama University
Okayama, Japan

Abraham Nyska
Laboratory of Experimental Pathology
National Institute of Environmental Health
  Sciences
Research Triangle Park, North Carolina

Kazuichi Okazaki
The Third Department of Internal Medicine
Kansai Medical University
Osaka, Japan

Denise Orzech
Toxicology Operation Branch
National Institute of Environmental Health
  Sciences
Research Triangle Park, North Carolina

Krishan K. Pandey
Department of Biochemistry and Molecular
  Biology
The Eppley Institute for Research in Cancer
  and Allied Diseases
University of Nebraska Medical Center
Omaha, Nebraska

Parviz M. Pour
UNMC Eppley Cancer Center and
  Department of Pathology and
  Microbiology
University of Nebraska Medical Center
Omaha, Nebraska
Department of General Surgery
University of Heidelberg
Heidelberg, Germany

Bogdan Prokopczyk
Penn State Cancer Institute
Pennsylvania State University
University Park, Pennsylvania

M. Sambasiva Rao
Department of Pathology
Northwestern University
Feinberg School of Medicine
Chicago, Illinois

Janardan K. Reddy
Department of Pathology
Northwestern University
Feinberg School of Medicine
Chicago, Illinois

Lawrence Rosenberg
Division of Surgical Research
McGill University
Montreal, Canada

Alexander Schneider
Department of Medicine II
  (Gastroenterology, Hepatology, Infectious
  Diseases)
University Hospital of Heidelberg at
  Mannheim
Mannheim, Germany

Donald M. Sells
Battelle Columbus Laboratories
Columbus, Ohio

Manfred V. Singer
Department of Medicine II
  (Gastroenterology, Hepatology, Infectious
  Diseases)
University Hospital of Heidelberg at
  Mannheim
Mannheim, Germany

Ajay P. Singh
Department of Biochemistry and Molecular
  Biology
The Eppley Institute for Research in Cancer
  and Allied Diseases
University of Nebraska Medical Center
Omaha, Nebraska

Jens Standop
Klinik Und Poliklinik Fuer Allgemein-,
  Viszeral-, Thorax- Und Gefäßchirurgie
  Rheinische Friedrich-Wilhelms-Universität
Bonn, Germany
Department of Surgery
University of Bonn
Bonn, Germany

Michael L. Steer
Department of Surgery Tufts–New England
  Medical Center
Professor of Surgery, Anatomy, and Cellular
  Biology
Tufts University School of Medicine
Boston, Massachusetts

Kazunori Takeda
Department of Human Morphology,
  Functional Physiology, Biophysiological
  Science
Okayama University Graduate School of
  Medicine and Dentistry
Okayama, Japan

Masao Tanaka
Department of Surgery and Oncology
Graduate School of Medical Sciences
Kyushu University
Fukuoka, Japan

Alexis Ulrich
Klinik Und Poliklinik Fuer Allgemein-,
  Viszeral-, Thorax- Und Gefäßchirurgie
  Rheinische Friedrich-Wilhelms-
  Universität
Bonn, Germany
Department of Surgery
University of Heidelberg
Heidelberg, Germany

Nigel J. Walker
Laboratory of Computational Biology and
  Risk Analysis
National Institute of Environmental Health
  Sciences
Research Triangle Park, North Carolina

Yu Wang
David Geffen School of Medicine at UCLA
University of California at Los Angeles
Los Angeles, California

F. Ulrich Weiss
Division of Gastroenterology, Endocrinology
  and Nutrition
Ernst-Moritz-Arndt-Universität Greifswald
Greifswald, Germany

Michael E. Wyde
Toxicology Operation Branch
National Institute of Environmental Health
  Sciences
Research Triangle Park, North Carolina

Mehmet Yalniz
The Eppley Institute for Research in Cancer
  and Allied Diseases
University of Nebraska Medical Center
Omaha, Nebraska

Koji Yamaguchi
Department of Surgery and Oncology
Graduate School of Medical Sciences
Kyushu University
Fukuoka, Japan

CONTENTS

1

DEVELOPMENTAL BIOLOGY OF THE VERTEBRATE PANCREAS

John W. Lin and Steven D. Leach

CONTENTS

1.1 INTRODUCTION

The pancreas remains an organ of considerable interest in medicine and biology. Understanding pancreatic epithelial differentiation will fundamentally modify our approach to its major diseases. A major thrust of investigators in diabetes mellitus is the search for a renewable source of islet tissue for use as cell replacement therapy.[1] Identification of pancreatic stem cells and understanding the events required to promote islet cell differentiation may provide an ideal source for cell replacement therapy. In addition, pancreatic ductal adenocarcinoma remains an essentially fatal disease, despite advances in early detection and surgical therapy (see Chapter 20 through Chapter 25). The 2001 American Cancer Society report estimates 29,000 new cases in the United States in that year alone[2]; most new cases are surgically unresectable at the time of diagnosis and carry a median survival of less than 12 months.[3] Although the earliest events in pancreatic tumorigenesis remain unclear, there is growing evidence that premalignant tumor initiation involves the reactivation of dormant transcription factors known to play pivotal roles in normal pancreatic development.[4] Finally, knowledge regarding normal pancreatic growth and differentiation during development will inform ongoing studies of pancreatic regeneration following surgical pancreatectomy and toxic injury.[5]

1.2 PHYLOGENY OF VERTEBRATE PANCREAS

Consideration of invertebrate and lower vertebrate organisms provides enormous insight into the evolution of the pancreas.[6] Although no islets of Langerhans exist in any invertebrate, the insulin signaling pathway itself is highly conserved across a broad phylogenetic range. A range of invertebrates, from arthropods[7] to worms[8] to mollusks,[9] have been found to express genes orthologous to mammalian insulins; in these invertebrates, insulin orthologs are typically expressed within cells of the nervous system (Figure 1.1). The *Drosophila melanogaster* genome, for example, contains seven insulin-like peptides and one identified insulin receptor. Loss of function mutations in the *Drosophila* insulin receptor produces a phenotype similar to that observed in flies whose neuronal insulin producing cells have been ablated through a cytotoxic transgene.[7,10] Both groups of flies experience growth retardation, as measured by the length of the larval stage and wing size, and carbohydrate levels in transgenic *Drosophila* hemolymph are elevated during the brief fasting period of larval development, suggesting a conserved role in glucostasis for *Drosophila* insulin-like peptides.[7] Insulin signaling has been recognized in *Caenorhabditis elegans* since the late 1990s and has been shown to play roles in

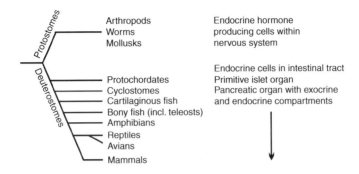

Figure 1.1 Schematic phylogenetic tree illustrating the evolutionary versions of the modern vertebrate pancreas. Animal phylogeny above coelenterate organisms is broadly divided into protostomian and deuterostomian lines. The protostomes include the arthropods, worms, and mollusks and possess endocrine hormone producing cells within their nervous systems (see text). In protochordates, primitive members of the deuterostomian line, endocrine hormone producing cells are no longer found in the nervous system, but are instead scattered throughout the intestinal tract. Cyclostomes, the most primitive vertebrates, possess a primitive islet organ but no exocrine parenchyma. A modern pancreas, with specialized exocrine and endocrine cells and a ductal system, is first seen in holocephan cartilaginous fish.

glucostasis and longevity. As in *Drosophila*, insulin-like ligands are produced within neuronal cell types in *C. elegans*.[8]

The evidence for conserved functions in these two organisms suggests that other invertebrate examples of endocrine hormone-producing cells may have functions analogous to their mammalian counterparts. The ancient colocalization of endocrine-like cells and neuronal tissue is echoed today in the observation that many transcription factors controlling pancreatic differentiation (e.g., *ngn3*, *neuroD*, *Ptf1a*) are also expressed, at least transiently, in the developing nervous system. The recent identification of short-lived *Pdx1*-regulated somatostatin expressing cells in the developing rat brain provides further evidence for this relationship.[11]

As one conceptually ascends the phylogenetic tree to protochordates, the immediate precursors to true vertebrates, these endocrine-like cells are no longer predominantly located in brain, but instead are found scattered throughout the intestinal mucosa.[6] The first islet organ appears in hagfish and lampreys, primitive jawless fish of the cyclostome class. In both organisms, endocrine cells are primarily concentrated in a single organ attached to the bile duct (in hagfish) or gut (in lampreys), and tissue expressing exocrine zymogens are enmeshed within the liver.[12] These organisms also bear the first recognizable thyroid and pituitary glands, suggesting a general trend toward condensation of endocrine tissue to form specialized endocrine organs. A modern pancreas is found in the next higher order of vertebrates, the holocephan cartilaginous fish. These possess a pancreas with an exocrine architecture, a main pancreatic duct, and organized islets.[6] Ascending the phylogenetic tree further, one finds surprising conservation of embryonic pancreatic morphogenesis among a wide variety of higher vertebrates, from fish[13] and reptiles[14] to avians[15] and mammals (Figure 1.1).[5]

In this review, we compare and contrast pancreatic development in a variety of vertebrate organisms. In so doing, we emphasize molecular regulatory factors required for normal cellular differentiation and morphogenesis. Even though details of morphogenesis may vary from species to species in association with evolutionary changes in body plan, the molecular mechanisms governing pancreatic development appear to be highly conserved.

1.3 PANCREATIC ORGANOGENESIS IN MAMMALIAN MODELS

Modern understanding of pancreatic organogenesis is derived in large part from the seminal reports of Wessels and Cohen and Pictet and Rutter, who reported detailed analyses of pancreatic morphogenesis in the mouse and rat.[16,17] The pancreas forms as dorsal and ventral evaginations of early foregut endoderm (Figure 1.2), which subsequently fuse into a single organ during the rotation of the intestinal tract. In contrast to the human pancreas, the ducts of Santorini and Wirsung in the rat do not typically fuse and drain independently into the intestinal lumen.[17] These dorsal and ventral buds begin as simple cylindrical evaginations of simple cuboidal epithelium and later elaborate digitations to produce a complex branching tree of folded epithelial sheets.[16] These folds and branches are reminiscent of the branching ductal tree seen in adult pancreas, but no true ductal epithelium exists at this point as evidenced by the lack of ultrastructural features of ductal epithelium and the lack of mature duct-specific markers, such as mucins or the cystic fibrosis transmembrane regulator.[17–19]

In the mouse, the dorsal bud initially appears as a bulge in the intestinal tube at E9.5 (embryonic days post coitus). Classic works describe a single ventral bud that appears later at E10.5 and fuses by day E12.5.[17] Recent work, guided by modern markers of pancreatic endoderm (*Pdx1*, see below), demonstrates two symmetric ventral pancreatic bulges in the E9.5 mouse embryo in intimate association with each of the paired vitelline veins. The left vitelline vein and the left ventral pancreatic bud regress as the mouse embryo develops, leaving the right (portal) vitelline vein and a single ventral pancreatic bud by E10.5.[20] Similar observations of three initial pancreatic buds have also been made in *Xenopus*[21] and

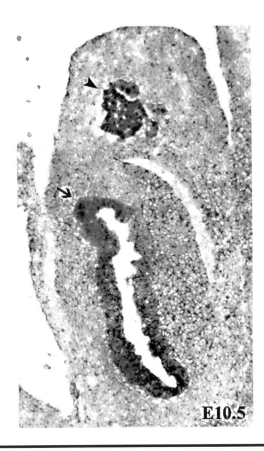

Figure 1.2   Dorsal pancreatic anlage in the E10.5 embryonic mouse. Immuno-histochemical staining for PDX-1 identifies developing intestine (arrow) and dorsal pancreatic bud (arrowhead). Image courtesy Dr. Farzad Esni.

chick,[15] suggesting that human ventral pancreas may first appear in a similar manner.

Pictet and Rutter noted the appearance of glucagon producing cells as early as 20 to 22 somites (E9.0). These cells appeared to be an integral part of the epithelial sheet, joined at the apical surface by tight junctions to neighboring protodifferentiated cells. Between 20 and 35 somites (E9.5 to E10.25), these glucagon-positive cells were seen to separate and cluster, such that they no longer contacted foregut lumen but continued, like all epithelia, to remain bounded from adjacent mesenchyme by basal lamina.[17] Although this work inferred the presence of glucagon based on characteristic granule appearance on electron microscopy, recent work confirms these early conclusions. Reverse transcription polymerase chain reaction (RT-PCR) has demonstrated the appearance of glucagon, as well as other endocrine hormones, at this early stage. Somatostatin messenger RNA (mRNA) is detectable as early as E8.5 (prior to dorsal bud formation), and insulin and glucagon mRNA appear at E9.5.[22] These results were also confirmed by antibody-based detection of cells that express either glucagon alone or glucagon and insulin together in early pancreatic buds.[23] The initial, reasonable hypothesis was that these cells were direct precursors to the islet forms seen in newborn pups. However, it can be dangerous to infer cell lineage merely from the expression of a few markers. As discussed below, more recent work suggests that these early endocrine cells are not the true precursors to mature islets.[24,25]

Beginning at about E13.5, there is an exponential increase in measurable digestive enzyme activity and insulin content in the pancreatic rudiments. Pictet and Rutter termed this period the "secondary transition."[17] Electron microscopy of pancreatic sections at this time reveals the accumulation of rough endoplasmic reticulum and the first appearance of zymogen granules. A population of insulin and glucagon positive cells emerge and gradually accumulate in clusters adjacent to the epithelial sheet.[26] Mature islet-like clusters of endocrine cells do not appear until late in gestation at about E18.5. In contrast to the early appearance of endocrine hormones by RT-PCR analysis, amylase mRNA cannot be detected until E12.5.[22] The convoluted folded epithelial sheet of the pancreatic buds acquires a distinct acinar morphology during the secondary transition. By E14.5, these acinar structures are lined by columnar cells bearing apical zymogen granules immunoreactive for amylase, trypsin, and other zymogens.

1.3.1 Transcriptional Machinery in Pancreatic Development

The embryonic events of pancreatic development can be conceptually resolved into three phases.[27] First, a restricted portion of multipotential foregut endodermal epithelium is specified to become pancreatic anlagen. Second, the cell fates of these multipotential epithelial cells are determined in a regulated manner. Third, proliferation and organization of these pancreatic precursors ultimately leads to specialized islets of Langerhans and the extensively arborized epithelial tree of the adult pancreas. Advances in molecular biologic and genetic techniques over the past 15 years have yielded insight into the transcriptional machinery that regulates each of these developmental steps.

1.3.1.1 Early Patterning of Pancreatic Anlagen

A number of molecules play important roles in inducing or specifying foregut endoderm to a pancreatic fate, including the *Pdx1*, *Hlxb9*, and *Ptf1a* transcription factors and the sonic hedgehog (*Shh*) signaling pathways. The parahox homeodomain transcription factor *Pdx1* (pancreas duodenum homeobox-1) was the first gene identified in which inactivation by germline knockout specifically disrupted pancreatic development while sparing most endodermal and all mesenchymal derivatives.[28] Orthologs of *Pdx1* in humans, mice, rats, frogs, and fish are highly conserved in both sequence and function.[29,30,31] In the adult mouse and rat, *Pdx1* expression is largely restricted to the β- and δ-cells of the adult islets of Langerhans, where it contributes to the expression of insulin and somatostatin.[29] In the developing mouse, *Pdx1* is expressed throughout the entire exocrine and endocrine pancreas (including the pancreatic ducts), as well as in the proximal common bile duct and cystic ducts, the pyloric glands of the stomach, and the duodenal epithelium.[32]

The important role of *Pdx1* in pancreatic development was demonstrated in the mid–1990s, when *Pdx1⁻/⁻* null mice were generated and found to display aborted pancreatic morphogenesis and a lack of differentiated endocrine and exocrine cell types.[28] This requirement for *Pdx1* is cell autonomous to endoderm: recombination of *Pdx1⁻/⁻* endoderm with wild type mesenchyme fails to generate pancreas, and recombination of wild type endoderm with *Pdx1⁻/⁻* mesenchyme leads to normal pancreatic growth.[33] Early work in the ectopic expression of *Pdx1* involved *in ovo* electroporation of anterior chick endoderm; in this model, ectopic introduction of *Pdx1* leads to downregulation of nonpancreatic transcription factors such as *Hex* and *CdxA,* but does not lead to bud formation or pancreatic cytodifferentiation.[34] In contrast, more recent work involving adenoviral delivery of rat *Pdx1* to adult mouse liver has found endocrine hormone expression within liver parenchyma and restoration of glucostasis following streptozotocin ablation of pancreatic β-cells.[35,36] In addition, transgenic expression in *Xenopus* of a fusion protein combining a potent *VP16* transactivation domain with *Pdx1* results in expression of endocrine and exocrine markers within hepatic anlagen.[37] Therefore, overexpression of *Pdx1* in embryonic and mature liver is able to activate pancreas-specific gene expression, although neither experimental model generates characteristic pancreatic structural elements such as islets, acini, or ductal trees.

Although clearly essential to pancreatic organogenesis and eventual cytodifferentiation, initial hopes that *Pdx1* would be a "master gene" in pancreatic development have not been realized. Importantly, even though the pancreatic buds normally form within the domain of *Pdx1* positive endoderm, the absence of functional *Pdx1* does not prevent this initial bud formation.[28,32,33] The earliest events in specifying the pancreatic anlagen must, therefore, involve events upstream or independent of *Pdx1* function. Recent evidence suggests that the homeodomain transcription factor *Hlxb9* may play such a role for the dorsal anlagen. *Hlxb9⁻/⁻* mice fail to form a dorsal pancreatic bud and do not express *Pdx1* in dorsal foregut endoderm. Interestingly, ventral bud pancreas continues to develop in *Hlxb9*[-/-] mice, albeit with a moderate reduction in insulin positive cells and subtle perturbation of islet architecture.[38,39]

The patterning of a restricted domain of foregut to a pancreatic fate is also dependent upon *Shh* signaling. Although *Shh* is widely expressed in intestinal endoderm of the mouse,[40] *Shh* expression is conspicuously absent in dorsal and ventral pancreatic anlagen.[41] Misexpression of *Shh* in pancreatic anlagen leads to a mixed pancreatic-duodenal architecture in pancreatic buds,[41] and abrogation of *Shh* signaling through cyclopamine treatment of chick embryos leads to heterotopic pancreas throughout the foregut.[42] This repression of *Shh* depends on signals from notochord.[43,44] The inductive, propancreatic effect of notochord can be reproduced *in vitro* by soluble activin-B, a member of the transforming growth factor-family, and fibroblast growth factor 2 (FGF2).[43] Of note, even though the notochord spans the entire axial length of the embryo, only anterior endoderm can be induced to express pancreatic markers by notochord.[15] These observations may be unified by the observed anteroposterior restriction of activin receptor expression and function to an area of anterior endoderm that ultimately contributes to the dorsal pancreatic bud.[45]

The Class II basic helix-loop-helix (bHLH) transcription factor *Ptf1a* (pancreas transcription factor 1a, also referred to as *Ptf1a-p48* or *p48*) has also been recently implicated in early dorsal and ventral pancreatic development. *Ptf1a*[-/-] mice fail to develop a pancreas, although they do possess a dorsal pancreatic duct remnant similar to that seen in *Pdx1*[-/-] knockouts. However, in contrast to *Pdx1*[-/-] mice, differentiated endocrine cells are present in *Ptf1a*[-/-] mice, but are mislocated initially in the pancreatic mesentery and later in the spleen. This phenotype contributed to the initial view of *Ptf1a* as a key transcription factor in exocrine, but not endocrine, development.[46] Recent work has suggested an alternate view, in which *Ptf1a* plays an early role in the normal development of both endocrine and exocrine lineages. MacDonald, Wright, and colleagues crossed transgenic mice expressing Cre recombinase in the endogenous *Ptf1a* locus onto a Rosa26R background. In these mice, cells that activate the *Ptf1a* promoter are permanently labeled by the Cre-mediated genomic excision of loxP flanked sites in the Rosa26R line, an excision that results in a functional *lacZ* reporter gene; furthermore, all daughter cells are labeled as well, allowing accurate lineage tracing.[47] In heterozygous transgenics (*Ptf1a*[+/Cre]), lacZ activity was observed in both endocrine and exocrine cell types, but not in duodenum or other endodermal derivatives. By breeding these transgenics to homozygosity (*Ptf1a*[Cre/Cre]), the investigators were able to trace the fate of these cells in the absence of functional PTF1a protein. Surprisingly, a broad region of duodenal epithelium adjacent to the pancreatic duct remnant expressed lacZ, indicating that pancreatic progenitors, in the absence of PTF1a, revert to an intestinal cell fate.[47] These results provide evidence that PTF1a commits foregut endoderm to a pancreatic fate, and that in its absence, this endoderm defaults to intestinal differentiation.

1.3.1.2 Patterning of Dorsal vs. Ventral Anlagen

Although the dorsal and ventral pancreatic buds eventually fuse to form a single organ, they appear to be histologically (and perhaps evolutionarily) distinct. In developing *Xenopus*, insulin expression is restricted to the dorsal pancreatic bud and its derivatives[21]; however, in rats[48] and chicks,[15] the dorsal bud is enriched in glucagon positive cells. The induction of the dorsal and ventral anlagen also appears to occur via distinct molecular pathways. This fact is not surprising in light of the vastly different mesenchymal environments encountered by the dorsal and ventral buds. The dorsal anlagen develops in intimate contact first with notochord and later with dorsal aorta and pancreatic mesenchyme, and the ventral anlagen develops in contact with hepatic precursors, cardiac mesoderm, and septum transversum.[49,50]

As mentioned above, *Hlxb9*[-/-] mice completely fail to form a dorsal, but not ventral, pancreatic bud.[38,39] Further work focusing on ventral pancreatic endoderm in the mouse suggests a distinct role for hedgehog signaling as well. Naked ventral endoderm harvested from E8 to E8.5 mice grown in collagen gels, unlike dorsal endoderm, thrives in the absence of mesenchymal signals, does not express *Shh,* and acquires *Pdx1* expression. If exposed to FGF2 or cocultured with cardiac mesenchyme, ventral endoderm cultures do not acquire *Pdx1* expression, but do express *Shh* and albumin (a marker of hepatic cell fate). These results propose a model in which ventral endoderm is bipotential and that FGF2 signaling from cardiac mesenchyme patterns the anterior portion toward liver, rather than toward a default of pancreas.[50] Thus FGF2 in ventral endoderm appears to block pancreatic differentiation and upregulate *Shh*, and FGF2 signaling in the dorsal chick endoderm appears to promote pancreatic differentiation and downregulate *Shh*.[43]

1.3.1.3 Endocrine Cytodifferentiation

A number of transcription factors important to endocrine differentiation have been identified, and schematic cascades of transcription factor activation have been proposed to unify existing data and explain the formation of four specialized cell types from foregut endodermal precursors.[51] Although these schemata depict linear cascades, the actual interplay is likely to be more complex and modified by as yet unidentified molecules.

Neurogenin3 (*ngn3*) is a Class II bHLH transcription factor that plays an important role in endocrine differentiation. Targeted knockout of *ngn3* leads to a complete absence of islets or differentiated endocrine cells.[52] Overexpression of *ngn3* in transgenic mice leads to hypomorphic, poorly branched pancreatic buds composed principally of endocrine cells, suggesting that *ngn3* commits a precursor population to an endocrine fate.[53,54] *Ngn3* transcripts are first detected in E9.0 embryonic pancreatic buds and peak at E15.5, coincident with the "second wave" of cytodifferentiation. *Ngn3* is undetectable in newborn and adult mouse pancreas and cannot be detected in embryonic cells positive for insulin, glucagon, somatostatin, or peptide YY, suggesting that *ngn3*-expressing cells represent a population of committed but still undifferentiated endocrine precursors.[54]

Other transcription factors have been identified to be important in endocrine pancreas development, although none have a null phenotype as dramatic and specific as that seen in *ngn3* knockouts. *NeuroD* (also known as BETA2, beta cell E-box transactivator-2) is inducible by *ngn3*,[55] enhances insulin promoter activity,[51] and its null phenotype exhibits marked reduction of all endocrine cell types but with preserved *ngn3* expression.[56] Inactivation of the paired domain homeobox gene *Pax4* leads to an absence of mature β- and δ-cells, with a preponderance of α-cells and preservation of exocrine development; this phenotype suggests a role for *Pax4* as a switch between α and β/δ cell fates.[57] Knockout of the related homeobox gene *Pax6* results in a pancreas with few or no α-cells, in addition to dramatic reduction of β-, δ-, and PP-cells.[58] *Nkx2.2*[59] and *Nkx6.1*[60] are members of the NK class of homeodomain proteins, both of which are broadly expressed in the early (E9.5/E10.5) pancreatic bud and then progressively restricted to mature endocrine cells by E15.5. Knockout of either disrupts β-cell development, and analysis of double knockout mice suggests that *Nkx6.1* lies downstream of *Nkx2.2*.[60]

1.3.1.4 Exocrine Cytodifferentiation

In contrast to endocrine pancreatic differentiation, few regulatory components responsible for exocrine differentiation have been identified.[61] As discussed above, *Ptf1a* plays an early role in pancreatic development and is required for the appearance of exocrine cell types. *Ptf1a* is the only known transcription factor specific to the adult exocrine pancreas.[62] The trimeric transcriptional activator complex PTF1 and its cognate binding sequence was initially identified through DNA footprint analysis of the promoters of rat amylase, elastase, and trypsin genes.[63] Subsequent analysis demonstrated PTF1 to be composed of three subunits, of which only one, PTF1a-p48, is tissue specific. Like several genes important in endocrine pancreatic differentiation (e.g., *ngn3*, *neuroD*), *Ptf1a* is a Class

II bHLH protein.[64] Transcripts are first detectable at E9.5 in the pancreatic buds, although transcripts can be transiently found at E8.5 in the developing hindbrain.[47,65]

Mist1 is a recently identified Class II bHLH transcription factor that is only expressed in acinar cells of the pancreas, although it is expressed in a variety of secretory extrapancreatic tissues, such as salivary acini, and serous cells of the stomach, prostate, and seminal vesicles.[66,67] In the embryonic pancreas, Mist1 is detectable early (E10.5) in the dorsal pancreatic bud; at E14, when acinar structures are histologically distinct, Mist1 expression is confined to acinar cells. Knockout mice null for Mist1 survive and at birth are reported to be grossly undistinguishable from control littermates. Examination of Mist1-/- mice later in life reveals defects in acinar cell organization and loss of acinar cell polarity; similar defects are also found in salivary and seminal vesicle epithelia.[68] The bHLH domain of Mist1 is highly similar to that found in the recently characterized Drosophila transcription factor dimmed, which is required for amplified levels of secretory activity in Drosophila neuroendocrine cells.[69] Given the restriction of Mist1 to secretory epithelial cell types in mammals, dimmed and Mist1 may ultimately prove to be functional orthologs that both promote cellular machinery necessary to maintain a secretory cell type.

Broadly speaking, the exocrine pancreas encompasses both the specialized acinar cells, which secrete digestive zymogens, the ductal epithelium, and the centroacinar cells at the junction of acinar and ductal elements. Even though many studies consider acinar and ductal differentiation under the umbrella of exocrine pancreas, it may be erroneous to link ductal and acinar differentiation, as demonstrated by a recent study also employing Cre labeling techniques. Melton and colleagues created a transgenic mouse using the Pdx1 promoter to drive expression of a tamoxifen-inducible Cre recombinase. By administering a tamoxifen pulse to pregnant mothers at varying stages of gestation and later examining the embryos for Cre reporter activity, the investigators noted early divergence between ductal labeling and acinar or islet labeling. If given tamoxifen between E9.5 and E11.5, all three cell types were labeled. However, if tamoxifen was administered earlier (E8.5) or later (E12.5) only acinar and islet cell types were labeled.[70] Furthermore, by creating a transgenic in which Pdx1 expression could be inhibited by administration of tetracycline to pregnant mothers, MacDonald and colleagues demonstrated that inhibition of Pdx1 after E12.5 prevents acinar and islet differentiation, but does not prevent main pancreatic duct formation.[71] These results suggest that ductal precursors may diverge from a common acinar/islet precursor during early development of the pancreatic buds.

1.3.1.5 Regulation of Endocrine and Exocrine Differentiation

Several lines of evidence point to a common progenitor cell for endocrine and acinar cells. All epithelium in the pancreatic buds express *Pdx1* at early stages, but with cytodifferentiation acinar cell types lose high-level *Pdx1* expression. Conversely, *Ptf1a* is only found in acinar cells of the adult pancreas, but Cre-loxP lineage tracing demonstrates that a large proportion of islet cells express *Ptf1a* at some point in their ontogeny. The events that govern this determination toward exocrine or endocrine cell fates is especially relevant to the goal of directing an as-yet-unidentified pancreatic stem cell to adopt β-cell fates for cell replacement therapy in diabetes.

A recent report regarding the "transcriptome" of dorsal bud pancreatic cells provides some insight into the cascade of events that diverge to become endocrine or exocrine cytodifferentiation.[72] These investigators trypsinized dorsal pancreatic buds from E10.5 mice to yield single epithelial cells. The mRNA isolated from an individual epithelial cell was then amplified and hybridized on a custom microarray containing cDNAs of transcription factors previously implicated in pancreatic development. By analyzing dozens of individual cell "transcriptomes," several consistent patterns emerged. Although ordered relationships of precursors and progeny cannot be established using this method, conceptually ordering the transcriptomes based on stepwise accumulation or loss of a transcription factor provides a logical and reasonable estimate of ordered relationships. In this manner, Chiang and Melton proposed that a population of *Pdx1*⁺ cells (which ubiquitously coexpress *Nkx2.2* and *Nkx6.1*) first acquire *Ptf1a* expression and then diverge into two paths:

1. An exocrine path, in which *amylase* and *trypsin* expression is initiated, followed by later loss of *Nkx2.2* and *Nkx6.1*
2. An endocrine path, in which *ngn3* is activated and *Ptf1a* is lost[24,72]

Notch signaling appears to play an important role in regulating the commitment of this common progenitor to endocrine and exocrine cell fates. In the developing nervous system, Notch signaling mediates the phenomenon of lateral inhibition, in which a differentiated cell instructs its neighbors to maintain an undifferentiated state. Interruption of Notch signaling through targeted knockout of its intracellular mediator *RBP-J* or the Notch ligand *Dll1* (Delta-like ligand 1) led to apparent expansion of *ngn3* positive precursors in the early foregut endoderm.[53] Mice null for the Notch downstream effector *Hes1* have a hypoplastic pancreas, with a predominance of endocrine tissue.[73] Finally, although HES-1 is uniformly detected throughout the early pancreatic buds, by E9.5 a HES-1 negative population emerges; these HES-1 negative cells, in which Notch signaling is presumably inactive, acquire markers of early endocrine differentiation like *ngn3, neuroD,* and *Pax6.* These early observations were unified in a model in which active Notch signaling is required to prevent a population of pancreatic precursors from undergoing early endocrine differentiation, and thus be available for expansion and later exocrine differentiation. In this model, ablation of Notch signaling results in precocious and early endocrine differentiation, depleting the pool of pancreatic precursors and resulting in a hypoplastic, endocrine-dominant pancreas.[73] This model is informed by more recent work involving transgenic misexpression of the constitutively active Notch1 intracellular domain (NotchICD) in *Pdx1*-positive cells.[74] The pancreata of these mice are severely deficient in both differentiated endocrine and exocrine cell types, consistent with a role for active Notch signaling in maintaining a precursor-like state in pancreatic anlagen.

1.3.2 Epithelial-Mesenchymal Signaling

The seminal experiments of Golosow and Grobstein and Wessels and Cohen indicate an important role for endodermal-mesenchymal interactions in pancreatic development.[16,75] Golosow and Grobstein dissected intact dorsal pancreatic buds from E11 mouse embryos and cultured them in specialized Eagle's media. Pancreatic buds with intact, attached mesenchyme grow and eventually differentiate into exocrine and endocrine cell fates with appropriate acinar architecture. Removal of the mesenchyme, however, results in pancreatic buds that regress in size and show no histologic evidence of differentiation. Surprisingly, recombining these naked buds even with heterotopic (nonpancreatic) mesenchyme restores growth and differentiation. Finally, Golosow and Grobstein demonstrated that separating a pancreatic bud and salivary mesenchyme by a Millipore filter did not eliminate this trophic effect, suggesting that soluble factors are involved as opposed to direct ligand-receptor interactions.[75] This work was expanded upon by Wessels and Cohen, who tested the ability of pre-E11 endoderm to thrive in culture.[16] Through exhaustive tissue recombination experiments they noted that foregut endoderm from early embryos (3 to 13 somites, E8.0) develops acini in tissue culture if recombined with pancreatic mesenchyme, but fail to do so if recombined with nonpancreatic mesenchyme. Endoderm taken from later stage embryos (14 somites, ~ E9.0), in contrast, is able to form pancreas when provided with nonpancreatic mesenchyme.

These experiments demonstrated three principal concepts:

1. That mesenchyme is required for growth and differentiation of the pancreatic bud at all stages.

2. Mesenchymal trophic factors are likely permissive (and not instructive) for endoderm beyond 14 somites of age, as nonpancreatic mesenchyme serves equivalently well.
3. As a corollary to the previous conclusion, the instructive signals specifying these endodermal buds to become pancreas have already taken place by the time of bud formation at E9.5.

A recent demonstration of the permissive role of mesenchyme comes from mice null for the LIM homeodomain gene *Isl1*; these knockouts failed to form dorsal pancreatic mesenchyme, but retain a normal ventral mesenchymal environment (septum transversum). In turn, *Isl1*[-/-] mice exhibited severe failure of dorsal, but not ventral, pancreatic development. *In vitro* tissue recombination experiments further demonstrated that the defect is intrinsic to the mesenchyme, as recombination of dorsal foregut endoderm from *Isl1*[-/-] mice with wildtype mesenchyme led to normal pancreatic development.[76]

Efforts to purify the responsible "mesenchymal trophic factor" in the 1960s and 1970s were unsuccessful, partly due to the purification technology available at the time and partly due to the likely fact that this "factor" must in fact comprise multiple molecules.[5,77] The recent recognition of notochord repression of *Shh* in early foregut endoderm, via soluble factors such as activin-βB or FGF2, suggests a likely molecular correlate for the early instructive signals identified in the early experiments of Wessels and Cohen. Elegant experiments in the mouse have also demonstrated a required role for endothelial tissue in pancreatic specification.[20] Isolated foregut endoderm from E8.5 embryos grown *in vitro* fail to differentiate into pancreas and develop an intestinal-like epithelium. Coculturing this endoderm with notochord, however, led to expression of *Pdx1*; coculturing with isolated dorsal aortae or other sources of endothelial tissue led to the expression of *Pdx1* and insulin.

Other experiments have also revealed a role for mesenchymal factors in biasing pancreatic epithelia toward exocrine or endocrine pancreas. Whereas pancreatic buds with intact mesenchyme survive and generate acinar, ductal, and islet cells when transplanted underneath renal capsule, buds transplanted without their mesenchyme fail to form acinar cells, but do form dense clusters of insulin and glucagon positive cells.[78] *In vitro* experiments in which E12.5 pancreatic buds are grown in collagen gels with mesenchyme demonstrate predominantly acinar epithelium by Day 7 of culture, with a 5 to 1 ratio of amylase-positive to insulin-positive cells. In contrast, naked buds grown without mesenchyme, while smaller, have approximately equal ratios of amylase- and insulin-positive cells. Addition of follistatin to cultures of naked buds restored the acinar predominance.[79] Follistatin is a soluble protein known to bind to and inactivate activin and other members of the transforming growth factor-β (TGF-β) family (see also Chapter 13), suggesting that active TGF signaling may favor endocrine over exocrine cell types. In fact, treatment of *in vitro* pancreatic bud cultures with TGF-β1 leads to increased endocrine and decreased acinar cell mass, with little effect on ductal cell mass.[80] Transgenic expression of a dominant negative TGF-β receptor leads to increased acinar cell proliferation and apoptosis.[81] Together, these results suggest that pancreatic mesenchyme, at least *in vitro*, may negatively regulate endocrine pancreas development through inhibition of soluble TGF-β family members.

Other growth factors have been investigated in *in vitro* pancreatic bud cultures. E11.5 rat dorsal pancreatic buds cultured for 7 days in collagen gels fail to grow or develop exocrine differentiation in the absence of mesenchyme. Treatment with FGF1, FGF7, or FGF10, however, leads to marked growth and expansion of the exocrine, but not endocrine cell mass[82]; misexpression of FGF10 in pancreatic epithelia (under control of the *Pdx1* promoter) leads to a hyperplastic pancreas deficient in differentiated endocrine and exocrine cell types.[83,84] Treatment of naked E13.5 rat dorsal pancreatic buds cultured in collagen gels with epidermal growth factor (EGF) leads to a similar increase in epithelial cell mass, but with an apparent downregulation of amylase, insulin, and glucagon expression. Withdrawal of EGF in these cultures is followed by the appearance of insulin expression throughout the enlarged bud.[85] Knockout mice lacking functional EGF receptor have smaller pancreas with reduced endocrine cell mass; at birth, their pancreas lack islets, but have organized islet-like streaks of endocrine tissue along pancreatic ducts.[86] All *in vitro* experiments must be interpreted with due caution, however, given the necessarily artificial conditions. For example, culturing pancreatic buds in collagen gels tends to give a predominantly acinar epithelium, and culturing buds in a basement membrane-like matrix (Matrigel™) results in a predominantly endocrine epithelium.[80] *In vivo* experiments, such as tissue-specific overexpression of cytokines or their receptors, will help to dissect the role of known cytokine families in pancreatic development. The effects of the TGF and EGF family on the adult pancreas are described in Chapter 13.

1.3.3 The Lineage of Endocrine Cell Types

Given the medical importance of the pancreatic islet and the β-cell, interest has been intense in the endocrine stem cell and in the factors controlling endocrine cytodifferentiation (see also Chapter 28). Surprisingly, the cell of origin for the mature islet is not clear. As described above, previous work in the early pancreatic bud (circa E9.5) had demonstrated the presence of cells expressing glucagon alone, cells coexpressing glucagon and either insulin or pancreatic polypeptide, or multihormonal cells expressing all three endocrine hormones.[24] Logically, these cells were thought to give rise to the wave of differentiated endocrine cells arising after E13.5, but several lines of evidence suggest that these early cells give rise only to the α-cell mantle of mature islets and not to mature β- or PP-cells.

These early glucagon-positive endocrine cells do not coexpress *Pdx1* or *Nkx6.1*, two markers usually found in developing endocrine cells, and during the secondary transition they come to surround the developing wave of insulin-positive cells.[24] Disruption of pancreas formation by knockout of *Pdx1*,[33] knockout of *Ptf1a*,[46] or by overexpression of *Hlxb9*[87] does not prevent the formation of these early glucagon-positive endocrine cells. Finally, Herrera has reported two suggestive lineage tracing studies. In the first, transgenic expression of diphtheria toxin under a glucagon promoter eliminated α-cells, but not β-cells, and expression under an insulin promoter reduced α-cell populations, but eliminated β-cells from mature islets. Questions regarding the cell autonomy of the toxic effect led to a second experiment, in which lineage tracing was performed using Cre-loxP mediated genomic recombination events. These results suggested that the ontogeny of mature β-cells does not include activation of the glucagon promoter.[25] Taken together, these data suggest that these early multihormonal cells contribute to the α-cells of mature islets, and a distinct group of progenitors contribute to mature β-cells.

1.4 PANCREATIC ORGANOGENESIS IN NONMAMMALIAN MODELS

Studies employing rodent animal models have contributed a wealth of data allowing us to dissect the biology of pancreatic development. However, recent investigators have increasingly turned toward other models of pancreatic development. There are several major motives for this trend. First, many experimental designs that are impractical or impossible in the mouse are readily achieved in other animal models. Second, the explosion of genomic information available for nonmammalian species has facilitated the identification of orthologous genes, facilitating the translation of work from one species to another. Third, mechanisms of pancreatic development found to be conserved across various species are likely to represent fundamental, ancient programs of organogenesis; these programs will be of principal interest in the applied goals of pancreatic developmental biology, such as islet cell generation and pancreatic carcinogenesis, as well as in the general application to programming of other organ systems.

Excepting mammals, three higher vertebrates have been of principal interest in pancreatic development — the zebrafish, *Danio rerio*, an advanced teleost fish; the amphibian *Xenopus laevis*; and the avian chick. The state of knowledge regarding pancreatic development in these organisms will be discussed in some detail, to better understand their relationship with mammalian development. In addition, the experimental advantages unique to each organism will be described, to better interpret the contribution of a particular model to our understanding of pancreatic organogenesis.

1.4.1 Pancreatic Development in the Zebrafish (*Danio rerio*, Formerly *Brachydanio rerio*)

Though the freshwater zebrafish (*Danio rerio*) is native to the rivers of India, *Danio* variants are commonly seen worldwide in household aquariums. The zebrafish possesses a pancreas largely composed of exocrine tissue organized in acini that ultimately drain into a main pancreatic duct.[13,88,89] Throughout its early development, the pancreas contains a single, large islet in the anterior most portion of the pancreas (the pancreatic "head"). As the adult zebrafish contains several secondary islets throughout its mass, stem cells competent to form islets are felt to persist beyond the embryonic period.[90] The principal zebrafish islet closely resembles the human islet on histologic examination, with specialized cells expressing orthologs of all four mammalian endocrine pancreatic hormones. As in mammals, insulin producing β-cells occupy the center of the islet with glucagon producing α-cells and pancreatic polypeptide producing PP-cells in the periphery. Unlike mammalian islets, somatostatin secreting δ-cells are predominantly localized to the core of the zebrafish islet.[91] Brockmann bodies, clumps of endocrine tissue devoid of exocrine cells, are also occasionally found scattered along the intestine of the adult zebrafish, as in other adult teleost fishes.[92]

In contrast to mammalian pancreas, the zebrafish pancreas does not form exclusively as a bud from the intestinal tube. In fact, the intestinal tube in zebrafish forms relatively late in development, well after somitogenesis is complete. Early morphogenesis of the zebrafish does not include an in-folding of the lateral embryo to form a hollow intestinal tube. In zebrafish, the marginal blastomeres, which comprise the endodermal progenitors, involute early in gastrulation and migrate medially to form a monolayer of cells on the ventral embryo; the advancing wave of cells appears to reach the ventral midline at about the 18 somite stage (18 hpf, hours post fertilization).[93] On histologic sections, the zebrafish endoderm appears to be an unpatterned sheet of cells up through 24 somites (21 hpf), at which point, a row of endoderm begins to radially organize into a tube. Over the next 12 hours, this pattern extends rostrally, the cells polarize, and then an intestinal lumen is generated through cavitation.[89]

In spite of relatively late formation of an intestinal tube, pancreatic markers are identifiable quite early in the zebrafish embryo; compared to the timing of somitogenesis, the onset of pancreatic markers loosely parallels that observed in other vertebrates. *Pdx1* and insulin-positive cells appear in bilateral rows of endoderm early in development at 10 and 12 somites of age, respectively.[90] In mouse endoderm, *Pdx1* expression is also first detectable at E8.5, which corresponds approximately to 10 somites of age (Table 1.1).[90,94] Another teleost fish, the medaka goldfish, also begins to express *Pdx1* at 10 somites of age; in contrast to zebrafish, though, medaka do form an early intestinal tube and have a single dorsal and paired ventral pancreatic remnants as in other vertebrates.[95] Furthermore, orthologs to *Isl1*, *Nkx2.2*, *Pax6*, and *neuroD*, transcription factors important in mammalian pancreatic development, are expressed in conserved spatiotemporal patterns in zebrafish embryos.[90,96]

In parallel with the medial movement of the endodermal progenitors, the *Pdx1* and insulin-positive cells converge together in the midline by the 18 somite stage (18 hpf). By 24 hpf, somitogenesis is complete, a definitive insulin-positive islet is visible, and glucagon-positive cells first appear.[90] The exocrine anlage is first detectable at 32 hpf, as a spatially distinct domain of *ptf1a*-positive cells within ventral, *pdx1*-positive endoderm (Lin and Leach, personal observations). By 48 hpf, transcripts of both the exocrine-specific transcription factor *ptf1a* and the digestive enzyme *trypsin* are detectable in a ring of cells surrounding the islet; at this stage, *pdx1* transcripts are present in both a lateral stripe of intestinal-fated endoderm and in the principal islet itself (Lin and Leach, personal observations) (Figure 1.3A and Figure 1.3B). The segregated origin of endocrine and exocrine anlagen and their ultimate convergence is elegantly observable in the gutGFP transgenic zebrafish, which expresses green fluorescent protein (GFP) throughout early endoderm.[102]

Like *Xenopus* and chick, the zebrafish embryo is amenable to classic techniques in developmental biology, including cell transplantation and label-injection fate mapping experiments. Due to recent technical advances and several specific advantages, utilization of the zebrafish in developmental biology has become increasingly widespread. First, zebrafish development occurs on a more rapid time scale than many vertebrate systems, with major organ patterning complete by 24 hpf and definitive pancreatic and hepatic structures evident by 72 hpf. Second, overexpression or loss of function phenotypes are robustly generated by microinjection of single-cell stage embryos with mRNA or antisense oligonucleotides. Finally, zebrafish are easily and cost-effectively raised in large numbers. This last fact has made the zebrafish an ideal vertebrate organism for forward

Table 1.1 Comparative View of Approximate Developmental Stages in Vertebrate Models of Pancreatic Development

Stage	Homo sapiens	Mus musculus	Rattus norvegicus	Chick	Danio rerio
Gastrulation	Carnegie Stage 7 15–17 dpc	Theiler Stage 9 E7.0	Witschi Stage 12 E8.5	H-H Stage 2 6 hpf	50% epiboly 5.25 hpf
First somite	Carnegie Stage 9 19 dpc	Theiler Stage 12 E8.0	Witschi Stage 14 E9.5	H-H Stage 7 24 hpf	10.5 hpf
10 somites	Carnegie Stage 10 23 dpc	Theiler Stage 13 E8.5	Witschi Stage 15 E10.0	H-H Stage 10 34 hpf	14 hpf
20 somites	Carnegie Stage 11 26 dpc	Theiler Stage 14 E9.5	Witschi Stage 17 E11.0	H-H Stage 13 50 hpf	19 hpf
30 somites	Carnegie Stage 13 30 dpc	Theiler Stage 16 E10.0	Witschi Stage 19 E11.75	H-H Stage 18 3 dpf	24 hpf
Hatching or birth	9 months	E19.0	E22.0	20 dpf	(3 dpf)*

Gestational ages are approximate and adapted from staging systems described for human,[97] mouse,[98] rat,[99] chick,[97,100] and zebrafish[101] embryos.

Abbreviations: dpc, days post conception; E, embryonic days post conception; hpf, hours post fertilization; dpf, days post fertilization.

*Zebrafish embryos develop completely externally and typically hatch from their chorions at 3 dpf.

Source: Adapted from Hamburger and Hamilton,[100] Kimmel et al.,[101] O'Rahilly,[97] Theiler,[98] and Witschi.[99]

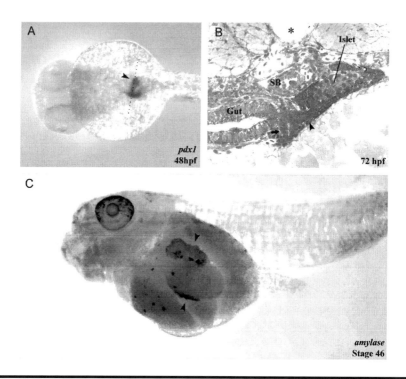

Figure 1.3   Pancreata of *Danio rerio* and *Xenopus laevis*. (A) Embryonic *pdx1* expression in the 48 hpf zebrafish. Whole-mount *in situ* hybridization identifies *pdx1* transcripts in the single principal islet (arrowhead) and a lateral stripe of gut-fated endoderm. (B) Coronal section of a 72 hpf embryo at the level of the dotted line in (A), counterstained with methylene blue and azure II, demonstrates the right-sided embryonic zebrafish islet surrounded by exocrine parenchyma; the gut tube is visible to the left, and the swim bladder dorsally. The embryonic main pancreatic duct (arrowhead) and gallbladder (arrow) are also identifiable. (C) Whole-mount *in situ* hybridization for amylase transcripts in stage 46 *Xenopus* embryo outlines the dorsal and ventral pancreatic anlagen. SB, swim bladder; asterisk in (B) marks midline notochord.

genetic, large-scale mutagenesis screens, in which random point mutations are induced by alkylating agents and the resulting progeny screened for interesting mutations.[103] The nearly completed zebrafish genome sequencing effort also makes this organism highly attractive for a forward genetics approach (http://www.ensembl.org).

Generating knockouts, or targeted loss of function phenotypes, in the mouse is a powerful technique for dissecting the role of specific genes in development. Although the technology to do so is now widely available, it still remains relatively time-consuming and expensive. In contrast, loss of function phenotypes in the zebrafish are easily generated by the microinjection of single cell embryos with antisense morpholinos targeted to a mRNA transcript of interest.[104] These single-stranded morpholino compounds resemble mRNA in structure, but are highly resistant to ribonuclease (RNAse) digestion by virtue of their morpholine ring backbone. The binding of a morpholino oligonucleotide to a stretch of mRNA near the initiation codon prevents ribosome assembly, resulting in inhibition of translation. This process is referred to as gene *knockdown*, and loss of function phenotypes generated by morpholino injection are known as *morphants*. For example, injection of a morpholino targeted against *pdx1* leads to a reduction in endocrine and exocrine cell types, similar to the phenotype seen in the mouse *Pdx1* knockout.[105]

In contrast to gene knockdown, gain-of-function phenotypes can also be generated in zebrafish by the injection of capped mRNA for a gene of interest. Such experiments can confirm, for example, epistatic relationships between members of a signaling cascade by demonstrating that injection of a downstream mRNA can rescue the phenotype generated by knocking down an upstream member of the cascade. As chip-based analysis of tissue-specific complementary DNAs (cDNAs) (the "transcriptome") becomes more common, dozens of genes are likely to be differentially expressed in tissues of interest and thus be earmarked for investigation. The zebrafish will be an ideal organism to quickly and easily generate both loss- and gain-of-function phenotypes to test these candidate genes on a large-scale basis.

The zebrafish's large clutch size and ease of rearing in large numbers has also made it an ideal organism for forward genetics, the identification of interesting phenotypic mutations followed by cloning and characterization of these loci. Typically, screening of the mutated fry is performed either by visual inspection, taking advantage of the transparency of the zebrafish embryo, or by analysis of large-scale screens using antibodies or RNA antisense probes to label structures of interest.[103] Loss of function mutations in two important signaling pathways implicated in mammalian pancreatic development have already been well characterized, namely sonic hedgehog (*sonic you, syu*)[106] and Delta-Notch (*mindbomb, mib*).[107] A number of other interesting mutant strains have been observed to specifically affect the foregut while sparing general embryonic development, including a number in which exocrine, but not endocrine, pancreas fails to form.[88] Work is ongoing to positionally clone and characterize these loci.

The role of *Shh* has been extended in a surprising direction by recent studies employing the zebrafish developmental model. Based on previous work in mouse and chick models, one would predict that zebrafish bearing a loss-of-function mutation in the *Shh* ortholog, *syu*, would develop an expanded pancreatic domain. Instead, *syu⁻/⁻* zebrafish exhibit a severe defect in endocrine pancreas formation, with preservation of exocrine differentiation and *pdx1* expression. No islet is seen on histologic sections, and insulin expression is either absent or greatly reduced and found in ectopic areas of endoderm.[108,109] A similar effect is seen when treating early (pregastrulation) zebrafish embryos with cyclopamine. Delaying the onset of cyclopamine treatment until early somitogenesis, however, restores islet development and leads to ectopic insulin expression similar to that seen in previous chick experiments.[42,108] This combination of experiments suggest that *syu* has distinct early (pregastrulation) and late (early-somitogenesis) roles: in early, pregastrulation stages, hedgehog is required for endocrine pancreatic differentiation, but in later stages hedgehog negatively regulates pancreatic differentiation.

A similar experiment has not been conducted in chicks, where embryos are accessible only after somitogenesis has begun, or in mice, where complete interruption of *Shh* signaling through combined knockout of hedgehog family members is uninformative due to early lethality. The zebrafish embryo, in contrast, is accessible immediately after fertilization and can survive for much of embryogenesis on simple diffusion of oxygen and nutrients, so that even mutations with profound cardiovascular defects can maintain viability throughout the embryonic period.[91]

It is also interesting to speculate on the correlation between zebrafish pancreatic development and either the dorsal or ventral buds of other vertebrates. Experimental evidence is not yet clear as to whether zebrafish pancreas is patterned by signals from dorsal mesenchyme, cardiac mesenchyme, or both. Of note, the adult zebrafish does possess extrapancreatic foci of both exocrine and endocrine cell types. Brockmann bodies, clusters of specialized endocrine hormone cells, are scattered along the length of the intestine. The proximal gut of adult zebrafish is also studded with pyloric caeca, pouches of secretory epithelium of indeterminate function.[30] The pyloric caeca also stain for a variety of exocrine pancreatic enzymes (trypsin, lipase, amylase), endocrine pancreatic hormones (glucagon, somatostatin, cholecystokinin), and *pdx1* transcripts.[92] It may be that the zebrafish represents a transitional evolutionary step, with both the remnants of an ancient preholocephan enteroendocrine system and a rudimentary program for the organized pancreas seen in higher vertebrates.

1.4.2 Pancreatic Development in *Xenopus laevis*

The South African claw-toed frog *Xenopus laevis* has been employed for generations in developmental biology. It was the organism in which Nieuwkoop performed his classic animal cap assay experiments, demonstrating that mesoderm was induced by signals from the vegetal pole. The reliance upon this organism can be largely explained by three advantages of the *Xenopus* model system. Like zebrafish, *Xenopus* are relatively easy to raise and keep in a laboratory environment. Both species have embryos that develop externally, allowing continuous, noninvasive observation of development. Finally, and in contrast to zebrafish, *Xenopus* embryos are large and tolerant of extensive tissue manipulation. Their size facilitates many techniques, from blastomere grafting to microinjection.[110]

Recently, however, an exciting new technique for generating transgenic frogs ensures growing interest in this organism.[111] Standard transgenesis techniques in both mouse and zebrafish require the use of F1 progeny for analysis, as the incorporation of the transgene in the F0 population is typically extremely mosaic; this generation time leads to a several month delay, waiting for the F0 population to reach sexual maturity. In *Xenopus*, decondensed sperm nuclei and the transgene of interest are incubated with a dilute restriction enzyme, allowing high frequency integration of the transgene into sperm genomic DNA. These nuclei are then microinjected into unfertilized oocytes. As the incorporation of the transgene occurs prior to fertilization, the F0 population is nonmosaic. The efficiency of transgene incorporation and the relative ease of oocyte injections allows rapid large-scale of F0 nonmosaic transgenics in *Xenopus*.[111]

As with birds and mammals, early morphogenesis of the *Xenopus* embryo leads to the formation of a hollow intestinal tube lined by a monolayer of endoderm. The pancreas forms as three buds from the anterior portion of this tube, a process that has been described in detail.[21] The dorsal pancreatic bud is the first to form, in close proximity to the dorsal aorta, at Nieuwkoop Stage 35/36 (50 hpf). Paired, symmetric ventral buds are visible at Stage 37/38 (53 hpf). These three pancreatic buds rotate in conjunction with the maturing, coiling intestinal tube to fuse into a single pancreas by Stage 41 (76 hpf).

At early stages, the endocrine hormones insulin, glucagon, and somatostatin are not only expressed in the developing pancreas but also in scattered cells of the intestinal epithelium.[21] As the embryo matures, expression of these hormones is ultimately restricted to the pancreatic buds in scattered cells. Islet-like structures are first discernible at Stage 50 (8 dpf, days post fertilization),[21] though these structures appear to represent clusters of α- or β-cells and not true Islets of Langerhans.[112] As seen in zebrafish, chicks and mammals, the appearance of exocrine enzymes lags behind that of endocrine hormones in *Xenopus*. Whereas insulin is detectable in Stage 35 dorsal buds, amylase is only first detected at Stage 41 (76 hpf) and carboxypeptidase A at Stage 44 (92 hpf) (Figure 1.3C).[21,112]

After being initially expressed in a narrow band of foregut endoderm, the expression of *XlHbox8*, the *Xenopus* ortholog of *Pdx1*, is restricted to endocrine cells of the pancreas and a short portion of gut adjacent to the pancreas.[110] The transcription factor *Sox9* is also expressed throughout dorsal and ventral anlagen, but, in contrast to *XlHbox8*, is not expressed in adjacent duodenal epithelia.[113] Beyond this descriptive data, though, the molecular events governing *Xenopus* pancreatic development have not been specifically studied. Fate mapping experiments of neurula stage embryos (Stage 14) show that distinct areas of neurula endoderm contribute to the dorsal and ventral pancreatic buds.[114] Whether this early determination is simply due to varying migrational cues or to cell autonomous molecular signals is unclear.

A recent report by Slack and colleagues highlights the potential power of *Xenopus* transgenesis in studying development.[37] Theorizing that previous attempts at inducing pancreatic differentiation by overexpression of *Pdx1* had failed due to a lack of cooperative DNA binding partners (e.g., *Pbx*, *Mrg1*),[29,115] these authors created a fusion protein joining *XlHbox8* with the VP16 activation domain from herpes simplex virus. Applying the sperm nuclear transplantation transgenesis technique, these investigators efficiently generated *Xenopus* lines that expressed an *XlHbox8-VP16* fusion gene in response to a liver-specific transthyretin promoter. In these transgenics, the liver primordium was completely replaced by pancreatic cell types, as determined by RNA *in situ* hybridization for insulin, glucagon, and amylase transcripts. Although insulin expression might be explained by direct activation by excess *XlHbox8*, the glucagon and amylase promoters are not known to be regulated by *Pdx1* or *XlHbox8*.[29] An interesting point is that once the hepatic precursors adopt pancreatic gene expression, the transthyretin promoter is no longer activated. Thus the conversion of hepatic precursors to pancreatic cell types appears to be permanent, despite predicted loss of inappropriate *XlHbox8-VP16* expression.[37] It should be noted, however, that *Pdx1* is known to positively regulate its own promoter, raising the possibility that transient transgene expression may induce sustained activation of endogenous *Pdx1*.[29] It remains to be tested whether the modified *XlHbox8-VP16* gene can induce pancreatic differentiation in intestinal or other nonhepatic tissues. The observation of hepatocellular differentiation in models of pancreatic regeneration (see Chapter 18 for additional information), though, suggest that hepatic and pancreatic stem cells likely share many common features.[5,116]

Although long generation times and the impracticality of keeping large numbers of *Xenopus* adults limit the organism's suitability for forward genetics, it is a highly suitable organism for a number of reverse genetic approaches. As in zebrafish, gain or loss of function phenotypes are straightforward to generate through the microinjection of single cell embryos. The recent development of sperm nuclear transgenesis is certain to bring about a renaissance as well. The easy creation of multiple transgenic lines can lead to an unprecedented ability to express regulatory or effector proteins with a variety of tissue-specific or inducible promoters. The Joint Genome Institute is currently hosting an ongoing sequencing effort for the *Xenopus tropicalis* genome, with a targeted completion in 2005 (http://genome.jgi-psf.org). This effort should facilitate the identification of orthologs to mammalian genes known to play important roles in pancreatic differentiation.

1.4.3 Pancreatic Development in the Chick

In comparison to teleost fish and amphibians, avians are phylogenetically close relatives to mammals. Both avians and mammals are amniotes and warm blooded. The basic body plan of the early human and chick embryos are remarkably similar. Although the externally developing embryos of zebrafish and *Xenopus* allow unparalleled access at all stages of development, the postimplantation mammalian embryo is essentially inaccessible to many forms of manipulation, as removal from the mother arrests development. Chick embryos represent an intermediate, but nevertheless possess remarkable accessibility. The developing embryo can be accessed through a small window in the eggshell, manipulated *in situ*, and then replaced in an incubator, thus allowing a number of *in vivo* experimental designs. For example, a recent method was developed to introduce transgenes into endoderm through electroporation: the embryo is accessed through a small window, the transgene injected between the embryo and yolk, and then electroporation electrodes applied across the thickness of the embryo. *In vitro* culture methods have been developed as well, and chick endodermal plugs grow remarkably well on collagen/agar plugs. Like *Xenopus*, the chick embryo heals well after grafting or excision experiments, making it amenable to surgical manipulation.[15]

Historically, the chick model organism has played a significant role in establishing the endodermal origin of pancreatic islet cells. Given the histologic and functional similarity between islet cells and neuroendocrine cells of neural crest origin, there was initial speculation that pancreatic endocrine cells may be of ectodermal origin. This speculation increased in the early 1980s as antisera- and RNA hybridization-based techniques discovered catecholamine processing enzymes in islet cells.[1] Two independent experiments in the 1970s, however, demonstrated via quail-chick chimeras that pancreatic endocrine cells were of endodermal origin.[117,118] The chick, owing to its tolerance of the surgical manipulation of notochord deletions and notochord grafts, has also contributed much to our knowledge regarding *Shh* and notochord induction of pancreas.

Pancreatic embryology in chicks closely parallels that in mice of equivalent somite age.[15] As in mice, the first appearance of a pancreatic bud is on the dorsal endoderm of the intestinal tube at the 26-somite stage (Hamburger and Hamilton Stage 15). Glucagon expressing cells appear first in the dorsal bud at Stage 17 (30 somites), followed shortly thereafter by insulin positive cells at Stage 19 (38 somites). Paired ventral buds first appear at Stage 21 and Stage 22, both of which ultimately fuse with the dorsal pancreas.[21]

Even sensitive RT-PCR techniques do not detect chick *Pdx1* mRNA until Stage 15 (26 somites), relatively late in comparison to its onset at 10 somites in the mouse and zebrafish.[44] The expression domain of *Pdx1* recapitulates that seen in other vertebrates.[15] At this stage, insulin mRNA is also detectable, and insulin protein is detectable by immunohistochemistry at Stage 19. As in other models of pancreatic development, differentiated exocrine products lag those of endocrine cells. Carboxypeptidase mRNA is detectable by RT-PCR at Stage 19 and protein at Stage 22.[44]

1.5 PANCREAS EMBRYOLOGY IN HUMANS

Early cell division of blastomeres in the implanted human embryo leads to the formation of a primitive sheet of cells, the epiblast.[119] Between the 2nd and 3rd week, cells of the epiblast gastrulate in a process remarkably similar to that seen in avian embryos. Gastrulation results in the three classical embryonic germ layers. Cranial-caudal and lateral in-folding of the embryonic layers by the 4th week forms a tubular gut, lined by endoderm. The middle portion of this tubular gut is initially floorless, remaining in open connection with the yolk sac cavity. The cranial and caudal boundaries of this floorless portion are termed the anterior and posterior intestinal portals, and these two portals divide the endoderm into the foregut, midgut, and hindgut. By this time, the earliest precursors of the liver and pancreas have already appeared.

As seen in other higher vertebrates, the human pancreas develops as two distinct evaginations or buds. The dorsal pancreatic bud forms from the dorsal endoderm of the duodenum at Carnegie Stage 12.[120] The ventral pancreatic bud forms slightly later (Carnegie Stage 14) from the endoderm of the ventrally located hepatic diverticulum. Both buds are organized around a central duct-like structure, the dorsal duct communicating directly with the duodenal lumen and the ventral duct communicating with the developing bile duct. During the rotation of the intestinal contents the duodenum acquires its adult C-loop configuration, while the ventral pancreatic bud rotates dorsal to the duodenum to fuse with the dorsal bud; this fusion is complete by Carnegie Stage 17.[120] The ventral bud contributes to the posterior head and uncinate process of the adult pancreas, and the dorsal bud contributes to the remainder.[119] The proximal portion of the ventral duct and the distal portion of the dorsal duct fuse to form the main pancreatic duct (duct of Wirsung) (see also Chapter 2). The proximal dorsal duct usually regresses, but occasionally persists as an accessory duct (duct of Santorini). Little is known about the ontogeny of the pancreatic innervation or pancreatic vascular and lymphatic systems,[17] although the adult pancreas is richly vascularized from both foregut (celiac) and midgut (superior mesenteric) vascular trees.[5] (For additional information, please refer to Chapter 2 and Chapter 4.)

1.5.1 Pancreatic Development and Human Disease

Some common anomalies of human pancreatic anatomy may be related to known developmental molecular events. Mice deficient in hedgehog signaling are consistently noted to develop an annular pancreas (encircling of the duodenum by the pancreatic head) and are occasionally noted to have defects in dorsal and ventral duct fusion.[121] Loss of function mutations in hedgehog pathway components may contribute to the anatomic anomaly of annular pancreas.[122,123] A more common anatomic anomaly is heterotopic pancreas. These nests of organized pancreatic exocrine and endocrine tissue (often with a central papilla) can be found throughout the gastrointestinal tract, although they are most frequent in the stomach and duodenum. These ectopic foci recall examples of extrapancreatic exocrine and endocrine cells seen in animal models: the pyloric caeca and Brockmann bodies of teleost fish; the enteroendocrine system of lower vertebrates; the anterior expansion of pancreatic anlagen seen after cyclopamine inhibition of hedgehog signaling in chick or zebrafish embryos. Drawing direct analogies between animal and human phenotypes can be misleading, however, as evidenced by the apparent lack of a pancreatic phenotype in Currarino syndrome patients, who are deficient in *HLXB9*.[124]

The major transcription factors involved in pancreatic development are infrequently a direct cause of human disease, and many mutations are likely unrecognized due to embryonic lethality. Nevertheless, a homozygous frameshift mutation in the human *Pdx1* locus has been reported to be the cause of pancreatic agenesis in a female infant that survived to 18 days of age.[125] Examination of the pedigrees for both parents revealed heterozygous mutations and an autosomal dominant pattern of diabetes, leading to the establishment of *Pdx1* mutations as the MODY-4 class (maturity onset diabetes of the young, Type 4).[29] Hypomorphic point mutations in *Pdx1* have also been identified as the cause of pancreatic agenesis in a newborn infant.[126] Mutations in *NeuroD* have also been associated with Type 2 diabetes mellitus in humans, and these mutations form the MODY-6 class.[127]

1.6 EPILOGUE

Expanded understanding of pancreatic development will likely have a great impact on human disease in two indirect ways. First, a robust understanding of pancreatic development will guide the search for an endocrine stem cell population to use as cell replacement therapy in diabetes.[1] A large number of transcription factors have been demonstrated to play key roles in endocrine pancreatic development, but even these have not yet yielded a viable stem cell candidate in either embryonic or adult pancreas. The second impact of pancreatic developmental biology will occur in our understanding of the earliest events in pancreatic carcinogenesis.[4]

Analysis of human specimens of ductal adenocarcinoma reveal expression of a variety of cell-type markers, including those typically associated with acinar and ductal cells[128] as well as those associated with epithelial cells of the mature stomach, duodenum, and colon,[129] suggesting the potential reacquisition of developmental pluripotency in these cancers. Metaplastic ductal epithelium in transgenic mice overexpressing TGF- demonstrates reactivation of *Pdx1*, *Pax6*, and Notch pathway components, regulatory pathways normally extinguished after exocrine differentiation is complete.[130,131] Hedgehog signaling is also reactivated in PanIN lesions (pancreatic intraepithelial neoplasia), the proposed precursor lesion to invasive pancreatic cancer,[132] and the growth of pancreatic cell lines is efficiently inhibited by cyclopamine or a neutralizing antibody targeting hedgehog ligands.[133,134] A more complete understanding of the regulatory machinery involved in pancreatic development will likely implicate additional candidate pathways for the study of pancreatic proliferation in both normal (e.g., injury repair and regeneration) and abnormal (e.g., neoplasia) contexts.

REFERENCES

1. Edlund H. 2002. *Nat. Rev. Genet.* 3:524–532.
2. American Cancer Society. 2001. *CA. Cancer J. Clin.* 51:15–36.
3. Lillemoe K.D., Yeo C.J., and Cameron J.L. 2000. *CA. Cancer J. Clin.* 50:241–268.
4. Meszoely I.M., Means A.L., Scoggins C.R., and Leach S.D. 2001. *Cancer J.* 7:242–250.
5. Slack J.M. 1995. *Development* 121:1569–1580.
6. Falkmer S. 1985. In *The diabetic pancreas*. Volk B.W. and Arquilla E.R. Eds., New York: Plenum.
7. Rulifson E.J., Kim S.K., and Nusse R. 2002. *Science* 296:1118–1120.
8. Nelson D.W. and Padgett R.W. 2003. *Genes Dev.* 17:423–428.
9. Smit A.B., Vreugdenhil E., Ebberink R.H.M., Geraerts W.P.M., Kluotwijk J., and Joosse J. 1988. *Nature* 331:535–538.
10. Tatar M., Kopelman A., Epstein D., Tu M.P., Yin C.M., and Garofalo R.S. 2001. *Science* 292:107–110.
11. Perez-Villamil B., Rivera A., Moratall R., and Vallego M. 2000. *J. Biol. Chem.* 275:19106–19114.

12. Falkmer S. and Wilson S. 1964. In *The structure and metabolism of the pancreatic islets.* Brolin S.E., Hellman B., and Knutson, H. Eds. Oxford, U.K.: Pergamon Press. pp. 17–32.
13. Youson J.H. and Al-Mahrouki A.A. 1999. *Gen. Comp. Endocr.* 116:303–335.
14. Jackintell L.A. and Lance V.A. 1994. *Gen. Comp. Endocr.* 94:244–260.
15. Kim S.K., Hebrok M., and Melton D.A. 1997b. *Cold Spring Harbor Symposia on Quantitative Biology* 62:377–383.
16. Wessels N.K. and Cohen J.H. 1967. *Dev. Biol.* 15:237–270.
17. Pictet R. and Rutter W.J. 1972. In *Handbook of physiology*, Vol. 1, Sect. 7. Steiner D.F. and Frenkel N. Eds. Washington, DC: Williams and Wilkins. pp. 25–66.
18. Foulkes A.G. and Harris A. 1993. *Pancreas* 8:3–6.
19. Chambers J.A., Hollingsworth M.A., Trezise A.E.O., and Harris A. 1994. *J. Cell. Sci.* 107:413–424.
20. Lammert E., Cleaver O., and Melton D. 2001. *Science* 294:564–567.
21. Kelly O.G. and Melton D.A. 2000. *Dev. Dyn.* 218:615–627.
22. Gittes G.K. and Rutter W.J. 1992. *Proc. Natl. Acad. Sci.* 89:1128–1132.
23. Teitelman G., Albert S., Polak J., Martinez A., and Hanahan D. 1993. *Development* 118:1031–1039.
24. Larsson L.I. 1998. *Microsc. Res. Tech.* 43:284–291.
25. Herrera P.L. 2000. *Development* 127:2317–2322.
26. Herrera P.L., Huarte J., Sanvito F., Meda P., Orci L., and Vassalli J.D. 1991. *Development* 113:1257–1265.
27. Edlund H. 1998. *Diabetes* 47:1817–1823.
28. Jonsson J., Carlsson L., Edlund T., and Edlund H. 1994. *Nature* 371:606–609.
29. Hui H. and Perfetti R. 2002. *Eur. J. Endocr.* 146:129–141.
30. Milewski W.M., Duguay S.J., Chan S.J., and Steiner D.F. 1998. *Endocrinology* 139:1440–1449.
31. Wright C.V., Schnegelsberg P., and De Robertis E.M. 1988. *Development* 104:787–794.
32. Offield M.F., Jetton J.L., Labosky P.A., Ray M., Stein R.W., Magnuson M.A., Hogan B.L.M., and Wright C.V.E. 1996. *Development* 122:983–995.
33. Ahlgren U., Jonsson J., and Edlund H. 1996. *Development* 122:1409–1416.
34. Grapin-Botton A., Majithia A.R., and Melton D.A. 2001. *Genes Dev.* 15:444–454.
35. Ber I., Shternhall,K., Perl S., Ohanuna Z., Goldberg I., Barshack I., Benvenisti-Zarum L., Meivar-Levy I., and Ferber S. 2003. *J. Biol. Chem.* 278:31950–31957.
36. Ferber S., Halkin A., Cohen H., Ber I., Einav Y., Goldberg I., Barshack I., Seijffers R., Kopolovic J., Kaiser N., and Karasik A. 2000. *Nat. Med.* 6:568–572.
37. Horb M.E., Shen C., Tosh D., and Slack J.M. 2003. *Curr. Biol.* 13:105–115.
38. Harrison K.A., Thaler J., Pfaff S., Gu H., and Kehrl J.H. 1999. *Nat. Genet.* 23:71–75.
39. Li H., Arber S., Jessell T.M., and Edlund H. 1999. *Nat. Genet.* 23:67–70.
40. Bitgood M.J. and McMahon A.P. 1995. *Dev. Biol.* 172:126–138.
41. Apelqvist A., Ahlgren U., and Edlund H. 1997. *Curr. Biol.* 7:801–804.
42. Kim S.K. and Melton D.A. 1998. *Proc. Natl. Acad. Sci.* 95:13036–13041.
43. Hebrok M., Kim S.K., and Melton D.A. 1998. *Genes Dev.* 12:1705–1713.
44. Kim S.K., Hebrok M., and Melton D.A. 1997a. *Development* 124:4243–4252.
45. Kim S.K., Hebrok M., Li E., Oh S.P., Schrewe H., Harmon E.B., Lee J.S., and Melton D.A. 2000. *Genes Dev.* 14:1866–1871.

46. Krapp A., Knöfler M., Ledermann B., Bürki K., Berney C., Zoerkler N., Hagen-büchle O., and Wellauer P.K. 1998. *Genes Dev.* 12:3752–3763.
47. Kawaguchi Y., Cooper B., Gannon M., Ray M., MacDonald R.J., and Wright C.V.E. 2002. *Nat. Genet.* 32:128–134.
48. Spooner B.S., Walther B.T., and Rutter W.J. 1970. *J. Cell. Biol.* 47:235–246.
49. Wells J.M. and Melton D.A. 2000. *Development* 127:1563–1572.
50. Deutsch G., Jung J., Zheng M., Lora J., and Zaret J.S. 2001. *Development* 128:871–881.
51. Schwitzgebel V.M. 2001. *Mol. Cell Endocr.* 185:99–108.
52. Gradwohl O., Dierich A., LeMeur M., and Guillemot F. 2000. *Proc. Natl. Acad. Sci.* 97:1607–1611.
53. Apelqvist A., Li H., Sommer L., Beatus P., Anderson D.J., Honjo T., Hrabe de Angelis M.U.L., and Edlund H. 1999. *Nature* 400:877–881.
54. Schwitzgebel V.M., Scheel D.W., Conners J.R., Kalamaras J., Lee J.E., Anderson D.J., Sussel L., Johnson J.D., and German M.S. 2000. *Development* 127:3533–3542.
55. Huang H.P., Liu M., El-Hodiri H.M., Chu K., Jarnrich M., and Tsai M.J. 2000. *Mol. Cell. Biol.* 20:3292–3307.
56. Naya F.J., Huang H.P., Qiu Y., Mutoh H., DeMayo F.J., Leiter A.B., and Tsai M.J. 1997. *Genes Dev.* 11:2323–2324.
57. Sosa-Pineda B., Chowdhury K., Torres M., Oliver G., and Gruss P. 1997. *Nature* 386:399–402.
58. St. Onge L., Sosa-Pineda B., Chowdhury K., Mansouri A., and Gruss P. 1997. *Nature* 387:406–409.
59. Sussel L., Kalamaras J., Hartigan-O'Connor D.J., Meneses J.J., Pedersen R.A., Rubenstein J.L., and German M.S. 1998. *Development* 125:2213–2221.
60. Sander M., Sussel L., Conners J.R., Scheel D.W., Kalamaras J., De la Cruz F., Schwitzgebel V.M., Hays-Jordan A., and German M.S. 2000. *Development* 127:5533–5540.
61. Means A.L. and Leach S.D. 2001. *Pancreatology* 1:587–596.
62. Rose S.D., Swift G.H., Peyton M.J., Hammer R.E., and MacDonald R.J. 2001. *J. Biol. Chem.* 276:44018–44026.
63. Cockell M., Stevenson B.J., Strubin M., Hagenbuchle O., and Wellauer P.K. 1989. *Mol. Cell. Biol.* 9:2464–2476.
64. Krapp A., Knofler M., Frutiger S., Hughes G.J., Hagenbuchle O., and Wellauer P.K. 1996. *EMBO. J.* 15:4317–4329.
65. Obata J., Yano M., Mimura H., Goto T., Nakayama R., Mibu Y., Oka C., and Kawaichi M. 2001. *Genes Cells* 6:345–360.
66. Lemercier C., To R.Q., Carrasco R.A., and Konieczny S.F. 1998. *EMBO. J.* 17:1412–1422.
67. Pin C.L., Bonvissuto A.C., and Konieczny S.F. 2000. *Anat. Rec.* 259:157–167.
68. Pin C.L., Rukstalis J.M., Johnson C., and Konieczny S.F. 2001. *J. Cell Biol.* 155:519–530.
69. Hewes R.S., Park D., Gauthier S.A., Schaefer A.M., and Taghert P.H. 2003. *Development* 130:1771–1781.
70. Gu G., Dubauskaite J., and Melton D.A. 2002. *Development* 129:2447–2457.
71. Holland A.M., Hale M.A., Kagami H., Hammer R.E., and MacDonald R.J. 2002. *Proc. Natl. Acad. Sci.*
72. Chiang M. and Melton D.A. 2003. *Dev. Cell.* 4:383–393.

73. Jensen J., Pedersen E.E., Galante P., Hald J., Heller R.S., Ishibashi M., Kageyama R., Guillemot F., Serup P., and Madsen O.D. 2000. *Nat. Genet.* 24:36–44.

74. Hald J., Hjorth J.P., German M.S., Madsen O.D., Serup P., and Jensen J. 2003. *Dev. Biol.* 260:426–437.

75. Golosow N. and Grobstein C. 1962. *Dev. Biol.* 4:242–255.

76. Ahlgren U., Pfaff S., Jessell T.M., Edlund T., and Edlund H. 1997. *Nature* 385:257–260.

77. Ronzio R.A. and Rutter W.J. 1973. *Dev. Biol.* 30:307–320.

78. Gittes G.K., Galante P.E., Hanahan D., Rutter W.J., and Debas H.T. 1996. *Development* 122:439–447.

79. Miralles F., Czernichow P., and Scharfmann R. 1998. *Development* 125:1017–1024.

80. Sanvito F., Herrera P.L., Huarte J., Nichols A., Montesano R., Orci L., and Vassalli J.D. 1994. *Development* 120:3451–3462.

81. Bottinger E.P., Jakubczak J.L., Roberts I.S.D., Mumy M., Hemmati P., Bagnall K., Merlino G., and Wakefield L.M. 1997. *EMBO. J.* 16:2621–2633.

82. Miralles F., Czernichow P., Ozaki K., Itoh N., and Scharfmann R. 1999. *Proc. Natl. Acad. Sci.* 96:6267–6272.

83. Hart A., Papadopoulou S., and Edlund H. 2003. *Dev. Dyn.* 228:185–193.

84. Norgaard G.A., Jensen J.N., and Jensen J. 2003. *Dev. Biol.* 264:323–338.

85. Cras-Meneur C., Elghazi L., Czernichow P., and Scharfmann R. 2001. *Diabetes* 50:1571–1579.

86. Miettinen P.J., Huotari M., Koivisto T., Ustinov J., Palgi J., Rasilainen S., Lehtonen E., Keski-Oja J., and Otonkoski T. 2000. *Development* 127:2617–2627.

87. Li H. and Edlund H. 2001. *Dev. Biol.* 240:247–253.

88. Wallace K.N. and Pack M. 2003. *Development* 255:12–29.

89. Biemar F., Argenton F., Schmidtke R., Epperlein S., Peers B., and Driever W. 2001. *Dev. Biol.* 230:189–203.

90. Ober E.A., Field H.A., and Stainier D.Y. 2003. *Mech Dev.* 120:5–18.

91. Moyle P.B. and Cech Jr. J.J. 1988. *Fishes. An introduction to ichytology.* New York: Prentice Hall.

92. Warga R.M. and Kimmel C.B. 1990. *Development* 108:569–580.

93. Guz Y., Montminy M.R., Stein R., Leonard J., Gamer L.W., Wright C.V., and Teitelman G. 1995. *Development* 121:11–18.

94. Assouline B., Nguyen V., Mahé S., Bourrat F., and Scharfmann R. 2002. *Mech. Dev.* 117:299–303.

95. Korzh V., Sleptsova I., Liao J., He J., and Gong Z. 1998. *Dev. Dyn.* 213:92–104.

96. O'Rahilly R. 1979. *Eur. J. Obstet. Gynecol. Reprod. Biol.* 9:273–280.

97. Pack M., Solnica-Krezel L., Malicki J., Neuhauss S.C.F., Schier A.F., Stemple D.L., Driever W., and Fishman M.C. 1996. *Development* 123:321–328.

98. Theiler K. 1989. *The house mouse: atlas of embryonic development.* New York: Springer-Verlag.

99. Witschi, E. 1962. In *Growth including reproduction and morphological development* Altman P.L. and Dittmer D.S. Eds. Washington, DC: Federation of American Societies for Experimental Biology. pp. 304–314.

100. Hamburger V. and Hamilton H.L. 1951. *J. Morphol.* 88:49–92.

101. Kimmel C.B., Ballard W.W., Kimmel S.R., Ullmann B., and Schilling T.F. 1995. *Dev. Dyn.* 203:253–310.

102. Field H.A., Dong P.D., Beis D., and Stainier D.Y. 2003. *Dev. Biol.* 261:197–208.

103. Driever W., Solnica-Krezel L., Schier A.F., Neuhauss S.C.F., Malicki J., Stemple D.L., Stainier D.Y., Zwartkruis F., Abdelilah S., Rangini Z., Belak J., and Boggs C. 1996. *Development* 123:37–46.

104. Nasevicius A. and Ekker S.C. 2000. *Nat. Genet.* 26:216–220.

105. Yee N.S., Yusuff S., and Pack M. 2001. *Genesis* 30:137–140.

106. Schauerte H.E., van Eeden F.J., Fricke C., Odenthal J., Strahle U., and Haffter P. 1998. *Development* 125:2983–2993.

107. Itoh M., Kim C.H., Parlardy G., Oda T., Jiang Y.J., Maust D., Yeo S.Y., Lorick K., Wright G.J., Ariza-McNaughton L., Weissman A.M., Lewis J., Chardrasekharappa S.C., and Chitnis A.B. 2003. *Dev. Cell.* 4:67–82.

108. diIorio P.J., Moss J.B., Sbrogna J.L., Karlstrom R.O., and Moss L.G. 2002. *Dev. Biol.* 244:75–84.

109. Roy S., Qiao T., Wolff C., and Ingham P.W. 2001. *Curr. Biol.* 11:1358–1363.

110. Chalmers A.D. and Slack J.M.W. 1998. *Dev. Dyn.* 212:509–521.

111. Kroll K.L. and Amaya E. 1996. *Development* 122:3173–3183.

112. Horb M.E. and Slack J.M. 2002. *Mech. Dev.* 113:153–157.

113. Lee Y.H. and Saint-Jeannet J.P. 2003. *Int. J. Dev. Biol.* 47:459–462.

114. Chalmers A.D. and Slack J.M. 2000. *Development* 127:381–392.

115. Dutta S., Gannon M., Peers B., Wright C., Bonner-Weir S., and Montminy M. 2001. *Proc. Natl. Acad. Sci.* 98:1065–1070.

116. Scarpelli D.G. and Rao M.S. 1981. *Proc. Natl. Acad. Sci.* 78:2577–2581.

117. Andrew A. 1976. *J. Embryol. Exp. Morphol.* 35:577.

118. Fontaine J. and Le Douarin N.M. 1977. *J. Embryol. Exp. Morphol.* 41:209–222.

119. Carlson B.M. 1994. *Human embryology and developmental biology.* St. Louis, MO: Mosby.

120. Park H.W., Chae Y.M., and Shin T.S. 1992. *Yonsei Med. J.* 33:104–108.

121. Hebrok M., Kim S.K., St. Jacques B., McMahon A.P., and Melton D.A. 2000. *Development* 127:4905–4913.

122. Hendricks S.K. and Sybert V.P. 1991. *Clin. Genet.* 39:383–385.

123. MacFadyen U.M. and Young I.D. 1987. *Am. J. Med. Genet.* 27:987–989.

124. Hagan D.M., Ross A.J., Strachan T., Lynch S.A., Ruiz-Perez V., Wang Y.M., Scambler P., Custard E., Reardon W., Hassan S., Nixon P., Papapetrou C., Winter R.M., Edwards Y., Morrison K., Barrow M., Cordier-Alex M.P., Correia P., Galvin-Parton P.A., Gaskill S., Gaskin K.J., Garcia-Minaur S., Gereige R., Hayward R., and Homfray T. 2000. *Am. J. Hum. Genet.* 66:1504–1515.

125. Stoffers D.A., Zinkin N.T., Stanojevic V., Clarke W.L., and Habener J.F. 1997. *Nat. Genet.* 15:106–110.

126. Schwitzgebel V.M., Mamin A., Brun T., Ritz-Laser B., Zaiko M., Maret A., Jornayvaz F.R., Theintz G.E., Michielin O., Melloul D., and Philippe J. 2003. *J. Clin. Endocr. Metab.* 88:4398–4406.

127. Malecki M.T., Jhala U.S., Antonellis A., Fields L., Doria A., Orban T., Saad M., Warram J.H., Montminy M.R., and Krolewski A.S. 1999. *Nat. Genet.* 23:323–328.

128. Kim J.H., Ho S.B., Montgomery C.K., and Kim Y.S. 1990. *Cancer* 66:2134–2143.

129. Sessa F., Bonato M., Frigerio B., Capella C., Solcia E., Prat M., Bara J., and Samloff I.M. 1990. *Gastroenterology* 98:1655–1665.

130. Song S.Y., Gannon M., Washington M.K., Scoggins C.R., Meszoely I.M., Goldenring J.R., Marino C.R., Sandgren E.P., Coffey Jr. R.J., Wright C.V., and Leach S.D. 1999. *Gastroenterology* 117:1416–1426.

131. Miyamoto Y., Maitra A., Ghosh B., Zechner U., Argani P., Iacobuzio-Donahue C.A., Sriuranpong V., Iso T., Meszoely I.M., Wolfe M.S., Hruban R.H., Ball D.W., Schmid R.M., and Leach S.D. 2003. *Cancer Cell* 3:565–576.
132. Hruban R.H., Wilentz R.E., Goggins M., Offerhaus G.J., Yeo C.J., and Kern S.E. 1999. *Ann. Oncol.* 10:S9–S11.
133. Thayer S.P., Di Magliano M.P., Heiser P.W., Nielsen C.M., Roberts D.J., Lauwers G.Y., Qi Y.P., Gysin S., Fernandez-Del Castillo C., Yajnik V., Antoniu B., McMahon M., Warshaw A.L., and Hebrok M. 2003. *Nature* 425:851–856.
134. Berman D.M., Karhadkar S.S., Maitra A., Montes De Oca R., Gerstenblith M.R., Briggs K., Parker A.R., Shimada Y., Eshleman J.R., Watkins D.N., and Beachy P.A. 2003. *Nature* 425:846–851.

2

GROSS ANATOMY OF THE PANCREAS

Koji Yamaguchi, Masahiko Kawamoto, and Masao Tanaka

CONTENTS

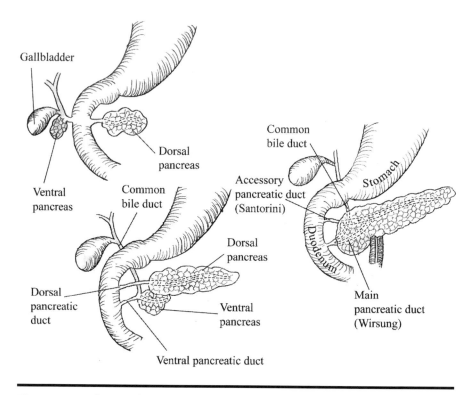

Figure 2.1 Schema of pancreatic development.

2.1 INTRODUCTION

The development of a complicated tissue, such as the pancreas, which consists of a mixture of various types of exocrine and endocrine cells, requires a meticulously planned and engineered assembly line during embryogenesis and organogenesis. An error in any stage of the assembly line can cause an immediate malfunction after birth or render the organ susceptible to toxicological insults during the life. Hence, the injury to the pancreas can start at its inception.

2.2 DEVELOPMENT

For pancreatic development, please see Chapter 1.

2.3 GENERAL TOPOGRAPHY

The pancreas is a retroperitoneal, yellowish-white lobulated organ, which weighs approximately 100 g in adult males and 85 g in adult females.[1] It is enveloped by retroperitoneal connective tissue, which forms a poorly defined capsule, and the peritoneum of the posterior wall of the lesser sac covers the anterior surface of the pancreas. The dimension of the pancreas is generally reported as between 15 to 25 cm in length, 1.4 to 4 cm in thickness, and 3 to 9 cm in height. Surgical access to the anterior aspect of the pancreas requires a dissection of the greater omentum to enter the lesser sac because the pancreas lies transversely in the retro-peritoneum behind the stomach and transverse colon. The head of the pancreas is inserted into the duodenal C loop and the tail extends to the hilus of the spleen. The body of the gland lies between the celiac trunk above and the superior mesenteric artery below.

The pancreas is divided into four parts — head, neck, body, and tail — but the exact anatomic demarcation of these parts is blurred. The head tends to be the thickest part of the organ and borders the duodenum, the common bile duct, the pancreatico-duodenal vessels, the vena cava, and the right renal artery and vein. The uncinate process of the head is part of the lower pole of the head, which extends upward from the lower border and passes behind the neck and the superior mesenteric vessels. It inserts itself between the portal vein and the inferior vena cava.[2] Carcinoma of the uncinate process of the pancreas easily invades these vessels at an early stage because of its close anatomical proximity.[3]

The superior mesenteric vessels sometimes form an indentation in the uncinate process to form the pancreatic notch. The notch between the head and the body accommodates the terminal part of the superior mesenteric vein and the beginning of the portal vein. The splenic vein runs from the left to right from the hilum of the spleen to join the superior mesenteric vein. Its course is behind the pancreas and is closely attached to the posterior surface of the pancreas. It forms no sulcus in the gland. The close relationship between the pancreas and the splenic and portal veins makes them vulnerable to invasion by cancer to the wall of the portal vein and it may play a decisive role in determining the resectability or other clinical aspects of a pancreatic tumor. A dissection of the pancreas from the splenic vein is sometimes difficult because there are many small branches of the splenic vein to the pancreas. This is the reason why the preservation of the spleen is difficult in cases undergoing a distal pancre-atectomy. The body and tail of the organ are elongated and extend to the left anterior to the aorta, left kidney, and left adrenal gland, and posterior to the stomach from which they are separated by the lesser peritoneal sac. The relationship of the pancreas to the lesser sac of the peritoneum is of particular importance in the formation of pancreatic pseudocysts, which sometimes develop following inflammation or injury to the pancreas.

2.3.1 Developmental Abnormalities

2.3.1.1 Agenesis

Agenesis of the pancreas is a rare condition and occurs due to the embryonic failure of both pancreatic buds to develop. Agenesis in one of the pancreatic buds is also rare. Agenesis of the dorsal primordium occurs due to an absence of the dorsal duct, accessory papilla, or the body and tail of the pancreas. Pancreatography shows no Santorini duct. Most of the islets of Langerhans are situated in the tail of the pancreas and the absence of the body and tail of the pancreas often leads to the development of diabetes. The term "pancreatic defect" is used to indicate a clinical uncertainty as to the reason for the absence of the body and tail of the gland, whether resulting from aplasia, hypoplasia, a malformation, or an acquired condition (atrophy).

2.3.1.2 Short Pancreas

A *congenital short pancreas* is a term that has been applied to a short pancreas that contains both dorsal and ventral ducts, but lacks a body and tail. It is believed to represent a partial agenesis of the pancreas.

2.3.1.3 Hypoplasia

Hypoplasia of the pancreas is a rare condition in which glandular pancreatic tissue is present at the normal location, but is greatly reduced in size and volume. The major ducts are intact but reduced in number and there is diminished duct branching. Therefore, there is a generalized reduction in the pancreatic size and volume, but there are no localized defects as seen in pancreatic agenesis.

2.3.1.4 Accessory Pancreas

Pancreatic tissue may develop in various sites apart from the pancreas. This condition has been referred to as an ectopic, aberrant, or accessory pancreas. Most instances of an accessory pancreas have been found in the wall of the stomach or the duodenum close to the opening of the major and minor duodenal papillae. Less frequent sites are the esophagus, small bowel, Meckel's diverticulum, colon, spleen, biliary tract, liver, and mesentery.

The clinical incidence of an accessory pancreas in several large series has been reported to range from 1 to 13%. Because the majority of such accessory pancreas cases are asymptomatic, it is probably more common than clinical incidence would suggest.

Accessory pancreatic nodules in the gastrointestinal tract are usually well-circumscribed, smooth masses that are located in the submucosal layer. The nodules vary in diameter from a few millimeters to several centimeters, and their tops often harbor a central umbilication corresponding to the orifice of a ductal structure.

Because they contain normal exocrine tissues in all and islet cell tissue in one-third of subjects, they may be subject to the usual pancreatic diseases, such as pancreatitis and pancreatic neoplasms including pancreatic carcinoma and islet cell tumors.

2.3.1.5 Annular Pancreas

An *annular pancreas* is a circular enclosure, partial or complete, of the second part of the duodenum by pancreas tissue that is in continuity with the normal pancreas. The anomaly was named by von Ecker in 1862 and was first corrected surgically by Vidal in 1905. Over 350 cases of annular pancreas had been described by 1967.

One hypothesis regarding the anomaly is that it is the result of the persistence of the left portion of the ventral anlage (which normally regresses) and its failure to accompany the rotation of the duodenum, which carries the right ventral anlage laterally and posteriorly to fuse with the dorsal bud of the pancreas. This anomaly may thus be subject to all the inflammatory and neoplastic conditions that occur in the normal pancreas.

The symptoms or signs are a duodenal obstruction, which usually occurs shortly after birth, although sometimes appears later in life. Annular pancreas appears to occur with increased frequency in patients with Down's syndrome. Resection of the duodenum or bypass operation, such as a gastrojejunostomy, is the treatment of choice.

2.4 BLOOD SUPPLY

2.4.1 Arteries

The main arterial blood supply comes from the celiac trunk and the superior mesenteric artery. The anterior and posterior pancreatico-duodenal arcades, formed by the anastomosis of the anterior and posterior branches of the superior and inferior pancreatico-duodenal arteries, supply the head of the pancreas and the duodenum. The neck, uncinate process, and body are supplied by the dorsal pancreatic artery, which may be a branch of the celiac, hepatic, splenic, or superior mesenteric artery, and by the inferior pancreatic artery, which is a continuation of the left branch of the dorsal pancreatic artery. In addition, the splenic artery supplies the distal portion of the pancreas with as many as nine branches. A large branch of the splenic artery, the great pancreatic (pancreatica magna) artery, contributes, within the pancreas, a left and right branch often oriented along the course of the Wirsung duct. The tail of the pancreas is supplied by branches of the splenic artery or left gastroepiploic artery; these consist of the caudal pancreatic arteries. (For detailed blood supply of the pancreas, please refer to Chapter 4.)

2.4.2   Veins

Pancreatic venous drainage is drained via the hepatic portal vein, which forms the superior mesenteric and splenic veins. The veins of the pancreas correspond to the relative arteries and flow the same course.[4] They are particularly abundant along the large pancreatic ducts. The splenic vein courses to the right in a groove on the posterior surface of the pancreas and, in its course, receives many small collaterals draining the gland. The pancreatico-duodenal veins pass ventral to the uncinate process and join the splenic vein behind the neck of the pancreas to form the hepatic portal vein. Because the major veins and arteries of the pancreas course posteriorly to the Wirsung duct, the ducts can be opened surgically through an anterior approach, thereby avoiding serious hemorrhaging when a pancreatojejunostomy is performed.[2] In the anterior aspect of the portal vein just beneath the pancreas neck, there are few veins of the portal vein, and the pancreas and the portal vein are easily dissected and taped for a pancreatic resection "tunneling of the pancreas." The intrinsic blood supply of the pancreas is well developed and ensures a functional relationship between the endocrine and exocrine tissues of the pancreas.

2.4.3   Lymphatic Drainage

The peripheral lymphatic plexi of the pancreas are formed by periacinar and perilobular lymphatic capillaries coursing next to blood capillaries in the interstitial connective tissue. At the surface of the pancreas, they drain into valved channels that run through the interlobular septa.[5] Thus, superficial lymphatics can be subdivided into five main collecting trunks and lymph node groups[6,7] including superior, inferior, anterior, posterior, and splenic.

1. Superior — These afferent vessels drain the lymphatics from the upper part of the pancreas into the suprapancreatic lymph nodes situated at the upper part of the head and body. These are called the superior head (SH) and superior body (SB) lymph nodes, respectively.
2. Inferior — These lymphatic channels drain the lower half of the head and body of the pancreas and end in the inferior pancreatic

group of the lymph nodes that are mainly located at the inferior border of the head (IH) and body (IB) of the gland. They may also be connected with the superior mesenteric and left lateroaortic lymph nodes.

3. Anterior — These collecting trunks course along the anterior surface of the head of the pancreas and terminate in the pyloric (Py) nodes, anterior pancreaticoduodenal (APD) nodes, and some mesenteric lymph nodes near the jejunum.

4. Posterior — These collecting lymphatic channels run along the posterior aspect of the head of the pancreas. They lead into the posterior pancreaticoduodenal (PPD) lymph nodes, the common bile duct (CBD) nodes, the right lateroaortic nodes, and some lymph nodes at the origin of the superior mesenteric artery. The PPD group drains most of the lymph flowing from the major papilla (ampulla of Vater) and the common bile duct.

5. Splenic — The lymphatic drains the lymph from the tail of the pancreas to the splenic group of lymph nodes (formed by the superior lymph nodes of the tail of the pancreas), the phrenicolienal ligament nodes, and those at the hilus of the spleen.

A recent study based on the frequency of lymph node involvement in cases of carcinoma of the head of the pancreas demonstrated that the main lymphatic pathway from the head of the pancreas to the paraaortic lymph nodes is through the lymph nodes around the superior mesenteric artery.[8]

2.4.4 Nerve Supply

The pancreas is innerved by both the vagal and sympathetic nerves. In addition, the origin is supplied by afferent fibers that course either with the sympathetic or vagal branches. This innervation of the head of the pancreas is greater than that in the tail of the pancreas.

The vagal fibers are part of the regulatory system that controls exocrine and endocrine activity of the pancreas as well as the capillary blood flow of the pancreas. Vagal efferent fibers originate from cell bodies located in the vagal dorsal nucleus. The fibers are carried by both right and left vagi to the celiac plexus and then to the interlobular septa of the pancreas where they synapse with these ganglia. The postganglionic nerve forming these ganglia have numerous nerve endings that are distributed to the acinar cells, islet cells, and smooth muscle cells around the ducts.[9]

Sympathetic innervation originates from neurons located in the lateral grey column of Segment 5 to Segment 10 of the thoracic spinal cord. The sympathetic fibers reach the celiac ganglia through the greater splanchnic nerve and synapse with these ganglia. Postganglionic fibers course along the hepatic, splenic, and superior mesenteric arteries and innervate the pancreatic arteries and arterioles, as well as the pancreatic veins. The direct vasomotor control of these vessels and arteriolar vessels regulate the blood flow of the pancreas.

A major portion of the pancreatic fibers of the vagus nerve are visceral afferent. These fibers are implicated in the duodenal-pancreatic reflex mechanisms, but are not involved in the transmission of pain from the pancreas, which occurs along the afferent sympathetic fibers. Afferent sympathetic fibers enter the spinal cord at the same levels as the efferent nerve. As a result, uncontrollable back pain is one of the frequent treatment targets of advanced pancreatic cancer.

2.5 PANCREATIC DUCT

2.5.1 Anatomy

The pancreas develops from both ventral and dorsal pancreatic anlages, which arise from the endodermal epithelium of the duodenum. The ventral anlage becomes the dorsocaudal portion and uncinate process of the head of the pancreas, and the dorsal anlage becomes the ventrocephalic portion of the head and body and tail of the gland.

The duct of the ventral anlage and of the dorsal anlage distal to the junction of both ducts serve as the main pancreatic duct (duct of Wirsung). The duct of the dorsal anlage proximal to the junction becomes the accessory pancreatic duct (duct of Santorini). The pancreatic duct system is arranged on the main channel principle, that is, 15 to 30 side branches of about equal thickness draining from both above and below into a main channel, namely the Wirsung duct.[10] Secondary ducts enter the main pancreatic duct in a herringbone pattern.

The duct of Wirsung is the main pancreatic duct and opens into the duodenum at the papilla of Vater (major papilla). It drains the major part of the gland and begins in the tail by the convergence of several small ducts (secondary ducts). The main pancreatic duct begins at the terminus of the tail of the gland. It runs closer to the posterior surface and midway between the superior and inferior margins through to the tail and body and runs nearer to the dorsal and superior surfaces of the gland than to the ventral and inferior surface in the head. Its course is horizontal in the tail and body and it moves posteriorly and inferiorly to form an arch (convex to right) in the head. It often receives secondary ducts that drain the posteroinferior part of the head of the pancreas (ventral pancreas), including the uncinate process. The diameter of the duct of Wirsung averages 3 mm[11] and is 4.8 mm in the head, 3.5 mm in the body, and 2.4 mm in the tail, respectively.

The accessory pancreatic duct of Santorini, which embryologically corresponds to the head part of the dorsal duct, is present in 99% of all humans. This duct drains the anterosuperior part of the head and may open into the duodenum at the minor papilla, about 2 cm cephalad to the papilla of Vater (major papilla). It also anastomoses with the duct of Wirsung near the neck of the pancreas. When pancreatitis is mainly located in the pancreas drained by the Santorini duct, such pancreatitis is termed *groove pancreatitis.*

In the head of the pancreas, the lower 3 to 5 cm part of the CBD runs in a groove in the posterior surface of the head of the pancreas and may seem to be embedded in the pancreatic tissue. In all but a few of the cases of the latter type, however, the duct lies in a cleft in the pancreas and is overlapped by the pancreatic tissue. The bile duct usually passes down close to, or runs in direct contact with, the left border of the second part of the duodenum; but it may be as much as 2 cm from the duodenal wall before it begins to incline to the right to reach and then pass through the wall of the second part of the duodenum. Its progress to the right may be in a straight line or toward the duodenum almost at a right angle.

In the head, the main pancreatic duct inclines caudally and dorsally and passes to the left caudal side of the intrapancreatic portion of the CBD, with which it usually unites while running obliquely through the duodenal wall. It finally opens into the major duodenal papilla (the papilla of Vater), a prominence located on the posteromedial wall of the second portion of the duodenum. Because of the anatomical intimacy of the distal bile duct and the pancreas head, the distal CBD tends to be at high risk of invasion by pancreatic carcinoma, which develops in the head of the pancreas and often produces obstructive jaundice.

2.5.2 Ampulla of Vater

The ampulla of Vater is the vase shaped expansion of the duodenal wall through which the CBD and the main pancreatic duct enter the duodenum. The duodenal mucosa and musculature (the muscle of Oddi) form the outer wall of the ampulla and project into the duodenum as a papilla, the major papilla. The ampullary inner surface is lined with a biliary ductal epithelium. There is considerable controversy regarding the opening of the ampulla of Vater, the openings of the CBD and pancreatic duct into the complex. There is even doubt as to the existence of an ampulla in some cases.

In about one-third of all patients, the common bile and main pancreatic ducts enter into a common ampulla, from 3 to 14 mm, from the apex of the duodenal orifice, so that a true "common channel" is formed. When carcinoma arises in the papilla of Vater, the CBD and main pancreatic duct are affected together, thus producing a "double duct sign" on images.

2.5.3 Variations of Main and Accessory Pancreatic Ducts

During the embryologic development of the main and accessory pancreatic ducts, several variations may occur.[12,13] The usual pattern of the pancreatic duct is that the accessory pancreatic duct is connected with the main pancreatic duct and then opens into the duodenum. The accessory duct is smaller in caliber than the main duct. There are hypoplasia and aplastic variations of the accessory duct and nonunion of both ducts (Figure 2.2). In these patterns, the main pancreatic duct carries most or all of the pancreatic secretion. In less common patterns, the accessory pancreatic duct carries most or all of the secretion. The variation is pancreas divisum and the details of this condition are described in Section 2.5.4.

A communication of the accessory and main ducts is present in 90% of patients and the patency of the accessory duct orifice is reported in 35

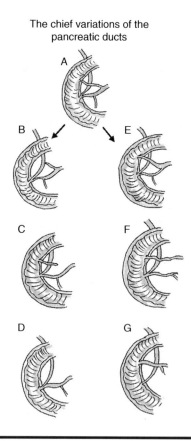

The chief variations of the
pancreatic ducts

Figure 2.2 Main and accessory pancreatic ducts.

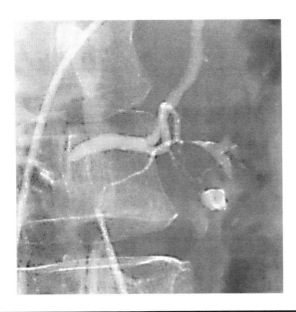

Figure 2.3   Prominent Santorini duct. The Santorini duct is larger than the Wirsung duct in diameter.

to 60% of cases.[13] The frequency of the variations of the ducts is 60% in the usual configuration, 30% in the suppression of the accessory duct, and 10% in the suppression of the main duct.[14] There seems to be at least some agreement that in about 90% of all pancreases, the sole or main excretory channel is the duct (of Wirsung) that opens at the major duodenal papilla. In about 50 to 70% of pancreases, an accessory (of Santorini) is present, thus resulting in the transmission of various amounts of secretion, from none up to all the gland's output into the duodenum at the minor duodenal papilla (Figure 2.3).

Two ducts of Santorini with two minor papilla in the duodenum and with a communication with the major duct system were found in a single case by Berman et al.[15] in their series of 130 specimens.

It has already been mentioned that variations in the course of the main pancreatic duct through the head of the pancreas are by no means uncommon. One striking variation in the course of the main pancreatic duct through the head of the pancreas is a complete loop of the duct (Figure 2.4). The looped type of duct was first recognized by Baldwin[16] and Rienhoff and Pickrell[17] and it was also demonstrated radiographically in autopsy materials by Newman et al.[18] They also showed that the duct of Wirsung may occasionally form a loop around the CBD and such a loop is sometimes referred to as the formation of an ansa pancreatica.[18] Some of the numerous variations in the course of the duct are illustrated.

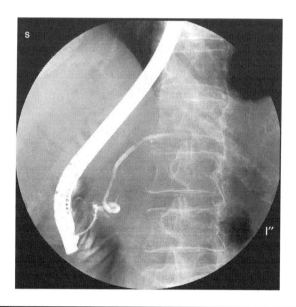

Figure 2.4   Loop of main pancreatic duct. Distal portion of the main pancreatic duct shows a loop.

2.5.4   Pancreas Divisum

A Santorini system and a Wirsung system may coexist with separate duodenal orifices and without communications within the gland (*pancreas divisum*) (Figure 2.5). This condition is termed a pancreas divisum (complete and incomplete). In this situation, the duct of Wirsung, normally the

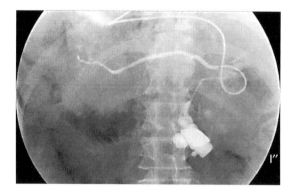

Figure 2.5   Pancreas divisum. Santorini and Wirsung ducts are separated. Intraductal papillary mucinous tumor is seen in a branch of the Santorini duct.

main pancreatic duct, may be small, namely no more than 1 to 2 cm in length, and the duct of Santorini, draining into the duodenum at the minor papilla, provides the main ductal system. When the duct of Santorini is well developed, its course through the head of the gland is in a straight line without the angles or loops that are usually observed in the course of the duct of Wirsung.

It has been postulated by certain workers that, in pancreas divisum, inadequate drainage via Santorini's duct may predispose to pancreatitis. This condition is called *dorsal pancreatitis*. When pancreatitis is confined to the ventral pancreas, such pancreatitis is named ventral pancreatitis. The anomaly was found in 24 patients by Cotton and Kizu[19] by endoscopic pancreatography and episodes of pancreatitis was said to have occurred in 14 of these cases, and 6 others had had episodes of abdominal pain. A failure of fusion of the ducts was found, also by endoscopic retrograde pancreatography, by Mitchell et al.[20] in 21 out of 449 (4.7%) successful pancreatograms. Pancreas divisum may be one cause of pancreatitis in patients with idiopathic pancreatitis. A papilllotomy, tube stenting, and sphincterotomy of the minor papilla may effectively prevent a relapse of acute pancreatitis. When the communication of the Santorini and Winslow ducts is incomplete, pancreas head carcinoma should be differentiated from this condition on pancreatography because incomplete communication mimics stenosis by pancreatic cancer.

2.5.5 Termination of the Common Bile Duct and the Main Pancreatic Duct

The main pancreatic and common bile ducts unite in the duodenal wall in 85% of cases to form a common channel that has been named the ampulla of Vater[13] after Vater who first noted it. The common channel, 1 to 14 mm in length, opens on the major duodenal papilla with a single opening (Figure 2.6). In rare types of the termination of both ducts, both ducts open on the major duodenal papilla via separate orifices or open into the duodenum at the separate points. Investigations that relied mainly upon dissection showed that between 20 and 30% of cases may have separate biliary and pancreatic orifices on the duodenal papilla. Rienhoff and Pickrell[17] found that in 24% of their 250 dissections there was no junction of the pancreatic and bile ducts, but Singh[21] found that in 30 of the 100 dissections he carried out the two ducts remained separate. However, injection methods followed by radiography in Millbourn's studies[22] showed it was only exceptionally (about 1 in 20 cases) that there were separate entrances. Somewhat similar results were obtained by Berman et al.[15] who made vinyl acetate casts of the ducts of the pancreases obtained at routine postmortem examinations and then digested away the

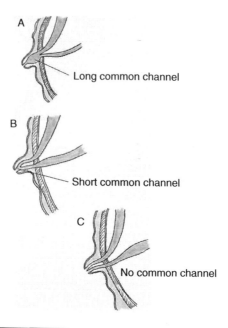

A — Long common channel

B — Short common channel

C — No common channel

Figure 2.6 Termination of the common bile duct and the main pancreatic duct.

glandular tissue. They found that only 6.2% of their 130 specimens displayed the main pancreatic duct (duct of Wirsung) and the CBD opening independently into the duodenum. Such variations in the results obtained by different workers means that there must be many cases in which it is difficult to discriminate between a common orifice and two ducts that open separately but close together.

2.5.6 Pancreatobiliary Maljunction

A *pancreatobiliary maljunction* (PBM) is a congenital anomaly that can be defined as a union of the pancreas and biliary ducts that is located outside of the duodenal wall and is beyond the influence of the sphincter of Odii (Figure 2.7 and Figure 2.8). The common channel measures 15 mm or greater in length. PBM is uncommon with the incidences of 1.0% in endoscopic retrograde cholangiopancreatography (ERCP) and 3.3% in biliary surgery in Japan. The frequency of PBM in Western countries is lower than that in Japan in contrast to pancreas divisum. PBM is frequently seen in Asian people, especially in women. PBM is closely related to a congenital bile duct dilatation.[23]

Because the function of the sphincter muscle of the duodenal papilla does not extend the full length of the common channel, it results in the regurgitation of the pancreatic juice and bile. In most cases, because the

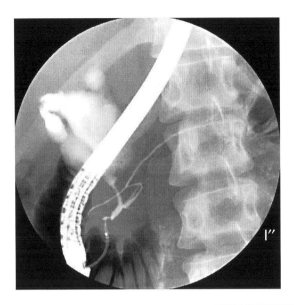

Figure 2.7　Pancreatobiliary maljunction (B-P type). The distal bile duct drains into the main pancreatic duct forming a long common channel. Congenital choledochal cyst is identified.

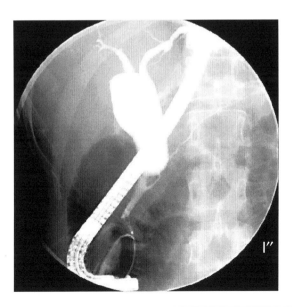

Figure 2.8　Pancreatobiliary maljunction (P-B type). The distal common bile duct drains into the main pancreatic duct forming a long common channel. Congenital choledochal cyst is present.

intrapressure of the pancreatic duct is higher than the intrapressure of the biliary tract, the pancreatic juice regurgitates into the biliary tract continuously and the biliary mucosa is continuously susceptible to damage as a result of a continued presence of infected bile and activated pancreatic enzymes. Regurgitated phospholipase A2 of the pancreatic juice activates lisolecithin in the bile, which injures the bile duct and evokes bile duct cancer. This eventually causes cancer to occur in the biliary mucosa and other pathological changes including gallstones (28.5 to 38.1%) or acute pancreatitis (23.4 to 26.6%). When PBM is associated with a congenital choledochal cyst,[24] bile duct cancer is often seen in the choledochal cyst (15.6 to 36.0%).[25] When PBM is not associated with congenital choledochal cyst, gallbladder carcinoma is sometimes seen due to a condensation of the bile containing pancreatic juice in the gallbladder.[26]

2.5.7 Choledochocele

Choledochocele occurs in patients older than those with PBM (Figure 2.9). Choledochocele has sometimes been reported to show dysmotility by sphincter of Oddi manometry and it is considered to be an acquired disorder produced either by a sphincter of Oddi dysfunction or stenosis. Others say that choledochocele is a congenital condition due to the presence of a duodenal mucosa lining on both sides of choledochocele. Ohtsuka et al.[27] reported the malignant potential of choledochocele by the bile with stagnating pancreatic juice in the choledochocele to be the same mechanism as that observed in PBM.

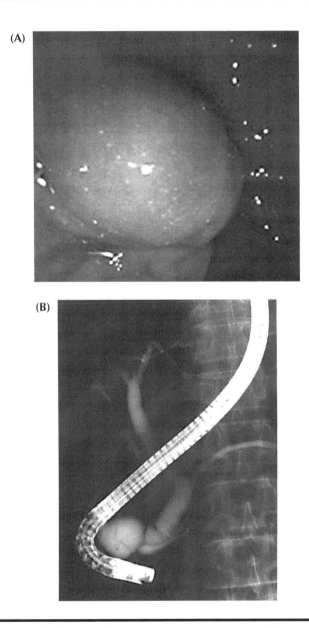

Figure 2.9 Choledochocele. The upper portion of the ampulla of Vater is enlarged (A). Endoscopic retrograde cholangiopancreatography shows a cystic dilatation of the distal end of the CBD (B).

REFERENCES

1. Bockman D. Anatomy of the pancreas. In: Go VLW, Gardner JD, Brooks FP, Lebenthal E, DiMagno EP, and Scheele GA, Eds. The exocrine pancreas: biology, pathobiology and diseases New York: Raven Press, 1993:1–8.
2. White T. Surgical anatomy of the pancreas. In: Carey LC, Ed. The pancreas St. Louis: CV Mosby, 1973:3–16.
3. Yamaguchi K. Carcinoma of the uncinate process of the pancreas with a peculiar manifestation. *Am J Gastroenterol* 1992;87:1046–50.
4. Falconer CWA, Griffiths E. The anatomy of the blood vessels in the region of the pancreas. *Brit J Surg* 1950;37:334–44.
5. Evans BP, Ochsner A. The gross anatomy of the lymphatics of the human pancreas. *Surgery* 1954;36:177–91.
6. Cubilla AL, Fitzgerald PJ. Tumor of the pancreas Atlas of tumor of pathology, ed. 2nd Edition Washington DC: Armed Forces Institute of Pathology, 1984:34–36.
7. Cubilla AL, Fortner J, Fitzgerald PJ. Lymph node involvement in carcinoma of the pancreas area. *Cancer* 1978;41:880–87.
8. Kayahara M, Nagakawa T, Kobayashi H, et al. Lymphatic flow in carcinoma of the head of the pancreas. *Cancer* 1992;70:2061–66.
9. Singh M, Webster PD. Neural hormonal control of pancreatic secretion. *Gastroenterology* 1978;74:294–309.
10. Anacker H. Radiological anatomy of the pancreas. In: Anacker H, Ed. Efficiency and limits of radiologic examination of the pancreas, 3rd Ed. Stuttgurt: Georg Thieme Pub, 1973:29–42.
11. Birnsting M, Study of pancreatography. *Br J Surg* 1959;47:128–39.
12. Cubilla AL, Fitzgerald PJ. Tumors of the exocrine pancreas, 2nd series, Fascicle 19. Washington: Armed Forces Institue of Pathology, 1984.
13. Skandalakis LJ, Gray SW, Skandalakis JE. Surgical antomy of the pancreas. In: Howard JM, Jordan JR, and Reber HA, Eds. Surgical diseases of the pancreas Philadelphia: Lea and Febriger, 1987:11–36.
14. Gray SW, Skandalakis JE, Skandalakis LJ. Enbryology and congenital anamolies of the pancreas. In: Howard JM, Jordan JR, and Reber HA, Eds. Surgical disease of the pancreas Philadelphia: Lea and Febriger, 1987:37–45.
15. Berman LG, Prior JT, Abramow SM, Ziegler DD. A study of the pancreatic duct system in man by the use of vinyl acetate casts of postmortem preparations. *Surg Gynecol Obstet* 1960;110:391–403.
16. Baldwin WM. The pancreatic ducts in man, together with a study of the microscopical structure of the minor duodenal papilla. *Anat Rec* 1911;5:197–228.
17. Rienhoff, Jr., WF, Pickrell KL. Pancreatitis: an anatomic study of the pancreatic and extrapancreatic biliary systems. *Arch Surg* 1945;51:205–19.
18. Newman HF, Weinberg SB, Newman EB, Northup JD. The papilla of Vater and distal portion of the common bile duct and duct of Wirsung. *Surg Gynecol Obstet* 1958;106:687–94.
19. Cotton PB. Progress report. ERCP. *Gut* 1977;18:316–41.
20. Mitchell CJ, Lintott DJ, Ruddell WS, Losowsky MS, Axon AT. Clinical relevance of an unfused pancreatic duct system. *Gut* 1979;20:1066–71.

21. Singh I. Observation on the mode of termination of the bile and pancreatic ducts: anatomical factors in pancreatitis. *J Anat Soc India* 1956;5:54–60.

22. Millbourn E. On the excretory ducts of the pancreas in man with special reference to their relations to each other to the common bile duct and to the duodenum. *Acta Anatom* 1950;9:1–34.

23. Babbitt DP. Congenital choledochal cysts: new etiological concept based on anomalous relationships of the common bile duct and pancreatic duct. *An Radiol* 1968;12:231–40.

24. Babbitt DP, Starshak RJ, Clemett AR. Choledochal cyst: a concept of etiology. *Am Roentgenol* 1973;119:57–62.

25. Todani T, Watanabe Y, Kobuchi K. Carcinoma arising in the wall of congenital bile duct cysts. *Jpn J Pedriatr Surg* 1977;9:77–83.

26. Kinoshita H, Nagata E, Hirohashi K. Carcinoma of the gallbladder with an anomalous connection between the choledochus and the pancreatic duct: report of 10 cases and review of the literature in Japan. *Cancer* 1984;54:762–69.

27. Ohtsuka T, Inoue K, Ohuchida J, Nabae T, Takahata S, Niiyama H, et al. Carcinoma arising in choledochocele. *Endoscopy* 2001;33:614–19.

3

FINE STRUCTURE OF THE PANCREAS

Dale E. Bockman, Tomoko Inagaki, and Toshio Morohoshi

CONTENTS

3.1   INTRODUCTION

In the absence of disease, the pancreas develops and functions silently, providing hormones and digestive enzymes in a regulated fashion. These products are delivered to the proper place at the appropriate time in the right form. Potentially damaging enzymes must be synthesized, stored, released, and transported in a manner that preserves the integrity of all of the components. Accomplishing this process requires not only the specialization of different types of cells, but also the restriction of products, at various stages, to specific compartments to avoid premature activation with resultant injury to the tissues. Transmission of material to and from the cell is closely regulated. Membranes separate intracellular contents. Material in the lumen is separated from the extracellular matrix by the organization of the epithelial components, which make up the acini and ducts.

Blood vessels provide nutrients to the cells of the pancreas and serve as the route for substances that affect the cells as well as for substances released from the pancreas. The relationship of blood vessels to the islets is important for hormonal distribution. A common vascular supply serves the exocrine and endocrine components.

Pancreatic nerves contribute to the normal function of the organ. *Efferent nerves* regulate secretion and circulation. *Sensory nerves* contribute to the regulatory process, carrying on their role silently in the absence of pancreatic disease.

3.2   DEVELOPMENT

Information on the development of the pancreas is described in Chapter 1 and Chapter 2.

3.3   GENERAL ORGANIZATION OF THE PANCREAS

The normal pancreas is rich in epithelial components that carry on their functions surrounded by a relatively sparse extracellular matrix. Blood vessels and nerves traverse the extracellular matrix to serve the epithelial components as well as the cellular elements within the extracellular matrix itself. The single organ is composed of two functional systems: the endocrine gland consisting of a variety of epithelial cells arranged as dispersed aggregates and the exocrine gland with cells specialized to synthesize and release digestive enzymes into the ductal system, which in turn modifies and conducts these products into the duodenum.

The most common arrangement for the delivery of pancreatic juice into the duodenum is for the main pancreatic duct to join with the bile duct at the major papilla and for an accessory pancreatic duct to empty into the duodenum, slightly anteriorly at the minor papilla. There are variations to this arrangement, the most common being the absence of a patent accessory duct. However, it is clear that exocrine secretions must be conducted from their point of secretion from cells through a system of channels that eventually will exit into the duodenum. The endocrine products are released directly into the bloodstream. Therefore, the delivery route is completely unrelated to that of the exocrine pancreas. However, the endocrine cells are intimately associated with cellular components of the exocrine system.

3.4   COMPARTMENTALIZATION

The normal pancreas maintains a system of larger and smaller compartments that provide regions for vital reactions to take place, for modulation of the transfer of substances from one component to another, and to protect regions from damage that might result from improper quantities of the wrong substance coming into contact with a part of the gland.

A component that can be a marker for the delineation of some compartments is seen in the basal lamina, which is found at the interface of epithelial cells and extracellular matrix. The basal lamina is observed by electron microscopy as a gray line paralleling the basal membrane of epithelial cells. It is part of the basement membrane, which can be revealed by histochemical staining. The basal lamina provides a continuous sheet separating epithelial cells from extracellular matrix beginning in the duodenum and continuing along the large ducts, over the ductal tree, and around the acini. Tight junctions between the epithelial cells where they join each other adjacent to the lumen contribute to the separation between the lumenal compartment and the extracellular matrix. The contents of the lumenal compartment differ from those of the extracellular matrix. Normal function relies on the compartments remaining separate. Breaching the barrier between the two compartments may lead to the development of abnormal products within the lumen and in the surrounding tissue.[1,2]

3.5   PANCREATIC ACINI

Acinar cells, which are responsible for generation and release of digestive enzymes into the lumenal system, are grouped together in spheroidal and tubular arrangements.[3,4] Sections through these groupings are referred to as acini (Figure 3.1 and Figure 3.2). Acinar cells are not identical in content, and they undergo changes with accumulation of intracellular products, stimulation, and secretion. However, they are similar in overall fine structure.[5–7]

Acinar cells are typical of glandular cells that synthesize, store, and secrete proteinaceous materials. They are the model most commonly used to delineate the steps that are necessary for this type of cell to carry out that process.[8]

Acinar cells are polarized. Their apical pole is smaller, bordering on a lumen that is surrounded by acinar cells or by a combination of acinar cells and ductular (centroacinar) cells (Figure 3.1 and Figure 3.2). The apical membrane projects into the lumen as microvilli. Mostly unactivated

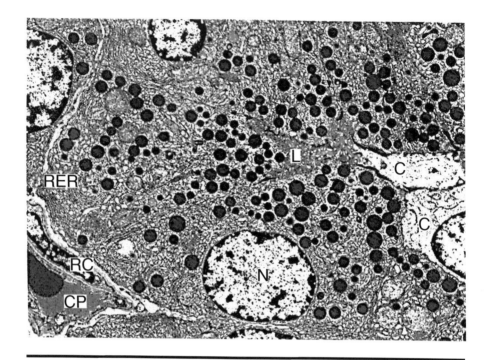

Figure 3.1   Electron micrograph of human pancreas. Acinar cells occupy most of the field. Zymogen granules (Z) are more numerous toward the lumen (L). Rough endoplasmic reticulum (RER), present throughout the cell, is prominent toward the base of acinar cells. Two centroacinar cells (C), a capillary (CP), and a pericyte (PC) are present. ×5600. (From Morohoshi and Bockman.[5] Used with permission.)

digestive enzymes (e.g., trypsinogen) are stored within the cells as spherical zymogen granules approximately 1 μm in diameter. A membrane that provides a barrier between the cytosol and the contained zymogen surrounds each zymogen granule. Zymogen granules tend to concentrate in the region of the cells closer to the lumen. Secretion involves the fusion of the zymogen granule membrane with the lumenal membrane of the acinar cell, releasing the contents of the zymogen granule into the lumen.

The broader basal pole of acinar cells forms the interface with extracellular matrix. Thus, the acini are one cell thick. Rough endoplasmic reticulum is prominent in the basal region of the acinar cell, providing an indication where the assembly of amino acids into proteins (digestive enzymes and lysosomal enzymes) begins. The nucleus occupies a more central region. The Golgi apparatus lies in general between the nucleus and the apical pole.

The acinar cells are immunohistochemically positive for anti-alpha-amylase, alpha-1-antitrypsin, alpha-1-antichymotrypsin, trypsin, and chy-

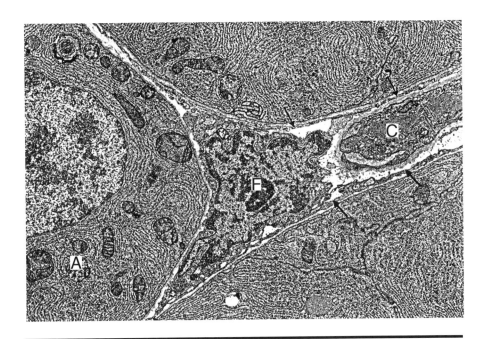

Figure 3.2 Acinar cells (A) lying on a basal lamina. At the outside of the membrane, thin periacinar mesenchymal cells (pancreatic stellate cell) with thin elongated cytoplasm are present (arrows). Fibroblastic cells (F) and capillary (C) are seen between the acini. ×7200.

motrypsin. These digestive enzymes and lysosomal enzymes are synthesized in the rough endoplasmic reticulum. These protein products are transported to the Golgi apparatus where they are sorted into separate compartments — zymogen granules for the digestive enzymes and lysosomes for the others. Under normal conditions, these compartments remain separate. Under unusual conditions, zymogens and lysosomal enzymes may become colocalized; acute pancreatitis is one situation in which colocalization and activation of trypsin is observed.[9–11]

3.6 PANCREATIC DUCTS

The smallest pancreatic ducts (*ductules*) begin in close association with acini. They are composed of a single layer of cells without the prominent endoplasmic reticulum and zymogen granules of acinar cells (Figure 3.3). They are characterized by a plain appearance and are recognized in part by the lack of acinar characteristics. In some places, ductular cells join with acinar cells to complete the tube of epithelial cells surrounding the lumen. In this location, the ductular cells may be referred to as centroacinar cells. Electronmicroscopically, centroacinar cells are relatively clear

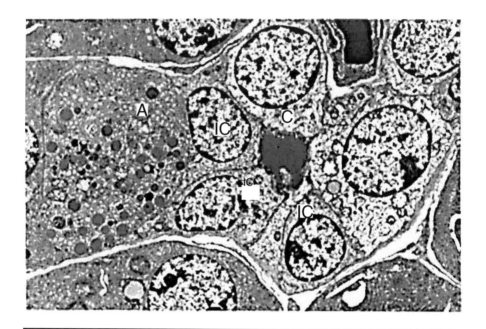

Figure 3.3 Section through the beginning of an intralobular ductule. Three smaller (IC) and two larger (C) ductular cells form the wall around the ductular lumen. An acinar cell (A) is intimately associated with the ductule, but most other acinar cells are separated by sparse intercellular matrix. ×3800. (From Morohoshi and Bockman.[5] Used with permission.)

cells joined to adjacent acinar cells by junctional complexes (Figure 3.1 and Figure 3.2). Their lateral membranes form interdigitations. Centroacinar cells contain little rough endoplasmic reticulum and fewer ribosomes. The luminal surface is smooth with a few microvilli (see Chapter 5 for additional information).

Intercalated ductules are formed by single flat to low cuboidal epithelium with clear cytoplasm. They are morphologically similar to centroacinar cells. One or two modified cilia project into the lumen from intercalated duct cells (Figure 3.4).[12] The cilia do not have the classical "9 + 2" arrangement of microtubules associated with motility. Rather, the number and arrangement of microtubules is variable, consistent with sensory cilia in other situations. It has been suggested that the cilia are kinocilia, which mix the pancreatic juice and help move along the ductal system or serve a sensory function; lack of the "9 + 2" arrangement makes this unlikely.[13] The lumen of the ductules is quite small in the normal condition, but can enlarge under some pathological conditions.

Intra- and interlobular ducts, some of which have mucin granules[5] are lined with a single layer of cuboidal or low columnar cells on a

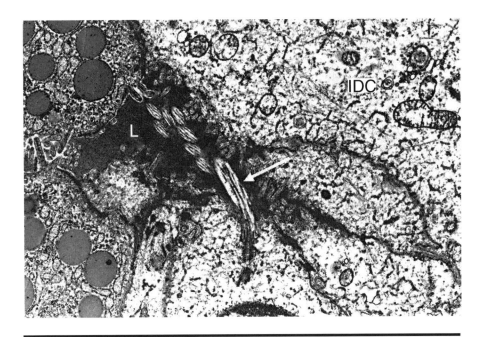

Figure 3.4   Acinar, centoacinar, and intercalated duct cells (IDC) have microvilli projecting into the lumen (L). Intercalated duct cells also have one or two cilia (arrow). ×17,200.

basement membrane. They secrete mucins into the lumen. Mucin-containing glands in the connective tissue wall empty into the large ducts. The mucins are mainly sulfated and stain strongly with Alcian blue at pH 1.0 as well as with high iron diamine.[14] They possess round or oval nuclei, some rough endoplasmic reticulum, mitochondria, and a well-developed Golgi apparatus. Numerous microvilli project into the lumen. In the intercalated duct epithelium, some scattered goblet are present. The goblet cells have mucin granules in the apical cytoplasm. The mucin granules are discharged by exocytosis.

The main pancreatic duct is lined by a single layer of columnar epithelium with oval nuclei in the basal region (Figure 3.5). The epithelium expands basally to provide mucous glands in some areas. The mucins have fewer sulphomucins and more neutral mucins and sialomucins. The epithelial cells have a large proportion of mucus-secreting cells that have an abundance of mucin granules in the apical region. Scattered endocrine cells may also be found in the basal region of the epithelium.

Ductal cells acquire characteristics associated with their function. These include cytokeratins, carbonic anhydrase, and secretin receptors.[15,16] Cytokeratins (e.g., 7 and 19) are useful as markers to identify duct cells and make them more easily recognizable.

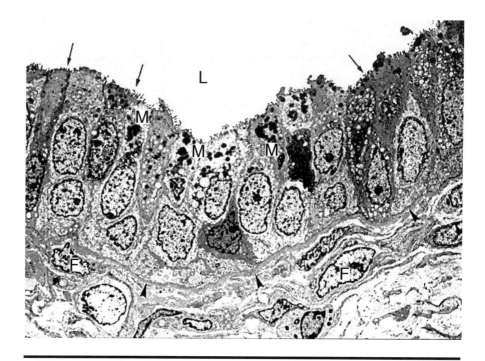

Figure 3.5 Fine structure of the main pancreatic duct. Columnar epithelial cells containing mucin granules (M) lie on a basal lamina (arrowheads). Microvilli (arrows) border the lumen (L). The extracellular matrix at the bottom contains cells, including fibroblasts (F). ×2400. (From Morohoshi and Bockman.[5] Used with permission.)

The importance of maintaining the lumen as a separate compartment by the continuous layer of epithelium joined by tight junctions, plus the underlying basement membrane, is evident from observations of breaches associated with pancreatic disease.[2,17] The ducts from patients with chronic pancreatitis display areas in which the barrier is missing, with unusual accumulations in the lumen as well as changes in the surrounding tissue. The number of capillaries increases and inflammatory cells accumulate within an altered extracellular matrix.

3.7   ISLETS

Most of the endocrine cells in the pancreas are clustered into the islets of Langerhans (see Chapter 28) and contribute to about 1 to 2% of the volume of the adult pancreas. There are also single cells that are interposed between the cells of the exocrine pancreas, mainly in the ducts. With hematoxylin and eosin stain, the islet cells appear as clusters of pale-

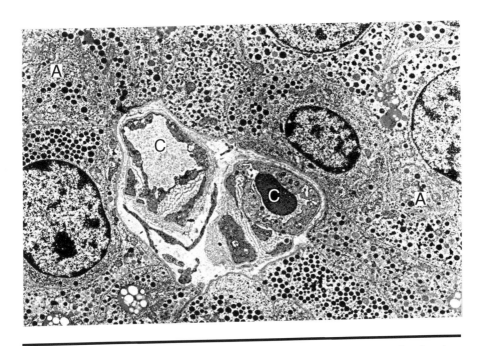

Figure 3.6 Fine structure of pancreatic islet composed of four types of endocrine cells (A-, B-, D-, PP-cell). Islet cells have numerous neuroendocrine granules diffusely distributed in cytoplasm. C: capillary; P: pericyte; A: Alpha cells; B: Beta cells. ×5600.

stained cells surrounded by more intensely stained pancreatic acini. The pancreatic islet is spheroidal, measuring 52 to 210 μm in diameter.[18] The cells within the islets are arranged as irregular cords. It is not possible to identify the various cell types within islets with routine methods. Pancreatic endocrine components consist of several cell types (Figure 3.6) producing different hormones. The four main cell types include A-cells secreting glucagon, B-cells secreting insulin, D-cells secreting somatostatin, and PP-cells secreting pancreatic polypeptide. There is a tendency for B-cells to be in the center part on the islet and non-B cells to occupy the periphery.

Granular size and morphology are relatively specific for each cell type. A-cell (alpha cell) granules measure 200 to 300 nm and contain an eccentrically located dense core within a less dense outer region that is separated from the limiting membrane by a thin halo.[19] B-cells (beta cells) contain 225- to 375-nm granules with a dense polyhedral core surrounded by a wide, electron-lucent halo underlying the limiting membrane.[19] D-cell (delta cell) granules are slightly smaller (170 to 220 nm) than A-cell granules and have a uniformly dense core.[19] The PP-cells of the ventrally derived portion of the pancreas have 180- to 220-nm granules of variable shape and density, whereas granules of other PP-cells are smaller and more homogeneous.[19]

The islets are highly vascular. The detailed patterns of these vessels are described in Chapter 4. The extracellular matrix within islets is sparse. Upon appropriate stimulation, the hormones, which are stored as granules, are released from the cells, taken up by capillaries in the islet, and distributed via the vascular system.

Like zymogen granules, each granule in islet cells is surrounded by a membrane, designating a functional compartmentalization of the stored material as distinct from the cytosol. Hormone release involves fusion of the granule membrane with the cell membrane. Islet cells containing insulin or glucagon are identifiable by the structure of the stored granules as observed by routine preparation for electron microscopy. Granules in insulin-containing cells (beta cells) characteristically have dense, angular, crystal-like structures separated from their membrane by light areas. Granules in glucagon-containing cells (alpha cells) have a dense, spherical core with a surrounding light space immediately inside the membrane. However, identification of these, and of cells producing pancreatic polypeptide and somatostatin, is more easily accomplished by using immunocytochemistry.

3.8 DEVELOPMENT AND PLASTICITY

Exocrine and endocrine cells of the pancreas are derived from the same source and blood vessels are intimately involved in the developmental process. In the mature state, however, the major vessels supplying the pancreas do not parallel the major ducts. Blood vessels enter the pancreas from its periphery, achieving association with the more central ducts secondarily.

In early development, the aorta lies immediately dorsal to the posterior foregut and the vitelline veins lie immediately ventrolateral. At this stage of development, the aorta, veins, and gut are simple tubes one cell layer thick; the vessels consist of one layer of endothelial cells. The vessels induce the gut epithelium to proliferate, producing buds that eventually will form the pancreas.[20] The dorsal and ventral pancreatic buds grow into the dorsal mesogastrium and eventually fuse. The epithelium continues to expand as primitive ductules that branch and anastomose.

Some of the cells begin to synthesize pancreatic hormones. More primitive cells may synthesize more than one hormone in the same cell.[21] Mature cells usually synthesize and store one type of hormone. Other cells begin to synthesize exocrine zymogens. Islets and acini are produced by accumulations of endocrine and acinar cells. Ducts and ductules are derived from the primitive ductules that are not induced to form islet cells or acinar cells. Ductal cells synthesize specific cytokeratins. Upon achieving maturity, the normal pancreas maintains this condition. Differentiated islet cells and acinar cells are maintained in approximately the same quantities. Cell proliferation and differentiation are modest in the mature pancreas.

Under a variety of conditions, acinar and islet cells are induced to transdifferentiate. This process tends to cause the previously differentiated cells to regress into an earlier developmental stage. Acini and islets revert to the morphology of ductules, complete with the markers that characterize them. Cells, which at one time possessed the granules typical of stored insulin or hydrolytic zymogens, take on the relatively dull fine structural characteristics of duct cells and cytokeratins typical of duct cells. This kind of conversion is observed under a variety of conditions, including occlusion of the pancreatic ducts, chronic pancreatitis, experimental manipulation *in vivo* and *in vitro*, and development of pancreatic cancer.[22–36] Because of the architectural arrangement of the pancreas,[3,4,37] when acinar cells transdifferentiate into ductular cells, the appearance of lobules is altered to appear as a collection of tubules. The collections thus produced are referred to as tubular complexes.[24]

3.9 THE EXTRACELLULAR MATRIX

The extracellular matrix in the normal pancreas is distributed throughout the gland, but is modest in volume. It is more prominent around the larger ducts and blood vessels. With the development of pancreatic disease, particularly chronic pancreatitis, the extracellular matrix expands to prominence and alters its characteristics.[17]

The extracellular matrix is composed of fibers and macromolecules with associated cells. Fibroblasts constitute a major cell type. In the normal situation, small numbers of macrophages, neutrophils, and lymphocytes may lie scattered at random. Mast cells, plasma cells, and fat cells may be present. Collagen fibers and to a lesser extent elastic fibers are distributed in the matrix.

A multitude of molecules constitute the relatively permanent part of the amorphous extracellular matrix or represent substances passing through the matrix. The amorphous intercellular ground substance, rich in glycoproteins and proteoglycans, is a viscous fluid that fills all of the space between the formed elements (cells and fibers). Fibronectin binds cells and collagen; laminin, in basal laminae, helps in the adherence of epithelial cells. Proteoglycan molecules, with a protein core to which are connected linear polysaccharides, are quite large. They bind water avidly and form connections to formed elements. Substances passing from blood vessels to pancreatic epithelial cells, or in the opposite direction, pass through the ground substance that intervenes between the vasculature and exocrine or endocrine cells.

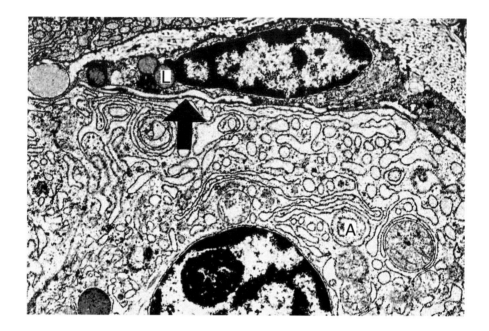

Figure 3.7 Spindle-shaped periacinar cell (pancreatic stellate cell) often shows myofibroblastic characters (arrow) with a few lipid droplets (L) in the cytoplasm and collagen fibrils at the periphery. ×17,200.

Pancreatic stellate cells (Figure 3.7) are identified in the normal organ by their shape (stellate), their content of lipid vesicles rich in retinol and retinyl esters, and the presence in their cytosol of the intermediate filament protein vimentin.[38] Like hepatic stellate cells, these cells, which are thought to be a homology of the so-called Ito cells,[39,40] are presumed to be activated to produce fibrosis, characteristic of cirrhosis and chronic pancreatitis. Activation of pancreatic stellate cells causes loss of the lipid accumulations and accumulation of smooth muscle alpha actin, producing cells with the features of myofibroblasts.[38] Stellate extensions of pancreatic stellate cells and the presence of lipid droplets may be demonstrated by electron microscopy. The actin filaments in activated stellate cells may be similarly demonstrated.

3.10 BLOOD VESSELS

3.10.1 Arteries

Three important groups of arteries supply the pancreas. The first group consists of branches of the gastroduodenal artery, including the superior pancreaticoduodenal artery. The second consists of branches of the inferior pancreaticoduodenal artery, which is derived from the superior mesenteric artery. The third consists of branches from the splenic artery; these supply mainly the body and tail of the pancreas from the posterior-superior surface. The pancreaticoduodenal artery supplies primarily the head, forming arcade-like shunts there.

Branches directly or secondarily from these give off smaller arteries that form arcades. Interlobular arteries branch from the arcades. As they traverse the connective tissue between lobules, they provide arterioles that supply the microvasculature of the lobule.[41]

The interacinar capillary network receives its blood supply from the intralobular arterioles. The capillaries that form this plexus display a wall composed of endothelial cells of varying thickness. Some "windows" or fenestrae are found in the capillaries and are closed by diaphragms. Transport vesicles are common in the cytoplasm of endothelial cells. A basal lamina surrounds each capillary, providing a boundary between endothelial cell and extracellular matrix (Figure 3.8).

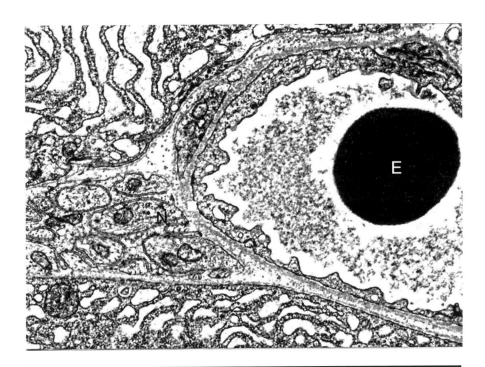

Figure 3.8   An erythrocyte (E) lies in the lumen of a capillary in the pancreas. Nerve fibers (N) lie close to the endothelial wall of the capillary, which is thin and contains vesicles. Rough endoplasmic reticulum of acinar cells is in the upper left and bottom. ×17,200. (From Morohoshi and Bockman.[5] Used with permission.)

The pattern of pancreatic microvasculature is made more complex by the insertion of islets in the circulation. Islets are supplied by a glomerulus of capillaries that provide a rich blood supply to the component cells. The islet glomerulus is supplied by one or more arterioles. The fenestrated capillaries branch and anastomose as they come into close approximation to the islet cells. The islet glomerulus is continuous with the interacinar plexus through capillaries that pass through the layer of connective tissue surrounding the islet. Thus, the capillary bed of the islet becomes continuous with the capillary bed surrounding acini.

3.10.2   Veins

After circulating through the interacinar capillary network, blood drains through venules and veins that lie alongside the interlobular arteries (see Chapter 4). These veins are tributaries of the hepatic portal system. Lobular circulation drains into small veins in the interlobular connective tissue. These are continuous with a venous plexus that surrounds larger ducts. The capillary plexus supplying the ducts is derived from the venous plexus.[42] Thus, ducts are supplied in part with venous blood that has been through the capillaries around islets and acini, and in part with arterial blood from branches of the interlobular arteries.

Veins join to form larger vessels that eventually empty into the splenic and superior mesenteric veins before they join to form the hepatic portal vein. In the liver, blood that has drained from capillaries in the pancreas traverses the hepatic sinusoids. It is clear that endocrine products from pancreatic islets are present in high concentrations within the microvasculature of the pancreas and of the liver.

3.11   LYMPHATICS

The lymphatic system of the pancreas is anything but prominent. However, it does provide a route for the drainage of some amount of fluid during normal functioning. Under conditions of pancreatic disease, this route may become more important. Pancreatic cancer can spread to lymph nodes near the pancreas and then proceed to more distant nodes. Cellular debris, toxic substances, and cells can enter lymphatics and be conducted away from the pancreas when inflammation causes damage to the gland.

Lymphatic capillaries are formed by thin endothelial cells that surround a lumen variable in size and shape. The edges of the endothelial cells tend to overlap loosely rather than be joined by tight junctions as might be expected in blood capillaries. A basal lamina is present around the lymphatic capillary, but it may not be continuous. Thin microfilaments may connect to the outside of the lymphatic endothelial cells from the surrounding connective tissue, providing a mechanism for enlarging the capillary under conditions of edema. Lymph drainage from the pancreas head reaches the anterior and posterior pancreaticoduodenal nodes (Node 13 and Node 17). The lymph flow from the anterior pancreaticoduodenal nodes reaches the inferior head lymph nodes (Node 14 and Node 15) or flows directly into the juxtaaortic and paraaortic lymph nodes via the intestinal lymph trunks. Most of the lymphatic flow from the posterior pancreaticoduodenal nodes reaches the juxtaaortic and paraaortic nodes. The lymphatics from the pancreatic body and tail flow into the juxtaaortic and paraaortic nodes via the splenic and superior body lymph nodes (Node 7 and Node 11). Red blood cells and increased levels of pancreatic enzymes appear in the thoracic duct during hemorrhagic pancreatitis.

3.12 NERVES

The pancreas is supplied with sympathetic, parasympathetic, and visceral sensory nerves. The pancreatic nerves are mainly unmyelinated (Figure 3.9), although some myelinated fibers are present. Bundles of unmyelinated nerves that travel through the interlobular connective tissues innervate the acini and ducts. Islets are innervated by both sympathetic and parasympathetic fibers.

Sympathetic fibers mainly travel the greater (originating from the T4 to T10 sympathetic ganglia) and lesser splanchnic nerves (originating from the T9 to L2 sympathetic ganglia). These nerves pass through the diaphragmatic crura to enter the celiac plexus and ganglia and then pass into the pancreatic parenchyma. Parasympathetic fibers originate from the vagus nerve. The vagus nerve fibers related to the pancreas pass through the celiac plexus or reach the pancreas directly via the hepatic and gastric rami of the vagus nerve.

Visceral sensory fibers accompany the efferent fibers in both parasympathetic and sympathetic contributors. The nerve fibers serving the pancreas join with each other to form a plexus around the arteries that supply the organ. The nerves distribute with the blood vessels and then leave them to achieve their destinations throughout the gland.

Sympathetic, parasympathetic, and visceral sensory fibers run within each nerve. The multiple nerve fibers (axons) within each nerve are wrapped within extensions of Schwann cells, usually several fibers per Schwann cell. A small amount of extracellular matrix occupies the area between the units of Schwann cells with their contained axons. Many units are bundled together to form a nerve. The periphery of the nerve is formed by the perineurium, which is a wall, commonly multilayered, of flattened epithelioid cells joined by tight junctions in each layer. A basal lamina is present on both the external and internal sides of each layer. Thus, a

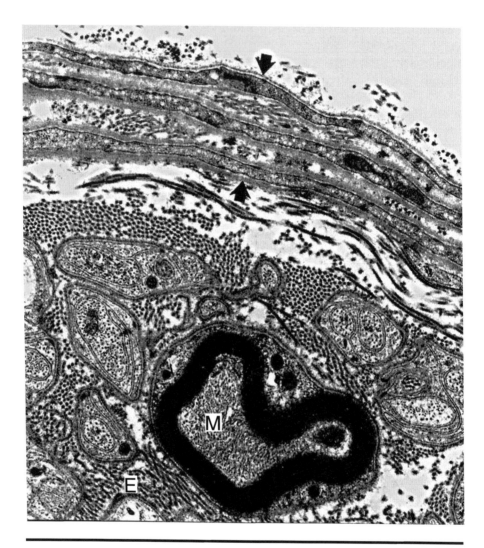

Figure 3.9 Cross-section of a large pancreatic nerve observed by electron microscopy. Multiple units of unmyelinated fibers, consisting of axons and Schwann cells, surround one myelinated axon (M). The endoneurium is separated from the general extracellular matrix outside the nerve (top of micrograph) by the perineurium (total of structures between the arrows). The perineurium consists of three layers of thin epithelioid cells, each of which has a basal lamina along each surface. Extracellular matrix lies between the layers of perineurium. The perineurium provides a barrier allowing a specialized compartment inside. ×20,880. (From Bockman et al.[43] Used with permission of the American Gastro-enterological Association.)

perineurium consisting of three layers would have six basal laminae. In addition, a thin layer of extracellular matrix separates layers. The intact perineurium provides a distinct barrier that enables the interior of the nerve to possess a unique microenvironment. The compartment within the nerve is separated from the open compartment outside the perineurium, which is the generalized extracellular matrix of the pancreas.[43]

Nerve cell bodies within the substance of the pancreas usually are grouped together as ganglia. The intrapancreatic ganglia are collections of secondary neurons of the parasympathetic system; they receive terminals from the vagus nerve and extend their fibers through the extracellular matrix. Parasympathetic and sympathetic fibers within the pancreas mediate pancreatic secretion and regulate blood flow, mainly by affecting the smooth muscle of blood vessels.

The visceral afferent fibers of the pancreas normally function silently. Although they may be involved in reflexes necessary for secretion and regulation of blood flow, their activity does not reach consciousness. However, pancreatic disease frequently causes pain that is mediated by the afferent fibers. The fibers may respond to edema, lack of oxygen, the presence of noxious substances, and direct damage.[43] In the case of direct damage, the internal compartment is no longer separate from the external, admitting any number of noxious substances into the nerve, potentially initiating pain, secretion, and alteration of vascular flow due to aberrant stimulation of affected nerves.

REFERENCES

1. Bockman D.E. 1984. Pathomorphology of pancreatitis: regressive changes in an acutely or chronically damaged epithelial organ. In *Pancreatitis — concepts and classification*. Gyr K.E., Singer M.V., and Sarles H. Eds. Amsterdam: Elsevier, pp. 11–15.

2. Bockman D.E., Müller M., Büchler M.W., Friess H., and Beger H.G. 1997. Pathological changes in pancreatic ducts from patients with chronic pancreatitis. *Int. J. Pancreatology* 21:119–126.

3. Bockman D.E. 1976. Anastomosing tubular arrangement of the exocrine pancreas. *Am. J. Anat.* 147:113–118.

4. Akao S., Bockman D.E., Lechene de la Porte P., and Sarles H. 1986. Three dimensional pattern of ductuloacinar associations in normal and pathological human pancreas. *Gastroenterology* 90:661–668.

5. Morohoshi T. and Bockman D.E. 1994. Microanatomy and Fine Structure of the Pancreas. In *Atlas of exocrine pancreatic tumors*. Pour P.M., Konishi Y., Klöppel G., and Longnecker D.S. Eds. Tokyo: Springer. pp. 17–30.

6. Bockman D.E. 1996. Microanatomy of the pancreas. In *Cellular interrelationships in the pancreas — implications for islet transplantation*, Rosenberg L. and Duguid W.P. Eds. Austin, TX: R.G. Landes. pp. 9–27.

7. Bockman D.E. 1998. Histology and fine structure. In *The pancreas*, Beger H.G., Warshaw A.L., Büchler M.W., Carr-Locke D.L., Neoptolemos J.P., Russell C., and Sarr M.G. Eds. Oxford, U.K.: Blackwell Science. pp. 19–26.

8. Palade G. 1975. Intracellular aspects of the process of protein synthesis. *Science* 189:347–358.

9. Luthen R., Niederau C., Niederau M., Ferrell L.D., and Grendell J.M. 1995. Influence of ductal pressure and infusates on activity and subcellular distribution of lysosomal enzymes in the rat pancreas. *Gastroenterology* 109:573–581.

10. Steer M.L. 1997. Pathogenesis of acute pancreatitis. *Digestion* 58 Suppl. 1:46–49.

11. Müller M.W., McNeil P.L., Büchler M.W., Friess H., Beger H.G., and Bockman D.E. 1999. Membrane wounding and early ultrastructural findings. In *Acute pancreatitis: novel concepts in biology and therapy*. Büchler M.W., Uhl W., Friess H., and Malfertheiner P. Eds. Berlin: Blackwell. pp. 27–34.

12. Nagata A. and Monno S. 1984. Ultrastructure of pancreatic duct and pancreatic ductal cells [in Japanese]. *The Cell* 16:397–402.

13. Bockman D.E., Büchler M., and Beger H.G. 1986. Structure and function of specialized cilia in the exocrine pancreas. *Int. J. Pancreatol.* 1:21–28.

14. Solcia E., Capella C., and Klöppel G. 1992. Tumors of the exocrine pancreas. Gross anatomy. In *Atlas of tumor pathology*. Rosai J. and Sobin L.H. Eds. 3rd series, Washington DC: AFIP.

15. Bouwens L. 1998. Cytokeratins and cell differentiation in the pancreas. *J. Pathol.* 184:234–239.

16. Ulrich A.B., Schmied B.M., Matsuzaki H., El-Metwally T., Moyer M.P., Ricordi C., Adrian T.E., Batra S.K., and Pour P.M. 2000. Establishment of human pancreatic ductal cells in a long-term culture. *Pancreas* 21:358–368.

17. Kennedy R.H., Bockman D.E., Uscanga L., Choux R., Grimaud J.-A., and Sarles H. 1987. Pancreatic extracellular matrix alterations in chronic pancreatitis. *Pancreas* 2:61–72.

18. Klöppel G. 1984. Anatomy and physiology of the endocrine pancreas. In *Pancreatic pathology*. Klöppel G. and Heitz P.U. Eds. London: Churchill Livingstone, pp.133–153.

19. Klimstra D.S. 1997. Pancreas. In *Histology for pathologists*, Sternberg S.S. Ed. 2nd ed. Philadelphia: Lippincott-Raven, pp. 613–647.

20. Lammert E., Cleaver O., and Melton D. 2001. Induction of pancreatic differentiation by signals from blood vessels. *Science* 294:564–567.

21. Polak M., Bouchareb-Banaei L., Scharfmann R., and Czernichow P. 2000. Early pattern of differentiation in the human pancreas. *Diabetes* 49:225–232.

22. Arias A.E. and Bendayan M. 1993. Differentiation of pancreatic acinar cells into duct-like cells in vitro. *Lab. Invest.* 69:518–530.

23. Bockman D.E., Black O., Mills L.R., and Webster P.D. 1978. Origin of tubular complexes developing during induction of pancreatic carcinoma by 7,12-dimethylbenz(a)anthracene. *Am. J. Pathol.* 90:645–658.

24. Bockman D.E., Boydston W.R., and Anderson M.C. 1982. Origin of tubular complexes in human chronic pancreatitis. *Am. J. Surg.* 144:243–249.

25. Bockman D.E., Guo J., Büchler P., Müller M.W., Bergmann F., and Friess H. 2003. Origin and development of the precursor lesions in experimental pancreatic cancer in rats. *Lab. Invest.* 83:853–859.

26. Bockman D.E. and Merlino G. 1992. Cytological changes in the pancreas of transgenic mice overexpressing transforming growth factor alpha. *Gastroenterology* 103:1883–1892.

27. DeLisle R.C. and Logsdon C.D. 1990. Pancreatic acinar cells in culture: expression of acinar and ductal antigens in a growth-related manner. *Eur. J. Cell. Biol.* 51:64–75.

28. Hall P.A. and Lemoine N.R. 1992. Rapid acinar to ductal transdifferentiation in cultured human exocrine pancreas. *J. Pathol.* 166:97–103.

29. Isaksson G., Ihse I., and Lundquist I. 1983. Influence of pancreatic duct ligation on endocrine and exocrine rat pancreas. *Acta. Physiol. Scand.* 117:281–286.

30. Rooman I., Heremans Y., Heimberg H., and Bouwens L. 2000. Modulation of rat pancreatic acinoductal transdifferentiation and expression of PDX-1 in vitro. *Diabetologia* 43:907–914.

31. Sandgren E.P., Luetteke N.C., Palmiter R.D., Brinster R.L., and Lee D.C. 1990. Overexpression of TGFα in transgenic mice: induction of epithelial hyperplasia, pancreatic metaplasia, and carcinoma of the breast. *Cell* 61:1121–1135.

32. Schmied B.M., Matsuzaki A., Ding H., Ding X., Ricordi C., Weide L., Moyer M.P., Batra S.K., Adrian T.E., and Pour P.M. 2001. Transdifferentiation of human islet cells in a long-term culture. *Pancreas* 23:157–171.

33. Wagner M., Greten F.R., Weber C.K., Koschnick S., Mattfeldt T., Deppert W., Kern H., Adler G., and Schmid R.M. 2001. A murine tumor progression model for pancreatic cancer recapitulating the genetic alterations of the human disease. *Genes Develop.* 15:286–293.

34. Wang R., Li J., and Rosenberg L. 2001. Factors mediating the transdifferentiation of islets of Langerhans to duct epithelial-like structures. *J. Endocrinol.* 171:309–318.

35. Wang R.N., Kloppel G., and Bouwens L. 1995. Duct to islet-cell differentiation and islet growth in the pancreas of duct-ligated adult rats. *Diabetologia* 38:1405–1411.

36. Yuan S., Rosenberg L., Paraskevas S., Agapitos D., and Duguid W.P. 1996. Transdifferentiation of human islets to pancreatic ductal cells in collagen matrix culture. *Differentiation* 61:67–75.

37. Bockman D.E., Boydston W.R., and Parsa I. 1983. Architecture of human pancreas: implications for early changes in pancreatic disease. *Gastroenterology* 85:55–61.

38. Bachem M.G., Schneider E., Gross H., Weidenbach H., Schmid R.M., Menke A., Siech M., Beger H., Grunert A., and Adler G. 1998. Identification, culture, and characterization of pancreatic stellate cells in rats and humans. *Gastroenterology* 115:421–432.

39. Watari N. 1984. Ultrastructural studies on the connective tissues in the pancreas [in Japanese]. *The Cell* 166:402–408.

40. Morohoshi T. and Kanda M. 1985. Periacinar fibroblastoid cell — its action on early stage of alcoholic pancreatitis [in Japanese]. *Tann to Sui* 6:1205–1211.

41. Bockman D.E. 1992. Microvasculature of the pancreas: relation to pancreatitis. *Int. J. Pancreatol.* 12:11–21.

42. Lifson N. and Lassa C.V. 1981. Note on the blood supply of the ducts of the rabbit pancreas. *Microvasc. Res.* 22:171–176.

43. Bockman D.E., Büchler M., Malfertheiner P., and Beger H.G. 1988. Analysis of nerves in chronic pancreatitis. *Gastroenterology* 94:1459–1469.

4

VASCULAR ANATOMY OF THE PANCREAS

Yukio Mikami, Kazunori Takeda, Takuro Murakami, and Seiki Matsuno

CONTENTS

4.1 INTRODUCTION

Most injuries to the pancreas are caused by chemicals that reach the pancreas through arteries. Consequently, knowledge of the vascular structure of the pancreas is of immense importance in understanding the site and the extent of the damage.

4.1.1 Arterial Anatomy

The pancreas receives its blood supply from branches of the celiac artery (CEA) and the superior mesenteric artery (SMA) (Figure 4.1). The gastroduodenal artery (GDA) generally departs from the common hepatic artery

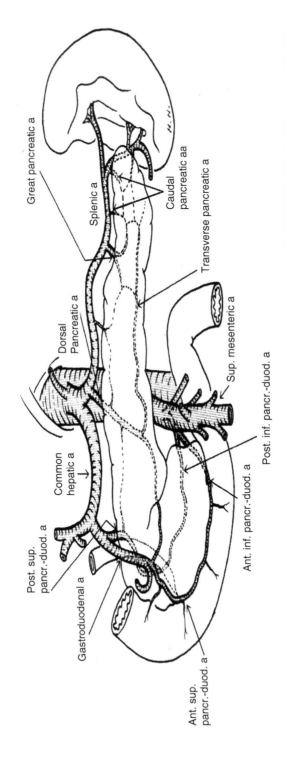

Figure 4.1 Pancreatic arterial anatomy. a, Artery; ant., anterior; sup., superior; post., posterior; inf., inferior. (From Pour et al.,[5] Used with permission.)

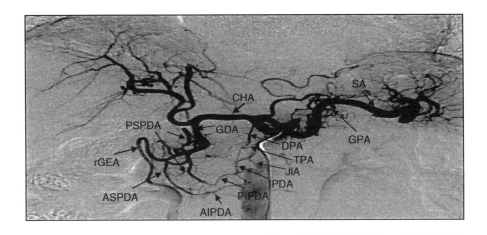

Figure 4.2   Selective celiac arteriogram. SA, splenic artery; GDA, gastroduodenal artery; CHA, common hepatic artery; PSPDA, posterior superior pancreaticoduodenal artery; rGEA, right gastroepiploic artery; ASPDA, anterior posterior pancreaticoduodenal artery; AIPDA, anterior posterior pancreaticoduodenal artery; PIPDA, posterior inferior pancreaticoduodenal artery; IPDA, inferior pancreaticoduodenal artery; J1A, first jejunal artery; DPA, dorsal pancreatic artery; TPA, transverse pancreatic artery; GPA, great pancreatic artery.

(CHA) and first gives off the posterior superior pancreaticoduodenal artery (PSPDA) near the upper border of the pancreas (Figure 4.2). The GDA then gives off the right gastroepiploic artery (rGEA) and turns into the anterior posterior pancreaticoduodenal artery (ASPDA). The ASPDA descends on the anterior surface of the head of the pancreas to join the anterior posterior pancreaticoduodenal artery (AIPDA). The PSPDA runs in front of the common bile duct (CBD) from left to right and descends on along the right side of the CBD on the posterior aspect of the pancreas. The PSPDA then runs behind the CBD from right to left to join the posterior inferior pancreaticoduodenal artery (PIPDA). The superior and inferior pancreaticoduodenal arteries (ASPDA, PSPDA, AIPDA, and PIPDA) form the anterior and posterior arterial arcades in the head of the pancreas. The ASPDA and PSPDA are consistent in origin, but the AIPDA and PIPDA arise separately or have a common trunk as the inferior pancreaticoduodenal artery (IPDA) from the SMA. And the IPDA also has variations, arising independently from the right side of the SMA or having a common artery composed of the IPDA and the first jejunal artery (J1A) that arises from the left side of the SMA. After branching off the J1A, the IPDA runs behind the SMA toward the right side and divides into the AIPDA and the PIPDA.[1] These variations in the IPDA have been described in detail by Murakami and coworkers (Figure 4.3).[2] The IPDA was found in 80% of 125 autopsy subjects, a common artery composed of the IPDA and the J1A in 56%, and

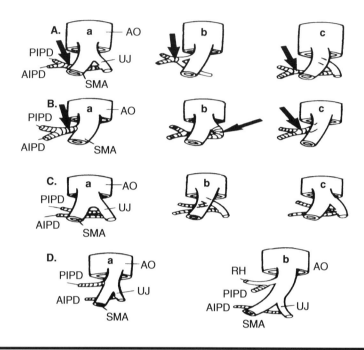

Figure 4.3 Variations in the topographical relationships of the orgin of the inferior pancreaticoduodenal arteries. Anterior aspects. Four types (A, B, C, and D) and 11 subtypes (a, b, and c) were classified in 125 specimens examined. Every incidence described below was estimated in the total specimens. Type A (55.6%) has the inferior pancreaticoduodenal artery (IPD, arrow) arising from (forming a common trunk with) the upper jejunal artery (UJ). Subtypes a (48.4%), b (6.4%), and c (0.8%) (a + b + c = A) of Type A represent differences in the topographical relationships of arteries. Type B (24.2%) is a "typical" pattern seen in many textbooks, in which the IPD arises directly from the superior mesenteric artery (SMA). This type is also divided into three subtypes: a (17.8%), b (4.8%), and c (1.6%) (a + b + c = B). The anterior posterior pancreaticoduodenal artery (AIPD) and posterior inferior pancreaticoduodenal artery (PIPD) originate from the SMA independently in Type C (3.3%) and its subtypes, a (1.6%), b (0.9%), and c (0.8%) (a + b + c = C). Type D (16.9%) consists of other patterns. In Type D a (11.3%), the SMA issues the PIPD, whereas the AIPD arises from (forms a common trunk with) the UJ. In Type D b (5.6%), the PIPD issues from the right (accessory) hepatic artery (RH) arising from the SMA. This type of RH was seen in 12.7% of the specimens examined; 44.1% of the specimens with an RH showed the Type D Ao, Aorta. (From Murakami et al.[2] Used with permission.)

the IPDA arose independently from the SMA in 24% of the subjects. The posterior arcade passes behind the CBD and is farther from the duodenum and in a more cephalad position than the anterior arcade. The head of the pancreas and duodenum are supplied with blood mainly from these arcades.

The dorsal pancreatic artery (DPA) lies behind the neck of the pancreas, arising from the splenic artery (SA), the CEA, the CHA, or the SMA. The GDA and the DPA give off branches and form the arcade along the superior margin of the pancreas. The branch from the GDA is called the superior pancreatic branch and the branch from the DPA is called the suprapancreatic branch.[3] The DPA then runs downward to the lower border of the pancreas and divides into the left and right branches. The DPA provides the main blood supply to the neck and body of the pancreas.

The transverse pancreatic artery (TPA) has the left branch of the DPA in 90% of the cases,[4] and it departed from the GDA near the point where the GDA divides into rGEA and the ASPDA. The TPA runs along the inferior margin of the pancreas to anastomize with the great pancreatic artery (GPA) and the caudal pancreatic arteries (CPAs) to form the arcade. This arcade is called the prepancreatic arcade.

The GPA is the greatest artery among the branches of the SA that course along the superior margin of the body and tail of the pancreas. It usually arises around the border between the body and tail of the pancreas and divides into the left and right branches to anastomize with the TPA, the DPA, and the CPAs.

The CPAs are small branches of the SA or the left gastroepiploic artery (lGEP). The TPA, the GPA, and the CPAs supply blood to the body and tail of the pancreas.[5]

In relation to the arterial anatomy of the pancreas, it is important to understand the variations of the CHA. The CHA sometimes departs from the SMA and divides into the left and right branches in the hilus of the liver, and the right hepatic artery sometimes departs from the SMA. The CHA has many other variations as shown in Figure 4.4.[6]

4.1.2 Venous Anatomy

The venous blood of the pancreas drains into the portal system around the pancreas; the splenic vein (SV), the superior mesenteric vein (SMV), the inferior mesenteric vein (IMV), and the portal vein (PV) (Figure 4.5). In general, the pancreatic veins run parallel to the arteries and lie superficial to them.

The SV runs inferior to the splenic artery along the posterior aspect of the pancreas to join the SMV, which passes anterior to the inferomedial aspect of the uncinate process to form the PV behind the neck of the pancreas.[5]

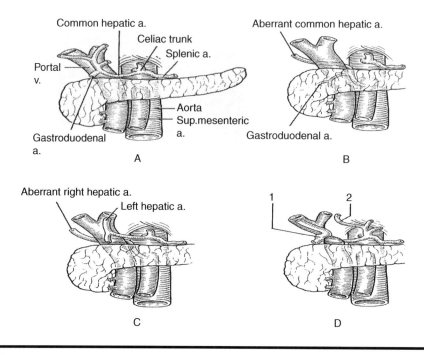

Figure 4.4 Some variations of the hepatic arteries in relation to the pancreas. A: Normal configuration. B: Aberrant common hepatic artery. C: Aberrant right hepatic artery. D: 1, common hepatic artery looping around the portal vein from behind (causing compression of the vein); 2, aberrant left hepatic artery arising from left gastric artery. (From Trede and Carter.[6] Used with permission.)

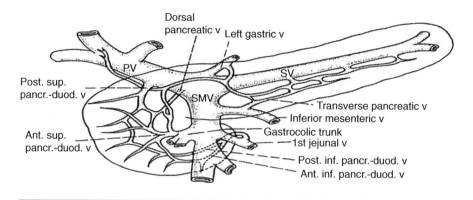

Figure 4.5 Pancreatic venous anatomy. PV, Portal vein; SMV, superior mesenteric vein; SV, splenic vein. (From Pour et al.[5] Used with permission.)

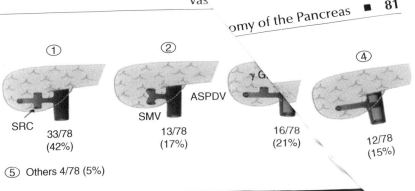

Figure 4.6 **Variations and their incidence in the gastrocolic trunk of H̖ SRC, Superior right colic vein; GEV, gastroepiploic vein; ASPDV, anterior supe. pancreaticoduodenal vein. (From Kimura et al.[7] and Gillot et al.[8] Used with permission.)**

The anterior superior pancreaticoduodenal vein (ASPDV) generally terminates in the SMV via the gastrocolic trunk. The gastrocolic trunk is called the gastrocolic trunk of Henle or Henle's trunk, because Henle reported a common trunk of the superior right colic vein (SRCV) and the right gastroepiploic vein (rGEV) in 1868.[7] In 1964, Gillot et al. reported that a gastrocolic trunk was found in about 60% of 78 subjects (Figure 4.6).[8] In 1912, Descomps and DeLalaubie reported that the ASPDV may join the gastrocolic trunk in some subjects.[9]

The anterior inferior pancreaticoduodenal vein (AIPDV) drains into the first jejunal vein or the SMV.

The posterior superior pancreaticoduodenal vein (PSPDV) terminates in the right posterior wall of the PV. The ASPDV and the AIPDV form an arcade on the anterior surface of the pancreas. However, Takamuro et al. reported that the PSPDV and the posterior inferior pancreaticoduodenal vein (PIPDV) sometimes formed arcades and at other times did not.[10] The PIPDV enters the first jejunal vein (J1V) frequently to form a common trunk with the AIPDV.

Sometimes a large vein running from the posterior aspect of the pancreas to the junction of the SMV and the PV is observed and this vein is called the dorsal pancreatic vein (DPV).

The transverse pancreatic vein (TPV) branches off the small veins together with the DPV and it terminates in the SMV, the IMV, or occasionally, the SV or the gastrocolic trunk. The SV receives some short pancreatic venous branches and these branches anastomize the TPV and drain the blood from the body and tail of the pancreas.

In relation to the venous anatomy of the pancreas, it is important to understand the variations in the IMV, SV, and SMV. Kimura et al. reported that the IMV joined the SV in 34%, the SMV in 42%, and the confluence of the SV and the SMV in 24% of 38 autopsy subjects.[1]

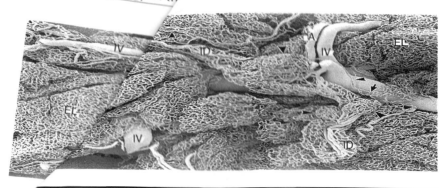

Figure 4.7 Overview of a replicated blood vascular bed of the human pancreas (a caudal segment exposed by dissection, 25-year-old woman). Note that the blood vascular plexuses of the exocrine lobules (EL) and secretory ducts (interlobular ducts, ID) are thoroughly reproduced together with their connecting interlobular arteries (IA) and veins (IV) and that the exocrine lobules (large arrowheads) closely associated with the ducts are smaller than the other lobules. The small arrowheads indicate the interlobular blood vascular plexuses, some capillaries of which continue into the lobular capillaries (arrows). ×40. EI, extralobular endocrine islet (islet of Langerhans) or its vascular plexus; EL, exocrine lobule or its vascular plexus; IA, interlobular artery; ID, interlobular duct or its vascular bed; II, interlobular endocrine islet or its vascular plexus; IV, interlobular vein; LA, lobular artery; LD, lobular duct or its vascular bed; LV, lobular vein; PA, periductal artery; PV, periductal vein; a, afferent vessel of the islet; e, efferent vessel of the islet; s, surface capillary network of the extralobular islet; v, venous efferent vessel (emissary vein) of the extralobular islet; la, branch of the lobular artery; lv, branch of the lobular vein. Abbreviations summarized here are used in Figure 4.7 through Figure 4.13. (From Murakami et al.[11] Used with permission.)

4.2 MICROCIRCULATION OF THE PANCREAS

The microcirculation system of the pancreas is studied by electron microscopic examination of vascular casts of human pancreas. The results are schematically illustrated in Figure 4.13.[11]

4.2.1 Lobular Vascular Bed

The vascular bed of the exocrine lobules (lobular plexus) consists of fine capillaries (Figure 4.7) that receive one or more afferent vessels (lobular arteries) from the interlobular arteries and issue one or more efferent vessels (lobular veins) continuous with the interlobular veins (Figure 4.8 and Figure 4.9). The lobular plexus occasionally possess insignificant fine connections with the interlobular or periductal plexus (Figure 4.7 and Figure 4.8).

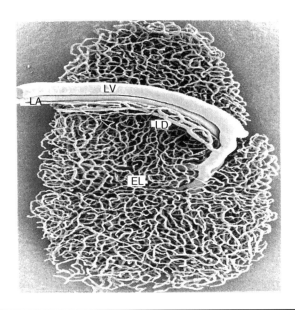

Figure 4.8 A lobular blood vascular plexus isolated together with its connecting lobular artery (LA), lobular vein (LV), and ductal plexus (LD). Note that the lobule is fairly independent. The ductal plexus drains at the hilus of the lobule into a branch of the lobular vein (large arrowhead). The small arrowhead indicates a rare fine capillary connection between the lobular and ductal plexuses. ×80. (From Murakami et al.[11] Used with permission.)

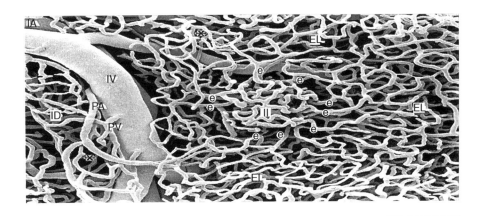

Figure 4.9 An intralobular islet (II) exposed in the lobular surface. Note that the islet emits marked efferent vessels (e) that continue, as the insulo-acinar portal vessels, into adjacent lobular capillaries (EL). * Injection defects. ×200. (From Murakami et al.[11] Used with permission.)

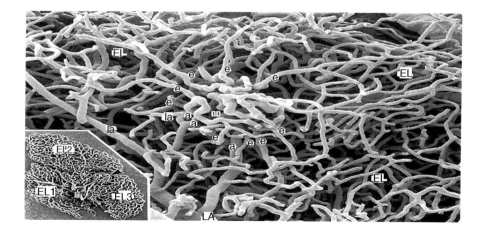

Figure 4.10 An intralobular islet (II) with three afferent vessels (a). The afferent vessels on the left hand run from the superficial aspect into the insular capillaries, and the right-hand one runs deep into the islet. This islet also emits many insulo-acinar portal vessels (e) continuous with the adjacent lobular capillaries (EL). Inset shows an isolated cluster of three lobules (EL1 to EL3). Note in this inset that only the EL3 lobule contains an islet (arrowhead). ×250. Inset: ×40. (From Murakami et al.[11] Used with permission.)

The size of the lobular plexus varies widely. Large lobules measuring more than 0.5 mm in length contain numerous fine capillaries (lobular capillaries), and small ones 500 μm or less in length contain a small number of lobular capillaries. Large lobules are typically located in the superficial layers of the pancreas, whereas smaller ones are in the deeper layers of the organ or in close association with the interlobular ducts (Figure 4.7).

4.2.2 Intralobular Islets and Their Blood Vessels

The vascular network in the islets of Langerhans (insular plexus) consists of thicker (sinusoidal) capillaries conglomerated into a globular mass, measuring 30 to 250 μm (usually, 100 to 150 μm) in diameter (Figure 4.9 to Figure 4.11). The peripheral or cortical capillaries of the intralobular islets issue numerous efferent vessels that radiate into the capillary network in the surrounding exocrine tissues (Figure 4.9 to Figure 4.11). These efferent vessels of the intralobular islets are relatively long, straight, or gently winding capillaries. However, as these vessels connect the intra-lobular islets and the lobular capillaries covering the exocrine acini and intralobular ducts, they therefore should be described as insulo-acinar portal vessels.[12–14] Some portal vessels arise deep in the islets (Figure 4.11),

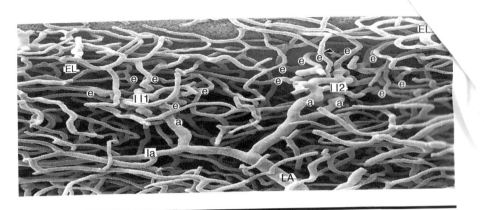

Figure 4.11 Two intralobular islets (II1, II2) as found in the same lobule. The II2 islet receives two afferent vessels (a) and one of its efferent vessels (arrowhead) arises deep in the islet. Even in these islets, all of the efferent vessels (e) (including that indicated by the arrowhead) continue, as the insulo-acinar portal vessels, into the adjacent lobular capillaries (EL). ×280. (From Murakami et al.[11] Used with permission.)

others more superficially. In humans, the intralobular islets issue no efferent vessels directly draining into the veins. The portal vessels are characteristically slender, being never thicker than the capillaries in the islets and as thick as or slightly thicker than the lobular capillaries (Figure 4.9 to Figure 4.11).

The number of the portal vessels varies widely among islets. Generally, larger islets possess a larger number of portal vessels. Larger islets exceeding 200 μm in diameter issue 30 or more portal vessels, whereas a small islet consisting of a few capillary loops issue 3 to 7 portal vessels. Usually, a part of the lobular capillary network is supplied with the portal vessels of the islets, and the remaining parts directly receive lobular arteries; both portions of the lobular capillaries are drained by the lobular veins. On rare occasions, the entire extent of the lobular capillary network is supplied by the portal vessels. In these latter cases, the lobular artery or arteries take the exclusive role as the afferent vessels of the islets.

The islets identified by this characteristic feature in the vascular casts were usually located intralobularly (Figure 4.9 to Figure 4.11), embedded in the general capillary network of the exocrine tissue (Figure 4.11). Only rarely did an intralobular islet expose its body to the lobular surface (Figure 4.9 and Figure 4.10).

By surveying many lobules with a light microscope, we are able to find a clearly definable insular plexus in one among seven lobules (Figure 4.10 inset). When a lobule revealed an islet, it is usually single, but occasionally several islets can be found in a lobule (Figure 4.11). Thicker

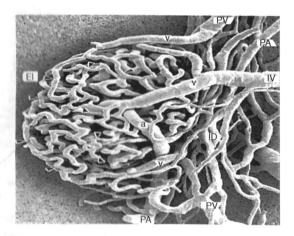

Figure 4.12 An extralobular islet (EI). Note that this islet is provided with a set of fine capillaries (arrowheads), which receives the sinusoidal capillaries of the islet and confluence into the emissary veins (v) finally continuous with the periductal (PV) or interlobular (IV) veins. ×250. (From Murakami et al.[11] Used with permission.)

lobules likely possess their islets more consistently. Nevertheless, it is reasonable to say that a considerable number of lobules in the human pancreas are devoid of any islets. Moreover, the range of the portal vessels is limited and rarely covers the entire exocrine lobules. Thus, it is suggested that, in humans, the insular control over the exocrine pancreas generally is valid in restricted areas of the lobule.

In humans, it is rare for an islet to be located interlobulary (extralobulary) or periductally (i.e., between the lobules or along the interlobular duct) (Figure 4.12). Species differences in this regard are conspicuous. In the mouse and rat, many islets are located interlobularly along the excretory ducts and drained via their surface network of fine capillaries into the interlobular or periductal veins.[15,16]

The intralobular islet receives one to three afferent vessels (insular arterioles) from the lobular artery (Figure 4.9 to Figure 4.11). These afferent vessels enter deep into the islet and form a conglomeration of sinusoidal capillaries. In some other islets, the afferent vessels divide superficially on one pole of the islet and continue into the sinusoidal capillaries. In typical cases, the afferent vessels break up into superficial and deep branches, which supply the islets both from the superficial and deep aspects. When the islet receives two or more afferent vessels, one often runs deep into the insular plexus and the other splits into its superficial aspects (Figure 4.10). In certain animal species, the pattern of insular microcirculation is known to be regular.[12,13,17,18] However, in human islets,

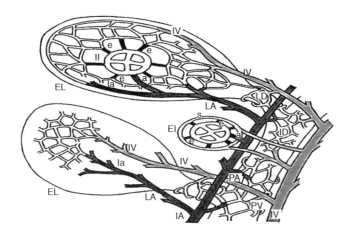

Figure 4.13 **A diagram showing the vascular arrangements of the human pancreas. From the top to the bottom are shown a lobule containing an islet, an extralobular (periductal) islet, and a lobule lacking in an islet. An interlobular duct is illustrated on the right hand. (From Murakami et al.[11] Used with permission.)**

no rule can be found as to whether afferent vessels are primarily connected to the deep or superficial portion of the islet, and A-, B-, and D-cells are rather irregularly intermingled within the islets.

4.2.3 Interlobular Islets and Their Vascular Connections

The human pancreas only occasionally reveals islets located in the interlobular connective tissue. The interlobular (extralobular) islets receive one or more afferent vessels from the interlobular or periductal arteries (Figure 4.12). The afferent arterioles penetrate deep into the islets to form a conglomeration of sinusoidal capillaries. This deep capillary plexus is surrounded and drained by a thin network of fine capillaries (outer capillary meshwork) (Figure 4.12). This marginal network, in turn, issues efferent vessels that are directly continuous with the interlobular or periductal veins (Figure 4.12). One idea proposed is that this capsular network of the interlobular islets is homologous with the lobular capillary bed for the intralobular islets. Thus, in the interlobular islets, the portal vessels may be said to exist between the core plexus of sinusoidal capillaries and the capsular network of thin capillaries.

4.2.4 Periductal Vascular Plexus

The vascular networks surrounding the interlobular and lobular ducts (periductal plexuses) are supplied with periductal arteries and veins that are derived from the interlobular arteries and veins, respectively (Figure 4.7 to Figure 4.9). The terminal segments of the periductal plexus (ductal plexus surrounding the lobular ducts) consist of several capillaries that drain into the lobular veins in or outside the lobular plexus (Figure 4.8). Few capillary connections can be recognized between the lobular and ductal plexuses (Figure 4.8). This indicates that the capillary plexuses of the exocrine lobules and extralobular secretory ducts are independent of each other in terms of the blood supply. Within the lobule, in contrast, the exocrine acini and their connecting intercalated and intralobular secretory ducts are commonly supplied.

REFERENCES

1. Kimura W. 2000. Surgical anatomy of the pancreas for limited resection. *J. Hepatobiliary Pancreat. Surg.* 7:473–479.
2. Murakami G., Hirata K., Takamuro T., Mukaiya M., Hata F., and Kitagawa S. 1999. Vascular anatomy of the pancreaticoduodenal region: a review. *J. Hepatobiliary Pancreat. Surg.* 6:55–68.
3. Bertelli E., Di Gregorio F., Bertelli L., Orazioli D., and Bastianini A. 1997. The arterial blood supply of the pancreas: a review. IV. The anterior inferior and posterior pancreaticoduodenal aa., and minor sources of blood supply for the head of the pancreas. An anatomical review and radiologic study. *Surg. Radiol. Anat.* 19:203–212.
4. Freeny P.C. and Lawson T.L. 1982. *Radiology of the pancreas.* New York: Springer Verlag. pp. 51–97.
5. Pour P.M., Konishi Y., Kloppel G., and Longnecker D.S. 1994. *Atlas of exocrine pancreatic tumors — morphology, biology and diagnosis with an international guide for tumor classification.* New York: Springer Verlag. pp. 1–15.
6. Trede M. and Carter D.C. 1993. *Surgery of the pancreas.* New York: Churchill Livingstone. pp. 23–25.
7. Henle J. 1868. *Handbuch der systematischen Anatomie des Menschen.* Braunschweig: Druck und Verlag von Friedrich Vieweg und Sohn. p. 391 (cited by Gillot et al.).
8. Gillot C., Hureau J., Aaron C., Martini R., and Thaler G. 1964. The superior mesenteric vein. *J. Int. Coll. Surg.* 41:339–369.
9. Descomps P. and De Lalaubie G. 1912. Les veines mesenteriques. *J. de l'Ant. Et physiol. Norm. et path. De l'homme et des animaux* 48:337–376 (cited by Gillot et al.).
10. Takamuro T., Oikawa I., Murakami G., and Hirata K. 1998. Venous drainage from the posterior aspect of the pancreatic head and duodenum. *Okajimas Folia Anat. Jpn.* 75:1–8.
11. Murakami T., Fujita T., Taguchi T., Nonaka Y., and Orita K. 1992. The blood vascular bed of the human pancreas, with special reference to the insulo-acinar portal system. Scanning electron microscopy of corrosion casts. *Arch. Histol. Cytol.* 55:381–395.
12. Fujita T. 1973. Insulo-acinar portal system in the horse pancreas. *Arch. Histol. Jpn.* 35:161–171.

13. Fujita T. and Murakami T. 1973. Microcirculation of monkey pancreas with special reference to the insulo-acinar portal system. A scanning electron microscope study of vascular casts. *Arch. Histol. Jpn.* 35:255–263.

14. Ohtani O. and Fujita T. 1981. Insulo-acinar portal system of the pancreas. A scanning electron microscope study of corrosion casts. *Prog. Clin. Biol. Res.* 59B:111–120.

15. Murakami T., Fujita T., Miyake T., Ohtsuka A., Taguchi T., and Kikuta A. 1993. The insulo-acinar portal and insulo-venous drainage systems in the pancreas of the mouse, dog, monkey and certain other animals: a scanning electron microscopic study of corrosion casts. *Arch. Histol. Cytol.* 56:127–147.

16. Murakami T. and Fujita T. 1992. Microcirculation of the rat pancreas, with special reference to the insulo-acinar portal and insulo-venous drainage systems: a further scanning electron microscope study of corrosion casts. *Arch. Histol. Cytol.* 55:453–476.

17. Bonner-Weir S. and Orci L. 1982. New perspectives on the microvasculature of the islets of Langerhans in the rat. *Diabetes* 31:883–889.

18. Ohtani O., Ushiki T., Kanazawa H., and Fujita T. 1986. Microcirculation of the pancreas in the rat and rabbit with special reference to the insulo-acinar portal system and emissary vein of the islet. *Arch. Histol. Jpn.* 49:45–60.

5

CENTROACINAR CELLS —
A NEGLECTED ESSENTIAL
ELEMENT OF THE PANCREAS

Parviz M. Pour

CONTENTS

5.1   INTRODUCTION

The pancreas is the most complex structure of the body. The presence of a variety of exocrine and endocrine cells has hampered our understanding of the function of the individual cell component and especially their interaction. The available data point to a complex dialogue between the individual exocrine cells, between the endocrine cells, and between the exocrine and endocrine cells.[1,2] The key element involved in a coordinated function of this heterogeneous cell population has remained a mystery. The transdifferentiation ability of individual cells into different exocrine or endocrine phenotypes adds to the complexity of this tissue. Strikingly, one of the most important cellular elements in the regulation of enzyme secretion, the centroacinar cells, has been largely ignored. The

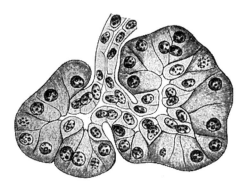

Figure 5.1 Schematic representation of centroacinar cells. Interestingly, two of the centroacinar cells in the drawing show the cytoplasmic extension between the acinar cells. This is hard to see in routinely prepared slides. Most probably, serial sectioning of the tissues was used for this drawing. There are three acinar cells showing chromatin clumps. Whether they present mitotic figures, which under normal conditions are hard to find (the labeling index of the acinar cells does not exceed 0.5%) or reflect nuclear fragmentation as a sign of toxicity is unclear. (From Zimmermann.[8] Used with permission.)

reason may be that their inconspicuous structure with almost transparent cytoplasm, has escaped the attention of anatomic histologists. Limited data have been presented by electron microscopy,[1,3–7] a method that is presently not widely used.

5.2 WHAT ARE CENTROACINAR CELLS?

The histological appearance of the centroacinar cells have been briefly described in the literature and were presented in a form of a drawing,[8] as show in Figure 5.1. In some handbooks for histology, only references to earlier publications have been made.[9] Their presentation in drawings reflects the difficulty in visualizing the centroacinar cells in their natural three-dimensional organization, which can only be achieved by serial sectioning of the tissue and reconstruction. In hematoxyllin- and eosin-stained sections, they appear as nearly transparent cytoplasm and round nuclei, which look lighter than those of the acinar cells do. Almost invisible in the routinely prepared sections, they can be demonstrated by the Lewis a (Le^a) antibody, which selectively recognized the centroacinar cells and the terminal ductular cells (Figure 5.2).

The centroacinar cells form the end portion "Endstücke"[8] of the intercalary duct portion. The cells are attached to the luminal surface of the acinar cells (Figure 5.1). However, it is unclear whether every acinar cell has its own centroacinar cell. It is equally unclear whether all acinar units

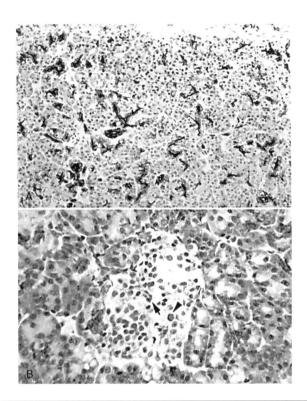

Figure 5.2 Human pancreas. A: Visualization of centroacinar cells in histological specimens with anti-Le^a antibody. The antibody immuno reacts only with centroacinar and terminal ductular cells. Based on the reactivity, centroacinar cells seem to be absent in some acini. This could also be due to the uneven cut through the actually existing centroacinar cells. ABC method ×50. B: The presence of centroacinar cells within some islets can be demonstrated with anti-Le^a antibody (arrowhead). A mitotic figure is present within the islet (arrow). ABC ×120.

of the pancreas have centroacinar cells. This possibility was raised when we discovered that there are two types of acinar in the human pancreas.[10]

Immunohistochemically, one group expresses the blood group antigen of the host (Type 1), whereas the other group (Type 2) does not (Figure 5.3). Detailed information is available in our earlier communication[10] and is summarized in Figure 5.4. The ratio between the two types shows a great variation. In individuals with blood group Type A, Type B, or Type AB, there is a larger number of compatible blood group expression, whereas in the O blood group type people the situation is reversed (Figure 5.3). Remarkably, there is a reverse relationship between the expression of the A, B, or AB antigen and the blood group Type Le^b antigen in that acinar cells expressing the A, B, or AB antigen do not react with the anti-

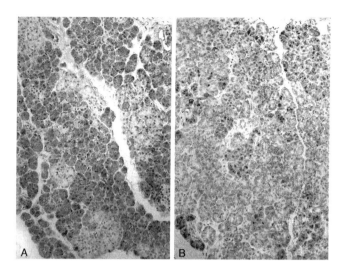

Figure 5.3 Immunoreactivity of anti-blood group A antibody (A) and anti-H antibody (B) with acinar cells of the human pancreas. A: Although most acinar cells show diffuse cytoplasmic staining with anti-A antibody (Type 1), a small group of acini (Type 2) do not show staining. B: The reverse correlation between the stained and unstained acini exist for H-antigen expression in the same specimen. Moreover, the staining with anti-H is of Golgi type. In people with blood group Type O, the entire acinar cells stain with the anti-H antibody. ABC method ×50.

Le[b] antibody.[8,10] Whether the differing expression of these antigens is related to the secretory status of the patient is not known. Nevertheless, the findings suggested differences in the functions of these two acini types and raised the question whether one of the types lack centroacinar cells. Although immunohostiochemically the distribution of centroacinar cells is absent in some areas (Figure 5.2), no correlation was found between the expression of Le[a] and the two types of acinar cells. We do not yet know whether individuals lacking the Lewis gene express the antigen in their centroacinar cells. Nevertheless, it appears that not all acini have centroacinar cells, as was proposed by Zimmermann.[8]

5.3 TOPOGRAPHY AND FINE STRUCTURE OF CENTROACINAR CELLS

The ultrastructural organization of centroacinar cells was described in humans by Kern and Ferner,[3] in rats by Elkholm and Endlund[5] and in other species by Ziegel[6] and Ichikawa.[7] The cells have poor organelles

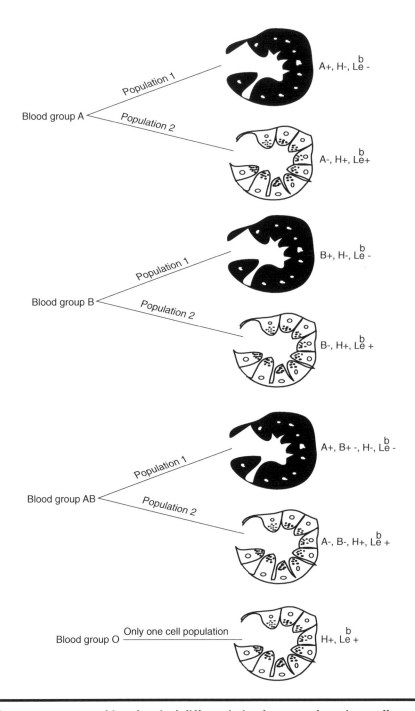

Figure 5.4   Immunohistochemical differentiation between the acinar cell types. For details see Uchida et al.[10]

and diffusely distributed polyribosomes. The usually small mitochondria are located supranuclear toward the luminal surface, where the Golgi apparatus also lies. Characteristic is the presence of 6 to 8 μ long ciliae (9 peripheral and 2 central) that begins near the nucleus and travels through the microvilli to open into the lumen.[7] These cilia that look identical to those occurring in ductal and ductular cells[3,11] have been observed in the pancreas of several species including humans,[3] mouse, guinea pig, Chinese hamsters, chicken,[6] and dogs.[7] The nucleus is round or oval with marginal indentation. Variation in nuclear size and shape, as well as in the density of the cytoplasm occurs in both humans and Syrian hamsters (Figure 5.5 and Figure 5.6). Numerous tonofibrillary structures that travel between the desmosomes of the neighboring centroacinar cells[3] reflect the striking mobility of these cells as described below. Refer to Kern and Ferner[3] and Elkhlom and Endlund[5] for detailed information on the subcellular organelles. Because most of these cytological and histological studies were performed in normal and possibly young tissues, variation in their topography and size in relation to the age, digestive periods, and diseases have remained unrecognized.

In the rat, some centroacinar cells present long cytoplasmic processes that extend between the proximal acinar cells in a complex manner.[1] We found similar cytoplasmic processes in the pancreas of hamsters and humans. In hamsters, the centroacinar cells are either aligned along the luminal face of the acinar cells or are positioned between the distal portion (close to the lumen) of acinar cells, pushing them apart (Figure 5.5B). In the aged hamsters, tiny cytoplasmic processes of the centroacinar cells covering a small or larger portion of the luminal surface of the neighboring acinar cells can be seen occasionally (Figure 5.5B and Figure 5.6B). These tiny cytoplasmic extensions, invisible in the routinely prepared sections, can be visualized in the hamster and human pancreas with the anti-blood group A antibody, and with the anti-Le[a] antibody, which selectively bind to the centroacinar cells of hamsters[12] and humans,[10] respectively (Figure 5.2). As described below, the frequency and the extent of these cytoplasmic processes (pseudopodiae) increase in some pathological conditions.

Confirming the results of a previous study,[1,3] we found a close contact of centroacinar cells with the acinar cells as well as with the endocrine cells (Figure 5.7).

Immunohistochemically, the presence of centroacinar cells within some islets can be seen in the normal pancreas by the Le[a] antibody (Figure 5.2B), indicating that their presence is not restricted to acinar glands (acini). Conversely, a single cell or a small group of islet cells are found intermingled with centroacinar cells (Figure 5.7). In such cases, it is difficult to decide whether the endocrine-like cells present the transdifferentiated centroacinar cells or vice versa are an independent contribution of each

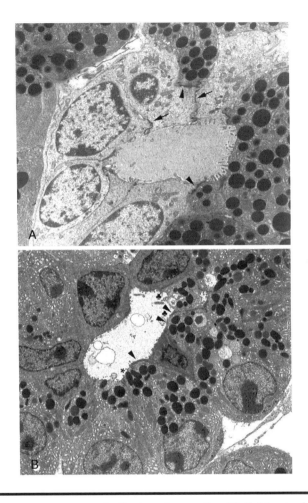

Figure 5.5   Human pancreas. A: Centroacinar cells with light cytoplasm contain-ing a few or no mitochondriae. Tight junctions are seen between the centroacinar cells (arrows) and between these and acinar cells (arrowheads). ×4400. B: Five centroacinar cells with dark cytoplasm in an acinus. One of the cells, sandwiched between the acinar cells, shows narrow and short cytoplasmic processes covering the surface of the neighboring acinar cells (arrowheads). Note that the affected acinar cells show signs of zymogen lysis (♣) and their zymogens are faced toward the unobstructed luminal surface (✳). ×2400.

cell type. The intimate dialogue between the centroacinar and acinar cells, centroacinar cells and islet cells, as well as islet cells and acinar cells are indicated by the presence of the tight junctions between them. Hence, the process of secretion appears to be coordinated by the input of these different cells.

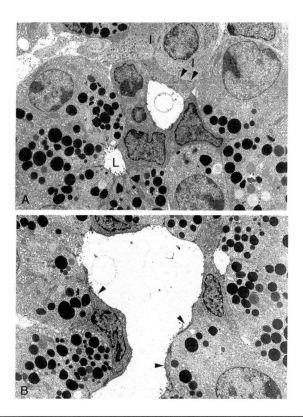

Figure 5.6   Centroacinar cells of the hamster pancreas. A: A small centroacinar cell with a dark cytoplasm surrounds the lumen and shows an intimate connection to islet cells (I). A short, tight junction between the centroacinar and islet cells (arrowheads). Two of the acinar cells have a reduced number of zymogens showing variation in size. In the lumen of another acinus (L) there is only one centroacinar cell, but the acinar cells seem to open freely into the lumen. Whether it is an acinus without centroacinar cells or we are looking at the unobstructed surface of acinar cells is not clear. ×2400. B: An acinus with distended lumen lined by centroacinar cells, the cytoplasmic extensions of which cover part of the luminal surface of acinar cells (arrowheads). Note the reduced number of zymogens in the affected acinar cells. ×2400.

5.4   PATHOLOGY OF CENTROACINAR CELLS

Striking changes in the number, size, and topography of the centroacinar cells occur during pancreatic carcinogenesis in the hamster model. In the normal pancreas of hamsters, the centroacinar cells are confined within the acinar lumen, although in aged hamsters, occasionally, tiny cytoplasmic processes covering a small part of the neighboring acinar cells can be

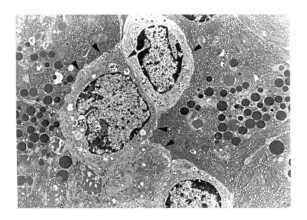

Figure 5.7   Centroacinar-like cells within an acinus of a hamster pancreas containing a few endocrine granules and several lysosomes. Narrow gaps are present between these and the acinar cells (arrowheads). It is difficult to decide whether the cells present degenerating endocrine cells or the transforming centroacinar cells. Both cells with the secretory granules without contact to the lumen show the pattern of extra-isular islet cells. Probably, these cells are functionally in a resting stage. ×4400.

observed (Figure 5.5A and Figure 5.6B). In animals treated with a pancreatic carcinogen, however, a wide variation in the number and size of the centroacinar cells occurs focally or multifocally depending on the dose of the carcinogen used.[2] Near the normal centroacinar cells, those with long cytoplasmic processes cover most or the entire luminal surface of acinar cells. The blockage of the entire luminal surface of the acinar cells by these centroacinar cells is evident by the reduced number of zymogen granules, signs of zymogen digestion, and the increased number of lysosomes or complete necrosis of the affected acinar cells (Figure 5.8 and Figure 5.9). The long and wide processes virtually separate the acinar cells all along their opposing lateral surface. Some of these processes are found to extend along the base of the acinar cells, separating the basal membrane of the acinar cells from the supplying blood vessel (Figure 5.8). The affected acinar cells show signs of degeneration and necrosis and are found expelled into the lumen (Figure 5.9 and Figure 5.10). The original position of the necrotic acinar cells is occupied by the now plump and duct-like appearing centroacinar cells (Figure 5.10). In advanced cases, centroacinar cells replacing the entire acinar cells of an acini give rise to a pseudoductular structure.[2] In the histological sections, the cytoplasmic extension of the centroacinar cells can be visualized by an anti-blood group A antibody, which selectively stains the hyperplastic centroacinar (Figure 5.8B and Figure 5.9B).

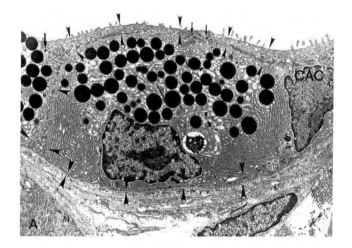

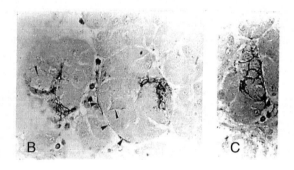

Figure 5.8 Pancreas of a hamster treated with pancreatic carcinogen. A: An acinar cell is "wrapped" with the long processes of the centroacinar cells. Arrowheads mark the extension of the processes, which are also present in the basal portion of the acinar cell (lower part). Tight junctions are present between the processes and the acinar cells (arrows). A large lysosome lies near the nucleous of the acinar cells. CAC — centroacinar cell. ×4400. B: A representative section of the pancreas stained with the anti-blood group A antibody. The distribution of the cytoplasmic processes of the centroacinar cells is demonstrated between the acinar cells (long arrowheads) and along the base of acinar cells (short arrowheads). C: A chain of the processes can be seen in many affected areas of the pancreas. ABC method. ×120.

The findings ultimately point to the regulatory role of centroacinar cells in the secretory process of acinar cells in the normal and pathological conditions and highlight the complicated machinery of the secretion. It appears that, in the healthy tissue, the centroacinar cells, due to their tonofibrillary architectures,[3] act as a valve regulating the amount of secreted zymogen granules. Although the function of the cilia in these as

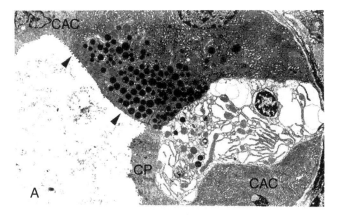

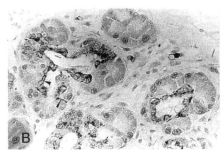

Figure 5.9 A section from the pancreas of a hamster treated with pancreatic carcinogen. A: Thick cytoplasmic process (CP) of centroacinar cells (CAC) is seen over a necrotic acinar cell. A long and narrow process covers the whole surface of the adjacent acinar cell (arrowheads). The blocking of the surface by the processes is apparently incomplete as the affected acinar cell does not show signs of damage. ×2400. B: Histological section of the pancreas of the same hamster. Hyperplastic centroacinar cells and their cytoplasmic extensions between the acinar cells (long arrowhead) and along the luminal surface of an acinar cell (short arrowhead) is visualized by an anti-A antibody. Many stained centroacinar cells covering the extended lumen of an acinus (right middle portion). ABC methods. ×120.

well as ductular cells is unknown, it may well be working as a sensory element that may trigger the function of the cytoplasmic processes. It is possible that the activated secreting acinar are unable, *per se*, to shut off the extrusion of the zymogens at once. The activation of centroacinar cells by humoral, neurogenic impulses or through the cilia, and the formation of cytoplasmic processes may guarantee the optimal and economical usage of the secretion. The intimate interaction between centroacinar cells and islet cells further point to the complexity of the regulatory process of the enzyme secretion.

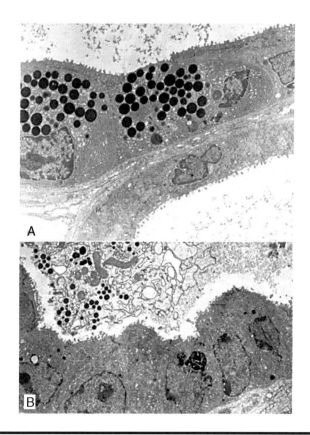

Figure 5.10 The pancreas of a carcinogen-treated hamster. A: A low magnification of the specimen shown in Figure 5.8. Cytoplasmic extensions cover the
surface of the acinar cells. The zymogens of the acinar cell are shifted toward
the free luminal surface of the cell. In the lower half of the figure, the original
acinus is replaced by duct-like cells, which are similar to centroacinar cell. ×2400.
B: A complete reconstruction of an acinus by centroacinar cells in advanced
stages of carcinogenesis in the hamster. Cylindrical cells make up the duct-like
structure irregular in shape and size differing from the original centroacinar cells.
A necrotic acinar cell expelled into the lumen is still present. These lesions
ultimately resulted in invasive carcinomas. ×2400.

The abnormal pseupodiae formation in carcinogen-treated animals
could be interpreted in two ways. Either the reaction of the centroacinar
cells is to remove acinar cells that are damaged by the carcinogen, or,
most likely, the centroacinar cells are the target of the carcinogen or other
toxic substances. This important question requires detailed cellular biological studies, which are presently suffering from the shift of interest
from cellular biology into molecular biology. The undisputable importance

of such studies is highlighted by the fact that the described changes also occur in the human pancreas.

Focal proliferation of centroacinar cells can be observed in some elderly individuals. In most cases, focal acinar cell atrophy or nesidioblastosis is evident. In these foci, the centroacinar cells are larger and fill the lumen of the acini. In chronic pancreatitis specimens, however, multifocal proliferation of these cells and the production of long cytoplasmic processes extending between the acinar cells can be detected ultrastructurally (Figure 5.11) or, more conveniently, by the use of an anti-Le[a] antibody, which selectively stains the centroacinar cells and their cytoplasmic processes (Figure 5.11B). A striking diffuse or focal hyperplasia of the centroacinar cells is seen in cases of infantile hyperinsulinemic hypoglycemia (Figure 5.12). In this abnormality, a mixture of islet cells and centroacinar cells can be found frequently. (Detailed morphological, immunohostochemical and ultrastructural findings of the disease can be found in the excellent contribution of Klöppel and Heitz[13].) Thus, it appears that, in this disease, the etiological factor causing the abnormality affects both centroacinar and islet cells. Whether the islet cells in this condition are primarily derived from transdifferentiated centroacinar cells remains to be investigated.

Based on our data, the role of the centroacinar cells in the normal physiology and pathology is indisputable. In pancreatic research, much attention has been paid to acinar and ductal cells. One should bear in mind that centroacinar cells with their tide attachment to acinar cells are the essential elements of the acini and cannot be separated from acinar cells. It is striking that in most *in vitro* studies, the presence of the centroacinar cells has been largely overlooked. Therefore, whether the reported transformation of acinar cells into ductular cells is a reality or an illusion remains a vital question.

5.5   SUGGESTED PHYSIOLOGICAL FUNCTION OF CENTROACINAR CELLS

Our findings, supported, in part, by earlier studies, strongly suggest that the centroacinar cells play a crucial role in the secretion process of pancreatic enzymes. We hypothesize the following scenario. The two types of acinar cells may indicate that the secretion process in acinar cells is random (i.e., a certain percentage of acinar cells function simultaneously at a given time, while the other group rest or recover). It also questions the possible differences in the machinery of enzyme production, in terms of the quality of the individual enzymes, in individuals with different blood group types. Equally, it is possible that a minor portion of acini is engaged in the enzyme secretion during the resting phase, and others function upon stimulation. Beside the secretion of bicarbonate, the centroacinar

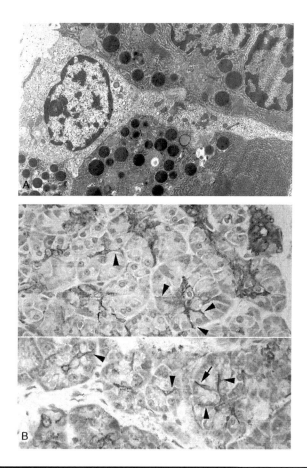

Figure 5.11 Specimen from the pancreas of a patient with chronic pancreatitis. A: Hyperplastic centroacinar cells were present in some areas of the pancreas. Some of the centroacinar cells with light cytoplasm, similar to those in hamsters, show long cytoplasmic extensions separating the acinar cells from each other. The ensuing loss of communication between the affected acinar cells may be responsible for acinar cell damage as reflected by the signs of zymogen digestion. ×4500. B: The immunoreactivity of anti-Le^a with the centroacinar cells in chronic pancreatitis patients. Proliferation of centroacinar cells is evident in the right upper corner. Cytoplasmic extension of the centroacinar cells between the acinar cells is seen in several areas (arrowheads). A centroacinar cell with its usual oval nuclei and cytoplasmic processes lies between two acinar cells (arrow). ABC ×120.

cells seem to regulate the excretion of zymogens from acinar cells by their cytoplasmic processes, which act as valves. Their intimate relationship with the acinar cells and islet cells reflect the coordinated effort in controlling the enzyme secretion. The presence of gaps between the centroacinar and acinar cells indicate that the centroacinar cells do not

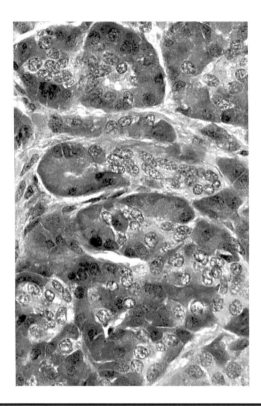

Figure 5.12 **Proliferation of centroacinar cells intermingled with islet cells in an infant with infantile hyperinsulinemic hypoglycemic disease. H&E ×120.**

totally block the excretion of zymogens in the normal condition. Their cytoplasmic processes become active to shut off the secretion from the acinar cells, which apparently do not have the ability to interrupt the secretion abruptly. The trigger of the activation of the centroacinar cells is probably as complicated as the question of whether this event is mainly of economical reason (in preventing the waste of the enzyme) or other more important reasons. Nevertheless, the apparent lack of centroacinar cells in some acini indicates that the secretion in these acini is uncontrolled. Theoretically, these acini may be engaged in the continuous enzyme secretion, possibly during the pre- and postprandial stage.

Centroacinar cells also seem to be the target of toxic substances. Their proliferation in some conditions, including inflammatory and neoplastic processes, counteracts their regulatory physiological function and leads to severe consequences, such as tissue loss and malignancy.[2] Reasons for their greater susceptibility to toxic substances, which may be excreted by acinar cells or are introduced directly by blood circulation, could be the presence of various drug-metabolizing enzymes in these cells, especially of CYP 2E1 and 3A4, which are involved in the metabolism of a large group of toxins and carcinogens[14] (see Chapter 7 and Chapter 8). Although chronic pancreatitis is suggested to be the end stage of repeated attacks of acute pancreatitis,[15,16] their development, at least in part, due to pathological function of the centroacinar cells in response to the etiological factor (alcohol) cannot be ruled out.

Our observation was based on a study to investigate the subcellular alteration of pancreatic cells during pancreatic carcinogenesis in treated and untreated controls. Because animals were taken randomly at different stages of the animal's digestive process, we do not have information about a relationship between the patterns of centroacinar cells and the stage of digestion. The same applies to our findings on the presence of immunologically different acinar cell types. Do these acinar cells produce different combinations of digestive enzymes or reflect different stages of secretory activation? What is the role of centroacinar cells during digestion? Do their shape and cytoplasmic processes function differently in relation to the stage of digestion? What triggers their activation? Is their presence unequivocal for proper enzyme secretion? If so, why do some acini not have centroacinar cells? These and many other pertinent questions await exploration.

Another important message of our findings relates to transdifferentiation. In the last decade, the process of islet cell neogenesis from ductular or other cells is of a great scientific and clinical importance for the development of a cell replacement therapy in diabetes mellitus. The close association between the endocrine and centroacinar cells as demonstrated in Figure 5.6 and Figure 5.7 suggests that centroacinar cells also present a source for nesidioblastosis. Both cell types, the centroacinar and endocrine cells, show some close similarities in the morphological appearance such as the pale cytoplasm and the nuclear size or distribution pattern of the chromatin (Figure 5.7). After transdifferentiation, the rough endoplasmic reticulum and the Golgi-apparatus as a typical sign of protein biosynthesis are much more developed in the cytoplasm in cells with secretory granules. At present, there is no morphological or biochemical marker to characterize the shift from the ductular into an endocrine cell type. Especially the endocrine pancreas of sand rat (*psammomys obesus*) has a great regenerative potency by neogenesis after beta cell destruction through a high-energy diet.

Nonetheless, the results of our limited study indicate that, from physiological and pathological points, centroacinar cells present an essential element of the pancreas.

ACKNOWLEDGMENTS

Supported, in part, by the National Institute of Health, National Cancer Institute SPORE Grant No. P50CA72712, the National Cancer Institute

Laboratory Cancer Research Center Support Grant CA367127, and the American Cancer Society Special Institutional Grant.

REFERENCES

1. Leeson T.S. and Leeson R. 1986. Close association of centroacinar/ductular and insular cells in the rat pancreas. *Histol. Histopathol.* 1(1):33–42.
2. Pour P.M. 1988. Mechanism of pseudoductular (tubular) formation during pancreatic carcinogenesis in the hamster model. An electron-microscopic and immunohistochemical study. *Am. J. Pathol.* 130(2):335–344.
3. Kern H.F. 1971. Die Feinstruktur des exokrinen Pankreasgewebe. *Zeitschrift fur Zellforschung* 113:322–343.
4. Eckholm R. 1959. Ultrastructure of the exocrine pancreas. *J. Ultrastruct. Res.* 2:453–481.
5. Ekholm R. and Edlund Y. 1962. The ultrastructural organization of the rat exocrine pancreas. II. Centroacinar cells, intercalary and intralobular ducts. *J. Ultrastruct. Res.* 7:73–83.
6. Ziegel R. 1962. On the occurence of cilia in several types of the chick pancreas. *J. Ultrastruct. Res.* 7:286–292.
7. Ishikawa A. 1965. Fine structural changes in response to hormonal stimulation of the perfused canine pancreas. *J. Cell. Biol.* 24:369–385.
8. Zimmermann K. 1927. *Handbuch der mikroskopischen Anatomie des Menschen*. Berlin: Springer.
9. Bargmann W. 1967. *Pancreas*. Stuttgart: George Thieme Verlag.
10. Uchida E.S.Z., Mroczek E., Buechler M., Burnett D., and Pour P.M. 1986. Presence of two distinct acinar cell populations in human pancreas based on their antigenicity. *Int. J. Pancreatol.* 1:213–225.
11. Althoff J., Wilson R.B., Ogrowsky D., and Pour P. 1979. The fine structure of pancreatic duct neoplasm in Syrian golden hamsters. *Prog. Exp. Tumor Res.* 24:397–405.
12. Pour P.M., Uchida E., Burnett D.A., and Steplewski Z. 1986. Blood-group antigen expression during pancreatic cancer induction in hamsters. *Int. J. Pancreatol.* 1(5–6):327–340.
13. Kloeppel G.P.H. 1984. *Pancreatic pathology*. London: Churchill Livingstone.
14. Standop J., Schneider M.B., Ulrich A., Chauhan S., Moniaux N., Buchler M.W., Batra S.K., and Pour P.M. 2002. The pattern of xenobiotic-metabolizing enzymes in the human pancreas. *J. Toxicol. Environ. Health* A65(19):1379–1400.
15. Kloppel G. 1992. The morphological basis for the evolution of acute pancreatitis into chronic pancreatitis. *Virchows. Arch. A Pathol. Anat. Histopathol.* 420:1–4.
16. Kloppel G. 1986. *Pathomorphology of chronic pancreatitis*. Heidelberg: Springer Verlag.

6

PHYSIOLOGY OF EXOCRINE PANCREAS

William Y. Chey and Ta-min Chang

CONTENTS

6.1 INTRODUCTION

The exocrine pancreas consists of the ducts and acini with the former secreting water and electrolytes and the latter secreting enzymes. The bicarbonate- and enzyme-rich pancreatic juice enters the duodenum to neutralize gastric acid and digest food chyme emptied from the stomach. Until the end of the 19th century, pancreatic exocrine secretion (PES) was thought to be regulated by the vagus nerve and acid in the duodenum. The role of the vagus was advocated by Pavlov's group who considered that "the vagus is the secretory nerve of the pancreas."[1] In 1902, Bayliss and Starling[2] reported a historical observation that intravenous infusion of an intestinal acid extract to dogs with extrinsically denervated pancreata resulted in stimulation of PES. The discovery of this humoral substance, termed *secretin,* thus established a new hormonal concept for regulation of PES. The hormonal concept was strengthened by purification and structural determination of secretin and cholecystokinin (CCK) by Jorpes and Mutt.[3,4] Subsequently, several regulatory peptides from the gastrointestinal tract and the pancreas were identified to affect PES. The physiological actions of these regulatory peptides have become understood through immunoneutralization of the circulating hormones with the corresponding specific antisera and the use of receptor antagonists specific for the corresponding regulatory peptides. Pancreatic exocrine secretion is also regulated by a feedback mechanism[5] involving the release of two key intestinal hormones, secretin and CCK. The feedback regulatory process has become understood through the discovery of secretin- and CCK-releasing peptides during the last two decades. In addition, the vagus nerve plays an important role to regulate PES by interacting with the regulatory peptides. It has become increasingly evident that PES is an integrated physiological process that is regulated by hormonal-hormonal and neurohormonal interactions involving interactions among classic hormones, neuropeptides, neurotransmitters, and the secretin- and CCK-releasing peptides.

Table 6.1   Hormones or Peptides that Influence the Exocrine Pancreas

Stimulants	Inhibitors
Hormones	Somatostatin
Secretin	Pancreatic polypeptide
Cholecystokinin	Peptide YY
Neurotensin	Glucagon/oxyntomodulin
Motilin	Pancreastatin
Insulin	Thyrotropin-releasing hormone
Serotonin (5-HT)	
Secretin and CCK-releasing peptides	
Secretin-releasing peptide	
Pancreatic phospholipase A$_2$	
CCK-releasing peptides	
Monitor peptide	
Luminal CCK-releasing factor (LCRF)	
Diazepam binding inhibitor (DBI)	

6.2   ENTEROPANCREATIC HORMONES

Enteropancreatic hormones known to influence exocrine pancreatic secretion are listed in Table 6.1. Stimulatory hormones include secretin, cholecystokinin, neurotensin, insulin, and motilin; whereas SS, peptide YY, glucagon, oxyntomodulin, Met-enkephalin, pancreastatin, thyrotropin releasing hormone (TRH), and pancreatic polypeptide are inhibitors. Serotonin (5-hydroxytryptamine, 5-HT) released from enterochromaffin (EC) cells may act either as a stimulant or an inhibitor through its multiple receptor subtypes.

6.3   SECRETIN

Secretin is localized mainly in upper small intestinal mucosa and is the main enteric hormone released by gastric acid to stimulate pancreatic exocrine secretion. It has been well established that circulating level of secretin increases postprandially.[6,7] Suppression of gastric acid secretion with a histamine H$_2$ blocker inhibits postprandial secretin release[8] and PES. The hormonal role of secretin on pancreatic secretion was further established in the dog through inhibition of postprandial pancreatic secretion by immunoneutralization with a specific anti-secretin serum (Figure 6.1).[9] It is now clear that secretin is released not only by acid, but also by other luminal stimuli including digestive product of fat and protein, bile acids, and herbal extracts. Several forms of pro-secretin have been

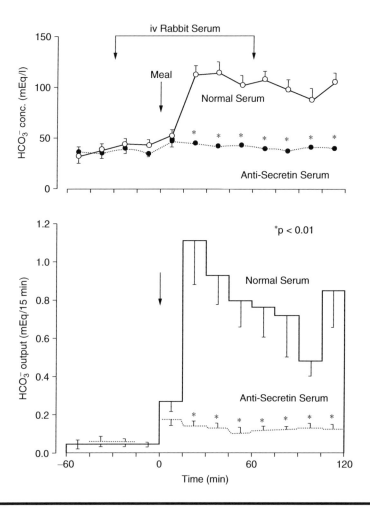

Figure 6.1   Effect of anti-secretin serum (Anti-S) or normal rabbit serum (NRS) on meal-stimulated pancreatic bicarbonate secretion. Anti-S suppressed meal-stumulated pancreatic bicarbonate secretion, but NRS did not. Source: Data from Chey et al.[9] Used with permission.

isolated from porcine intestinal extract and shown to stimulate PES with different potency. However, their physiological roles on PES have not been established.

6.4   CHOLECYSTOKININ

Purification and characterization of CCK has established that it possesses both cholecystokinin and pancreozymin activities.[4] As a candidate hormone, CCK has been shown to increase in the circulation after ingestion of a meal. The elevated plasma CCK concentration correlated well with the elevation in pancreatic enzyme secretion. The hormonal role of CCK was established through inhibition of pancreatic protein or enzyme secretion in response to a meal or intraduodenal infusion of a nutrient with a specific CCK receptor antagonist (see Chapter 14). For example, the CCK antagonist, proglumide inhibits intestinal fat-stimulated pancreatic secretion in the dog.[10] CCK also has a synergistic effect with secretin to stimulate pancreatic bicarbonate secretion. Thus, in both humans and dogs combining intravenous infusion of secretin and CCK in physiological doses stimulates pancreatic bicarbonate secretion to an extent greater than the sum of the effects of secretin and CCK given individually. Moreover, PES stimulated by combination of secretin and CCK in physiological doses was abolished by administration of proglumide.[11]

CCK also exists in multiple molecular forms and is localized predominantly in upper small intestinal mucosa. It is also present in the central and peripheral nervous systems including the vagus nerve. Thus, CCK may function as both a hormone and a neuropeptide to affect PES. The main circulation form of CCK is found to be CCK-58,[12] particularly when released upon intestinal stimulation.[13] On the other hand, electrical vagal stimulation in anesthetized dogs releases CCK-8 as the predominant form.[13] It is not clear at present whether or not CCK-8 released from the vagus nerve play an important role on PES under a physiological condition.

6.5 NEUROTENSIN

Neurotensin (NT) is present in enteroendocrine cells of small intestinal mucosa as well as the nervous systems. It is also elevated in the circulation after ingestion of a fatty meal in dogs and humans or intraduodenal infusion of fat or sodium oleate in dogs. Intravenous infusion of NT resulted in dose-dependent stimulation of pancreatic volume, bicarbonate, and protein secretion in dogs, humans, and rats. Infusion of NT at a dose producing the postprandial plasma NT concentration in dogs stimulated PES that was sensitive to atropine.[14] In dogs, the release of NT upon duodenal infusion of fat was also abolished by atropine.[14] Thus, both the release and action of endogenous NT are dependent on a cholinergic pathways. NT also potentiates with CCK to stimulate bicarbonate secretion and with secretin to stimulate enzyme secretion. In both rats and dogs, exogenous NT-stimulated PES was suppressed by the CCK-A receptor antagonist, L364,718, atropine, and neural blockade suggesting an action through a CCK-dependent cholinergic mechanism. NT-immunoreactive nerves were found in rat pancreas so that NT also may play a neurocrine role in the exocrine pancreas. This notion is supported by the observation that NT stimulates secretion from isolated rat pancreatic acini, lobules, and perfused pancreas. Immunoneutralization of circulating NT in dogs inhibits postprandial pancreatic protein secretion by 50%, suggesting that NT plays a significant hormonal role in postprandial PES.[15]

6.6   INSULIN

Local insulin in the pancreas appears to be important for hormonal actions of secretin and CCK. It has been shown in rats that upon immunoneutralization of endogenous insulin, postprandial PES is impeded and the physiological doses of secretin and CCK are unable to stimulate PES.[16] Similar effects of immunoneutralization of insulin have been observed in isolated and perfused rat and dog pancreata suggesting that insulin plays a local regulatory role. It is likely that endogenous insulin plays a permissive role for the actions of secretin and CCK (Section 6.17). For additional information, see Chapter 13 and Chapter 14.

6.7   MOTILIN

Motilin concentration in the circulation exhibits a cyclic change in association with the cyclic interdigestive migrating motor complex (MMC) of the upper gut. PES during interdigestive state in dogs also fluctuates with MMC. Exogenous motilin given intravenously in physiological doses evoked a transient stimulation of PES mimicking the cyclic secretion in dogs. Immunoneutralization of circulating motilin not only abolished MMC, but also suppressed the associated cyclic pancreatic secretion.[17] Thus, motilin appears to be a hormone regulating the cyclic secretion of pancreatic juice during interdigestive state. However, the underlying mechanism is not clear at present, although the cyclic increase of both circulating motilin and PES are suppressed by atropine suggesting that a cholinergic pathway is involved.

6.8   SOMATOSTATIN

Intravenous infusion of somatostatin (SS) inhibits the releases of secretin and CCK as well as PES. SS receptor is found in the exocrine pancreas. SS also inhibits exocrine secretion from isolated and perfused rat pancreas or acini. These observations indicate SS exerts direct action on the exocrine pancreas. In addition, CCK-stimulated protein secretion in anesthetized rats is augmented by administration of an anti-SS monoclonal antibody, indicating that endogenous SS participates in modulation of CCK-stimulated PES. Because SS is present in both the intestinal mucosa and the pancreas, it is likely that mucosal SS regulates the release of the gut hormones, and pancreatic SS regulates PES directly as a paracrine mes-

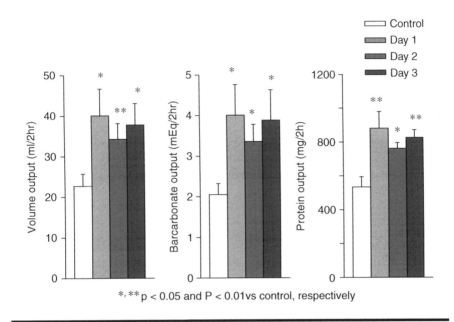

*, ** p < 0.05 and P < 0.01 vs control, respectively

Figure 6.2   Effect of antipancreatic polypeptide (anti-PP) on postprandial pancreatic exocrine secretion in 6 dogs. Intravenous administration of anti-PP in a large dose augmented postprandial PES for 3 days. Source: Data from Shiratori et al.[18] Used with permission.

senger. In addition, SS is also present in enteric nervous system and the vagus nerve, it cannot be ruled out that SS also functions as a neurocrine regulator of PES.

6.9   PANCREATIC POLYPEPTIDE

Pancreatic polypeptide (PP) given intravenously inhibits PES. Circulating PP also increases in a cyclic pattern in phase with MMC. Immunoneutralization of circulating PP in dogs resulted in a significant increase in PES of both fluids and protein in interdigestive as well as postprandial states (Figure 6.2).[18] Thus, PP appears to be another hormone that negatively regulates PES. However, PP inhibits agonist-stimulated PES *in vivo* but not *in vitro,* suggesting that it acts indirectly on the exocrine pancreas. In dogs with extrinsically denervated pancreas, PP remained effective as an inhibitor of PES, suggesting it may act through intrapancreatic neurons. Microinjection of PP in dorsal motor nucleus of anesthetized rats inhibited both 2-deoxy-glucose and CCK-8 stimulated PES in a dose-dependent manner. The inhibitory effect of PP on CCK-stimulated PES was abolished by vagotomy indicating the involvement of the vagus nerve. PP also inhibits PES in response to bile-pancreatic juice diversion that is mediated by the vagus nerve. However, vagotomy but not capsaicin-elicited afferent denervation prevented the inhibition of diversion elicited PES by PP, indicating the involvement of the vagal efferent pathway.

6.10 PEPTIDE TYROSINE TYROSYLAMIDE

Intravenous infusion of peptide tyrosine tyrosylamide (PYY) in dogs inhibits PES in response to exogenous secretin and CCK as well as to intraduodenal administration of sodium oleate, suggesting that PYY may be an inhibitory regulator of PES. Circulating PYY is increased after ingestion of a meal in dogs, humans, and rats. In rats, exogenous PYY infused to simulate postprandial concentration inhibited PES stimulated by physiological doses of secretin and CCK. Moreover, immunoneutralization of circulating PYY in rats with an anti-PYY serum resulted in augmentation of the stimulated PES[19] as well as PES stimulated by duodenal infusion of oleic acid, suggesting that endogenous PYY participates in regulation of PES. Both PYY and PP dose-dependently inhibited PES stimulated by 2-deoxy-glucose (2-DG), CCK, bethanechol, and electrical vagal stimulation in anesthetized rats.[20] Inhibition of 2-DG-stimulated pancreatic secretion, which is not mediated by secretin and CCK, was inhibited by PYY in a half maximal dose of 10 pmol/kg/h, indicating that circulating PYY may regulate PES stimulated via vagal efferent pathway. On the other hand, inhibition of CCK-stimulated secretion by PYY in a half maximal dose of 250 pmol/kg/h (a pharmacological dose) cannot be considered as a hormonal action. The different responses of 2-DG- and CCK-stimulated pancreatic secretion to PYY appear to indicate two different target sites for PYY. Because PYY does not inhibit CCK-8- and urecholine-stimulated amylase secretion from isolated dog and rat pancreatic acini, it has been postulated that its inhibitory effect is indirect and may be neural mediated. However, in isolated guinea pig pancreatic acini, PYY inhibited VIP- but not CCK- or bombesin-stimulated amylase secretion and exhibited specific receptor binding, indicating a direct action in this species. The results of some *in vitro* studies have indicated that both neuropeptide Y (NPY, an analog of PYY and PP) and PYY act by inhibiting the release of neurotransmitters, particularly acetylcholine and norepinephrine. In conscious dogs, PES stimulated either by secretin, CCK, or a meal was inhibited by PYY through the adrenergic pathway. PYY occurs mainly in the ileum and colon so that its release during the early postprandial stage has been considered to be neural mediated. PYY also occurs as a neuropeptide and may exert a neurocrine inhibitory effect. Thus, exogenous PYY at a high pharmacological dose may be required to attain its neurocrine concentration at the target neuron or cell. However, it has been observed[21] that in extrinsically denervated dog pancreas, PYY inhibited PES stimulated by secretin and CCK (in pharmacological doses), and the inhibition by PYY was reversed by simultaneous infusion of bethanechol with the two hormones. It was thus proposed that PYY also acts on an intrapancreatic cholinergic pathway. Similarly in isolated and perfused rat pancreas, PYY was able to inhibit CCK-stimulated amylase secretion. Thus, the neural target sites of PYY remain to be identified.

6.11 SEROTONIN

Although serotonin (5-HT) is elevated in circulation postprandially, it is now proposed that 5-HT released from EC cells in response to luminal stimuli functions as a local mediator that acts on enteric sensory nerve fibers.[22] Thus, through different receptor subtypes, 5-HT may act as a stimulator or an inhibitor of PES. It has become evident that 5-HT plays significant roles in regulation of PES. For instance, PES and secretin release in response to duodenal acidification in anesthetized rats are inhibited by an antagonist of 5-HT_3 receptor, ondansetron, and by the antagonist of 5-HT_2 receptor, ketanserin, in a dose-dependent manner.[23] Moreover, the secretion of fluid and bicarbonate stimulated by exogenous secretin in physiological doses is also inhibited significantly by ondansetron and ketanserin.[23] Thus, 5-HT appears to regulate acid-stimulated PES through modulation of both the release and action of secretin via the two receptor subtypes. The 5-HT_3 receptor subtype is also involved in postprandial pancreatic protein secretion. Thus, the protein secretion increased by intragastric administration of rodent chow in conscious rats[24] was suppressed 54% by a CCK-A receptor antagonist, L-364,718, and inhibited 94% by combination of the CCK-A receptor antagonist and the 5-HT_3 antagonist, ICS 205-930 (Figure 6.3).[24] This observation suggests that in the rat, two additive pathways involving CCK-A receptor and 5-HT_3 receptor, respectively, mediate stimulation of postprandial pancreatic protein secretion. However, 5-HT is also present in enteric, intrapancreatic, and vagus nerves and functions as a neurocrine regulator of PES as it is discussed in Section 6.17.

6.12 THYROTROPIN-RELEASING HORMONE

In addition to its presence in the central nervous system, TRH is widely distributed in the gastrointestinal tract. In the pancreas, TRH is found in the insulin producing cells in the islet of Langerhans. Intravenous administration of TRH inhibited secretin-, CCK-, carbachol-, or 2-deoxy-D-glucose-stimulated pancreatic secretion of volume, bicarbonate, and enzyme. In isolated pancreatic lobules or acini, TRH also inhibited CCK- or carba-

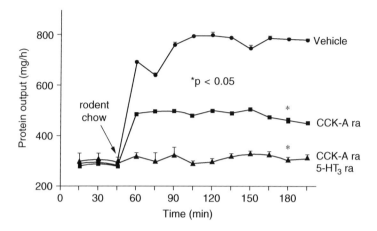

Figure 6.3 Effect of a CCK-A receptor antagonist (CCK-A ra) given alone or incombination with a 5-HT$_3$ receptor antagonist (5-HT$_3$ ra) on postprandial protein secretion in rats. CCK-A ra significantly suppressed postpradial protein secretion, and a combination of CCK ra with 5-HT$_3$ ra abolished the increase of postprandial protein secretion. The results indicated that postprandial increase in pancreatic protein secretion involves both CCK-A receptor- and 5-HT$_3$ receptor-dependent pathways in an additive manner. Source: Data from Li et al.[24] Used with permission.

chol-stimulated enzyme secretion, which indicates a direct effect on pancreatic acini. The stable metabolite of TRH, cyclo-His-Pro, has a similar spectrum of action on PES. The release of TRH from the pancreas is stimulated by 5-HT and inhibited by carbachol, suggesting that it is neural regulated. The physiological role of TRH on PES and its mechanism of action have not yet been established.

6.13 NEURAL REGULATION OF PANCREATIC EXOCRINE SECRETION

Neurohormonal regulation of PES has been known for decades. The role of acetylcholine on PES is established by inhibition of postprandial-, secretin-, or CCK-stimulated PES by intravenous (IV) administration of atropine.[25] The observations that atropine inhibits secretin-stimulated pancreatic fluid and bicarbonate secretion and CCK-stimulated enzyme secretion suggest that acetylcholine may either interact with the two gut hormones through a synergistic action or mediate the actions of the two hormones to stimulate PES. On the other hand, the ability of cholinergic agonists to stimulate ductal secretion in isolated pancreatic ducts and enzyme secretion from isolated pancreatic acini indicate that acetylcholine is directly involved in PES. However, it has become clear recently that the releases and actions of secretin and CCK, like NT, are neural mediated through cholinergic and noncholinergic pathways.

6.14 NEURAL REGULATION OF THE RELEASE AND ACTION OF SECRETIN AND CCK

In humans, dogs, and guinea pigs, PES stimulated by either exogenous or endogenous secretin or CCK is profoundly inhibited by atropine, suggesting the involvement of the muscarinic cholinergic pathway. In anesthetized rats, PES stimulated by duodenal acidification in the rat is substantially inhibited by tetrodotoxin, mucosal application of lidocaine, or IV infusion of a specific anti-gastrin-releasing polypeptide (GRP) serum, partially inhibited by a pituitary adenylase-cyclase activating polypeptide (PACAP) antagonist, $PACAP_{6-27}$, but not by administration of atropine or hexamethonium. These observations suggest that PES stimulated by duodenal acidification in the rat is mediated by a noncholinergic neural pathway that may involve the two neuropeptides. Perivagal capsaicin treatment in conscious rats inhibited PES in response to secretin at a physiological dose, but not at a pharmacological dose (Figure 6.4),[26] suggesting the involvement of vagal afferent pathway. Similarly, CCK-stimulated pancreatic enzyme secretion is also mediated through vagal afferent pathway in rats. Thus, PES in response to intraduodenal infusion of casein in rats was suppressed by perivagal capsaicin treatment, truncal vagotomy or a CCK-A receptor antagonist, L-364,718; whereas, chemical activation of vagal efferent pathway by 2-DG was not sensitive to capsaicin treatment. Perivagal capsaicin treatment, afferent rootlet section or mucosal application of capsaicin in the duodenum also resulted in suppression of PES in response to exogenous CCK-8 in physiological doses.[27] These observations clearly indicated that the actions of secretin and CCK on the exocrine pancreas are modulated via the vagal ascending pathway. However, it should be noted that, unlike in the dog, human, and guinea pig, secretin-stimulated pancreatic ductal secretion is not sensitive to atropine in the rat and thus must be mediated by noncholinergic neurotransmitters or neuropeptides. Vasoactive intestinal polypeptide (VIP), GRP, PACAP, and nitric oxide (NO) are possible candidates.

6.15 NEUROPEPTIDES AND NEUROTRANSMITTERS

Several neuropeptides and neurotransmitters listed in Table 6.2 are known to affect pancreatic exocrine secretion. Acetylcholine, NO, GRP, CCK, VIP, TRH PACAP, orexin A, and cocaine- and amphetamine-regulated transcript (CART) peptide are stimulants of PES. On the other hand, galanin, SS,

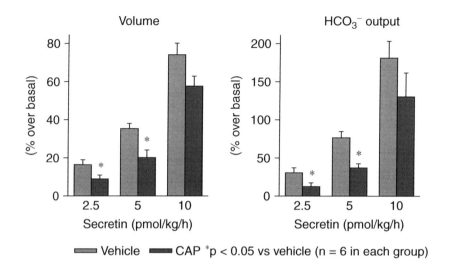

Figure 6.4 Effect of perivagal application of capsaicin (CS) on pancreatic secretion in response to exogenous secretin in conscious rats. Capsaicin treatment significantly inhibited significantly pancreatic secretion stimualted by secretin at physiological doses (2.5 and 5 pmol/kg/h), but not at a pharmacological dose (10 pmol/kg/h). The results indicate that action of secretin at a physiological dose on PES is dependent on vagal afferent pathway. Source: Unpublished data supporting an observation made by Li et al.[26]

PYY, calcitonin gene-related peptide (CGRP), NPY, and enkephalins are inhibitors of PES. Again neuronal 5-HT may function in both stimulation and inhibition of PES through its different receptor subtypes. Because the neuropeptides and neurotransmitters distribute widely in the enteric nervous system, the vagus nerve, and intrapancreatic neurons, their physiological roles and mechanism of actions in PES are not yet established.

6.16 NITRIC OXIDE

Nitric oxide synthase (NOS) inhibitors inhibit PES in response to a meal, a lumenal stimulus, or exogenously administered secretin or CCK, thereby establishing the role of NO in PES. Thus, the NOS inhibitor, N$^\gamma$-nitro-L-arginine, inhibits PES in response to duodenal infusion of dilute HCl, 15% casein, ingestion of a meal or IV infusion of physiological doses of secretin or CCK in anesthetized and conscious rats. The inhibition was reversed by the NOS substrate, L-arginine. The NOS inhibitor, however, had no effect on the release of secretin and CCK. Similar results were observed in the dog. Therefore, NO appears to mediate secretin- and CCK-stimulated pancreatic secretion.

Table 6.2 Neuropeptide or Neurotransmitters That Influence Exocrine Pancreas

Stimulants	Inhibitors
Pituitary adenylate cyclase activating polypeptide	Galanin
Gastrin releasing peptide/bombesin	Enkephalins
Vasoactive intestinal polypeptide	Neuropeptide Y/Peptide YY
Substance P	Calcitonin gene related peptide
Neurotensin	Somatostatin
Cholecystokinin	Serotonin (5-HT)
Thyrotropin-releasing hormone	Dopamine
Acetylcholine	
Nitric oxide	
γ-Aminobutyric acid (GABA)	
Serotonin (5-HT)	
Orexin A (Hypocretin I)	
Cocaine- and amphetamine-regulated transcript peptide	

6.17 PITUITARY ADENYLASE-CYCLASE ACTIVATING POLYPEPTIDE

Intravenous administration of PACAP-27 resulted in dose-dependent stimulation of PES and the release of secretin and CCK in rats.[28] In addition, PES stimulated by PACAP-27 is substantially inhibited by intravenous administration of a specific rabbit anti-secretin serum and the CCK-A receptor-specific antagonist, loxiglumide, and is abolished by combination of the anti-secretin serum and the CCK antagonist. These results indicate that PACAP stimulates PES through the release of secretin and CCK, a notion supported by the fact that PACAP stimulates hormone release from secretin- and CCK-producing cells. However, PACAP is also a potent stimulant of insulin release. Because insulin is needed for the action of secretin and CCK, it is likely that PACAP may also act through the release of insulin that permits the actions of secretin and CCK released by PACAP. Nevertheless, the physiological role of the endogenous PACAP remains to be defined.

6.18 VASOACTIVE INTESTINAL POLYPEPTIDE

VIP is widely distributed in central and peripheral nervous systems including the intrapancreatic nerves. Electrical vagal stimulation of isolated and perfused porcine pancreas resulted in the release of VIP and increased

PES both of which were inhibited by infusion of SS and galanin. Thus, VIP appears to mediate PES in response to vagal efferent stimulation. Intravenous infusion of VIP in rats resulted in increase PES of water and bicarbonate, but little protein. Nevertheless, the physiological role of VIP in PES, particularly protein secretion remains to be determined.

6.19 OREXIN A (HYPOCRETIN I)

Orexin A is hypothalamus-specific neuropeptide of secretin/glucagons/VIP superfamily. Intracerebroventricular (ICV), but not intravenous, administration of orexin A resulted in stimulation of pancreatic output of fluid and protein dose-dependently.[29] The stimulatory effect of orexin A was abolished by pretreatment with hexamethonium or atropine, but not by omeprazole, indicating neural mediation through the cholinergic pathway but not through stimulation of gastric acid secretion. Centrally administered orexin probably act through the vagal efferent pathway as it also stimulated vagal efferent nerves in anesthetized rats. The physiological role of orexin in the regulation of pancreatic exocrine secretion, however, remains to be determined.

6.20 GASTRIN RELEASING PEPTIDE/BOMBESIN

GRP or its analog bombesin given intravenously also stimulates the release of secretin and CCK as well as PES. In isolated and perfused rat pancreas, electrical field stimulation resulted in the release of GRP and potentiation of pancreatic fluid and bicarbonate secretion stimulated by secretin. In isolated rat pancreatic lobules, GRP stimulates dose-dependently [^{3}H]-acetylcholine and amylase release with the latter inhibited by tetrodotoxin and atropine. These observations suggest that intrapancreatic GRP-containing nerves participate in regulation of PES through the release of acetylcholine and thus appear to strengthen the latter view. Moreover, immunoneutralization of GRP in dogs inhibited postprandial PES and the release of NT but not CCK, thereby strongly suggesting that GRP does have a regulatory role on PES probably through the release of NT, which is present in pancreatic neurons and acts through a cholinergic pathway.

6.21 THYROTROPIN-RELEASING HORMONE

In contrast to the inhibition of PES by intravenous administration, intra-cereventricular injection of TRH in rats stimulates pancreatic juice flow and enzyme output. This effect is inhibited by central administration of SS that does not inhibit PES in response to the vagal efferent stimulation by 2-deoxy-D-glucose. The effect of TRH is completly abolished by VIP antagonist, [D-p-Cl-Phe6-Leu17]-VIP, strongly suggesting mediation by VIP.

The site of TRH action appears to be the dorsal vagal complex as microinjection of an active TRH analog in other brain stem sites had no effect. The central effect of TRH on PES is abolished by vagotomy or inhibited by atropine indicating mediation via a specific efferent vagal cholinergic pathway. The physiological role of TRH-mediated neural regulation of PES remains to be determined.

6.22   TACHYKININS

Tachykinins are a family of structurally related neuropeptides including substance P, neurokinin A (substance K), and neurokinin B (neuromedin K) as well as several nonmammalian peptides. Mammalian tachykinins are widely distributed in afferent nerve fibers, the enteric nervous system, and the intrinsic nerves of the pancreas. The tachykinin receptors, NK1, NK2, and NK3 are found in the exocrine pancreas suggesting that tachykinins may participate in regulation of PES. Intravenous of mammalian tachykinins in dogs stimulated exocrine secretion of protein and bicarbonate and increased pancreatic blood flow in a dose-dependent manner. The tachykinins-stimulated PES was abolished by cholinergic blockade with atropine, suggesting the involvement of a cholinergic pathway. In isolated rat pancreatic acini, the tachykinins stimulated dose-dependently enzyme secretion and potentiated the stimulatory effect of CCK-8 and urecholine. On the other hand, in isolated vascularly perfused rat pancreas, substance P concentration dependently inhibited CCK- and secretin-stimulated amylase and fluid secretion, respectively. The inhibitory effect of substance P was partially blocked by either NK1 or NK2 receptor antagonists, partially blocked by atropine and substantially blocked by tetrodotoxin. In addition, sensory stimulation by capsaicin in low concentration inhibited the pancreatic secretion stimulated by secretin and CCK, and the inhibition was reversed by the NK1 receptor antagonist, CP-96345. In the pancreas of pig, rat, and chicken, substance P immunoreactive nerve fibers are found around acini, ducts, blood vessels, and the islets. Infusion of capsaicin in isolated and perfused porcine pancreas stimulated the release of both substance P and neurokinin A, whereas electrical vagal stimulation did not. Infusion of substance P or neurokinin A stimulated exocrine secretion, which was blocked specifically by the NK1 receptor antagonist. Despite the species differences regarding the effect of tachykinins on PES, it appears that endogenous substance P/neurokinin A released from sensory nerves participates in regulation of pancreatic secretion through a neural mediated mechanism involving NK1 receptor. Nevertheless, the physiological stimulants and pathways that evoke stimulation of substance P/neurokinin A containing nerves to regulate exocrine pancreatic secretion remains to be determined.

6.23 CALCITONIN GENE-RELATED PEPTIDE

ICV injection of CGRP in anesthetized rats inhibited dose-dependently the vagally-dependent pancreatic protein secretion stimulated by 2-DG (Figure 6.5),[30] hyperosmotic saline, maltose, or CCK-8 in a physiological dose. ICV injection of CGRP had no effect on the pancreatic protein secretion stimulated by electrical vagal efferent stimulation or bethanechol. Although centrally administered CGRP is a potent stimulator of parasympathetic neurons and the release of norepinephrine, the inhibitory effects of ICV-administered CGRP on 2-DG-, maltose-, and CCK-8-stimulated pancreatic protein secretion were not affected by α-adrenergic blockade with phentolamine nor chemical sympathectomy with guanethidine. In addition, ICV CGRP had no effect on CCK-8-stimulated pancreatic protein secretion in chronically vagotomized rats that was dependent on a local enteropancreatic reflex mechanism adapted after vagotomy. Similarly, ICV injection of CGRP inhibited pancreatic secretion and CCK release during bile-pancreatic juice diversion and this effect was not reversed by intravenous infusion of either α– or β–adrenergic antagonists, phentolamine, or propranolol. In isolated guinea pig acini, CGRP stimulates amylase release and binds a specific receptor coupled to adenylate cyclase. Because CGRP-containing nerves are abundant in the pancreas, it is possible that CGRP released from intrapancreatic nerves participate in regulation of PES. The physiological significance of CGRP in neural regulation of PES and the pathway involved remains to be elucidated.

6.24 NEUROPEPTIDE TYROSYLAMIDE

NPY is another member of PP/PYY family of regulatory peptide. NPY is widely distributed in the central and peripheral nervous systems. Its effect on PES is similar to those of PP and PYY. For example, like PP and PYY, intravenous infusion of NPY inhibits PES stimulated by 2-deoxy-D-glucose. In addition, NPY containing neurons are abundant in the pancreas and thus may play a neurocrine regulatory role on PES. For example, in isolated pancreatic lobules, NPY inhibited enzyme secretion stimulated by K^+ deplorization.

6.25 COCAINE- AND AMPHETAMINE-REGULATED TRANSCRIPT PEPTIDE

CART peptide is a newly discovered neuropeptide that is widely distributed in the central and peripheral nervous systems including the vagus nerve and myenteric plexus. Intravenous infusion of CART peptide resulted in stimulation of pancreatic amylase secretion in rats.[31] The stimulatory effect

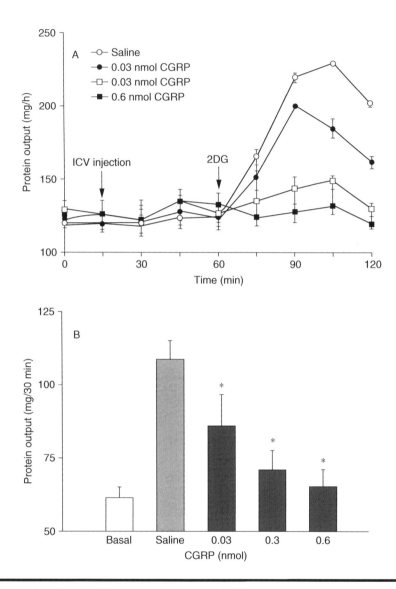

Figure 6.5 Effect of ICV administration of CGRP on 2-deoxy-D-glucose (2-DG)-stimulated pancreatic protein secretion in anesthetized rats. ICV adminstration of CGRP did not affect on basal pancreatic secretion but inhibited dose-dependently 2-DG-stimulated pancreatic protein secretion (panel A). The results of the last 30 min of each study are presented in panel B. * P < 0.01, significant difference vs. saline control. Source: Data reproduced from Li et al.[30] Used with permission.

of CART peptide was partially inhibited by atropine or the CCK-A receptor antagonist, L-364,718 and abolished by combination of the two inhibitors and vagotomy suggesting mediation through the vagus nerve involving both cholinergic- and CCK-dependent mechanisms. Thus, CART peptide is another candidate neuropeptide participating neural regulation of the exocrine pancreas.

6.26 SEROTONIN (5-HT)

Kirchgessner and Gershon[32] have provided histochemical evidence in both rats and guinea pigs that the pancreas is innervated by nerve fibers projected from the stomach and the small intestine. Many of these neurons contain 5-HT. Mucosal application of veratridine in the duodenum activated cytochrome oxidase activity in pancreatic neurons, islets, and acini that were abolished by tetrodotoxin and hexamethonium. Their observation thus confirmed the presence of a cholinergic-dependent enteropancreatic neural reflex mechanism proposed by Singer et al.[33] The enteropancreatic neurons are also abundant in axons that contain 5-HT immunoreactivity and $5-HT_{1A}$ and $5-HT_{1P}$ receptor subtypes. Application of veratridine to isolated rat pancreatic lobules stimulated amylase release and c-*fos* expression in pancreatic neurons, islets, and acinar cells that were inhibited by tetrodotoxin, atropine, 5-HT, and 5-hydroxyindalpine (5-OHIP, a $5-HT_{1P}$ agonist). The effects of 5-HT and 5-OHIP were blocked by the $5-HT_{1P}$ receptor antagonist, N-acetyl-5-hydroxytryptophyl-5-hydroxytryptophan amide (5-HTP-DP). These observations suggested that serotonergic nerves in the pancreas mediate inhibition of pancreatic enzyme secretion via presynaptic receptors on cholinergic neurons and that the inhibitory receptor is of $5-HT_{1P}$ subtype. The physiological role of this neural pathway is yet to be defined. It should be noted that 5-HT containing fibers were found in the walls of blood vessels, ducts, and in the periacinar and periinsular regions of rat pancreas. Administration of 5-HT increased insulin secretion while it inhibited glucagon secretion. Thus, intrapancreatic 5-HT-containing nerves may participate regulation of PES through regulation of insulin and glucagon release.

Li et al.[24] observed that pancreatic protein secretion in response to intraduodenal administration of maltose, a hypertonic saline or light mucosal stroking was abolished by vagal afferent rootlet resection. p-Chlorophenylalanine, an inhibitor of 5-HT synthesis, but not 5,7-hydroxytryptamine, a 5-HT neurotoxin, suppressed pancreatic secretion in response to these luminal stimuli that are not mediated by CCK. The $5-HT_3$ antagonist also markedly inhibited pancreatic secretory response to maltose and hypertonic saline; whereas combination of the $5-HT_3$ antagonist with ketanserin (a $5-HT_2$ antagonist) inhibited the response to

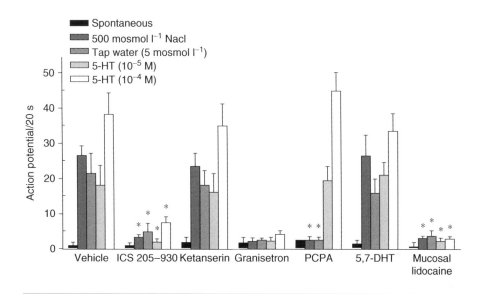

Figure 6.6 **Effect of various 5-HT receptor antagonists and other agents on discharge of nodose ganglia in response to intraluminal osmotic stimulation and to 5-HT.** The responses of nodose ganglia to luminal osmotic stimuli or 5-HT were abolished by mucosal lidocaine, 5-HT$_3$/5HT$_4$ receptor antagonist, ICS 205-930, and 5-HT$_3$ antagonist, granisetron, but not by 5-HT$_{2A}$ antagonist, ketanserin. The response to luminal osmotic stimuli was also abolished by the inhibitor of 5-HT synthesis, p-chlorophenylalanine (PCPA). The 5-HT neurotoxin, 5,7-DHT also had no effect on the responses to mucosal osmotic stimuli and 5-HT. The results suggest that 5-HT released from mucosal enterochromaffin cells stimulate a subset of nodose ganglia through mucosal afferent and 5-HT$_3$ receptors. Source: Data adopted from Zhu et al.[34] Used with permission.

mucosal stroking. A subsequent study demonstrated that luminal carbohydrate or hypertonic saline stimulated discharge of a subset of mucosal vagal afferent fibers recorded from nodose ganglia that were abolished by the 5-HT$_3$/5-HT$_4$ antagonist, tropisetron (ICS 205-930) and 5-HT$_3$ antagonist, granisetron (Figure 6.6).[34] The luminal stimuli-elicited vagal afferent discharges were also abolished by p-chlorophenylalanine, but not by 5,7-hydroxytryptamine. Intraluminal application of 5-HT also elicited vagal afferent discharge from nodose ganglia that was abolished by acute subdiaphragmatic vagotomy, mucosal application of lidocaine, or administration of the 5-HT$_3$ receptor antagonist. It was proposed that these luminal stimuli evoked 5-HT release from EC cells, which acts through 5-HT$_3$ receptor on vagal afferent fiber and a vagal reflex mechanism to stimulate pancreatic protein secretion. These observations together with the aforementioned effects of 5-HT antagonists on acid-stimulated secretin release and pancreatic secretions of fluid and bicarbonate appear to suggest that different luminal stimuli may evoke different serotonergic pathways to regulate pancreatic exocrine secretion.

6.27 ENDOCRINE–EXOCRINE AXIS OF THE PANCREAS

The pancreatic islets are interspersed throughout the pancreatic exocrine tissue, acini, by making a close contact through the islet-acinar portal system (see Chapter 4 and Chapter 9). It was proposed that islet hormones reach the acinar cells at relatively high concentration to influence acinar function. Exogenous insulin has been reported to potentiate CCK- and acetylcholine-stimulated amylase secretion. In conscious rats, immunoneutralization of circulating insulin resulted in a drastic inhibition of the pancreatic exocrine secretion of enzyme, bicarbonate, and fluid stimulated by ingestion of a meal or intravenous infusion of secretin and CCK.[16] In isolated and perfused rat pancreata, PES stimulated by secretin, CCK, or a combination of secretin and CCK was inhibited also by infusion of a specific rabbit anti-insulin serum. The infusion of anti-insulin serum also resulted in elevation of SS in portal venous effluent, and the inhibition of PES by the anti-insulin serum was partially reversed by coinfusion of an anti-SS serum. The same effects of anti-insulin serum were observed in isolated and perfused canine pancreata. In addition to SS, PP was also elevated in the portal effluent by the anti-insulin serum. The inhibition of fluid and bicarbonate secretion by the anti-insulin serum was abolished by coinfusion of both anti-SS and anti-PP serum. On the other hand, anti-SS and anti-PP only partially reversed the inhibition of protein secretion by anti-insulin serum. These results indicate that endogenous insulin as a local or paracrine messenger inhibits the release of both SS and PP to regulate their inhibitory action on secretin and CCK-stimulated PES. Moreover, another mediator (or mediators) in addition to SS and PP may be involved in action of endogenous insulin to regulate protein secretion.

6.28 FEEDBACK REGULATION OF PANCREATIC EXOCRINE SECRETION

In 1972, Green and Lyman[5] reported that diversion of pancreatic juice from the duodenum or intraduodenal infusion of soybean trypsin inhibitor resulted in an increase in pancreatic secretion of proteins. The increase was abolished by the return of pancreatic juice or trypsin. It was proposed that pancreatic proteases play a negative feedback regulatory role on pancreatic exocrine secretion. Subsequent studies have indicated that such feedback regulatory mechanism is operative in several species, including the rat, human, dog, pig, and guinea pig.[35] In rats, pigs, and humans, luminally administered proteases were shown to suppress PES both in the fasting state and the intestinal phase; whereas in dogs and guinea pigs, PES was suppressed only in the intestinal phase stimulated by emulsified oleic acid or sodium oleate or a meal (in dogs). In fasting rats, diversion of pancreatic juice from the duodenum or intraduodenal infusion of a trypsin inhibitor results in a significant increase in pancreatic secretion of protein and bicarbonate that was accompanied by elevation of both CCK and secretin in circulation. The increased pancreatic secretion was attributable to the increased release of CCK and secretin as it was abolished either by administration of a CCK receptor antagonist, an anti-CCK serum, or an anti-secretin serum. Diversion of pancreatic juice in dogs also increased plasma neurotensin level, which was suppressed by intraduodenal infusion of lipase. Moreover, diversion-evoked increase in PES was reduced by immunoneutralization of circulating neurotensin. Similarly, plasma PYY concentration in rats increased after pancreatic juice diversion. Immunoneutralization of circulating PYY in rats resulted in augmentation of PES during pancreatic juice diversion. Thus, the feedback regulatory mechanism appears to involve the release of neurotensin in dogs and PYY in rats as well. Subsequent studies have demonstrated that the effect of pancreatic protease was mediated through degradation of a CCK-releasing peptide activity secreted from the intestinal mucosa. Similarly a secretin-releasing peptide (SRP) activity was shown to mediate the release of secretin in response to duodenal acidification. It is now clear that intestinal acid perfusate of both rats and dogs contain the SRP activity. On the other hand, the inhibitory effect of lipase on neurotensin release is yet to be explained.

A CCK-releasing peptide, termed *monitor peptide,* was isolated from rat pancreatic juice.[36] Intraduodenal administration of monitor peptide in fasting rats with bile and pancreatic juice diversion resulted in a significant increase in pancreatic enzyme secretion and elevation of plasma CCK concentration. The peptide also stimulated the release of CCK from mucosal cells isolated from rat duodenum. The discovery of monitor peptide activity suggested that a positive feedback regulation of CCK release by pancreatic juice exists. Similarly, SRP activity was found in canine pancreatic juice. During the past decade, two CCK-RPs have been purified. Spannagel et al.[37] have purified a luminal CCK-releasing factor (LCRF) from rat intestinal secretion. LCRF is a peptide of 8136.5 Da, whose N-terminal 41 residues have been determined. Intraduodenal administration of LCRF stimulated pancreatic fluid and protein secretion that was inhibited by the CCK-A antagonist, MK-329 (Figure 6.7).[37] In addition, when a partially purified LCRF isolated from rat intestinal secretion was administered intraduodenally, pancreatic secretions of protein in recipient rats were elevated. The CCK-RP activity of the partially purified LCRF was

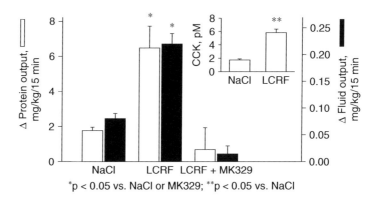

Figure 6.7 Effect of intraduodenal infusion of LCRF on pancreatic secretion and plasma CCK levels in conscious rats. Intraduodenal infusion of LCRF stimulated pancreatic secretion of fluid and protein. The effects of LCRF were abolished by the CCK-A receptor antagonist, MK329 indicating mediation through the release of CCK that was demonstrated by LCRF-stimulated increase in plasma CCK level over the saline control (inset). The results strongly suggest that LCRF is a luminal CCK releasing factor. Source: Data from Spannagel et al.[37] Used with permission.

abolished when it was percolated through an affinity column made of an anti-LCRF$_{1-6}$ serum, but not when that was made of a normal rabbit serum. Synthetic fragments of LCRF, LCRF$_{1-41}$, LCRF$_{1-35}$, and LCRF$_{11-25}$ but not LCRF$_{1-6}$ were also bioactive. Interestingly, intravenous administration of LCRF$_{1-35}$ also stimulated CCK release and pancreatic secretion. These results indicate that LCRF is an active form of CCK-RP in rat intestinal secretion. On the other hand, Herzig et al.[38] have isolated a peptide of 86 amino acid residues with CCK-RP activity from the extracts of porcine small intestine. The amino acid sequence of this peptide was identical to diazepam-binding inhibitor (DBI$_{1-86}$) previously purified by Mutt and coworkers.[39] Intraduodenal administration of a fragment of DBI$_{1-86}$, DBI$_{33-50}$, in anesthetized and atropinized rats resulted in a dose-dependent increase in pancreatic protein secretion and elevation of plasma CCK concentration, although the fragment was less potent than DBI$_{1-86}$. DBI was also shown to stimulate the release of CCK from mucosal I cell (CCK cell)-enriched preparation isolated from rat upper small intestine. Intraduodenal administration of an anti-DBI$_{33-50}$ serum suppressed pancreatic protein secretion and plasma CCK level in response to bile–pancreatic juice diversion as well as intraduodenal infusion of a peptone solution (Figure 6.8).[40] This observation suggests that DBI is also released into the intestinal lumen and contribute to the CCK-RP activity.

Two SRPs have been purified from canine pancreatic juice.[41] Mass spectral analysis indicated that they were both 14 kDa polypeptides.

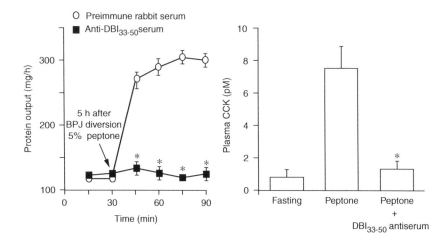

Figure 6.8 Effect of anti-DBI$_{33-50}$ antiserum on pancreatic secretion and plasma CCK concentration in response to duodenal infusion of peptone. Intraduodenal infusion of anti-DBI$_{33-50}$ together with peptone suppressed peptone-stimulated protein output (left panel) and increase of plasma CCK concentration (right panel). The results indicated that a DBI$_{33-50}$-like peptide mediates the stimulatory effect of peptone on pancreatic protein secretion and CCK release, thereby supporting the contention that DBI is a luminal CCK-releasing factor. Source: Data from Li et al.[40] Used with permission.

N-terminal amino acid sequence analysis indicated they were identical or highly homologous to canine pancreatic PLA$_2$. Both of these peptides stimulated the release of secretin from rat secretin-containing cells and murine neuroendocrine cell line, STC-1. Porcine pancreatic PLA$_2$ also occur in multiple molecular forms of 14 kDa and each was shown to stimulate secretin release from secretin-producing cells. Intravenous administration in anesthetized rats of a specific anti-PLA$_2$ serum produced against purified porcine pancreatic PLA$_2$ resulted in suppression of PES and the release of secretin in response to duodenal acidification. Moreover, duodenal acidification increased PLA$_2$-like immunoreactivity in the luminal perfusate. Preincubating the concentrate of acid perfusate (CAP) with anti-PLA$_2$ antiserum followed by removal of the high molecular weight antibody-antigen complex by ultrafiltration diminished the SRP activity of CAP; whereas normal rabbit serum-treated CAP retained the SRP activity (Figure 6.9).[42] These observations indicate that PLA$_2$ is a luminal SRP (LSRP) that participate regulation of secretin release and PES.

The release and action of LSRP are also neural regulated. Thus, SRP activity in CAP prepared from vagotomized or propanolol (an adrenergic antagonist)-treated donor rats was significantly reduced as compared with the corresponding control CAPs. In addition, the stimulation of PES and

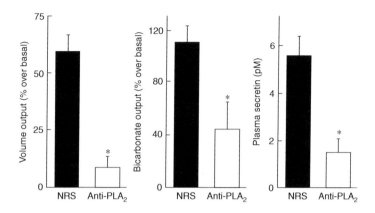

Figure 6.9 **Effect of NRS or anti-PLA2 serum on secretin-releasing peptide activity of a concentrate of duodenal acid perfusate (CAP) in recipient rats. Preincubation of the CAP preparation was preincubated with anti-PLA2 serum, but not with NRS inhibited the SRP activity of CAP. Source: Data from Li et al.[42] Used with permission.**

secretin release by CAP was diminished by pretreatment of the recipient rats with tetrodotoxin, perivagal capsaicin, or vagotomy.[26] Moreover, treatment of donor rats with methionine-enkephalin, which is known to inhibit PES and secretin release, also reduced the SRP activity of CAP. These observations suggest that the release and action of LSRP in response to duodenal acidification is neural mediated and that Met-ENK may be an inhibitory mediator of SRP release. Similarly, the release of CCK-RP is also neural regulated. The release and action of CCK-RP are inhibited by SS. The 5-HT$_3$ receptor antagonist, ISC 205-930, ketanserin (5-HT2 receptor antagonist), a substance P antagonist, CP 96,345, tetrodotoxin, atropine, hexamethonium, and mucosal lidocaine inhibited the release of CCK-RP activity by intraduodenally administered peptone solution. Based on the observation, it was proposed that peptone in the intestinal lumen activates the enterochromaffin cells to release 5-HT, which in turn activate substance-P-containing submucosal sensory neurons.[43] The signal was then transmitted through the interneurons in the myenteric plexus to relay stimulatory signal to the submucosal secretomotor cholinergic neurons to release acetylcholine that acts on the CCK-RP containing cells in the mucosal epithelium to release CCK-RP.

In summary, PES is regulated by neurohormonal regulatory mechanism, involving integrated actions of hormones, neuropeptides, neurotransmitters, and hormone-releasing peptides. Although the roles of several key elements are now known, the mechanisms and pathways of their regulatory processes remain to be elucidated.

6.29 EFFECTS OF TOXIC AGENTS TO THE EXOCRINE PANCREAS

There are some toxic agents that cause damage to the exocrine pancreas resulting in acute or chronic pancreatitis or cancer.[44] The toxic agents include toxic chemicals (see Chapter 22), therapeutic drugs, hormones, and bile salts. Chemicals toxic to the pancreas included alcohol (see Chapter 15), alloxan, streptozotocin, azaserine, dimethylbenzo[α]anthracene, ethionine, methanol, oleic acid, 4-hydroxyaminoquinoline-1-oxide, beta-oxidized derivatives of dipropylnitrosamine, and chloroform that can cause acute pancreatic injury and death or hyperplasia, metaplasia, and malignant transformation. Toxic therapeutic agents that are known to induce acute pancreatitis include azathioprine, estrogens, furosemide, methyldopa, pentamidineprocainamide, sulfonamides, and thiazide diuretics. The immunosuppressants FK506 and cyclosporine A also inhibit pancreatic secretory response to CCK and cause tissue damage in the exocrine pancreas. CCK and its analog cerulein are also toxic to the exocrine pancreas when administered in high dose leading to acute pancreatitis (see Chapter 14). Injection of a bile salt in high dose to the pancreatic duct also caused acute pancreatitis. A similar condition may arise if the common bile duct is obstructed due to tumor growth or blockage by a gallstone.

The mechanism of pancreatic tissue injury may vary with the toxic agents. The effect of ethanol, however, is studied most extensively (see Chapter 15). The cause of pancreatic damage by alcohol may be due to premature activation of digestive enzyme in the acinar cells. It has been shown that alcohol increases the synthesis of digestive enzyme in the pancreas[45] and increases the fragility of the zymogen granules[46] and lysosomes.[47] Premature activation of pancreatic protease by lysosomal cathepsin B may lead to the subsequent tissue damage. Long-term consumption of alcohol increases gene expression of pancreatic cholesterol esterase, ES-10 and fatty acid ethyl ester synthase III in rats[48] that may be the cause of accumulation of cholesteryl esters and fatty acid ethyl esters (FAEEs), ethanol metabolites that increase the fragility of the lysosomes, in the pancreas.[47,49] On the other hand, the fragility of zymogen granules is increased by alcohol through reduced synthesis of the granule membrane protein, GP2. Alcohol also increases production of reactive oxygen species or free radicals. It has been shown that acute alcohol consumption increases the level of lipid peroxidation products in rat pancreas, resulting from reaction of oxygen free radicals with the membrane components. Formation of alcohol toxic metabolites such as acetaldehyde and FAEEs in the pancreas can also increase tissue injury.

Intravenous infusion of ethanol at low dose significantly increased PES in dogs, whereas infusion of ethanol at high dose inhibited PES. The alcohol-stimulated pancreatic protein secretion in the rat appeared to be mediated through the release of a CCK-RP,[50] whereas the increased fluid and bicarbonate secretion was not mediated through the release of secretin. The effects of acute ethanol were abolished by atropine indicating involvement of a cholinergic component. Indeed it was reported that in alcoholic dogs, pancreatic protein secretion became resistant to bethanechol.[51] It was observed in mice that after prolonged (4 months) alcohol consumption, the majority of periacinar nerve terminal exhibited degenerative changes and there were slight decreases in the intensity of VIP and SP immomunoreactivity and near abolishment of PP-containing fibers.[52] The observation also suggested changes in neural regulation. In long-term alcohol-fed dogs, postprandial PES decreased as compared with nonalcoholic dogs. However, secretagogue-stimulated PES after long-term alcohol consumption may vary depending on the extent of tissue damage. The effect of other toxic agents may involve some similar mechanisms of tissue injury. On the other hand, drugs such as alloxan, streptozotocin, and cyclosporine that also affect the endocrine pancreas may also reduce secretin- and CCK-stimulated PES by abolishing insulin secretion (see Chapter 27). Endogenous CCK appears to increase the extent of tissue injury exerted by a toxic agent, as administration of a CCK receptor antagonist often lessens the injury. In addition, drugs abused together often caused more profound pancreatic tissue damage. The best examples are alcohol consumption–cigarette smoking and alcohol–cocaine abuse. It has been reported that cigarette smoke enhanced ethanol-induced pancreatic tissue damage in rats.[53] Nicotine appears to be one of the toxic agents in cigarette smoking as it is known to cause pancreatic tissue damage and inhibits CCK-stimulated amylase secretion from isolated rat pancreatic acini.[54] Cocaine inhibited CCK- and urecholine-stimulated amylase secretion and inhibited protein synthesis in isolated rat pancreatic tissue.[55] The inhibition of protein synthesis by cocaine was also observed *in vivo*. Combination of cocaine with ethanol resulted in a greater extent of inhibition of carbachol- and cerulein-stimulated amylase secretion from isolated guinea pig pancreatic lobules than that exerted by each drug alone. The metabolite of cocaine and ethanol, cocaethylene exhibited a more potent inhibitory effect on amylase secretion than its parent compounds combined.[56] Because cocaethylene is formed *in vivo* when cocaine and alcohol are abused together, it may contribute to pancreatic tissue damage caused by coadministration of the two drugs. Nevertheless, regardless of the mechanism of tissue injury, once pancreatitis is developed, there is a diminished secretory response to secretin and CCK.

REFERENCES

1. Babkin B.P. 1949. *Pavlov: a biography.* Chicago: Chicago University Press.
2. Bayliss M. and Starling B.H. 1902. Mechanism of pancreatic secretion. *J. Physiol. Lond.* 28:325–353.
3. Mutt V., Magnusson S., Jorpes J.E., and Dahl E. 1965. Structure of porcine secretin. I. Degradation with trypsin and thrombin, sequence of the tryptic peptides with C-terminal residue. *Biochemistry* 4:2358–2362.
4. Jorpes E. and Mutt V. 1966. Cholecystokinin and pancreozymin, one single hormone? *Acta Physiol. Scand.* 66:196–202.
5. Green G.M. and Lyman R.L. 1972. Feedback regulation of pancreatic enzyme as a mechanism for trypsin inhibitor-induced hypersecretion in rats. *Proc. Soc. Exp. Biol. Med.* 140:6–12.
6. Chey W.Y., Lee Y.H., Hendricks J.F., Rhodes R.A., and Tai H.H. 1978. Plasma secretin concentration in fasting and postprandial in man. *Am. J. Dig. Dis.* 23:981–988.
7. Schaffalitzky de Muckadel O.B. and Fahrenkrug J. 1978. Secretion pattern of secretin in man. Regulation of gastric acid. *Gut* 19:812–828.
8. Kim M.S., Lee K.Y., and Chey W.Y. 1979. Plasma secretin concentrations in fasting and postprandial states in dog. *Am. J. Physiol.* 236 (*Endocrinol. Metab. Gastrointest. Physiol.* 5):E539–E544.
9. Chey W.Y., Kim M.S., Lee K.Y., and Chang T.M. 1979. Effect of rabbit anti-secretin serum on postprandial pancreatic secretion in dog. *Gastroenterology* 77:1268–1275.
10. Hildenbrand P., Beglinger C., Gyr K., Jansen J.B.M.J., Rovati L., Zuercher M., Lamers C.B.H.W., Setnikar I., and Stalder G.A. 1990. Effect of a cholecystokinin receptor antagonist on intestinal phase of pancreatic biliary responses in man. *J. Clin. Invest.* 85:640–646.
11. Jo Y.H., Lee K.Y., Chang T.M., and Chey W.Y. 1991. Role of cholecystokinin in pancreatic bicarbonage secretion in dogs. *Pancreas* 6:197–201.
12. Eysselein V.E., Eberlein G.A., Hesse W.H., Singer M.V., Goebell H., and Reeve Jr. J.R. 1987. Cholecystokinin-58 is the major circulating form of cholecystokinin in canine blood. *J. Biol. Chem.* 262:214–217.
13. Chang T.M., Thagesen H., Lee K.Y., Roth F.L., and Chey W.Y. 2000. Canine vagus nerve stores cholecystokinin-58 and -8 but releases only cholecystokinin-8 upon electrical vagal stimulation. *Regul. Pept.* 87:1–7.
14. Feurle G.E., Baca I., and Knauf W. 1982. Atropine depresses release of neurotensin and its effect on the exocrine pancreas. *Regul. Pept.* 4:75–82.
15. Nustede R., Schmidt W.E., Kohler H., Folsch U.R., and Schafmayer A. 1993. Role of neurotensin in the regulation of exocrine pancreatic secretion in dogs. *Regul. Pept.* 44:25–32.
16. Lee K.Y., Zhou L., Ren X.S., Chang T.M., and Chey W.Y. 1990. An important role of endogenous insulin on exocrine pancreatic secretion in rats. *Am. J. Physiol.* 258 (*Gastrointest. Liver Physiol.* 21):G268–G274.
17. Lee K.Y., Shiratori K., Chen Y.F., Chang T.M., and Chey W.Y. 1986. A hormonal mechanism for the interdigestive pancreatic secretion in dogs. *Am. J. Physiol.* 251:G759–G764.

18. Shiratori K., Lee K.Y., Chang T.M., Jo Y.H., Coy D.H., and Chey W.Y. 1988. Role of pancreatic polypeptide in the regulation of pancreatic exocrine secretion in dogs. *Am. J. Physiol.* 255:G535–G541.

19. Jin H., Cai L., Lee K.Y., Chang T.M., Li P., Wagner D., and Chey W.Y. 1993. A physiological role of peptide YY on exocrine pancreatic secretion in rats. *Gastroenterology* 105:208–215.

20. Putman W.S., Liddle R.A., and Williams J.A. 1989. Inhibitory regulation of rat exocrine pancreas by peptide YY and pancreatic polypeptide. *Am. J. Physiol.* 256:G698–G703.

21. Demar A.R., Taylor I.L., and Fink A.S. 1991. Pancreatic polypeptide and peptide YY inhibit the denervated canine pancreas. *Pancreas* 6:419–426.

22. Gershon M.D. 1981. The enteric nervous system. *Ann. Rev. Neurosci.* 4:227–272.

23. Li J.P., Chang T.M., and Chey W.Y. 2001. Roles of 5-HT receptors in the release and action of secretin on pancreatic secretion in rats. *Am. J. Physiol. Gastrointest. Liver Physiol.* 280:G595–G602.

24. Li Y., Hao Y., Zhu J., and Owyang C. 2000. Serotonin released from intestinal enterochromaffin cells mediated luminal non-cholecystokinin-stimulated pancreatic secretion in rats. *Gastroenterology* 118:1197–1207.

25. You C.H., Rominger J.M., and Chey W.Y. 1982. Effects of atropine on the action and release of secretin in humans. *Am. J. Physiol.* 242:G608–G611.

26. Li P., Chang T.M., and Chey W.Y. 1995. Neuronal regulation of the release and action of secretin-releasing peptide and secretin. *Am. J. Physiol.* 269:G305–G312.

27. Li Y. and Owyang C. 1993. Vagal afferent pathway mediates physiological action of cholecystokinin on pancreatic enzyme secretion. *J. Clin. Invest.* 92:418–424.

28. Lee S.T., Lee K.Y., Li P., Coy D., Chang T.M., and Chey W.Y. 1998. PACAP stimulates pancreatic secretion via the release of secretin and cholecystokinin (CCK) in rats. *Gastroenterology* 114:490–497.

29. Miyasake K., Masuda M., Kanai S., Sato N., Kurosawa M., and Funakoshi A. 2002. Central Orexin-A stimulates pancreatic exocrine secretion via the vagus. *Pancreas* 25:400–404.

30. Li Y., Jiang Y.C., and Owyang C. 1998. Central CGRP inhibits pancreatic enzyme secretion by modulation of vagal parasympathetic outflow. *Am. J. Physiol.* 275:G957–G963.

31. Cowles R.A., Segura B.J., and Mulholland M.W. 2001. Stimulation of rat pancreatic exocrine secretion by cocaine- and amphetamine-regulated transcript peptide. *Regul. Pept.* 99:61–68.

32. Kirchgessner A.L. and Gershon M.D. 1990. Innervation of the pancreas by neurons in the gut. *J. Neurosci.* 10:1626–1642.

33. Singer M.V., Solomon T.E., Wood J., and Grossman M.I. 1980. Latency of pancreatic enzyme response to intraduodenal stimulants. *Am. J. Physiol.* 238:G23–G29.

34. Zhu J.X., Zhu X.Y., Owyang C., and Li Y. 2001. Intestinal serotonin acts as a paracrine substance to mediate vagal signal transmission evoked by luminal factors in the rat. *J. Physiol. Lond.* 530:431–442.

35. Chey W.Y. 1993. Hormonal control of pancreatic exocrine secretion. In *The pancreas: biology, pathobiology, and disease.* Go V.L.W et al. Eds. 2nd ed. New York: Raven Press. pp. 403–424.

36. Iwai K., Fukuoka S., Fushiki T., Tsujikawa M., Hirose M., Tsunasawa S., and Sakiyama F. 1987. Purification of a trypsin-sensitive cholecystokinin-releasing peptide from pancreatic juice. *J. Biol. Chem.* 262:8956–8959.

37. Spannagel A.W., Green G.M., Guan D., Liddle R.A., Faull K., and Reeve Jr. J.R. 1996. Purification and characterization of luminal cholecystokinin-releasing factor from rat intestinal secretion. *Proc. Natl. Acad. Sci. USA* 93:4415–4420.

38. Herzig K.H., Schon I., Tatemoto K., Ohe Y., Li Y., Folsch U.R., and Owyang C. 1996. Diazepam binding inhibitor is a potent cholecystokinin-releasing peptide in the intestine. *Proc. Natl. Acad. Sci. USA* 93:7927–7932.

39. Chen Z.W., Agerberth B., Gell K., Anderson M., Mutt V., Ostenson C.G., Efendic S., Barros-Sderling J., Persson B., and Jornvall H. 1988. Isolation and characterization of porcine diazepam-binding inhibitor. A polypeptide not only of cerebral occurrence but also common in intestinal tissues and with effects on regulation of insulin release. *Eur. J. Biochem.* 174:239–245.

40. Li Y., Hao Y., and Owyang C. 2000. Diazepambinding inhibitor mediate feedback regulation of pancreatic secretion and postprandial release of cholecystokinin. *J. Clin. Invest.* 105:351–359.

41. Chang T.M., Lee K.Y., Chang C.H., Li P., Song Y., Roth F.L., and Chey W.Y. 1999. Purification of two secretin-releasing peptides structurally related to phospholipase A$_2$ from canine pancreatic juice. *Pancreas* 19:401–405.

42. Li J.P., Chang T.M., Wagner D., and Chey W.Y. 2001. Pancreatic phospholipase A2 from the small intestine is a secretin-releasing factor in the rat. *Am. J. Physiol. Gastrointest. Liver Physiol.* 581:G526–G532.

43. Li Y. and Owyang C. 1996. Peptone stimulated CCK-releasing peptide secretion by activating intestinal submucosal cholinergic neurons. *J. Clin. Invest.* 97:1463–1470.

44. Scarpelli D.G. 1989. Toxicology of the pancreas. *Toxicol. Appl. Pharmacol.* 101:543–554.

45. Apte M.V., Wilson J.S., McCaughan G.W., Korsten M.A., Haber P.S., Norton I.D., and Pirola R.C. 1995. Ethanol-induced alterations in messenger RNA levels correlate with glandular content of pancreatic enzymes. *J. Lab. Clin. Med.* 125:634–640.

46. Haber P.S., Wilson J.S., Apte M.V., Korsten M.A., and Pirola R.C. 1994. Chronic ethanol consumption increases the fragility of rat pancreatic zymogen granules. *Gut* 35:1474–1478.

47. Wilson J.S., Apte M.V., Thomas M.C., Haber P.S., and Pirola R.C. 1992. The effects of ethanol, acetaldehyde and cholesteryl esters on pancreatic lysosomes. *Gut* 33:1099–1104.

48. Pfutzer R.H., Tadic S.D., Li H.S., Thompson B.S., Zhang J.Y., Ford M.E., Eagon P.K., and Witcomb D.C. 2002. Pancreatic cholesterol esterase, ES-10, and fatty acid ethyl ester synthase III gene expression are increased in the pancreas and liver but not in the brain or heart with long-term ethanol feeding in rats. *Pancreas* 25:101–106.

49. Haber P.S., Wilson J.S., Apte M.V., and Pirola R.C. 1993. Fatty acid ethyl esters increase rat pancreatic lysosomal fragility. *J. Lab. Clin. Med.* 121:759–764.

50. Saluja A.K., Lu L., Yamaguchi Y., Hofbauer B., Runzi M., Dawra R., Bhatia M., and Steer M.L. 1997. A cholecystokinin-releasing factor mediates ethanol-induced stimulation of rat pancreatic secretion. *J. Clin. Invest.* 99:506–512.

51. Schmidt D.N., Devaux M.A., Biedzinski T.M., and Sarles H. 1983. Cholinergic stimulation and inhibition of pancreatic secretion in alcohol-adapted dogs. *Scand. J. Gastroenterol.* 18:425–431.
52. Berger Z. and Feher E. 1997. Degeneration of intrapancreatic nerve fibers after chronic alcohol administration in mice. *Int. J. Pancreatol.* 21:165–171.
53. Hartwig W., Werner J., Ryschich E., Mayer H., Schmidt J., Gebhard M.M., Herfarth C., and Klar E. 2000. Cigarette smoke enhances ethanol-induced pancreatic injury. *Pancreas* 21:272–278.
54. Chowdhury P., Doi R., Tangoku A., and Rayford P.L. 1995. Structural and functional changes of rat exocrine pancreas exposed to nicotine. *Int. J. Pancreatol.* 18:257–264.
55. Hamel E. and Morisset J. 1978. Effects of cocaine on rat pancreatic enzyme secretion and protein synthesis. *Am. J. Dig. Dis.* 23:264–268.
56. Linari G., Antonille L., Nencini P., and Nucerito V. 2001. Ethanol combined with cocaine inhibits amylase release in guinea pig pancreatic lobules. *Pharmacol. Res.* 44:41–45.

7

DRUG-METABOLIZING
ENZYMES IN THE HUMAN
PANCREAS

Jens Standop, Alexis Ulrich, and Parviz M. Pour

CONTENTS

7.1   INTRODUCTION

Every living cell of the organism is exposed to a variety of substances, some of which are harmful by causing acute, subacute, or chronic functional failure, mutation, or teratogenesis. Although the toxicity of several compounds and natural products have been known for centuries and some, such as arsenic, have been used as poisons, the carcinogenicity of chemicals was first described in 1775 by Sir John Perivall Pott,[1] who observed scrotal cancer in chimneysweepers in the U.K. Since then, the tumor producing effect of many chemicals was shown primarily in laboratory animals. Tissues, which are mostly affected, include liver, esophagus, lung, nasal mucosa, urinary bladder, tongue, stomach, and pancreas.[2] A relationship has been found between the dose of the substance, its

139

chemical structure, and the affected tissue.[3] The simplest nitrosamine, nitrosodimethylamine (NDMA), for example, exhibits a particular affinity for the liver of animals and humans,[4] N-nitrosomethylbenzylamine (NMBA) is widely employed to study esophageal cancer in rats, and N-nitrosobis(2-oxopropyl)amine (BOP) is used for induction of pancreatic adenocarcinoma in hamsters.[2,5] Exposure to high concentrations of toxic and carcinogenic substances often occurs in working environments (e.g., in the rubber, leather, and metal industries), but can also occur by certain dietary practices (e.g., charbroiled red meat) and the use of cosmetics and tobacco.[2] The association between cigarette smoke with bronchial and pancreatic cancer has been well established.[6] Moreover, some therapeutic drugs, such as chemotherapeutic agents, can have considerable toxic and carcinogenic properties. Naturally occurring products, such as streptozotocin, isolated as an antibiotic and antitumor agent from Streptomyces achromogenes, has been shown to have a diabetogenic effect by selectively destroying insulin-producing islet cells, as well as neoplastic effects by inducing tumors of the pancreas, kidney, and liver.[7]

During their evolution, living cells and organisms have developed sophisticated defense mechanisms against potentially harmful substances, called *drug-metabolizing enzymes,* which are made up of a biological system for the deactivation of a wide variety of environmental chemicals (xenobiotics).[8] Drug-metabolizing enzymes are responsible for the oxidative, peroxidative, and reductive metabolic transformation of drugs, environmental chemicals, dietary products, and natural exogenic or endogenic (e.g., steroids) compounds. Many of these enzymes are inducible. An increasing supply of substrate is followed by a higher concentration of the metabolizing enzyme.[9] Although the enzymes generally convert xenobiotics to less toxic, more water-soluble products, the reactions frequently lead to the formation of reactive intermediates or allow the leakage of free radicals with toxic properties.[9–13] In mammals, Phase I drug-metabolizing enzymes (e.g., cytochrome P450 mono-oxygenases (CYP)) can be identified in nearly every tissue and more than 500 CYP enzymes have now been identified and characterized. The metabolism of synthetic chemicals and natural products, however, is mainly governed by about 20 CYP enzymes, which belong to CYP Family 1 through Family 4.[9,13–17] Exogenous (xenobiotics) and endogenous compounds that are substrates for the drug-metabolizing enzymes can increase the oxidative load of the tissue and induce chronic inflammation,[18,19] mutation, and malignant transformation.[8,20,21]

Prime examples for CYP-substrates are polycyclic aromatic hydrocarbons (CYP 1A1 and CYP 3A4), heterocyclic amines and caffeine (CYP 1A2), and the tobacco-specific carcinogen 4-(Methylnitrosamino)-1-(3-pyridyl)-1-butanone, better known as NNK (CYP 2D6); CYP 2B6 and 2E1 are among the most possible candidates for the metabolism of carcinogenic nitrosamines (Chapter 22).[13,22–25] The activation of anticancer drugs also depends on CYP-mediated metabolism (e.g., cyclophosphamide: CYP 2B6; paclitaxel: CYP 2C8/9; ifosfamide: CYP 3A4).[8,9,13,20] The NADPH cytochrome P450 oxido-reductase (NA-OR) is an indispensable cofactor for all forms of CYP enzymes involved in xenobiotic metabolism. Up until now, CYP 1A1, CYP 1A2, CYP 2B6, CYP 2C8/9/19, CYP 2D6, CYP 2E1, and CYP 3A4 were among the most frequently investigated Phase I drug-metabolizing enzymes of the human pancreas.[26–30]

CYP metabolism usually converts the parent compound into a form that is either more reactive or becomes a substrate for conjugation by the Phase II enzymes, which are composed of numerous transferases, including the super family of glutathione S-transferases (GSTs).[10,21,31] GSTs catalyze the conjugation of substrates, often the product of Phase I metabolism, with reduced glutathione (GSH) to give a glutathione-substrate (GS)-conjugate. The GS-conjugate is then excreted from the cell either passively or by one of the excretory mechanisms such as the multidrug resistance protein.[32–36] Four main classes of GST are known, α (A), μ (M), π (P), and θ (T), but more than 90% of the GST content is represented by GST-α, μ, and π.[21,37–39]

Studies on the expression and pathophysiological significance of drug-metabolizing enzymes have been mainly conducted in laboratory animals.[3,40] Recent research regarding human subjects, however, has concentrated on the CYP expression in organs particularly exposed to environmental pollutants such as skin, gut, respiratory tree, and the liver. The role of drug-metabolizing enzymes in the pathogenesis of pancreatic diseases, on the other hand, has received less attention. Although the anatomic position of the pancreas deep in the abdomen makes the immediate threat of toxins and carcinogens to this tissue unlikely, potentially harmful substances can reach the organ via two paths — the arterial blood stream and bile reflux.

Most pancreatic diseases, including pancreatitis and pancreatic cancer, both with fatal consequences, are believed to be caused by internal or external environmental toxins. Acute and chronic pancreatitis are linked to the consumption of alcohol and environmental substances,[18,41–43] and pancreatic cancer is strongly associated with the exposure to environmental carcinogens.[44–46] The susceptibility to pancreatic diseases, as well as their extent and severity, seem to depend on the integrity of the cellular detoxification process, which is governed by drug-metabolizing enzymes.

7.2 PHASE I METABOLIZING ENZYMES — CYTOCHROME P450s

CYPs belong to the Phase I biotransforming enzymes and metabolize lipophilic substances, depending on the nature of the parent substrate, to toxic or nontoxic products.[19,47] All of the presented enzymes have been linked to the activation of various nitrosamines and of a broad spectrum of xenobiotics, including carcinogens and therapeutic agents.[13,17,20,22,23,48] The expression of CYP enzymes has been studied immunohistochemically and with molecular biological methods in cultured human ductal and islet cells and in tissue specimens from the normal pancreas, chronic pancreatitis, and pancreatic cancer. In 1985, Acheson et al. examined the pharmacokinetics of antipyrine and of theophylline, which are validated probes for CYP activities, in a series of patients with chronic pancreatitis and pancreatic cancer.[49] The half-life of each drug was significantly lower, and its clearance faster, in patients than in controls and this pattern was detected in the subgroups with acute pancreatitis, chronic pancreatitis, or pancreatic cancer. Their data suggest an induction of CYPs in all forms of exocrine pancreatic diseases. The enzyme induction was unlikely to be secondary to pancreatic malfunction as there was no correlation between prevailing exocrine status, as assessed by secretin-pancreozymin tests, and the half-life or clearance of either drug. These results were supported by data from individuals with tropical chronic pancreatitis.[50]

As stated above, metabolic activation is a prerequisite for the carcinogenic effect of many carcinogens, and considerable interindividual variations exist in the metabolic capacity to activate the procarcinogens. These variations could be due to polymorphisms of CYPs responsible for the activation of carcinogens. For example, polymorphisms of the CYP 1A1, CYP 2D6, and possibly CYP 2E1 are known to be related to increased susceptibility to smoking related Kreyberg Type I lung cancer.[51] Therefore, in a study by Lee et al. the relationship of genetic polymorphisms of CYPs to the susceptibility to pancreatic cancer, another smoking-related malignancy,[6] was investigated.[52] However, no association of CYP polymorphism with increased susceptibility to pancreatic cancer was found. Two other groups of investigators confirmed that there was no evidence for an interaction between CYP 1A1 polymorphism and exocrine pancreatic cancer.[53,54] Conversely, according to Li et al., polymorphism of the CYP 1A1 gene in pancreatic cancer patients was significantly associated with the level of aromatic DNA adducts,[55] which is considered an early step in carcinogenesis.

Maruyama et al. examined genotype patterns of CYP 2E1, together with two alcohol dehydrogenases (see below), in patients with chronic alcoholic pancreatitis who were diagnosed in general hospitals all over Japan in comparison with chronic nonalcoholic pancreatitis patients or in alcoholics with normal pancreatic function. Additionally, the relationship between pancreatic fibrosis or pancreatitis and genotypes of alcohol-metabolizing enzymes in autopsy cases of alcoholics was assessed.[56,57] No significant differences in the distribution of genotypes of CYP 2E1, however, were found in patients with chronic pancreatitis, alcoholic pancreatitis, or pancreatic fibrosis. It has to be said that all specimens were obtained from the Asian population with a different expression of alcohol-metabolizing enzymes.

One of the first systematical investigations regarding the cellular localization of CYPs in the normal and diseased pancreas came from a research group in Manchester, U.K. In 1993, Foster et al. studied the expression of CYP 1A2, CYP 2E, CYP 3A1, and the NA-OR in liver and pancreas samples from organ donors compared with the enzyme expression in patients with chronic pancreatitis and pancreatic cancer.[30] In the donor specimens, they found comparable enzyme levels in hepatocytes and pancreatic acinar cells. In chronic pancreatitis and pancreatic cancer specimens, on the other hand, CYP enzyme levels were greater in both the liver and the pancreas than in the donors' samples. Remarkably, the islets of Langerhans showed high levels of CYP 1A2 in the donor group. Moreover, a clear induction of CYP 3A1 and NA-OR was found in islet cells of patients with chronic pancreatitis.[30]

The role of CYPs in the pathogenesis of chronic pancreatitis was also examined by Wacke et al., who compared the expression of CYP 1A1, CYP 1A2, CYP 2C9, CYP 2E1, and CYP 3A in the normal pancreas and specimens obtained from diseased patients by immunohistochemistry.[29] In the normal pancreas they found a weak to moderate expression of all five enzymes in up to 50% of ductal epithelia, acinar cells, and the islets of Langerhans. In contrast, in chronic pancreatitis cases with up to 100% upregulation was observed in ductal and acinar cells of some cases. This upregulation in diseased specimens was especially pronounced for CYP 2C9 and CYP 2E1. Contrary to Foster and colleagues, who used antibodies raised against enzymes from rat liver, Wacke used recombinant human CYP enzymes as the antigen source and, moreover, the capacity of microsomal preparations to metabolize verapamil was also assayed. Verapamil is among the synthetic substrates available for the assessment of cytochrome-mediated metabolism. Although lower than in liver microsomes, the oxidative capacity of microsomal preparations from chronic pancreatitis was higher than that of preparations obtained from control tissues.

Our more detailed studies on the CYP expression in the human pancreas was published recently.[26–28] In this study, normal human pancreatic specimens and cultured human ductal and islet cells were analyzed by immunohistochemistry, Western blot, or reverse transcription polymerase chain reaction (RT-PCR) for the expression of CYPs, which are known to be associated with the metabolism of various toxins and carcinogens.[26] Remarkable differences in the cellular distribution of CYP enzymes were found between the individuals and between different pancreatic cells in the same individual. Non-diabetics expressed more of the enzymes than diabetics, females more than males, younger more than older individuals, and organ donors (all young individuals) more than deceased autopsy specimens in deceased. CYP 2B6 was significantly more often expressed in the pancreas of organ donors than autopsies in deceased. Most of the enzymes were localized in islet cells and were often restricted to, or expressed in a higher concentration, in glucagon or pancreatic polypeptide cells. Furthermore, a different cellular localization of the enzymes was found in some individuals (e.g., cytoplasmic vs. Golgi pattern of staining and a frequent nuclear localization of CYP 2E1 in females).

The unexpected high concentration of many CYP enzymes in islet cells prompted us to compare the cellular localization of the enzyme in an anatomically different portion of the pancreas [i.e., tissues derived from the ventral Anlage (head: PP-rich) and dorsal Anlage (corpus and tail: PP-poor areas)].[28] Strikingly, more islets in the head region expressed CYP 2B6, CYP 2C8/9/19, CYP 2E1, and the NA-OR, than those in the body and tail. Moreover, the expression of CYP 2B6 and CYP 2E1 was restricted to the PP-cells, and the concentration of CYP 3A4 was stronger in PP-cells than in other islet cells. Based on the results, it was suggested that the predominant expression of these enzymes, especially of CYP 2E1 in the PP cells, which are frequently incorporated into the ductal epithelium, may explain the frequent development of pancreatic cancer in the head of the pancreas.

In another study,[27] the cellular localization of the above drug-metabolizing enzymes was analyzed by immunohistochemistry in tissue specimens from the normal pancreas, chronic pancreatitis, and pancreatic cancer. Compared to the normal pancreas, a higher expression of CYP 1A2, CYP 2B6, CYP 2C8/9/19, CYP 2D6, and NA-OR was noticed in chronic pancreatitis, and of CYP 1A1, CYP 2B6, CYP 2C8/9/19, CYP 2D6, CYP 2E1, CYP 3A4, and NA-OR in pancreatic cancer. On the other hand, a weaker expression of CYP 1A1 and CYP 2E1 was shown in specimens derived from chronic pancreatitis patients.

The expression of drug-metabolizing enzymes was also investigated by us in cultured human islet cells by using immunohistochemistry, confocal microscopy, Western blot, and RT-PCR.[26,58–60] The results were comparable with the *in vivo* studies. The antibody staining was generally of a moderate intensity and of a diffuse cytoplasmic type. CYP 2E1 was expressed in a moderate staining intensity in the cytoplasm but, as shown in whole tissue specimens, with a stronger intensity in the nuclei of most of the cells. The predominant nuclear-localization of this enzyme was confirmed by confocal microscopy.[26]

The capacity of islet cells to metabolize toxins and carcinogens was further investigated *in vitro* in a mutual study conducted at the Eppley Institute for Research in Cancer and Allied Diseases, Omaha, NE; the

American Health Foundation, Valhalla, NY; and the Department of Surgery, University of Ulm, Germany.[60] Because cigarette smoking is causatively associated with pancreatic cancer and significantly higher levels of the carcinogenic tobacco-specific nitrosamine, NNK, have been detected in the pancreatic juice of smokers as opposed to nonsmokers,[61] the ability of human pancreatic microsomes, cytosol, and islet and ductal cells to metabolize NNK was analyzed. In all preparations, metabolites derived from α-hydroxylation and from carbonyl reduction were detected. Among metabolites derived from α-hydroxylation, keto alcohol was the major metabolite produced by islet cells; this compound was not detected in NNK-treated ductal cells. These findings and the evidence for the presence of high concentrations of the aforementioned drug-metabolizing enzymes in the endocrine pancreatic cells[26–28,30,59,62] clearly suggest that islet cells are the primary sites of CYP-enzyme induction and prime targets of blood-borne toxins and carcinogens.

7.3 PHASE II METABOLIZING ENZYMES — GLUTATHIONE S-TRANSFERASES

The fate of the active intermediates of the Phase I reaction is governed by the Phase II detoxifying enzymes, like GSTs, in conjugation with glutathione or reduction reactions, resulting in less toxic and readily excreteable metabolites.[32,34,36] It is assumed that GSTs are one of the major detoxification systems that protect cells from cytotoxic and carcinogenic insults.[34,47] Therefore, the extent and duration of bioactivated toxic or carcinogenic compounds in an organ depend on the interplay of both biotransforming enzyme systems. Clearly, the yield of these metabolites would be higher if cells were rich in Phase I bioactivating enzymes, but poor on detoxifying Phase II enzymes.

As with Phase I metabolizing enzymes, genetic variation in Phase II enzymes metabolizing pancreatic carcinogens, such as heterocyclic and aromatic amines, could determine the risk of developing pancreatic diseases. Again, as with Phase I metabolizing enzymes, the reported data on polymorphisms and cancer susceptibility are contradictory. In a case control study by Bartsch et al., whole blood cell samples collected from Caucasian patients with pancreatic cancer, nonalcoholic and alcoholic pancreatitis, and from asymptomatic control subjects were analyzed for the effect of genetic polymorphism of different Phase II metabolizing enzymes on the risk of pancreatic diseases (cancer, pancreatitis).[63] A significant overrepresentation of the GSTM1 AB or GSTM1 B genotype was found in all pancreatic disease cases combined. Only a modest increase in susceptibility to pancreatic diseases could be found to be associated with the polymorphism of GSTM1 enzymes. On the contrary, a prospective study of 149 unselected incident cases of pancreatic adenocarcinoma and 146 ethnically matched controls by Liu et al. found no associations between GSTM1 polymorphism and pancreatic cancer susceptibility.[54] In a population-based case-control study, conducted in six San Francisco Bay area counties from 1994 to 2001, an association between homozygous deletions of two GST genes, GSTM1 and GSTT1; smoking; and pancreatic cancer was analyzed.[53] In this population, homozygous deletions of the GSTM1 and GSTT1 genes alone did not affect the risk of pancreatic cancer. An interaction between the GSTT1-null genotype and cigarette smoking was observed, especially among women. Therefore, in this subpopulation, the combination of a deletion polymorphism in GSTT1 and heavy smoking might be associated with an increased susceptibility of pancreatic cancer. It should be noted that subjects for this study were exclusively of the Caucasian population and that ethnic and genetic factors are known to substantially influence the expression pattern of drug-metabolizing enzymes.[10,63,64]

In 1994, Collier and coworkers systematically investigated the immunohistochemical expression of GST-α, GST-μ, and GST-π in specimens from the normal pancreas and from patients with pancreatic cancer.[65] Anti-GST-μ was not immunoreactive either in normal or cancer specimens. Malignant cells expressed GST-π alone in more than half of the cancer specimens. In only 8% of the malignancies, GST-π and GST-α were expressed in parallel; in another 8%, GST-α was exclusively expressed. In healthy control specimens, the anti-GST-π antibody was immunoreactive in intralobular ducts and centroacinar cells, and large ducts expressed GST-π and GST-α in parallel. Acinar cells expressed only GST-α. From these results, Collier et al. came to the conclusion that the GST expression could possibly reflect the cells of origin, suggesting that the tumor arises from the centroacinar cells or intralobular ducts, a possibility that was already discussed in the early 1980s.[66–68]

In one of the most recent studies, Verlaan et al. investigated whether polymorphisms in the GSTM1, GSTT1, and GSTP1 genes modify the risk for chronic pancreatitis.[69] In this study, the rates of GSTT1 and GSTP1 genotypes did not differ between chronic pancreatitis patients and healthy controls. GSTM1 null genotypes, on the other hand, were significantly less common in patients with chronic pancreatitis induced by alcohol as compared to alcoholics without pancreatitis and normal controls. The authors concluded that GSTM1 null alcohol users are less susceptible to chronic pancreatitis.

Furthermore, the cellular localization of the most frequently expressed GSTs — α, μ, and π — was studied in the normal pancreas and in tissue samples from primary chronic pancreatitis, secondary chronic pancreatitis resulting from ductal obstruction by cancer, and pancreatic cancer patients.[47,70] Ductal and ductular cells of all three groups as well as cancer cells expressed GST-π. Acinar cells were not stained in any specimen. A various number of islet cells was strongly immunoreactive. Significantly more islet cells expressed GST-π in chronic pancreatitis specimens than in the normal controls or pancreatic cancer specimen. Interestingly, the GST-π expression in islet cells was restricted to glucagon cells, however, not all glucagon cells expressed the enzyme. To examine whether the overexpression or deficiencies of GST-π are compensated for by other GST isozymes, the expression of GST-α and GST-μ was investigated in the same specimens.[70] Compared to the normal pancreas, the expression of all three GSTs was higher in the ductal, acinar, and islet cells of primary chronic pancreatitis, but conversely, it was lower in the secondary chronic pancreatitis. Unlike the GST-π, the distribution of GST-α and GST-μ within islets did not show marked differences between the three groups. The reactivity of cancer cells to GST-α and GST-π antibodies was similar to that of ductal cells in the normal pancreas, chronic pancreatitis, and secondary chronic pancreatitis, but 55% of cancers did not express GST-μ, suggesting an association between pancreatic cancer and the lack of GST-μ expression. The lack of GST-μ expression is in line with findings in other cancers.[71] Contrary to previous studies,[65,72,112] an expression of GST-α was found in islet cells. The reason for the differing results could be the small number of islet cells expressing GST-α (up to 10%).

Contrary to the findings on Phase I metabolizing enzymes, which were more frequently expressed in islet cells of the head, GST-μ and GST-π were expressed primarily in islets of the body and tail. GST-α was expressed in islet cells of one-half of the specimens, regardless of their anatomic origin.[28] As with GST-π, the expression of GST-α within all islets was restricted to α-cells; the GST-μ antibody had no preference for a special endocrine cell type.[47,70]

The fourth class of enzymes, GST-θ, was also investigated by Foster et al.[30] Contrary to the findings on CYPs, GST levels of the liver and the pancreas were similar between healthy individuals and pancreatic cancer patients. In specimens from patients with chronic pancreatitis, elevated GST-θ levels were found in the islets of Langerhans.

Recently Trachte et al. investigated abnormalities in gene expression in ductal adenocarcinoma of the pancreas in comparison with the normal pancreas by using cDNA arrays and immunohistochemistry.[74] While evaluating approximately 2000 genes, they found, contrary to the results of Ulrich et al., an upregulation of GST-π in all investigated adenocarcinomas.

7.4 MISCELLANEOUS DRUG-METABOLIZING ENZYMES

Although CYPs and GSTs belong to the most important drug-metabolizing enzymes,[13,21] some further drug-metabolizing enzymes with pathophysiological consequences for the pancreas should be briefly mentioned just to highlight the complexity of the drug-metabolizing system (refer to Chapter 8 and Chapter 22).

The microsomal epoxide hydrolase is involved in the first-pass metabolism of highly reactive epoxide intermediates and acts coordinately with CYPs (e.g., CYP 1A1 and CYP 1A2) to inactivate deleterious polycyclic hydrocarbon oxides and epoxides. The enzyme can also convert inactive diols to highly toxic and carcinogenic metabolites, thus exhibiting the same dual role of procarcinogen detoxification and activation found in CYPs.[21,75,76] Only a few studies on the expression of epoxide hydrolase in the human pancreas are published. In one study, the expression of epoxide hydrolase was found only in acinar cells.[77]

N-Acetyltransferases (NATs) make up another group of drug-metabolizing enzymes. NAT1 and NAT2 are polymorphic and have various xenobiotics with primary aromatic amine or hydrazine structures as their substrates.[78–80] As with GSTM1, Bartsch et al. also found a possible association of NAT1 polymorphism with a modest increase in susceptibility to pancreatic diseases.[63] Anderson et al. analyzed pancreatic tissue samples from smokers and nonsmokers for their content of a variety of drug-metabolizing enzymes. They detected low levels of epoxide hydrolase, cytosolic sulfotransferase, and of a variety of CYP enzymes (see Section 7.2). However, the O-acetyltransferase activity of pancreatic cytosols was high, about two-thirds of the levels measured in the human colon. Moreover, cytosols predominantly expressed NAT1.[81]

NAD(P)H:quinone oxido-reductase, an important enzyme in the activation and detoxification pathways of quinone compounds, sulfotransferases, uridine diphosphate, and the flavin-containing monooxygenases, which are polymorphic,[21] have been rarely or never described in the human pancreas.

7.5 PROBLEM AREAS

The pancreas is one of the most important tissues for mammals. The healthy tissue guarantees normal digestive function including the blood glucose stability. Its diseases have severe consequences including pain and death. Toxic substances, including alcohol and many yet unknown environmental xenobiotics, generally are the cause of these pancreatic diseases. Chronic ethanol consumption and exposure to environmental pollutants have been implicated in the etiology of acute and chronic pancreatitis. Numerous studies have found an increased risk for pancreatic cancer among smokers and smokeless tobacco users and certain occupational groups.[18,30,41,44–46,82–84] Various forms of the toxins require metabolic activation by CYPs; however, the CYP-mediated metabolism of some substances like ethanol (CYP 2E1) and caffeine (CYP 1A2) usually results in their detoxification or inactivation.[8,9,13,20,85] However, this detoxification depends on the availability and capability of the Phase II conjugating system, like GSTs. The depletion of these enzymes results in a higher concentration and prolonged effect of CYP-activated toxic substances.

The half-life of most of these activated intermediates is estimated to be 1 to 10 seconds.[86] Therefore, the majority of carcinogens are believed to be activated within the target cells by CYP enzymes, the regulation and substrate specificity of which differ markedly.[12,87] Consequently, adequate information about the tissue distribution of drug-metabolizing enzymes in the normal and diseased state can provide insight into the target cells of toxins and carcinogens, the susceptibility of the cells to these agents, and the etiology of pancreatic diseases. Moreover, polymorphisms of genes encoding drug-metabolizing enzymes could possibly determine a person's or a population's risk of chronic pancreatitis or pancreatic cancer.

Nevertheless, risk estimation based on an enzyme expression profile is more than pretentious, due to the vast number of drug-metabolizing enzymes and the numerous factors affecting their expression pattern. Individual differences found in the expression of these enzymes are attributed to sex, age, dietary and ethnic practices, genetics, lifestyle (e.g., smoking, alcohol, drugs), exposure to environmental substances, patho-physiological status, and rhythmic patterns in the activity of the enzymes.[10,88,89] Our knowledge on the spectrum of the drug-metabolizing enzyme is limited. Most studies on these enzymes are based on the use of antibodies for immunohistochemistry or Western blot analysis. The specificity of these antibodies is currently a problem as is the method of immunohistochemical procedures. The lack of standard procedures can explain differing published data. Another important variable is the tissue of origin. Although tissues from organ donors have been used as normal (control) samples, their real normality is questionable. The cause of death, the underlying diseases, the nutritional status at the time of death and hours before that, and the premortem treatment (e.g., survival procedures, drugs, infusions) all can affect the results. Moreover, the higher sensitivity of some enzymes for postmortem degradation, especially in the pancreas, which is prone to fast autolysis, could affect the results.

In most of the studies, the parent population from which the samples were derived was highly variable with respect to gender and age, which are also known to affect the CYP expression profile.[10] Also, information on the lifestyle of the patients, including tobacco use, occupation, dietary and ethnic practices, would be valuable for a more detailed comparison. Moreover, regional differences in the etiology of the disease, such as alcohol consumption in chronic pancreatitis cases[90] vs. occupational expo-sure to xenobiotics in Foster's cases,[30] are interacting with the expression

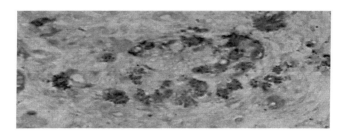

Figure 7.1   Expression of GST-alpha (black) and glucagon (gray) in an islet cell of the normal pancreas. ×100.

profiles of the enzymes. As yet, CYP 1A2, CYP 2B6, CYP 2C8/9/19, CYP 2D6, CYP 2E1, and CYP 3A1, the NA-OR, and GST-α, GST-μ, GST-π, and GST-θ have been found to be induced or higher expressed in chronic pancreatitis specimens compared to the normal pancreas.[26,27,30,47,70] In pancreatic cancer, a stronger immunoreactivity compared to the normal pancreas has been found for anti-CYP 1A1, CYP 1A2, CYP 2B6, CYP 2C8/9/19, CYP 2D6, CYP 2E1, CYP 3A1, CYP 3A4, NA-OR, and GST-π and GST-θ antibodies.[27,30,74] This apparent upregulation suggests a role of these xenobiotic-metabolizing enzymes in the pathogenesis of the diseases, although most of the differences were not statistically significant and not all authors came to a universal result. Remarkably, CYP 1A1, CYP 3A1, CYP 3A4, and GST-α and GST-π were especially associated with glucagon cells (Figure 7.1).[26,70] Because glucagon cells are more closely associated with the islet capillaries,[91,92] the high concentration of various CYP and GST enzymes in these cells highlights the function of these enzymes in the metabolism of blood-borne toxins and therefore in the defense mechanism of the pancreas. In this context, it is noteworthy that the number of glucagon cells is significantly greater in diabetic and pancreatic cancer tissues.[93] A similar zonal distribution of CYP enzymes is also known for the human liver.[94] The integrity of the defense mechanism of islet cells could, notwithstanding, determine the susceptibility of the organ to disease-causing toxins.

Results on gene polymorphisms are also ambiguous. No correlation between genetic polymorphisms of CYP 2E1 and the susceptibility to chronic pancreatitis has been found.[56,57] Polymorphisms of CYP 1A1 and the risk of pancreatic cancer also seem to be incoherent.[52–54] The results on polymorphisms of Phase II drug-metabolizing enzymes and an increased risk for pancreatic diseases are not clear and even contradictory. This might be particularly related to the dissimilar study populations. In one study, only a modest increase in susceptibility to pancreatic diseases was seemingly associated with the polymorphism of GSTM1 or NAT1.[63]

However, Verlaan et al. described a reduced risk for GSTM1-null geno-types for alcohol induced chronic pancreatitis.[69] Even though five μ-class genes (M1 to M5), located in tandem on Chromosome 1, exist,[95–97] there is an overlap in their substrates, implying that deficiencies in one gene may be compensated for by the expression of other genes in the family.[98] For GST-θ, which consists of the two genes T1 and T2, a deletion polymorphism has also been described.[21,53,99,100] In another study, an interaction between the GSTT1-null genotype, heavy smoking, and pan-creatic cancer especially in Caucasian women has been noted.[53] Taking the results on enzyme expression profiles and genetic polymorphisms together, many different factors seem to determine the individual's risk for pancreatic diseases.

Besides giving a risk estimation or prediction for pancreatic disease development, the investigation of drug-metabolizing enzymes might give some insight into the pathophysiological mechanisms and could improve the understanding of the biology or, even perhaps, the etiology of the diseases. Considering the blood supply of the pancreas, where a portion of arterial blood passes through to the islets before nourishing the exocrine pancreas,[101] the expression and apparently the higher concentration of most of these enzymes in islet cells in humans and animals (Chapter 8)[3,40,102,103] is understandable. The islet cells are the first cells that are exposed to any blood-borne carcinogens and toxins (Figure 7.2A and Figure 7.2B). Consequently, islets appear to be an important target of blood-borne toxins and, hence, are equipped with defensive enzymes. The toxins and carcinogens, which have been activated by islet cells, could reach the surrounding exocrine tissue and exert their deleterious effect, especially if the Phase II detoxifying enzymes are depleted.

Inter-individual differences exist in the cellular localization of CYP enzymes. In general, and in the majority of the cases, immunostaining of CYP enzymes is in a diffuse cytoplasmic pattern.[26–30,47,70] In some speci-mens, however, CYP enzymes are exclusively located in the Golgi region.[26] A frequent nuclear localization of CYP 2E1, especially in specimens obtained from female subjects, has also been noticed (Figure 7.2C and Figure 7.2D). Although contrary to the observations in animals (Chapter 8),[104,105] an intranuclear presence of CYPs in human pancreatic cells has not been reported earlier; the existence of this enzyme in the region of the cell where the prime target for mutagenesis and carcinogenesis, the DNA, is located[104] is noteworthy. GSTs, which play an important role in the detoxification process of CYP-activated compounds, have never been found in the nucleus. Hence, hypothetically, intranuclear CYP 2E1, which is involved in the metabolism of pancreatic carcinogens, could play a key role in carcinogenesis as has been suggested experimentally.[25] Because CYP 2E1 is one of the sex-hormone-dependent enzymes,[25] its primary

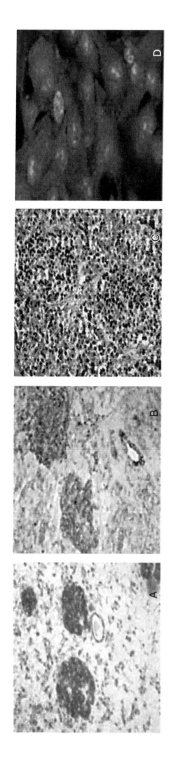

Figure 7.2   A: Strong staining of islet and ductular cells with the anti-CYP 2D6. The reactivity was found in all islet cells in a diffuse cytoplasmic pattern. ×75. B: Strong staining of islet and ductal cells with anti-NA-OR, compared to a weak immunoreactivity of acinar cells. ×75. C: Islet cells with a strong staining of most nuclei with the anti-CYP 2E1 antibody. Acinar cells were stained with a weak intensity. ×80. D: Immunoreactivity of cultured human islet cells with anti-CYP 2E1. The staining intensity was much stronger in the nuclei than in the cytoplasm. Confocal microscopy. ×1000.

presence in the nuclei of female individuals may indicate the role of female sex hormones in the induction of this enzyme. It is possible that some environmental agents, such as drugs and tobacco products, shift the cellular localization of this enzyme. Moreover, the interindividual differences in the cellular expression of the CYP enzymes could reflect different functions of the enzymes in different cells.

Around 70% of pancreatic adenocarcinomas and chronic pancreatitis cases develop in the head of the pancreas.[90,106,107] Although many hypotheses have been offered for this preferred localization of the diseases, none have been validated. The reflux of carcinogen-containing bile into the pancreatic duct as a possible mechanism[108] could not be confirmed experimentally.[109,110] Similarly, the increased viscosity of pancreatic fluid caused by alcohol as the etiology of chronic pancreatitis[18] cannot be considered valid as this would rather lead to an inflammation in the distal portion of the pancreas. For yet undisclosed reasons, the islets in the head region, which derive from the ventral anlage, contain more PP-cells and less glucagon cells compared to the areas derived from the dorsal anlage (i.e., the body and tail of the organ).[111] In a recent study, where the expression of selected CYP and GST enzymes was investigated between tissues from the head (including the uncinate process) and the body/tail region,[28] a higher concentration of the CYPs was found in islet cells than in exocrine cells, and a higher concentration in PP-rich areas than in PP-poor areas (Figure 7.3). Because most of these enzymes were localized within PP-cells, these cells seem to play a major role in metabolizing substances that present respective enzyme substrates. It can be further considered that these enzymes are also involved in the preparation of hormones, as was already shown for their key reactions in steroidogenesis.[112] Remarkably, a higher expression of the enzymes was also found in the exocrine cells derived from the ventral anlage than in tissues from the dorsal anlage. The higher expression of these enzymes in the PP-rich areas could be related to the paracrine effect of PP-cells on the exocrine cells with which they are so intimately associated. Contrary to the distribution of CYPs, the number of cells expressing GST-μ and GST-π was significantly lower than in other areas of the pancreas compared to the PP-rich areas. Because both GSTs play an important role in protecting cells from cytotoxic and carcinogenic agents,[113] the toxic load caused by CYP-mediated products might lead to a more rapid depletion of GSTs in the head than in the body or tail region, and consequently to a pronounced toxic effect. The greater content of CYP enzymes and a lower expression of GST enzymes in the head of the pancreas could render the tissue from the pancreas head more vulnerable to toxins and carcinogens and could be one of the reasons for the greater susceptibility of this area for inflammatory and malignant diseases.

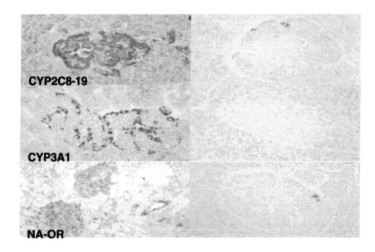

Figure 7.3   Expresssion of CYP-enzymes in pancreatic specimens derived from the ventral (left) and dorsal (right) anlage. ×80. Note that only one cell in the body of the pancreas is stained with the anti-CYP 2C8-19 (top right) or NA-OR (bottom right) antibody. No staining of the islet cells in the body of the pancreas with the CYP 3A1 antibody (middle right).

Diabetes mellitus is another unfortunate disease of the pancreas.[58] In one study, a remarkable decrease in the expression of the enzymes was found in diabetics compared to non-diabetics; CYP 1A2, CYP 2B6, and CYP 3A2 were not detectable in half of the specimens from diabetics.[26] It has been well established that intact β-cells, directly or indirectly, play a role in the normal function of the exocrine pancreas.[115] Therefore, the lack of enzymes in the diabetics might be due to degenerative alterations in the exocrine pancreas in response to an altered endocrine function. Whether the lack of certain CYP enzymes could also affect the susceptibility of the pancreas of diabetics to certain pancreatic diseases is unknown, although an association between diabetes and pancreatic cancer has been suggested.[115]

The expression of numerous drug-metabolizing enzymes in pancreatic cancer cells is also an important finding relative to the development of anticancer drugs. Chemotherapeutics that can be activated by the enzymes, which are present in cancer cells but are missing in normal pancreatic cells, could have desirable therapeutic effects. In a study by Lohr and coworkers,[116,117] human embryonic epithelial cells have been genetically modified to express the CYP 2B1 enzyme under the control of a CMV immediate-early promoter. As stated above, CYP 2B1 converts the chemotherapeutic oxazaphosphorines (ifosfamide or cyclophosphamide) to their active cytotoxic compounds. The cells were then injected in the pancreas of nude mice bearing preestablished human pancreatic cancer cell line tumors, following intraperitoneal administration of ifosfamide. Compared to untreated control animals, this treatment scheme, requiring CYP 2B1 activity, led to a partial or complete tumor ablation.

7.6 EPILOGUE

The human pancreas is abundantly equipped with many drug-metabolizing enzymes, making this gland the target of potentially harmful environmental substances. Therefore, the tissue is able to metabolize toxins and carcinogens directly. It seems that the pattern of the expression and genetic polymorphisms of individual drug-metabolizing enzymes and their particular distribution in acinar, ductal, and islet cells could determine the individual susceptibility for diseases. Studies on tissues from acute and chronic pancreatitis, pancreatic cancer patients, and especially from the members of hereditary chronic pancreatitis and familial pancreatic cancer families may shed some light onto the role of the xenobiotic-metabolizing enzymes in the etiology of these diseases and could help to develop preventive strategies. The establishment of more appropriate methods for the detection of enzymes is the prerequisite for gaining such information.

REFERENCES

(In some instances references are reviews, which should be consulted for citations of the original literature.)

1. Pott J.P. 1775. *Chirurgical observations relative to the cataract, the polypus of the nose, the cancer of the scrotum, the different kinds of ruptures, and the mortification of the toes and feet.* London: T.J. Carnegy.
2. Hecht S.S. 1997. Approaches to cancer prevention based on an understanding of N-nitrosamine carcinogenesis. *Proc. Soc. Exp. Biol. Med.* 216(2):181–191.
3. Ulrich A.B., Standop J., Schmied B.M., Schneider M., Lawson T.A., and Pour P.M. 2002. Expression of drug-metabolizing enzymes in the pancreas of hamster, mouse and rat, responding differently to pancreatic carcinogenicity of N-nitrosobis(2-oxopropyl)amine (BOP). *Pancreatology* 2(6):519–527.
4. Lijinsky W. 1984. Species differences in nitrosamine carcinogenesis. *J. Cancer Res. Clin. Oncol.* 108(1):46–55.
5. Pour P., Althoff J., Kruger F.W., and Mohr U. 1977. A potent pancreatic carcinogen in Syrian hamsters: N-nitrosobis(2-oxopropyl)amine. *J. Natl. Cancer Inst.* 58(5):1449–1453.
6. Silverman D.T. 2001. Risk factors for pancreatic cancer: a case-control study based on direct interviews. *Teratog. Carcinog. Mutagen.* 21(1):7–25.
7. Schoental R. and Connors T.A. 1981. Carcinogenic mycotoxins: VIII: Streptozotocin. In: *Dietary influences on cancer: traditional and modern.* Boca Raton, FL: CRC Press, Inc. p. 120.

8. Murray G.I. 2000. The role of cytochrome P450 in tumour development and progression and its potential in therapy. *J. Pathol.* 192(4):419–426.

9. Pelkonen O., Maenpaa J., Taavitsainen P., Rautio A., and Raunio H. 1998. Inhibition and induction of human cytochrome P450 (CYP) enzymes. *Xenobiotica* 28(12):1203–1253.

10. Lewis D.F., Ioannides C., and Parke D.V. 1998. Cytochromes P450 and species differences in xenobiotic metabolism and activation of carcinogen. *Environ. Health Perspect.* 106(10):633–641.

11. Lucier G.W., Thompson C.L., and Hoel D.G. 1992. Omeprazole, cytochrome P450, and chemical carcinogenesis. *Gastroenterology* 102(5):1823–1824.

12. Pelkonen O. and Raunio H. 1997. Metabolic activation of toxins: tissue-specific expression and metabolism in target organs. *Environ. Health Perspect.* 105 Suppl. 4:767–774.

13. Rendic S. and Di Carlo F.J. 1997. Human cytochrome P450 enzymes: a status report summarizing their reactions, substrates, inducers, and inhibitors. *Drug Metab. Rev.* 29(1–2):413–580.

14. Guengerich F.P. 1988. Roles of cytochrome P-450 enzymes in chemical carcinogenesis and cancer chemotherapy. *Cancer Res.* 48(11):2946–2954.

15. Nebert D.W., Nelson D.R., Coon M.J., Estabrook R.W., Feyereisen R., Fujii-Kuriyama Y. et al. 1991. The P450 superfamily: update on new sequences, gene mapping, and recommended nomenclature [published erratum appears in *DNA Cell. Biol.* 1991(Jun)10(5):397–398]. *DNA Cell. Biol.* 10(1):1–14.

16. Gonzalez F.J. and Gelboin H.V. 1994. Role of human cytochromes P450 in the metabolic activation of chemical carcinogens and toxins. *Drug Metab. Rev.* 26(1–2):165–183.

17. Guengerich F.P., Shimada T., Yun C.H., Yamazaki H., Raney K.D., Thier R. et al. 1994. Interactions of ingested food, beverage, and tobacco components involving human cytochrome P4501A2, 2A6, 2E1, and 3A4 enzymes. *Environ. Health Perspect.* 102 Suppl. 9:49–53.

18. Braganza J.M. 1998. A framework for the aetiogenesis of chronic pancreatitis. *Digestion* 59(Suppl 4):1–12.

19. Wallig M.A. 1998. Xenobiotic metabolism, oxidant stress and chronic pancreatitis. Focus on glutathione. *Digestion* 59(Suppl. 4):13–24.

20. Kaminsky L.S. and Spivack S.D. 1999. Cytochromes P450 and cancer. *Mol. Aspects. Med.* 20(1–2):70–84, 137.

21. Hirvonen A. 1999. Polymorphisms of xenobiotic-metabolizing enzymes and susceptibility to cancer. *Environ. Health Perspect.* 107 Suppl. 1:37–47.

22. Crespi C.L., Penman B.W., Gelboin H.V., and Gonzalez F.J. 1991. A tobacco smoke-derived nitrosamine, 4-(methylnitrosamino)-1-(3-pyridyl)-1-butanone, is activated by multiple human cytochrome P450s including the polymorphic human cytochrome P4502D6. *Carcinogenesis* 12(7):1197–1201.

23. Hasler J.A. 1999. Pharmacogenetics of cytochromes P450. *Mol. Aspects Med.* 20(1–2):12–24, 25–137.

24. Guengerich F.P. 1997. Comparisons of catalytic selectivity of cytochrome P450 subfamily enzymes from different species. *Chem. Biol. Interact.* 106(3):161–182.

25. Kazakoff K., Iversen P., Lawson T., Baron J., Guengerich F.P., and Pour P.M. 1994. Involvement of cytochrome P450 2E1-like isoform in the activation of N-nitrosobis(2-oxopropyl)amine in the rat nasal mucosa. *Eur. J. Cancer B. Oral Oncol.* 30B(3):179–185.

26. Standop J., Schneider M.B., Ulrich A., Chauhan S., Moniaux N., Buchler M.W. et al. 2002. The pattern of xenobiotic-metabolizing enzymes in the human pancreas. *J. Toxicol. Environ. Health A* 65(19):1379–1400.

27. Standop J., Schneider M., Ulrich A., Buchler M.W., and Pour P.M. 2003. Differences in immunohistochemical expression of xenobiotic-metabolizing enzymes between normal pancreas, chronic pancreatitis and pancreatic cancer. *Toxicol. Pathol.* In press.

28. Standop J., Ulrich A., Schneider M., Buchler M.W., and Pour P.M. 2002. Differences in the expression of xenobiotic-metabolizing enzymes between islets derived from the ventral and dorsal anlage of the pancreas. *Pancreatology* 2(6):510–518.

29. Wacke R., Kirchner A., Prall F., Nizze H., Schmidt W., Fischer U. et al. 1998. Up-regulation of cytochrome P450 1A2, 2C9, and 2E1 in chronic pancreatitis. *Pancreas* 16(4):521–528.

30. Foster J.R., Idle J.R., Hardwick J.P., Bars R., Scott P., and Braganza J.M. 1993. Induction of drug-metabolizing enzymes in human pancreatic cancer and chronic pancreatitis. *J. Pathol.* 169(4):457–463.

31. Yu M.W., Gladek-Yarborough A., Chiamprasert S., Santella R.M., Liaw Y.F., and Chen C.J. 1995. Cytochrome P450 2E1 and glutathione S-transferase M1 polymorphisms and susceptibility to hepatocellular carcinoma. *Gastroenterology* 109(4):1266–1273.

32. Ketterer B. 1988. Protective role of glutathione and glutathione transferases in mutagenesis and carcinogenesis. *Mutat. Res.* 202(2):343–361.

33. Booth J., Boyland E., and Sims P. 1961. An enzyme from rat liver catalysing conjugations with glutathione. *Biochem. J.* 79:516–524.

34. Jakoby W.B. 1978. The glutathione S-transferases: a group of multifunctional detoxification proteins. *Adv. Enzymol. Relat. Areas Mol. Biol.* 46:383–414.

35. Chasseaud L.F. 1979. The role of glutathione and glutathione S-transferases in the metabolism of chemical carcinogens and other electrophilic agents. *Adv. Cancer Res.* 29:175–274.

36. Boyer T.D. 1989. The glutathione S-transferases: an update. *Hepatology* 9(3):486–496.

37. Mannervik B., Alin P., Guthenberg C., Jensson H., Tahir M.K., and Warholm M. et al. 1985. Identification of three classes of cytosolic glutathione transferase common to several mammalian species: correlation between structural data and enzymatic properties. *Proc. Natl. Acad. Sci. USA* 82(21):7202–7206.

38. Meyer D.J., Coles B., Pemble S.E., Gilmore K.S., Fraser G.M., and Ketterer B. 1991. Theta, a new class of glutathione transferases purified from rat and man. *Biochem. J.* 274(Pt. 2):409–414.

39. Awasthi Y.C., Sharma R., and Singhal S.S. 1994. Human glutathione S-transferases. *Int. J. Biochem.* 26(3):295–308.

40. Ulrich A.B., Standop J., Schmied B.M., Schneider M.B., Lawson T.A., and Pour P.M. 2002. Species differences in the distribution of drug-metabolizing enzymes in the pancreas. *Toxicol. Pathol.* 30(2):247–253.

41. Braganza J.M., Jolley J.E., and Lee W.R. 1986. Occupational chemicals and pancreatitis: a link? *Int. J. Pancreatol.* 1(1):9–19.

42. Sandilands D., Jeffrey I.J., Haboubi N.Y., MacLennan I.A., and Braganza J.M. 1990. Abnormal drug metabolism in chronic pancreatitis. Treatment with antioxidants [see comments]. *Gastroenterology* 98(3):766–772.

43. Schulz H.U., Niederau C., Klonowski-Stumpe H., Halangk W., Luthen R., and Lippert H. 1999. Oxidative stress in acute pancreatitis. *Hepatogastroenterology* 46(29):2736–2750.

44. Pietri F. and Clavel F. 1991. Occupational exposure and cancer of the pancreas: a review. *Br. J. Ind. Med.* 48(9):583–587.

45. Muscat J.E., Stellman S.D., Hoffmann D., and Wynder E.L. 1997. Smoking and pancreatic cancer in men and women. *Cancer Epidemiol. Biomarkers Prev.* 6:15–19.

46. Tolbert P.E. 1997. Oils and cancer. *Cancer Causes Control* 8(3):386–405.

47. Ulrich A.B., Schmied B.M., Matsuzaki H., Lawson T.A., Friess H., Andren-Sandberg A. et al. 2001. Increased expression of glutathione S-transferase-pi in the islets of patients with primary chronic pancreatitis but not secondary chronic pancreatitis. *Pancreas* 22(4):388–394.

48. Shimada T., Iwasaki M., Martin M.V., and Guengerich F.P. 1989. Human liver microsomal cytochrome P-450 enzymes involved in the bioactivation of procarcinogens detected by umu gene response in Salmonella typhimurium TA 1535/pSK1002. *Cancer Res.* 49(12):3218–3228.

49. Acheson D.W., Rose P., Houston J.B., and Braganza J.M. 1985. Induction of cytochromes P-450 in pancreatic disease: consequence, coincidence or cause? *Clin. Chim. Acta.* 153(2):73–84.

50. Chaloner C., Sandle L.N., Mohan V., Snehalatha C., Viswanathan M., and Braganza J.M. 1990. Evidence for induction of cytochrome P-450I in patients with tropical chronic pancreatitis. *Int. J. Clin. Pharmacol. Ther. Toxicol.* 28(6):235–240.

51. Nebert D.W., Petersen D.D., and Puga A. 1991. Human AH locus polymorphism and cancer: inducibility of CYP1A1 and other genes by combustion products and dioxin. *Pharmacogenetics* 1(2):68–78.

52. Lee H.C., Yoon Y.B., and Kim C.Y. 1997. Association between genetic polymorphisms of the cytochromes P-450 (1A1, 2D6, and 2E1) and the susceptibility to pancreatic cancer. *Korean J. Intern. Med.* 12(2):128–136.

53. Duell E.J., Holly E.A., Bracci P.M., Liu M., Wiencke J.K., and Kelsey K.T. 2002. A population-based, case-control study of polymorphisms in carcinogen-metabolizing genes, smoking, and pancreatic adenocarcinoma risk. *J. Natl. Cancer Inst.* 94(4):297–306.

54. Liu G., Ghadirian P., Vesprini D., Hamel N., Paradis A.J., and Lal G. et al. 2000. Polymorphisms in GSTM1, GSTT1 and CYP1A1 and risk of pancreatic adenocarcinoma. *Br. J. Cancer* 82(10):1646–1649.

55. Li D., Firozi P.F., Zhang W., Shen J., DiGiovanni J., and Lau S. et al. 2002. DNA adducts, genetic polymorphisms, and K-ras mutation in human pancreatic cancer. *Mutat. Res.* 513(1–2):37–48.

56. Maruyama K., Takahashi H., Matsushita S., Nakano M., Harada H., and Otsuki M. et al. 1999. Genotypes of alcohol-metabolizing enzymes in relation to alcoholic chronic pancreatitis in Japan. *Alcohol Clin. Exp. Res.* 23(4 Suppl.):85S–91S.

57. Matsumoto M., Takahashi H., Maruyama K., Higuchi S., Matsushita S., Muramatsu T. et al. 1996. Genotypes of alcohol-metabolizing enzymes and the risk for alcoholic chronic pancreatitis in Japanese alcoholics. *Alcohol Clin. Exp. Res.* 20(9 Suppl.):289A–292A.

58. Schmied B.M., Ulrich A., Matsuzaki H., Ding X., Ricordi C., Moyer M.P. et al. 2000. Maintenance of human islets in long term culture. *Differentiation* 66:173–180.

59. Standop J., Ulrich A., Schneider M., Buchler M.W., and Pour P.M. 2002. Expression of xenobiotic-metabolizing enzymes in the human pancreas. *Pancreatology* 2:285, A166.

60. Prokopczyk B., Trushin N., Leder G., Ramadani M., Beger H.G., Siech M. et al. 2002. 4-(Methylnitrosamino)-1-(3-pyridyl)-1-butanone (NNK) and human pancreatic cancer: model studies. *Cancer Research: Proceedings of the AACR* A3415.

61. Prokopczyk B., Hoffmann D., Bologna M., Cunningham A.J., Trushin N., Akerkar S. et al. 2002. Identification of tobacco-derived compounds in human pancreatic juice. *Chem. Res. Toxicol.* 15(5):677–685.

62. Yokose T., Doy M., Taniguchi T., Shimada T., Kakiki M., Horie T. et al. 1999. Immunohistochemical study of cytochrome P450 2C and 3A in human non-neoplastic and neoplastic tissues. *Virchows. Arch.* 434(5):401–411.

63. Bartsch H., Malaveille C., Lowenfels A.B., Maisonneuve P., Hautefeuille A., and Boyle P. 1998. Genetic polymorphism of N-acetyltransferases, glutathione S-transferase M1 and NAD(P)H:quinone oxidoreductase in relation to malignant and benign pancreatic disease risk. The International Pancreatic Disease Study Group [see comments]. *Eur. J. Cancer Prev.* 7(3):215–223.

64. Lewis D. 1996. *Cytochromes P450: structure, function and mechanism.* London: Taylor & Francis.

65. Collier J.D., Bennett M.K., Hall A., Cattan A.R., Lendrum R., and Bassendine M.F. 1994. Expression of glutathione S-transferases in normal and malignant pancreas: an immunohistochemical study. *Gut* 35(2):266–269.

66. Pour P.M. 1980. Experimental pancreatic ductal (ductular) tumors. *Monogr. Pathol.* 21:111–139.

67. Pour P.M. 1994. Pancreatic centroacinar cells. The regulator of both exocrine and endocrine function. *Int. J. Pancreatol.* 15(1):51–64.

68. Pour P.M. and Wilson R.B. 1980. Experimental Tumors of the Pancreas. In *Tumors of the pancreas.* Moossa A. Ed. Baltimore: Williams and Wilkins. pp. 37–158.

69. Verlaan M., Te Morsche R.H., Roelofs H.M., Laheij R.J., Jansen J.B., Peters W.H. et al. 2003. Glutathione S-transferase mu null genotype affords protection against alcohol induced chronic pancreatitis. *Am. J. Med. Genet.* 120A(1):34–39.

70. Ulrich A., Schmied B.M., Standop J., Schneider M.B., Lawson T., Friess H. et al. 2002. Differences in the expression of glutathione S-transferases in normal pancreas, chronic pancreatitis, secondary chronic pancreatitis and pancreatic cancer. *Pancreas* 24(3):291–297.

71. McWilliams J.E., Sanderson B.J., Harris E.L., Richert-Boe K.E., and Henner W.D. 1995. Glutathione S-transferase M1 (GSTM1) deficiency and lung cancer risk. *Cancer Epidemiol. Biomarkers Prev.* 4(6):589–594.

72. Campbell J.A., Corrigall A.V., Guy A., and Kirsch R.E. 1991. Immunohistologic localization of alpha, mu, and pi class glutathione S- transferases in human tissues. *Cancer* 67(6):1608–1613.

73. Hayes J.D.P.C., Mantle T.J. Ed. 1990. *Glutathione S-transferases and drug resistance.* London: Taylor & Francis.

74. Trachte A.L., Suthers S.E., Lerner M.R., Hanas J.S., Jupe E.R., Sienko A.E. et al. 2002. Increased expression of alpha-1-antitrypsin, glutathione S-transferase pi and vascular endothelial growth factor in human pancreatic adenocarcinoma. *Am. J. Surg.* 184(6):642–647; discussion 647–648.

75. Oesch F. 1973. Mammalian epoxide hydrases: inducible enzymes catalysing the inactivation of carcinogenic and cytotoxic metabolites derived from aromatic and olefinic compounds. *Xenobiotica* 3(5):305–340.

76. Sims P., Grover P.L., Swaisland A., Pal K., and Hewer A. 1974. Metabolic activation of benzo(a)pyrene proceeds by a diol-epoxide. *Nature* 252(5481):326–328.

77. Coller J.K., Fritz P., Zanger U.M., Siegle I., Eichelbaum M., Kroemer H.K. et al. 2001. Distribution of microsomal epoxide hydrolase in humans: an immunohistochemical study in normal tissues, and benign and malignant tumours. *Histochem. J.* 33(6):329–336.

78. Blum M., Grant D.M., McBride W., Heim M., and Meyer U.A. 1990. Human arylamine N-acetyltransferase genes: isolation, chromosomal localization, and functional expression. *DNA Cell. Biol.* 9(3):193–203.

79. Hearse D.J. and Weber W.W. 1973. Multiple N-acetyltransferases and drug metabolism. Tissue distribution, characterization and significance of mammalian N-acetyltransferase. *Biochem. J.* 132(3):519–526.

80. Hein D.W., Doll M.A., Rustan T.D., Gray K., Feng Y., Ferguson R.J. et al. 1993. Metabolic activation and deactivation of arylamine carcinogens by recombinant human NAT1 and polymorphic NAT2 acetyltransferases. *Carcinogenesis* 14(8):1633–1638.

81. Anderson K.E., Hammons G.J., Kadlubar F.F., Potter J.D., Kaderlik K.R., Ilett K.F. et al. 1997. Metabolic activation of aromatic amines by human pancreas. *Carcinogenesis* 18(5):1085–1092.

82. Muscat J., Stellman S., Hoffmann D., Wynder E. 1997. Smoking and pancreatic cancer in men and women. *Cancer Epidemiol Biomarkers Prev* 6:15–19

83. Pour P.M., Runge R.G., Birt D., Gingell R., Lawson T., Nagel D. et al. 1981. Current knowledge of pancreatic carcinogenesis in the hamster and its relevance to the human disease. *Cancer* 47(6 Suppl.):1573–1589.

84. Mulder I., van Genugten M.L., Hoogenveen R.T., de Hollander A.E., and Bueno-de-Mesquita H.B. 1999. The impact of smoking on future pancreatic cancer: a computer simulation. *Ann. Oncol.* 10(Suppl. 4):74–78.

85. Shimizu M., Lasker J.M., Tsutsumi M., and Lieber C.S. 1990. Immunohistochemical localization of ethanol-inducible P450IIE1 in the rat alimentary tract. *Gastroenterology* 99(4):1044–1053.

86. Mirvish S.S. 1995. Role of N-nitroso compounds (NOC) in etiology of gastric, esophageal, nasopharyngeal and bladder cancer, and contribution to cancer of known exposures to NOC. *Cancer Lett.* 93:17–48.

87. Chen S.C., Wang X., Xu G., Zhou L., Vennerstrom J.L., Gonzalez F. et al. 1999. Depentylation of [3H-pentyl]methyl-n-amylnitrosamine by rat esophageal and liver microsomes and by rat and human cytochrome P450 isoforms. *Cancer Res.* 59(1):91–98.

88. Gibson G.G. and Skett P. 1994. *Introduction to drug metabolism.* 2nd ed. London: Chapman and Hall.

89. Pelkonen O. 1992. Carcinogen metabolism and individual susceptibility. *Scand. J. Work Environ. Health* 18(Suppl. 1):17–21.

90. Buechler M.W., Uhl W., and Malfertheiner P. 1996. *Pankreaserkrankungen: akute pankreatitis, chronische pankreatitis, tumore des pankreas.* Basel, Switzerland: Karger.

91. Chester-Jones I., Ingleton P.M., Phillips G. Eds. 1986. *Fundamentals of comparative vertebrate endocrinology.* New York: Plenum Publishing Corporation.

92. Ferner H. 1952. *Das inselorgan des pankreas.* Stuttgart, Germany: Thieme.

93. Schmied B.M., Ulrich A.B., Matsuzaki H., Li C., Friess H., and Buechler M.W. et al. 2000. Alteration of the langerhans islet in pancreatic cancer patients. *Int. J. Pathol.* 28(3):187–197.

94. Lindros K.O. 1997. Zonation of cytochrome P450 expression, drug metabolism and toxicity in liver. *Gen. Pharmacol.* 28(2):191–196.

95. Taylor J.B., Oliver J., Sherrington R., and Pemble S.E. 1991. Structure of human glutathione S-transferase class Mu genes. *Biochem. J.* 274(Pt. 2):587–593.

96. Zhong S., Spurr N.K., Hayes J.D., and Wolf C.R. 1993. Deduced amino acid sequence, gene structure and chromosomal location of a novel human class Mu glutathione S-transferase, GSTM4. *Biochem. J.* 291(Pt. 1):41–50.

97. Pearson W.R., Vorachek W.R., Xu S.J., Berger R., Hart I., Vannais D. et al. 1993. Identification of class-mu glutathione transferase genes GSTM1-GSTM5 on human chromosome 1p13. *Am. J. Hum. Genet.* 53(1):220–233.

98. Inskip A., Elexperu-Camiruaga J., Buxton N., Dias P.S., MacIntosh J., Campbell D. et al. 1995. Identification of polymorphism at the glutathione S-transferase, GSTM3 locus: evidence for linkage with GSTM1*A. *Biochem. J.* 312(Pt. 3):713–716.

99. Juronen E., Tasa G., Uuskula M., Pooga M., and Mikelsaar A.V. 1996. Purification, characterization and tissue distribution of human class theta glutathione S-transferase T1-1. *Biochem. Mol. Biol. Int.* 39(1):21–29.

100. Pemble S., Schroeder K.R., Spencer S.R., Meyer D.J., Hallier E., Bolt H.M. et al. 1994. Human glutathione S-transferase theta (GSTT1): cDNA cloning and the characterization of a genetic polymorphism. *Biochem. J.* 300(Pt. 1):271–276.

101. Murakami T., Hitomi S., Ohtsuka A., Taguchi T., and Fujita T. 1997. Pancreatic insulo-acinar portal systems in humans, rats, and some other mammals: scanning electron microscopy of vascular casts. *Microsc. Res. Tech.* 37(5–6): 478–488.

102. Zhang L., Weddle D.L., Thomas P.E., Zheng B., Castonguay A., Schuller H.M. et al. 2000. Low levels of expression of cytochromes P-450 in normal and cancerous fetal pancreatic tissues of hamsters treated with NNK and/or ethanol. *Toxicol. Sci.* 56(2):313–323.

103. Norton I.D., Apte M.V., Haber P.S., McCaughan G.W., Pirola R.C., and Wilson J.S. 1998. Cytochrome P4502E1 is present in rat pancreas and is induced by chronic ethanol administration. *Gut* 42(3):426–430.

104. Bresnick E., Boraker D., Hassuk B., Wayne L., and Thomas P.E. 1979. Intranuclear localization of hepatic cytochrome P448 by an immunochemical method. *Mol. Pharmacol.* 16:324–331.

105. Bresnick E., Hassuk B., Liberator P., Levin W., and Thomas P. 1980. Nucleolar cytochrome P-450. *Mol. Pharmacol.* 19:550–552.

106. Beger H.G., Schlosser W., Siech M., and Poch B. 1999. The surgical management of chronic pancreatitis: duodenum-preserving pancreatectomy. *Adv. Surg.* 32:87–104.

107. Pedrazzoli S., Beger H.G., Obertop H., Andren-Sandberg A., Fernandez-Cruz L., Henne-Bruns D. et al. 1999. A surgical and pathological based classification of resective treatment of pancreatic cancer. Summary of an international workshop on surgical procedures in pancreatic cancer. *Dig. Surg.* 16(4):337–345.

108. Braganza J. 1983. Pancreatic disease: a casualty of hepatic "detoxification"? *Lancet* 1000–1003.

109. Pour P.M. 1989. Experimental pancreatic cancer. *Am. J. Surg. Pathol.* 13(Suppl. 1):96–103.

110. Pour P. and Donnelly T. 1978. Effect of cholecystoduodenostomy and choledochostomy in pancreatic carcinogenesis. *Cancer Res.* 38(7):2048–2051.

111. Hazelwood R.L. 1993. The pancreatic polypeptide (PP-fold) family: gastrointestinal, vascular, and feeding behavioral implications. *Proc Soc Exp Biol Med* 202:44–63.

112. Graham-Lorence S. and Peterson J.A. 1996. P450s: structural similarities and functional differences. *Faseb. J.* 10(2):206–214.

113. Hayes J.D. and Pulford D.J. 1995. The glutathione S-transferase supergene family: regulation of GST and the contribution of the isoenzymes to cancer chemoprotection and drug resistance. *Crit. Rev. Biochem. Mol. Biol.* 30(6):445–600.

114. Saruc M. and Pour P.M. 2003. Diabetes and its relationship to pancreatic carcinoma. *Pancreas* 26(4):381–387.

115. Warren S. and Lecompte P. 1952. *The pathology of diabetes mellitus.* Philadelphia: Lea and Febiger.

116. Lohr M., Bago Z.T., Bergmeister H., Ceijna M., Freund M., Gelbmann W. et al. 1999. Cell therapy using microencapsulated 293 cells transfected with a gene construct expressing CYP2B1, an ifosfamide converting enzyme, instilled intra-arterially in patients with advanced-stage pancreatic carcinoma: a phase I/II study. *J. Mol. Med.* 77(4):393–398.

117. Lohr M., Muller P., Karle P., Stange J., Mitzner S., Jesnowski R. et al. 1998. Targeted chemotherapy by intratumour injection of encapsulated cells engineered to produce CYP2B1, an ifosfamide activating cytochrome P450. *Gene Ther.* 5(8):1070–1078.

8

DRUG-METABOLIZING
ENZYMES IN THE
PANCREAS OF ANIMALS

Alexis Ulrich, Jens Standop, and Parviz M. Pour

CONTENTS

8.1   INTRODUCTION

This chapter deals with the expression of drug-metabolizing enzymes (see Chapter 7) in laboratory animals.

8.2   MODELS FOR TOXICOLOGICAL TESTING

Due to ethical and rational reasons, laboratory animals are used to test the safety of new drugs and environment and to elucidate the metabolic pathways of pollutants and carcinogens.[1] *In vitro* studies on cultured cells, if available, have their limitations, because of the possible modification of the drug-metabolizing enzymes in artificial conditions. However, the bioassays in animals have created controversy about the extrapolation of the results to humans. For example, more than 400 known carcinogens exist in rodents, whereas only a small number of compounds (approximately 30) have been assumed to be carcinogenic in humans, indicating that small rodents might be more susceptible to carcinogenicity. In part, these differences were attributed to variations in the type or activity of the drug-metabolizing enzymes.[2] Many cytochrome P450 mono-oxygenase (CYP) genes and their corresponding enzyme activities have been identified in laboratory animals but not in humans, and vice versa.[3]

Different factors are believed to influence the activity of drug-metabolizing enzymes, hampering meaningful comparisons between xenobiotic metabolism of animals and humans. These include age, sex, environmental exposure, diet, pathophysiological status, drug–drug interactions, and genetics (interindividual differences in the activity level of various isoforms, polymorphisms).[2] Sex differences in the expression of CYPs within the same species have been shown for rat and mouse, whereas a sex specific human CYP isozyme has not yet been reported.[3] Concerning environmental exposure, there is even little comparison between rats and mice (and between strains of the same species) with respect to chemical carcinogenicity and in the pathways for both activation and detoxification.[2,4] Experimental models are kept in a controlled environment with a defined dose regimen and dietary restrictions, limiting the chance of drug–drug interactions; whereas in humans, drug-metabolizing enzymes are affected by medication, lifestyle, pathophysiological state, involuntary (and unknown) exposure to pollutants as well as possible inborn metabolic defects, causing genetic polymorphisms.[2,5,6] However, the selection of the appropriate animal model for each CYP isozyme might increase the comparability. For example, based on a review of Zuber et al.[1] all commonly used experimental animal models were appropriate for CYP 1A-mediated pathways, except dogs, which were most suitable for CYP 2D studies. The monkey (*Maccaucus rhesus*) might be a good representative for CYP 2C, the pig and minipig for CYP 3A isozymes.[1] Bioengineered animal models with CYP gene knockout mice that do not express either CYP 1A2, CYP 1A1, CYP 1B1, CYP 2E1, or both CYP 1A2 and CYP 2E1 might facilitate studies on chemical metabolism, toxicity, and carcinogenicity in the future.[7]

In the beginning of research on drug-metabolizing enzymes, the liver was regarded as the most important organ in drug metabolism. This has changed since it became known that many xenobiotics that are transformed into reactive metabolites by CYP enzymes do not damage all cells in a tissue, but rather a specific cell type or groups of cells located in selected areas of the tissue.[8] Hence, it appears that the susceptibility of cells to toxic substances is related to cell-specific differences in xenobiotic metabolism.[8] Since then, other organs, including the pancreas, became subjects of research interest.

8.3 THE PANCREAS AS A SITE OF DRUG METABOLISM

Studies on drug-metabolizing enzymes, which are known to be involved in the metabolism of a variety of substrates within the pancreas of laboratory animals (hamster, mouse, rat, rabbit, pig, dog, monkey), have shown a wide variation in the cellular localization and quantity of these enzymes (CYP 1A1, CYP 1A2, CYP 2B6, CYP 2C8, CYP 2C9, CYP 2C19, CYP 2D1, CYP 2E1, CYP 3A1, CYP 3A2, CYP 3A4, and GST-α, GST-π, and GST-μ) between the species.[9] (The specific substrates of these enzymes are mentioned in Chapter 7). Most of the enzymes were expressed in the pancreas of hamsters, mice, and monkeys, whereas rats, pigs, rabbits, and dogs were lacking several isozymes. In part, these findings were consistent with other reports.[10] Differences found between the relative studies could be due to more specific antibodies being available today than was available 16 years ago.[11–14]

8.4 THE ROLE OF ISLET CELLS IN DRUG METABOLISM

Interestingly, in all investigated species, the primary localization of the enzymes was the pancreatic islet.[9] In each species, an average of nine enzymes were expressed in the islets, eight enzymes within ductal cells, and seven within acinar cells. The exclusive expression of isozymes in the islet cells was seen in hamsters (CYP 2E1), mice (CYP 1A1, CYP 1A2, GST-α, GST-μ), rats (CYP 2C8, CYP 2C9, CYP 2C19), rabbits (CYP 1A2, CYP 2B6, GST-π), and pigs (CYP 1A1) (Figure 8.1). These findings, nevertheless, imply a greater importance of the islet cells in the metabolism of xenobiotics within the pancreas.[9] This is in line with the findings in humans (see Chapter 7). Strikingly, in all species, most of these enzymes were located within the α-cells as was ascertained by a multilabeling technique and serial section examination (Figure 8.2).[9] Considering the blood supply of the pancreas, the primary localization of the drug-metabolizing enzymes in islet cells makes sense. As presented in Chapter 7, generally, in human and other species part of the arterial blood passes through the islets first before nourishing the exocrine pancreas.[15,16] The islet–acinar portal system ensures, for example, that acinar cells are exposed to concentrations of islet peptide messengers (see also Chapter 9), which may be 20 times higher than those of peripheral blood.[17] Hence, it is conceivable that islet cells are the first to face bloodborne toxins. It appears that there are species differences in the spatial distribution of islet cells within the islet and, consequently, in their relation to blood vessels. In humans, α-, β-, and δ-

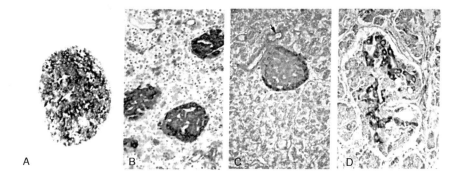

Figure 8.1   The immunoreactivity of anti-CYP 2E1 with the pancreas of hamster (A), rat (B), mouse (C), and, for comparison, human (D). In the hamster, the reactivity was restricted to the islet cells. In the rat and mouse, a weaker immunoreactivity was seen in ductal or ductular cells. Note that in the rat (B) and mouse the islet cells corresponding to the α-cells are stained stronger. (From Ulrich et al.[46] Used with permission.)

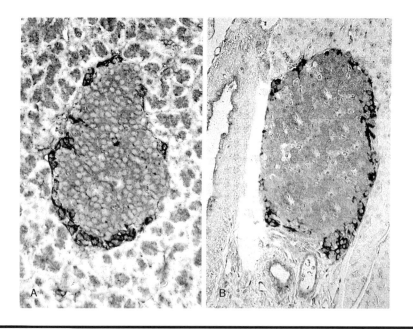

Figure 8.2   The expression of CYP 3A1 in the hamster (A) and rat pancreas (B). In the hamster, the α-cells (located in the periphery) are stained stronger than the remaining islet cells. Acinar cells are stained with a weaker intensity. In the rat (B), ductal cells are also stained. (From Ulrich et al.[46] Used with permission.)

cells show an inconsistent arrangement, whereas in rats and hamsters α-
and δ-cells are consistently located in the periphery and β-cells in the
center.[15,18] Accordingly, in the rat and possibly in hamster islets, the blood
flows from the α–δ-mantle to the β-core.[19,20] In monkeys, where the α-
and δ-cells are in the center, the afferent arteries run into the center first.[15]
In humans however, the microcirculation is indecisive.[15] The arrangement
of α-cells around the capillaries has already been reported by Ferner in
1942[21] and suggests that these cells have a closer contact to the blood
vessel. This could also explain why many of the drug-metabolizing enzymes
are located within the α-cells. The lack of drug-metabolizing enzymes in
acinar cells might be a reason for their susceptibility to toxic agents,[22]
which may have escaped the drug-metabolic capacity of islet cells or may
have been metabolically activated to more potent toxic substances within
the islets and secreted into the blood reaching the acini. Because the
capillary plexus of exocrine lobules and that of extralobular secretory ducts
(lobular ducts) have no connection with the vessels supplying the islet,[15]
the presence of enzymes in ductal cells appears self-explanatory.

8.5 DRUG METABOLISM IN PANCREATIC DISEASES

Based on the histology of most human pancreatic cancers, which are
generally detected at later stages due to their late clinical symptoms, it is
commonly believed that these carcinomas derive from the exocrine tissue,
especially ductal cells. Animal models, however, made it possible to
examine the histogenesis of the induced pancreatic tumors at all stages
of carcinogenicity. The Syrian Golden Hamster (SGH) is the only species
thus far that produces pancreatic tumors that mimic the human tumors
morphologically, clinically, and biologically. The carcinogen, N-nitrosobis
(2-oxopropyl)amine (BOP) has been effective in hamsters, but not in rats
or mice.[23–25] Because BOP needs metabolic activation for its carcinogenic
action, it was assumed that the reason for species differences is the
presence and cellular localization of drug-metabolizing enzymes. The lack
of potentially BOP-activating CYP-isozymes in the mouse and rat, but their
presence in the hamster, or the presence of the detoxifying enzymes,
glutathione S-transferases (GSTs), in the mouse and rat, but not in the
hamster could validate this possibility. The expression of CYP 1A1 in the
acinar cells of the hamster, but not of the mouse or rat, suggested that
the acinar cells could be the BOP-metabolizing site and, thus, tumor
progenitor cells.[26–29] However, several other studies have shown that in
hamsters, BOP alkylates the DNA in ductal cells, but not in acinar cells,[28]
supporting the notion that ductal cells are the BOP target in the hamster.
However, most cancers in the hamster model develop within the
islets.[24,30–32] Interestingly, the isozymes CYP 1A1, CYP 2B6, and CYP 2E1,

which are regarded as the most possible candidates of nitrosamine-metabolizing enzymes,[33–35] were expressed strongly in the islet cells of the hamster; although, the mouse and rat expressed these isozymes in almost similar patterns. Consequently, differences in pancreatic cell response to BOP could be related to other factors, including the rate of DNA repair in these species.[28,35,36] Rats lacked several drug-metabolizing enzymes compared to the mouse and hamster (GST-α in islets; CYP 2B6, CYP 2C8, CYP 2C9, CYP 2C19, and CYP 3A1 in acinar cells; CYP 3A4 in ductal cells).[37]

8.6 THE ROLE OF ENZYMES IN PANCREATIC DISEASES

Chronic pancreatitis appears to be the result of exposure to xenobiotics too. Finding an increased expression of the Phase II enzyme GST-π in the islets of human patients with chronic pancreatitis compared with normal pancreas and secondary chronic pancreatitis due to duct obstruction in pancreatic cancer supported this hypothesis (see Chapter 7).[38] Lacking an animal model of chronic pancreatitis, Rutishauser et al.[39] fed SGHs with a low or high fat diet that was supplemented with a prototype inducer of CYP 2 (Phenobarbitone) or CYP 1 (β-naphthoflavone) enzyme families, with or without a putative enzyme inhibitor (cimetidine). They concluded that drug modifiers of CYP magnified the deleterious effects of corn oil-enriched diets, comparable to those found in humans. Furthermore, these functional derangements were accompanied by pancreatic lipoatrophy.[39] Other groups fed rats with either ethanol or 3-methylcholanthrene (MC) and investigated their influence on the activity of CYP 2E1 and CYP 1A1 in rat pancreatic microsomes. CYP 2E1 is believed to play a major role in the metabolization of ethanol, whereas CYP 1A1 is linked to the metabolism of polycyclic aromatic hydrocarbons. The results demonstrated that in pancreatic microsomes, ethanol and MC exerted striking inductive effects on CYP 2E1 and CYP 1A1 activities. These findings suggested that these enzymes play a role in the pathogenesis of pancreatitis or pancreatic cancer.[40,41] Clarke et al. focused on the expression of CYP 1A-like proteins in the pancreatic islets after treatment with MC. They concluded that CYP 1A-protein was inducible in the islets and may have a role in the alteration of pancreatic β-cell responsiveness as it mediated the metabolism of arachidonic acid to metabolites with stimulating effects on insulin exocytosis.[42]

In rats, pancreatic carcinomas, generally of acinar appearance, could be obtained by injection of a single dose of azaserine.[43,44] In this model, two distinct types of foci of abnormal acini were described — basophilic and acidophilic foci. The latter, also termed atypical acinar cell nodules (AACN) were believed to have the potential to progress to carcinoma.[45,46] Daly et al. evaluated the GSTs α, μ, and π as early immunocytochemical markers for the development of the AACN compared to haematoxylin and eosin staining (H&E). They found an overexpression of GST-μ in all foci detected with H&E. However, 64% of the foci detected with GST-μ had not been identified with H&E as AACN during a prior examination. Reevaluation of these H&E sections revealed that some of these foci showed subtle morphological changes indicative of AACNs. They concluded that immunocytochemical staining for GST-μ is a more reliable and sensitive method than H&E for detecting the early stages of azaserine-induced foci.[47] An overexpression of GST-π was not found, in contrast to the reports of overexpression in pancreatic lesions induced in the hamster by nitrosamines[13] and in the rat by hydroxyaminoquinoline 1-oxide.[14] This might be explained by species and substrate differences.

8.7 THE ROLE OF CYP ENZYMES IN OTHER DISEASES

Nijhoff et al.[48] investigated the effect of the dietary, naturally occurring anticarcinogens flavone, coumarin, α-angelicalactone, and ellagic acid on the activity of GSTs in the esophagus, stomach, and pancreas of male Wistar rats. The treatment resulted in strong chemoprotective effects in the esophagus and stomach and, to a lesser extent, in the pancreas by enhancing the GST detoxification system. The lesser effect in the pancreas was explained with the shorter exposure time and intensity with the dietary substance.[48] These findings are in line with reports of GST enzyme induction in the pancreas of rat after a diet rich in cruciferous vegetables.[49]

8.8 EPILOGUE

There are many indications that both pancreatic cancer and chronic pancreatitis are the result of exposure to toxic and carcinogenic substances. Drug-metabolizing enzymes, such as CYPs and GSTs govern the deactivation of exogenous compounds, offering cellular protection against such substances. However, some substrates could be activated into more toxic compounds, instead. Because it is known that many xenobiotics that are transformed into reactive metabolites by CYP enzymes do not damage all cells in a tissue, rather than injuring a specific cell type, researchers were interested in the cellular localization of drug-metabolizing enzymes in the normal and diseased tissues. This way they hoped to get insight into the biology or, perhaps, etiology of the disease.

As ethical and rational reasons limited the ability to investigate the drug-metabolizing system in humans, animal models were used. However, due to marked differences found in the metabolism of xenobiotics by different species and the interindividuality in humans, the extrapolation of the animal studies to humans presented problems. A wide variation in the distribution and cellular localization of the selected drug-metabolizing enzymes was found in the pancreata of eight species on one hand and between the pancreases of humans on the other hand.[9] Even between rodents, such as rats and mice, marked differences in the drug-metabolizing machinery involved in activation and detoxification of xenobiotics exist.[2,4] These results question the validity of many other toxicological results. Most toxicological bioassays are conducted in rats. And yet, this species lacks many drug-metabolizing enzymes compared to mice and hamsters. Moreover, it appears that the metabolic capacity of the same tissue from different species varies considerably, as does the localization of the enzymes in different cells of the same tissue in the same species.[9]

However, the finding that islet cells express most drug-metabolizing enzymes in the pancreas of almost all species points to their potential role as the gatekeepers of the pancreas. In this context, the localization of most of these enzymes in α-cells, which has been claimed to have a closer contact to the blood vessels,[21] is striking.

Though various factors are believed to influence the activity of drug-metabolizing enzymes, experimental animals are tools for rough estimation of potentially hazardous substances to humans. The selection of the appropriate animal model for each CYP isozyme might increase the comparability.

REFERENCES

1. Zuber R., Anzenbacherova E., and Anzenbacher P. 2002. Cytochromes P450 and experimental models of drug metabolism. *J. Cell. Mol. Med.* 6(2):189–198.
2. Lewis D.F., Ioannides C., and Parke D.V. 1998. Cytochromes P450 and species differences in xenobiotic metabolism and activation of carcinogen. *Environ. Health Perspect.* 106(10):633–641.
3. Gonzalez F.J. and Nebert D.W. 1990. Evolution of the P450 gene superfamily: animal-plant 'warfare,' molecular drive and human genetic differences in drug oxidation. *Trends Genet.* 6(6):182–186.
4. Di Carlo F.J. 1984. Carcinogenesis bioassay data: correlation by species and sex. *Drug Metab. Rev.* 15(3):409–413.
5. Manjgaladze M., Chen S., Frame L.T., Seng J.E., Duffy P.H., Feuers R.J., Hart R.W., and Leakey J.E.A. 1993. Effects of caloric restriction on rodent drug and carcinogen metabolizing enzymes: implications for mutagenesis and cancer. *Mutat. Res.* 295:201–222.
6. Seng J.E., Gandy J., Turturro A., Lipman R., Bronson R.T., Parkinson A., Johnson W., Hart R.W., and Leakey J.E.A. 1996. Effects of caloric restriction on expression of testicular cytochrome P450 enzymes associated with the metabolic activation of carcinogens. *Biophys.* 335:42–52.
7. Ghanayem B.I., Wang H., and Sumner S. 2000. Using cytochrome P-450 gene knock-out mice to study chemical metabolism, toxicity, and carcinogenicity. *Toxicol. Pathol.* 28(6):839–250.

8. Baron J. 1991. In situ sites for xenobiotic activation and detoxication: implications for the differential susceptibility of cells to the toxic actions of environmental chemicals. In *Histo- and cytochemistry as a tool in environmental toxicology*. Graumann W. and Drukker J. Eds. New York: Fischer Verlag.

9. Ulrich A.B., Standop J., Schmied B.M., Schneider M.B., Lawson T.A., and Pour P.M. 2002. Species differences in the distribution of drug-metabolizing enzymes in the pancreas. *Toxicol. Pathol.* 30(2):247–253.

10. Clarke J., Flatt P.R., and Barnett C.R. 1997. Cytochrome P450 1A-like proteins expressed in the islets of Langerhans and altered pancreatic beta-cell secretory responsiveness. *Br. J. Pharmacol.* 121(3):389–394.

11. Hudson C.E., DeHaven J.E., Schulte B.A., and Norris J.S. 1999. Exogenous 17beta-estradiol blocks alpha and mu but not pi class glutathione S-transferase immunoreactivity in epithelium of Syrian hamster vas deferens. *J. Histochem. Cytochem.* 47(1):91–98.

12. March T.H., Jeffery E.H., and Wallig M.A. 1998. Characterization of rat pancreatic glutathione S-transferases by chromatofocusing, reverse-phase high-performance liquid chromatography, and immunohistochemistry. *Pancreas* 17(3):217–228.

13. Moore M.A., Bannasch P., Satoh T., Hacker H.J., and Ito N. 1987. Immunohistochemically demonstrated increase in glutathione S-transferase species in propylnitrosamine-induced focal proliferative and neoplastic Syrian hamster pancreatic lesions. *Virchows. Arch. B. Cell. Pathol. Incl. Mol. Pathol.* 52(6):479–488.

14. Moore M.A., Makino T., Tsuchida S., Sato K., Ichihara A., Amelizad Z., Oesch F., and Konishi Y. 1987. Altered drug metabolizing potential of acinar cell lesions induced in rat pancreas by hydroxyaminoquinoline 1-oxide. *Carcinogenesis* 8(8):1089–1094.

15. Murakami T., Fujita T., Taguchi T., Nonaka Y., and Orita K. 1992. The blood vascular bed of the human pancreas, with special reference to the insulo-acinar portal system. Scanning electron microscopy of corrosion casts. *Arch. Histol. Cytol.* 55(4):381–395.

16. Murakami T., Hitomi S., Ohtsuka A., Taguchi T., and Fujita T. 1997. Pancreatic insulo-acinar portal systems in humans, rats, and some other mammals: scanning electron microscopy of vascular casts. *Microsc. Res. Tech.* 37(5–6):478–488.

17. Lifson N., Kramlinger K.G., Mayrand R.R., and Lender E.J. 1980. Blood flow to the rabbit pancreas with special reference to the islets of Langerhans. *Gastroenterology* 79(3):466–473.

18. Pour P. 1978. Islet cells as a component of pancreatic ductal neoplasms. I. Experimental study: ductular cells, including islet cell precursors, as primary progenitor cells of tumors. *Am. J. Pathol.* 90(2):295–316.

19. Miyake T., Murakami T., and Ohtsuka A. 1992. Incomplete vascular casting for a scanning electron microscope study of the microcirculatory patterns in the rat pancreas. *Arch. Histol. Cytol.* 55(4):397–406.

20. Ohtani O., Ushiki T., Kanazawa H., and Fujita T. 1986. Microcirculation of the pancreas in the rat and rabbit with special reference to the insulo-acinar portal system and emissary vein of the islet. *Arch. Histol. Jpn.* 49(1):45–60.

21. Ferner H. 1942. Beitraege zur histobiologie der Langerhansschen inseln des Menschen mit besonderer Beruecksichtigung der Silberzellen und ihrer Beziehung zum pankreas-diabetes. *Virchows. Arch. Pathol. Anat.* 309:87–136.

22. Braganza J.M. 1998. A framework for the aetiogenesis of chronic pancreatitis, *Digestion* 59(Suppl. 4):1–12.

23. Longnecker D.S., Wiebkin P., Schaeffer B.K., and Roebuck B.D. 1984. Experimental carcinogenesis in the pancreas. *Int. Rev. Exp. Pathol.* 26:177–229.

24. Pour P.M. 1989. Experimental pancreatic cancer. *Am. J. Surg. Pathol.* 13(Suppl. 1):96–103.

25. Scarpelli D.G., Reddy J.K., and Longnecker D.S. 1986. *Experimental Pancreatic Carcinogenesis*. Boca Raton, FL: CRC Press.

26. Bockman D.E., Black O. Jr., Mills L.R., and Webster P.D. 1978. Origin of tubular complexes developing during induction of pancreatic adenocarcinoma by 7,12-dimethylbenz(a)anthracene. *Am. J. Pathol.* 90(3):645–658.

27. Kawabata T.T., Wick D.G., Guengerich F.P., and Baron, J. 1984. Immunohistochemical localization of carcinogen-metabolizing enzymes within the rat and hamster exocrine pancreas. *Cancer Res.* 44(1):215–223.

28. Lawson T., Kolar C., Garrels R., Kirchmann E., and Nagel D. 1989. The activation of 3H-labeled N-nitrosobis(2-oxopropyl)amine by isolated hamster pancreas cells. *J. Cancer Res. Clin. Oncol.* 115(1):47–52.

29. Scarpelli D.G., Kokkinakis D.M., Rao M.S., Subbarao V., Luetteke N., and Hollenberg P.F. 1982. Metabolism of the pancreatic carcinogen N-nitroso-2,6-dimethylmorpholine by hamster liver and component cells of pancreas. *Cancer Res.* 42(12):5089–5095.

30. Pour P.M. and Kazakoff K. 1996. Stimulation of islet cell proliferation enhances pancreatic ductal carcinogenesis in the hamster model. *Am. J. Pathol.* 149(3):1017–1025.

31. Pour P.M., Kazakoff K., and Carlson K. 1990. Inhibition of streptozotocin-induced islet cell tumors and N-nitrosobis(2-oxopropyl)amine-induced pancreatic exocrine tumors in Syrian hamsters by exogenous insulin. *Cancer Res.* 50(5):1634–1639.

32. Pour P.M., Weide L., Liu G., Kazakoff K., Scheetz M., Toshkov I., Ikematsu Y., Fienhold M.A., and Sanger W. 1997. Experimental evidence for the origin of ductal-type adenocarcinoma from the islets of Langerhans. *Am. J. Pathol.* 150(6):2167–2180.

33. Guengerich F.P. 1997. Comparisons of catalytic selectivity of cytochrome P450 subfamily enzymes from different species. *Chem. Biol. Interact.* 106(3):161–182.

34. Hasler J.A. 1999. Pharmacogenetics of cytochromes P450. *Mol. Aspects Med.* 20(1–2):12–24, 25–137.

35. Mori Y., Yamazaki H., Toyoshi K., Makino T., Obara T., Yokose Y., and Konishi Y. 1985. Mutagenic activation of carcinogenic N-nitrosopropylamines by rat liver: evidence for a cytochrome P-450 dependent reaction. *Carcinogenesis* 6(3):415–420.

36. Kokkinakis D.M., Scarpelli D.G., Subbarao V., and Hollenberg P.F. 1987. Species differences in the metabolism of N-nitroso(2-hydroxypropyl)(2-oxopropyl)amine. *Carcinogenesis* 8(2):295–303.

37. Ulrich A.B., Standop J., Schmied B.M., Schneider M.B., Lawson T.A., and Pour P.M. 2002. Expression of drug-metabolizing enzymes in the pancreas of hamster, mouse, and rat, responding differently to the pancreatic carcinogenicity of BOP. *Pancreatology* 2:519–527.

38. Ulrich A.B., Schmied B.M., Matsuzaki H., Lawson T.A., Friess H., Andren-Sandberg A., Buchler M.W., and Pour P.M. 2001. Increased expression of glutathione S-transferase-pi in the islets of patients with primary chronic pancreatitis but not secondary chronic pancreatitis. *Pancreas* 22(4):388–394.

39. Rutishauser S.C., Ali A.E., Jeffrey I.J., Hunt L.P., and Braganza J.M. 1995. Toward an animal model of chronic pancreatitis. Pancreatobiliary secretion in hamsters on long-term treatment with chemical inducers of cytochromes P450. *Int. J. Pancreatol.* 18(2):117–126.

40. Kessova I.G., DeCarli L.M., and Lieber C.S. 1998. Inducibility of cytochromes P-4502E1 and P-4501A1 in the rat pancreas. *Alcohol Clin. Exp. Res.* 22(2):501–504.

41. Norton I.D., Apte M.V., Haber P.S., McCaughan G.W., Pirola R.C., and Wilson J.S. 1998. Cytochrome P4502E1 is present in rat pancreas and is induced by chronic ethanol administration. *Gut* 42:426–430.

42. Zeldin D.C., Foley J., Boyle J.E., Moomaw C.R., Tomer K.B., Parker C., Steenbergen C., and Wu S. 1997. Predominant expression of an arachidonate epoxygenase in islets of Langerhans cells in human and rat pancreas. *Endocrinology* 138(3):1338–1346.

43. Longnecker D.S. 1984. Lesions induced in rodent pancreas by azaserine and other pancreatic carcinogens. *Environ. Health Perspect.* 56:245–251.

44. Standop J., Schneider M.B., Ulrich A., and Pour P.M. 2001. Experimental animal models in pancreatic carcinogenesis: lessons for human pancreatic cancer. *Dig. Dis.* 19(1):24–31.

45. Rao M.S., Upton M.P., Subbarao V., and Scarpelli D.G. 1982. Two populations of cells with differing proliferative capacities in atypical acinar cell foci induced by 4-hydroxyaminoquinoline-1-oxide in the rat pancreas. *Lab. Invest.* 46:527–534.

46. Roebuck B.D., Baumgartner K.J., and Longnecker D.S. 1984. Characterization of two populations of pancreatic atypical acinar cell foci induced by azaserine in the rat. *Lab. Invest.* 50:141–146.

47. Daly J.M., Tee L.B.G., Oates P.S., Morgan R.G.H., and Yeoh G.C.T. 1991. Glutathione S-transferas (μ class) as an early marker of azaserine-induced foci in the rat. *Carcinogenesis* 12(7):1237–1240.

48. Nijhoff W.A. and Peters W.H.M. 1994. Quantification of induction of rat oesophageal, gastric and pancreatic glutathione and glutathione S-transferases by dietary anticarcinogens. *Carcinogenesis* 15(9):1769–1772.

49. Wallig M.A., Kingston S., Staack R., and Jeffery E.H. 1998. Induction of rat pancreatic glutathione s-transferase and quinone reductase activities by a mixture of glucosinolate breakdown derivatives found in brussels sprouts. *Food and Chemical Toxicology* 36:365–373.

9

PANCREATIC EXOCRINE-ENDOCRINE INTERACTION

Vay Liang W. Go and Yu Wang

CONTENTS

9.1   INTRODUCTION

Disorders of the exocrine and endocrine pancreas constitute some of the most prevalent and life-threatening diseases that affect people today. Each year, 5 to 10 per 100,000 people are diagnosed with chronic pancreatitis[1–6]; 17 to 28 per 100,000 people with acute pancreatitis,[1–6] 30,000 cystic fibrosis cases,[7] 8 to 10 per 100,000 people with pancreatic cancer,[8–10] and 1.3 million new cases of diabetes occur in the United States.[11,12a,12b] Moreover,

175

exocrine pancreatic dysfunction often potentiates endocrine pancreatic dysfunction, and vice versa.

Traditionally, the endocrine and exocrine pancreas has been described as two distinct organs, each managed by separate medical specialists: the gastroenterologist over exocrine pancreatic diseases and the endocrinologist over endocrine disorders. In reality, the pancreas is an integrated organ. It is true that, functionally, the exocrine pancreas is involved with digestion and absorption of nutrients, and the endocrine pancreas is responsible for the regulation of glucose homeostasis and metabolism. However, during the past three decades, the reciprocal influences of the exocrine pancreas on the endocrine were established and the so-called "islet-acinar axis" was documented.[13] In addition, anatomically, there is an extensive body of evidence that a continuous insulo-acinar venous portal circulation connects the endocrine and exocrine tissue, and circulates hormone-rich blood from islets to acinar parenchyma.[14]

Recent finding that most islets, such as the acinar tissue, are connected to a duct system further established that the functional unit of the pancreas consists of an endocrine, exocrine, and ductal system.[15] Moreover, the components of this functional unit share a common embryological origin. Hence, an understanding of pancreatic embryogenesis and development becomes a frontier for clinical application and the relevance of genetics to the pathogenesis of pancreatic diseases, including diabetes mellitus, pancreatic cancer, and inflammatory disorders. In this chapter, we plan to elucidate the nature of these exocrine–endocrine interrelationships.

9.2 PANCREATIC MORPHOGENESIS

Although the pancreas is a complex endocrine and exocrine gland that is essential for life, man and mouse share a strikingly similar morphogenesis.[16] Thus, studies on mouse-model organogenesis are allowing a better understanding of the molecular mechanisms implicated in the development of the pancreas, as well as its diseases.

During embryogenesis (see Chapter 1), the pancreatic endoderm is sequentially exposed to distinct mesodermal cell populations (Figure 9.1).[16] The dorsal endoderm contacts the notochord, aorta, and pancreatic mesenchyme. The ventral endoderm lies adjacent to the septum transversum and cardiogenic mesoderm, and then contacts the ventral veins and mesenchyme. Pancreatic morphogenesis then begins in response to signals from these mesodermal tissues and is orchestrated by a cascade of transcription factors that are expressed by specific cells in a specific sequence (see Chapter 1 and Chapter 28).[17] Two essential homeodomains contain the transcription factors *Hlxb9* and *PDX1/Ipf1*, which are expressed before bud formation (approximately E9.5). However, *Hlxb9* expression becomes

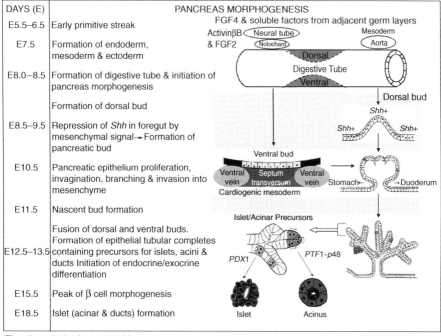

DAYS (E)	PANCREAS MORPHOGENESIS
E5.5–6.5	Early primitive streak
E7.5	Formation of endoderm, mesoderm & ectoderm
E8.0–8.5	Formation of digestive tube & initiation of pancreas morphogenesis
	Formation of dorsal bud
E8.5–9.5	Repression of *Shh* in foregut by mesenchymal signal→ Formation of pancreatic bud
E10.5	Pancreatic epithelium proliferation, invagination, branching & invasion into mesenchyme
E11.5	Nascent bud formation
E12.5–13.5	Fusion of dorsal and ventral buds. Formation of epithelial tubular completes containing precursors for islets, acini & ducts Initiation of endocrine/exocrine differentiation
E15.5	Peak of β cell morphogenesis
E18.5	Islet (acinar & ducts) formation

(E) embryonic day in mouse *shh+*

Figure 9.1   Morphogenesis of the mouse pancreas and the expression of key genes during embryonic development. (From Kandeel et al.[16] Used with permission.)

negative within 2 days of bud formation and then does not reappear until islet cell differentiation. *PDX1* expression, on the other hand, continues in duodenal and dorsal stomach epithelium (E11.5). In fact, the absence of *PDX1* will suspend growth and morphogenesis of the bud epithelia. The inductive signals required for development of the dorsal and ventral foregut endoderm into the pancreas are provided by adjacent mesodermal structures, including the notochord, dorsal aorta, cardiogenic mesoderm, and septum transversum. The expression of Sonic hedgehog (Shh) in a region of the foregut is repressed by the notochord signal, which causes the dorsal foregut endoderm to become the pancreas. The absence of this repressor signal leads to formation of the intestines.[16]

In the classic description of pancreatic morphogenesis, the molecular mechanisms underlying dorsal and ventral pancreatic development are similar: islet and acinar cells derive from ductal cells, and hormone-expressing cells initially detected in the nascent pancreatic islets become precursors for what later develops into endocrine cells.[18,19] The fusion of the dorsal and ventral pancreatic rudiments then leads to a variety of other possible fusions and rearrangements of the main pancreatic duct system by the postnatal pancreas.[20] Hence, the mammalian pancreas is composed of both endocrine and exocrine tissue. The endocrine pancreas is comprised of 1 to 2% of the organ's cellular mass, with four endocrine cell types located in the islets of Langerhans. The remaining is exocrine tissue organized into acini, which synthesizes and secretes digestive enzymes, and ducts that secrete a bicarbonate fluid to flush the acinar secretion into the intestines.

9.2.1 Gene Expression Cascades in Pancreatic Organogenesis

Pancreatic development is orchestrated by a cascade of transcription factors that are expressed in specific cell types and characteristic sequences.[21] The recent discovery of various transcriptional and signaling factors, and the results of several new studies, encourages a revision of fundamental concepts and formulation of a hypothesis on the underlying molecular and cellular mechanisms of pancreatic organogenesis. Zhang and Sarvetonik recently looked at these transcriptional factors and other biomarkers during cell differentiation at different stages of embryogenesis, from stem/progenitor cells to the exocrine and endocrine pancreas, as illustrated in Figure 9.2 and discussed in Chapter 1 and Chapter 28.[19]

The expression of key pancreatic transcription factors, such as *PDX1*, *PBX1*, *PTF1-p48,* and *Neurogenin 3* (*Ngn3*), is required throughout development of the pancreas. Although *PDX1* expression is not essential for the initiation of pancreatic budding from the endoderm, the expression of *PDX1* and *Ngn3* is necessary for regulating pancreatic morphogenesis and differentiation. *PBX1* is a member of the three-amino acid loop extension (TALE) class of homeodomain transcription factors that regulates the activity of other homeobox factors and is essential for normal pancreatic endocrine and exocrine development and function. Therefore, *PBX1* points to the concept that exocrine and endocrine tissues differentiate from a pluripotent progenitor derived from the putative ductal stem cells. *PTF1-p48* is the pancreas-specific subunit of the heteromeric bHLH protein complex *PTF1* that binds and activates transcriptional enhancers of the acinar hydrolytic enzyme genes. The major role of *p48* may be during acinar cell differentiation beginning at approximately E13.5. However, *p48* is expressed as early as E9.5 and has been shown to play a more fundamental role in endocrine and exocrine pancreas development.

9.2.2 Exocrine and Endocrine Cell

Pancreatic mesenchyme signals are key to stimulating pancreatic exocrine development and determining the exocrine to endocrine tissue ratio. In

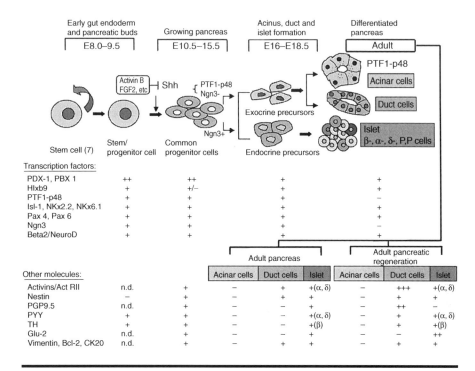

Early gut endoderm and pancreatic buds	Growing pancreas	Acinus, duct and islet formation	Differentiated pancreas
E8.0–9.5	E10.5–15.5	E16–E18.5	Adult

Transcription factors:

	Stem/progenitor cell	Common progenitor cells	Exocrine precursors / Endocrine precursors	Adult
PDX-1, PBX 1	++	++	+	+
Hlxb9	+	+/–	+	+
PTF1-p48	+	+	+	–
Isl-1, NKx2.2, NKx6.1	+	+	+	+
Pax 4, Pax 6	+	+	+	+
Ngn3	+	+	+	–
Beta2/NeuroD	+	+	+	+

Other molecules:

		Adult pancreas			Adult pancreatic regeneration		
		Acinar cells	Duct cells	Islet	Acinar cells	Duct cells	Islet
Activins/Act RII	n.d.	+	–	+	+(α, δ)	–	+++	+(α, δ)
Nestin	–	+	–	+	+	–	+	+
PGP9.5	n.d.	+	–	–	+	–	++	–
PYY	+	+	–	–	+(α, δ)	–	+	+(α, δ)
TH	+	+	–	–	+(β)	–	+	+(β)
Glu-2	n.d.	+	–	–	+	–	–	++
Vimentin, Bcl-2, CK20	n.d.	+	–	+	+	–	+	+

Figure 9.2   A schematic model for the process of pancreatic differentiation. The identified transcription factors and other molecules that are candidates for islet progenitor markers are shown. +++: high expression; ++: relative high expression; +: moderate expression; +/– : very low expression; n.e.: no expression. n.d.: not determined. (From Zhang and Sarvetnick.[19] Used with permission.)

in vitro pancreatic organ culture experiments, removal of large portions of the mesenchyme abolished the pancreatic exocrine cell development and promoted exclusive differentiation of the epithelium into endocrine cells.[22] Follistatin, an inhibitory factor from pancreatic mesenchyme, is one of the participating factors in the induction of pancreatic exocrine development. *p48* is the cell-specific acini component of the *PTF1* complex that binds to the promoters of the gene encoding the pancreatic digestive enzymes. Therefore, the *p48* protein is required for the formation of pancreatic acini and duct cells. Mutations of both *PTF1-p48* alleles will halt exocrine tissue development.

The islets of Langerhans are the functional units of the endocrine pancreas. β-cells form the core of the islets. They represent the majority of the endocrine cell population and account for 60 to 80% of the cell population that secrete insulin. The other endocrine cells are found at the periphery of the islets. These α-, δ-, and PP-cells secrete glucagon, somatostatin, and pancreatic polypeptide, respectively. A simplified model

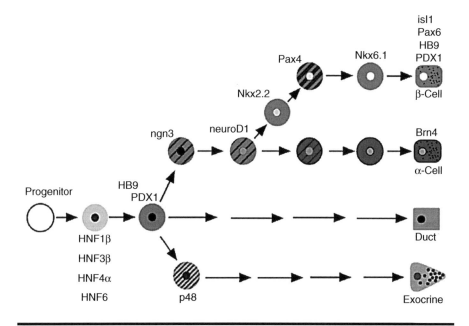

Figure 9.3   A simplified model for the role of islet transcription factors in endocrine differentiation in the developing pancreas. The proposed position for each transcription factor is based on its timing of expression, timing of predominant functional role, or both. Clearly some factors function at several steps, but a single step is shown for simplicity. (From Wilson et al.[23] Used with permission.)

for the role of islet transcription factors in endocrine differentiation of the developing pancreas has been proposed and reviewed by Wilson, Scheel, and German.[23] This model is summarized in Figure 9.3.[23] During differentiation, *PDX1* regulates the development of the endocrine cells, and plays a key role in the initiation and final differentiation of mature islet cells (see Chapter 1 and Chapter 28). After its early expression in the prepancreatic endoderm and early pancreatic buds, *PDX1* expression is down-regulated. Its expression is reactivated in most differentiated β-cells and some δ-cells, where it helps maintain the differentiated phenotypes. *Ngn3* then binds and activates *pax 4* gene promoter. After *Ngn3*, the late factors *pax 6, isl 1, PDX1,* and *brn 4* function in the final steps of islet cell differentiation. *brn 4* activation is restricted to non-β-cells to regulate glucagon gene expression. Many of the transcription factors discussed above and in Figure 9.3 have been found to be mutated in monogenic forms of diabetes,[23] causing mature onset of maturity onset diabetes of the young (MODY). In addition, mutations in the coding sequence of *isl 1* and *pax 4* have been implicated in families with late onset diabetes. Therefore, analysis of these mutations can give insight into the process by which β-cell dysfunction leads to the development of diabetes.[16,23]

9.3 INSULO-ACINAR PORTAL SYSTEM

The pancreas is part of the splanchnic circulation, where the celiac and superior mesenteric arteries supply blood by and venous drainage empties into the portal vein. The superior mesenteric artery supplies the head of the pancreas, while the celiac supplies the body and tail via the splenic artery.[14] The microcirculation of the pancreas usually divides into that of the acini and duct and the islets, but remains interconnected (see also Chapter 4). Wharton first described these microvascular connections between the islets and exocrine pancreas half a century ago.[24] Years later, this microvascular connection of capillary networks in the islets and acinar tissue came to be referred as the "insulo-acinar portal circulation."[25]

The hypothesis is that the islets of Langerhans were scattered throughout the exocrine pancreas and that this microcirculation helped coordinate hormonal and digestive function of the pancreas. As blood passes through the islets, it becomes enriched with a high concentration of islet hormones (see Chapter 4 and Chapter 6). These hormones then diffuse through highly permeable capillaries into the acinar tissue to regulate its function. Over the past two decades, the insulo-acinar portal system has been confirmed in rats, guinea pigs, dogs, and man.[26,27] Furthermore, the microvascular architecture within the islets appears to be organized to drain blood from α- and δ-cell areas to β-cell areas of the islets. In addition, there are three types of efferent vessels from the islets to acinar tissue:

1. Insulo-acinar portal vessels that radiate from the islets to join the capillary network
2. Emissary venules of the islets that lead directly into the systemic circulation
3. Insulo-ductal portal vessels that drain into the periductal capillary network

Therefore, the functional unit of the pancreas consists of the islets and acinar and ductal tissues, which are all connected by an insulo-acinar portal system.

9.4 REGULATION OF PANCREATIC SECRETION

Anatomically, the exocrine and endocrine pancreas each are innervated by the parasympathetic and sympathetic branches of the autonomic nervous system (see Chapter 2 and Chapter 6). Parasympathetic innervation via the vagus nerves potentiates hormone- and meal-stimulated exocrine pancreatic secretion and glucose-stimulated insulin and glucagon secretion. Sympathetic innervation via the splanchnic nerve inhibits glucose-stimulated insulin and somatostatin secretion, as well as basal- and hormone-stimulated exocrine pancreatic secretion, but enhances glucagon secretion. The central nervous system itself exerts control over insulin secretion, either directly via innervation to the endocrine pancreas, or indirectly via hormonal control from the gastrointestinal tract. Secretin, cholecystokinin (CCK), gastric inhibitory polypeptide (GIP), and other gastrointestinal hormones released from endocrine cells in the upper portion of the small intestine also regulate both exocrine and endocrine pancreatic functions. Moreover, pancreatic exocrine and endocrine secretions are controlled by the central and enteric nervous systems and gut hormones released after meal ingestion (Figure 9.4). Although most of the neurohormonal peptides are well defined and characterized, their physiologic roles and the magnitude of their interrelationships and interactions have yet to be fully determined.

Postprandial or nutrient-stimulated pancreatic secretion is triphasic. The cephalic, gastric, and intestinal phases of postprandial pancreatic secretion interact and overlap, often occurring simultaneously during ingestion of a meal. The sight, smell, taste, or thought of food stimulates the cephalic phase of gastric and pancreatic secretion, which is mediated by vagal cholinergic pathways. The gastric phase refers to gastric emptying and pancreatic secretion through regulation of duodenal loads of pancreatic stimulants. Apart from gastric emptying, the mechanisms involved in the regulation of duodenal nutrient loads of stimulants can activate enteric nervous system and gut hormonal secretion. Cholinergic pathways and hormones also mediate the intestinal phase, which accounts for most of the pancreatic response to a meal in humans. CCK and secretin are the primary hormonal mediators of pancreatic exocrine secretion. Together with GIP in concert with plasma glucose level, they stimulate insulin secretion. Intraluminal content of fatty acids and amino acids are potent stimulants for the release of these hormones.

The functional interrelationships between the endocrine and exocrine pancreas and between the gut and the pancreas have been coined the islet-acinar axis and enteropancreatic axis, respectively. The enteropancreatic axis is often separated into the enteroexocrine and enteroinsular axes to differentiate the separate interactions of the small intestine with the exocrine and endocrine pancreas. The islet-acinar axis is morphologically based on the existence of an islet-acinar portal blood system, formed by the dispersion of islet cells among acini in the exocrine tissue and by the presence of saturable insulin receptors, receptors for two insulin-like growth factors, IGF-I and IGF-II, and others on acini membranes.[13] These

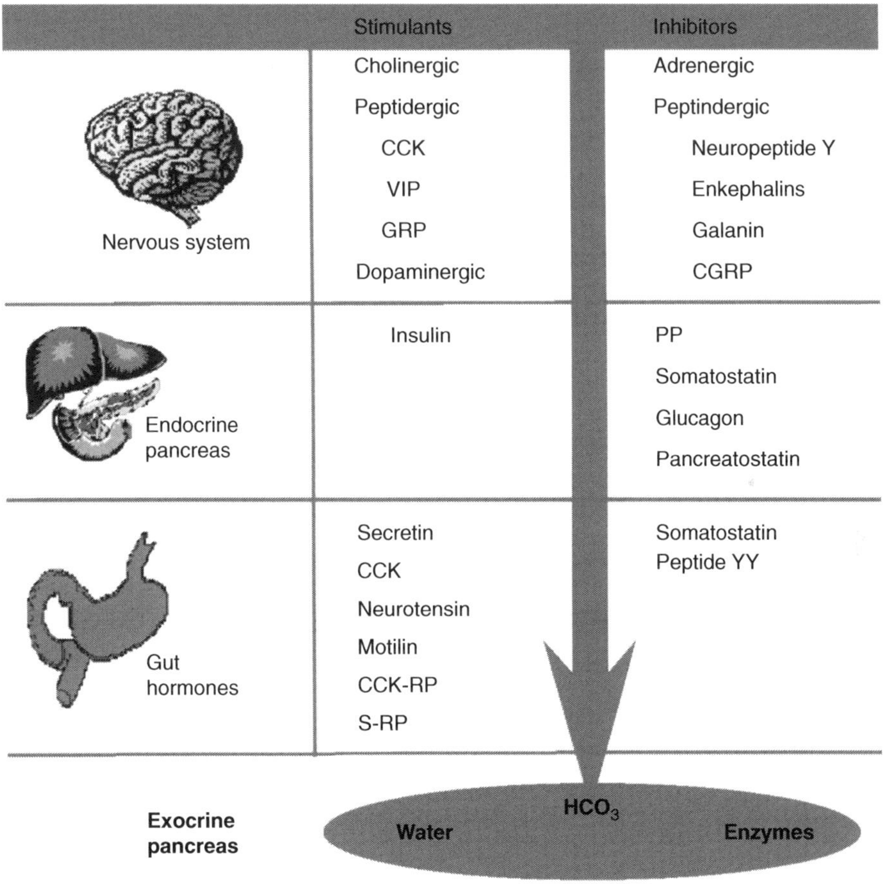

	Stimulants	Inhibitors
Nervous system	Cholinergic Peptidergic CCK VIP GRP Dopaminergic	Adrenergic Peptindergic Neuropeptide Y Enkephalins Galanin CGRP
Endocrine pancreas	Insulin	PP Somatostatin Glucagon Pancreatostatin
Gut hormones	Secretin CCK Neurotensin Motilin CCK-RP S-RP	Somatostatin Peptide YY
Exocrine pancreas	Water	HCO₃ Enzymes

Figure 9.4 Neurohormonal factors that influence pancreatic exocrine secretion. CCK = cholecystokinin; VIP = vasoactive intestinal peptide; GRP = gastrin-releasing peptide; CGRP = calcitonin gene-related peptide; PP = human pancreatic polypeptide; CCK-RP = cholecystokinin-releasing peptide; S-RP = secretin-releasing peptide.

structural arrangements facilitate functional interactions between endocrine islet hormones, such as insulin, glucagon, somatostatin, and pancreatic polypeptide and exocrine secretion. In general, insulin stimulates exocrine pancreatic secretion, whereas glucagon, somatostatin, and pancreatic polypeptide inhibit exocrine secretion. In addition, there are also intraislet hormonal interactions among the endocrine secretions of the islets. For example, somatostatin inhibits insulin secretion. Insulin is released from endocrine β-cells in response to rising glucose levels under the influence of cholinergic and gut hormones after meal intake (see Chapter 6).

The enteroinsular axis has also been characterized. GIP is also called glucose-dependent insulinotropic polypeptide. On absorption of glucose, galactose, sucrose, or fat (corn oil), the duodenum secretes GIP.[29–32b] GIP has been identified as a possible incretin, which is an endocrine factor from the gut with insulinotropic activity. The direct metabolic effects of GIP include antagonizing the lipolytic action of glucagon in fat cells, reducing glucagon-induced increase of cyclic adenosine monophosphate, and reducing hepatic glucose output without a concomitant rise in plasma insulin.[33] Incretins are released by nutrients and stimulate insulin secretion in the presence of elevated blood glucose levels. The connection between the gut and the pancreatic islets has been coined the enteroinsular axis. Because the enteroinsular axis acts as a feedback loop for suppression of pancreatic secretion, Isaksson and Ihse[34] have proposed its use in the treatment of pain induced by pancreatic hypersecretion during chronic pancreatitis. Six randomized trials have evaluated the administration of pancreatic enzymes as a method to provide pain relief. A meta-analysis of these trials, however, concluded that pancreatic enzyme therapy is ineffective in controlling pain.[28]

Gut hormones such as CCK and secretin, and nutrients such as glucose and amino acids stimulate the secretion of somatostatin, which reduces digestive functions and thus decreases the rate of nutrient absorption into the portal circulation.[35] In humans, somatostatin inhibits exocrine enzyme and bicarbonate secretion. Somatostatin significantly inhibits exocrine enzyme and protein production in response to CCK, CCK-octapeptide and cerulein, and enzyme secretion in response to electrical stimulation of the vagus nerve.[36] The site and mechanism of action of somatostatin-induced inhibition of exocrine secretion is not well studied, but it is proposed that the inhibitory effects of somatostatin play an important role in the physiologic regulation of pancreatic secretion. Ingestion of mixed meals and intragastric administration of glucose, fat, protein, and hydrochloric acid produces a rise in circulating somatostatin levels in the effluent gastric and pancreatic veins and the inferior vena cava.[37] Via the enteropancreatic axis, somatostatin inhibits CCK-induced or amino acid-induced exocrine secretion. As part of the enteroinsular axis, however, mixed meal ingestion also stimulates somatostatin release, which ultimately inhibits the pancreatic secretion initially induced by CCK. In addition, islet somatostatin inhibits insulin and glucagon secretion (see Chapter 6 and Chapter 14).

Pancreatic polypeptide inhibits exocrine pancreatic and biliary secretion. In humans, bovine pancreatic polypeptide inhibits basal as well as secretin-stimulated or CCK-stimulated pancreatic enzyme secretion.[38] The inhibitory effect of bovine pancreatic polypeptide is observed with infusion rates of bovine pancreatic polypeptide that produce plasma levels similar to postprandial levels, suggesting that pancreatic polypeptide also plays an important role in the physiologic regulation of pancreatic exocrine function.[38] In particular, the inhibitory effect of pancreatic polypeptide may involve a feedback loop between the enteropancreatic and islet-acinar axes. In this feedback loop, pancreaticobiliary secretions stimulate the release of pancreatic polypeptide, which, in turn, inhibits pancreatic and biliary secretions.

The interdigestive pancreatic secretion has been shown to cycle in temporal coordination with gastrointestinal motility. Specifically, pancreatic enzyme and bicarbonate secretion and antroduodenal motility fluctuate in tandem. Inhibition of gastric acid secretion by selective M1 receptor antagonism using telenzepine inhibits amylase, lipase, trypsin, and chymotrypsin outputs by 85 to 90% during interdigestive Phase I and Phase II, and by more than 95% during Phase III, pointing to cholinergic mediation of the interaction between pancreatic secretion and the gut. Studies demonstrating similar effects of atropine and telenzepine on pancreatic secretion also point to cholinergic regulation of pancreatic secretion.[39–41]

9.5 CLINICAL IMPLICATIONS OF EXOCRINE AND ENDOCRINE INTERACTIONS

Exocrine–endocrine pancreatic functional interrelationships play a significant role in the development of endocrine and exocrine disorders of the pancreas. Disorders of the exocrine pancreas, such as chronic and acute pancreatitis and pancreatic adenocarcinoma, can induce endocrine pancreatic disorders such as diabetes mellitus and islet cancer.[35] In turn, diabetes and glucose intolerance are often associated with exocrine pancreatic dysfunction, and may participate in pancreatic carcinogenesis. New epidemiological studies confirmed that glucose intolerance is a risk factor for pancreatic cancer, such that association cannot be accounted for by an adverse impact of early pancreatic cancer on β-cell function[42] and that insulin may also act as a promoter for pancreatic carcinogenesis. Recently, the nutritional factor link to pancreatic carcinogenesis became a main focus of interest.[43] These include dietary roles and interaction between various risk factors, including hyperinsulinemia, diabetes mellitus, chronic pancreatitis, obesity and energy balance, hereditary pancreatitis, and genetic polymorphisms that affect carcinogen metabolism of epigenetic mechanisms.[43,44] In addition, biomarker investigation has become an area of great promise. Genomic, proteomic, and metabolic profiling are now utilized in not only pancreatic cancer but also inflammatory and genetic diseases of the pancreas (see also Chapter 19, Chapter 20, and Chapters 23–25). In fact, pancreatic cancer and inflammatory diseases are now considered to be preventable.[45,46]

Currently, exocrine pancreatic insufficiencies are managed by using high-dose pancreatic minimicrospheres to achieve acceptable result by improving their nutrient absorptive function.[47] However, it has recently been demonstrated that chronic pancreatitis patients receiving this enzyme replacement show exacerbated pancreatic endocrine dysfunction,[48] because of their improved carbohydrate malabsorption, which led to a brittle nature of the blood glucose control in the malnourished insulin-dependent patients. Therefore, understanding the exocrine–endocrine functional interrelationships that underlie the development of numerous exocrine and endocrine pancreatic disorders gives way to important implications for clinical diagnosis and treatment of these diseases.

9.6 EPILOGUE

Great strides have been made in the diagnosis and treatment of pancreatic disorders. These strides are due in large part to advances made in the understanding of embryologic, organogenic, morphologic, physiologic, and functional relationships shared by the exocrine and endocrine pancreas. These interrelationships must be taken into consideration for the diagnosis and management of both exocrine and endocrine pancreatic disorders and diseases.

ACKNOWLEDGMENTS

This work is partially supported by the American Gastroenterological Association's Miles and Shirley Fiterman Foundation Clinical Research in Nutrition (Hugh R. Butt) Award and the NIH-funded UCLA Clinical Nutrition Research Unit (P01 CA 42710). The authors are also grateful for the graphic design by Manish Champaneria for Figure 9.4.

REFERENCES

1. Dufour M.C. and Adamson M.D. 2003. The epidemiology of alcohol-induced pancreatitis. *Pancreas* 27:286–290.
2. Gastroenterology Therapy Online. 2002. Pancreatitis workload "on the rise." Online: http://www.gastrotherapy.com/news/html/2002112812166.asp (accessed September 29, 2003).
3. National Digestive Diseases Information Clearinghouse. 2003. Digestive diseases topics list: pancreatitis. Publication No. 03-1596. NDDIC is a service of the National Institute of Diabetes and Digestive and Kidney Diseases, National Institutes of Health. Online: http://digestive.niddk.nih.gov/ddiseases/pubs/pancreatitis/ (accessed September 29, 2003).

4. Tinto A., Lloyd D.A., Kang J.Y., Majeed A., Ellis C., Williamson R.C., and Maxwell J.D. 2002. Acute and chronic pancreatitis — diseases on the rise: a study of hospital admissions in England 1989/90–1999/2000. *Aliment. Pharmacol. Ther.* 16:2097–2105.

5. The Society for Surgery of the Alimentary Tract (SSAT). 2004. SSAT Guidelines for Physicians. Online: http://www.ssat.com/cgi-bin/guidelines.cgi (accessed April 11, 2005).

6. eMedicine.com. 2005. Pancreatitis by Khoury G. and Deeba S. Online: http://www.emedicine.com/EMERG/topic354.htm (accessed April 11, 2005).

7. Cystic Fibrosis Foundation. 2005. About Cystic Fibrosis: What is CF? Online: http://www.cff.org/about_cf/what_is_cf/ (accessed April 11, 2005).

8. The Lustgarten Foundation for Pancreatic Cancer Research. 1999. *The epidemiology of pancreatic cancer.* Fall/Winter 1999 Newsletters. Online: http://www.lustgartenfoundation.org/LUS/Content/CDA/NewslettersDetail/0,2936,2-0_136,00.html (accessed September 29, 2003).

9. Surveillance, Epidemiology and End Results. 2002. Incidence: Pancreas Cancer, 2002 estimate. National Cancer Institute. Online: http://www.seer.cancer.gov/faststats/html/inc_pancreas.html (assessed September 29, 2003).

10. eMedicine.com. 2005. Pancreatic Cancer by Erickson R.A. Online: http://www.emedicine.com/med/topic1712.htm (accessed April 11, 2005).

11. 4woman.gov, National Women's Health Information Center. 2003. Diabetes: overview. NWHIC is a service of the U.S. Department of Health and Human Services Office on Women's Health. Online: http://www.4woman.gov/faq/diabetes.htm (accessed September 29, 2003).

12a. National Center for Chronic Disease Prevention and Health Promotion. 2005. National Diabetes Fact Sheet. Online: http://www.cdc.gov/diabetes/pubs/estimates.htm (accessed April 11, 2005).

12b. National Diabetes Information Clearinghouse. 2003. National Diabetes Statistics. A service of the National Institute of Diabetes and Digestive and Kidney Diseases, National Institutes of Cancer. Online: http://diabetes.niddk.nih.gov/dm/pubs/statistics/index.htm (accessed April 11, 2005).

13. William J.A. and Goldfine I.D. 1993. The insulin-acinar relationship. In *The pancreas: biology, pathobiology, and disease.* Go V.L.W., DiMagno E.P., Gardner J.D. et al. Eds. 2nd ed. New York: Raven Press Ltd. pp. 789–802.

14. Bonner-Weir S. 1993. The microvasculature of the pancreas, with emphasis on that of the islets of Langerhans: anatomy and functional implications. In *The pancreas: biology, pathobiology, and disease.* Go V.L.W., DiMagno E.P., Gardner J.D. et al. Eds. 2nd ed., New York: Raven Press Ltd. pp. 759–768.

15. Bertelli E., Regoli M., Orazioli D., and Bendayan M. 2001. Association between islets of Langerhans and pancreatic ductal system in adult rat. Where endocrine and exocrine meet together? *Diabetologia* 44:575–584.

16. Kandeel F., Smith C.V., Todorov I., and Mullen Y. 2003. Advances in islet cell biology: from stem cell differentiation to clinical transplantation. *Pancreas* 27:E63–E78.

17. Kim S.K. and MacDonald R.J. 2002. Signaling and transcriptional control of pancreatic organogenesis. *Curr. Opin. Genet. Dev.* 12:540–547.

18. Ashizawa S., Brunicardi F.C., and Wang X.-P. 2004. PDX-1 and the pancreas, *Pancreas* 28:109–120.

19. Zhang Y.Q. and Sarvetnick N. 2003. Development of cell markers for the identification and expansion of islet progenitor cells. *Diabetes Metab. Res. Rev.* 19:363–374.
20. Githens S. 1993. Differentiation and development of the pancreas in animals. In *The pancreas: biology, pathobiology, and disease*. Go V.L.W., DiMagno E.P., Gardner J.D. et al. Eds. 2nd ed. New York: Raven Press Ltd. pp. 21–55.
21. Schwitzgebel V.M. 2001. Programming of the pancreas. *Mol. Cell. Endocrinol.* 185:99–108.
22. Gittes G.K., Galante P.E., Hanahan D., Rutter W.J., and Debase H.T. 1996. Lineage-specific morphogenesis in the developing pancreas: role of mesenchymal factors. *Development* 122:439–447.
23. Wilson M.E., Scheel D., and German M.S. 2003. Gene expression cascades in pancreatic development. *Mechanisms Dev.* 120:65–80.
24. Wharton G.K. 1932. The blood supply of the pancreas, with special reference in that of Islet of Langerhans. *Anat. Rec.* 54:5581.
25. Fujita T. 1973. Insulo-acinar portal system in the horse pancreas. *Arch. Histol. Jpn.* 35:161–171.
26. Murakami T., Hitomi S., Ohtsuka A., Taguchi T., and Fujita T. 1997. Pancreatic insulo-acinar portal systems in humans, rats and some other mammals: scanning electron microscopy and vascular casts. *Microsc. Res. Tech.* 37:478–488.
27. Ohtani O. and Wang Q.X. 1997. Comparative analysis of insulo-acinar portal system in rats, guinea pigs, and dogs. *Microsc. Res. Tech.* 37:489–496.
28. Brown A., Hughes M., Tenner S., and Banks P.A. 1997. Does pancreatic enzyme supplementation reduce pain in patients with chronic pancreatitis: A meta-analysis. *Am. J. Gastroenterol.* 92:2032–2035.
29. Falko J.M., Crockett S.E., Cataland S., and Mazzaferri E.L. 1975. Gastric inhibitory polypeptide (GIP) stimulated by fat ingestion in man. *J. Clin. Endocrinol. Metab.* 41:260–265.
30. Falko J.M., Reynolds J.C., O'Dorisio T.M., Bossetti B., and Cataland S. 1982. The role of gastric inhibitory polypeptide in the augmented insulin response to sucrose. *Diabetes Care* 5:379–385.
31. Folsch U.R., Ebert R., and Creutzfeldt W. 1981. Response of serum levels of gastric inhibitory polypeptide and insulin to sucrose ingestion during long-term application of acarbose. *Scand. J. Gastroenterol.* 16:629–632.
32a. Morgan L.M., Wright J.W., and Marks V. 1979. The effect of oral galactose on GIP and insulin secretion in man. *Diabetologia* 16:235–239.
32b. Brown J.C., Dryburgh J.R., Ross S.A., and Dupre J. 1975. Identification and actions of gastric inhibitory polypeptide. *Recent Prog. Horm. Res.* 31:487–532.
33. Andersen D.K., Elahi D., Brown J.C., Tobin J.D., and Andres R. 1978. Oral glucose augmentation of insulin receptor secretion: interactions of gastric inhibitory polypeptide with ambient glucose and insulin levels. *J. Clin. Invest.* 62:152–161.
34. Isaksson G. and Ihse I. 1983. Pain reduction by an oral pancreatic enzyme preparation in chronic pancreatitis. *Dig. Dis. Sci.* 28:97–102.
35. Owyang C. 1993. Endocrine changes in pancreatic insufficiency. In *The pancreas: biology, pathobiology, and disease*. Go V.L.W., DiMagno E.P., Gardner J.D. et al. Eds. 2nd ed. New York: Raven Press. pp. 803–814.

36. Dollinger H.C., Raptis S., and Pfeiffer E.F. 1976. Effects of somatostatin on exocrine and endocrine pancreatic function stimulated by intestinal hormones in man. *Horm. Metab. Res.* 8:74–78.

37. Schusdziarra V., Harris V., Conlon J.M., Arimura A., and Unger R. 1978. Pancreatic and gastric somatostatin release in response to intragastric and intraduodenal nutrients and HCl in the dog. *J. Clin. Invest.* 62:509–618.

38. Greenberg G.R., McCloy R.F., Chadwick V.S., Adrian T.E., Baron J.H., and Bloom S.R. 1979. Effect of bovine pancreatic polypeptide on basal pancreatic and biliary outputs in man. *Dig. Dis. Sci.* 24:11–14.

39. DiMagno E.P., Hendricks J.C., Go V.L.W., and Dozios R.R. 1979. Relationship among canine fasting pancreatic and biliary secretions, pancreatic duct pressure and duodenal Phase III motor activity: Boldyreff revisited. *Dig. Dis. Sci.* 24:689–693.

40. Layer P., Chan A.T.H., Go V.L.W., and DiMagno E.P. 1988. Human pancreatic secretion during Phase II antral motility of the interdigestive cycle. *Am. J. Physiol.* 254:G249–G253.

41. Layer P., Chan A.T.H., Go V.L.W., Zinsmeister A.R., and DiMagno E.P. 1992. Adrenergic modulation of interdigestive pancreatic secretion in humans. *Gastroenterology* 103:990–993.

42. McCarty M.F. 2001. Insulin secretion as a determinant of pancreatic cancer risk. *Medical Hypotheses* 57:146–150.

43. Hine R.J, Srivastava S., Milner J.A., and Ross S.A. 2003. Nutritional links to plausible mechanisms underlying pancreatic cancer: a conference report. *Pancreas* 27:356–366.

44. Saruc M. and Pour P.M. 2003. Diabetes and its relationship to pancreatic carcinoma. *Pancreas* 26:381–387.

45. Boros L.G., Lee W.N., and Go V.L. 2002. A metabolic hypothesis of cell growth and death in pancreatic cancer. *Pancreas* 24:26–33.

46. Go V.L.W., Butrum R., and Wong D.A. 2003. Diet, nutrition, and cancer prevention: the postgenomic era. *J. Nutr.* 133:3830S–3836S.

47. Layer P. and Keller J. 2003. Lipase supplementation therapy: standards, alternatives, and perspectives. *Pancreas* 26:1–7.

48. O'Keefe S.J.D., Cariem A.K., and Levy M. 2001. The exacerbation of pancreatic endocrine dysfunction by potent pancreatic exocrine supplements in patients with chronic pancreatitis. *J. Clin. Gastroenterol.* 32:319–323.

10

NUTRITIONAL REQUIREMENTS OF THE PANCREAS

Daniel S. Longnecker

CONTENTS

10.1 INTRODUCTION

The pancreas plays a central role in nutrition and metabolism by virtue of the production of exocrine digestive enzymes and the secretion of islet polypeptide hormones. The digestive enzymes produced by the exocrine pancreas include trypsin, chymotrypsin, carboxypeptidase, elastase, lipase, RNase, DNase, and amylase. A deficiency of these enzymes in the small intestine results in digestive failure and malabsorption; however, the focus of this chapter is on the impact of nutritional deficiencies on the pancreas rather than on the effects of malabsorption on the host.

Because of the central role of the pancreas in maintaining normal nutrition, the selective advantage of having the organ be relatively immune to nutritional deficiencies is obvious. With the exception of kwashiorkor and tropical pancreatitis there are few clinically recognized syndromes of pancreatic dysfunction that are attributable to nutritional deficiency in humans.[1] The effects of nutritional factors on the pancreas are more easily recognized in the laboratory where attention can be focused on a single factor under controlled conditions.

The effect of malnutrition on the pancreas has been reviewed previously and additional references to older literature will be found in those reviews.[1–3]

10.2 EXOCRINE PANCREAS

10.2.1 Protein Energy Malnutrition

Protein energy malnutrition (protein–calorie malnutrition) includes nutritional diseases that have been called kwashiorkor and marasmus. These represent a spectrum of disease with varying clinical manifestations and the two diseases are not entirely separable on the basis of cause or the characteristics of the affected populations.[3] In this discussion, the focus is on deficiency of protein or specific amino acids rather than on starvation. The potential contribution of vitamin or trace element deficiency in humans with protein energy malnutrition is recognized[1] as is the potential for increased sensitivity to toxic agents.[2]

The recycling of amino acids probably represents a subtle but important function of the pancreas. Digestive enzymes are synthesized and secreted even during periods of fasting. It is reasonable to assume that these enzyme proteins are digested in the small intestine and their amino acids reabsorbed. The amount of zymogen stored in the pancreas varies greatly with cycles of feeding and fasting. This seems to represent an ideal system for short-term storage and release of amino acids. The normal pancreas constitutes about 1% of the body's total weight. More than 80% of its mass is acinar tissue providing the basis for synthesis and secretion of an estimated 6 to 20 g of zymogen protein per day.[3] The impact of the loss of exocrine pancreatic enzyme secretion on normal amino acid metabolism is not specifically defined but is certainly part of the basis for kwashiorkor.

10.2.1.1 Studies in Animals

The impact of protein deprivation and single amino acid deficiency on the pancreas has been demonstrated in experimental animals, thus providing experimental models for kwashiorkor. Pancreatic atrophy has been described in rats fed diets that were deficient in histidine, isoleucine, leucine, lysine, phenylalanine, tryptophan, threonine, or valine.[4] Most of these studies were for 1 week or less, and the principal histological change was a reduction in the size of acinar cells with reduced zymogen content. The pathological changes were similar regardless of which amino acid was restricted. Atrophy did not develop with deficiencies of methionine or arginine,[4] but it should be noted that methionine plays a role in both pancreatic development and transmethylation reactions as discussed below.

Atrophic changes were noted in rats that were force-fed several plant proteins of poor quality but not in rats that were allowed to eat these diets *ad libitum*.[1] This difference between forced feeding and *ad libitum.* consumption was also observed in studies of rats fed diets that were deficient in single amino acids.[4] In several instances, edema was noted in the pancreas when rats were force-fed amino acid-deficient diets. This finding suggests that there was mild pancreatitis although infiltration by leukocytes was not described. Studies of longer duration have utilized *ad libitum* feeding. Ultrastructural studies in lysine-deficient rats showed reduced zymogen, increased intracellular lipid deposits, and mitochondrial abnormalities during an experiment lasting 95 days. In the aggregate, these studies demonstrate the sensitivity of the acinar cell to protein deficiency and amino acid imbalances.

10.2.1.2 The Pancreas in Kwashiorkor

The clinical syndrome of kwashiorkor is complex and reflects the systemic effects of prolonged consumption of a protein-deficient diet.[5] It has been pointed out that the dietary deficiency in populations that develop kwashiorkor is more correctly described as protein deficiency with a relatively high-carbohydrate intake as contrasted with complete starvation. The pancreas is affected early and the loss of exocrine pancreatic function contributes to malabsorption, which further compounds the nutritional problem. Scrimshaw and Bdhar state that lipase, trypsin, and amylase activities in the duodenal secretions are lowered almost to zero in the acute stage of kwashiorkor.[6] Veghelyi et al., in describing observations of infants fed diets lacking animal protein, report that atrophic changes were present in pancreatic acinar cells as early as 17 days after the beginning of the deficient diet.[7] Patients with kwashiorkor suffer from steatorrhea, with stools that are bulky and soft and contain considerable quantities of undigested food.[6] At this stage, absorption of fat soluble vitamins is compromised.[3]

The activities of the pancreatic enzymes apparently return rapidly to normal when an adequate diet is given to patients in an acute phase of kwashiorkor and excellent recovery was reported in adults with a history of prolonged protein energy malnutrition when a high-protein diet was consumed for 12 to 14 weeks.[8] This sequence in the clinical picture reflects the dependence of the pancreas on dietary amino acids to maintain the high rate of protein synthesis required for production of the pancreatic exocrine digestive enzymes.

Minor differences in the clinical syndrome have been described in various geographical locations. It has been suggested that this might reflect relatively greater degrees of deficiency of specific amino acids in the diets in various regions and population groups, but the possibility that there are deficiencies of vitamins or trace elements should be considered as well. The disease develops following weaning and has its peak prevalence between the ages of 1 and 5, but it can develop in children of all ages and in adults who consume an inadequate amount of dietary protein.[3]

The pathological changes in the pancreas include atrophy of acinar cells with a striking decrease in their zymogen content at an early stage. In an ultrastructural study carried out in infants dying of kwashiorkor, autophagy was a conspicuous process in the acinar cells during the acute phase.[9] After 1 month of protein deficiency, the acini are markedly dilated and lined by flat epithelium. When kwashiorkor is prolonged, there is perilobular and periductal fibrosis. In advanced cases, the acinar tissue virtually disappears and may be replaced by fibrous tissue. Davies and Brist describe an advanced case in a 24-year-old man in whom the pancreas had been almost completely replaced by fibrous tissue.[10] In advanced cases, restoration of a normal diet does not reverse all structural changes or restore normal function.[3] Islets may survive for months to years, but the number of islets is decreased in advanced cases. Calcification of duct contents has been found in some cases (fewer than half) — a change regarded as characteristic of the process[3] although this change is more specifically associated with tropical pancreatitis.

10.2.1.3 Tropical (Nutritional) Pancreatitis

A subset of patients with protein–calorie malnutrition has been identified by Pitchumoni and others under the diagnosis of "nutritional pancreatitis."[2] The chronic pancreatitis patients in Uganda described by Shaper[11] have been included in this category. The characteristic difference between this group and patients with typical kwashiorkor is the consistent presence of extensive intrapancreatic calcification (calculi) and diabetes mellitus.[12] Pancreatic atrophy and fibrosis were also present. Many such patients have been studied in the state of Kerala, India, whereas only a few cases have been reported from other parts of India. The peak prevalence occurred between the ages of 16 and 20, usually in patients with a history of episodes of upper-abdominal pain during childhood. Chronic alcoholism was excluded as a cause in the adolescent patients, but the disease was not attributed solely to protein malnutrition. It has been suggested that some toxin, perhaps of dietary origin, initiates the process, or that a trace element deficiency plays a role. The focused geographical distribution and characteristic clinical picture are consistent with the view that tropical pancreatitis may be a separate disease from kwashiorkor in which protein malnutrition operates synergistically with an additional dietary deficiency[12] or toxin.[3]

10.2.2 Methyl Deficiency

Studies in a variety of experimental systems document that the pancreas is affected by dietary deficiency of methyl donors and folate. Disturbed methylation is implicated in abnormalities of growth, differentiation, and function of the exocrine pancreas as well as its response to toxins and carcinogens. The supporting experimental data have been reviewed.[13] The endocrine cells of the pancreas (islets) are less affected than the acinar tissue by methyl deficiency.

The effect of methyl deficiency on pancreatic growth and differentiation was studied in organ culture of fetal pancreas.[14,15] Morphologic indices of acinar cell differentiation showed a general correlation with biochemical assays of the activity of amylase, lipase, and chymotrypsin. These studies indicate that there is a basal methionine requirement for growth of the exocrine pancreas and that higher methionine levels are required to achieve partial or full differentiation of acinar cells.

The importance of methylation as the critical mechanism was supported by the observation that a supplement of S-adenosylmethionine (SAM) was more effective than methionine in reversing the effect of methionine-restricted culture media. The requirement for methyl groups could be partially met by choline supplementation in the presence of the basal level of methionine, but a high level of choline failed to overcome the requirement for methionine completely. A supplement of homocysteine enhanced the effectiveness of the choline supplement in the presence of the basal level of methionine suggesting that this enhanced the formation of SAM.[16]

Additional evidence for a role of methylation pathways in maintaining normal differentiation within the pancreas comes from studies of the formation of foci of hepatocytes within the pancreas (see Chapter 18). Such foci have been observed after a variety of experimental manipulations in rats and hamsters including the feeding of a severely methyl-group-deficient diet to rats[17] and in regimens that include administration of ethionine.

Other investigators have demonstrated that inhibitors of methylation directly inhibit amylase secretion *in vitro,*[18] and that combinations of ethionine treatment and choline deficiency have a similar effect *in vivo.*[19]

10.2.3   Choline Deficiency

A model of acute hemorrhagic pancreatitis in mice implicates choline deficiency as an important factor in predisposing the pancreas to catastrophic injury.[20,21] It has long been known that ethionine is toxic for the pancreas in rats. Although injections of large doses (1 g/kg) induced pancreatitis,[22] injections or feeding of lower levels caused severe, albeit reversible, pancreatic atrophy. A conspicuous degeneration of acinar cells occurred in such rats. Subsequently, Lombardi has shown that female mice fed a choline-deficient semipurified diet containing ethionine develop a fatal acute hemorrhagic pancreatitis over a 4-to-5-day course, and mice fed a choline supplemented version of the diet do not develop pancreatitis.[20]

A series of studies concerned with the pathogenesis of the disease in this model have implicated the intracellular activation of digestive enzymes with autodigestion of the pancreatic tissues. Increased cathepsin β 1 activity in the pancreas and the presence of this enzyme in the cytosol have been demonstrated.[23] It is known that cathepsin β 1 can activate trypsinogen. The hypothesis has been advanced that an alteration of intracellular membrane permeability or stability resulting from altered phospholipid metabolism in the choline-deficient, ethionine-fed mice is important in the pathogenesis of the acute autodestruction of the acinar cells.[20] Such changes in the membrane give the lysosomal cathepsin access to the normally segregated zymogens. Integrity of intracellular compartmentation is thought to be influential in preventing the intracellular activation of pancreatic enzymes. The significance of choline in phospholipid metabolism, and thus in membrane synthesis, is established.

Ethionine seems to play a major pancreatoxic role in this model, because mice fed the choline-deficient diet do not develop pancreatitis, nor has pancreatitis been a feature of experimental choline deficiency in other species. The delicacy of the pathogenetic mechanism is emphasized by the fact that male mice fed the choline-deficient, ethionine-containing diet do not develop acute pancreatitis unless they are also treated with estrogen.[24] The importance of choline deficiency as a predisposing factor for pancreatitis in humans is unknown.

10.2.4 Trace Element Deficiency

10.2.4.1 Selenium

Pancreatic atrophy and fibrosis have been described in chicks maintained on a selenium-deficient diet.[1,3] Whereas several lesions noted in chicks maintained on such diets were attributable to vitamin E deficiency resulting from fat malabsorption, pancreatic lesions developed even when the model was manipulated to prevent this vitamin deficiency. Selenomethionine was highly effective in preventing such pancreatic atrophy, presumably because of its great affinity for the pancreas. The effect on function consisted of progressive loss of exocrine pancreatic secretion with decreased production of lipase, trypsin, and chymotrypsin. Although food intake and body weight both decreased, forced-feeding experiments have supported the view that this loss of pancreatic function is specifically the result of selenium deficiency. Investigation of the biochemical mechanism of this deficiency in the chick pancreas has shown depletion of glutathione peroxidase activity suggesting that free radical–induced injury may play a role.[3]

No corresponding clinical deficiency state is known in man, but it is of interest that low blood selenium levels were reported in children with kwashiorkor[1] and in chronic alcoholics.[3] This raises the possibility that selenium deficiency may play a role induction of pancreatic lesions in some cases of kwashiorkor and chronic pancreatitis.[25] Concerning the effects of long-term treatment of selenium please see Chapter 21.

10.2.4.2 Copper

Atrophy of acinar tissue and its replacement by fat and reduced secretion of exocrine enzymes have been reported in rats that were maintained on copper-free diets for prolonged periods.[26,27] The rate of development of clinically significant copper deficiency was accelerated by feeding a chelating agent such as penicillamine to speed the depletion of tissue stores. Islet and ductal tissues were little affected and the atrophic changes were not accompanied by significant inflammation or fibrosis. The periinsular acinar cells were relatively resistant to the injury and survived longer than the bulk of acinar cells.

The biochemical basis of the effect of copper deficiency in the pancreas is not known, but deficient activity of one or more metalloenzymes seems likely. Increased storage of iron in the pancreas leading to peroxidative damage has also been proposed as a mechanism.[26] Severe depletion of cytochrome oxidase activity, a copper-dependent enzyme, was reported in the acinar tissue of copper-deficient rats. Ultrastructural studies in these rats revealed enlargement, abnormal configuration, and degenerative changes in the mitochondria in acinar cells. Vesiculation of rough endoplasmic reticulum, autophagy, and reduced zymogen content were also noted.

Hamsters are resistant to the pancreatic injury induced by copper deficiency, and the effect of this deficiency on the pancreas in humans is unknown.

10.2.4.3   Zinc

Long-term dietary zinc deficiency has long been shown to cause acinar cell abnormalities and diminish exocrine pancreatic function in rats, specifically reducing secretion of carboxypeptidase A and carboxypeptidase B, enzymes that contain zinc.[1,3] More recently, dietary zinc levels were shown to affect the response of the pancreas to exogenous toxic agents. Zinc deficiency enhanced ethanol-induced injury in the pancreas whereas a zinc-supplemented diet was protective against ethionine induced pancreatitis.[3]

10.2.4.4   Magnesium

One study of rats fed a low magnesium diet showed retention of zymogen granules in acinar cells suggesting impaired secretory ability.[3]

10.2.5   Vitamins

Although it would seem inevitable that vitamin deficiencies affect pancreatic function, there are few reports of pancreatic dysfunction or lesions associated with vitamin deficiencies in humans although case reports and series have shown exocrine abnormalities associated with vitamin A deficiency and pellagra.[3] Squamous metaplasia of the interlobular pancreatic ductal epithelium has been described in rats maintained on a vitamin A-free diet for several months[28]; squamous metaplasia also has been found in the pancreatic ducts of patients with cystic fibrosis, some of whom were known to be deficient in vitamin A.[29] Pancreatic exocrine enzyme content or secretion was reduced in rats fed pyridoxine- or riboflavin-deficient diets.[3]

10.2.5.1   Folate Deficiency

The pancreas contains high folate levels, second only to the liver.[30] Glycine-N-methyltransferase, which requires folate coenzymes, is abundant in the liver and pancreas of rats.[31] The ratio of SAM to SAH is significantly reduced in rats fed a folate deficient diet.[32] The pancreas of rats fed a folate-deficient diet contained more immature secretory granules than the pancreas of controls[30] and pancreatic amylase secretion was reduced in folate deficient rats.[33] In the context of the studies of Parsa et al. cited above, these studies suggest that pancreatic exocrine function was reduced as a result of disturbed methyl metabolism secondary to dietary folate deficiency.

One epidemiologic study reports an inverse relationship of risk for carcinoma of the pancreas and serum folate levels among a group of male smokers with a significant p value for trend.[34] Studies of the effect of choline deficiency combined with ethionine treatment on pancreatic carcinogenesis in the hamster model[35,36] provides additional support for an enhanced risk of pancreatic carcinoma among populations with a "hypomethylation" state.

10.2.6 Dietary Imbalances and Toxins

10.2.6.1 Chronic Alcoholism

Although the etiology of chronic pancreatitis in humans is often unknown, chronic alcoholism is one of the major known causes of this disease and one of the apparent causes of acute pancreatitis. The role of free radical-induced cell injury in chronic pancreatitis is widely discussed, and the potential role of vitamin A, vitamin E, and vitamin C in defense of such injury is recognized.[3] Even so, the pancreatic disease of chronic alcoholism may be more correctly classed as an example of toxic cellular injury than as a nutritional disease (see Chapter 15).[3]

10.2.6.2 Trypsin Inhibitors

The fact that the ratio of digestive enzymes produced by the pancreas changes in response to long-term changes in the composition of the diet is well established.[3] Feeding raw soy flour diet to rats and chickens induces hyperplasia of the pancreas. Diets containing several other natural or chemical trypsin inhibitors have a similar effect. It seems likely that the mechanisms by which pancreatic enzyme production is adapted to the composition of the diet are similar to those that mediate the effects of diet on pancreatic hypertrophy and hyperplasia. These effects can be regarded as adaptive physiological responses rather than pathological.

Pancreatic enzyme secretion is stimulated by cholecystokinin (CCK) that is made in the mucosa of the proximal small intestine (see also Chapter 6 and Chapter 14). The release of CCK appears, at least in part, to be controlled by a negative-feedback mechanism such that the presence of active trypsin in the intestine suppresses CCK release (see also Chapter 11) by destroying signaling peptides secreted by the pancreas and by the intestinal mucosa.[37] Dietary proteins bind the trypsin in the duodenal lumen allowing the signal peptides to survive and stimulate secretion of CCK, which in turn stimulates further pancreatic secretion of trypsin and other exocrine enzymes. Trypsin inhibitors, such as the one contained in raw soy flour (soy bean trypsin inhibitor, SBTI), also bind trypsin in the lumen and prolong CCK release (see also Chapter 11).

CCK is a trophic hormone for the rat pancreas. Prolonged administration of exogenous CCK or of closely related peptides such as cerulein that stimulate pancreatic enzyme secretion has been shown to stimulate pancreatic growth (see Chapter 14).[38] Both hypertrophy and hyperplasia of the pancreatic tissue have been reported. The hypothesis that consumption of a diet containing a trypsin inhibitor, such as SBTI, leads to hyperplasia because of sustained CCK secretion led to evaluation of a derivative hypothesis that such hyperplasia might enhance the process of carcinogenesis in the pancreas. It was first shown that rats fed raw soy flour diet during exposure to a chemical carcinogen, azaserine, developed a higher frequency of pancreatic carcinomas than a control group that was fed heated soy flour.[39] SBTI is inactivated by heat. In a subsequent study, the same trend was shown in rats treated with another carcinogen, N-nitrosobis(2-hydroxypropyl)amine.[40] During the course of these studies, an increased prevalence of hyperplastic foci and adenomas in the pancreas was observed in control groups that were fed raw soy flour but not treated with carcinogen. Finally, in long-term studies, a low prevalence of acinar cell neoplasms has been observed in groups of rats fed raw soy flour diet without prior carcinogen treatments[41] and in rats fed SBTI isolate.[42]

Species differences in response to dietary trypsin inhibitors are dramatic and the effects of these agents on the pancreas in rats and some other species seem not to pertain in the human. No epidemiological data have been presented that implicates consumption of diets containing trypsin inhibitors as a risk factor for pancreatic carcinoma in humans.

10.2.6.3 High-Fat Diets

Epidemiological studies have shown a general correlation of national levels of dietary fat consumption with the prevalence of carcinoma of the pancreas (see Chapter 21).[43] Enhanced progression of pancreatic carcinomas has been demonstrated in azaserine-treated rats and in N-nitrosobis(2-oxopropyl)amine-treated hamsters fed several high fat diets following exposure to the carcinogen (see Chapter 21).[38,44] However, other recent epidemiologic studies have failed to identify high fat diets as a risk factor for carcinoma of the pancreas in the United States,[45] whereas high body mass index and caloric intake was associated with increased risk (see Chapter 20).[46] The latter finding was consistent by both race and gender, whereas those with elevated values for one of these factors but not the other experienced no increased risk. The authors suggest that energy balance may play a major role in pancreatic carcinogenesis. These recent studies may shift emphasis from dietary fat to total caloric intake and physical activity.

10.3 ENDOCRINE PANCREAS

The effect of nutritional factors on the islet cells is less clear than are the nutritional relationships in the exocrine pancreas. The responsiveness of the β-cells to blood glucose levels is well established. It has been postulated that persistent hyperglycemia in the mild-moderate Type 2 diabetic stresses the β-cells and leads to a progressive decline in the functional capacity to secrete insulin. This somewhat mystical pathogenetic scheme seems to neglect the more important consideration of the underlying cause of the islet dysfunction (see Chapter 9, Chapter 27, and Chapter 28).

Trace elements, e.g., zinc, are known to complex with insulin *in vitro,* but deficiency states are not recognized clinically to affect islet function. Dietary chromium deficiency has been identified as a cause or contributing factor in the glucose intolerance of infants with protein–calorie malnutrition, of adults with maturity-onset diabetes, and of certain other patient groups.[47] Impaired glucose tolerance has been reported in chromium-deficient rats. These observations suggest that chromium may be an essential trace element for normal β-cell function.

Islet function may suffer as a result of diseases of the exocrine tissue, e.g., in advanced stages of chronic pancreatitis, especially tropical pancreatitis, and the advanced stages of kwashiorkor. Rats fed a protein-deficient diet for 8 weeks were reported to have reduced levels of plasma insulin, although their glucagon levels were similar to those in control rats.[48] Thus nutritional deficits may be an indirect cause of diabetes under special circumstances.

10.4 EPILOGUE

Although kwashiorkor is the only clinically important pancreatic disease known to be due entirely to nutritional deficiency, this statement requires a caveat. Even though the pancreatic lesions of kwashiorkor are usually attributed entirely to nutritional imbalance with deficient dietary protein, the possible contribution of some trace element deficiency or increased vulnerability to toxins in food should be recognized — especially in diseases such as tropical pancreatitis. Furthermore, the etiology of many cases of pancreatitis, pancreatic carcinoma, and diabetes remains unknown. Until the causes of these diseases are better understood, the possibility that nutritional factors play at least a contributory role in some

Pathways linking nutritional deficiency and pancreatic disease

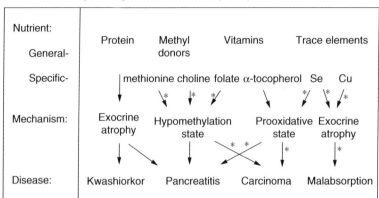

Figure 10.1 This general scheme links nutritional deficits and exocrine pancreatic disease based on clinical observations in human and animal studies. Pathways marked with an asterisk (*) are supported primarily by studies in animals and should be regarded as tentative for the human. See the text for discussion of detail, additional specific deficiencies, and involvement of the islets.

patients merits consideration (Figure 10.1). An example is the possible role of folate deficiency in the development of some cases of carcinoma. It should also be recognized that enzyme polymorphisms may modify individual response to nutritional deficiency or toxins.

REFERENCES

1. Longnecker D. 1985. Nutritionally induced pancreatic disease. In *Nutritional pathology*. Sidransky H. Ed. New York: Marcel Dekker. pp. 115–126.
2. Pitchumoni C. 1973. Pancreas in primary malnutrition disorders. *Am. J. Clin. Nutr.* 26:374–379.
3. Pitchumoni C. and Scheele G. 1993. Interdependence of nutrition and exocrine pancreatic function. In *The exocrine pancreas: biology, pathobiology, and diseases*. Go V., DiMagno E., Gardner J., Lebenthal E., Reber H., and Scheele G. Eds. 2nd ed. New York: Raven Press.. pp. 449–473.
4. Sidransky H. 1972. Chemical and cellular pathology of experimental acute amino acid deficiency. *Methods Achiev. Exp. Pathol.* 6:1.
5. Viteri F. and Torun B. 1980. Protein-calorie malnutrition. In *Modern nutrition in health and disease*. Goodhart R. and Shils M. Eds. 6th ed. Philadelphia: Lea & Febiger. p. 697.
6. Scrimshaw N. and Bdhar M. 1961. Protein malnutrition in young children. *Science* 133:2039.
7. Veghelyi P., Kemény T., Pozsonyi J., and Sos J. 1950. Dietary lesions of the pancreas. *Am. J. Dis. Child.* 79:658.

8. Tandon B., Banks P., George P., Sama S., Ramachandran K., and Gandhi P. 1970. Recovery of exocrine pancreatic function in adult protein-calorie malnutrition. *Gastroenterology* 58:358–362.

9. Blackburn W. and Vinijhaikul K. 1969. The pancreas in kwashiorkor; an electron microscopy study. *Lab. Invest.* 20:305–318.

10. Davies J. and Brist M. 1948. The essential pathology of kwashiorkor. *Lancet* 1:317.

11. Shaper A. 1960. Chronic pancreatic disease and protein malnutrition. *Lancet* 1:1223.

12. Mohan V., Nagalotimath S., Yajnik C., and Tripathy B. 1998. Fibrocalculous pancreatic diabetes. *Diabetes-Metab. Rev.* 14(2):153–170.

13. Longnecker D. 2002. Abnormal methyl metabolism in pancreatic toxicity and diabetes. *J. Nutr.* 132:2373S–2376S.

14. Parsa I., Marsh W., and Fitzgerald P. 1970. Pancreas acinar cell differentiation. 3. Importance of methionine in differentiation of pancreas anlage in organ culture. *Am. J. Pathol.* 59(1):1–22.

15. Parsa I., Marsh W., and Fitzgerald P. 1972. Pancreas acinar cell differentiation. V. Significance of methyl groups in morphologic and enzymatic development. *Exptl. Cell Res.* 73(1):49–56.

16. Parsa I., Marsh W., and Fitzgerald P. 1972. Pancreas acinar cell differentiation. VI. Effects of methyl donors and homocysteine. *Fed. Proc.* 31(1):166–175.

17. Hoover K. and Poirier L. 1986. Hepatocyte-like cells within the pancreas of rats fed methyl-deficient diets. *J. Nutr.* 116(8):1569–1575.

18. Capdevila A., Decha-Umphai W., Song K., Borchardt R., and Wagner C. 1997. Pancreatic exocrine secretion is blocked by inhibitors of methylation. *Arch. Biochem. Biophys.* 345(1):47–55.

19. Gilliland L. and Steer M. 1980. Effects of ethionine on digestive enzyme synthesis and discharge by mouse pancreas. *Am. J. Physiol.* 239(5):G418–G426.

20. Lombardi B. 1976. Influence of dietary factors on the pancreatotoxicity of ethionine. *Am. J. Pathol.* 84(3):633–648.

21. Niederau C., Luthen R., Niederau M., Grendell J., and Ferrell L. 1992. Acute experimental hemorrhagic-necrotizing pancreatitis induced by feeding a choline-deficient, ethionine-supplemented diet. Methodology and standards. *Eur. Surg. Res.* 24 Suppl.(1):40–54.

22. Farber E. and Popper H. 1950. Production of acute pancreatitis with ethionine and its prevention by methionine. *Proc. Soc. Exptl. Biol. Med.* 74:838–840.

23. Rao K., Zuretti M., Baccino F., and Lombardi B. 1980. Acute hemorrhagic pancreatic necrosis in mice: The activity of lysosomal enzymes in the pancreas and the liver. *Am. J. Pathol.* 98:45–59.

24. Rao K., Eagon P., Okamura K., Van Thiel D., Gavaler J., Kelly R. et al. 1982. Acute hemorrhagic pancreatic necrosis in mice. Induction in male mice treated with estradiol. *Am. J. Pathol.* 109:8–14.

25. Bowrey D., Morris-Stiff G., and Puntis M. 1999. Selenium deficiency and chronic pancreatitis: disease mechanism and potential for therapy. *HPB Surg.* 11(4):207–215.

26. Fields M. and Lewis C. 1997. Impaired endocrine and exocrine pancreatic functions in copper-deficient rats: the effect of gender. *J. Am. Coll. Nutr.* 16(4):346–351.

27. Rao M. and Reddy J. 1995. Hepatic transdifferentiation in the pancreas. *Semin. Cell Biol.* 6(3):151–156.

28. Raica Jr. N., Stedham M., Herman Y., and Sauberlich H. 1969. Vitamin A deficiency in germ-free rats. In *The fat-soluble vitamins.* DeLuca H. and Suttie J. Eds. Madison: University of Wisconsin Press. p. 283.

29. Blackfan K. and May C. 1938. Inspissation of secretion, dilatation of the ducts and acini, atrophy and fibrosis of the pancreas in infants. *J. Pediatr.* 13:627.

30. Balaghi M., Horne D., Woodward S., and Wagner C. 1993. Pancreatic one-carbon metabolism in early folate deficiency in rats. *Am. J. Clin. Nutr.* 58(2):198–203.

31. Yeo E. and Wagner C. 1994. Tissue distribution of glycine N-methyltransferase, a major folate-binding protein of liver. *Proc. Natl. Acad. Sci. USA* 91(1):210–214.

32. Balaghi M. and Wagner C. 1992. Methyl group metabolism in the pancreas of folate-deficient rats. *J. Nutr.* 122(7):1391–1396.

33. Balaghi M. and Wagner C. 1995. Folate deficiency inhibits pancreatic amylase secretion in rats. *Am. J. Clin. Nutr.* 61(1):90–96.

34. Stolzenberg-Solomon R., Albanes D., Nieto F., Hartman T., Tangrea J., Rautalahti M. et al. 1999. Pancreatic cancer risk and nutrition-related methyl-group availability indicators in male smokers. *J. Natl. Cancer Inst.* 91(6):535–541.

35. Mizumoto K., Tsutsumi M., Denda A., and Konishi Y. 1988. Rapid production of pancreatic carcinoma by initiation with N-nitroso-bis(2-oxopropyl)amine and repeated augmentation pressure in hamsters. *J. Natl. Cancer Inst.* 80(19):1564–1567.

36. Mizumoto K., Tsutsumi M., Kitazawa D., Denda A., and Konishi Y. 1990. Usefulness of rapid production model for pancreatic carcinoma in male hamsters. *Cancer Lett.* 49:211–215.

37. Spannagel A.W., Green G.M., Guan D., Liddle R.A., Faull K., and Reeve Jr. J. 1996. Purification and characterization of a luminal cholecystokinin-releasing factor from rat intestinal secretion. *Proc. Natl. Acad. Sci. USA* 93(9):4415–4420.

38. Longnecker D.S. 1993. Experimental models of exocrine pancreatic tumors. In *The exocrine pancreas: biology, pathobiology, and diseases.* Go V.L.W., DiMagno E.P., Gardner J.D., Lebenthal E., Reber H.A., and Scheele G.A. Eds. 2nd ed. New York: Raven Press. pp. 551–564.

39. Morgan R., Levinson D., Hopwood D., Saunders J., and Wormsley K. 1977. Potentiation of the action of azaserine on the rat pancreas by raw soya bean flour. *Cancer Lett.* 3:87–90.

40. Levison D.A., Morgan R.G.H., Brimacombe J.S., Hopwood D., Coghill G., and Wormsley K.G. 1979. Carcinogenic effects of di(2-hydroxypropyl)nitrosamine (DHPN) in male Wistar rats: Promotion of pancreatic cancer by a raw soya flour diet. *Scand. J. Gastroenterol.* 14:217–224.

41. McGuinness E., Morgan R., Levison D., Frape D., Hopwood D., and Wormsley K. 1980. The effects of longterm feeding of soya flour on the rat pancreas. *Scand. J. Gastroenterol.* 15:497–502.

42. Gumbmann M.R., Spangler W.L., Dugan G.M., Rackis J.J., and Liener I.E. 1985. The USDA trypsin inhibitor study, IV. The chronic effects of soy flour and soy protein isolate on the pancreas in rats after two years. *Qual. Plant Foods Human Nutr.* 35:275–314.

43. Ghadirian P., Thouez J.P., and PetitClerc C. 1991. International comparisons of nutrition and mortality from pancreatic cancer. *Cancer Detection Prev.* 15(5):357–362.
44. Longnecker D.S. 1998. Clues from Experimental Models. In *Pancreatic cancer: pathogenesis, diagnosis, and treatment.* Reber H.A. Ed. Totowa, NJ: Humana Press. pp. 53–70.
45. Michaud D., Giovannucci E., Willett W., Colditz G., and Fuchs C. 2003. Dietary meat, dairy products, fat, and cholesterol and pancreatic cancer risk in a prospective study. *Am. J. Epidemiol.* 157(12):1115–1125.
46. Silverman D. 2001. Risk factors for pancreatic cancer: a case-control study based on direct interviews. *Teratogenesis, Carcinogenesis, & Mutagenesis* 21(1):7–25.
47. Hambidge K. 1974. Chromium nutrition in man. *Am. J. Clin. Nutr.* 27:505–514.
48. Anthony L. and Faloona G. 1974. Plasma insulin and glucagon levels in protein-malnourished rats. *Metab. Clin. Exptl.* 23:303–306.

11

ACTIVITY OF PANCREATIC SECRETORY TRYPSIN INHIBITOR IS ESSENTIAL AGAINST PANCREATOTOXIC FACTORS

Masahiko Hirota

CONTENTS

11.1   INTRODUCTION

Inappropriate activation of trypsinogen in the pancreas leads to the development of pancreatitis. Once activated, trypsin is capable of activating many other digestive proenzymes in the pancreas and enhances autodigestion of the pancreas. Trypsin also activates cells via trypsin receptor, which has recently become known as one of the protease activated receptors, namely PAR-2. Both acinar and duct cells express abundant PAR-2.[1]

Trypsin activity is thought to be predominantly controlled by pancreatic secretory trypsin inhibitor (PSTI), which is also known as serine protease inhibitor Kazal Type 1 (SPINK1). PSTI is synthesized in acinar cells of the pancreas and acts as a potent natural inhibitor of trypsin to prevent the occurrence of pancreatitis. When trypsinogen is converted into trypsin in the pancreas, the PSTI immediately binds to trypsin to prevent further activation of pancreatic enzymes. The PSTI also blocks the further activation of pancreatic cells via trypsin receptor, PAR-2.

Several gene mutations in trypsinogen have been identified and have been presumed to be pathogenic in patients with hereditary pancreatitis through the enhancement of intrapancreatic trypsin activity. The mutations lead to an 80% likelihood of developing pancreatitis. Although gene mutations in trypsinogen have been identified and are presumed to be pathogenic in patients with hereditary pancreatitis, no causative gene mutation was found in about 50% of the patients. Subsequently, we proposed the hypothesis that, if the function of PSTI is impaired by its genetic mutation, trypsin may easily promote autodigestion causing acute or chronic pancreatitis. Mutation of PSTI gene may promote a predisposition to pancreatitis, by lowering the function of inhibiting trypsin activity. Five independent groups, including ours, started the mutational analysis of PSTI gene at approximately the same time and reported several mutations of the gene in patients with pancreatitis.[2–7]

11.2 MUTATIONAL ANALYSIS OF PSTI GENE IN FAMILIAL AND JUVENILE PANCREATITIS IN JAPAN

All 4 exons of the PSTI gene and their flanking intronic regions were sequenced for 37 familial pancreatitis patients (24 families), 15 juvenile pancreatitis patients, 22 sporadic pancreatitis patients (15 acute and 7 chronic), and 33 healthy volunteers. Three types of exonic mutations in the PSTI gene were observed. N34S was found in 6 familial pancreatitis patients (3 families) and 1 juvenile pancreatitis patient, and R67C was found in 1 familial pancreatitis patient and 1 juvenile pancreatitis patient. It should be noted that the N34S mutation was cosegregated with 2 intronic mutations, specifically, IVS1-37T>C and IVS3-69insTTTT (Table 11.1).[8] Cosegregation of this set of mutations (N34S + IVS1-37T>C + IVS3-69insTTTT) was also observed in other countries.

There are some supports for the idea that N34S mutation leads to the development of pancreatitis:

1. The frequency of the N34S mutation in pancreatitis patients was considerably higher than that in nonpancreatitis subjects.

Table 11.1 Summary of Mutational Analysis of PSTI in Japan

Mutation	Familial Pancreatitis	Juvenile Pancreatitis	Sporadic Pancreatitis	Healthy Volunteer
IVS1-37T>C	8/74	1/30	0/44	0/66
Exon 3: N34S	8/74	1/30	0/44	0/66
IVS3-69insTTTT	8/74	1/30	0/44	0/66
Exon 4: R67C	1/74	1/30	0/44	0/66
Exon 4: 272C>T	2/74	4/30	6/44	5/66

2. The rate of association of pancreatitis in subjects with the homozygous N34S mutation was assumed to be high based on the data collected from recent reports (98%, 49/50). This high rate of association of pancreatitis in homozygous N34S subjects suggests that this mutation may be a recessive inherited trait.

The R67C mutation has not previously been discussed in reports from other countries. Hence, R67C may be a uniquely Japanese mutation. The mature PSTI protein has been reported to contain three intrachain disulfide bonds: ^{32}C-^{61}C, ^{39}C-^{58}C, and ^{47}C-^{79}C. The R67C mutation potentially forms a novel disulfide bridge between ^{67}Cys and any of the other Cys residues. ^{67}Cys may also form an intermolecular disulfide bond, such as PSTI homodimer or PSTI-albumin complex (Table 11.2). Alternatively, ^{67}Cys may easily be oxidized, producing a modified molecular form or causing the destruction of acinar cells through endoplasmic reticulum stress.

Table 11.2 Predicted Molecular Forms of R67C

PSTI-^{67}Cys-SH
PSTI-^{67}Cys-S-S-^{67}Cys-PSTI: Homodimer
PSTI-^{67}Cys-S-S-Albumin
PSTI-^{67}Cys-S-S-another protein
PSTI-^{67}Cys-S-S-X
PSTI-^{67}Cys-Sox
PSTI-^{67}Cys-S-X

$$\overset{\displaystyle S}{\underset{\displaystyle SH}{\overset{/ \ \backslash}{PSTI\text{-}^{67}Cys\text{-}S}}}$$

We also found a 272C>T mutation in the 3í untranslated region of exon 4 in one patient with familial pancreatitis, four with juvenile pancreatitis, three with sporadic acute pancreatitis, three patients with sporadic acute pancreatitis, and three patients with sporadic chronic pancreatitis. This mutation, however, has been reported with the same frequency even in healthy volunteers and apparently indicates a normal polymorphism (Table 11.1).

11.3 GENERATION AND ANALYSIS OF PSTI KNOCKOUT MICE

Mutations in the PSTI gene seem to promote a predisposition to pancreatitis, possibly by lowering the threshold of pancreatitis caused by pancreatotoxic factors. To validate this hypothesis, we have generated a mouse bearing a null allele of the gene encoding PSTI.[9] The basic structure of the pancreas, including acinar cells, was once constructed before birth. The rapid and synchronized onset of acinar cell death occurred within a few days after birth and resulted in the disappearance of all acinar cells. There was no apparent degeneration of the ductal epithelium and islet cells.

The heterozygously knockout mice ($Psti^{+/-}$) were healthy, fertile, and indistinguishable from wild type. To produce homozygous knockout mice ($Psti^{-/-}$), male and female heterozygous mice were intercrossed. The ratio of living newborn mice matched the Mendelian rate of 1:2:1, suggesting that PSTI deficiency does not cause embryonic lethality. In $Psti^{-/-}$ mice, PSTI mRNA and protein were not detected at all. At birth, $Psti^{-/-}$ mice were indistinguishable from wild type ($Psti^{+/+}$) and fed milk. However, they did not gain weight, died day by day, and never survived more than 14 days. At autopsy of $Psti^{-/-}$ mice just before birth, all organs were present and normal. The pancreas also looked normal, suggesting normal embryonic development of the pancreas. However, the pancreas progressively disappeared after birth. The pancreatic bed was quite small, transparent, and the size of the spleen was also reduced. Thus, atrophy of the pancreas and deficiency of exocrine functions were considered to be the main cause of severe growth retardation and death of the $Psti^{-/-}$ mice.

The process of atrophy of the pancreas was examined with histological analysis. At 15.5 days postcoitum (dpc), when acinar cells have not been formed, there is no difference in the pancreatic morphology between the $Psti^{+/+}$ and $Psti^{-/-}$ mice. One day later, the acinar cells can be recognized clearly. $Psti^{-/-}$ embryo already showed mild vacuolization of acinar cells. This represents that degeneration of the acinar cells begins at the same time with the appearance of the zymogen granule, probably due to intracellular trypsin activation. At 18.5 dpc, just before birth, vacuolization of acinar cells in $Psti^{-/-}$ embryo became more apparent and severe, although

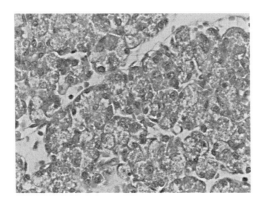

Figure 11.1 Histological appearance of the pancreas. *Psti*/- pups at 0.5 dpp showed marked acinar cell vacuolization.

the acinar cells functioned as shown by the production of amylase. Acinar cell degeneration in *Psti*/- mice peaked at 0.5 days postpartum (dpp), just after birth. Almost all acinar cells showed massive cytoplasmic vacuolization (Figure 11.1). One day later, normal structure of acinar cells was not detected anymore, and only duct cells forming tubular complex and endocrine cells remained.

The mechanism of the acinar cell degeneration in PSTI/SPINK1 knockout mice was analyzed as a next step. We checked whether apoptosis occurred or not, however, there was no increase in the number of apoptotic cells detected by terminal deoxynucleotidyl transferase-mediated dUTP nick-end labeling (TUNEL) staining. Apoptosis did not seem to be involved in this process. Then, we performed electron microscopic analysis of the pancreas. At 18.5 dpc, just before birth, the cytoplasm of acinar cells in *Psti*+/+ mice was filled with zymogen granules. On the other hand, in *Psti*/- mice, considerable number of vacuoles were observed. At 0.5 dpp, just after birth, *Psti*/- acinar cells showed more marked vacuolization. In higher magnification, some vacuoles contained cellular organelles, indicating that these vacuoles are autophagic vacuoles. A huge vacuole containing digested organelle was also present. Any nuclear changes or apoptotic body formation were not observed, suggesting that this acinar cell death is not apoptosis. Also, the basement membrane was preserved and there was no scattering of cellular contents, indicating that it is not necrosis. The appearance of numerous cytoplasmic vacuoles indicates that the acinar cell death in *Psti*/- mice is via an autophagic cell death mechanism. Autophagic cell death is one type of major physiological cell death, characterized by the formation of autophagic vacuoles in the cytoplasm. On the other hand, the histological appearance of the pancreas of *Psti*+/- mice was quite normal, suggesting that one functional allele is sufficient to maintain normal structure and function of the pancreas under unstimulated condition.

The results of *Psti$^{-/-}$* mice can be summarized as follows:

1. PSTI activity is indispensable to maintain exocrine integrity of the pancreas.
2. A lack of PSTI activity induces massive autophagic cell death of pancreatic acinar cells, by itself.
3. One functional PSTI allele is sufficient to maintain normal structure and function of the pancreas under unstimulated condition.

11.4   FUNCTIONAL ANALYSIS OF RECOMBINANT PSTI PROTEINS WITH AMINO ACID SUBSTITUTION

We hypothesized that mutation of the PSTI gene may promote predisposition to pancreatitis, possibly by lowering the function of inhibiting trypsin activity. Based on the hypothesis, we developed a producing system of PSTI recombinant proteins and performed a biochemical analysis.

Trypsin inhibitory activity of recombinant protein was analyzed using human and bovine trypsin.[10] The activity of PSTI protein with a point mutation of the most common type, N34S, was compared to that of the wild type. The function of the N34S PSTI remained unchanged under both usual alkali and acidic conditions as compared to the wild type PSTI (Figure 11.2). Calcium concentration did not affect the activity of recombinant PSTI. Trypsin susceptibility of the N34S protein did not increase either.

The interaction of recombinant N34S with human and bovine trypsin was also analyzed by using a surface-plasmon-resonance (SPR) biosensor technique.[11] The binding kinetics of N34S PSTI protein to trypsin was compared to the wild type. The binding kinetics of N34S PSTI did not decrease as compared to the wild type PSTI (Figure 11.3). These results along with enzymatic functional analysis suggest that other mechanisms than the conformational change of PSTI by amino acid substitution may possibly underlie the predisposition to pancreatitis in patients with N34S. Enhanced digestion of N34S by enzymes other than trypsin and abnormal splicing may be applicable.

N34S is usually associated with two intronic mutations (i.e., IVS1-37T>C and IVS3-69insTTTT). Mutations in the intronic polypyrimidine tract, such as IVS3-69insTTTT that converts consecutive T5 to a T9 structure, sometimes result in a splicing abnormality. Hence, intronic mutations rather than N34S itself may be associated with the decreased function of PSTI. The failure to exhibit function loss in the recombinant N34S PSTI protein supports this possibility.

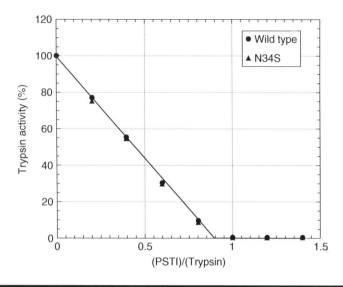

Figure 11.2 Inhibitory activity of recombinant PSTI proteins for human trypsin. The function of N34S PSTI (triangles) remained unchanged compared with wild type PSTI (circles).

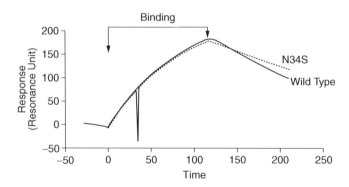

Figure 11.3 Binding affinity of recombinant PSTI proteins to human trypsin. The interaction of recombinant N34S with human trypsin was analyzed using an SPR biosensor technique. The binding kinetics of N34S PSTI did not decrease compared with wild type PSTI.

As for R67C, there are many isoforms produced in the recombinant protein producing system, as suggested above. All these isoforms of R67C recombinant PSTI protein lost their reactivities with anti-PSTI (wild type) antibody, suggesting the massive conformational alterations. As a result of the above-mentioned reasons, we could not purify recombinant R67C PSTI. R67C is possibly associated with the predisposition to pancreatitis.

The results of recombinant protein experiments are summarized as follows:

1. Trypsin inhibitory activity of N34S recombinant PSTI protein was maintained *in vitro*, suggesting that one of the accompanying intronic mutations, rather than N34S itself, may be associated with the predisposition to pancreatitis.
2. R67C may diminish the biological activity of PSTI due to its conformational alteration.

11.5 EPILOGUE

Figure 11.4 represents the intrapancreatic balance of trypsin and PSTI activity. Lack of PSTI activity can promote the massive pancreatic acinar cell death by itself, as shown in the knockout mice experiment. One

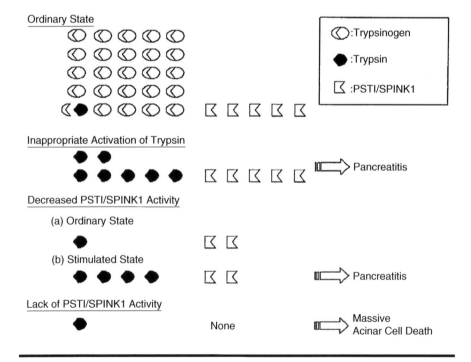

Figure 11.4 Intrapancreatic balance of trypsin and PSTI. Genetic mutations in the PSTI gene seem to promote a predisposition to pancreatitis, possibly by lowering the threshold for pancreatitis.

functional PSTI allele is sufficient to maintain normal structure of the pancreas under an unstimulated condition. However, the mutations in the PSTI gene seem to promote a predisposition to pancreatitis, possibly by lowering the threshold for pancreatitis. Some pancreatotoxic factors, such as alcohol, may easily activate trypsin in patients with PSTI gene mutation due to the decreased trypsin-inhibitory capacity.

REFERENCES

1. Maeda K., Hirota M., Kimura Y., Ichihara A., Ohmuraya M., Sugita H., Ogawa M. Proinflammatory role of trypsin and protease-activated receptor-2 in a rat model of acute pancreatitis. *Pancreas* 2005 (in press).
2. Chen J.M., Mercier B., Audrezet M.P., Raguenes O., Quere I., and Ferec C. 2001. Mutations of the pancreatic secretory trypsin inhibitor (PSTI) gene in idiopathic chronic pancreatitis. *Gastroenterology* 120:1061–1063.
3. Kaneko K., Nagasaki Y., Furukawa T., Mizutamari H., Sato A., Masamune A., Shimosegawa T., and Horii A. 2001. Analysis of the human pancreatic secretory trypsin inhibitor (PSTI) gene mutations in Japanese patients with chronic pancreatitis. *J. Hum. Genet.* 46:293–297.
4. Kuwata K., Hirota M., Sugita H., Kai M., Hayashi M., Nakamura M., Matsuura T., Adachi N., Nishimori I., and Ogawa M. 2001. Genetic mutations in exons 3 and 4 of the pancreatic secretory trypsin inhibitor in patients with pancreatitis. *J. Gastroenterol.* 36:612–618.
5. Ogawa M., Hirota M., and Kuwata K. 1999. Genetic mutations in pancreatic secretory trypsin inhibitor (PSTI) gene in patients with pancreatitis. Reported at the Annual Meeting of Research Committee for Intractable Pancreatic Diseases in Japan on September 9, 1999.
6. Pfutzer R.H., Barmada M.M., Brunskill A.P., Finch R., Hart P.S., Neoptolemos J., Furey W.F., and Whitcomb D.C. 2000. SPINK/PSTI polymorphisms act as disease modifiers in familial and idiopathic chronic pancreatitis. *Gastroenterology* 119:615–623.
7. Witt H., Luck W., Hennies H.C., Classen M., Kage A., Lass U., and Landt O. 2000. Mutations in the gene encoding the serine protease inhibitor, Kazal Type I are associated with chronic pancreatitis. *Nature Genet.* 25:213–216.
8. Kuwata K., Hirota M., Nishimori I., Otsuki M., and Ogawa M. 2003. Mutational analysis of the pancreatic secretory trypsin inhibitor (*PSTI*) gene in familial and juvenile pancreatitis in Japan. *J. Gastroenterol.* 38:365–370.
9. Ohmuraya M., Hirota M., Araki M., Mizushima N., Matsui M., Mizumoto T., Haruna K., Kume S., Takeya M., Ogawa M., Araki K., and Yamamura K. Lack of pancreatic secretory trypsin inhibitor induces autophagic cell death of pancreatic acinar cells. Submitted.
10. Kuwata K., Hirota M., Shimizu H., Nakae M., Nishihara S., Takimoto A., Mitsushima K., Kikuchi N., Endo S., Inoue M., and Ogawa M. 2002. Functional analysis of recombinant pancreatic secretory trypsin inhibitor proteins with amino acid substitution. *J. Gastroenterol.* 37:928–934.
11. Hirota M., Kuwata K., Ohmuraya M., and Ogawa M. 2003. From acute to chronic pancreatitis: the role of mutations in the pancreatic secretory trypsin inhibitor gene. *J. Pancreas* 4:83–88.

12

PATHOLOGICAL PERSPECTIVES OF PANCREATIC MUCINS

Mahefatiana Andrianifahanana, Nicolas Moniaux, and Surinder K. Batra

CONTENTS

12.1   INTRODUCTION

Historically, the term *mucin* (MUC for human and Muc for other species) was coined in reference to members of a family of large glycoproteins representing the major structural components of the mucus.[1–3] Thus far, a total of nineteen human mucins have been identified and more are likely awaiting discovery. These include MUC1, MUC2, MUC3A, MUC3B, MUC4, MUC5AC, MUC5B, MUC6, MUC7, MUC8, MUC9, MUC11, MUC12, MUC13, MUC15, MUC16, MUC17, MUC19, and MUC20.

Due to the unique characteristics and multifunctional properties of mucins, there has been a renewed interest in an improved understanding of their intricate biology, as evidenced by the plethora of literature that has accumulated over the past decade.* Among the many facets of mucins, aspects pertaining to mechanisms underlying the regulation of their production carry a particular biological importance. Under normal physiological conditions, mucins feature a well-defined profile of tissue-, time-, and developmental state–specific expression. Nonetheless, this tightly regulated homeostatic expression is often compromised by a variety of insults that affect cellular integrity, thereby leading to pathological conditions. Deregulated mucin expression has been considered as one of the most prominent characteristics of numerous types of inflammatory diseases and cancers.[1,4]

In this chapter, we will attempt to summarize the past findings and discuss recent advances related to mucin expression, with a particular emphasis on pathophysiological implications in the pancreas.

12.2 STRUCTURE AND CLASSIFICATION OF MUCINS

The basic structure of a mucin molecule reveals a protein backbone, termed "apomucin," decorated with a large number of O-linked oligosaccharides and a few N-glycan chains. Additional modifications including sialylation or sulfation are also commonly observed on mature mucin glycoproteins. A hallmark of mucins is the occurrence of specific tandemly repeated motifs (often referred to as the "tandem repeats" [TRs]) that occupy a central position in the protein backbones.[1,3,5] Due to the fluctuating sizes of their TR regions, mucins frequently exhibit a polymorphism of the variable number of tandem repeat (VNTR) type, both within and between individuals. Of note, TRs are particularly rich in serine, threonine, and proline residues, which represent potential sites for extensive O-glycosylation.[6,7] The sugar moieties constitute up to 80% of the mucin's total mass and confer an elongated and rigid structure upon the molecule, while contributing to the viscoelastic properties of mucus secretions.[1,3,8,9]

On the basis of their structural characteristics and physiological fates, mucins have been categorized into three classes — the secreted/gel-forming, membrane-bound, and soluble mucins (Table 12.1, Figure 12.1). The main difference between members of these classes lies with the

---

* http://www.ncbi.nlm.nih.gov/entrez/query.fcgi?db=PubMed (*keyword:* "mucin").

Table 12.1 Main Characteristics of the Human Mucin Genes

MUC gene	Subfamily	Locus	RNA	Splice Form Identified	Repeat Size aa	Apomucin Molecular Weight	Experimental Molecular Weight	Specific domain
MUC1	Membrane	1q21	4.4 kb	yes	20	122 kDa	250 to 1000 kDa	β-catenin, SEA
MUC2	Secreted, gel forming	11p15.5	15.7 kb	no	23	540 kDa	ND	pVW
MUC3	Membrane	7q22	ND	yes	17	ND	ND	EGFs, SEA
MUC4	Membrane	3q29	26.5 kb	yes	16	930 kDa	ND	EGFs
MUC5AC	Secreted, gel forming	11p15.5	17.5 kb	no	8	641 kDa	ND	pVW
MUC5B	Secreted, gel forming	11p15.5	17 kb	no	29	620 kDa	ND	pVW
MUC6	Secreted, gel forming	11p15.5	ND	ND	169	ND	ND	pVW
MUC7	Secreted	4q13-q21	2.3 kb	no	23	41.5 kDa	150 to 200 kDa	
MUC8	Secreted	12q24-3	ND	ND	13 and 41	ND	ND	ND
MUC9	Secreted	1p13	2.3 kb	ND	15	83 kDa	120 kDa	ND
MUC11	ND	7q22	ND	ND	28	ND	ND	ND
MUC12	Membrane	7q22	ND	ND	28	ND	ND	EGFs, SEA
MUC13	Membrane	3q13.3	2.8 kb	yes	27	54.7 kDa	175 kDa	EGFs, SEA
MUC15	Membrane	11p14.3	3.4 kb	yes	none	33.3 kDa	110 kDa	
MUC16	Membrane	19q13.3	35 kb	no	156	>1000 kDa	ND	SEA
MUC17	Membrane	7q22	ND	ND	59	ND	ND	EGFs, SEA
MUC19	Secreted, gel forming	12q12	ND	no	ND	ND	ND	pVW
MUC20	Membrane	3q29	2.4 kb	yes	18	55 to 79 kDa	ND	

Abbreviations used: aa, amino acid; ND, not determined; EGF, epidermal growth factor; SEA, sea-urchin sperm protein, enterokinase and agrin; pVW, pro von Willebrand.

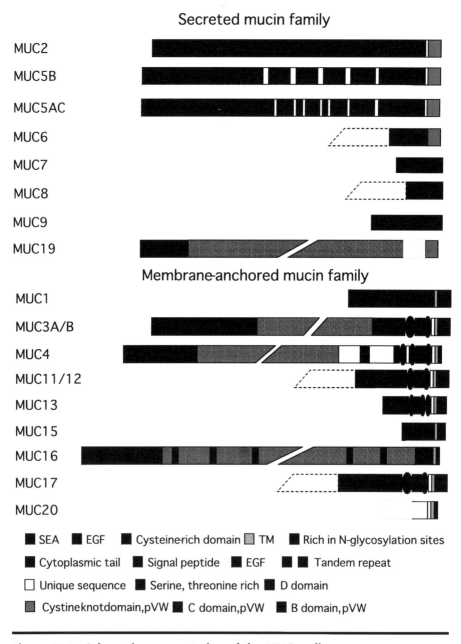

Figure 12.1 Schematic representation of the MUC coding sequences.

presence of a transmembrane domain, which serves to anchor membrane-bound mucins to the plasmalemma.[10–15] Moreover, a distinctive-property of secreted mucins is their ability to form a network of multimeric complexes as a result of intra- and intermolecular associations via cysteine-containing domains in their protein sequences.[3,16] Two mucins, MUC8 and MUC11, have only been partially characterized and, therefore, remain unclassified. Furthermore, due to the structural features of MUC9 and MUC15, their assignment to the mucin family has been subject to controversy. For example, although classification of MUC15 among membrane-bound mucins has been proposed,[17] the lack of TR may link this glycoprotein to the family of mucinlike molecules (e.g., endomucin-1 and endomucin-2).[18]

12.2.1 Secreted Mucins

The supramucosal mucus consists of a secreted gel layer and a membrane-associated glycocalyx.[19,20] Secreted mucins are produced by specialized structures generally referred to as the goblet cells (GC), from the surface epithelium, and the mucous cells, from the submucosal glands.[21] Members of this class of mucins are the major components of the supramucosal insoluble gel-like mucus layer. Genes encoding the classical gel-forming mucins, including MUC2, MUC5AC, MUC5B, and MUC6 (Table 12.1) are clustered within a 400-kb long, GC-rich region of band p15.5 on Chromosome 11.[16] These genes present a large number of symmetrical and repeated structures, suggesting the occurrence of duplication events and a common ancestral origin.[16,22,23] In a recent study, a novel secreted mucin, MUC19, has been identified. Unlike the 11p15 MUC genes, MUC19 localizes to Chromosome 12 in the q12 region.[24] From a structural standpoint, the apomucin moiety of secreted mucins is typically characterized by the presence, at their distal ends, of cysteine-rich regions termed CK (Cystine Knot) and D domains, which exhibit extensive similarities to corresponding regions of the blood-clotting pre-pro von Willebrand factor (vWF).[23,25,26] While the CK domains carry sites for mucin dimerization within the rough endoplasmic reticulum (RER),[27–29] the D domains facilitate their subsequent oligomerization.[30]

12.2.2 Membrane-Bound Mucins

This group comprises a total of nine mucins, including MUC1, MUC3A, MUC3B, MUC4, MUC12, MUC13, MUC16, MUC17, and MUC20 (Table 12.1). The gene coding for MUC1 maps to Chromosome 1, whereas MUC3, MUC12, and MUC17 are located within Band q22 of Chromosome 7.[2,3] The recent characterization of MUC20[31] has allowed the localization of this gene to Chromosome 3q29 near the MUC4 locus, suggesting the existence of yet another cluster of mucin genes. From amino- to carboxyl-terminal ends, the overall structure of a membrane-bound mucin exposes three main regions:

1. An extracellular domain, or ectodomain, which carries the central TR region flanked by unique sequences
2. A single-pass transmembrane domain
3. A cytoplasmic tail

The ectodomain protrudes into the glycocalyx and is organized into modules that harbor at least three types of putatively functional domains, including the epidermal growth factor (EGF)-like domains that are present in MUC3,[32] MUC4,[3] MUC12,[13] MUC13,[15] and MUC17[33]; the sea-urchin sperm protein, enterokinase, and agrin (SEA) module, which has been reported for MUC1,[34] MUC13,[15] and MUC17[33]; and the cysteine-rich domains, as observed with MUC4.[35] The carboxyl-proximal portion of the molecule consists of a short cytoplasmic tail, which may contain potential serine or tyrosine phosphorylation sites.[14,32,33,35-37] These sites are thought to play a role in signal transduction.[37,38] Despite their membrane-bound nature, these mucins may also yield soluble products as a result of proteolytic cleavage events[14,33,35,39-43] or via generation of alternatively spliced gene products.[12,14,44-49] The functional relevance of these variants, however, remains unclear.

12.2.3 Soluble Mucin

Thus far, only one soluble mucin, MUC7, has been identified.[50] MUC7 is a relatively small glycoprotein (39-kDa apomucin) that does not share the common characteristics of other secreted mucins, with the exception of its high contents in serine and threonine residues and the presence of the TR region.[51] Moreover, MUC7 is not known to multimerize and, therefore, is incapable of forming a gel. Nonetheless, dimeric species of MUC7 have been recovered from saliva secretions.[52]

12.3 MUCIN BIOSYNTHESIS

The biosynthesis of mucins is channeled via separate pathways depending on their ultimate location in different cellular compartments (i.e., secreted vs. membrane-bound). Following transcription of the MUC genes, translation occurs along polysomal structures, where *N*-glycosylation takes place early, concomitant with the internalization of nascent peptides into RERs. *N*-glycosylation occurs at the consensus peptide sequence asparagine-X-serine/threonine (NXS/T, where X represents any amino acid), located mostly outside the TR regions[1,8,35,53,54] and often associated with a β-turn structure.[55] For secreted mucins, dimerization subsequently ensues, a process that is facilitated by specific N-glycan chains, as observed with MUC2. Moreover, N-glycans (e.g., MUC2) are recognized by the ER lectins, calnexin and calreticulin, which may modulate mucin synthesis during folding and oligomerization in the ER.[56] These events are followed by O-glycosylation, oligomerization, and storage of mature mucins in secretory granules for further release into the extracellular environment. It is believed that dimerization and multimerization take place in distinct subcellular compartments.[23,27,29,57] O-glycosylation of mucin monomers and dimers takes place during their transit through the Golgi cisternae. This occurs via well-defined pathways, dictated by the substrate specificity of individual glycosyltransferases, which, in turn, are responsible for the sequential addition of monosaccharides to the growing oligosaccharide chain along the mucin peptide.[27,29,57–60] The amino acid sequence, XTPXP, has been proposed to serve as a signal for protein O-glycosylation.[61]

The biosynthesis of the membrane-bound mucin, MUC1, features a unique two-step mechanism. Following its neosynthesis, the partially glycosylated MUC1 molecule is shuttled to the cell membrane and subsequently reinternalized toward the trans-Golgi, where O-glycosylation is completed. The resulting MUC1 glycoprotein is then retranslocated to the cell membrane in its mature form.[62,63] On the other hand, synthesis of the rat homologue to MUC4, sialomucin complex (SMC), involves early proteolytic cleavage of the SMC precursor prior to substantial O-glycosylation and during its transit from the ER to the Golgi. The resulting subunits, ascites sialoglycoprotein 1 (ASGP-1) and ASGP-2, subsequently reassociate via stable but non-covalent linkages to yield the mature SMC molecule.[41,64]

12.4 EXPRESSION PROFILES OF HUMAN MUCINS IN THE PANCREAS

Mucins are produced by a variety of epithelial cells. They are commonly encountered at, although not restricted to, the interface between epithelial tissues lining ducts or lumens and the adjoining external milieu.[1,3] Mucins are characterized by a well-defined expression pattern, which may be disrupted in response to a variety of environmental insults, or following intrinsic changes (e.g., epigenetic modifications) that alter gene expression or secretion rate. Different organs serving specific functions display distinct profiles of mucin expression. Nonetheless, a certain number of mucins share an overlapping expression among different organs, although their biochemical characteristics (e.g., glycosylation pattern) are not necessarily identical depending on the type of organ, developmental stage, or phys-

Table 12.2 Mucin Gene Expression under Different Pancreatic Physiologic Conditions

Gene	Fetus	Normal Pancreas	Pancreatitis	IPMT	PanIN	Adenocarcinoma Tissue	Adenocarcinoma Cell Line
MUC1	-/+	+	++	+	++	+++	+++
MUC2	-	-	-	+++	-	-	-/++
MUC3	-	±	±	±	ND	±	-/+
MUC4	-	-	-	++	+ to +++	+++	+++
MUC5AC	-	+	-/+	++	ND	-/+	-/+
MUC5B	-	+	++	++	ND	+/++	-/++
MUC6	+	++	++/+++	++	ND	++/+++	-/+++
MUC11/12	ND	-/+	ND	ND	ND	ND	ND
MUC16	ND	ND	ND	ND	ND	++	ND
MUC17	ND	ND	ND	ND	ND	ND	+

Abbreviations used: IPMT, intraductal papillary-mucinous tumors; PanIN, pancreatic intraepithelial neoplasia; -, negative; +, low; ++, moderate; +++, intense; ND, none determined; -/+, from negative to positive depending on the sample; ±, negative or positive depending on the study.

iological state (i.e., normal vs. diseased) of the cells.[65,66] The major types of mucin-producing organs are generally assigned to three main spheres according to their localization and functional traits — the airway/respiratory system, the digestive tract, and the genitourinary system. Mucins are also expressed in other locales such as the mammary glands as well as the auditory and ocular systems.

Several mucins are commonly expressed in the normal pancreas (Table 12.2), including MUC1, MUC5AC, MUC5B, MUC6, MUC11, MUC12, and MUC20, whereas MUC2, MUC3, MUC4, MUC7, and MUC17 usually remain undetectable.[31,33,67–70] Low levels of MUC13 have also been observed in pancreatic acinar and ductal cells.[15] Some studies, however, have reported a heterogeneous expression of MUC3 in epithelial cells of interlobular pancreatic ducts[68,71] and the lack of detection of MUC5AC, both in ductal and acinar cells.[68,70] It is likely that these discrepancies may be linked to variations in the sensitivity or specificity of the technical procedures used in these studies or to the heterogeneity of the clinical material analyzed.[67] The expression profiles of pancreatic mucins are altered during inflammation and carcinogenesis. Chronic pancreatitis, an inflammatory disorder of the pancreas, is characterized by a downregulation of MUC5AC and, to a lesser extent, MUC1.[67] In pancreatic ductal adenocarcinoma, the prevalent type of exocrine pancreatic malignancy, the most frequently observed changes include the upregulation of MUC1 and aberrant/ectopic expression of MUC4.[67,72–77] We have also reported the upregulation of MUC2 in several pancreatic tumor cell lines,[67] although in clinical specimens, its ectopic expression is mostly observed in non-invasive intraductal papillary mucinous tumors of the villous dark-cell type, but not in invasive ductal carcinomas.[70,78] Moreover, a downregulation of MUC5AC in a large number of pancreatic tumors and its expression to varying levels in cell lines have been reported.[67] A high percentage of intraductal papillary mucinous tumors and a few cases of invasive ductal carcinomas were shown to express this mucin. Importantly, expression of MUC5AC has been correlated with a better survival prognosis in pancreatic cancer patients.[70] Recently, upregulation of MUC17 has been reported in several pancreatic cancer cell lines.[33]

MUC gene expression has not been extensively studied in intraductal papillary-mucinous tumors (IPMTs) of the pancreas. MUC2 and MUC5AC levels were shown to be high in IPMT compared to adenocarcinomas.[79–83] Recently, an altered expression of MUC genes (overexpression of MUC2 and MUC5AC) in IPMTs was reported, and the expression profile correlated with types and lesions of IPMTs.

To recapitulate the accrued knowledge about genetic and morphological changes associated with different stages of pancreatic tumors, a progression model for pancreatic carcinogenesis has been proposed.* This model highlights the development of infiltrating tumors from histologically identifiable precursor lesions known as pancreatic intraepithelial neoplasias (PanINs), where a progression from a low-grade PanIN1A to high-grade PanIN3 is defined according to the severity of dysplasia.[82,83] Interestingly, the expression profiles of MUC3,[71] MUC4,[71,77] MUC5AC,[84] and MUC6,[84] and of the mucin-associated tumor antigens, sialyl Tn[84] and sialyl Le[x],[71] suggest a potential role of these mucins in the progression of pancreatic cancer.

In the hepatobiliary system, large bile ducts of normal livers exhibit high levels of MUC3 and low amounts of MUC5B, contrasting with small bile ducts, which show no detectable expression of any mucins. An occasional expression of MUC1, MUC2, MUC5AC, and MUC6 also occurs in these structures, whereas MUC4 and MUC7 usually remain undetectable. Moreover, hepatocytes display a weak and diffuse expression of MUC3.[68] The normal gall bladder features a high level of MUC3, moderate amounts of MUC5B and MUC6, and low expression of MUC1, MUC2, and MUC5AC.[68,85] Increased mucus secretion by gall bladder epithelial cells has been implicated in gallstone formation.[86] Similar to the pancreas, the normal gall bladder does not express MUC4; however, a strong upregulation of this mucin has been observed in gall bladder adenocarcinomas.[68]

---

* http://pathology.jhu.edu/pancreas_panin.

Although mucin expression has been canonically attributed to cells of epithelial origin, certain mucins are also detected in other cell types. For example, expression of the membrane-bound mucin, MUC1, has been observed in several types of resting and activated immune cells.[87–90] A variant form of MUC1, MUC1/Z, is also expressed by non-epithelial tumor cell lines, including neuroblastomas, astrocytomas, and melanomas.[91] Moreover, we have recently reported the detection of MUC4 transcript in the peripheral blood mononuclear cells (PBMCs) of pancreatic cancer patients.[69] Another membrane-bound mucin, MUC13, has also been iden-tified in mature lymphoid cells.[15] Thus far, the physiological significance of these nonclassical mucin occurrences or their biochemical attributes has remained obscure and merits further investigation. Nonetheless, it has been proposed that MUC1 may serve as a negative regulator of T-cell activation.[89]

12.5 ABNORMAL GLYCOSYLATION OF MUCINS IN PANCREATIC CANCER

Aberrant glycosylation is common in many pathological conditions includ-ing cancers. Many glycosyl epitopes constitute tumor-associated antigens (TAAs) and have been used as a target for immunotherapy and diagnosis. Aberrant glycosylation as a result or a cause of cancer remains enigmatic. Many recent studies indicate that some, if not all, cases of aberrant glycosylation are the results of initial oncogenic transformation and rep-resent key events in the induction of invasion and metastasis. The mucin-type O-glycans are attached to the glycoproteins through O-glycosidic linkages between N-acetylgalactosamine and serine or threonine amino acid residues in the apomucin moieties (Figure 12.2). O-glycans are assembled by a series of reactions catalyzed by glycosyltransferases and sulfotransferases in the Golgi compartment. In cancer cells, the expression pattern of these enzymes is dysregulated and leads to tumor-specific glycanic epitopes.[92,93] Depending on the glycosyltransferases expressed, hundreds of different O-glycan structures have been described and clas-sified based on their cores. At least eight distinct cores have been char-acterized in mammalian species. All the cores have the N-acetylgalactosamine linked to the residues of serine or threonine that represents the Tn antigen in common. When sialylated, this antigen forms sialyl-Tn (sTn) epitope, reported to be expressed at the surface of mucin cores. Indeed, sTn is detected at the surface of MUC1 overexpressed by pancreatic adenocarcinoma cells.[94] It has been proposed that the appear-ance of the sTn antigen might result from MUC1 overexpression. The abundance of acceptor sites may saturate the posttranslational modifica-tion.[94] The sTn antigen is recognized by the mouse monoclonal antibody

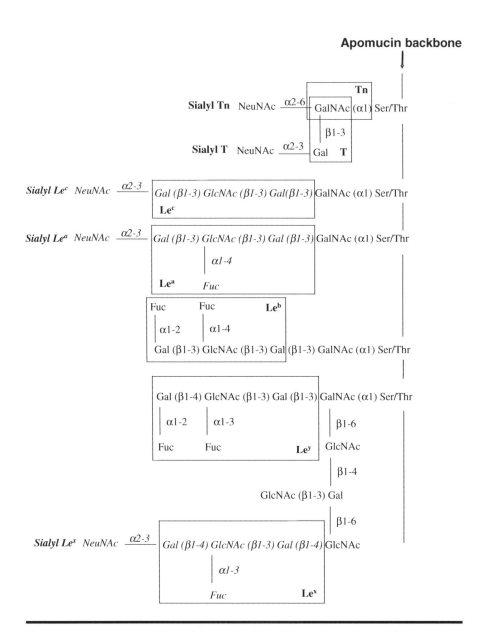

Figure 12.2   Representation of the main glycanic structure expressed on mucins.
NeuNAc, acid sialic or neuraminic acid; Gal, galactose; GalNac, *N*-acetylgalac-
tosamine; GLC, glucose; GlcNAc, *N*-acetylglucosamine; Fuc, fucose; Ser/Thr,
serine or threonine. The epitopes carried by mucins in pathological conditions
are boxed and their sialylated forms are indicated in italics.

CC49, which is extensively being investigated for the development of radioimmunotherapeutic applications.[95]

Alternatively the Tn antigen could be elongated to form one of the eight distinct cores and be terminated with sulfate, sialic acid, fucose, galactose, N-acetylglucosamine, and N-acetylgalactosamine. Depending on the physiologic conditions, the peripheric sugar may vary and give rise to tumor-associated antigens such as Le[a] (Lewis[a]), Le[b], Le[x], Le[y], sLe[a], sLe[c], and sLe[x]. The CA 19-9 previously described recognizes the sLe[a] and the DUPAN-2 is the sLe[c] antigen.

12.6 FUNCTIONAL ASPECTS OF MUCINS AND PATHOPHYSIOLOGICAL IMPLICATIONS

Classically, mucins are known to play a central role in the protection, lubrication, and moisturizing of the external surfaces of epithelial tissue layers lining the intricate network of ducts and lumens within the human body (e.g., References 7 and 20). However, several lines of evidence suggest the implication of mucins in other important and more sophisticated biological processes, such as epithelial cell renewal and differentiation, cell signaling, and cell adhesion.[37,96–98]

As previously described, both the glycosylation and the expression of mucins are dysregulated in the development and progression of pancreatic adenocarcinoma. The two main mucins that present the most deregulated expression pattern are MUC1 and MUC4. These two mucins belong to the membrane-anchored mucin subfamily. Therefore, this chapter will focus on the known functions of MUC1 and MUC4. Both mucins play a key role in the development of pancreatic cancer as well as for its progression (i.e., invasion and metastasis). The high incidence of metastases in pancreatic cancer is the major cause for its lethality. MUC1 and MUC4 are implicated in almost all of the steps associated with the development of metastases.

First of all, MUC1 and MUC4 possess antiadhesive properties. The negative charge of the O-glycosidic chains carried by their central repetitive domain provides an extended conformation to MUC1 and MUC4, with a size up to 500 nm[99] and 2.12 μm,[35] respectively. Moreover, MUC1 and MUC4 are not only upregulated in pancreatic cancer, but their upregulation is associated with the loss of their strictly apical localization. The steric hindrance caused by the overexpression of these two extended proteins disturbs the cell–cell or cell–matrix interactions.[100] This physical mechanism favors the release of the cancer cells into the circulation. MUC1 also interacts directly with β-catenin via an SXXXXXSSL motif (where X represents any amino acid residues) in its cytoplasmic tail.[101] β-catenin plays important functions in the formation of cell–cell junctions via interaction with E-cadherin.[102,103] β-catenin also binds to APC (Adenomatous Polyposis Coli),[102–104] its partner in the Wingless/Wnt-1 signalling pathway.[105] The Wnt-1 intracellular pathway is directly implicated in the development of the central nervous system.[106] Of note, the binding of β-catenin to any of its partners is a mutually exclusive process.[107] APC overexpression reduces the level of free cytoplasmic β-catenin, and thus reduces the number of β-catenin/E-cadherin complexes as well as intercellular adherence.[108,109] The formation of these complexes is regulated by the phosphorylation of the cytoplamic tail of each partner by the GSK3β.[110] After phosphorylation, β-catenin is degraded. GSK3β is also able to phosphorylate the β-catenin binding site on the cytoplasmic tail of MUC1.[111] The more the cytoplasmic tail of MUC1 is phosphorylated, the less MUC1 interacts with β-catenin.[112] The relative level of MUC1, E-cadherin, β-catenin, GSK3β, and APC seem to be critical to maintaining the epithelium integrity. Overexpression of MUC1 leads to an increase in the interaction between MUC1 and β-catenin, thereby inhibiting β-catenin/E-cadherin interaction and disrupting cell–cell contacts, which facilitates the release of the tumor cell from the tissue.

The expression of the mucin-associated carbohydrate antigens, sLe^x and sTn, appears in higher-grade PanIN lesions[71] and are thought to result from a shift in the expression of glycosyltransferases. These mucin-associated antigens are known to bind endothelial cells through the P-, E-selectins, and ICAM-1.[113,114] The P-selectin ligand (PSGL-1)[115] presents a structure similar to MUC1 and carries within its repetitive domain O-glycosidic chains with terminal sugars comparable to those found on MUC1 and MUC4 in tumors. Therefore, in addition to favoring the release of pancreatic cancer cells from tumor masses, MUC1 and MUC4 also facilitate their extravasation to the circulation.[94,116] MUC4 is also known to inhibit tumor killing by lymphokine-activated killer cells.[117]

The role of MUC1 and MUC4 on the growth of the primary pancreatic tumors has remained obscure until recently. Both MUC1 and MUC4 have been reported as potential modulators of receptor tyrosine kinase pathways. Indeed, MUC1 has been referred to as a heterodimeric partner for EGFR/HER1[118,119] and MUC4 for HER2.[120,121] Upon interaction of MUC1 with EGFR, the MUC1 cytoplasmic tail becomes phosphorylated and leads to the activation of c-Src.[118] Moreover, formation of the complex activates the mitogenic MAP kinase pathway.[119] For MUC4, its membrane-bound subunit acts as a ligand for HER2, activating this growth factor receptor, and in synergy with neuregulin, potentiates the phosphorylation of HER2 and HER3. Phosphorylation of HER2 occurs at residue Tyr1248 after binding with MUC4 and leads to the upregulation of the cell-cycle inhibitor p27^{kip}.[120] However, in the presence of neuregulin, MUC4/HER2 activation downregulates p27^{kip} but activate the protein kinase B/Akt via HER3. Therefore, MUC4 appears to be a modulator of cell proliferation. Inhibition of MUC4 expression using antisense or short-interfering RNA (siRNA) oligonucleotides specific to MUC4 resulted in inhibition of tumorigenecity and the dissemination of the cancer cells.[122] We believe that MUC4 is implicated in tumor growth and metastasis by directly altering the tumor-cell properties (adhesion/aggregation and motility), and, in part, via modulating HER2 expression.

12.7 MUCINS IN DIAGNOSIS AND THERAPY

Several TAAs have been shown to have many properties consistent with their association with mucins. Tumor-associated epitopes on mucins (Figure 12.2) have been implicated in the pathogenesis of many cancers including pancreatic cancer. Several clinically useful antibodies that recognize epitopes on mucins were initially generated against tumors of secretory epithelial cell origin from the pancreas, breast, colon, and lung, and include (but are not limited to) DUPAN2, HMFG2, SM-3, and CA 19-9. Some of the antibodies recognize epitopes that are aberrantly expressed on tumor mucin but not on corresponding normal cell populations (SM-3), whereas others recognize epitopes expressed on mucin from both normal and malignant cells from a given organ site (DUPAN2 and CA 19-9). These and other mucin-reactive antibodies generally show distinct patterns of tissue reactivity by immunohistochemical analysis.

There are no tumor-specific markers for pancreatic cancer; markers such as serum CA 19-9, DUPAN-2, and CA125 that are being used as potential targets have low specificity.[123–126] For over two decades, the structure of these antigens (CA 19-9, DUPAN2, or CA 125) has been heavily investigated for the development of serum-based immunoassays to detect early cancers. These antigens contain oligosaccharide structures that are being recognized by monoclonal antibodies. Further, these saccharide epitopes are present on the O-glycosidic chains harbored by the apomucin backbones. For example, MUC1 carries CA19-9[94,127,128] and DUPAN-2[75,127] epitopes, and MUC16 carries the CA 125 epitope.[129]

At this time, CA19-9 remains the only Food and Drug Administration (FDA)-approved marker to monitor pancreatic cancer. The CA 19-9 recognizes a mucin-type glycoprotein sialosyl Lewis A/B antigen.[130] Because 5% of the general population are Lewis A/B negative, the maximum sensitivity of this marker is 95%.[131,132] Patients (65%) with pancreatic cancer will have CA 19-9 levels greater than 120 U/L, whereas only 2% of the cases of pancreatitis will have levels this high. The specificity for pancreatic cancer increases with high levels of CA 19-9; however, most of these cancers will be unresectable.

A unique TAA is TAG-72, a pan-adenocarcinoma antigen that is expressed by majority of human adenocarcinoma of pancreas, colon, ovary, prostate, lung, and esophagus and is absent in most normal tissues. TAG-72 has been identified by its immunoreactivity with the monoclonal antibody (MAb) B72.3, a murine MAb, which was developed by the immunization of mice with a membrane-enriched fraction of human metastatic breast carcinoma tissue. The epitope recognized by MAb B72.3 is sialyl-Tn, a unique disaccharide present in multiple copies on the tumor-associated mucin TAG-72.

A retrospective analysis of 25 primary adenocarcinomas of the pancreas, 16 metastatic pancreatic tumors, 8 cases of chronic pancreatitis, and 3 adult normal pancreas for the expression of TAG-72 antigen using MAb B72.3 was performed.[133] Out of 25 malignant primary tumors, 21 (83%) were reactive, and all 16 metastatic sites expressed the B72.3 antigen. In contrast, all cases of pancreatitis and normal pancreas were either weakly reactive or nonreactive. Interestingly, 10 malignant and 2 benign pancreatic fine-needle aspirates showed results similar to those seen with fixed tissues. Another study examined the incidence and expression of TAG-72 along with other tumor-associated antigens CA19.9, DU-PAN2, and CA125 in serum and tissues of patients with pancreatic cancer.[134] Eighty-three percent of tumor tissues demonstrated the expression of 3 or more antigens. The least commonly detected antigen was CA125. Expression of TAG-72 was rare in the normal pancreas but was commonly expressed in ductal cells of chronic pancreatitis. Serologic coexpression of elevated antigen levels was less common because 30% of the patients showed increased levels of 3 or more antigens.

Nevertheless, the unique epitopes provided by these antigens are being used as potential targets for active and passive immunotherapy of cancer. Radioimmunoconjugates against TAG-72 of B72.3 and CC49 monoclonal antibodies have been tested in clinical trials for colorectal, pancreatic, and other cancers. Results so far in Phase I and Phase II trials (systemic administration) of these agents have not produced cures for several reasons.[135] A new generation of genetically engineered immunoconjugates, with desired pharmacokinetics and biodistribution properties, has been developed to address these issues and are in the process of being evaluated in preclinical and clinical trials.[136]

The immune system has the potential to recognize such TAA structures as foreign and to mount specific immune responses against them, so as to reject tumor cells. This provides the basis for the development of active specific immunotherapeutic agents (tumor vaccines). The first mucin cDNA to be cloned, MUC1, is highly overexpressed and differentially glycosylated by pancreatic adenocarcinomas and is a cell surface-associated mucin with a structure that is remarkably similar to some of the selectins and selectin ligands. There have been studies that evaluated the effect of immunizations with MUC1 on immunity to tumors in animal models. Vaccine strategies against MUC1 include DNA, peptide, glycopeptide, fusion proteins, and recombinant vaccines delivered by adenovirus, vaccinia virus, and other viral vectors, administered directly or *in vivo* to manipulated cells.

In one study, the vaccinia vector expressing MUC1 (epithelial tumor antigen) provided protection in a murine model against tumors expressing ETA; however, the specificity of the protective response for MUC1 was not demonstrated unequivocally.[137] In another study, major histocompatibility complex (MHC)-restricted MUC1-specific CTLs were produced in C57BL/6 mice immunized with a MUC1-mannan fusion protein.[138] CTLs that specifically recognize carbohydrate structures on specific peptides in association with MHC Class I molecules have been described.[139]

Human T-lymphocytes that recognize epitopes on the MUC1 core protein have been produced by *in vitro* stimulation with MUC1 from patients with breast, pancreatic, and ovarian carcinomas.[140,141] These *in vitro* stimulated T-lymphocytes are unusual in that they show some cytotoxic activity against human tumor cells but they are not MHC-restricted in their specificity. The fact that they are derived from patients with progressive disease suggests that some mechanism prevents these cells from eradicating tumors *in situ*. There is an anecdotal report, however, of a case in which a long-term breast cancer survivor has relatively high levels of CTL precursors with MHC restricted specificity for MUC1 and a clinical history that is consistent with the development of anti-MUC1 responses *in situ*. A Phase I trial was conducted by Finn and collaborators[142] in which 3 doses of 100 mg of 105 amino acid peptide corresponding to 5 tandem repeat units of MUC1 mixed with BCG were administered over a 9-week period. Persistent recall DTH (delayed type hypersensitivity) responses were noted in over half of the immunized patients. No significant side effects were observed, and no objective responses were reported from this trial. Most of the studies on MUC1 utilized exclusively the tandem repeat moiety of the protein. Because this portion of the molecule is secreted and found on the extracellular domain of tumor cells, it may not be the best target for immunization procedures that will produce conventional cellular immune responses (i.e., to peptides associated with MHC molecules). Recent studies are investigating the use of other domains of MUC1 such as the signal sequence or cytoplasmic tail.

12.8 EPILOGUE

Mucins are members of an expanding family of large multifunctional glycoproteins. Pancreatic mucins have important biological roles in the lubrication and protection of normal ducts. The functional roles of these mucins in normal and pathological conditions are highly complex. The study of the mucins structures as well as the relationship between structure and function show that mucins possess several other important functions including the regulation of differentiation and proliferation of normal and malignant cells, through processes of ligand–receptor interactions and morphogenetic signal transduction. Under normal physiological conditions, the production (synthesis or secretion) of mucins is maintained by a host of elaborate and coordinated regulatory mechanisms, thereby affording a well-defined pattern of tissue-, time-, and developmental state-specific distribution. Mucin homeostasis may be disrupted by the action of environmental or intrinsic factors that affect cellular integrity. The net result is an altered cell behavior, often culminating into a variety of pathological conditions. Deregulated mucin production is indeed a hallmark of numerous types of inflammatory and malignant disorders of the pancreas and other organs. Differential glycosylation, aberrant expression, and alternative splice variants of mucins are being investigated in pancreatic cancer. MUC1 and MUC4 are finding unique functions in the early diagnosis and progression of the disease. The potential for using these mucins for targeted therapy of pancreatic and other cancers is under active investigation.

ACKNOWLEDGMENTS

The authors of this chapter were supported by the grants from the National Institutes of Health (RO1 CA78590 and PO CA CA72712) and the Nebraska Research Initiative. Ms. Kristi L.W. Berger, editor, Eppley Institute, is greatly acknowledged for editorial assistance.

REFERENCES

1. Gendler S.J. and Spicer A.P. 1995. Epithelial mucin genes. *Ann. Rev. Physiol.* 57:607–634.
2. Hollingsworth M.A. and Swanson B.J. 2004. Mucins in cancer: protection and control of the cell surface. *Natl. Rev. Cancer* 4:45–60.
3. Moniaux N., Escande F., Porchet N., Aubert J.P., and Batra S.K. 2001. Structural organization and classification of the human mucin genes. *Front. Biosci.* 6:D1192–D1206.
4. Devine P.L. and McKenzie I.F. 1992. Mucins: structure, function, and associations with malignancy. *Bioessays* 14:619–625.
5. Griffiths B., Matthews D.J., West L., Attwood J., Povey S., Swallow D.M., Gum J.R., and Kim Y.S. 1990. Assignment of the polymorphic intestinal mucin gene (MUC2) to chromosome 11p15. *Ann. Hum. Genet.* 54:277–285.
6. Corfield A.P., Myerscough N., Gough M., Brockhausen I., Schauer R., and Paraskeva C. 1995. Glycosylation patterns of mucins in colonic disease. *Biochem. Soc. Trans.* 23:840–845.
7. Strous G.J. and Dekker J. 1992. Mucin-type glycoproteins. *Crit. Rev. Biochem. Mol. Biol.* 27:57–92.

8. Gum Jr. J.R. 1992. Mucin genes and the proteins they encode: structure, diversity, and regulation. *Am. J. Respir. Cell. Mol. Biol.* 7:557–564.

9. Kim Y.S. and Gum Jr. J.R. 1995. Diversity of mucin genes, structure, function, and expression. *Gastroenterology* 109:999–1001.

10. Gendler S.J., Burchell J.M., Duhig T., Lamport D., White R., Parker M., and Taylor-Papadimitriou J. 1987. Cloning of partial cDNA encoding differentiation and tumor-associated mucin glycoproteins expressed by human mammary epithelium. *Proc. Natl. Acad. Sci. USA* 84:6060–6064.

11. Gendler S.J., Lancaster C.A., Taylor-Papadimitriou J., Duhig T., Peat N., Burchell J., Pemberton L., Lalani E.N., and Wilson D. 1990. Molecular cloning and expression of human tumor-associated polymorphic epithelial mucin. *J. Biol. Chem.* 265:15286–15293.

12. Moniaux N., Escande F., Batra S.K., Porchet N., Laine A., and Aubert J.P. 2000. Alternative splicing generates a family of putative secreted and membrane-associated MUC4 mucins. *Eur. J. Biochem.* 267:4536–4544.

13. Williams S.J., McGuckin M.A., Gotley D.C., Eyre H.J., Sutherland G.R., and Antalis T.M. 1999. Two novel mucin genes down-regulated in colorectal cancer identified by differential display. *Cancer Res.* 59:4083–4089.

14. Williams S.J., Munster D.J., Quin R.J., Gotley D.C., and McGuckin M.A. 1999. The MUC3 gene encodes a transmembrane mucin and is alternatively spliced. *Biochem. Biophys. Res. Commun.* 261:83–89.

15. Williams S.J., Wreschner D.H., Tran M., Eyre H.J., Sutherland G.R., and McGuckin M.A. 2001. Muc13, a novel human cell surface mucin expressed by epithelial and hemopoietic cells. *J. Biol. Chem.* 276:18327–18336.

16. Pigny P., Guyonnet-Duperat V., Hill A.S., Pratt W.S., Galiegue-Zouitina S., d'Hooge M.C., Laine A., Van Seuningen I., Degand P., Gum J.R., Kim Y.S., Swallow D.M., Aubert J.P., and Porchet N. 1996. Human mucin genes assigned to 11p15.5: identification and organization of a cluster of genes. *Genomics* 38:340–352.

17. Pallesen L.T., Berglund L., Rasmussen L.K., Petersen T.E., and Rasmussen J.T. 2002. Isolation and characterization of MUC15, a novel cell membrane-associated mucin. *Eur. J. Biochem.* 269:2755–2763.

18. Kinoshita M., Nakamura T., Ihara M., Haraguchi T., Hiraoka Y., Tashiro K., and Noda M. 2001. Identification of human endomucin-1 and -2 as membrane-bound O-sialoglycoproteins with anti-adhesive activity. *FEBS Lett.* 499:121–126.

19. Hilkens J., Ligtenberg M.J., Vos H.L., and Litvinov S.V. 1992. Cell membrane-associated mucins and their adhesion-modulating property. *Trends Biochem. Sci.* 17:359–363.

20. van Klinken B.J., Dekker J., Buller H.A., and Einerhand A.W. 1995. Mucin gene structure and expression: protection vs. adhesion. *Am. J. Physiol.* 269:G613–G627.

21. Davis C.W. and Randell S.H. 2001. Airway goblet and mucous cells: identical, similar, or different? In *Cilia and mucus. From development to respiratory defense.* Salathe M. Ed. New York: Marcel Dekker. p. 195–210.

22. Desseyn J.L., Buisine M.P., Porchet N., Aubert J.P., Degand P., and Laine A. 1998. Evolutionary history of the 11p15 human mucin gene family. *J. Mol. Evol.* 46:102–106.

23. Perez-Vilar J. and Hill R.L. 1999. The structure and assembly of secreted mucins. *J. Biol. Chem.* 274:31751–31754.

24. Chen Y., Zhao Y.H., Kalaslavadi T.B., Hamati E., Nehrke K., Le A.D., Ann D.K., and Wu R. 2003. Genome-wide search and identification of a novel gel-forming mucin MUC19/Muc19 in glandular tissues. *Am. J. Respir. Cell Mol. Biol.* 30(2):155–165.

25. Mayadas T.N. and Wagner, D.D. 1991. von Willebrand factor biosynthesis and processing. *Ann. NY Acad. Sci.* 614:153–166.

26. Mayadas, T.N. and Wagner D.D. 1992. Vicinal cysteines in the prosequence play a role in von Willebrand factor multimer assembly. *Proc. Natl. Acad. Sci. USA* 89:3531–3535.

27. Asker N., Axelsson M.A., Olofsson S.O., and Hansson G.C. 1998. Human MUC5AC mucin dimerizes in the rough endoplasmic reticulum, similarly to the MUC2 mucin. *Biochem. J.* 335:381–387.

28. Asker N., Baeckstrom D., Axelsson M.A., Carlstedt I., and Hansson G.C. 1995. The human MUC2 mucin apoprotein appears to dimerize before O-glycosylation and shares epitopes with the 'insoluble' mucin of rat small intestine. *Biochem. J.* 308:873–880.

29. van Klinken B.J., Einerhand A.W., Buller H.A., and Dekker J. 1998. The oligomerization of a family of four genetically clustered human gastrointestinal mucins. *Glycobiology* 8:67–75.

30. Offner G.D., Nunes D.P., Keates A.C., Afdhal N.H., and Troxler R.F. 1998. The amino-terminal sequence of MUC5B contains conserved multifunctional D domains: implications for tissue-specific mucin functions. *Biochem. Biophys. Res. Commun.* 251:350–355.

31. Higuchi T., Orita T., Nakanishi S., Katsuya K., Watanabe H., Yamasaki Y., Waga I., Nanayama T., Yamamoto Y., Munger W., Sun H.W., Falk R.J., Jennette J.C., Alcorta D.A., Li H., Yamamoto T., Saito Y., and Nakamura M. 2004. Molecular cloning, genomic structure, and expression analysis of MUC20, a novel mucin protein, up-regulated in injured kidney. *J. Biol. Chem.* 279:1968–1979.

32. Gum Jr. J.R., Ho J.J., Pratt W.S., Hicks J.W., Hill A.S., Vinall L.E., Roberton A.M., Swallow D.M., and Kim Y.S. 1997. MUC3 human intestinal mucin. Analysis of gene structure, the carboxyl terminus, and a novel upstream repetitive region. *J. Biol. Chem.* 272:26678–26686.

33. Gum Jr. J.R., Crawley S.C., Hicks J.W., Szymkowski D.E., and Kim Y.S. 2002. MUC17, a novel membrane-tethered mucin. *Biochem. Biophys. Res. Commun.* 291:466–475.

34. Bork P. and Patthy L. 1995. The SEA module: a new extracellular domain associated with O-glycosylation. *Protein Sci.* 4:1421–1425.

35. Moniaux N., Nollet S., Porchet N., Degand P., Laine A., and Aubert J.P. 1999. Complete sequence of the human mucin MUC4: a putative cell membrane-associated mucin. *Biochem. J.* 338:325–333.

36. Yin B.W. and Lloyd K.O. 2001. Molecular cloning of the CA125 ovarian cancer antigen: identification as a new mucin, MUC16. *J. Biol. Chem.* 276:27371–27375.

37. Zrihan-Licht S., Baruch A., Elroy-Stein O., Keydar I., and Wreschner D.H. 1994. Tyrosine phosphorylation of the MUC1 breast cancer membrane proteins. Cytokine receptor-like molecules. *FEBS Lett.* 356:130–136.

38. Baruch A., Hartmann M., Yoeli M., Adereth Y., Greenstein S., Stadler Y., Skornik Y., Zaretsky J., Smorodinsky N.I., Keydar I., and Wreschner D.H. 1999. The breast cancer-associated MUC1 gene generates both a receptor and its cognate binding protein. *Cancer Res.* 59:1552–1561.

39. Boshell M., Lalani E.N., Pemberton L., Burchell J., Gendler S., and Taylor-Papadimitriou J. 1992. The product of the human MUC1 gene when secreted by mouse cells transfected with the full-length cDNA lacks the cytoplasmic tail. *Biochem. Biophys. Res. Commun.* 185:1–8.

40. Hanisch F.G. and Muller S. 2000. MUC1: the polymorphic appearance of a human mucin. *Glycobiology* 10:439–449.

41. Komatsu M., Arango M.E., and Carraway K.L. 2002. Synthesis and secretion of Muc4/sialomucin complex: implication of intracellular proteolysis. *Biochem. J.* 368:41–48.

42. Patton S., Gendler S.J., and Spicer A.P. 1995. The epithelial mucin, MUC1, of milk, mammary gland and other tissues. *Biochem. Biophys. Acta* 1241:407–423.

43. Wang R., Khatri I.A., and Forstner J.F. 2002. C-terminal domain of rodent intestinal mucin Muc3 is proteolytically cleaved in the endoplasmic reticulum to generate extracellular and membrane components. *Biochem. J.* 366:623–631.

44. Choudhury A., Moniaux N., Ringel J., King J., Moore E., Aubert J.P., and Batra S.K. 2001. Alternate splicing at the 3′-end of the human pancreatic tumor-associated mucin MUC4 cDNA. *Teratog. Carcinog. Mutagen.* 21:83–96.

45. Choudhury A., Moniaux N., Winpenny J.P., Hollingsworth M.A., Aubert J.P., and Batra S.K. 2000. Human MUC4 mucin cDNA and its variants in pancreatic carcinoma. *J. Biochem. (Tokyo)* 128:233–243.

46. Crawley S.C., Gum Jr. J.R., Hicks J.W., Pratt W.S., Aubert J.P., Swallow D.M., and Kim Y.S. 1999. Genomic organization and structure of the 3′ region of human MUC3: alternative splicing predicts membrane-bound and soluble forms of the mucin. *Biochem. Biophys. Res. Commun.* 263:728–736.

47. Ligtenberg M.J., Vos H.L., Gennissen A.M., and Hilkens J. 1990. Episialin, a carcinoma-associated mucin, is generated by a polymorphic gene encoding splice variants with alternative amino termini. *J. Biol. Chem.* 265:5573–5578.

48. Smorodinsky N., Weiss M., Hartmann M.L., Baruch A., Harness E., Yaakobovitz M., Keydar I., and Wreschner D.H. 1996. Detection of a secreted MUC1/SEC protein by MUC1 isoform specific monoclonal antibodies. *Biochem. Biophys. Res. Commun.* 228:115–121.

49. Wreschner D.H., Hareuveni M., Tsarfaty I., Smorodinsky N., Horev J., Zaretsky J., Kotkes P., Weiss M., Lathe R., and Dion A. 1990. Human epithelial tumor antigen cDNA sequences. Differential splicing may generate multiple protein forms. *Eur. J. Biochem.* 189:463–473.

50. Bobek L.A., Tsai H., Biesbrock A.R., and Levine M.J. 1993. Molecular cloning, sequence, and specificity of expression of the gene encoding the low molecular weight human salivary mucin (MUC7). *J. Biol. Chem.* 268:20563–20569.

51. Bobek L.A., Liu J., Sait S.N., Shows T.B., Bobek Y.A., and Levine M.J. 1996. Structure and chromosomal localization of the human salivary mucin gene, MUC7. *Genomics* 31:277–282.

52. Mehrotra R., Thornton D.J., and Sheehan J.K. 1998. Isolation and physical characterization of the MUC7 (MG2) mucin from saliva: evidence for self-association. *Biochem. J.* 334:415–422.

53. Desseyn J.L., Aubert J.P., Van S.I, Porchet N., and Laine A. 1997. Genomic organization of the 3′ region of the human mucin gene MUC5B. *J. Biol. Chem.* 272:16873–16883.

54. Toribara N.W., Ho S.B., Gum E., Gum Jr. J.R., Lau P., and Kim Y.S. 1997. The carboxyl-terminal sequence of the human secretory mucin, MUC6. Analysis of the primary amino acid sequence. *J. Biol. Chem.* 272:16398–16403.

55. Aubert J.P., Biserte G., and Loucheux-Lefebvre M.H. 1976. Carbohydrate-peptide linkage in glycoproteins. *Arch. Biochem. Biophys.* 175:410–418.

56. McCool D.J., Okada Y., Forstner J.F., and Forstner G.G. 1999. Roles of calreticulin and calnexin during mucin synthesis in LS180 and HT29/A1 human colonic adenocarcinoma cells. *Biochem. J.* 341:593–600.

57. Sheehan J.K., Thornton D.J., Howard M., Carlstedt I., Corfield A.P., and Paraskeva C. 1996. Biosynthesis of the MUC2 mucin: evidence for a slow assembly of fully glycosylated units. *Biochem. J.* 315:1055–1060.

58. Brockhausen I. 1999. Pathways of O-glycan biosynthesis in cancer cells. *Biochem. Biophys. Acta* 1473:67–95.

59. Feizi T., Gooi H.C., Childs R.A., Picard J.K., Uemura K., Loomes L.M., Thorpe S.J., and Hounsell E.F. 1984. Tumour-associated and differentiation antigens on the carbohydrate moieties of mucin-type glycoproteins. *Biochem. Soc. Trans.* 12:591–596.

60. Van den S.P., Rudd P.M., Dwek R.A., and Opdenakker G. 1998. Concepts and principles of O-linked glycosylation. *Crit. Rev. Biochem. Mol. Biol.* 33:151–208.

61. Yoshida A., Suzuki M., Ikenaga H., and Takeuchi M. 1997. Discovery of the shortest sequence motif for high level mucin-type O-glycosylation. *J. Biol. Chem.* 272:16884–16888.

62. Litvinov S.V. and Hilkens J. 1993. The epithelial sialomucin, episialin, is sialylated during recycling. *J. Biol. Chem.* 268:21364–21371.

63. Pimental R.A., Julian J., Gendler S.J., and Carson D.D. 1996. Synthesis and intracellular trafficking of Muc-1 and mucins by polarized mouse uterine epithelial cells. *J. Biol. Chem.* 271:28128–28137.

64. Sheng Z.Q., Hull S.R., and Carraway K.L. 1990. Biosynthesis of the cell surface sialomucin complex of ascites 13762 rat mammary adenocarcinoma cells from a high molecular weight precursor. *J. Biol. Chem.* 265:8505–8510.

65. Lan M.S., Bast Jr. R.C., Colnaghi M.I., Knapp R.C., Colcher D., Schlom J., and Metzgar R.S. 1987. Co-expression of human cancer-associated epitopes on mucin molecules. *Int. J. Cancer* 39:68–72.

66. Shimizu M. and Yamauchi K. 1982. Isolation and characterization of mucin-like glycoprotein in human milk fat globule membrane. *J. Biochem. (Tokyo)* 91:515–524.

67. Andrianifahanana M., Moniaux N., Schmied B.M., Ringel J., Friess H., Hollingsworth M.A., Buchler M.W., Aubert J.P., and Batra S.K. 2001. Mucin (MUC) gene expression in human pancreatic adenocarcinoma and chronic pancreatitis: a potential role of MUC4 as a tumor marker of diagnostic significance. *Clin. Cancer Res.* 7:4033–4040.

68. Buisine M.P., Devisme L., Degand P., Dieu M.C., Gosselin B., Copin M.C., Aubert J.P., and Porchet N. 2000. Developmental mucin gene expression in the gastroduodenal tract and accessory digestive glands. II. Duodenum and liver, gallbladder, and pancreas. *J. Histochem. Cytochem.* 48:1667–1676.

69. Ringel J. and Lohr M. 2003. The MUC gene family: their role in diagnosis and early detection of pancreatic cancer. *Mol. Cancer* 2:9.

70. Yonezawa S., Horinouchi M., Osako M., Kubo M., Takao S., Arimura Y., Nagata K., Tanaka S., Sakoda K., Aikou T., and Sato E. 1999. Gene expression of gastric type mucin (MUC5AC) in pancreatic tumors: its relationship with the biological behavior of the tumor. *Pathol. Int.* 49:45–54.

71. Park H.U., Kim J.W., Kim G.E., Bae H.I., Crawley S.C., Yang S.C., Gum Jr. J., Batra S.K., Rousseau K., Swallow D.M., Sleisenger M.H., and Kim Y.S. 2003. Aberrant expression of MUC3 and MUC4 membrane-associated mucins and sialyl lex antigen in pancreatic intraepithelial neoplasia. *Pancreas* 26:E48–E54.

72. Balague C., Gambus G., Carrato C., Porchet N., Aubert J.P., Kim Y.S., and Real F.X. 1994. Altered expression of MUC2, MUC4, and MUC5 mucin genes in pancreas tissues and cancer cell lines. *Gastroenterology* 106:1054–1061.

73. Choudhury A., Singh R.K., Moniaux N., El Metwally T.H., Aubert J.P., and Batra S.K. 2000. Retinoic acid-dependent transforming growth factor-beta 2-mediated induction of MUC4 mucin expression in human pancreatic tumor cells follows retinoic acid receptor-alpha signaling pathway. *J. Biol. Chem.* 275:33929–33936.

74. Hollingsworth M.A., Strawhecker J.M., Caffrey T.C., and Mack D.R. 1994. Expression of MUC1, MUC2, MUC3 and MUC4 mucin mRNAs in human pancreatic and intestinal tumor cell lines. *Int. J. Cancer* 57:198–203.

75. Khorrami A.M., Choudhury A., Andrianifahanana M., Varshney G.C., Bhatta-charyya S.N., Hollingsworth M.A., Kaufman B., and Batra S.K. 2002. Purification and characterization of a human pancreatic adenocarcinoma mucin. *J. Biochem. (Tokyo)* 131:21–29.

76. Sirivatanauksorn V., Sirivatanauksorn Y., and Lemoine N.R. 1998. Molecular pattern of ductal pancreatic cancer. *Langenbecks Arch. Surg.* 383:105–115.

77. Swartz M.J., Batra S.K., Varshney G.C., Hollingsworth M.A., Yeo C.J., Cameron J.L., Wilentz R.E., Hruban R.H., and Argani P. 2002. MUC4 expression increases progressively in pancreatic intraepithelial neoplasia. *Am. J. Clin. Pathol.* 117:791–796.

78. Balague C., Audie J.P., Porchet N., and Real F.X. 1995. In situ hybridization shows distinct patterns of mucin gene expression in normal, benign, and malignant pancreas tissues. *Gastroenterology* 109:953–964.

79. Yonezawa S., Sueyoshi K., Nomoto M., Kitamura H., Nagata K., Arimura Y., Tanaka S., Hollingsworth M.A., Siddiki B., Kim Y.S., and Sato E. 1997. MUC2 gene expression is found in noninvasive tumors but not in invasive tumors of the pancreas and liver: its close relationship with prognosis of the patients. *Hum. Pathol.* 28:344–352.

80. Yonezawa S., Taira M., Osako M., Kubo M., Tanaka S., Sakoda K., Takao S., Aiko T., Yamamoto M., Irimura T., Kim Y.S., and Sato E. 1998. MUC-1 mucin expression in invasive areas of intraductal papillary mucinous tumors of the pancreas. *Pathol. Int.* 48:319–322.

81. Terada T. and Nakanuma Y. 1996. Expression of mucin carbohydrate antigens (T, Tn and sialyl Tn) and MUC-1 gene product in intraductal papillary-mucinous neoplasm of the pancreas. *Am. J. Clin. Pathol.* 105:613–620.

82. Hruban R.H., Goggins M., Parsons J., and Kern S.E. 2000. Progression model for pancreatic cancer. *Clin. Cancer Res.* 6:2969–2972.

83. Hruban R.H., Adsay N.V., Albores-Saavedra J., Compton C., Garrett E.S., Goodman S.N., Kern S.E., Klimstra D.S., Kloppel G., Longnecker D.S., Luttges J., and Offerhaus G.J. 2001. Pancreatic intraepithelial neoplasia: a new nomenclature and classification system for pancreatic duct lesions. *Am. J. Surg. Pathol.* 25:579–586.

84. Kim G.E., Bae H.I., Park H.U., Kuan S.F., Crawley S.C., Ho J.J., and Kim Y.S. 2002. Aberrant expression of MUC5AC and MUC6 gastric mucins and sialyl Tn antigen in intraepithelial neoplasms of the pancreas. *Gastroenterology* 123:1052–1060.

85. Ho S.B., Niehans G.A., Lyftogt C., Yan P.S., Cherwitz D.L., Gum E.T., Dahiya R., and Kim Y.S. 1993. Heterogeneity of mucin gene expression in normal and neoplastic tissues. *Cancer Res.* 53:641–651.

86. Kuver R., Savard C., Oda D., and Lee S.P. 1994. PGE generates intracellular cAMP and accelerates mucin secretion by cultured dog gallbladder epithelial cells. *Am. J. Physiol.* 267:G998–G1003.

87. Agrawal B., Krantz M.J., Parker J., and Longenecker B.M. 1998. Expression of MUC1 mucin on activated human T cells: implications for a role of MUC1 in normal immune regulation. *Cancer Res.* 58:4079–4081.

88. Brugger W., Buhring H.J., Grunebach F., Vogel W., Kaul S., Muller R., Brummendorf T.H., Ziegler B.L., Rappold I., Brossart P., Scheding S., and Kanz L. 1999. Expression of MUC-1 epitopes on normal bone marrow: implications for the detection of micrometastatic tumor cells. *J. Clin. Oncol.* 17:1535–1544.

89. Chang J.F., Zhao H.L., Phillips J., and Greenburg G. 2000. The epithelial mucin, MUC1, is expressed on resting T lymphocytes and can function as a negative regulator of T cell activation. *Cell Immunol.* 201:83–88.

90. Wykes M., MacDonald K.P., Tran M., Quin R.J., Xing P.X., Gendler S.J., Hart D.N., and McGuckin M.A. 2002. MUC1 epithelial mucin (CD227) is expressed by activated dendritic cells. *J. Leukoc. Biol.* 72:692–701.

91. Oosterkamp H.M., Scheiner L., Stefanova M.C., Lloyd K.O., and Finstad C.L. 1997. Comparison of MUC-1 mucin expression in epithelial and non-epithelial cancer cell lines and demonstration of a new short variant form (MUC-1/Z). *Int. J. Cancer* 72:87–94.

92. Brockhausen I. 1999. Pathways of O-glycan biosynthesis in cancer cells. *Biochem. Biophys. Acta* 1473:67–95.

93. Hakomori S. 1989. Aberrant glycosylation in tumors and tumor-associated carbohydrate antigens. *Adv. Cancer Res.* 52:257–331.

94. Burdick M.D., Harris A., Reid C.J., Iwamura T., and Hollingsworth M.A. 1997. Oligosaccharides expressed on MUC1 produced by pancreatic and colon tumor cell lines. *J. Biol. Chem.* 272:24198–24202.

95. Colcher D., Pavlinkova G., Beresford G., Booth B.J., and Batra S.K. 1999. Single-chain antibodies in pancreatic cancer. *Ann. NY Acad. Sci.* 880:263–280.

96. Carraway K.L., Price-Schiavi S.A., Komatsu M., Idris N., Perez A., Li P., Jepson S., Zhu X., Carvajal M.E., and Carraway C.A. 2000. Multiple facets of sialomucin complex/MUC4, a membrane mucin and erbb2 ligand, in tumors and tissues (Y2K update). *Front Biosci.* 5:D95–D107.

97. Wesseling J., van der Valk S.W., Vos H.L., Sonnenberg A., and Hilkens J. 1995. Episialin (MUC1) overexpression inhibits integrin-mediated cell adhesion to extracellular matrix components. *J. Cell. Biol.* 129:255–265.

98. Wesseling J., van der Valk S.W., and Hilkens J. 1996. A mechanism for inhibition of E-cadherin-mediated cell-cell adhesion by the membrane-associated mucin episialin/MUC1. *Mol. Biol. Cell.* 7:565–577.

99. Jentoft N. 1990. Why are proteins O-glycosylated? *Trends Biochem. Sci.* 15:291–294.

100. Komatsu M., Carraway C.A., Fregien N.L., and Carraway K.L. 1997. Reversible disruption of cell-matrix and cell-cell interactions by overexpression of sialomucin complex. *J. Biol. Chem.* 272:33245–33254.

101. Yamamoto M., Bharti A., Li Y., and Kufe D. 1997. Interaction of the DF3/MUC1 breast carcinoma-associated antigen and beta-catenin in cell adhesion. *J. Biol. Chem.* 272:12492–12494.

102. Hulsken J., Birchmeier W., and Behrens J. 1994. E-cadherin and APC compete for the interaction with beta-catenin and the cytoskeleton. *J. Cell. Biol.* 127:2061–2069.

103. Hulsken J., Behrens J., and Birchmeier W. 1994. Tumor-suppressor gene products in cell contacts: the cadherin-APC-armadillo connection. *Curr. Opin. Cell. Biol.* 6:711–716.

104. Munemitsu S., Albert I., Souza B., Rubinfeld B., and Polakis P. 1995. Regulation of intracellular beta-catenin levels by the adenomatous polyposis coli (APC) tumor-suppressor protein. *Proc. Natl. Acad. Sci. USA* 92:3046–3050.

105. Peifer M. 1996. Regulating cell proliferation: as easy as APC. *Science* 272:974–975.

106. Bhat R.V., Baraban J.M., Johnson R.C., Eipper B.A., and Mains R.E. 1994. High levels of expression of the tumor suppressor gene APC during development of the rat central nervous system. *J. Neurosci.* 14:3059–3071.

107. Rubinfeld B., Souza B., Albert I., Munemitsu S., and Polakis P. 1995. The APC protein and E-cadherin form similar but independent complexes with alpha-catenin, beta-catenin, and plakoglobin. *J. Biol. Chem.* 270:5549–5555.

108. Peifer M. 1993. Cancer, catenins, and cuticle pattern: a complex connection. *Science* 262:1667–1668.

109. Burchill S.A. 1994. The tumour suppressor APC gene product is associated with cell adhesion. *Bioessays* 16:225–227.

110. Rubinfeld B., Albert I., Porfiri E., Fiol C., Munemitsu S., and Polakis P. 1996. Binding of GSK3beta to the APC-beta-catenin complex and regulation of complex assembly. *Science* 272:1023–1026.

111. Li Y., Bharti A., Chen D., Gong J., and Kufe D. 1998. Interaction of glycogen synthase kinase 3beta with the DF3/MUC1 carcinoma-associated antigen and beta-catenin. *Mol. Cell. Biol.* 18:7216–7224.

112. Quin R.J. and McGuckin M.A. 2000. Phosphorylation of the cytoplasmic domain of the MUC1 mucin correlates with changes in cell-cell adhesion. *Int. J. Cancer* 87:499–506.

113. Regimbald L.H., Pilarski L.M., Longenecker B.M., Reddish M.A., Zimmermann G., and Hugh J.C. 1996. The breast mucin MUC1 as a novel adhesion ligand for endothelial intercellular adhesion molecule 1 in breast cancer. *Cancer Res.* 56:4244–4249.

114. Majuri M.L., Mattila P., and Renkonen R. 1992. Recombinant E-selectin-protein mediates tumor cell adhesion via sialyl-Le(a) and sialyl-Le(x). *Biochem. Biophys. Res. Commun.* 182:1376–1382.

115. Sako D., Chang X.J., Barone K.M., Vachino G., White H.M., Shaw G., Veldman G.M., Bean K.M., Ahern T.J., and Furie B. 1993. Expression cloning of a functional glycoprotein ligand for P-selectin. *Cell* 75:1179–1186.

116. Sawada T., Ho J.J., Chung Y.S., Sowa M., and Kim Y.S. 1994. E-selectin binding by pancreatic tumor cells is inhibited by cancer sera. *Int. J. Cancer* 57:901–907.

117. Komatsu M., Yee L., and Carraway K.L. 1999. Overexpression of sialomucin complex, a rat homologue of MUC4, inhibits tumor killing by lymphokine-activated killer cells. *Cancer Res.* 59:2229–2236.

118. Li Y., Ren J., Yu W., Li Q., Kuwahara H., Yin L., Carraway III K.L., and Kufe D. 2001. The epidermal growth factor receptor regulates interaction of the human DF3/MUC1 carcinoma antigen with c-Src and beta-catenin. *J. Biol. Chem.* 276:35239–35242.

119. Schroeder J.A., Thompson M.C., Gardner M.M., and Gendler S.J. 2001. Transgenic MUC1 interacts with epidermal growth factor receptor and correlates with mitogen-activated protein kinase activation in the mouse mammary gland. *J. Biol. Chem.* 276:13057–13064.

120. Jepson S., Komatsu M., Haq B., Arango M.E., Huang D., Carraway C.A., and Carraway K.L. 2002. Muc4/sialomucin complex, the intramembrane ErbB2 ligand, induces specific phosphorylation of ErbB2 and enhances expression of p27(kip), but does not activate mitogen-activated kinase or protein kinaseB/Akt pathways. *Oncogene* 21:7524–7532.

121. Carraway K.L., Perez A., Idris N., Jepson S., Arango M., Komatsu M., Haq B., Price-Schiavi S.A., Zhang J., and Carraway C.A. 2002. Muc4/sialomucin complex, the intramembrane ErbB2 ligand, in cancer and epithelia: to protect and to survive. *Prog. Nucleic Acid Res. Mol. Biol.* 71:149–185.

122. Singh A.P., Moniaux N., Chauhan S.C., Meza J.L., and Batra S.K. 2004. Inhibition of MUC4 expression suppresses pancreatic tumor cell growth and metastasis. *Cancer Res.* 64:622–630.

123. Rhodes J.M. 1999. Usefulness of novel tumour markers. *Ann. Oncol.* 10:118–121.

124. McLaughlin R., O'Hanlon D., Kerin M., Kenny P., Grimes H., and Given H.F. 1999. Are elevated levels of the tumour marker CA19-9 of any clinical significance? — an evaluation. *Ir. J. Med. Sci.* 168:124–126.

125. Fabris C., Malesci A., Basso D., Bonato C., Del Favero G., Tacconi M., Meggiato T., Fogar P., Panozzo M.P., Ferrara C., Scalon P., and Naccarato R. 1991. Serum DU-PAN-2 in the differential diagnosis of pancreatic cancer: influence of jaundice and liver dysfunction. *Br. J. Cancer* 63:451–453.

126. Hamori J., Arkosy P., Lenkey A., and Sapy P. 1997. The role of different tumor markers in the early diagnosis and prognosis of pancreatic carcinoma and chronic pancreatitis. *Acta Chir. Hung.* 36:125–127.

127. Ho J.J., Norton K., Chung Y.S., and Kim Y.S. 1993a. Expression of CA19-9, DU-PAN-2, and SPan-1 antigens on two types of normal salivary mucins. *Oncol. Res.* 5:347–356.

128. Ho J.J. and Kim Y.S. 1994. Serological pancreatic tumor markers and the MUC1 apomucin. *Pancreas* 9:674–691.

129. Yin B.W., Dnistrian A., and Lloyd K.O. 2002. Ovarian cancer antigen CA125 is encoded by the MUC16 mucin gene. *Int. J. Cancer* 98:737–740.

130. Sell S. 1990. Cancer-associated carbohydrates identified by monoclonal antibodies. *Hum. Pathol.* 21:1003–1019.

131. Lamerz R. 1999. Role of tumour markers, cytogenetics. *Ann. Oncol.* 10 Suppl. 4:145–149.

132. Orntoft T.F., Vestergaard E.M., Holmes E., Jakobsen J.S., Grunnet N., Mortensen M., Johnson P., Bross P., Gregersen N., Skorstengaard K., Jensen U.B., Bolund L., and Wolf H. 1996. Influence of Lewis alpha1-3/4-L-fucosyltransferase (FUT3) gene mutations on enzyme activity, erythrocyte phenotyping, and circulating tumor marker sialyl-Lewis a levels. *J. Biol. Chem.* 271:32260–32268.

133. Lyubsky S., Madariaga J., Lozowski M., Mishriki Y., Schuss A., Chao S., and Lundy J. 1988. A tumor-associated antigen in carcinoma of the pancreas defined by monoclonal antibody B72.3. *Am. J. Clin. Pathol.* 89:160–167.

134. Tempero M., Takasaki H., Uchida E., Takiyama Y., Colcher D., Metzgar R.S., and Pour P.M. 1989. Co-expression of CA 19-9, DU-PAN-2, CA 125, and TAG-72 in pancreatic adenocarcinoma. *Am. J. Surg. Pathol.* 13 Suppl. 1:89–95.

135. Batra S.K., Goel A., Pavlinkova G., and Colcher D. 2003. Monoclonal antibody targeted radionuclide therapy. In *Targeted therapy for cancer.* Syrigos K. and Harrington K. Eds. London: Oxford Press.

136. Batra S.K., Jain M., Wittel U.A., Chauhan S.C., and Colcher D. 2002. Pharmacokinetics and biodistribution of genetically engineered antibodies. *Curr. Opin. Biotechnol.* 13:603–608.

137. Hareuveni M., Gautier C., Kieny M.P., Wreschner D., Chambon P., and Lathe R. 1990. Vaccination against tumor cells expressing breast cancer epithelial tumor antigen. *Proc. Natl. Acad. Sci. USA* 87:9498–9502.

138. Apostolopoulos V., Loveland B.E., Pietersz G.A., and McKenzie I.F. 1995. CTL in mice immunized with human mucin 1 are MHC-restricted. *J. Immunol.* 155:5089–5094.

139. Haurum J.S., Arsequell G., Lellouch A.C., Wong S.Y., Dwek R.A., McMichael A.J., and Elliott T. 1994. Recognition of carbohydrate by major histocompatibility complex class I-restricted, glycopeptide-specific cytotoxic T lymphocytes. *J. Exp. Med.* 180:739–744.

140. Barnd D.L., Lan M.S., Metzgar R.S., and Finn O.J. 1989. Specific, major histocompatibility complex-unrestricted recognition of tumor-associated mucins by human cytotoxic T cells. *Proc. Natl. Acad. Sci. USA* 86:7159–7163.

141. Jerome K.R., Barnd D.L., Bendt K.M., Boyer C.M., Taylor-Papadimitriou J., McKenzie I.F., Bast Jr. R.C., and Finn O.J. 1991. Cytotoxic T-lymphocytes derived from patients with breast adenocarcinoma recognize an epitope present on the protein core of a mucin molecule preferentially expressed by malignant cells. *Cancer Res.* 51:2908–2916.

142. Finn O.J., Jerome K.R., Henderson R.A., Pecher G., Domenech N., Magarian-Blander J., and Barratt-Boyes S.M. 1995. MUC-1 epithelial tumor mucin-based immunity and cancer vaccines. *Immunol. Rev.* 145:61–89.

13

GROWTH FACTORS IN PANCREATIC DISEASES

Ahmed Guweidhi, Jörg Kleeff, Parviz M. Pour,
Markus W. Büchler, and Helmut Friess

CONTENTS

13.1   INTRODUCTION

There are three major pathological conditions of the pancreas — acute pancreatitis (AP), chronic pancreatitis (CP), and pancreatic cancer (PC). AP is an acute inflammatory disease of the pancreas typically presenting with abdominal pain and increased pancreatic enzymes in blood or urine.[1] Clinically mild and severe forms of AP are differentiated depending on the development of pancreatic or peripancreatic necrosis. Mild AP is morphologically characterized by interstitial edema, vacuolization of acinar cells, and infiltrates of polymorphonuclear leukocytes in the pancreas, with rapid recovery and an uncomplicated clinical course. In contrast, significant local and systemic complications are associated with severe AP, including extensive peripancreatic and intrapancreatic fat necrosis, parenchymal necrosis, hemorrhage, and in the later course the development of fibrosis.[2–5]

CP is an inflammatory disease primarily of the exocrine pancreas, characterized by irreversible and progressive destruction of the whole organ resulting in severe exocrine and endocrine insufficiency.[1,6,7] Heavy alcohol consumption is the main etiologic factor of CP in Western industrialized countries.[7–9] The morphological changes of CP include acinar cell degeneration, dilatation of the duct system with intraductal protein plugs and stones, necrosis, and replacement of the lost parenchyma by dense fibrous tissue. The destruction of pancreatic parenchyma is accompanied by infiltration of different subsets of inflammatory cells and followed by tissue remodeling, with ductal cell proliferation and ductular hyperplasia.[1,6,10,11]

Carcinoma of the pancreas is the fourth to fifth leading cause of cancer-related mortality in the Western world.[12] It is a devastating disease with a poor prognosis, an overall 5-year survival rate of less than 1%, and a median survival time after diagnosis of approximately 5 to 6 months. The incidence of PC seems to have increased slightly over the past decades, because of better diagnostic options and a general increase in the life expectancy.[12,13] At the time of diagnosis, between 75 and 85% of PC patients have tumors that are deemed unresectable, and these patients can only be offered palliative surgical options.[12,14,15] Furthermore, PC patients typically have metastatic lesions at the time of diagnosis, and most patients undergoing potentially curative tumor resection die within the first 2 postoperative years due to local recurrence or distant metastasis.[12,14]

In recent decades, many laboratories have focused on understanding the molecular alterations that occur in pancreatic diseases, with the hope that this will ultimately lead to improved diagnostic and therapeutic modalities. Modern molecular biology techniques provide important clues for clarifying morphological changes and pathophysiological aspects of pancreatic diseases.

Growth factors are produced by many different cell types and exert their effects via autocrine or paracrine mechanisms. The molecular knowledge acquired regarding changes in the expression of growth factors in pancreatic diseases has the potential to improve diagnostic and therapeutic treatment strategies in the near future. In this review, we will summarize our current understanding of the role of growth factors and growth factor receptors in pancreatic diseases.

13.2 INSULIN IN PANCREATIC DISEASES

Insulin is an anabolic hormone with powerful metabolic effects. It is synthesized by the beta cells of the islets of Langerhans as a single chain precursor called proinsulin. Insulin consists of 2 dissimilar polypeptide chains, an A chain with 21 amino acids and a B chain with 30 amino acids, which are linked by 2 disulfide bonds. Chain A and Chain B are derived from a 1-chain precursor, proinsulin. Proinsulin is converted to insulin by the enzymatic removal of a segment that connects the amino end of the A chain to the carboxyl end of the B chain. This segment is called the connecting C peptide.[16–18] Like other growth factors, insulin uses phosphorylation and the resultant protein–protein interactions as essential tools to transmit and compartmentalize its signal. Insulin initiates its wide variety of growth and metabolic effects by binding to the insulin receptor. The insulin receptor belongs to the large family of growth factor

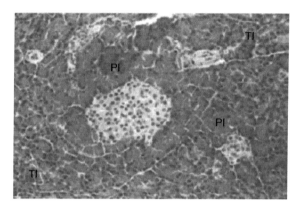

Figure 13.1 Normal human (rat) pancreas. Note the larger size of the acinar cells in the peri-insular (PI) area compared to those in the teleinsular (TI) region.

receptors with intrinsic tyrosine kinase activity.[18] Following insulin binding, the receptor undergoes autophosphorylation on multiple tyrosine residues. This results in activation of the receptor kinase and tyrosine phosphorylation of a variety of docking proteins including insulin receptor substrate (IRS) proteins.[18–20] Phosphorylated IRS proteins serve as docking proteins between the insulin receptor and a complex network of intracellular signaling molecules containing Src homology 2 (SH2) domains.[19,20] Four members (IRS-1, IRS-2, IRS-3, IRS-4) of this family have been identified and they play different and specific roles *in vivo*.[21,22] Activation of SH2 domain proteins initiates a cascade of a biochemical reactions and activation of intracellular pathways, including regulation of cell differentiation, growth, survival, and metabolism that ultimately transmit the insulin signal.

13.2.1 Insulin in the Normal Pancreas

Insulin is the hormone that is produced solely in the pancreas of mammals (see Chapter 1, Chapter 9, and Chapter 28). It is an important growth factor not only for the normal physiological function of the pancreas, but also for the repair of toxicological injuries. The effect of insulin on the pancreas is both through blood circulation and paracrine. Based on the insulo-acinar vessel architecture of the pancreas, as described in Chapter 4, the insulin released from the β-cells reaches the exocrine tissue via efferent branches of the arterial complex within the islet. Because the concentration of insulin is much higher in the tissue immediately surrounding the islets, peri-insular acini are larger and contain more protein and enzyme than acinar cells remote from the islets

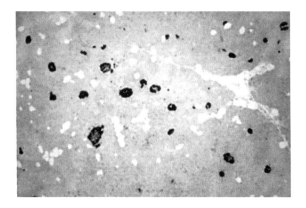

Figure 13.2 Human pancreas. Random distribution of islets (black in photo) of various size. Anti-insulin antibody, Avidin-Biotin Complex Method ×30.

(tele-insular acini).[23,24] This paracrine function of the islet cells can be demonstrated histologically, particularly in the rat pancreas, where peri-insular acinar cells distinguish themselves from the tele-insular acinar cells by their more intense esosinophilic color giving rise to the so-called "halo" phenomenon (Figure 13.1).[23,25] A difference in the enzyme content of the peri-insular and tele-insular acini has been shown.[26,27] The random distribution of islets within the pancreas occupying every 1.1 mm of the exocrine tissue (Figure 13.2)[28] reflects the importance of the paracrine pathway of insulin for the normal digestive function and the repair of toxicological or mechanical injuries. This topographical insulin concentration may explain also the reason that in the hamster model most cancers develop within or immediately around the islets.[3,29] The reason for the fast growth and expansion of tumors arising from within or around the islets could well be related to a suitable environment within or in the vicinity of islets, where high concentrations of growth factors are present.[29]

13.2.2 Insulin in AP

The importance of insulin in the repair of the damaged pancreas has been shown by several studies. Pancreatic endocrine function impairment following AP is associated with decreased plasma levels of both basal and glucose-stimulated insulin and it is more common after severe than after mild AP.[31] Overall, the incidence of diabetes mellitus in patients after AP is about 15.8%. Alcoholic pancreatitis is more often complicated with impaired glucose tolerance and diabetes mellitus than other causes of pancreatitis. In addition, the severity of exocrine insufficiency in AP patients correlates strongly with the severity of concomitant pancreatic endocrine insufficiency, the severity of which has been categorized according to insulin dependency.[32]

In the rat's AP model, both basal and glucose-stimulated insulin secretory response of pancreatic islets isolated from rats with AP is markedly decreased although pancreatic islets appear histologically intact.[33,34] Moreover, the production of inducible nitric oxide (iNOS) by β-cells of isolated islet cells is markedly increased. Increased iNOS activity leads to an elevated nitric oxide (NO) production[33] and high levels of NO are involved in pancreatic β-cell dysfunction and apoptosis. Therefore, it was suggested that the reduced insulin secretory response of islets in AP to glucose is a consequence of the toxic action of NO produced by iNOS in the β-cells.[33] Recently, it was discovered in patients with AP that the patient's insulin is progressively degraded by gelatinase B (MMP-9), a key regulator in the pathophysiology of autoimmune diseases. In AP, gelatinase B is produced by the inflammatory cells that are mainly localized next to the insulin secreting β-cells in the islets suggesting that gelatinase B cleaves insulin, secreted by β-cells in close proximity, and this degradation of insulin could lead to the development of overt diabetes mellitus. Therefore, MMP inhibitors may have a beneficial effect on insulin levels in the treatment of diabetes mellitus.[3]

13.2.3 Insulin in CP

Diabetes, usually of insulin-dependent type, eventually develops in 30 to 50% of CP patients.[36] In CP, a reduction in the number of β-cells induces β-cell neogenesis in extrainsular tissue compartments to compensate for the loss of insulin in islets. In CP patients, the extrainsular endocrine cells are the predominant insulin cell type.[37] Despite the compensatory β-cell neogenesis, insulin secretion deficiency and development of diabetes mellitus still present a common complication in CP. More recently, it was found that the ductular epithelial cells in CP express considerable amounts of gelatinase B (MMP-9) and within the islets some cells stain positively for gelatinase B. It was furthermore demonstrated that insulin, which serves as an important substrate for gelatinase B, is progressively degraded by gelatinase B.[35] This could explain the reduction of insulin secretion in early stages of CP despite histological and ultrastructural evidence for intact islets of Langerhans.[38] In severe cases of CP associated with the atrophy of the islet cells and especially of β-cells,[39] with the reduced insulin synthesis and discharge, the regenerative capability of the pancreas is reduced or abolished. Consequently, the damaged pancreas requires other growth factors for repair, although apparently noninsulin hormones do not have the same pancreatic-specific effects of insulin.

13.2.4 Insulin in PC

Abnormal glucose tolerance and overt diabetes mellitus develop in up to 80% of PC patients.[37,40] The development of diabetes or altered glucose tolerance generally occurs shortly before the clinical manifestation of the disease.[40] Thus, diabetes mellitus is considered to be an early symptom rather than the cause of PC.[40] The cellular growth promoting effect of insulin has an unwanted effect on the cancer cells. Unfortunately, the growth of PC cells benefits greatly from the paracrine secretory ability of insulin. Most PC cell lines require insulin for their maintenance and growth *in vitro*. However, contrary to the normal tissue, cancer cells, which eventually destroy the surrounding insulin-providing islets, in the absence of insulin, acquire the ability to produce their own growth factors, including neuropeptides, except for insulin. For yet unknown reasons, the synthesis of insulin is restricted to the islet cells and tumor cells derived from them. Many investigators have shown that the majority of well-differentiated ductal adenocarcinomas contain a few or a conspicuous number of different types of islet cells including β-cells.[4] Whether these cells serve as an energy provider for tumor cells or simply present bystanders is unknown. Moreover, these endocrine cells are found both in the invasive part of cancers[42] and in their metastasis,[41] which indicates that these endocrine cells are an integral part of the malignant exocrine tissue.

In PC patients, islets in the vicinity of cancer cells show a decreased number of β-cells and increased number of α-cells.[37] It has been suggested that, the depletion of β-cells in PC is related to the effect of substances, such as amylin, released by cancer cells. In that case, suppression of all islets within the pancreas is expected. However, this has not been shown.[37] It has been shown that the alterations of islets occur only in the vicinity of tumors perhaps through cancer cell derived substances that reach the islets by a paracrine pathway.[37] Nevertheless, the alteration of β-cells in PC patients is mainly restricted to the endocrine cells within the islets and there is no compensatory proliferation of β-cells in other pancreatic areas.[37] It has also been hypothesized that pancreatic ductal adenocarcinoma arises from within islets, most probably from undifferentiated cells or transdifferentiated cells, which loose the insulin secretory ability during carcinogenesis.

13.3 THE INSULIN-LIKE GROWTH FACTOR AND ITS RECEPTOR IN PANCREATIC DISEASES

Insulin-like growth factors (IGFs) have been implicated as regulators of cell differentiation and cell proliferation in a number of cell systems and have been reported to play an important role in growth regulation of human tumors.[43,44] The IGF family includes IGFs Type I and Type II (IGF-I and IGF-II), which are structurally related to proinsulin. IGF-I is a mitogenic factor that has the ability to bind and activate the insulin receptor as well as the IGF-I receptor.[45–48] The IGF-I receptor binds IGF-I with higher affinity than insulin, and in general the mitogenic effects of both factors are mediated by the IGF-I receptor. Ligand binding to the IGF-I receptor results in activation of the intracellular receptor kinase domains and transphosphorylation.[45–48] Following receptor phosphorylation, adapter molecules (IRS-1 and IRS-2) associate with the receptor and are phosphorylated. The presence of a variety of phosphorylation motifs on each of the two IRS proteins enables them to interact with a variety of downstream signaling molecules, which allows them to transmit IGF-I signals as well as to contribute to signal transduction of several cytokine receptors.[49–54]

13.3.1 IGF in AP

In experimental models of AP, the expression of IGF-I and its receptor (IGF-IR) was increased in a temporal fashion. Pancreatic IGF-I levels increased over 50-fold during regeneration, between 1 and 3 days after induction of pancreatitis, reaching a maximum at Day 2.[55,56] IGF-I is localized in fibroblasts within the areas of interstitial tissue and is expressed in primary cultures of pancreatic fibroblasts, but not in cultured pancreatic acinar cells, and IGF-I receptors are expressed in cultured acinar cells.[55] This indicates that acinar cell proliferation during regeneration from pancreatitis is mediated at least in part by paracrine release of IGF-I from fibroblasts.

13.3.2 IGF in PC

In PC there is an increase in IGF-I receptor expression in comparison with the normal pancreas and also a marked increase in IGF-I.[44,57] IGF-I is expressed in the stroma elements in the normal pancreas, but in PC, IGF-I is abundantly expressed in both the cancer cells and in the surrounding stroma in PC tissues.[44] Furthermore, the receptor substrates IRS-1 and IRS-2 are also overexpressed in human PC, predominantly in the cancer cells.[58,59] Increased IGF-I serum levels were found in 10% of patients with PC, and this was inversely correlated with fasting serum glucose levels, which may indicate that serum IGF-I level is related to altered glucose metabolism.[60,61]

In vitro experiments in cultured human PC cell lines have revealed that several cell lines express the IGF-I receptor.[44] Furthermore, IGF-I stimulates the growth of cultured PC cell lines, and this effect is blocked by a specific anti-IGF-I receptor antibody, suggesting that this mitogenic effect is mediated via the IGF-I receptor.[44] In addition, the growth of cultured human PC cells is inhibited by IGF-I receptor antisense oligonucleotides that specifically block IGF-I receptor synthesis.[44] Together, these observations suggest that the IGF-I receptor mediates mitogenic signaling in PC cells. Furthermore, it has recently been shown that IRS-2 is expressed in cultured PC cells and that IGF-I enhances IRS-2 phosphorylation and PI3-kinase activation in PC cells, suggesting that IRS-2 is an important mediator of mitogenic signaling in PC cells, and that this effect is mediated at least in part via PI3-kinase.[59] It has also been demonstrated that IGF-I enhances phosphorylation of IRS-1 in PC cells.[58] The *in vitro* and *in vivo* data collectively suggest that IGF-I — acting in a paracrine or autocrine manner via the IGF-I receptor — together with the concomitant overexpression of IGF-I, IGF-I receptor, IRS-1, and IRS-2 may participate in aberrant pathways to enhance PC cell growth *in vivo*.

13.4 THE TRANSFORMING GROWTH FACTOR β FAMILY IN PANCREATIC DISEASES

Transforming growth factor (TGF-β) was originally discovered as a secreted factor that induces malignant transformation *in vivo*.[62] Now it is known that the TGF-β superfamily of ligands consists of a large family of polypeptide growth factors that exert a wide range of biological effects, including cell growth and differentiation, angiogenesis, cell invasion, extracellular matrix composition, local immune function, and apoptosis.[63,64]

The mammalian TGF-β superfamily includes the TGF-β family itself (TGF-β1, TGF-β2, and TGF-β3), the activin/inhibin family, the bone morphogenetic protein (BMP) family, the Vg-1 family, and Müllerian inhibitory substance. These growth factors are usually synthesized as precursors that undergo proteolytic cleavage to yield biologically active proteins.[63,64]

TGF-β family members initiate their cellular action by binding to receptors with intrinsic serine/threonine kinase activity. This receptor family consists of two subfamilies, Type I and Type II receptors, which are structurally similar, with small cysteine-rich extracellular regions and intracellular parts consisting mainly of the kinase domains. Type I receptors, but not Type II receptors, have a region rich in glycine and serine residues (GS domain) in the juxtamembrane domain. Each member of the TGF-β superfamily binds to a characteristic combination of Type I and Type II receptors, both of which are needed for signaling.[63,64]

The Type II TGF-β receptor (TβRII) binds ligands in the absence of the Type I TGF-β receptor (TβRI). Upon ligand binding, TβRII forms a heteromeric complex with TβRI, which cannot bind ligands in the absence of the TβRII.[63,64] A number of different TβRI receptors have been described; however, the principal Type I receptor for TGF-β seems to be ALK5.[62] Following heterodimerization of Type I and Type II TGF-β receptors, activated TβRII then transphosphorylates the glycine and serine (GS)-rich domain of the Type I receptor kinase, thereby activating TβRI.[63,64] Activated TβRI then transiently associates with Smad2 or Smad3. Phosphorylated Smad2 or Smad3 then separately form heteromeric complexes with Smad4, the common mediator of the TGFβ superfamily signaling pathway.[65,66] These complexes then translocate to the nucleus, where they can function as transcriptional activators.[65,67]

The Smad family has been divided into three subgroups. The first group — Smad1, Smad2, Smad3, and Smad8 — encodes pathway-specific signal transducing Smads, which mediate TGF-β, activin, or BMP signaling.[65,67] The second subgroup contains Smad4, which binds to all pathway-specific Smads. The third subgroup — Smad6 and Smad7 — is composed of inhibitory Smads, which antagonize the function of pathway-specific Smads.[68]

In addition to the three mammalian TGF-βs, the TGF-β superfamily also comprises the activins/inhibins and BMPs.[64] Both activins/inhibins and BMPs signal through specific Type I and Type II transmembrane serine-threonine kinase receptors. Two Type I receptors for activin/inhibin (actRI and actRIb) and two Type II receptors specific for activin/inhibin (actRII and actRIIb) have been described.[64] Furthermore, two Type I receptors specific for BMPs (BMPR-Ia and BMPR-Ib) as well as a specific Type II receptor for BMP (BMPR-II) have been identified.[64]

13.4.1 The TGF-β Family in AP

In AP, the regenerative response of the damaged pancreas is assumed to be determined by a balance between newly synthesized and deposited extracellular matrix (ECM) and degradation of ECM. Pancreatic tissues in animal models are almost completely restored to normal after an attack of AP, once the cause of the disease is removed.[69] TGF-β1 plays an important role in the process of wound healing. TGF-β1 directs the migration of monocytes and fibroblasts and increases the synthesis and secretion of ECM components.[70,71] TGF-βs have been shown to increase the synthesis of collagen and fibronectin by fibroblasts.[71–73] Furthermore, neutralizing TGF-β1 antibodies reduces the expression of procollagen Type I and Type III and fibronectin in the pancreas after induction of pancreatitis using cerulean.[74] These observations suggest that TGF-β1 is a major regulator of ECM synthesis during regeneration after AP induced by various models. Moreover, these data support the view that ECM creates a platform for pancreatic regeneration.[75]

In humans, all three TGF-β isoforms and their receptors Type I and Type II as well as collagen Type I are overexpressed in acute necrotizing pancreatitis as compared with the samples from a control group. Expression of TGF-β and its receptor is restricted to the remaining acinar and ductal cells in most of the necrotizing pancreatitis samples.[76]

The expression of TGF-β1 during AP is biphasic, with an initial increase probably related to pancreatic damage and inhibition of cell proliferation and with a later increase accompanied by the stimulation of the synthesis of extracellular matrix components, thereby contributing to tissue remodeling and repair.[77] Recently it was shown that TGF-β1 serum levels are significantly reduced in patients with severe pancreatitis and multiple organ failure and that there was a negative correlation with MMP-1 levels in these patients. It was hypothesized that in patients with severe AP, TGF-β1 levels decrease and MMP-1 levels remain high, resulting in persisting severe inflammation and therefore no regeneration of the pancreas.[78]

Regeneration after cerulein-induced AP is also associated with the activation of TGF-β-dependent pathways. It has been demonstrated in a rat model of cerulein-induced AP that the injection of neutralizing antibodies against TGF-β1 results in a significant reduction in pancreatic levels of Collagen I and Collagen III.[79]

13.4.2　The TGF-β Family in CP

TGF-βs and their receptors are involved in tissue destruction and fibrogenesis in CP. Animal models also support the concept of the involvement of TGF-β in extracellular matrix remodeling in pancreatic diseases. For example, transgenic mice overexpressing TGF-β1 develop severe pancreatic fibrosis.[80] Strong immunostaining for TGF-β1 has been localized in the majority of ductal cells in CP, but also in mononuclear inflammatory cells and fibroblasts.[81,82] In addition, CP tissues also overexpress TβRI and TβRII, which are predominantly localized in ductal cells and atrophic acinar cells.[83] Moreover, connective tissue growth factor (CTGF), which is one of the downstream effectors of TGF-β and a strong stimulator of fibrogenesis,[84] is overexpressed in degenerating acinar cells and primarily in fibroblasts surrounding areas of tissue destruction in CP.[83] CTGF messenger RNA (mRNA) expression levels positively correlate with the degree of fibrosis in CP tissues.[83]

13.4.3　The TGF-β Family in PC

Since epithelial cells are usually growth inhibited by TGF-βs,[64] the observation of overexpression of these ligands in some types of cancer is

TGF-β1-pancratic cancer

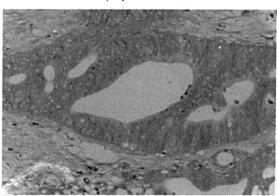

Figure 13.3   Transforming growth factor beta 1 (TGF-β1) immunohistochemistry in PC. Many PC cells exhibit intense TGF-β1 immunoreactivity.

surprising.[85–87] However, it is becoming increasingly clear that these cancer cells have lost their ability to respond to the growth suppressive effects of TGF-βs.[87–89] It has been hypothesized that TGF-βs may also act to increase the expression of adhesion molecules and extracellular matrix components including fibronectin, collagen, and laminin, thereby enhancing the metastatic potential of cancer cells. Furthermore, it has been suggested that TGF-βs may act to stimulate angiogenesis and to suppress cancer-directed immune mechanisms.

Human PCs overexpress all three TGF-β ligands (Figure 13.3), and the overexpression of any of these isoforms is associated with a worse prognosis.[86] Furthermore, TβRII is also expressed at increased levels in the cancer cells of human pancreatic tumors in comparison with the normal pancreas.[90,91] Despite expressing high levels of TGF-βs and TβRII, cultured human PC cell lines are usually resistant to the growth inhibitory effects of TGF-βs.[88,92] The resistance to TGFβ-mediated growth inhibition in PC might be due to several genetic and epigenetic alterations. Thus, a subgroup of PCs express low levels of TβRI,[88] and it has been shown that the responsiveness to TGF-β in T3M4 human PC cells can be restored by transfecting these cells with a full-length TβRI.[92] Deletions and point mutations of the tumor suppressor gene Smad4 (also called DPC4, for deleted in pancreatic carcinoma), the common mediator of the TGF-β superfamily signaling pathway, have been identified in about 50% of pancreatic adenocarcinomas, which suggests that Smad4/DPC4 may have a specific role in pancreatic tumorigenesis.[66,87,93,94]

It has also been previously demonstrated that deletion or mutational inactivation of the Smad4/DPC4 gene correlates with a loss of responsiveness to TGF-β-induced growth inhibition and TGF-β-inducible p21[wafl] expression, and that Smad4/DPC4 can reestablish TGF-β-inducible reporter gene activity in a Smad4/DPC4-null human pancreatic adenocarcinoma cell line.[89]

The TGF-β signaling inhibitors Smad6 and Smad7 are also markedly overexpressed in PCs.[95,96] Smad6 and Smad7 expression is localized in the cancer cells within the tumor mass and also to a lesser extent in the endothelial cells and CP-like areas adjacent to the tumor.[95,96] When the TGF-β-responsive PC cell line COLO-357 was transfected with Smad6 and Smad7, a complete abrogation of the growth inhibitory effects of TGF-β occurred.[95,96] However, in the Smad6 and Smad7 overexpressing cancer cells, TGF-β was still able to induce the expression of PAI-1, which functions to enhance tumor invasion and metastasis.[95,96] Smad6 and Smad7 transfected cells also displayed enhanced anchorage-independent growth rates in nude mice.[95,96] These *in vitro* and *in vivo* data suggest that PC cells have redundant barriers to TGF-β signaling that may allow the cancer cells to escape the TGF-β-induced growth inhibition while still allowing for the expression of metastasis-promoting genes. In addition, tumor cell-derived TGF-β may act in a paracrine manner to enhance angiogenesis and suppress cancer-directed immune mechanisms.

PCs also markedly overexpress the activin/inhibin βA subunit and to a lesser extent the βB subunit, whereas the α subunit is not increased in PC samples.[97] Activin/inhibin βA expression is present in the diffuse infiltrating as well as in the duct-like cancer cells within the pancreatic tumor mass.[97] Furthermore, activin/inhibin Receptor I, Receptor Ib, and Receptor II were also overexpressed in these tissues, and the expression was also localized within the cancer cells. Both the ligand and its receptors were often coexpressed in these samples.[97] Interestingly, *in vitro* experiments in cultured human PC cells revealed that Activin A stimulates the growth of some PC cell lines,[97] suggesting that autocrine or paracrine effects of Activin A in some PC tissues may contribute to PC cell growth *in vivo*.

Recently it has also been demonstrated that a member of the BMP family — BMP2 — is overexpressed in PC, that this overexpression is localized within the tumor cells, and that enhanced expression of this polypeptide correlates with decreased survival of PC patients.[98] Concomitantly, these cancers also express high levels of the Type Ia and the Type II BMP receptor.[98] It could be shown that BMP2 inhibits the growth of COLO-357 PC cells, which express a normal Smad4 gene,[99] whereas BMP2 stimulates the growth of ASPC-1 and CAPAN-1 cells, which both express a truncated Smad4 gene.[94,98] Transfection of COLO-357 cells with a dominant-negative Smad4 gene abrogates BMP-2-induced growth inhibition in these cells, whereas transfection of a wild type Smad4 gene into ASPC-1 and CAPAN-1 cells leads to blockage of the mitogenic effects of BMP-2 in these cells.[98] These observations suggest that in the presence of a mutated Smad4 gene, BMP-2 might act as a mitogen. Together with the *in vivo* data showing that enhanced BMP-2 expression correlates with decreased survival, these findings point to an important role of BMP-2 in the pathobiology of PC. The ability of BMP-2 to act as a mitogenic factor in PC cells might also explain in part the propensity of PC to harbor Smad4 mutations.

The number of genetic and epigenetic changes observed in the TGF-β superfamily system of ligands and their receptors in PC points to an important role of this system in pancreatic tumorigenesis.

13.5 THE EGF RECEPTOR FAMILY AND ITS LIGANDS IN PANCREATIC DISEASES

The epidermal growth factor (EGF) receptor (EGFR) is a 170 kilodalton (kDa) glycosylated phosphoprotein. EGF has a profound effect on the differentiation of specific cells *in vivo* and is a potent mitogenic factor for a variety of cultured cells of both ectodermal and mesodermal origin.[100] In addition to having mitogenic effects, EGF contributes to several non-mitogenic responses, such as the inhibition of gastric acid secretion, wound repair, milk production, as well as some functions in the male reproductive system.[101,102] The EGFR is one of four homologous transmembrane proteins that mediate the actions of a family of growth factors. The EGF family of

EGFR (HER-1) – pancreatic cancer

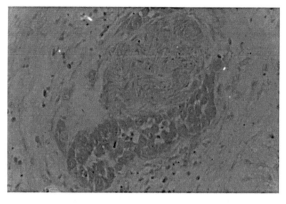

Figure 13.4 Epidermal growth factor receptor (EGFR, HER-1) immunohistochemistry in PC. The picture shows intense EGFR immunoreactivity in pancreatic cancer cells closely located to a pancreatic nerve.

receptors includes human EGF receptors Type 1 (HER1/EGFR/ErbB1), Type 2 (HER2/ErbB2/Neu), Type 3 (HER3/ErbB3), and Type 4 (HER4/ErbB4).[103-106] These four growth factor receptors contain an extracellular ligand-binding domain, a transmembrane domain and an intracellular domain with tyrosine kinase activity.[103-106]

A number of HER-specific ligands, which can bind to and activate EGF receptors, have been identified. Those ligands have been divided into three groups according to their EGF-like domain. The first group includes EGF, amphiregulin (AR), and TGF-α, which bind specifically to HER1. The second group contains betacellulin (BTC), heparin-binding EGF (HB-EGF), and epiregulin (EPR), which exhibit dual specificity in that they bind HER1 and HER4. The third group is composed of the neuregulins (NRGs), which form two subgroups based upon their capacity to bind HER3 and HER4 (NRG-1 and NRG-2) or only HER4 (NRG-3 and NRG-4). No direct ligand for HER2 has been discovered.[107]

13.5.1 The EGF Receptor Family and Its Ligands in the Normal Pancreas

In the normal human pancreas, HER1, HER2, HER3, and HER4 mRNA are present at different levels. The distribution of those receptors differs with respect to the various cell types in the normal human pancreas. Moderate HER1 and strong HER3 immunoreactivity are present predominantly in the cytoplasm of acinar cells and to a lesser extent in the ductal cells (Figure 13.6). Whereas, strong HER2 immunoreactivity is present in the islet cells (Figure 13.5) and HER4 is predominantly present in the cell membrane and cytoplasm of the ductal and acinar cells and at a much lower level in islet cells.[108,109]

13.5.2 The EGF Receptor Family and Its Ligands in AP

In AP, there are no significant differences in the levels of HER1, HER2, and HER3 as compared to the levels in the normal pancreas. However, some AP patients exhibit a significant increase in HER1 levels. Furthermore, the distribution of the receptors differs with respect to the various cell types in the human pancreas. There is a marked increase in HER1 immunoreactivity in acinar and ductal-like cells; whereas, HER3 immunoreactivity is less prominent in acini and increased in ductal-like cells. HER2 immunoreactivity is again mainly evident in islet cells, but is also present in the ductal-like cells. These findings indicate that there is altered distribution of HER receptors in the pancreas following AP and raise the possibility that HER receptors may be involved in the process of pancreatic regeneration during recovery from AP.[109]

ErbB2 (HER-2)

| normal | chronic pancreatitis |

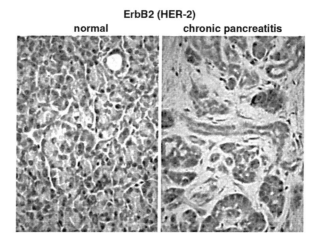

Figure 13.5 ErbB2 (HER-2) immunohistochemistry in normal pancreas and in CP. In the normal pancreas, some acinar cells exhibited HER-2 immunoreactivity. In CP moderate to strong HER-2 immunoreactivity was present in degenerating acinar cells and in acinar cells dedifferentiating into tubular complexes.

In pancreatic acinar cells, EGF has been found to directly increase DNA synthesis and stimulate cell growth *in vivo* and *in vitro*.[110,111] *In vitro*, high doses of EGF stimulate amylase release in isolated rat pancreatic acini, whereas low concentrations of EGF inhibit cholecystokinin-stimulated amylase release from acinar cells.[112] After the induction of cerulein-induced pancreatitis in rats, expression of EGF shows an increase during the initial 5 days, probably limiting the extent of AP and enhancing pancreatic repair and regeneration.[113] Moreover, treatment with EGF reduces the severity of pancreatic damage evoked by cerulein-induced pancreatitis and accelerates tissue repair. The beneficial effects of EGF appear to depend on the improvement of pancreatic blood flow, as well as on an increase of pancreatic cell growth.[114,115] EGF treatment is also beneficial in preventing septic complications in AP. EGF minimizes intestinal damage and decreases bacterial translocation from the gastrointestinal tract to mesenteric lymph nodes and bacterial spread to distant organs.[11]

13.5.3 The EGF Receptor Family and Its Ligands in CP

In CP, the receptors HER1, HER2, and HER3 are overexpressed both at the mRNA and the protein level (Figures 13.5 and 13.6). HER1 and HER3 are overexpressed in the pancreatic acinar and ductal cells.[117] Interestingly, HER2 expression is increased only in CP associated with pancreatic head enlargement.[118] It was also found that EGF and TGF-β1 are

ErbB3 (HER-3)

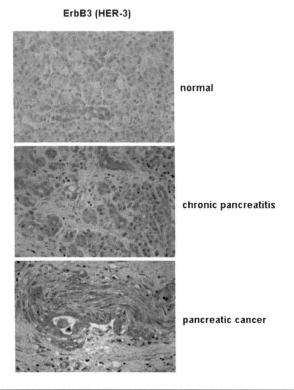

normal

chronic pancreatitis

pancreatic cancer

Figure 13.6 ErbB3 (HER-3) immunohistochemistry in normal pancreas, CP and PC. In the normal pancreas, some acinar and ductal cells exhibited HER-3 immunoreactivity. In CP moderate to strong HER-3 immunoreactivity was present in degenerating acinar cells and in acinar cells dedifferentiating into tubular complexes. Most PC cells exhibited strong HER-3 immunoreactivity. Also, some pancreatic nerves showed HER-3 immunostaining.

expressed at high levels in CP.[119] CRIPTO, an EGF-like growth factor that does not bind to EGFR, is also overexpressed in CP as compared to the normal pancreas.[120]

13.5.4 The EGF Receptor Family and Its Ligands in PC

In vivo studies demonstrate that human PCs overexpress EGF, TGF-α, and HB-EGF. *In vitro* experiments with human PC cell lines show that HB-EGF, EGF, TGF-α, and AR enhance the proliferation of these cells.[121–123] Human PCs usually exhibit high levels of the EGF receptor, HER2, and HER3 (Figures 13.4 and 13.6).[119,121,124,125] In contrast, quantitative PCR analysis shows that the expression of the fourth member of the EGF receptor family (HER4) is not changed (in advanced tumors) or even decreased (in early tumor stages) in PC.[108] EGF receptor expression progressively increases in the transition from acinar to duct-like to transformed cells.[126] Recently, a novel negative regulator of EGFR, termed EGF receptor related protein (ERRP), was discovered. ERRP is usually expressed in benign ductal epithelium, but not in ductal adenocarcinoma. ERRP expression appears to attenuate EGFR activation. Low levels of ERRP are associated with poor clinical outcome, suggesting that progressive loss of ERRP, a negative regulator of EGFR, may partly stimulate aggressive tumor cell growth in pancreatic adenocarcinoma.[127,128]

In addition, increased expression of HER3 has been shown to be associated with advanced tumor stage and significantly reduced postoperative survival,[119] while enhanced HER2 expression in PC does not appear to be associated with tumor progression, but rather with a better differentiated tumor phenotype.[124] Recently, it was reported that the lack of HER4 expression might increase the metastatic capacity of PC cells.[129]

The concomitant overexpression of EGF receptor and EGF or TGF-α in most PC cells indicates that autocrine and paracrine mechanisms of this receptor-ligand system play a crucial role in the pathogenesis of PC cell growth.[130–132] This hypothesis is supported by the observation that coexpression of the EGF receptor and either EGF or TGF-α or AR is associated with decreased survival in patients with PC, and up-regulation of EGF and the EGF receptor occurs more frequently in metastatic lesions than in primary PCs.[122,130] Further evidence of the importance of the EGF system of ligands and receptors comes from experiments in transgenic mice overexpressing TGF-α in the exocrine pancreas. In these animals, dysplastic changes of the pancreas, resembling those frequently seen in human PC, were found in 50% of the TGF-α transgenic mice.[126] In addition, malignant pancreatic tumors were observed in approximately 20% of TGF-α transgenic mice older than 1 year.[126]

Based on these findings, new treatment possibilities have evolved for PC. For example, transfection of a truncated EGFR lacking the tyrosine kinase domain into PANC-1 PC cells leads to dominate negative abrogation of EGF receptor-dependent signaling pathways and decreased anchorage-independent growth of these cancer cells.[133] In addition, AR antisense oligonucleotides act to decrease AR protein levels in T3M4 PC cells, decrease AR release into the medium, and inhibit cell growth in a dose-dependent manner.[133]

More recently, treatment of human PC cell lines and xenografts with combinations of ERBITUX™ anti-EGFR antibody, gemcitabine, and radiation was examined. Treatment of cells with a combination of ERBITUX and gemcitabine and radiation produced the highest induction of apoptosis and inhibition of proliferation *in vitro*. Moreover, combination treatment resulted in complete regression of xenograft tumors for more than 250 days.[134]

Taken together, these observations suggest that PC derives an important growth advantage through autocrine or paracrine mechanisms of EGF and EGF-related ligands, and that a therapeutic approach aimed at abrogating signaling of these ligands might be promising in the treatment of PC in the future.

13.6 THE HEPATOCYTE GROWTH FACTOR AND MET RECEPTOR IN PANCREATIC DISEASES

Hepatocyte growth factor (HGF) is another important polypeptide growth factor that signals through tyrosine kinase receptors encoded by the c-met oncogene. Upon ligand binding and subsequent receptor phosphorylation, cells are stimulated to proliferate or invade.[135-137] In the normal human pancreas, HGF is expressed in the stromal cells and c-met is expressed in ductal cells.[138,139] Activation of c-met stimulates motility, mitosis, and morphogenesis, and these processes are involved in organ regeneration or progression of malignancies.

13.6.1 HGF and MET in AP

In experimental and clinical AP, overexpression and elevated plasma HGF levels have been reported.[140-142] HGF mRNAs are up-regulated in the regeneration phase of AP and markedly increased after 48 hours.[143] In one study in rats with severe pancreatitis, HGF levels significantly increased in injured organs like liver, kidney, and lung.[140] When anti-HGF neutralizing antibodies were administered in severe pancreatitis, liver dysfunction worsened and apoptotic cells increased in the kidney. These results suggest that HGF is produced in injured organs and may function as an organotrophic and antiapoptotic factor against organ injuries in AP.[140]

C-met, the HGF receptor, is significantly up-regulated in experimental pancreatitis as compared to the normal pancreas[143,144] and is localized in areas of regenerating tissue.[144] Furthermore, in cultured pancreatic epithelial cells, HGF stimulates the expression of c-met in an autocrine manner.[143] Thus, enhanced c-met expression after AP suggests that HGF/c-met has an important role in pancreatic regeneration.[144] Serum HGF levels were found to be positive in 92.1% of patients with AP and were significantly elevated in the patients with higher Ranson scores, higher APACHE II scores, or higher computed tomography grades.[142] Furthermore, serum HGF levels in patients with organ dysfunction (liver, kidney, or lung) were significantly higher than in patients without organ dysfunction. Moreover, serum HGF levels on admission were significantly higher in nonsurvivors than in survivors. These results suggest that serum human HGF levels may reflect the severity, organ dysfunction, and prognosis in AP.[142]

Serum HGF levels were also elevated in experimental pancreatitis. The degree of serum HGF elevation correlated with the severity of AP and the associated organ dysfunctions.[140] Moreover, treatment with HGF during induction of cerulein-induced pancreatitis in rats attenuates pancreatic damage and is closely correlated with the increase in production of interleukin-10, the reduction in release of interleukin-1beta and interleukin-6, and the improvement of pancreatic blood flow.[145] Therefore, HGF has been targeted as a possible new treatment for AP. In trypsin-induced AP, pretreatment with a deleted form of hepatocyte growth factor (dHGF) increased survival rates.[14]

13.6.2 HGF and MET in CP

In experimental CP, HGF expression increased until Days 7 to 14 followed by a decrease to control levels, while c-met expression was constantly increased. HGF was localized in mesenchymal cells whereas c-met was present in epithelial cells.[147] Serum HGF levels in patients with CP are higher than those in normal individuals.[148] The significant increase of HGF and c-met expression suggests an essential role of this growth factor in the morphological changes during the development of CP.

13.6.3 HGF and MET in PC

In PC samples there is enhanced expression of HGF presumably derived from the stromal cells and not from the cancer cells.[138] In contrast, c-met is expressed strongly in the duct-like cancer cells in human PCs.[138,139] HGF enhances met receptor phosphorylation in cultured PC cell lines and stimulates cell growth and movement. In cell lines overexpressing the met receptor, receptor phosphorylation occurs in the absence of endogenously or exogenously added ligand.[139] *In vitro* and *in vivo* data suggest that increased expression of the met receptor and HGF in human PC leads to aberrant activation of met-dependent signaling pathways that have the potential to contribute to PC cell growth.

13.7 THE FIBROBLAST GROWTH FACTOR RECEPTORS AND THEIR LIGANDS IN PANCREATIC DISEASES

Fibroblast growth factors (FGFs) are a family of heparin-binding polypeptide growth factors that activate transmembrane tyrosine kinase receptors. They are involved in mitogenesis, cell differentiation, and angiogenesis and presently consist of more than 20 members.[149–151] The best known members of the family are acidic FGF (FGF1 or aFGF) and basic FGF (FGF2 or bFGF), the prototypes of this growth factor family. This family also includes int-2 (FGF3), hst (FGF4 or Kaposi FGF), FGF5, FGF6, keratinocyte growth factor (KGF or FGF7), androgen-induced growth factor (FGF8), glia-activating factor (FGF9), and FGF10.[152,153] Signaling by FGFs is mediated by a dual-receptor system consisting of four high-affinity transmembrane tyrosine-kinase FGF receptors (FGFRs) that function as signaling molecules to transmit the effects of FGFs, as well as by low-affinity heparin sulfate-proteoglycans (HSPGs) that are devoid of signaling capabilities but enhance ligand presentation to FGFRs.[149–151] The extracellular domain of the four high-affinity FGFRs (FGFR1 to FGFR4) consists of three immunoglobulin (Ig)-like domains, including a stretch between Domain 1 and Domain 2. Adjacent to the transmembrane region, these receptors consist of a split tyrosine-kinase domain. Alternative splicing results in several receptor isoforms with different ligand-binding affinities. These splice variants may exert important functions that influence the biological actions of various FGFs in different cell types. After ligand binding, FGFRs undergo homo- and heterodimerization, resulting in auto- and transphosphorylation of tyrosine residues within their cytoplasmic domains. As with the EGF receptor, mitogenic signaling is mediated via various pathways, including the ras/raf/MAP kinase cascade.[149–151]

13.7.1 FGFs and Receptors in the Normal Pancreas

In the normal pancreas, FGF1, FGF2, FGF5, and FGF7 and the four high-affinity FGF receptors are present.[154–157] FGF1 and FGF2 are expressed in the cytoplasm of acinar and ductal cells in the normal pancreas. However, FGF1 is expressed preferentially in ductal cells and FGF2 is more frequently found in acinar cells.[154,155] In addition, FGF5 is present in ductal and islet cells as well as in fibroblasts in the normal pancreas.[156] KGF receptor (KGFR) is also localized in islet cells, ductal cells, and centroacinar cells in the normal pancreas.[158]

13.7.2 FGFs and Receptors in AP

In human AP, the expression and distribution of FGFs and their receptors has two patterns. In regions that have undergone necrosis, there is complete loss of FGF1, FGF2, and FGFR1, while in the regenerating areas, FGF1 and FGF2 are readily evident in exocrine-type cells, in association with a marked increase in FGFR1. These findings indicate that FGF/receptor expression may be involved in the process of pancreatic exocrine regeneration during recovery from AP.[159] In cerulein-induced AP in rats, the expression of FGF1 and FGF2 is markedly increased. Thus, FGF1 is up-regulated between Day 3 and Day 5 after induction of pancreatitis, and FGF2 has a pronounced transcriptional elevation between 1 and 3 days after induction of pancreatitis.[143] Moreover, expression of FGF10 is induced and is localized in the vascular smooth muscle cells (vsmc), and the expression of KGFR is lost from centroacinar cells and is detected in some acinar cells and vsmc in addition to islet cells. However, total levels of KGFR are decreased after induction of pancreatitis.[158] These findings suggest that FGF10 contributes to the regeneration and differentiation of acinar cells and angiogenesis in AP through KGFR.

13.7.3 FGFs and Receptors in CP

In human CP samples there is an increase of FGF1, FGF2, and FGF5 expression in comparison to the normal pancreas. FGF1 and FGF2 are present in degenerating acinar and ductal cells and in areas exhibiting tubular complexes. Areas of CP tissues with minor damage show FGF1 and FGF2 expression in acinar and ductal cells that is only slightly increased compared to normal controls.[155] In contrast, FGF5 is predominantly localized in the periductal fibroblasts and endocrine islet cells, as well as in the atrophic acinar and ductal cells.[157]

13.7.4 FGFs and Receptors in PC

In PC, it has been demonstrated that FGF1, FGF2, FGF5, and FGF7 are overexpressed (Figure 13.7).[151,154,156] Overexpression of both FGF1 and FGF2 occurs within the cancer cells and is associated with a more advanced tumor stage, and the presence of FGF2 in the cancer cells is associated with shorter postoperative patient survival, whereas FGF1 does not seem to have a strong influence on survival.[154] FGF5 localizes predominantly in the stromal fibroblasts, infiltrating and surrounding the cancer cells, and to a lesser extent in the cancer cells.[156] In addition, FGF7 is expressed predominantly in the cancer cells.[16] *In vitro* studies have shown that FGF1, FGF2, FGF3, FGF4, FGF5, and FGF7 are expressed at variable levels in cultured PC cell lines and are coexpressed with members of the FGFR family.[160–163] Furthermore, FGF5 is expressed and released by cultured PC cells, and FGF1, FGF2, and FGF5 enhance the growth of PC cells *in vitro*.[151,156,159,164] Human PCs express disproportionately high levels of the 2-Ig loop form of FGFR1 (Figure 13.7), whereas the 3-Ig loop form dominates in the normal pancreas.[161] Furthermore, it has been shown that FGFR2 and its splice variant KGFR are also expressed at high levels in PC cells *in vivo*.[45] Regarding the role of the low-affinity HSPG receptor, it seems that glypican-1, a GPI-anchored protein, is the

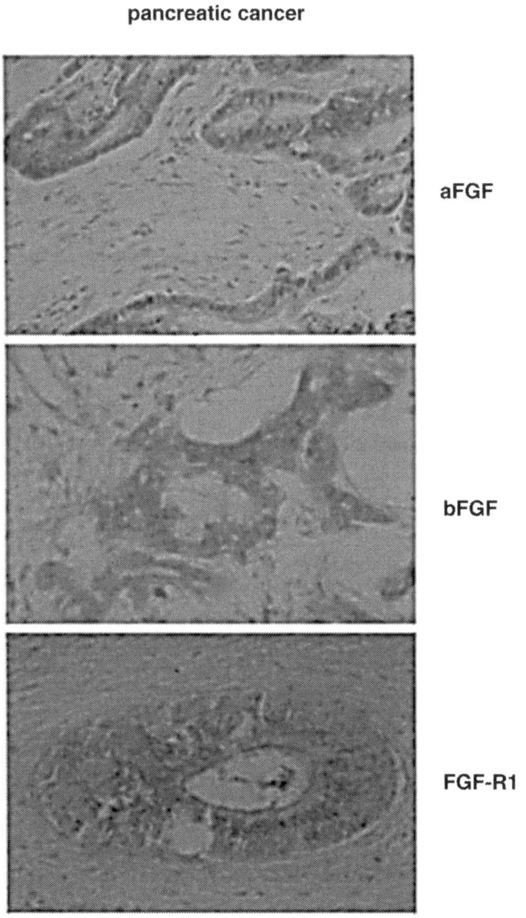

pancreatic cancer

aFGF

bFGF

FGF-R1

Figure 13.7 **Acidic fibroblast growth factor (aFGF), basic fibroblast growth factor (bFGF), and fibroblast growth factor Receptor 1 (FGFR1) immunohistochemistry in PC. Most PC cells exhibited moderate to strong aFGF, bFGF, and FGFR1 immunoreactivity.**

most important coreceptor for heparin-binding growth factors in PC.[164] Thus, the HSPG glypican-1 is overexpressed in a large proportion of PCs, and its expression occurs predominantly in the cancer cells and in the fibroblasts surrounding the tumor mass.[164] Decreased glypican-1 expression *in vitro* via a glypican antisense mRNA approach decreases the responsiveness of PC cells to heparin-binding growth factors.[164] Together these observations suggest that there is a potential for aberrant activation of FGF-dependent signaling in human PC. This is supported by recent observations that transfection of a truncated FGFR1 into PC cells leads to dominant-negative blockade of FGFR-dependent pathways and decreased tumor growth *in vivo* in nude mice.[165]

13.8 NEUROTROPHINS IN PANCREATIC DISEASES

The neurotrophins (NTs) are a family of proteins that are essential for the development of the vertebrate nervous system and were originally identified as neuronal survival factors. The NT family also includes nerve growth factor (NGF), brain-derived neurotrophic factor (BDNF), neurotrophin-3 (NT-3), and neurotrophin-4/5 (NT-4/5).[166–168]

Each NT binds to two different classes of transmembrane receptor proteins, the tropomyosin receptor kinases (Trks) and the NT receptor p75. This dual system allows the transduction of different signals following ligand binding, which can be as different as cell death signaling through p75 or cell survival through the Trk receptors.[169]

The p75 receptor is a transmembrane glycoprotein that binds to all NTs (pan-NT receptor) and also acts as a coreceptor for Trk receptors.[169,170] The Trk receptors belong to the family of transmembrane receptor tyrosine kinases, and three Trk genes have been identified in mammals. Each receptor can be activated by one or more of four NTs. Different NTs show binding specificity for particular receptors. Thus NGF binds preferentially to tyrosine receptor kinase A (TrkA); BDNF and NT4/5 to TrkB; and NT3 to TrkC. These specificities are not absolute, and NT3 is also a ligand for TrkA and TrkB in some cell types.[170–173]

Binding of the NTs to the Trk receptors leads to receptor tyrosine phosphorylation and activation of intercellular signaling pathways. NT-mediated activation of Trk receptors regulates cell survival, proliferation, the fate of neural precursors, axon and dendrite growth and patterning, and the expression and activity of functionally important proteins, such as ion channels and neurotransmitter receptors.[170–172,174–176]

13.8.1 NTs in AP

In acute necrotizing pancreatitis, up-regulation of NTs is a critical component of the response to pancreatic injury. The development of acute necrotizing pancreatitis leads to a significant increase in NGF production. Toma et al. observed two phases of NGF production: an early release from pancreatic islets at 2 and 6 hours and a later increase at 3 to 5 days together with maximum parenchymal necrosis.[17] In addition, the levels of BDNF, NT3, and NT4 in the inflamed pancreas reached a peak at 1 week.[17] In the normal pancreas, NGF, BDNF, NT3, and NT4 expression was observed in the islets.[17,177] In acute necrotizing pancreatitis at 2 to 6 hours, NGF showed widespread distribution in the parenchyma and was observed

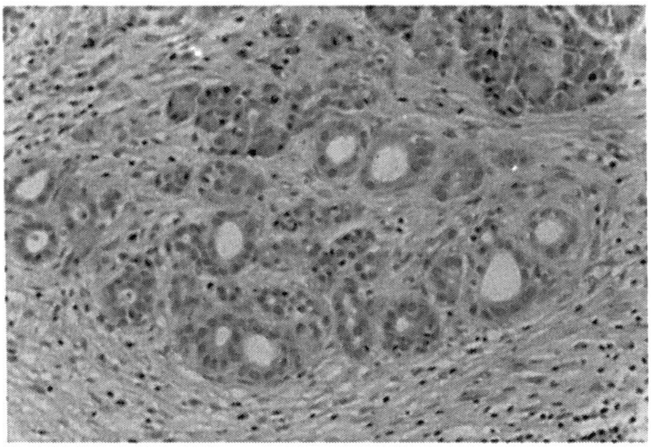

NGF – chronic pancreatitis

Figure 13.8 Nerve growth factor (NGF) immunohistochemistry in CP. Moderate to strong NGF immunoreactivity was present in degenerating acinar cells and in acinar cells dedifferentiating into tubular complexes.

in the cytoplasm of exocrine pancreatic tissues, including acinar and ductal cells, while a marked reduction of NGF level in the islets was observed. The same pattern was observed for the other NTs. There was marked reduction in the expression of BDNF, NT3, and NT4 in the islets at 2 and 6 hours in the acutely inflamed pancreas. In contrast, acinar and ductal cells, inflammatory cells, and neural elements displayed the expression of BDNF, NT3, and NT4 in the inflamed pancreas from 2 hours to 2 weeks.[17]

These findings demonstrate a shift in the localization of NTs from the endocrine to the exocrine pancreas. It is possible that NGF mediates peripheral sensitization and contributes to the generation of pain in acute necrotizing pancreatitis.

13.8.2 NTs in CP

In CP, recurrent abdominal pain, which occurs in approximately 80 to 90% of the patients, is the dominant clinical symptom. It is reported that the NGF/TrkA pathway is activated in CP and that this activation might influence nerve growth and the pain syndrome. NGF and its high-affinity TrkA receptor are increased in CP samples in comparison to the normal pancreas.[179] In CP, enhanced NGF expression is present in metaplastic ductal cells, in degenerating acinar cells, and in acinar cells dedifferentiating into tubular structures (Figure 13.8). TrkA is intensely present in the perineurium of most enlarged pancreatic nerves in CP. In addition, enhanced NGF and TrkA signals are also present in intrapancreatic ganglia cells in CP samples. These findings indicate that these factors are synthesized in these cellular elements. Furthermore, a significant correlation between NGF mRNA levels and pancreatic fibrosis and acinar cell damage and between TrkA mRNA and pain intensity has been observed.[179] These findings suggest that nerve changes in CP might be influenced by activation of the NGF/TrkA pathway through paracrine mechanisms. NGF released from degenerating acinar cells, dedifferentiating acinar cells, and metaplastic ductal cells might interact with TrkA located in the perineurium of pancreatic nerves. The presence of NGF and TrkA in intrapancreatic ganglia cells suggests that the NGF/TrkA pathway is also activated in intrinsic neural structures. These observations are of clinical interest because changes in neural morphology are associated with pain in CP. However, the interaction of NGF and TrkA in CP might not be limited to influencing nerve growth, but may be more directly tied to pain generation. For example, NGF might directly influence chronic pain by the regulation of transcription and the synthesis of substance P and calcitonin gene-related peptide (CGRP), as well as through the release of histamine.[180] Blocking of NGF by specific anti-NGF antibodies produces a sustained thermal and chemical hypoalgesia and a down-regulation of substance P and CGRP in rat experiments.[182] The sensory neurotransmitters substance P and CGRP are increased in enlarged pancreatic nerves in CP, indicating a potential regulatory interaction between both mediator systems in pain generation.[183]

BDNF was also found to be overexpressed in human CP compared to its expression in the normal pancreas.[184] BDNF is present in most of the ductal cells in the normal pancreas. In CP, BDNF expression is present in most cells of ductular complexes and in the perineurium of enlarged nerves. BDNF is found in degenerating acinar cells and islet cells and intrinsic pancreatic ganglia cells in CP samples. The expression level of BDNF was positively correlated with the pain intensity and frequency of CP patients.[184] Association of BNDF with pain suggests that it functions as a peripheral and central pain modulator, as reported previously in other inflammatory disorders.[185] Therefore, besides increased NGF levels, enhanced levels of BDNF in pancreatic nerves and degenerating acinar cells and ductular complexes may also play a role in nerve repair, regeneration, and progression of CP.[179]

Recently, overexpression of p75, the low-affinity receptor of NTs that mediates apoptosis, was reported in human CP.[186] p75 expression is present in some ductal cells in the normal pancreas, whereas in CP moderate p75NTR expression is present in acinar cells next to fibrosis, ductal cells, and cells of ductular structures as well as in some islet cells.[1] The expression of p75 positively correlates with the apoptotic index in the exocrine and endocrine pancreas. Thus the overexpression of p75 contributes to the apoptotic process of the exocrine and endocrine pancreas in CP.[1]

13.8.3 NTs in PC

In PC, the frequent infiltration of PC cells in pancreatic nerves has been noted for a long time. Perineural invasion extending to the extrapancreatic nerve plexus is a histopathologic characteristic in PC that leads to retropancreatic tumor extension, precludes curative resection, promotes local recurrence, and finally influences the prognosis of the patients negatively.[187–189] However, the mechanisms contributing to the invasion of pancreatic nerves and to the spread of cancer cells along nerves are poorly understood. NGF has been suggested to stimulate tumor growth, cancer cell invasion, and the formation of metastases in neuronal and nonneuronal tumors like lung cancer, prostate cancer, carcinoid tumors, medullary thyroid carcinoma, Wilms tumor, melanoma, and glioblastomas.[190–196] Therefore, it has been hypothesized that aberrant expression of the NTs or their receptors may contribute to the malignant phenotype of pancreatic ductal adenocarcinoma through autocrine or paracrine interactions. Therefore, the role of NTs and their receptors in PC was extensively studied in the past decade.

NGF and TrkA are significantly increased in PC tissues compared to the normal pancreas. NGF is strongly present in the cytoplasm of PC cells. Further, ductal and acinar cells, as well as neural tissue and cancer cells, express NGF.[19]

TrkA is intensely present in the perineurium of pancreatic nerves.[198] In addition, TrkA expression is also found in blood vessels, the neural tissue, duct cells, scattered islet cells, and cancer cells. TrkB has been found in the α-cells of the islet of normal and cancerous samples, and TrkC staining was similar to that of TrkA.[197] The low-affinity receptor p75NTR was expressed in the neural tissue and in scattered duct cells of the normal pancreas only.[19] NT3 expression has been noted in capillary endothelia and erythrocytes in both normal and cancerous tissues, and NT4 shows strong cytoplasmic staining of duct cells and heterogeneous staining of cancer cells.[19] Overexpression of TrkA correlates significantly with cancer proliferation, and TrkC overexpression has a significant correlation with cancer invasion, including venous and perineural invasion.[199]

There is no difference in NGF and TrkA expression between early and advanced tumor stages and between well or moderately differentiated and poorly differentiated tumors. Interestingly, tumors with high NGF and TrkA expression levels exhibit more frequent perineural invasion in histopathologic analysis. Furthermore, increased NGF and TrkA expression levels are associated with a higher degree of pain.[198]

In vitro studies have shown that NGF, TrkA, and p75 are expressed in PC cell lines.[200] In addition, exogenous NGF stimulates the growth of PC cells.[200] Thus, NGF–TrkA interactions are important factors influencing cell growth and spread in this malignancy[200] Moreover, stable transfection of NGF in PC cells results in enhanced anchorage-dependent growth and in an increase in anchorage-independent cell growth and cell invasion. Furthermore, stably transfected cells have shown enhanced tumorigenicity in nude mice.[201.]

Therefore, the disruption of NTs' Trk signaling in pancreatic carcinomas is a potential therapeutic target. Recently, the antitumor efficacy of administration of NT neutralizing antibodies on the growth of human pancreatic ductal adenocarcinoma xenografts in nude mice was examined. Tumor-bearing nude mice were treated with a mixture of NT neutralizing antibodies composed of anti-NGF, anti-BNDF, anti-NT3, and anti-NT4/5. Treatment with the antibody mixture significantly inhibited the growth of PC cell line xenografts as compared with IgG-treated controls.[202]

In conclusion, NGF has the capacity to act in a paracrine or an autocrine manner in PC. It enhances cancer cell growth and perineural invasion *in vivo* and *in vitro*, thereby contributing to the aggressiveness and poor prognosis of PC, and may contribute to the pain syndrome in human PC. Furthermore, targeting of NTs and their receptor pathways holds potential for future treatment of PC.

13.9 EPILOGUE

Molecular research over the past years has significantly contributed to a better understanding of the pathophysiological changes in pancreatic diseases. Increased expression of a number of growth factors and growth factor receptors in diseased pancreatic tissues contributes to the progression of both AP and CP and to the malignant phenotype of PC cells. Our expanding knowledge of both normal and abnormal pancreatic cell biology, which consists of complex interaction among both known, and probably to a major extent still unknown, systems makes it difficult to choose molecular targets that are amenable for therapeutic intervention. Nonetheless, future innovations regarding effective treatment options for PC must apply our present molecular knowledge of this disease.

REFERENCES

1. Sarner M. and Cotton P.B. 1984. Definitions of acute and chronic pancreatitis. *Clin. Gastroenterol.* 13(3):865–870.
2. Ranson J.H. 1984. Acute pancreatitis: pathogenesis, outcome and treatment. *Clin. Gastroenterol.* 13(3):843–863.
3. Pitchumoni C.S., Agarwal N., and Jain N.K. 1988. Systemic complications of acute pancreatitis. *Am. J. Gastroenterol.* 83(6):597–606.
4. Steer M.L. 1989. Classification and pathogenesis of pancreatitis. *Surg. Clin. N. Am.* 69(3):467–480.
5. Sarner M. and Cotton P.B. 1984. Classification of pancreatitis. *Gut* 25(7):756–759.

6. Sarles H., Bernard J.P., and Johnson C. 1989. Pathogenesis and epidemiology of chronic pancreatitis. *Ann. Rev. Med.* 40:453–468.

7. DiMagno E.P. 1993. A short, eclectic history of exocrine pancreatic insufficiency and chronic pancreatitis. *Gastroenterology* 104(5):1255–1262.

8. Adler G. and Schmid R.M. 1997. Chronic pancreatitis: still puzzling? *Gastroenterology* 112(5):1762–1765.

9. Steer M.L., Waxman I., and Freedman S. 1995. Chronic pancreatitis. *N. Engl. J. Med.* 332(22):1482–1490.

10. Malfertheiner P., Buchler M., Stanescu A., and Ditschuneit H. 1986. Exocrine pancreatic function in correlation to ductal and parenchymal morphology in chronic pancreatitis. *Hepatogastroenterology* 33(3):110–114.

11. Malfertheiner P., Buchler M., Stanescu A., and Ditschuneit H. 1987. Pancreatic morphology and function in relationship to pain in chronic pancreatitis. *Int. J. Pancreatol.* 2(1):59–66.

12. Jemal A., Murray T., Samuels A., Ghafoor A., Ward E., and Thun M.J. 2003. Cancer statistics, 2003. *CA Cancer J. Clin.* 53(1):5–26.

13. Friess H., Kleeff J., Silva J.C., Sadowski C., Baer H.U., and Buchler M.W. 1998. The role of diagnostic laparoscopy in pancreatic and periampullary malignancies. *J. Am. Coll. Surg.* 186(6):675–682.

14. Warshaw A.L. and Fernandez-del Castillo C. 1992. Pancreatic carcinoma. *N. Engl. J. Med.* 326(7):455–465.

15. Beger H.G., Buchler M.W., and Friess H. 1994. Surgical results and indications for adjuvant measures in pancreatic cancer. *Chirurg.* 65(4):246–252.

16. Steiner D.F., Cunningham D., Spigelman L., and Aten B. 1967. Insulin biosynthesis: evidence for a precursor. *Science* 157(789):697–700.

17. Steiner D.F., Clark J.L., Nolan C., Rubenstein A.H., Margoliash E., Aten B. et al. 1969. Proinsulin and the biosynthesis of insulin. *Recent Prog. Horm. Res.* 25:207–282.

18. De Meyts P. and Whittaker J. 2002. Structural biology of insulin and IGF1 receptors: implications for drug design. *Nat. Rev. Drug Discov.* 1(10):769–783.

19. Virkamaki A., Ueki K., and Kahn C.R. 1999. Protein-protein interaction in insulin signaling and the molecular mechanisms of insulin resistance. *J. Clin. Invest.* 103(7):931–943.

20. White M.F. 1997. The insulin signalling system and the IRS proteins. *Diabetologia* 40 Suppl. 2:S2–S17.

21. Sesti G., Federici M., Hribal M.L., Lauro D., Sbraccia P., and Lauro R. 2001. Defects of the insulin receptor substrate (IRS) system in human metabolic disorders. *Faseb. J.* 15(12):2099–2111.

22. Giovannone B., Scaldaferri M.L., Federici M., Porzio O., Lauro D., Fusco A. et al. 2000. Insulin receptor substrate (IRS) transduction system: distinct and overlapping signaling potential. *Diabetes Metab. Res. Rev.* 16(6):434–441.

23. Mossner J. 1985. Insulin — a regulator of exocrine pancreas function? *Z. Gastroenterol.* 23(12):694–702.

24. Bendayan M. 1993. Pathway of insulin in pancreatic tissue on its release by the B-cell. *Am. J. Physiol.* 264(2 Pt 1):G187–G194.

25. von Schonfeld J., Goebell H., and Muller M.K. 1994. The islet-acinar axis of the pancreas. *Int. J. Pancreatol.* 16(2–3):131–140.

26. Kramer M.F. and Tan H.T. 1968. The peri-insular acini of the pancreas of the rat. *Z. Zellforsch. Mikrosk. Anat.* 86(2):163–170.

27. Bendayan M. and Ito S. 1979. Immunohistochemical localization of exocrine enzymes in normal rat pancreas. *J. Histochem. Cytochem.* 27(6):1029–1034.
28. Pour P. 1978. Islet cells as a component of pancreatic ductal neoplasms. I. Experimental study: ductular cells, including islet cell precursors, as primary progenitor cells of tumors. *Am. J. Pathol.* 90(2):295–316.
29. Pour P.M. 1997. The role of Langerhans islets in pancreatic ductal adenocarcinoma. *Front. Biosci.* 2:D271–D282.
30. Pour P., Mohr U., Cardesa A., Althoff J., and Kruger F.W. 1975. Pancreatic neoplasms in an animal model: morphological, biological, and comparative studies. *Cancer* 36(2):379–389.
31. Malecka-Panas E., Gasiorowska A., Kropiwnicka A., Zlobinska A., and Drzewoski J. 2002. Endocrine pancreatic function in patients after acute pancreatitis. *Hepatogastroenterology* 49(48):1707–1712.
32. Boreham B. and Ammori B.J. 2003. A prospective evaluation of pancreatic exocrine function in patients with acute pancreatitis: correlation with extent of necrosis and pancreatic endocrine insufficiency. *Pancreatology* 3(4):303–308.
33. Qader S.S., Ekelund M., Andersson R., Obermuller S., and Salehi A. 2003. Acute pancreatitis, expression of inducible nitric oxide synthase and defective insulin secretion. *Cell. Tissue Res.* 313(3):271–279.
34. Abe N., Watanabe T., Ozawa S., Masaki T., Mori T., Sugiyama M. et al. 2002. Pancreatic endocrine function and glucose transporter (GLUT)-2 expression in rat acute pancreatitis. *Pancreas* 25(2):149–153.
35. Descamps F.J., Van den Steen P.E., Martens E., Ballaux F., Geboes K., and Opdenakker G. 2003. Gelatinase B is diabetogenic in acute and chronic pancreatitis by cleaving insulin. *Faseb. J.* 17(8):887–889.
36. Lankisch P.G., Lohr-Happe A., Otto J., and Creutzfeldt W. 1993. Natural course in chronic pancreatitis. Pain, exocrine and endocrine pancreatic insufficiency and prognosis of the disease. *Digestion* 54(3):148–155.
37. Schmied B.M., Ulrich A.B., Friess H., Buchler M.W., and Pour P.M. 2001. The patterns of extrainsular endocrine cells in pancreatic cancer. *Teratog. Carcinog. Mutagen.* 21(1):69–81.
38. Yeo C.J., Bastidas J.A., Schmieg Jr. R.E., Walfisch S., Couse N.F., Olson J.L. et al. 1989. Pancreatic structure and glucose tolerance in a longitudinal study of experimental pancreatitis-induced diabetes. *Ann. Surg.* 210(2):150–158.
39. Schmied B.M., Ulrich A.B., Matsuzaki H., Li C., Friess H., Bochler M.W. et al. 2000. Alteration of the Langerhans islets in pancreatic cancer patients. *Int. J. Pancreatol.* 28(3):187–197.
40. Wang F., Herrington M., Larsson J., and Permert J. 2003. The relationship between diabetes and pancreatic cancer. *Mol. Cancer* 2(1):4.
41. Eusebi V., Capella C., Bondi A., Sessa F., Vezzadini P., and Mancini A.M. 1981. Endocrine-paracrine cells in pancreatic exocrine carcinomas. *Histopathology* 5(6):599–613.
42. Pour P.M., Permert J., Mogaki M., Fujii H., and Kazakoff K. 1993. Endocrine aspects of exocrine cancer of the pancreas. Their patterns and suggested biologic significance. *Am. J. Clin. Pathol.* 100(3):223–230.
43. Lee A.V. and Yee D. 1995. Insulin-like growth factors and breast cancer. *Biomed. Pharmacother.* 49(9):415–421.

44. Bergmann U., Funatomi H., Yokoyama M., Beger H.G., and Korc M. 1995. Insulin-like growth factor I overexpression in human pancreatic cancer: evidence for autocrine and paracrine roles. *Cancer Res.* 55(10):2007–2011.

45. Lee J. and Pilch P.F. 1994. The insulin receptor: structure, function, and signaling. *Am. J. Physiol.* 266(2 Pt 1):C319–C334.

46. Knutson V.P. 1991. Cellular trafficking and processing of the insulin receptor. *Faseb. J.* 5(8):2130–2138.

47. McKinnon P., Ross M., Wells J.R., Ballard F.J., and Francis G.L. 1991. Expression, purification and characterization of secreted recombinant human insulin-like growth factor-I (IGF-I) and the potent variant des(1-3) IGF-I in Chinese hamster ovary cells. *J. Mol. Endocrinol.* 6(3):231–239.

48. Butler A.A., Yakar S., Gewolb I.H., Karas M., Okubo Y., and LeRoith D. 1998. Insulin-like growth factor-I receptor signal transduction: at the interface between physiology and cell biology. *Comp. Biochem. Physiol. B Biochem. Mol. Biol.* 121(1):19–26.

49. Ogawa W., Matozaki T., and Kasuga M. 1998. Role of binding proteins to IRS-1 in insulin signalling. *Mol. Cell. Biochem.* 182(1–2):13–22.

50. He W., Craparo A., Zhu Y., O'Neill T.J., Wang L.M., Pierce J.H. et al. 1996. Interaction of insulin receptor substrate-2 (IRS-2) with the insulin and insulin-like growth factor I receptors. Evidence for two distinct phosphotyrosine-dependent interaction domains within IRS-2. *J. Biol. Chem.* 271(20):11641–11645.

51. Kasuga M., Izumi T., Tobe K., Shiba T., Momomura K., Tashiro-Hashimoto Y. et al. 1990. Substrates for insulin-receptor kinase. *Diabetes Care* 13(3):317–326.

52. Kerouz N.J., Horsch D., Pons S., and Kahn C.R. 1997. Differential regulation of insulin receptor substrates-1 and -2 (IRS-1 and IRS-2) and phosphatidylinositol 3-kinase isoforms in liver and muscle of the obese diabetic (ob/ob) mouse. *J. Clin. Invest.* 100(12):3164–3172.

53. Myers Jr. M.G., Sun X.J., and White M.F. 1994. The IRS-1 signaling system. *Trends Biochem. Sci.* 19(7):289–293.

54. Sun X.J., Wang L.M., Zhang Y., Yenush L., Myers Jr. M.G., Glasheen E. et al. 1995. Role of IRS-2 in insulin and cytokine signalling. *Nature* 377(6545):173–177.

55. Ludwig C.U., Menke A., Adler G., and Lutz M.P. 1999. Fibroblasts stimulate acinar cell proliferation through IGF-I during regeneration from acute pancreatitis. *Am. J. Physiol.* 276(1 Pt 1):G193–G198.

56. Gomez G., Lee H.M., He Q., Englander E.W., Uchida T., and Greeley Jr. G.H. 2001. Acute pancreatitis signals activation of apoptosis-associated and survival genes in mice. *Exp. Biol. Med. (Maywood)* 226(7):692–700.

57. Ishiwata T., Bergmann U., Kornmann M., Lopez M., Beger H.G., and Korc M. 1997. Altered expression of insulin-like growth factor II receptor in human pancreatic cancer. *Pancreas* 15(4):367–373.

58. Bergmann U., Funatomi H., Kornmann M., Beger H.G., and Korc M. 1996. Increased expression of insulin receptor substrate-1 in human pancreatic cancer. *Biochem. Biophys. Res. Commun.* 220(3):886–890.

59. Kornmann M., Maruyama H., Bergmann U., Tangvoranuntakul P., Beger H.G., White M.F. et al. 1998. Enhanced expression of the insulin receptor substrate-2 docking protein in human pancreatic cancer. *Cancer Res.* 58(19):4250–4254.

60. Meggiato T., Plebani M., Basso D., Panozzo M.P., and Del Favero G.. 1999. Serum growth factors in patients with pancreatic cancer. *Tumour Biol.* 20(2):65–71.

61. Basso D., Plebani M., Fogar P., Panozzo M.P., Meggiato T., De Paoli M. et al. 1995. Insulin-like growth factor-I, interleukin-1 alpha and beta in pancreatic cancer: role in tumor invasiveness and associated diabetes. *Int. J. Clin. Lab. Res.* 25(1):40–43.

62. Bloom B.B., Humphries D.E., Kuang P.P., Fine A., and Goldstein R.H. 1996. Structure and expression of the promoter for the R4/ALK5 human type I transforming growth factor-beta receptor: regulation by TGF-beta. *Biochem. Biophys. Acta.* 1312(3):243–248.

63. Heldin C.H., Miyazono K., and ten Dijke P. 1997. TGF-beta signalling from cell membrane to nucleus through SMAD proteins. *Nature* 390(6659):465–471.

64. Massague J. 1998. TGF-beta signal transduction. *Ann. Rev. Biochem.* 67:753–791.

65. Derynck R., Zhang Y., and Feng X.H. 1998. Smads: transcriptional activators of TGF-beta responses. *Cell* 95(6):737–740.

66. Hahn S.A., Schutte M., Hoque A.T., Moskaluk C.A., da Costa L.T., Rozenblum E. et al. 1996. DPC4, a candidate tumor suppressor gene at human chromosome 18q21.1. *Science* 271(5247):350–353.

67. Kretzschmar M. and Massague J. 1998. SMADs: mediators and regulators of TGF-beta signaling. *Curr. Opin. Genet. Dev.* 8(1):103–111.

68. Whitman M. 1997. Signal transduction. Feedback from inhibitory SMADs. *Nature* 389(6651):549–551.

69. Singer M.V., Gyr K., and Sarles H. Revised classification of pancreatitis. 1985. Report of the Second International Symposium on the Classification of Pancreatitis in Marseille, France, March 28–30, 1984. *Gastroenterology* 89(3):683–685.

70. Gillessen A., Voss B., Rauterberg J., and Domschke W. 1993. Distribution of collagen types I, III, and IV in peptic ulcer and normal gastric mucosa in man. *Scand. J. Gastroenterol.* 28(8):688–689.

71. Ignotz R.A. and Massague J. 1986. Transforming growth factor-beta stimulates the expression of fibronectin and collagen and their incorporation into the extracellular matrix. *J. Biol. Chem.* 261(9):4337–4345.

72. Ignotz R.A., Endo T., and Massague J. 1987. Regulation of fibronectin and type I collagen mRNA levels by transforming growth factor-beta. *J. Biol. Chem.* 262(14):6443–6446.

73. Varga J., Rosenbloom J., and Jimenez S.A. 1987. Transforming growth factor beta (TGF beta) causes a persistent increase in steady-state amounts of type I and type III collagen and fibronectin mRNAs in normal human dermal fibroblasts. *Biochem. J.* 247(3):597–604.

74. Menke A., Yamaguchi H., Gress T.M., and Adler G. 1997. Extracellular matrix is reduced by inhibition of transforming growth factor beta1 in pancreatitis in the rat. *Gastroenterology* 113(1):295–303.

75. Kihara Y., Tashiro M., Nakamura H., Yamaguchi T., Yoshikawa H., and Otsuki M. 2001. Role of TGF-beta1, extracellular matrix, and matrix metalloproteinase in the healing process of the pancreas after induction of acute necrotizing pancreatitis using arginine in rats. *Pancreas* 23(3):288–295.

76. Friess H., Lu Z., Riesle E., Uhl W., Brundler A.M., Horvath L. et al. 1998. Enhanced expression of TGF-betas and their receptors in human acute pancreatitis. *Ann. Surg.* 227(1):95–104.

77. Riesle E., Friess H., Zhao L., Wagner M., Uhl W., Baczako K. et al. 1997. Increased expression of transforming growth factor beta s after acute oedematous pancreatitis in rats suggests a role in pancreatic repair. *Gut* 40(1):73–79.

78. Nakae H., Endo S., Inoue Y., Fujino Y., Wakabayashi G., Inada K. et al. 2003. Matrix metalloproteinase-1 and cytokines in patients with acute pancreatitis. *Pancreas* 26(2):134–138.

79. Muller-Pillasch F., Menke A., Yamaguchi H., Elsasser H.P., Bachem M., Adler G. et al. 1999. TGFbeta and the extracellular matrix in pancreatitis. *Hepatogastroenterology* 46(29):2751–2756.

80. van Laethem J.L., Robberecht P., Resibois A., and Deviere J. 1996. Transforming growth factor beta promotes development of fibrosis after repeated courses of acute pancreatitis in mice. *Gastroenterology* 110(2):576–582.

81. Slater S.D., Williamson R.C., and Foster C.S. 1995. Expression of transforming growth factor-beta 1 in chronic pancreatitis. *Digestion* 56(3):237–241.

82. van Laethem J.L., Deviere J., Resibois A., Rickaert F., Vertongen P., Ohtani H. et al. 1995. Localization of transforming growth factor beta 1 and its latent binding protein in human chronic pancreatitis. *Gastroenterology* 108(6):1873–1881.

83. di Mola F.F., Friess H., Martignoni M.E., Di Sebastiano P., Zimmermann A., Innocenti P. et al. 1999. Connective tissue growth factor is a regulator for fibrosis in human chronic pancreatitis. *Ann. Surg.* 230(1):63–71.

84. Igarashi A., Okochi H., Bradham D.M., and Grotendorst G.R. 1993. Regulation of connective tissue growth factor gene expression in human skin fibroblasts and during wound repair. *Mol. Biol. Cell.* 4(6):637–645.

85. Naef M., Ishiwata T., Friess H., Buchler M.W., Gold L.I., and Korc M. 1997. Differential localization of transforming growth factor-beta isoforms in human gastric mucosa and overexpression in gastric carcinoma. *Int. J. Cancer* 71(2):131–137.

86. Friess H., Yamanaka Y., Buchler M., Ebert M., Beger H.G., Gold L.I. et al. 1993. Enhanced expression of transforming growth factor beta isoforms in pancreatic cancer correlates with decreased survival. *Gastroenterology* 105(6):1846–1856.

87. Hata A., Shi Y., and Massague J. 1998. TGF-beta signaling and cancer: structural and functional consequences of mutations in Smads. *Mol. Med. Today* 4(6):257–262.

88. Baldwin R.L., Friess H., Yokoyama M., Lopez M.E., Kobrin M.S., Buchler M.W. et al. 1996. Attenuated ALK5 receptor expression in human pancreatic cancer: correlation with resistance to growth inhibition. *Int. J. Cancer* 67(2):283–288.

89. Grau A.M., Zhang L., Wang W., Ruan S., Evans D.B., Abbruzzese J.L. et al. 1997. Induction of p21waf1 expression and growth inhibition by transforming growth factor beta involve the tumor suppressor gene DPC4 in human pancreatic adenocarcinoma cells. *Cancer Res.* 57(18):3929–3934.

90. Friess H., Yamanaka Y., Buchler M., Beger H.G., Kobrin M.S., Baldwin R.L. et al. 1993. Enhanced expression of the type II transforming growth factor beta receptor in human pancreatic cancer cells without alteration of type III receptor expression. *Cancer Res.* 53(12):2704–2707.

91. Lu Z., Friess H., Graber H.U., Guo X., Schilling M., Zimmermann A. et al. 1997. Presence of two signaling TGF-beta receptors in human pancreatic cancer correlates with advanced tumor stage. *Dig. Dis. Sci.* 42(10):2054–2063.

92. Wagner M., Kleeff J., Lopez M.E., Bockman I., Massaque J., and Korc M. 1998. Transfection of the type I TGF-beta receptor restores TGF-beta responsiveness in pancreatic cancer. *Int. J. Cancer* 78(2):255–260.

93. Lagna G., Hata A., Hemmati-Brivanlou A., and Massague J. 1996. Partnership between DPC4 and SMAD proteins in TGF-beta signalling pathways. *Nature* 383(6603):832–836.

94. Schutte M., Hruban R.H., Hedrick L., Cho K.R., Nadasdy G.M., Weinstein C.L. et al. 1996. DPC4 gene in various tumor types. *Cancer Res.* 56(11):2527–2530.

95. Kleeff J., Maruyama H., Friess H., Buchler M.W., Falb D., and Korc M. 1999. Smad6 suppresses TGF-beta-induced growth inhibition in COLO-357 pancreatic cancer cells and is overexpressed in pancreatic cancer. *Biochem. Biophys. Res. Commun.* 255(2):268–273.

96. Kleeff J., Ishiwata T., Maruyama H., Friess H., Truong P., Buchler M.W. et al. 1999. The TGF-beta signaling inhibitor Smad7 enhances tumorigenicity in pancreatic cancer. *Oncogene* 18(39):5363–5372.

97. Kleeff J., Ishiwata T., Friess H., Buchler M.W., and Korc M. 1998. Concomitant over-expression of activin/inhibin beta subunits and their receptors in human pancreatic cancer. *Int. J. Cancer* 77(6):860–868.

98. Kleeff J., Maruyama H., Ishiwata T., Sawhney H., Friess H., Buchler M.W. et al. 1999. Bone morphogenetic protein 2 exerts diverse effects on cell growth in vitro and is expressed in human pancreatic cancer in vivo. *Gastroenterology* 116(5):1202–1216.

99. Kleeff J. and Korc M. 1998. Up-regulation of transforming growth factor (TGF)-beta receptors by TGF-beta1 in COLO-357 cells. *J. Biol. Chem.* 273(13):7495–7500.

100. Carpenter G. and Cohen S. 1979. Epidermal growth factor. *Ann. Rev. Biochem.* 48:193–216.

101. Carpenter G. 1980. Epidermal growth factor is a major growth-promoting agent in human milk. *Science* 210(4466):198–199.

102. Tsutsumi O., Kurachi H., and Oka T. 1986. A physiological role of epidermal growth factor in male reproductive function. *Science* 233(4767):975–977.

103. Sarkar F.H., Ball D.E., Li Y.W., and Crissman J.D. 1993. Molecular cloning and sequencing of an intron of Her-2/neu (ERBB2) gene. *DNA Cell. Biol.* 12(7):611–615.

104. Prigent S.A. and Lemoine N.R. 1992. The type 1 (EGFR-related) family of growth factor receptors and their ligands. *Prog. Growth Factor Res.* 4(1):1–24.

105. Kraus M.H., Issing W., Miki T., Popescu N.C., and Aaronson S.A. 1989. Isolation and characterization of ERBB3, a third member of the ERBB/epidermal growth factor receptor family: evidence for overexpression in a subset of human mammary tumors. *Proc. Natl. Acad. Sci. USA* 86(23):9193–9197.

106. Carraway 3rd K.L. and Cantley L.C. 1994. A neu acquaintance for erbB3 and erbB4: a role for receptor heterodimerization in growth signaling. *Cell* 78(1):5–8.

107. Holbro T., Civenni G., and Hynes N.E. 2003. The ErbB receptors and their role in cancer progression. *Exp. Cell. Res.* 284(1):99–110.

108. Graber H.U., Friess H., Kaufmann B., Willi D., Zimmermann A., Korc M. et al. 1999. ErbB-4 mRNA expression is decreased in non-metastatic pancreatic cancer. *Int. J. Cancer* 84(1):24–27.

109. Ebert M., Friess H., Buchler M.W., and Korc M. 1995. Differential distribution of human epidermal growth factor receptor family in acute pancreatitis. *Dig. Dis. Sci.* 40(10):2134–2142.

110. Logsdon C.D. 1986. Stimulation of pancreatic acinar cell growth by CCK, epidermal growth factor, and insulin in vitro. *Am. J. Physiol.* 251(4 Pt 1):G487–G494.

111. Dembinski A., Gregory H., Konturek S.J., and Polanski M. 1982. Trophic action of epidermal growth factor on the pancreas and gastroduodenal mucosa in rats. *J. Physiol.* 325:35–42.

112. Stryjek-Kaminska D., Piiper A., Stein J., Caspary W.F., and Zeuzem S. 1995. Epidermal growth factor receptor signaling in rat pancreatic acinar cells. *Pancreas* 10(3):274–280.

113. Konturek P.C., Dembinski A., Warzecha Z., Ihlm A., Ceranowicz P., Konturek S.J. et al. 1998. Comparison of epidermal growth factor and transforming growth factor-beta1 expression in hormone-induced acute pancreatitis in rats. *Digestion* 59(2):110–119.

114. Dembinski A., Warzecha Z., Konturek P.C., Ceranowicz P., Stachura J., Tomaszewska R. et al. 2000. Epidermal growth factor accelerates pancreatic recovery after caerulein-induced pancreatitis. *Eur. J. Pharmacol.* 398(1):159–168.

115. Warzecha Z., Dembinski A., Konturek P.C., Ceranowicz P., and Konturek S.J. 1999. Epidermal growth factor protects against pancreatic damage in cerulein-induced pancreatitis. *Digestion* 60(4):314–323.

116. Liu Q., Djuricin G., Nathan C., Gattuso P., Weinstein R.A., and Prinz R.A. 1997. The effect of epidermal growth factor on the septic complications of acute pancreatitis. *J. Surg. Res.* 69(1):171–177.

117. Korc M., Friess H., Yamanaka Y., Kobrin M.S., Buchler M., and Beger H.G. 1994. Chronic pancreatitis is associated with increased concentrations of epidermal growth factor receptor, transforming growth factor alpha, and phospholipase C gamma. *Gut* 35(10):1468–1473.

118. Friess H., Yamanaka Y., Buchler M., Hammer K., Kobrin M.S., Beger H.G. et al. 1994. A subgroup of patients with chronic pancreatitis overexpress the c-erb B-2 protooncogene. *Ann. Surg.* 220(2):183–192.

119. Friess H., Yamanaka Y., Kobrin M.S., Do D.A., Buchler M.W., and Korc M. 1995. Enhanced erbB-3 expression in human pancreatic cancer correlates with tumor progression. *Clin. Cancer Res.* 1(11):1413–1420.

120. Friess H., Yamanaka Y., Buchler M., Kobrin M.S., Tahara E., and Korc M. 1994. Cripto, a member of the epidermal growth factor family, is over-expressed in human pancreatic cancer and chronic pancreatitis. *Int. J. Cancer* 56(5):668–674.

121. Korc M., Chandrasekar B., Yamanaka Y., Friess H., Buchier M., and Beger H.G. 1992. Overexpression of the epidermal growth factor receptor in human pancreatic cancer is associated with concomitant increases in the levels of epidermal growth factor and transforming growth factor alpha. *J. Clin. Invest.* 90(4):1352–1360.

122. Ebert M., Yokoyama M., Kobrin M.S., Friess H., Lopez M.E., Buchler M.W. et al. 1994. Induction and expression of amphiregulin in human pancreatic cancer. *Cancer Res.* 54(15):3959–3962.

123. Kobrin M.S., Funatomi H., Friess H., Buchler M.W., Stathis P., and Korc M. 1994. Induction and expression of heparin-binding EGF-like growth factor in human pancreatic cancer. *Biochem. Biophys. Res. Commun.* 202(3):1705–1709.

124. Yamanaka Y., Friess H., Kobrin M.S., Buchler M., Kunz J., Beger H.G. et al. 1993. Overexpression of HER2/neu oncogene in human pancreatic carcinoma. *Hum. Pathol.* 24(10):1127–1134.

125. Lemoine N.R., Lobresco M., Leung H., Barton C., Hughes C.M., Prigent S.A. et al. 1992. The erbB-3 gene in human pancreatic cancer. *J. Pathol.* 168(3):269–273.

126. Wagner M., Luhrs H., Kloppel G., Adler G., and Schmid R.M. 1998. Malignant transformation of duct-like cells originating from acini in transforming growth factor transgenic mice. *Gastroenterology* 115(5):1254–1262.

127. Yu Y., Rishi A.K., Turner J.R., Liu D., Black E.D., Moshier J.A. et al. 2001. Cloning of a novel EGFR-related peptide: a putative negative regulator of EGFR. *Am. J. Physiol. Cell. Physiol.* 280(5):C1083–C1089.

128. Feng J., Adsay N.V., Kruger M., Ellis K.L., Nagothu K., Majumdar A.P. et al. 2002. Expression of ERRP in normal and neoplastic pancreata and its relationship to clinicopathologic parameters in pancreatic adenocarcinoma. *Pancreas* 25(4):342–349.

129. Thybusch-Bernhardt A., Beckmann S., and Juhl H. 2001. Comparative analysis of the EGF-receptor family in pancreatic cancer: expression of HER-4 correlates with a favourable tumor stage. *Int. J. Surg. Investig.* 2(5):393–400.

130. Yamanaka Y., Friess H., Kobrin M.S., Buchler M., Beger H.G., and Korc M. 1993. Coexpression of epidermal growth factor receptor and ligands in human pancreatic cancer is associated with enhanced tumor aggressiveness. *Anticancer Res.* 13(3):565–569.

131. Korc M. 1998. Role of growth factors in pancreatic cancer. *Surg. Oncol. Clin. N. Am.* 7(1):25–41.

132. Lemoine N.R. and Hall P.A. 1990. Growth factors and oncogenes in pancreatic cancer. *Baillieres Clin. Gastroenterol.* 4(4):815–832.

133. Wagner M., Cao T., Lopez M.E., Hope C., van Nostrand K., Kobrin M.S. et al. 1996. Expression of a truncated EGF receptor is associated with inhibition of pancreatic cancer cell growth and enhanced sensitivity to cisplatinum. *Int. J. Cancer* 68(6):782–787.

134. Buchsbaum D.J., Bonner J.A., Grizzle W.E., Stackhouse M.A., Carpenter M., and Hicklin D.J. et al. 2002. Treatment of pancreatic cancer xenografts with ERBITUX (IMC-C225) anti-EGFR antibody, gemcitabine, and radiation. *Int. J. Radiat. Oncol. Biol. Phys.* 54(4):1180–1193.

135. Paul S.R., Merberg D., Finnerty H., Morris G.E., Morris J.C., Jones S.S. et al. 1992. Molecular cloning of the cDNA encoding a receptor tyrosine kinase-related molecule with a catalytic region homologous to c-met. *Int. J. Cell. Cloning* 10(5):309–314.

136. Chan A., Rubin J., Bottaro D., Hirschfield D., Chedid M., and Aaronson S.A. 1993. Isoforms of human HGF and their biological activities. *Exs.* 65:67–79.

137. Cioce V., Csaky K.G., Chan A.M., Bottaro D.P., Taylor W.G., Jensen R. et al. 1996. Hepatocyte growth factor (HGF)/NK1 is a naturally occurring HGF/scatter factor variant with partial agonist/antagonist activity. *J. Biol. Chem.* 271(22):13110–13115.

138. Ebert M., Yokoyama M., Friess H., Buchler M.W., and Korc M. 1994. Coexpression of the c-met proto-oncogene and hepatocyte growth factor in human pancreatic cancer. *Cancer Res.* 54(22):5775–5778.

139. Di Renzo M.F., Poulsom R., Olivero M., Comoglio P.M., and Lemoine N.R. 1995. Expression of the Met/hepatocyte growth factor receptor in human pancreatic cancer. *Cancer Res.* 55(5):1129–1138.

140. Ueda T., Takeyama Y., Hori Y., Shinkai M., Takase K., Goshima M. et al. 2000. Hepatocyte growth factor increases in injured organs and functions as an organotrophic factor in rats with experimental acute pancreatitis. *Pancreas* 20(1):84–93.

141. Ueda T., Takeyama Y., Hori Y., Nishikawa J., Yamamoto M., and Saitoh Y. 1997. Hepatocyte growth factor in assessment of acute pancreatitis: comparison with C-reactive protein and interleukin-6. *J. Gastroenterol.* 32(1):63–70.

142. Ueda T., Takeyama Y., Toyokawa A., Kishida S., Yamamoto M., and Saitoh Y. 1996. Significant elevation of serum human hepatocyte growth factor levels in patients with acute pancreatitis. *Pancreas* 12(1):76–83.

143. Menke A., Yamaguchi H., Giehl K., and Adler G. 1999. Hepatocyte growth factor and fibroblast growth factor 2 are overexpressed after cerulein-induced acute pancreatitis. *Pancreas* 18(1):28–33.

144. Otte J.M., Kiehne K., Schmitz F., Folsch U.R., and Herzig K.H. 2000. C-met protooncogene expression and its regulation by cytokines in the regenerating pancreas and in pancreatic cancer cells. *Scand. J. Gastroenterol.* 35(1):90–95.

145. Warzecha Z., Dembinski A., Konturek P.C., Ceranowicz P., Konturek S.J., Tomaszewska R. et al. 2001. Hepatocyte growth factor attenuates pancreatic damage in caerulein-induced pancreatitis in rats. *Eur. J. Pharmacol.* 430(1):113–121.

146. Arisawa H., Fukui K., Imai E., Yamashita Y., Iga Y., and Masunaga H. 2001. Pretreatment with a deleted form of hepatocyte growth factor (dHGF) prevents the mortality of plasma-loss-induced hypovolemic shock in rats. *Shock* 16(6):438–443.

147. Otte J.M., Schwenger M., Brunke G., Sparmann G., Emmrich J., Schmitz F. et al. 2001. Expression of hepatocyte growth factor, keratinocyte growth factor and their receptors in experimental chronic pancreatitis. *Eur. J. Clin. Invest.* 31(10):865–875.

148. Matsuno M., Shiota G., Umeki K., Kawasaki H., Kojo H., and Miura K. 1997. Clinical evaluation of hepatocyte growth factor in patients with gastrointestinal and pancreatic diseases with special reference to inflammatory bowel disease. *Res. Commun. Mol. Pathol. Pharmacol.* 97(1):25–37.

149. Basilico C. and Moscatelli D. 1992. The FGF family of growth factors and oncogenes. *Adv. Cancer Res.* 59:115–165.

150. Givol D. and Yayon A. 1992. Complexity of FGF receptors: genetic basis for structural diversity and functional specificity. *Faseb. J.* 6(15):3362–3369.

151. Kornmann M., Beger H.G., and Korc M. 1998. Role of fibroblast growth factors and their receptors in pancreatic cancer and chronic pancreatitis. *Pancreas* 17(2):169–175.

152. Friess H., Kleeff J., Gumbs A., and Buchler M.W. 1997. Molecular versus conventional markers in pancreatic cancer. *Digestion* 58(6):557–563.

153. Klagsbrun M. 1989. The fibroblast growth factor family: structural and biological properties. *Prog. Growth Factor Res.* 1(4):207–235.

154. Yamanaka Y., Friess H., Buchler M., Beger H.G., Uchida E., Onda M. et al. 1993. Overexpression of acidic and basic fibroblast growth factors in human pancreatic cancer correlates with advanced tumor stage. *Cancer Res.* 53(21):5289–5296.
155. Friess H., Yamanaka Y., Buchler M., Beger H.G., Do D.A., Kobrin M.S. et al. 1994. Increased expression of acidic and basic fibroblast growth factors in chronic pancreatitis. *Am. J. Pathol.* 144(1):117–128.
156. Kornmann M., Ishiwata T., Beger H.G., and Korc M. 1997. Fibroblast growth factor-5 stimulates mitogenic signaling and is overexpressed in human pancreatic cancer: evidence for autocrine and paracrine actions. *Oncogene* 15(12):1417–1424.
157. Ishiwata T., Kornmann M., Beger H.G., and Korc M. 1998. Enhanced fibroblast growth factor 5 expression in stromal and exocrine elements of the pancreas in chronic pancreatitis. *Gut* 43(1):134–139.
158. Ishiwata T., Naito Z., Lu Y.P., Kawahara K., Fujii T., Kawamoto Y. et al. 2002. Differential distribution of fibroblast growth factor (FGF)-7 and FGF-10 in L-arginine-induced acute pancreatitis. *Exp. Mol. Pathol.* 73(3):181–190.
159. Ebert M., Yokoyama M., Ishiwata T., Friess H., Buchler M.W., Malfertheiner P. et al. 1999. Alteration of fibroblast growth factor and receptor expression after acute pancreatitis in humans. *Pancreas* 18(3):240–246.
160. Siddiqi I., Funatomi H., Kobrin M.S., Friess H., Buchler M.W., and Korc M. 1995. Increased expression of keratinocyte growth factor in human pancreatic cancer. *Biochem. Biophys. Res. Commun.* 215(1):309–315.
161. Kobrin M.S., Yamanaka Y., Friess H., Lopez M.E., and Korc M. 1993. Aberrant expression of type I fibroblast growth factor receptor in human pancreatic adenocarcinomas. *Cancer Res.* 53(20):4741–4744.
162. Beauchamp R.D., Lyons R.M., Yang E.Y., Coffey Jr. R.J., and Moses H.L. 1990. Expression of and response to growth regulatory peptides by two human pancreatic carcinoma cell lines. *Pancreas* 5(4):369–380.
163. Leung H.Y., Gullick W.J., and Lemoine N.R. 1994. Expression and functional activity of fibroblast growth factors and their receptors in human pancreatic cancer. *Int. J. Cancer* 59(5):667–675.
164. Kleeff J., Ishiwata T., Kumbasar A., Friess H., Buchler M.W., Lander A.D. et al. 1998. The cell-surface heparan sulfate proteoglycan glypican-1 regulates growth factor action in pancreatic carcinoma cells and is overexpressed in human pancreatic cancer. *J. Clin. Invest.* 102(9):1662–1673.
165. Wagner M., Lopez M.E., Cahn M., and Korc M. 1998. Suppression of fibroblast growth factor receptor signaling inhibits pancreatic cancer growth in vitro and in vivo. *Gastroenterology* 114(4):798–807.
166. Levi-Montalcini R. 1987. The nerve growth factor 35 years later. *Science* 237(4819):1154–1162.
167. Levi-Montalcini R., Skaper S.D., Dal Toso R., Petrelli L., and Leon A. 1996. Nerve growth factor: from neurotrophin to neurokine. *Trends Neurosci.* 19(11):514–520.
168. Barbacid M. 1995. Neurotrophic factors and their receptors. *Curr. Opin. Cell. Biol.* 7(2):148–155.
169. Barbacid M. 1995. Structural and functional properties of the TRK family of neurotrophin receptors. *Ann. NY Acad. Sci.* 766:442–458.
170. Bibel M. and Barde Y.A. 2000. Neurotrophins: key regulators of cell fate and cell shape in the vertebrate nervous system. *Genes Dev.* 14(23):2919–2937.

171. Huang E.J. and Reichardt L.F. 2001. Neurotrophins: roles in neuronal development and function. *Ann. Rev. Neurosci.* 24:677–736.

172. Huang E.J. and Reichardt L.F. 2003. TRK Receptors: Roles in Neuronal Signal Transduction. *Ann. Rev. Biochem* 72:609–642. *EPub* March 27, 2003 Review.

173. Klein R., Jing S.Q., Nanduri V., O'Rourke E., and Barbacid M. 1991. The trk proto-oncogene encodes a receptor for nerve growth factor. *Cell* 65(1):189–197.

174. Poo M.M. 2001. Neurotrophins as synaptic modulators. *Nat. Rev. Neurosci.* 2(1):24–32.

175. Miller F.D. and Kaplan D.R. 2001. Neurotrophin signalling pathways regulating neuronal apoptosis. *Cell. Mol. Life Sci.* 58(8):1045–1053.

176. Miller F.D. and Kaplan D.R. 2002. Neurobiology. TRK makes the retrograde. *Science* 295(5559):1471–1473.

177. Toma H., Winston J., Micci M.A., Shenoy M., and Pasricha P.J. 2000. Nerve growth factor expression is up-regulated in the rat model of L-arginine-induced acute pancreatitis. *Gastroenterology* 119(5):1373–1381.

178. Toma H., Winston J.H., Micci M.A., Li H., Hellmich H.L., and Pasricha P.J. 2002. Characterization of the neurotrophic response to acute pancreatitis. *Pancreas* 25(1):31–38.

179. Friess H., Zhu Z.W., di Mola F.F., Kulli C., Graber H.U., Andren-Sandberg A. et al. 1999. Nerve growth factor and its high-affinity receptor in chronic pancreatitis. *Ann. Surg.* 230(5):615–624.

180. McMahon S.B., Bennett D.L., Priestley J.V., and Shelton D.L. 1995. The biological effects of endogenous nerve growth factor on adult sensory neurons revealed by a trkA-IgG fusion molecule. *Nat. Med.* 1(8):774–780.

181. McMahon S.B. 1996. NGF as a mediator of inflammatory pain. *Philos. Trans. R. Soc. Lond. B Biol. Sci.* 351(1338):431–440.

182. Aloe L., Probert L., Kollias G., Micera A., and Tirassa P. 1995. Effect of NGF antibodies on mast cell distribution, histamine and substance P levels in the knee joint of TNF-arthritic transgenic mice. *Rheumatol. Int.* 14(6):249–252.

183. Buchler M., Weihe E., Friess H., Malfertheiner P., Bockman E., Muller S. et al. 1992. Changes in peptidergic innervation in chronic pancreatitis. *Pancreas* 7(2):183–192.

184. Zhu Z.W., Friess H., Wang L., Zimmermann A., and Buchler M.W. 2001. Brain-derived neurotrophic factor (BDNF) is upregulated and associated with pain in chronic pancreatitis. *Dig. Dis. Sci.* 46(8):1633–1639.

185. Oddiah D., Anand P., McMahon S.B., and Rattray M. 1998. Rapid increase of NGF, BDNF and NT-3 mRNAs in inflamed bladder. *Neuroreport* 9(7):1455–1458.

186. Zhu Z., Friess H., Shi X., Wang L., di Mola F.F., Wirtz M. et al. 2003. Up-regulation of p75 neurotrophin receptor (p75NTR) is associated with apoptosis in chronic pancreatitis. *Dig. Dis. Sci.* 48(4):717–725.

187. Pour P.M., Egami H., and Takiyama Y. 1991. Patterns of growth and metastases of induced pancreatic cancer in relation to the prognosis and its clinical implications. *Gastroenterology* 100(2):529–536.

188. Gudjonsson B. 1987. Cancer of the pancreas. 50 years of surgery. *Cancer* 60(9):2284–2303.

189. Bockman D.E., Buchler M., and Beger H.G. 1994. Interaction of pancreatic ductal carcinoma with nerves leads to nerve damage. *Gastroenterology* 107(1):219–230.

190. Oelmann E., Sreter L., Schuller I., Serve H., Koenigsmann M., Wiedenmann B. et al. 1995. Nerve growth factor stimulates clonal growth of human lung cancer cell lines and a human glioblastoma cell line expressing high-affinity nerve growth factor binding sites involving tyrosine kinase signaling. *Cancer Res.* 55(10):2212–2219.

191. Bold R.J., Ishizuka J., Rajaraman S., Perez-Polo J.R., Townsend Jr. C.M., and Thompson J.C. 1995. Nerve growth factor as a mitogen for a pancreatic carcinoid cell line. *J. Neurochem.* 64(6):2622–2628.

192. Geldof A.A., De Kleijn M.A., Rao B.R., and Newling D.W. 1997. Nerve growth factor stimulates in vitro invasive capacity of DU145 human prostatic cancer cells. *J. Cancer Res. Clin. Oncol.* 123(2):107–112.

193. McGregor L.M., McCune B.K., Graff J.R., McDowell P.R., Romans K.E., Yanco-poulos G.D. et al. 1999. Roles of trk family neurotrophin receptors in medullary thyroid carcinoma development and progression. *Proc. Natl. Acad. Sci. USA* 96(8):4540–4545.

194. Djakiew D., Pflug B.R., Delsite R., Onoda M., Lynch J.H., Arand G. et al. 1993. Chemotaxis and chemokinesis of human prostate tumor cell lines in response to human prostate stromal cell secretory proteins containing a nerve growth factor-like protein. *Cancer Res.* 53(6):1416–1420.

195. Donovan M.J., Hempstead B., Huber L.J., Kaplan D., Tsoulfas P., Chao M. et al. 1994. Identification of the neurotrophin receptors p75 and trk in a series of Wilms' tumors. *Am. J. Pathol.* 145(4):792–801.

196. Marchetti D., McQuillan D.J., Spohn W.C., Carson D.D., and Nicolson G.L. 1996. Neurotrophin stimulation of human melanoma cell invasion: selected enhancement of heparanase activity and heparanase degradation of specific heparan sulfate subpopulations. *Cancer Res.* 56(12):2856–2863.

197. Schneider M.B., Standop J., Ulrich A., Wittel U., Friess H., Andren-Sandberg A. et al. 2001. Expression of nerve growth factors in pancreatic neural tissue and pancreatic cancer. *J. Histochem. Cytochem.* 49(10):1205–1210.

198. Zhu Z., Friess H., di Mola F.F., Zimmermann A., Graber H.U., Korc M. et al. 1999. Nerve growth factor expression correlates with perineural invasion and pain in human pancreatic cancer. *J. Clin. Oncol.* 17(8):2419–2428.

199. Sakamoto Y., Kitajima Y., Edakuni G., Sasatomi E., Mori M., Kitahara K. et al. 2001. Expression of Trk tyrosine kinase receptor is a biologic marker for cell proliferation and perineural invasion of human pancreatic ductal adenocarci-noma. *Oncol. Rep.* 8(3):477–484.

200. Zhu Z.W., Friess H., Wang L., Bogardus T., Korc M., Kleeff J. et al. 2001. Nerve growth factor exerts differential effects on the growth of human pancreatic cancer cells. *Clin. Cancer Res.* 7(1):105–112.

201. Zhu Z., Kleeff J., Kayed H., Wang L., Korc M., Buchler M.W. et al. 2002. Nerve growth factor and enhancement of proliferation, invasion, and tumorigenicity of pancreatic cancer cells. *Mol. Carcinog.* 35(3):138–147.

202. Miknyoczki S.J., Wan W., Chang H., Dobrzanski P., Ruggeri B.A., Dionne C.A. et al. 2002. The neurotrophin-trk receptor axes are critical for the growth and progression of human prostatic carcinoma and pancreatic ductal adenocarci-noma xenografts in nude mice. *Clin. Cancer Res.* 8(6):1924–1931.

14

TOXICOLOGY OF NATURAL HORMONES AND HORMONELIKE CHEMICALS

Michael L. Steer

CONTENTS

14.1 INTRODUCTION

Recently, considerable progress has been made in achieving a molecular understanding of the events that underlie many of the toxic effects of hormones and hormonelike compounds on the pancreas. For the most part, these toxic effects appear to be primarily exerted on the exocrine rather than the endocrine pancreas. Among the large number of hormones and hormonelike factors that are known to affect pancreatic function, only the toxic effects of a selected few have been studied to any depth. In those studies, which have mostly been performed using experimental animals, the toxic effects of cholecystokin, cholecystokinin analogs, acetylcholine, secretin, and the administration of a soya protein-enriched diet (which exerts its toxic effects via release of endogenous cholecystokinin) have been evaluated. Furthermore, studies utilizing experimental laboratory animals have explored the molecular events that underlie the development of these toxic effects. In some cases, toxicity is manifest by an inflammatory process (i.e., pancreatitis), and in others, toxicity reflects the growth-promoting effects of the agent and it results in hypertrophy, hyperplasia, and neoplasia. In this chapter, I will review the mechanisms by which these hormones and hormonelike factors regulate the pancreas under physiological conditions and current concepts regarding the mechanisms by which these agents exert toxic effects on the pancreas.

14.2 PHYSIOLOGICAL AND CELL BIOLOGICAL CONSIDERATIONS

14.2.1 CCK-Related Stimulus-Secretion Coupling

See Figure 14.1.

14.2.1.1 CCK and CCK Receptors

Cholecystokinin (CCK) is a 58-amino acid peptide hormone that is secreted by I-cells that are located within the mucosa of the duodenum and upper jejunum. Following secretion, CCK-58 is proteolytically cleaved to yield

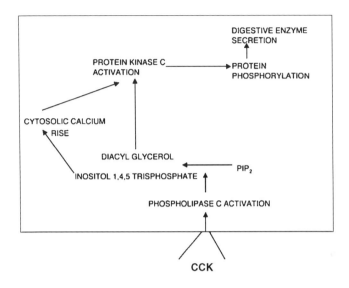

Figure 14.1 **CCK-induced stimulus-secretion coupling in pancreatic acinar cells.**
Cholecystokinin or its analogs binds to G protein-coupled receptors on the baso-
lateral surface of the plasma membrane and activates phospholipase C. As a result,
the membrane phospholipid PIP2 is hydrolyzed liberating inositol 1,4,5 trispho-
sphate and diacyl glycerol. Together these second messengers increase cytosolic
free calcium levels and activate protein kinase C leading to downstream protein
phosphorylation and an accelerated rate of digestive enzyme secretion from the
luminal surface of the cell.

smaller but still active products, the most abundant of which are the 33
and 8 amino acid species (i.e., CCK-33 and CCK-8). Considerable controversy has surrounded the question of which form of CCK is physiologically
dominant but, from a practical standpoint, fragments as small as the 8
amino acid form (CCK-8) are fully active agonists. Cerulein, the 10 amino
acid analog of CCK originally isolated from amphibian skin, has been
extensively utilized in experimental studies that have examined toxic
inflammatory and neoplastic effects of CCK.

CCK is a powerful exocrine pancreatic secretagogue for rodents and
for humans (see Chapter 6). Full-length CCK, and its biologically active
fragments including cerulein, bind to 7-transmembrane G-protein coupled
receptors on the cell surface of target cells. In the rodent pancreas, these
CCK receptors are located on the surface of acinar cells, but their presence
on human acinar cells has been questioned and, in humans, CCK may
act via binding to CCK receptors that are located on cholinergic vagal
efferent nerve fibers.

Stimulation of membrane-bound phospholipase C and hydrolysis of
membrane phosphatidyl inositol-4,5 bisphosphate (PIP_2) leading to the

generation of the second messengers inositol 1,4,5-trisphosphate (IP_3) and diacyl glycerol (DAG) appear to be the earliest change that follows CCK binding to its acinar cell receptor. Subsequently, IP_3 binds to IP_3 receptors located on the surface of the endoplasmic reticulum and, possibly, on zymogen granules and mitochondria. This causes the opening of Ca^{2+} channels and results in Ca^{2+} release into the cytoplasmic compartment from intracellular storage pools located within the endoplasmic reticulum, zymogen granule, and mitochondrial compartments. DAG triggers the activation of protein kinase C and, by mechanisms that remain to be elucidated, the combination of protein kinase C activation and elevated cytoplasmic Ca^{2+} levels leads to up-regulated fusion of zymogen granules with the apical plasmalemma and discharge of secretory protein (i.e., digestive enzymes and zymogens) into the acinar lumen (i.e., ductal space). The net result of this stimulus-secretion cascade is the CCK-induced acceleration of pancreatic acinar cell protein secretion into pancreatic juice.[1–3]

14.2.1.2 CCK-Mediated Secretion in Humans

The mechanisms by which CCK and its analogs regulate pancreatic secretion of protein in humans is not entirely clear (see Chapter 6). Although the above description may be valid for the human condition, the purported absence of CCK receptors on human acinar cells would make that seem unlikely.[4] Rather, it would suggest that CCK stimulates acinar cell protein secretion via an indirect mechanism (i.e., by causing acetylcholine release from vagal efferents and, thereby, triggering acetylcholine-stimulated acinar cell secretion of digestive enzymes (see Section 14.2.3).

14.2.2 Regulation of Acinar Cell Growth by CCK

In addition to acting as a powerful pancreatic acinar cell secretagogue, CCK also plays an important role in regulating acinar cell growth and replication.[5,6] The earliest events involved in this growth response appear to be identical to those involved in CCK-mediated protein secretion (i.e., activation of phospholipase C, generation of DAG and IP_3, activation of protein kinase C, and a rise in cytoplasmic Ca^{2+} levels). Downstream events involved in the growth response, however, have not been well defined but they appear to involve phospholipase D and tyrosine kinase.[7]

14.2.3 Acetylcholine-Mediated Acinar Cell Secretion

Pancreatic acinar cells of most species, including rodents and humans, express cell surface receptors for the neurotransmitter acetylcholine and respond to the local release of acetylcholine from intrapancreatic vagal efferent nerve endings. In contrast to the extensive number of studies that have explored CCK-related stimulus-secretion coupling in the pancreas, acinar cell events involved in cholinergically stimulated secretion have not been as well studied. Is is clear, however, that cholinergic stimulation of acinar cell protein secretion is mediated via muscarinic receptors and by an intracellular cascade that is similar or identical to that involved in CCK-mediated secretion (i.e., stimulation of phospholipase C, generation of DAG and IP$_3$, and a rise in cytosolic Ca^{2+}). For the complex interaction between CCK and other intestinal hormones and neural regulation, see Chapter 6.

14.2.4   The Effects of Secretin

14.2.4.1   Duct Cells

Secretin is produced and secreted by mucosal cells lining the duodenum. The dominant effect of pancreatic stimulation with secretin is the stimulation of fluid and bicarbonate secretion by cells lining the pancreatic ducts. Secretin binds to 7-transmembrane G-protein coupled receptors on the surface of those duct cells and causes activation of duct cell adenylate cyclase. The cyclic AMP (cAMP) that is generated at the cytoplasmic face of the basolateral membrane subsequently stimulates cAMP regulated chloride channels in the apical membrane (the cystic fibrosis transmembrane rectifier or CFTR) leading to active apical chloride secretion. That secreted chloride is then exchanged for intracellular bicarbonate via an apical membrane chloride/bicarbonate exchanger. The net result of these events is secretin-stimulated, cAMP-mediated, secretion of bicarbonate (and accompanying fluid) into the ductal system.[8]

14.2.4.2   Acinar Cells

In addition to pancreatic duct cells, the acinar cells of the pancreas also express secretin receptors. Secretin is a weak stimulant of acinar cell digestive enzyme secretion, but the mechanisms by which secretin stimulates acinar cell protein secretion are not entirely clear. In an overall sense, the process is mediated by secretin-stimulated cAMP formation and it does not appear to involve an elevation in cytosolic Ca^{2+} levels. Synergistic stimulation of digestive enzyme secretion by acinar cells follows exposure of those cells to CCK along with any intervention that leads to increased acinar cell cAMP levels.

14.2.5   The CCK Feedback Loop

CCK is released into the circulation by I-cells that are scattered throughout the duodenal and proximal jejunal mucosa. I-cell secretion of CCK is,

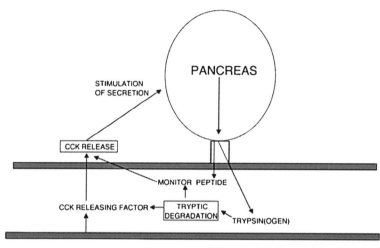

DUODENUM AND PROXIMAL JEJUNUM

Figure 14.2 The CCK feedback loop. Monitor peptide (secreted by the pancreas in pancreatic juice) and CCK releasing factor (secreted by mucosal cells in the duodenum and upper jejunum are trypsin-sensitive lumenally active proteins that stimulate CCK release from intestinal I cells into the circulation. In the presence of free active trypsin, the lumenally active proteins are degraded and CCK release is down-regulated. In the presence of undigested protein, trypsin inhibitors, or diminished trypsin secretion, the lumenally active proteins escape degradation and CCK release is up-regulated.

itself, stimulated by the interaction of I-cells with lumenally active CCK-releasing factors. Two types of CCK-releasing factors have been identified — one secreted by the pancreas in pancreatic juice ("monitor peptide") and the other secreted by mucosal cells in the duodenum and proximal jejunum. Both types of CCK-releasing factors can be proteolytically degraded if exposed to proteases within the duodenum and, once degraded, they are no longer biologically active.

To an extent that appears to be species-specific, CCK release (and, thus, acinar cell protein secretion) is regulated by a feedback loop (see Figure 14.2) involving the CCK-releasing factors.[9–11] When intraduodenal protease levels are high, as would be the case when secreted proteases are not quenched by being bound to ingested proteins or protease inhibitors within the duodenal lumen, the releasing factors are degraded and CCK secretion is down-regulated. After a high protein meal or ingestion of protease inhibitors, on the other hand, secreted proteases are bound to the ingested proteins or protease inhibitors and, as a result, the releasing factors escape degradation and they are free to stimulate CCK release.

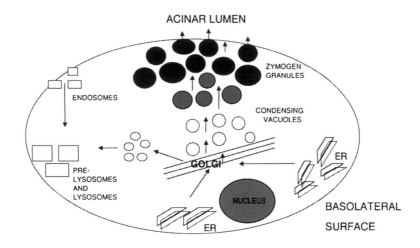

ACINAR LUMEN

Figure 14.3 Protein synthesis, intracellular transport, and secretion in pancreatic acinar cells. Newly synthesized proteins are assembled in the cisternae of the endoplasmic reticulum (ER) and vectorially transported to the Golgi stacks where lysosome-bound enzymes are posttranslationally glycosylated, 6-mannose phosphorylated, bound to mannose-6-phosphate receptors, and transported to the lysosomal compartment. Digestive enzymes are carried through the Golgi and packaged in condensing vacuoles that evolve into zymogen granules as they move toward the lumenal cell surface where they fuse with the lumenal plasmalemma and discharge their contents into the lumenal (i.e., ductal) space.

Raw soya flour contains a potent trypsin inhibitor, but the capacity of roy soya protein to inhibit trypsin is lost after the flour is heated. Both pancreatic acinar cell hypertrophy and hyperplasia have been observed when chickens or rodents are fed a diet of raw, but not heated, soya flour. This growth response appears to be stimulated by CCK and to involve the CCK feedback loop (i.e., ingestion of the trypsin inhibitor contained in raw soya flour inhibits intralumenal trypsin activity and, as a result, prevents degradation of the trypsin-sensitive lumenally-active CCK releasing factors). This leads to chronic hypercholecystokininemia and the stimulation of pancreatic acinar cell growth or replication.

14.2.6 Protein Synthesis, Transport, and Secretion

The acinar cell of the pancreas is believed to be the body's most active protein-synthesizing cell and roughly 90% of the protein synthesized by the pancreatic acinar cells is digestive enzyme protein that is destined to be secreted, along with pancreatic juice, into the intestine (see Figure 14.3). Newly synthesized proteins are assembled within the cisternae of the endoplasmic reticulum and vectorially transported, in small transport vesicles, to the Golgi complex. As the newly synthesized proteins traverse the Golgi complex, those destined for transport to the lysosomal compartment are posttranslationally modified by glycosylation, phosphorylated at the 6-position of mannose residues, and bound to mannose-6-phosphate receptors. These receptor-bound proteins are shuttled to the lysosomal compartment where, under acidic conditions, they dissociate from the mannose-6-phosphate receptors and act, within the acidic lysosomal compartment as intracellular digestive enzymes. On the other hand, most of the newly synthesized pancreatic proteins are not mannose-6-phosphorylated in the Golgi and, as a result, they pass through the Golgi stacks to the transsurface where they are packaged into membrane-bounded condensing vacuoles. The condensing vacuoles become increasingly dense as they migrate toward the luminal surface of the cell and evolve into zymogen granules. At the lumenal cell surface, and in response to secretagogue stimulation, the zymogen granule membrane fuses with the lumenal plasmalemma (fusion/fission) creating a small opening, or fusion pore, that establishes communication between the granule interior and the lumenal (or ductal) space. Granule contents, including digestive enzymes and their zymogens, are discharged from the cell through this opened fusion pore while the zymogen granule membrane becomes incorporated into the lumenal plasmalemma. It is subsequently internalized, transported to lysosomes, and degraded.[12,13]

14.3   CCK-INDUCED PANCREATITIS

Hormones that act as acinar cell secretagogues by the phospholipase $C/IP_3/Ca^{2+}$ cascade have been shown to induce acute pancreatitis in both mice and rats and it is likely that they can induce pancreatitis in many other species as well. Perhaps the earliest report of hormone-induced pancreatitis was that of Solcia and coworkers who, in 1972, noted that animals given high doses of the CCK analog cerulein developed evidence of pancreatic injury.[14] In 1975, Lampel and Kern brought this phenomenon to the attention of the pancreatitis research community in a frequently referenced manuscript[15] that described the characteristics of cerulein-induced acute pancreatitis in rats. Subsequently, Kern and his coworkers showed that high doses of an acetylcholinelike agent (carbamylcholine) could also induce acute pancreatitis when given to rats,[16] and other groups noted that cerulein could also induce pancreatitis in mice. Because of its ease of application, relatively low cost, and high degree of reproducibility, the cerulein-induced models of pancreatitis, in rats and mice, are by far the most widely employed experimental models of pancreatitis and numerous studies designed to elucidate the cellular events that underlie cell injury in these models have been performed.

14.3.1 Pathology

Cerulein-induced pancreatitis in rats is characterized by extensive acinar cell vacuolization and pancreatic edema with relatively little acinar cell necrosis. There is evidence of a mild to moderate parenchymal inflammatory reaction with sequestration of neutrophils within the pancreas. Furthermore, there is extensive acinar cell apoptosis. Cerulein-induced pancreatitis in rats is associated with mild and transient pancreatitis-associated lung injury, characterized by neutrophil sequestration within the lung, death of Type-2 pneumocytes, and increased permeability of the alveolar/capillary membrane barrier. This is comparable to the changes noted in the Adult Respiratory Distress Syndrome in humans. In rats, both cerulein-induced pancreatitis and its associated lung injury are transient events and, after the termination of cerulein administration, all of the above-described changes regress, full morphologic recovery usually occurs, and there is little or no mortality.

Cerulein-induced pancreatitis in mice is much more severe than the disease in rats.[17] Pancreatic changes in mice include extensive acinar cell necrosis and intrapancreatic hemorrhage in addition to acinar cell vacuolization, pancreatic edema, and a pancreatic parenchymal inflammatory reaction that includes chemoattraction of neutrophils to the pancreas and sequestration of those neutrophils within the pancreatic parenchyma. Cerulein-induced pancreatitis in mice is also associated with lung injury that, in this species, is also more severe than it is in rats. As with the rat cerulein-induced pancreas and lung injury, in mouse cerulein-induced pancreatitis there is little or no associated mortality. In contrast to the disease in rats, however, several studies have suggested that cerulein-induced pancreatitis in mice is associated with chronic changes, including pancreatic fibrosis, that may persist even after the cerulein administration has been terminated. (See Figure 14.4 and Table 14.1.)

14.3.2 Evolution of Cerulein-Induced Pancreatitis

Cerulein-induced pancreatitis, in both rats and mice, evolves rapidly and with a similar time dependence (Table 14.2).[18] The earliest changes are observed within the pancreas where edema, acinar cell vacuolization, and parenchymal inflammation can be detected within 3 hours after the start of cerulein administration. Biochemical changes including intrapancreatic activation of digestive enzyme zymogens such as trypsinogen, chymotrypsinogen, and procarboxypeptidase as well as a rise in circulating amylase/lipase levels can be detected even before the morphological changes of pancreatitis are noted. Intrapancreatic activation of digestive enzyme zymogens occurs within 15 to 30 minutes of cerulein administration and neutrophil sequestration as well as edema are observed somewhat

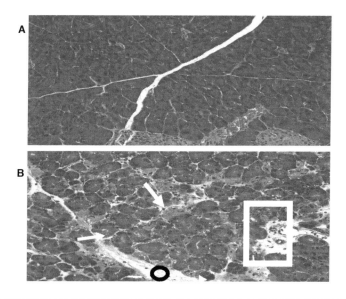

Figure 14.4 Morphologic features of cerulein-induced pancreatitis. Mice were given injections of either saline (Panel A) or cerulein (50 µg/kg) at hourly intervals for 12 hours and then the pancreas was examined in hemotoxylin/eosin stained samples. In Panel B, the large arrow indicates an area of necrosis; the small arrow indicates acinar cell vacuolization; the black circle indicates an area of intralobular edema; and the white box indicates an area of inflammation.

Table 14.1   Sequence of Events in Cerulein-Induced Pancreatitis

Event	Time
Rise in cytoplasmic calcium	<3 min
Digestive zymogen/lysosomal hydrolase colocalization	10 to 15 min
Activation of digestive zymogens	10 to 15 min
NF-κB and AP-1 activation	15 min
Proinflammatory transcription factor activation	15 min
Neutrophil sequestration in pancreas	30 min
Hyperamylasemia/hyperlipasemia	30 to 60 min
Pancreatic edema	30 to 60 min
Acinar cell injury/necrosis	30 to 60 min

Table 14.2   Elements of Secretagogue-Induced Acute Pancreatitis and Pancreatitis-Associated Lung Injury

Pathology of Secretagogue-Induced Acute Pancreatitis
Acinar cell vacuolization
Pancreatic and peripancreatic edema
Acinar cell injury/necrosis
Sequestration of neutrophils in pancreas and lung
Alveolar/capillary membrane thickening
Alveolar/capillary membrane leak
Type II pneumocyte injury and Acute Respiratory Distress Syndrome

later. Cerulein-induced pancreatitis-associated lung injury can be first detected approximately 6 hours after the start of cerulein administration, after the pancreatic injury is clearly established. Lung injury has been observed to progress and become even more severe if cerulein administration is continued for 12 hours. Longer periods of cerulein administration have not been evaluated in detail.

14.3.3   CCK Receptor Affinity State and Cerulein-Induced Pancreatitis

Rodent acinar cells express CCK-A receptors that exist in two affinity states — high and low affinity. The high affinity state of the CCK receptor mediates the stimulation of CCK-induced protein secretion from acinar cells while low affinity CCK receptors appear to mediate inhibition of protein secretion. As a result of this relationship, the dose/response for CCK or cerulein stimulation of acinar cells is biphasic with stimulation of protein secretion at low CCK/cerulein concentrations and inhibition of secretion at high concentrations of CCK or cerulein (Figure 14.5). In practice, concentrations of CCK or cerulein that exceed those causing maximal rates of protein secretion are referred to as being either "supramaximally stimulating" or "supraphysiological." These concentrations of CCK and cerulein bring about the inhibition of protein secretion and it is these supramaximally stimulating concentrations that cause pancreatitis by interacting with low affinity CCK-A receptors.[19]

14.3.4   Acinar Cell Events during Cerulein-Induced Pancreatitis

Studies reported by several different investigative groups have defined a number of acinar cell events, coupled to the occupancy of the low affinity CCK receptors, that occur when acinar cells are exposed to supramaximally

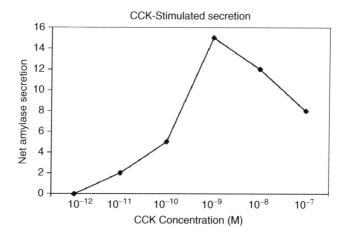

Figure 14.5 Dose-response relationship for CCK-stimulated amylase secretion from pancreatic acinar cells. Acinar cells were incubated with varying concentrations of CCK or cerulein and amylase secretion into the medium was quantitated over 30 minutes.

stimulating concentrations of cerulein or CCK. Most of the evidence suggests that these low affinity CCK-A receptors are merely an altered state of the high affinity receptors (i.e., 7-transmembrane G protein-coupled receptors that are normally coupled to activation of phospholipase C, generation of DAG and IP_3, activation of protein kinase C, and release of Ca^{2+} from intracellular storage pools). The rise in intracellular Ca^{2+} levels that occurs after supramaximal cerulein stimulation differs from that observed when maximally or submaximally stimulating doses of the CCK analog are used. In the presence of maximally or submaximally stimulating CCK concentrations, oscillating rises and falls in cytoplasmic Ca^{2+} levels are observed and the frequency as well as the magnitude of those oscillations depend on the concentration of cerulein to which the cells are exposed. In contrast, a large rise in cytoplasmic Ca^{2+} levels followed by a slow decline to a sustained, but persistently elevated, plateau level of Ca^{2+} is observed when acinar cells are exposed to a supramaximaly stimulating concentration of cerulein. This rise in Ca^{2+} levels appears to be essential for the subsequent downstream events leading to pancreatic injury because interventions that abort the cerulein-induced rise in cytoplasmic Ca^{2+} can prevent those downstream events and prevent acinar cell injury.[20]

In addition to changes in cytoplasmic Ca^{2+}, exposure of acinar cells to supramaximally stimulating doses of CCK or cerulein results in a number of other changes. Those changes can be grouped according to their potential role in the evolution of pancreatitis. Thus, a series of changes including inhibition of secretion, cytoskeletal reorganization, redistribution of digestive enzyme zymogens and lysosomal hydrolases, intracellular activation of digestive enzyme zymogens, and acinar cell vacuolization appear to be critical events that couple supramaximal stimulation with acinar cell injury; whereas, other changes including mitogen activated protein kinase (MAPK) activation, NF-κB activation, and acinar cell generation of proinflammatory chemo/cytokines appear to be critical determinants of pancreatitis severity.[21] Recently reported studies indicate that supramaximal stimulation of acinar cells with cerulein inhibits digestive enzyme synthesis by acinar cells,[22] but the relationship between this response and either cell injury or pancreatitis is not clear.

14.3.4.1 Events Related to Acinar Cell Injury

Supramaximal stimulation of acinar cells with cerulein, or CCK, results in the inhibition of acinar cell digestive enzyme secretion. The mechanisms responsible for this inhibition of secretion have not been established. It is associated with the spike or sustained elevation of cytoplasmic calcium levels that is observed when acinar cells are exposed to supramaximally stimulating concentrations of cerulein; but it is not clear whether or not it is this pattern of calcium change that actually causes the inhibition of secretion that is observed. Supramaximal stimulation of acinar cells with cerulein is also associated with profound morphological changes involving, primarily, the cytoskeleton. A rapid and reversible disruption of the subapical filamentous actin web occurs and filamentous actin is redistributed from the apical to the basal-lateral surface of the cell.[23] Evidence has been presented that suggests that the subapical actin web may play a critical permissive role in secretagogue-induced digestive enzyme secretion. Thus, the disruption of the subapical actin web that occurs in the presence of supramaximally stimulating concentrations of cerulein may account for the inhibition of enzyme secretion that follows.[24] Unfortunately, the actual mechanisms by which supramaximally stimulating concentrations of cerulein bring about disruption of the subapical actin web and cause filamentous actin redistribution are not known.

Exposure of acinar cells to supramaximally stimulating concentrations of cerulein, or CCK, has a profound effect on the intracellular trafficking and localization of lysosomal hydrolases (Figure 14.6). Normally, lysosomal hydrolases are sorted away from the exported digestive enzyme zymogens as both types of newly synthesized proteins traverse the Golgi stacks. In the presence of supramaximally stimulating concentrations of cerulein, however, sorting of lysosomal hydrolases is defective and increased amounts of lysosomal hydrolases remain in the secretory path-

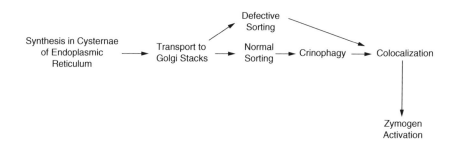

Figure 14.6   Digestive enzyme zymogen and lysosomal hydrolase colocalization leads to intraacinar cell zymogen activation. During the early stages of CCK-induced pancreatitis, a combination of defective protein sorting in the Golgi and postsorting crinophagy causes digestive enzyme zymogens and lysosomal hydrolases to become colocalized within the same intracellular organelle. Within that organelle, the lysosomal hydrolase cathepsin B and, possibly, other lysosomal hydrolases activates trypsinogen and trypsin activates the other zymogens.

way where they are packaged, along with digestive enzymes, in condensing vacuoles and zymogen granules. In addition to this defect in sorting, supramaximal stimulation with cerulein also causes zymogen granules and lysosomes to fuse by a process referred to as *crinophagy*. The ultimate result of defective sorting and crinophagy is the redistribution of lysosomal hydrolases from the lysosomal compartment to the secretory compartment and the colocalization of lysosomal hydrolases with digestive enzyme zymogens within the same intracellular organelles. Those organelles appear to be cytoplasmic vacuoles that arise in the region of the trans-Golgi network shortly after the onset of supramaximal stimulation with cerulein. Intraacinar cell activation of digestive enzyme zymogens has been shown to occur within those vacuoles and considerable evidence has been presented indicating that that activation is mediated by the lysosomal hydrolase cathepsin B.[25,26] Other lysosomal hydrolases may also contribute to this activation phenomenon.

The mechanisms responsible for digestive enzyme/lysosomal hydrolase colocalization after supramaximal stimulation with cerulein have not been established with certainty but preliminary studies have suggested that the phenomenon may be mediated by phosphoinositide-3-kinase (PI3K). Pharmacologic inhibition of PI3K has been shown to prevent the colocalization phenomenon, prevent intraacinar cell activation of trypsinogen, and reduce the severity of cerulein-induced pancreatitis.[27] The class of PI3K that is responsible for the colocalization phenomenon and intraacinar cell zymogen activation is not entirely clear and evidence has been presented that it may be either a Class I PI3K (regulated by G protein-coupled receptors and linked to downstream activation of protein kinase B (Akt)) or a Class III PI3K (a constitutively active, non-Akt activating enzyme).

14.3.4.2 Proinflammatory Events

Supramaximal stimulation of pancreatic acinar cells with cerulein or CCK leads to rapid activation of at least two proinflammatory transcription factors — NF-κB and AP-1[28] and the early generation of chemokines and other proinflammatory factors.[29] It is likely that other transcription factors regulating synthesis of inflammatory factors are also activated by supra-maximal stimulation with cerulein but, to date, they have not been identified. Activation of NF-κB appears to be downstream to the cerulein-induced rise in cytoplasmic calcium levels and activation of protein kinase C because it can be prevented by either preventing the rise in calcium levels or by inhibiting protein kinase C.[27,30] NF-κB activation does not depend on intracellular activation of digestive enzyme zymogens or cell injury.[31] Once activated, NF-κB is translocated to the nucleus, where it functions to regulate the synthesis of a number of inflammatory mediators.

Pancreatic acinar cells have been shown to elaborate a number of inflammatory factors as a result of supramaximal stimulation with either cerulein or CCK. The list of acinar cell generated factors includes the following: TNFα, ICAM-1, substance P, MOB-1, IL-1, MIP-1, MCP-1, RANTES, KC, and LIX. The cell signaling events and transcription factors involved in regulating the synthesis of these, and other, mediators of inflammation after supramaximal stimulation with cerulein have not been defined, but they are the subject of on-going studies.

14.3.4.3 Sequence of Acinar Cell Events

As noted above, supramaximal stimulation of acinar cells with CCK or cerulein brings about a number of acinar cell events but the cause–effect relationship between many of these various events is not entirely clear (Figure 14.7). For example, considerable controversy has surrounded the question of whether the colocalization of digestive zymogens with lysos-omal hydrolases *leads* to acinar cell injury and pancreatitis or whether that colocalization is the *result* of acinar cell injury. Most investigators believe that intracellular zymogen activation *causes* acinar cell injury that leads to inflammatory cell activation and sequestration of neutrophils in the pancreas but some have argued that zymogen activation is actually *mediated* by neutrophils that have previously been activated and seques-tered in the pancreas. Similarly, some have questioned whether or not the activation of proinflammatory transcription factors and the cytoskeletal changes of supramaximal stimulation with cerulein *lead* to cell injury or

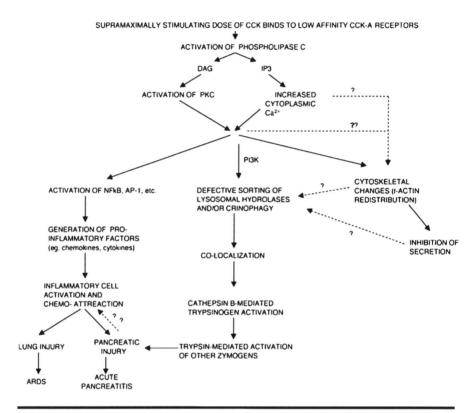

Figure 14.7 **Sequence of events in CCK-induced pancreatitis. Supramaximally stimulating concentrations of CCK or cerulein bind to low affinity CCK-A receptors triggering activation of membrane bound phospholipase C and generation of both DAG and IP3. Calcium is released from intracellular storage pools and protein kinase C is activated. Subsequently, digestive enzyme secretion is inhibited, proinflammatory transcription factors are activated, and intracellular transport of newly synthesized proteins is perturbed leading to the colocalization of digestive enzyme zymogens with lysosomal hydrolases. PI3k may play an important role in bringing about the colocalization phenomenon. Colocalization leads to zymogen activation and acinar cell injury. Inflammatory cells are activated and chemoattracted to the pancreas in response to acinar cell injury and acinar cell generation of proinflammatory factors. Inflammatory cell activation and chemoattraction in the lung leads to lung injury and Acute Respiratory Distress Syndrome and acinar cell injury and pancreatic inflammation result in acute pancreatitis.**

result from acinar cell injury. A number of studies focusing on the characteristics of these intracellular events have been performed in an attempt to resolve these issues. In a general sense, two types of studies have been undertaken:

1. Those that define the temporal relationships between events in order to determine which may be cause and which are effect.
2. Those in which an event has been aborted and the effects of aborting that event on other phenomena are determined.

The time-dependence studies reported to date have, for the most part, indicated that the earliest changes following supramaximal stimulation involve proinflammatory transcription factor activation[29] and digestive zymogen/lysosomal hydrolase colocalization.[32,33] These changes appear to occur prior to evidence of cell injury, pancreatic edema, or hyperamylasemia and long before macroscopic evidence of pancreatitis can be detected. At roughly the same time as colocalization is detected (i.e., 15 minutes after initial exposure to a supramaximally stimulating dose of cerulein), zymogen activation within the acinar cell can be noted.[34] Shortly thereafter, acinar cell vacuolization, pancreatic edema, subapical actin redistribution, and evidence of acinar cell injury can be observed. Neutrophil sequestration within the pancreas occurs at a still later time, after intraacinar cell activation of zymogens such as trypsinogen has clearly occurred.[35,36] These findings would suggest that the initial events that occur as a result of supramaximal stimulation with cerulein involve up-regulation of proinflammatory transcription factor generation and redistribution of lysosomal hydrolases. The former leads to up-regulated generation, by acinar cells, of proinflammatory factors; the latter leads to intraacinar cell zymogen activation and, subsequently, to acinar cell injury, acinar cell vacuolization, and acinar cell cytoskeletal changes including redistribution of subapical f-actin. These studies would argue strongly against the hypotheses that acinar cell injury leads to zymogen activation as zymogen activation appears to precede cell injury. Similarly, these studies would argue against the notion that neutrophil sequestration within the pancreas plays a critical role in zymogen activation because zymogen activation appears to be nearly complete prior to the sequestration of neutrophils within the pancreas.

A variety of inhibitors, with greater or lesser degrees of specificity, have been employed to abort certain events and examine the effects of that intervention. Inhibitors of PI3K have been shown to prevent the colocalization phenomenon, prevent intraacinar cell activation of zymogens, and reduce the severity of cerulein induced pancreatitis.[27] This has provided further support to the concept that colocalization leads to zymogen activation and zymogen activation leads to cell injury. Inhibitors of protein kinase C and interventions that prevent the rise in acinar cell cytoplasmic Ca^{2+} levels that follows supramaximal cerulein stimulation have been shown to prevent the colocalization phenomenon, prevent zymogen activation, and reduce the extent of cerulein-induced acinar cell injury.[37] In addition to supporting the concept that colocalization leads to zymogen activation, these findings suggest that colocalization is, itself, caused by the combined effects of protein kinase C activation and an increase in cytoplasmic Ca^{2+}. Inhibitors of trypsin and other proteases have been noted to prevent acinar cell injury, but to not prevent the colocalization phenomenon.[38] These observations further support the conclusion that colocalization leads to zymogen activation and cell injury and they argue against the suggestion that zymogen activation and cell injury lead to colocalization.[19] Inhibiting the activity of lysosomal hydrolases including cathepsin B has been shown to prevent zymogen activation, but not colocalization.[39] In addition to supporting the conclusion that colocalization is the cause and not the result of zymogen activation, these studies indicate that zymogen activation may be mediated by lysosomal hydrolases including cathepsin B.[36] Similar changes have been noted when genetically altered mice that lack cathepsin B have been given supramaximally stimulating doses of cerulein (i.e., intrapancreatic zymogen activation and the severity of pancreatitis were both reduced).

Although the acinar cell generation of proinflammatory factors is a relatively late phenomenon that occurs during evolution of the pancreatic inflammatory reaction, the activation of proinflammatory transcription factors such as NF-κB occurs shortly after the start of supramaximal stimulation. This has led to the conclusion that acinar cell generation of proinflammatory factors probably contributes to the evolution of pancreatic inflammation, but that it may also be, at least in part, a result of pancreatic injury. The relationship between zymogen activation, the intraacinar cell presence of activated proteases such as trypsin, and the activation of acinar cell proinflammatory transcription factors has also been examined. Most of the evidence indicates that the intraacinar cell activation of NF-κB, and probably other proinflammatory transcription factors, occurs in parallel with intraacinar cell activation of zymogens,[29] but that it is neither mediated by activated zymogens nor the cause of zymogen activation.[31]

14.3.5 Pancreatitis-Related Effects of Acetylcholine and Secretin

14.3.5.1 Acetylcholine

Like CCK, acetylcholine interacts with cell surface G protein-coupled receptors that are linked to phospholipase C activation and a rise in cytoplasmic free calcium levels. Some have reported that concentrations of acetylcholine that exceed those causing maximal rates of digestive enzyme secretion from acinar cells can cause inhibition of enzyme secretion and that supramaximally stimulating doses of acetylcholine analogs such as carbamylcholine can cause pancreatitis in experimental animals.[16] This form of secretagogue-induced pancreatitis has proven difficult to study because animals exposed to these high levels of cholinergic agonists develop severe bronchorrhea leading to respiratory distress and, not infrequently, the animal's death. For this reason, little is known about the cellular events involved in supramaximal cholinergic stimulation-induced pancreatitis.

14.3.5.2 Secretin

Toxic pancreatic effects of secretin have not been observed when the hormone is administered alone even in supramaximally stimulating doses. On the other hand, recent studies[40] have suggested that secretin may sensitize the acinar cells of the pancreas to CCK-induced injury. This conclusion is based on experiments in which rat pancreatic acini were incubated with either a maximally or a supramaximally stimulating dose of cerulein with or without secretin. When secretin was given alone, intraacinar cell activation of zymogens did not occur. Similarly, intraacinar cell activation of zymogens did not occur when maximally or submaximally stimulating doses of cerulein were given without secretin, but activation did occur when secretin was given along with maximally stimulating doses of cerulein. As expected, supramaximally stimulating doses of cerulein caused zymogen activation even in the absence of secretin. The sensitizing effects of secretin in these experiments appeared to be mediated by the cAMP, which is generated in response to secretin stimulation.

Taken together, these studies raise the interesting possibility that secretin sensitizes the acinar cells of the pancreas to CCK-induced toxic injury. The cellular mechanisms underlying this sensitization remain to be established as does the relationship between the mechanisms underlying the evolution of pancreatitis and the combined effects of secretin and CCK stimulation.

14.4 HORMONE-INDUCED GROWTH AND NEOPLASIA

Rats, hamsters, and chickens develop an enlarged pancreas when they are given a diet containing raw soya protein.[41] Many studies have shown that raw soya protein is a potent trypsin inhibitor and it is likely that the pancreaticotrophic effects of administering the raw soya protein diet reflect the ability of that diet to inhibit trypsin within the upper gastrointestinal tract. Indeed, pancreatic hypertrophy has also been observed when synthetic short-chain trypsin inhibitors have been fed to experimental animals.

The mechanisms responsible for the growth promoting effects of lumenally active trypsin inhibitors are only partially understood. The available evidence suggests that administration of a lumenally active trypsin inhibitor prevents trypsin-mediated degradation of lumenally active CCK releasing factors (see Figure 14.2) and, as a result, CCK release from intestinal I-cells into the circulation is chronically stimulated. This chronic hypercholecysto-kininemia exerts a trophic effect on the acinar cells of the pancreas promoting both hypertrophy and hyperplasia of the gland.[42] It is not at all clear why feeding raw soya protein to rats, hamsters, and chickens, but not to mice or other animals, leads to promotion of pancreatic growth or neoplasia. These observations, however, would certainly be compatible with the hypothesis that the cholecystokinin feedback loop might be more active or less sensitive to down-regulation in certain animals, but not in others.

In the 1980s, a series of studies by Wormsley's group as well as others indicated that administering a diet rich in trypsin inhibitors could also lead to pancreatic neoplasia. Rats fed the raw soya protein diet developed hyperplastic pancreatic nodules, adenomas, and cancers. They also showed that feeding rodents a raw soya protein diet sensitized the pancreas to carcinogen-induced pancreatic cancers.[43,44] It is likely that these neoplasia-related effects of trypsin inhibitor administration are the result of the chronic hypercholecystokininemia that is caused by intralumenal trypsin inhibition.

14.5 EPILOGUE

Several naturally occurring hormones and hormonelike agents have been shown to exert toxic effects on the pancreas. For the most part, those toxic effects are observed in the exocrine, rather than the endocrine, pancreas and, to date, those toxic effects appear to specifically involve the acinar cells of the exocrine pancreas. CCK and its analogs have been the most well-studied of the pancreaticotoxic hormones but there is also evidence that acetylcholine and secretin can have toxic effects as well. To a great extent, the acute toxic effects of CCK and its analogs involve the induction of changes that lead to acute pancreatitis, but neoplastic changes are associated with chronic hypercholecystokininemia.

This chapter has focused on the cellular events and mechanisms that underlie the induction of pancreatitis by toxic doses of CCK or its analogs, particularly the decapeptide analog cerulein. Cerulein-induced acute pancreatitis is the most widely utilized experimental model of pancreatitis and, for that reason, many studies have been performed that probe the cellular mechanisms underlying this toxic effect of CCK. Cerulein-induced pancreatitis follows exposure of the pancreas to doses of the CCK analog that exceed those causing a maximal rate of digestive enzyme secretion. A series of intracellular events are triggered by administration of supra-maximally stimulating doses of cerulein that eventually lead to intraacinar cell activation of digestive enzyme zymogens and to activation of proin-flammatory transcription factors. Taken together, these changes lead to the acinar cell injury and intrapancreatic inflammatory reaction that are characteristic of acute pancreatitis.

Less is known about the neoplasia-related toxic effects of CCK. For the most part, the studies evaluating this issue have involved the chronic administration of diets containing trypsin inhibitors to experimental animals. This triggers chronic hypercholecystokininemia, probably by mechanisms involving the CCK feedback loop and lumenally active CCK releasing factors. The chronically elevated circulating levels of CCK that occur under these conditions appear to exert a trophic effect on the pancreas that can be translated into hypertrophy and hyperplasia, as well as neoplasia. At its extreme, pancreatic adenomas and carcinomas have been associated with the chronic administration of diets containing trypsin inhibitors. The cellular mechanisms involved in these neoplastic toxic effects of CCK have not been elucidated and their relevance to pancreatic cancer in humans is not known.

REFERENCES

1. Williams J.A. and Hootman S.R. 1986. Stimulus-secretion coupling in pancreatic acinar cells. In *The exocrine pancreas: biology, pathobiology, and diseases*. Go V.L.W., Gardner J.D., Brooks F.P., Leventhal E., DiMagno E.P., and Scheele G.A. Eds. New York: Raven Press. pp. 123–129.
2. Case R.M. 1998. Pancreatic exocrine secretion: mechanisms and control. In *The pancreas*. Beger H.G., Warshaw A.L., Buchler M.W., Carr-Locke D.I., Neoptolemos J.P., Russell C., and Sarr M.G. Eds. Oxford, U.K.: Blackwell Sciences. pp. 63–100.
3. Williams J.A. 2001. Intracellular signaling mechanisms activated by cholecystokinin-regulating synthesis and secretion of digestive enzymes in pancreatic acinar cells. *Ann. Rev. Physiol.* 63:77–97.
4. Ji B., Bi Y., Simeone D., Mortensen R.M., and Logsdon C.D. 2001. Human pancreatic acinar cells lack functional responses to cholecystokinin and gastrin. *Gastroenterology* 121:1380–1390.
5. Mainz D.L., Black O., and Webster P.D. 1973. Hormonal control of pancreatic growth. *J. Clin. Invest.* 52:2300–2304.
6. Dawra R., Saluja A., Lerch M.M., Saluja M., Logsdon C., and Steer M.L. 1993. Stimulation of pancreatic growth by cholecystokinin is mediated by high affinity receptors on rat pancreatic acinar cells. *Biochem. Biophys. Res. Commun.* 193:814–820.
7. Rivard N., Rydzewska G., and Morrisset J. 1994. Cholecystokinin-induced pancreatic growth involves the high-affinity CCK receptor and concomitant activation of tyrosine kinase and phospholipase D. *Ann. NY Acad. Sci.* 713:422–423.

8. Case R.M. and Argent B.E. 1993. Pancreatic duct cell secretion: control and mechanisms of transport. In *The pancreas: biology, pathobiology, and diseases.* Go V.L.W., DiMagno E.P., Gardner J.D., Lebenthal E., Reber H.A., and Scheele G.A. Eds. 2nd ed. New York: Raven Press. pp. 301–350.

9. Iwai K., Fushiki T., and Fokuoka S. 1988. Pancreatic enzyme secretion mediated by novel peptide: monitor peptide hypothesis. *Pancreas* 3:720–728.

10. Miyasaka K., Guan D.F., Liddle R.A., and Green G.M. 1989. Feedback regulation by trypsin: evidence for intraluminal CCK-releasing peptide. *Am. J. Physiol.* 257:G175–G181.

11. Lu L., Louie D., and Owyang C. 1989. A cholecystokinin releasing peptide mediates feedback regulation of pancreatic secretion.. *Am. J. Physiol.* 256:G430–G435.

12. Palade G. 1975. Intracellular aspects of the process of protein secretion. *Science* 189:347–358.

13. Kornfield S. 1986. Trafficking of lysosomal enzymes in normal and disease states. *J. Clin. Invest.* 77:1–6.

14. Chieli T., Bertazzoli C., Ferni G., Dell'oro I., Capella C., and Solcia E. 1972. Experimental toxicology of caerulein. *Toxicol. Appl. Pharmacol.* 23:480–491.

15. Lampel M. and Kern H.F. 1977. Acute interstitial pancreatitis in the rat induced by excessive doses of a pancreatic secretagogue. *Virchows. Archiv. A. Patholog. Anat. Histol.* 373:97–117.

16. Adler G., Gerhards G., Schick J., Rohr G., and Kern H.F. 1983. Effects of in vivo cholinergic stimulation of rat exocrine pancreas. *Am. J. Physiol.* 244:G626–G629.

17. Kaiser A., Saluja A., Sengupta A., Saluja M., and Steer M.L. 1995. Relationship between severity, necrosis and apoptosis in five models of experimental acute pancreatitis. *Am. J. Physiol.* 38:C1295–C1304.

18. Steer M.L. 1998. The early intraacinar cell events which occur during acute pancreatitis. The Frank Brooks Memorial Lecture. *Pancreas* 17:31–37.

19. Saluja A.K., Saluja M., Printz H., Zavertnik A., Sengupta A., and Steer M.L. 1989. Experimental pancreatitis is mediated by low-affinity cholecystokinin receptors that inhibit digestive enzyme secretion. *Proc. Natl. Acad. Sci. USA* 86:8968–8971.

20. Saluja A.K., Bhagat L., Lee H.S., Bhatia M., Frossard J.L., and Steer M.L. 1999. Secretagogue-induced digestive enzyme activation and cell injury in rat pancreatic acini. *Am. J. Physiol.* 276:G835–G842.

21. Steinle U., Weidenbach H., Wagner M., Adler G., and Schmid R. 1999. NFkB/Rel activation in cerulein pancreatitis. *Gastroenterology* 116:420–430.

22. Sans M.D., DiMagno M.J., D'Alecy L.G., and William J.A. 2003. Cerulein-induced acute pancreatitis inhibits protein synthesis in mouse pancreas through effects on eucaryotic initiation factors 2B and 4F. *Am. J. Physiol. (GI Physiol.)* 285:G517–28.

23. O'Konski M.S. and Pandol S.J. 1990. Effects of caerulein on the apical cytoskeleton of the pancreatic acinar cell. *J. Clin. Invest.* 86:1649–1657.

24. Muallem S., Kwiatowska K., XU X., and Yin H.L. 1995. Actin filament disassembly is a sufficient final trigger for exocytosis in nonexcitable cells. *J. Cell. Biol.* 128:589–598.

25. Van Acker G.J.D., Saluja A.K., Bhagat L., Singh V., Song A.M., and Steer M.L. 2001. Cathepsin B inhibition prevents trypsinogen activation and reduces pancreatitis severity. *Am. J. Physiol.* 283:G794–G800.

26. Singh V.P., Saluja A., Bhagat L., Song A., van Acker G.J.D., Soltoff S., Cantley L., and Steer M.L. 2001. Inhibition of phosphoinositide-3-kinase prevents trypsinogen activation and reduces the severity of acute pancreatitis. *J. Clin. Invest.* 108:1387–1395.

27. Gukovsky I., Gukovskaya A.S., and Pandol S.J. 1998. Cerulein activated NF-kB and AP-1 in pancreatic acinar cells. *Gastroenterology* 114:A465.

28. Grady T., Liang P., Ernst S.A., and Logsdon C.D. 1997. Chemokine gene expression in rat pancreatic acinar cells is an early event associated with acute pancreatitis. *Gastroenterology* 113:1966–1975.

29. Han B. and Logsdon C.D. 2000. CCK stimulates MOB-1 expression and NF-kappaB activation via protein kinase C and intracellular Ca(2+). *Am. J. Physiol.* 278:C344–C351.

30. Singh V.P., Saluja A.K., Bhagat L., van Acker G.J.D., Song A.M., Mykoniatis A., and Steer M.L. 2001. Secretagogue (Caerulein)-induced intrapancreatic activation of trypsinogen is mediated by protein kinase C (PKC), phosphoinositide-3-kinase (PI-3K), and tyrosine kinases (TKs). *Gastroenterology* 120:A1735.

31. Hietaranta A.J., Saluja A.K., Bhagat L., Singh V.P., Song A.M., and Steer M.L. 2001. Relationship between NF-kB and trypsinogen activation in rat pancreas after supramaximal caerulein stimulation. *Biochem. Biophys. Res. Commun.* 280:388–395.

32. Saluja A.K., Hashimoto S., Saluja M., Powers R.E., Meldolesi J., and Steer M.L. 1987. Subcellular redistribution of lysosomal enzymes during caerulein-induced pancreatitis. *Am. J. Physiol.* 251:G508–G516.

33. Grady T., Saluja A., Kaiser A., and Steer M. 1996. Pancreatic edema and intrapancreatic activation of trypsinogen during secretagogue-induced pancreatitis preceeds glutathione depletion. *Am. J. Physiol.* 271:G20–G26.

34. Steer M.L. 1992. How and where does acute pancreatitis begin? *Arch. Surg.* 127:1350–1353.

35. Van Acker G., Saluja A., Bhagat L., Hietaranta A., Singh V., Song A., Mykoniatis A., Pan A., and Steer M. 2000. Relationship between intrapancreatic activation of trypsinogen and intrapancreatic sequestration of neutrophils following supramaximal stimulation with caerulein in mice. *Pancreas* 21:487.

36. Van Acker G., Saluja A.K., Singh V.P., Bhagat L., Song A.M., Mykoniatis A., and Steer M.L. 2001. Intrapancreatic activation of trypsinogen precedes and is required for neutrophil sequestration in the pancreas during secretagogue-induced pancreatitis in mice. *Gastroenterology* 120:A2742.

37. Saluja A.K., Donovan E.A., Yamanaka K., Yamaguchi Y., Hofbauer B., and Steer M.L. 1997. Caerulein-induced in vitro activation of trypsinogen in rat pancreatic acini is mediated by cathepsin B. *Gastroenterology* 113:304–310.

38. Hofbauer B., Saluja A.K., Lerch M., Bhagat L., Bhatia M., Lee H.S., Frossard J.L., Adler G., and Steer M.L. 1998. Intra-acinar cell activation of trypsinogen during caerulein-induced pancreatitis in rats. *Am. J. Physiol.* 275:G352–G362.

39. Bhagat L., Saluja A.K., van Acker G.J.D., Singh V.P., Song A.M., Mykoniatis A., and Steer M.L. 2001. Lysosomal hydrolase/digestive zymogen co-localization and intra-acinar cell activation of trypsinogen: which is the horse and which is the cart? *Gastroenterology* 120:A2748.

40. Kolodecik T., Lu Z., Nyce M., and Gorelick F. 2003. Cyclic AMP agonists sensitize isolated rat pancreatic acini to cerulein induced zymogen activation. *Gastroenterology* 122:S1794.

41. Oates P.S. and Morgan G.H. 1982. Pancreatic growth and cell turnover in the rat fed raw soya flour. *Am. J. Pathol.* 108:217–224.

42. Smith J.C., Wilson F.D., Allen P.V., and Berry D.L. 1989. Hypertrophy and hyperplasia of the rat pancreas produced by short-term dietary administration of soya-derived protein and soybean trypsin inhibitor. *J. App. Toxicol.* 9:175–179.

43. McGuinness E.E., Morgan R.G.H., Levison D.A., Hopwood D., and Wormsley K.G. 1981. Interaction of azaserine and raw soya flour on the rat pancreas. *Scand. J. Gastroenterol.* 16:49–56.

44. McGuinness E.E., Hopwood D., and Wormsley K.G. 1982. Further studies of the effects of raw soya flour on the rat pancreas. *Scand. J. Gastroenterol.* 17:273–277.

15

ALCOHOL AND PANCREATITIS

Stephan L. Haas, Alexander Schneider, and Manfred V. Singer

CONTENTS

15.1 INTRODUCTION

Chronic pancreatitis represents a disease that is characterized by repeated episodes of acute pancreatitis, abdominal pain, and fibrotic destruction of the organ leading to exocrine and endocrine insufficiency. The investigation of the pathophysiological mechanisms underlying acute and chronic pancreatitis remains a major scientific and clinical challenge. Although alcohol appears to be one of the leading causes of pancreatitis, there exists no generally accepted theory on the development of alcoholic chronic pancreatitis.

Histological diagnosis of suspected chronic pancreatitis is seldom available during the early course of chronic pancreatitis, and most of the histological data are from specimens obtained during surgical interventions in the late phase of the disease. Due to difficulties in obtaining reliable data from patients with alcoholic chronic pancreatitis, it remains difficult to draw conclusions on the exact impact of alcohol consumption on pancreatic pathology. Animal models that satisfactorily represent the human disease have been difficult to develop, but revealed certain aspects of human alcoholic pancreatitis. The recent identification of stellate cells provided further insights into the mechanisms of pancreatic fibrosis (see Chapter 16).

The major aims of this chapter are to summarize the epidemiology of alcoholic chronic pancreatitis, to describe briefly the natural course of the disease, and to review the numerous effects of ethanol on the pancreas that may contribute to the development of alcoholic chronic pancreatitis.

15.2 EPIDEMIOLOGY OF ALCOHOLIC CHRONIC PANCREATITIS

A large number of studies have shown that alcohol consumption is the leading cause of chronic pancreatitis in industrialized countries. Based on these studies, approximately 70% of all cases of chronic pancreatitis can be attributed to this etiologic factor in industrialized countries with a variation from 38 to 94% in different investigations.[1–8] However, the proportion of patients with alcoholic chronic pancreatitis varies considerably between the different studies. For an adequate interpretation of these results, several aspects have to be taken into account: Precise information regarding alcohol consumption is frequently difficult to obtain. Furthermore, the amount of alcohol consumption that was used to define the disease as "alcoholic" pancreatitis was variable in the different investigations. The indistinguishable clinical picture of alcoholic and nonalcoholic chronic pancreatitis contributes to a potentially incorrect diagnosis. Of note, the International Classification of Diseases (ICD)-10 may help to improve future epidemiological data of chronic pancreatitis, because for the first time alcoholic pancreatitis is represented by a separate code.

Although excessive alcohol consumption is frequently observed in patients with chronic pancreatitis, clinically overt alcoholic pancreatitis is observed in only about 5% of heavy drinkers. A prospective study of patients with alcoholic chronic pancreatitis demonstrated an incidence of 8.2 cases per year per 100,000 individuals and an overall prevalence of 27.4 cases within a population of 100,000 individuals.[4] Further information regarding the frequency of subclinical pancreatic damage in patients with excessive alcohol consumption was obtained in autopsy studies. These studies revealed that about 20% of all alcoholics exhibit histological features of chronic pancreatitis without suffering signs of chronic pancreatitis prior to death. In an investigation of 99 alcoholic patients who died from alcohol-related disease, 18% presented with perilobular sclerosis as evidence of alcohol-induced pancreatic damage.[9] In another study of 783 patients that had died of alcoholic liver cirrhosis, 20% of these patients demonstrated histological signs of chronic pancreatitis.[10]

Several studies aimed to answer the question whether the consumption of specific types of alcoholic beverages is associated with an increased risk of alcoholic pancreatitis, but an association of pancreatitis with the type of beverage was not found.[2,11,12] Patients with alcoholic chronic pancreatitis usually present an ethanol-intake exceeding 80 g per day for several years before the first clinical onset of the disease.[2,5,7,8,13] In 1997, an international conference on alcoholic chronic pancreatitis agreed to define alcoholic chronic pancreatitis as chronic pancreatitis that occurs after a daily intake of ethanol equal to or greater than 80 g per day for several years.[14] However, there appears to be no defined threshold of alcohol consumption below which no increased risk of alcoholic pancreatitis exists.[2] The risk of alcoholic chronic pancreatitis increased logarithmically with higher amounts of alcohol consumption.[2] Of note, an increased risk has also been reported in patients with moderate amounts of alcohol consumption such as 20 g per day.[2]

15.3 NATURAL COURSE OF ALCOHOLIC CHRONIC PANCREATITIS

The majority of patients with alcoholic chronic pancreatitis are diagnosed between 35 and 40 years of age.[15] Alcoholic chronic pancreatitis usually presents with an early phase of recurrent attacks of acute pancreatitis, followed by the late phase of the disease characterized by the development of chronic pain, pancreatic calcifications, and exocrine and later endocrine insufficiency. The pancreas demonstrates a large functional capacity so that exocrine and endocrine insufficiency develops with a considerable delay during the course of alcoholic chronic pancreatitis.

The interval between the start of continuous alcohol abuse and the clinical manifestation of alcohol-induced chronic pancreatitis usually requires between 13 to 21 years.[2,5] Then, the interval between onset of the disease and the occurrence of exocrine insufficiency and calcifications averages 4.8 +/- 5.5 years in alcoholic chronic pancreatitis.[16] Another study reported a median time of 13.1 years to pancreatic exocrine insufficiency after onset of the disease, whereas endocrine insufficiency was diagnosed after a median time of 19.8 years.[6] Data regarding the progression of pancreatic insufficiency are conflicting. Several investigations did not show a progressive deterioration of exocrine pancreatic insufficiency or revealed even slight improvements in pancreatic insufficiency over time.[6,17,18] In contrast, other reports show progressive deterioration of pancreatic function during a median follow-up of 10.4 years in patients with alcoholic chronic pancreatitis.[19] The reasons for these differences are not clarified, but may result from differences in study designs or sensitivities of the pancreatic function tests used in the different investigations.[18] Although the progression toward pancreatic insufficiency is not completely halted by cessation of alcohol abuse, abstinence may diminish the progression to exocrine and endocrine insufficiency.[20] Moreover, a decrease in pain or a decrease in the frequency of episodes of acute pancreatitis has been shown in patients who stop consuming alcohol.[21]

The relationship between acute and chronic alcoholic pancreatitis remains controversial (see Chapter 16).[22] Studies in patients with an initial episode of acute pancreatitis revealed that these patients already presented histological signs of chronic pancreatitis.[6,9,23] However, several long-term clinical studies,[14,16,19] pathologic studies,[24] studies in patients with hereditary pancreatitis,[25,26] and recent experimental studies[27] provide strong evidence that recurrent attacks of acute pancreatitis can result in chronic pancreatitis. Indeed, the concept that acute pancreatitis represents the first recognition of chronic pancreatitis appears to be valid in about half of the 247 alcoholic patients who died of acute pancreatitis, but not in the other half of patients who demonstrated no signs of chronic pancreatic damage.[10]

The mortality rate in patients with alcoholic chronic pancreatitis is approximately one-third higher than that in an age and sex matched general population.[18] About 50% of patients with the disease die within 20 years after onset of the disease with a median age of 54 years.[16] However, mortality is not related to chronic pancreatitis and its complications in 80% of the patients.[6,16,28] The most common causes of death remain malignancies and cardiovascular diseases that are rather associated with smoking and ongoing alcohol abuse than with chronic pancreatitis.[29]

Table 15.1 Major Findings in Studies with Humans and Ethanol-Fed Animals on Pancreatic Exocrine Secretion

Acute Ethanol Administration

Oral and intragastric ethanol administration increases pancreatic bicarbonate and protein secretion.

Intravenous ethanol administration reduces basal and hormonally stimulated pancreatic bicarbonate and protein secretion.

Nonalcoholic constituents of beer may increase pancreatic secretion.

Chronic Ethanol Administration

In human alcoholics, the basal pancreatic enzyme secretion is increased.

In human alcoholics, the viscosity of the pancreatic juice is enhanced.

In human alcoholics, the pancreatic juice contains a higher concentration of proteins.

In human alcoholics, the pancreatic bicarbonate secretion is decreased.

In human alcoholics, an enhanced ratio of trypsinogen levels to pancreatic secretory trypsin inhibitor levels is present in the pancreatic juice.

In ethanol-fed animals, a diet rich in fat and protein increases the concentrations of enzymes in the pancreatic juice.

15.4 EFFECTS OF ETHANOL ON THE PANCREAS

Ethanol consumption results in various effects on pancreatic exocrine secretion (Table 15.1) (reviewed in Reference 30), and ethanol and its metabolites have additional damaging effects on the pancreas (Table 15.2) (reviewed in Reference 26). Thus, these different effects of ethanol on the pancreas have to be considered separately.

15.4.1 Alcohol and Pancreatic Exocrine Secretion

The pancreatic secretory response to intestinal nutrients is mediated by a highly complex interplay of neural and humoral responses. Studies addressing the effect of alcohol on pancreatic secretion yielded contradictory results. Therefore, several reasons have to be considered:

- The effects of ethanol vary depending on the route of administration (intravenous, intragastric, intrajejunal).
- The effects depend on the secretory state of the pancreas (stimulated, unstimulated).

Table 15.2   Major Findings with Animal Models of Acute and Chronic Ethanol Administration on Pancreatic Morphology

Animal Models of Acute Ethanol Administration

Ethanol administration (intragastrically, intraperitoneally, intravenously) with physiological stimulation (CCK, secretin) and obstruction of the pancreatic duct results in acute pancreatitis.

Ethanol administration enhances the vulnerability of the pancreas to develop acute pancreatitis and limits pancreatic regeneration from acute pancreatitis.

Ethanol administration selectively reduces pancreatic blood-flow and microcirculation.

Cigarette smoke enhances ethanol-induced pancreatic ischemia.

Ethanol administration increases free oxygen radical generation in the pancreas.

Ethanol metabolites directly damage the pancreas.

Animal Models of Chronic Ethanol Administration

Dietary fat potentiates ethanol-induced pancreatic injury.

Ethanol administration increases free oxygen radical generation in the pancreas.

Ethanol administration increases pancreatic acinar cell expression and glandular content of digestive and lysosomal enzymes.

Ethanol administration decreases the number of muscarinic receptor sites.

Ethanol administration limits pancreatic regeneration after temporary obstruction of the pancreatic duct and further aggravates already induced pancreatic damage.

Ethanol administration sensitizes pancreatic acinar cells to endotoxin-induced injury.

Ethanol administration enhances the vulnerability of the pancreas to pancreatitis caused by CCK octapeptide.

- There are variations in ethanol effects in different species.
- Acute and chronic effects of ethanol have to be differentiated.
- Different experimental conditions may provide contradictory results.

15.4.1.1   Effects of Acute Ethanol Administration In Vivo

In humans, cats, and pigs, oral or intragastric administration of ethanol causes a weak stimulation of pancreatic bicarbonate and protein secretion.[31–35] When the gastric content was prevented to reach the duodenum, intragastric, intraduodenal, or intrajejunal instillation of ethanol inhibited or did not effect pancreatic exocrine secretion in humans, dogs, and rats.[36–40] Therefore, it was suggested that the ethanol-induced pancreatic output depends on the secretion of gastric acid. The passage of the gastric acid to the duodenum is required to release secretin or other hormones to activate pancreatic secretion. In addition, enteropancreatic neural reflexes that are transmitted by gastric acid may be involved.

A further explanation for the induction of pancreatic secretion by ethanol is the release of gastrin. In dogs with partial gastrectomy, intragastric instillations of even high concentrations of ethanol were not accompanied by an increased pancreatic enzyme output suggesting that gastrin liberation is crucial for the stimulation of pancreatic secretion.[41] Although ethanol has been shown to stimulate pancreatic secretion in humans, neither a significant increase of gastrin release nor an increased gastric acid output has been demonstrated after ingestion of ethanol, indicating that in humans, contrary to animals, other still undefined physiological processes are involved.

Only a few experiments have been undertaken to address the question how ethanol modulates pancreatic secretion after food ingestion. In one study, an inhibition of postprandial enzyme secretion was observed with intragastric application of ethanol.[42] However, another study reported a mild decrease in the early postprandial period followed by a significant increase of enzyme secretion.[43]

Intravenous administration of ethanol is a rather artificial route of alcohol administration, but has the advantage to gain more insights in direct effects of ethanol on the pancreas. By this route of administration, ethanol administration resulted in a dose-dependent inhibition of the basal and hormonally stimulated pancreatic bicarbonate and enzyme output in humans and in different animal species.[30]

To delineate the contribution of vagal innervation on the observed inhibitory effect of ethanol on pancreatic secretion, several investigators blocked vagal input by truncal vagotomy or atropine administration. Using this approach, no inhibitory effect of ethanol was demonstrated in rats and dogs.[41,44,45] It was concluded that the inhibitory effect of ethanol is mediated by vagal activity. Others have argued that the blockage of parasympathetic nerves provides such a high inhibitory effect on pancreatic secretion that the demonstration of an additional inhibitory effect of ethanol is not feasible.[46]

15.4.1.2 Effects of Acute Ethanol Administration In Vitro

Studying the effects of ethanol on acinar cells proved to be an attractive alternative to *in vivo* experiments because additionally interfering humoral and neural effects are excluded within this experimental setting. The application of ethanol to isolated pancreatic acini led to an increase of basal amylase secretion, but decreased the cholecystokinin (CCK)-mediated amylase release.[47–49] These investigations have shown that acute ethanol administration modulated pancreatic secretion through the activation of cyclic adenosine monophosphate (cAMP), an increase of cytosolic-free calcium concentrations and inhibition of calcium efflux. The detected increase of the intracellular calcium concentration in the acinar cell may represent a pathway by which ethanol increases the secretion of amylase.

The inhibition of the CCK-receptor binding may result in the inhibition of CCK-induced amylase secretion. However, the exact mechanism still needs to be defined, but may be ascribed to unspecific cell membrane alterations with subsequent modifications of the receptor–ligand interaction. Others have attributed this inhibitory effect to an inhibition of the calcium pump activity of the pancreatic acinar cell membranes. There is growing evidence that ethanol may impair the microtubular function that is implicated in the secretion of pancreatic enzymes.[30]

15.4.1.3 Effects of Acute Administration of Alcoholic Beverages

Most investigators focused their interest on the effects of pure ethanol on the pancreas. However, alcoholic beverages like beer and wine contain numerous additional substances that might be involved in the development of pancreatic inflammation. However, data regarding the effects of the nonalcoholic constituents of alcoholic beverages on pancreatic secretion are limited.

The intragastric administration of beer in a dose (250 ml) that does not result in elevated plasma ethanol concentrations caused a significant stimulation of the basal pancreatic enzyme output, whereas the administration of pure ethanol with the same alcohol concentration had no effect on the pancreatic enzyme secretion.[38,46] This stimulatory effect may result from the liberation of CCK and gastrin by nonalcoholic constituents. A characterization of the involved substances is still lacking, but there is some evidence that fermentation products of glucose might be involved.[38]

In another study, the pancreatic enzyme output was determined after intragastric administration of amounts of beer (850 ml) or wine (400 ml) that lead to an elevation of plasma ethanol concentrations, but the basal pancreatic enzyme output remained unchanged.[39] It was suggested that the direct inhibitory effect of the circulating ethanol in the blood neutralized the stimulatory effect of the nonalcoholic components of beer and wine.[38,39]

Additional studies revealed that the meal-stimulated pancreatic enzyme output was inhibited by intragastric administration of beer, white wine, and gin. In these studies, the amount of alcohol also resulted in elevated ethanol plasma levels (reviewed in Reference 30).[42] Again, the inhibitory effects of circulating ethanol might have neutralized a possible stimulatory effect of beer and wine on the secretion of pancreatic enzymes.[42]

15.4.1.4   Effects of Chronic Ethanol Administration In Vivo

In human alcoholics, the basal pancreatic enzyme output was increased compared with the basal enzyme output in nonalcoholic individuals.[50–53] The viscosity of the pancreatic juice was enhanced in alcoholic subjects, and this rise was correlated with increased concentrations of proteins.[54] The volume of pancreatic juice was similar in controls and alcoholics, thereby indicating a true hypersecretion of pancreatic proteins in subjects with chronic excessive alcoholic consumption.[52] In addition, the pancreatic bicarbonate secretion was significantly lower in human alcoholics than in nonalcoholic individuals.[51,52] Of note, the basal plasma concentrations of secretin, CCK, and gastrin remained unchanged between alcoholic and nonalcoholic subjects.[30,55,56] An enhanced ratio of trypsinogen levels to pancreatic secretory trypsin inhibitor levels was found in the pancreatic juice from chronic alcoholics.[57] This alteration of the normal ratio in favor of trypsinogen may facilitate premature activation of pancreatic proenzymes within the pancreas resulting in an increased risk of subsequent pancreatic autodigestion.

15.4.2   Alcohol and Pancreatic Enzyme Synthesis

The effects of chronic alcohol consumption on pancreatic gene expression and glandular content of pancreatic enzymes were studied in rats.[18] Messenger RNA levels for lipase, trypsinogen, chymotrypsinogen, and cathepsin B were elevated in chronically ethanol fed rats suggesting that chronic ethanol consumption increases the capacity of the pancreatic acinar cell to synthesize digestive and lysosomal enzymes and that these changes might lead to an elevated susceptibility of the pancreas to enzyme-related damage.[18,58] Thus, these studies suggest that chronic alcohol consumption results in changes of the synthesis of pancreatic enzymes, which again may increase the risk of premature zymogen activation.

15.4.3   Metabolism of Alcohol in the Pancreas

The metabolism of ethanol is well-characterized in the liver and occurs via two major pathways: an oxidative pathway generating acetaldehyde and a nonoxidative pathway generating fatty acid ethyl esters. Recent experiments with pancreatic acinar cells demonstrated that alcohol is metabolized via both oxidative and nonoxidative pathways in the pancreas.

15.4.3.1 Oxidative Pathway

Haber et al. performed cell culture experiments with pancreatic acinar cells to identify the enzymes that are responsible for the oxidation of ethanol by acinar cells. These studies suggested that the alcohol dehydrogenase (ADH) activity within the pancreatic acinar cell is represented by ADH3, which is a nonsaturable form of ADH with a high K_m for ethanol.[59] Recent experiments also used isolated pancreatic acini and confirmed these findings.[60] However, the intrapancreatic generation of acetaldehyde has several pathophysiological implications.

Acetaldehyde is a reactive intermediate that forms covalent adducts with various cellular proteins. Animal studies have shown that acetaldehyde causes morphological damage to the pancreas.[61,62] Acetaldehyde also inhibited the stimulated secretion of pancreatic enzymes from isolated pancreatic acini.[48] This impairment of the pancreatic secretory function by acetaldehyde was suggested to result from microtubular dysfunction that may interfere with the exocytosis from pancreatic acinar cells.[48] Furthermore, the conversion of ethanol to acetaldehyde by ADH has profound effects on the acinar cell metabolism by altering the intracellular redox state reflected by a reduced NAD/NADH (nicotinamide adenine dinucleotide) ratio.

Finally, it was shown that cytochrome P450-expression is detectable in the pancreas and that the enzymatic activity is enhanced in pancreatic tissue of patients with chronic pancreatitis.[63] Norton et al. showed that the amount of cytochrome P450 2E1 (CYP 2E1) in the pancreas of chronically alcohol-fed rats increased five-fold compared to control animals.[64] The induction of pancreatic cytochrome P450-activity is associated with a concomitant generation of reactive oxygen species that may facilitate cellular damage by peroxidation of membrane systems, enzymes, structural proteins, and DNA. Therefore, the generation of oxidative stress via ethanol oxidation may also contribute to the development of pancreatic damage and will be discussed below.

15.4.3.2 Nonoxidative Pathway

A nonoxidative pathway of ethanol metabolism was also described in the pancreas and results in the formation of fatty acid ethyl esters (FAEEs). The formation of FAEE is catalyzed by FAEE-synthase. An autopsy study in individuals who died from acute alcohol intoxication revealed that FAEE and FAEE-synthase were found predominantly in organs damaged by ethanol, with the highest levels in the pancreas.[65] Interestingly, FAEE appear to have a variety of deleterious effects. FAEE induced mitochondrial dysfunction with subsequent inefficient mitochondrial energy production.[66] After incubation with FAEE, isolated pancreatic lysosomes showed an increased fragility, which may represent another reason for ethanol induced pancreatic damage. In addition, the intravenous infusion of FAEE in rats resulted in pancreatic injury with edema and intrapancreatic trypsinogen activation.[67]

15.4.4 Alcohol and Oxidative Stress

In general, oxidative stress represents an imbalance between the production of reactive oxygen species (ROS) and the antioxidant defense mechanisms within the cell such as glutathione, the enzymes glutathione peroxidase, superoxide dismutase and catalase, and their cofactors such as selenium or Vitamin E. All aerobic biological systems continuously generate ROS, and approximately 2 to 5% of the daily oxygen intake of humans is converted to ROS.[68] In low concentrations, these oxygen species are important for normal cellular function, but the increased generation of oxygen radicals is highly toxic for the cell through the modification of proteins, lipids, and DNA.

Acute and chronic ethanol consumption results in oxidative stress within the pancreas. Therefore, ethanol-induced oxidative stress was suggested to play a fundamental role in alcoholic chronic pancreatitis.[69] In humans, increased lipid peroxidation products were detected in pancreatic tissue from patients with chronic pancreatitis and in their pancreatic juice.[70] Furthermore, patients with alcoholic, but also patients with hereditary and idiopathic, chronic pancreatitis revealed a decreased antioxidative capacity.[71,72] In rats, acute alcohol ingestion resulted in an increase of lipid peroxidation and of protein oxidation and led to the depletion of glutathione.[73] In chronically ethanol-fed rats, a depletion of mitochondrial glutathione was observed together with an increase of oxidative stress markers.[74] Chronic alcohol consumption led to the accumulation of fat in the pancreatic acinar cell, which again appears to increase the susceptibility to lipid peroxidation.[75,76] Free oxygen radicals may also be implicated in the development of chronic pancreatitis in male Wistar-Bonn/Kobori (WBN/Kob) rats, which represent an inbred strain of Wistar rats, who develop spontaneously chronic pancreatitis-like lesions and a diabetic syndrome.[77] The inflammatory infiltrate, which frequently occurs in acute and chronic pancreatitis has also been suggested to contribute to the generation of oxidative stress. In a recent study, infiltrating neutrophils appeared to generate ROS. This oxidative burst was suggested to contribute to the intrapancreatic activation of trypsin with subsequent development of autodigestion and pancreatitis.[78] Another study with isolated pancreatic acinar cells showed that ROS activate the transcription factor NF-κB, which again results in the up-regulation of inflammatory cytokine expression.[79] Thus, the concept of oxidative stress in ethanol-induced pancreatic damage remains extremely interesting and requires further investigation.

15.4.5 Alcohol and Pancreatic Microcirculation

In several animal models of acute pancreatitis, the role of impaired pancreatic microcirculation has been emphasized. In the early stages of acute pancreatitis, a vasodilation and hyperperfusion was observed, whereas in the later stages of the disease, and especially in severe acute pancreatitis, a reduction of pancreatic microcirculation predominated.

Alcohol *per se* has profound effects on pancreatic microcirculation and oxygenation. Administration of ethanol to dogs and rats reduced pancreatic blood flow and pancreatic hemoglobin saturation significantly.[80,81] Various factors like platelet activating factor, nitric oxide, prostaglandins, and endothelin are implicated in the regulation of tissue microcirculation. Several lines of evidence suggest that Endothelin-1 (ET-1) represents one of the major mediators of impaired pancreatic perfusion via ethanol.[82] ET-1 is not only a powerful vasoconstrictor and binds to specific ET-1 receptors, but increases capillary permeability thereby facilitating capillary leakage.[83] Thus, activated pancreatic enzymes may gain access to the interstitial space and may contribute to pancreatic autodigestion. The reduction of pancreatic perfusion by ethanol may result in pancreatic hypoxia and may also induce oxidative stress.[84] Data regarding pancreatic perfusion in human alcoholic chronic pancreatitis are difficult to generate. Schilling and coworkers have shown a marked reduction of microcirculation in patients with alcoholic chronic pancreatitis.[85] This finding confirmed the earlier observation that pancreatic blood flow is decreased in cats with experimentally induced chronic pancreatitis. In the same animal model with cats, the intragastric and intravenous administration of ethanol reduced the pancreatic blood flow and the pancreatic interstitial pH.[86] Of note, a study with rats showed that smoking potentiated the impairment of pancreatic capillary perfusion. Therefore, smoking may aggravate and accelerate the development of chronic pancreatitis in alcohol abusers.[87]

15.4.6 Alcoholic Pancreatitis and Nutrition

The investigation of different dietary regimes and their impact on the development of ethanol-induced pancreatic damage remains an extremely important issue in both experimental investigations with chronically ethanol-fed animals and in clinical studies with human alcoholics.

In 1987, Ramo et al. studied the effects of different diets on rodents that were fed with alcohol for 12 weeks. The different diets were rich in fat, rich in protein, or rich in carbohydrates. Subsequent histologic examinations revealed no major differences except for an inflammatory infiltrate in 9 of 24 rats that received the high-fat diet.[88] Tsukamoto et al. continuously administered high amounts of ethanol through an intragastrically placed catheter for up to 160 days. In addition, the rats received a liquid diet with increasing amounts of unsaturated fats (corn oil). Up to 30% of the rats that received the high-fat diet and an extra-high-fat diet developed focal fibrotic lesions with mononuclear cell infiltration, ductal dilatations, and proteinaceous plugs. Therefore, this animal model showed pathological changes that are usually observed in human chronic pancreatitis.[89] Using a similar model, Kono et al. fed rats with increasing doses of ethanol in combination with either saturated or unsaturated fat.[90] The stepwise increase of the dosage of ethanol resulted in much higher blood ethanol concentrations compared with other studies. The combination of ethanol with unsaturated fat led to an increase in amylase and lipase serum levels. After 8 weeks, pancreatic fibrosis with an increased mRNA expression of collagen was detected. In addition, a significant increase in the formation of free radicals was observed. These parameters were only moderately elevated in alcohol-fed rats that received a diet with saturated medium-chain triglycerides. Thus, it was concluded that unsaturated fats aggravate alcohol-induced pancreatic damage. It was also suggested that the total amount of ethanol represents the important factor in this animal model.[90]

In human alcoholics, clinical studies that investigated the role of dietary factors in alcoholic pancreatitis produced conflicting results. Sarles et al. reported that patients with chronic alcoholic pancreatitis consumed more fat and protein compared with nonalcoholics without chronic pancreatitis.[91] Pitchumoni et al. observed a higher consumption of fat in patients with alcoholic pancreatitis compared with patients with alcoholic liver cirrhosis.[92] However, another study detected no dietary differences between patients with alcoholic pancreatitis and alcoholic liver cirrhosis.[11] Thus, further studies that focus on nutritional parameters in patients with alcoholic chronic pancreatitis are warranted.

15.4.7 Alcohol and Pancreatic Stellate Cells

The deposition of extracellular matrix proteins and the development of pancreatic fibrosis represent key characteristics of chronic pancreatitis. In 1998, two groups isolated pancreatic stellate cells from the pancreas and demonstrated that these cells are the main source of extracellular matrix in pancreatic fibrosis.[93,94] In the normal pancreas, these pancreatic stellate cells remain in a quiescent state and contain Vitamin A lipid droplets in the cytoplasm. These cells are morphologically similar to hepatic stellate cells that play a dominant role in hepatic fibrosis.

The role that pancreatic stellate cells play during the development of pancreatic fibrosis (see Chapter 16) has been investigated with two approaches. First, *in vivo* studies used either pancreatic specimens obtained from patients with chronic pancreatitis or pancreatic tissue obtained from animal models with experimental pancreatitis. Second, *in vitro* studies with cultured pancreatic stellate cells provided deep insights into the underlying mechanisms of pancreatic fibrosis.

Recent *in vivo* studies with tissue samples of patients with alcoholic chronic pancreatitis demonstrated an increased expression of lipid peroxidation products in acinar cells that were adjacent to areas of pancreatic fibrosis.[95] These findings suggest that oxidative stress plays an important role in the development of pancreatic fibrosis. In addition, the immunohistochemical staining of pancreatic tissue sections provided evidence that activated pancreatic stellate cells represent the main source of extracellular matrix during pancreatic fibrosis and that transforming growth factor-beta is secreted by activated pancreatic stellate cells.

Indeed, recent *in vitro* investigations have shown that pancreatic stellate cells are activated early in the course of pancreatic damage. Once activated, these cells loose their cytoplasmic lipid droplets, express alpha-smooth muscle actin (SMA), and synthesize extracellular matrix proteins such as Collagen I and Collagen III, fibronectin, and laminin. In addition, activated pancreatic stellate cells exhibit an enhanced proliferation rate and have the capacity to synthesize chemokines, adhesion molecules, and proinflammatory cytokines such as transforming growth factor-beta1.

Various stimuli like proinflammatory cytokines, growth factors, and oxidative stress activate pancreatic stellate cells. *In vitro* studies revealed that ethanol in clinically relevant concentrations activated SMA expression and Type I collagen synthesis in pancreatic stellate cells.[96] The ethanol-induced activation of pancreatic stellate cells was blunted by inhibiting the conversion of ethanol to acetaldehyde thereby indicating that stellate cells metabolize ethanol to acetaldehyde. It was further concluded that the ethanol-induced pancreatic stellate cell activation is mediated by acetaldehyde.[96]

Based on these findings, it was speculated that there exist a necroinflammatory and a nonnecroinflammatroy pathway of alcohol-induced pancreatic fibrogenesis. According to this concept, ethanol-induced acinar cell injury may lead to the generation of proinflammatory cytokines that subsequently activate pancreatic stellate cells (necroinflammatory pathway). In addition, ethanol and acetaldehyde may directly activate pancreatic stellate cells resulting in the development of pancreatic fibrosis (nonnecroinflammatory pathway).

15.5 EPILOGUE

The exact mechanisms underlying alcoholic chronic pancreatitis are not yet clarified. It is likely that several mechanisms act together and increase the risk to develop alcoholic chronic pancreatitis. Alcohol directly damages the pancreas through the effects of toxic ethanol metabolites and possibly by limiting pancreatic regeneration. Acute ethanol administration selectively reduces pancreatic blood-flow and microcirculation. These effects might also be important for the progression of underlying chronic pancreatitis. Chronic ethanol consumption significantly decreases the pancreatic bicarbonate secretion. In addition, chronic ethanol consumption increases the basal pancreatic enzyme output, the protein concentration, and the viscosity of the pancreatic juice. Thus, changes in pancreatic exocrine secretion patterns may contribute to the development of pancreatic damage. Chronic ethanol administration increases pancreatic acinar cell expression of digestive and lysosomal enzymes, increases the glandular content of these enzymes, and may facilitate the activation of pancreatic enzymes. Chronic ethanol administration leads to an enhanced ratio of trypsinogen to pancreatic secretory trypsin inhibitor. This distortion of the normal ratio in favor of trypsinogen may facilitate premature activation of pancreatic proenzymes within the pancreas thereby rendering the pancreas more susceptible to an enzyme-related injury. In animal models, obstruction of the pancreatic duct results in morphological changes similar to obstructive human chronic pancreatitis, and further alcohol application suppresses pancreatic regeneration. The development of pancreatic duct plugs within the course of alcoholic chronic pancreatitis likely contributes to the progression of the disease. Pancreatic stellate cells represent the main source of extracellular matrix in pancreatic fibrosis. These cells are activated by growth factors, inflammatory cytokines, alcohol, its metabolite acetaldehyde, and oxidative stress. Future research efforts must identify the mechanisms of progression toward chronic pancreatitis and must establish more effective means of early detection, early diagnosis, and early intervention.

REFERENCES

1. Gastard J., Joubaud F., Farbos T., Loussouarn J., Marion J., Pannier M., Renaudet F., Valdazo R., and Gosselin M. 1973. Etiology and course of primary chronic pancreatitis in Western France. *Digestion* 9:416–428.
2. Durbec J.P. and Sarles H. 1978. Multicenter survey of the aetiology of pancreatic diseases, relationship between the relative risk of developing chronic pancreatitis and alcohol, protein and lipid consumption. *Digestion* 18:337–350.
3. Goebell H., Hotz J., and Hoffmeister H. 1980. Hypercaloric nutrition as aetiological factor in chronic pancreatitis. *Zeitschrift für Gastroenterologie* 18:94–97.
4. Copenhagen Pancreatitis Study. 1981. An interim report from a prospective multicentre study. *Scand. J. Gastroenterol.* 16:305–312.
5. Dani R., Penna F.J., and Nogueira C.E. 1986. Etiology of chronic calcifying pancreatitis in Brazil: a report of 329 consecutive cases. *Intl. J. Pancreatol.* 1:399–406.
6. Layer P., Yamamoto H., Kalthoff L., Clain J.E., Bakken L.J., and DiMagno E.P. 1994. The different courses of early and late-onset idiopathic and alcoholic chronic pancreatitis. *Gastroenterology* 107:1481–1487.

7. Singer M.V. and Müller M.K. 1995. Epidemiologie, Ätiologie und Pathogenese der chronischen Pankreatitis. In *Erkrankungen des exkretorischen Pankreas*. Mössner J., Adler G., Fölsch U.R., Singer M.V. Eds. Jena, Germany: Gustav Fischer Verlag, pp. 313–324.

8. Worning H. 1998. Alcoholic chronic pancreatitis. In *The Pancreas*. Beger H.G., Warshaw A.L., Buchler M.W. et al. Eds. Malden, MA: Blackwell Science, pp. 672–682.

9. Suda K., Shiotsu H., Nakamura T., Akai J., and Nakamura T. 1994. Pancreatic fibrosis in patients with chronic alcohol abuse: correlation with alcoholic pancreatitis. *Am. J. Gastroenterol.* 89:2060–2062.

10. Renner I.G., Savage W.T., Pantoja J.L., and Renner V.J. 1985. Death due to acute pancreatitis: a retrospective analysis of 405 autopsy cases. *Dig. Dis. Sci.* 30:1005–1018.

11. Wilson J.S., Bernstein L., McDonald C., Tait A., McNeil D., and Pirola R.C. 1985. Diet and drinking habits in relation to the development of alcoholic pancreatitis. *Gut* 26:882–887.

12. Stigendal L. and Olsson R. 1984. Alcohol consumption pattern and serum lipids in alcoholic cirrhosis and pancreatitis. A comparative study. *Scand. J. Gastro-enterol.* 19:582–587.

13. Marks I.N. and Bank S. 1963. The etiology, clinical features and diagnosis of pancreatitis in the South Western Cape: a review of 243 cases. *South Afr. Med. J.* 37:1039–1053.

14. Ammann R.W. 1997. A clinically based classification system for alcoholic chronic pancreatitis: Summary of an international workshop on chronic pancreatitis. *Pancreas* 14:215–221.

15. Ammann R.W., Muellhaupt B., Meyenberger C., and Heitz P.U. 1994. Alcoholic nonprogressive chronic pancreatitis: prospective long-term study of a large cohort with alcoholic acute pancreatitis (1976–1992). *Pancreas* 9:365–373.

16. Ammann R., Heitz P., and Klöppel G. 1996. Course of alcoholic chronic pancreatitis: a prospective clinicomorphological long-term study. *Gastroenter-ology* 111:224–231.

17. Lankisch P.G., Seidensticker F., Lohr-Happe A., Otto J., and Creutzfeldt W. 1995. The course of pain is the same in alcohol- and nonalcohol-induced chronic pancreatitis. *Pancreas* 10:338–341.

18. Apte M.V. and Wilson J.S. 2003. Alcohol-induced pancreatic injury, *Best Pract. Res. Clin. Gastroenterol.* 17:593–612.

19. Ammann R.W. and Muellhaupt B. 1994. Progression of alcoholic acute to chronic pancreatitis. *Gut* 35:552–556.

20. Gullo L., Barbara L., and Labo G. 1988. Effect of cessation of alcohol use on the course of pancreatic dysfunction in alcoholic pancreatitis. *Gastroenterology* 95:1063–1068.

21. Strum W.B. 1995. Abstinence in alcoholic chronic pancreatitis. Effect on pain and outcome. *J. Clin. Gastroenterol.* 20:37–41.

22. Hanck C. and Singer M.V. 1997. Does acute alcoholic pancreatitis exist without preexisting chronic pancreatitis? *Scand. J. Gastroenterol.* 32:625–626.

23. Pitchumoni C.S., Glasser M., Saran R.M., Panchacharam P., and Thelmo W. 1984. Pancreatic fibrosis in chronic alcoholics and nonalcoholics without clinical pancreatitis. *Am. J. Gastroenterol.* 79:382–388.

24. Klöppel G. and Maillet B. 1991. Chronic pancreatitis: evolution of the disease. *Hepatogastroenterology* 38:408–412.

25. Whitcomb D.C., Gorry M.C., Preston R.A., Furey W., Sossenheimer M.J., Ulrich C.D., Martin S.P., Gates L.K., Jr, Amann S.T., Toskes P.P., Liddle R., McGrath K., Uomo G., Post J.C., and Ehrlich G.D. 1996. Hereditary pancreatitis is caused by a mutation in the cationic trypsinogen gene. Nat. Genet. 14:141–145.

26. Schneider A., Whitcomb D.C., and Singer M.V. 2002. Animal models in alcoholic pancreatitis — what can we learn? *Pancreatology* 2:189–203.

27. Sparmann G., Behrend S., Merkord J., Kleine H.D., Graser E., Ritter T., Liebe S., and Emmrich J. 2001. Cytokine mRNA levels and lymphocyte infiltration in pancreatic tissue during experimental chronic pancreatitis induced by dibutyltin dichloride. *Dig. Dis. Sci.* 46:1647–1656.

28. Lowenfels A.B., Maisonneuve P., Cavallini G., Ammann R.W., Lankisch P.G., Andersen J.R., DiMagno E.P., Andren-Sandberg A., Domellof L., and Di Francesco V. 1994. Prognosis of chronic pancreatitis: an international multi-center study. International Pancreatitis Study Group. *Am. J. Gastroenterol.* 89:1467–1471.

29. Ammann R.W., Akovbiantz A., Largiader F., and Schueler G. 1984. Course and outcome of chronic pancreatitis. Longitudinal study of a mixed medical-surgical series of 245 patients. *Gastroenterology* 86:820–828.

30. Niebergall-Roth E., Harder H., and Singer M.V. 1998. A review: acute and chronic effects of ethanol and alcoholic beverages on the pancreatic exocrine secretion in vivo and in vitro. *Alcoholism Clin. Exp. Res.* 22:1570–1583.

31. Demol P., Singer M.V., Hotz J., Hoffmann U., Hanssen L.E., Eysselein V.E., and Goebell H. 1986. Action of intragastric ethanol on pancreatic exocrine secretion in relation to the interdigestive gastrointestinal motility in humans. *Arch. Intl. Physiol. Biochimie* 94:251–259.

32. Tiscornia O.M., Gullo L., Sarles H., Mott C.B., Brasca A., Devaux M.A., Palasciano G., and Hage G. 1974. Effects of intragastric and intraduodenal ethanol on canine exocrine pancreatic secretion. *Digestion* 10:52–60.

33. Tsai J., Kobari M., Takeda K., Miyashita E., Rahman M.M., and Matsuno S. 1991. Changes of duodenal pH and pancreatic exocrine function after upper G-I intraluminal ethanol administration. *Tohoku J. Exp. Med.* 164:81–91.

34. Llanos O.L., Swierczek J.S., Teichmann R.K., Rayford P.L., and Thompson J.C. 1977. Effect of alcohol on the release of secretin and pancreatic secretion. *Surgery* 81:661–667.

35. Wheatley I.C., Barbezat G.O., Hickman R., and Terblanche J. 1975. The effect of acute ethanol administration on the exocrine pancreatic secretion of the pig. *Br. J. Surg.* 62:707–712.

36. Mott C., Sarles H., Tiscornia O., and Gullo L. 1972. Inhibitory action of alcohol on human exocrine pancreatic secretion. *Am. J. Dig. Dis. Sci.* 17:902–910.

37. Marin G.A., Ward N.L., and Fischer R. 1973. Effect of ethanol on pancreatic and biliary secretions in humans. *Am. J. Dig. Dis. Sci.* 18:825–833.

38. Chari S.T., Harder H., Teyssen S., Knodel C., Riepl R.L., and Singer M.V. 1996. Effect of beer, yeast-fermented glucose, and ethanol on pancreatic enzyme secretion in healthy human subjects. *Dig. Dis. Sci.* 6:1216–1224.

39. Hajnal F., Flores M.C., and Valenzuela J.E. 1989. Effect of alcohol and alcoholic beverages on nonstimulated pancreatic secretion in humans. *Pancreas* 4:486–491.

40. Brooks F.P. and Thomas J.E. 1953. The effect of alcohol on canine external pancreatic secretion. *Gastroenterology* 23:36–39.

41. Tiscornia O.M., Hage G., Palasciano G., Brasca A.P., Devaux M.A., and Sarles H. 1973. The effects of pentolinium and vagotomy on the inhibition of canine exocrine pancreatic secretion by intravenous ethanol. *Biomedicine* 18:159–163.

42. Hajnal F., Flores M., Radley S., and Valenzuela J.E. 1990. Effect of alcohol and alcoholic beverages on meal-stimulated pancreatic secretion in humans. *Gastroenterology* 98:191–196.

43. Jian R., Cortot A., Ducrot F., Jobin G., Chayvialle J.A., and Modigliani R. 1986. Effect of ethanol ingestion on postprandial gastric emptying and secretion, biliopancreatic secretions, and duodenal absorption in man. *Dig. Dis. Sci.* 31:604–614.

44. Alonso R., Alvarez C., Garcia L.J., Calvo J.J., and Lopez M.A. 1992. Cholincergic mechanisms involved in the effect of intraduodenal ethanol on exocrine pancreatic secretion. *Zeitschrift für Gastroenterologie* 30:117–120.

45. Tiscornia O.M., Brasca A.F., Hage G., Palasciano G., Devaux M.A., and Sarles H. 1972. Les effets de látropine, du penthonium, de la vagotomie et du fluothane sur la sécrétion pancréatique du chien. *Biologie et Gastroenterologie* 5:249–256.

46. Kölbel C.B.M., Singer M.V., Möhle T., Heinzel C., Eysselein V., and Goebell H. 1986. Action of intravenous ethanol and atropine on the secretion of gastric acid, pancreatic enzymes, and bile acids and the motility of the upper gastroduodenal tract in nonalcoholic humans. *Pancreas* 1:211–218.

47. Uhlemann E.R., Robberecht P., and Gardner J.D. 1979. Effects of alcohols on the actions of VIP and secretin on acinar cells from guinea pig pancreas. *Gastroenterology* 76:917–925.

48. Ponnappa B.C., Hoek J.B., Waring A.J., and Rubin E. 1987. Effect of ethanol on amylase secretion and cellular calcium homeostasis in pancreatic acini from normal and ethanol-fed rats. *Biochemical Pharmacology* 36:69–79.

49. Nakamura T., Okabayashi Y., Fujii M., Tani S., Fujisawa T., and Otsuki M. 1991. Effect of ethanol on pancreatic exocrine secretion in rats. *Pancreas* 6:571–577.

50. Planche N.E., Palasciano G., Meullenet J., Laugier R., and Sarles H. 1982. Effect of intravenous alcohol on pancreatic and biliary secretion in man. *Dig. Dis. Sci.* 27:449–453.

51. Renner I.G., Rinderknecht H., Valenzuela J.E., and Douglas A.P. 1980. Studies of pure pancreatic secretions in chronic alcoholics without pancreatic insufficiency. *Scand. J. Gastroenterol.* 15:41–44.

52. Sahel J. and Sarles H. 1979. Modifications of pure human pancreatic juice induced by chronic alcohol consumption. *Dig. Dis. Sci.* 24:897–905.

53. Brugge W.R., Burke C.A., Brand D.L., and Chey W.Y. 1985. Increased interdigestive pancreatic trypsin secretion in alcoholic pancreatic disease. *Dig. Dis. Sci.* 30:431–439.

54. Harada H., Takeda M., Yabe H., Hanafusa E., Hayashi T., Kunichika K., Kochi F., Mishima K., Kimura I., and Ubuga T. 1980. The hexosamine concentration and output in human pure pancreatic juice in chronic pancreatitis. *Gastroenterologia Japonica* 15:520–526.

55. Singer M.V. and Goebell H. 1985. Acute and chronic actions of alcohol on pancreatic exocrine secretion in humans and animals. In *Alcohol releated diseases in gastroenterology*. Seitz H.K. and Kommerell B. Eds. New York: Springer. pp. 376–414.

56. Singh M. and Simsek H. 1990. Ethanol and the pancreas. Current status. *Gastroenterology* 98:1051–1062.

57. Rinderknecht H., Stace N.H., and Renner I.G. 1985. Effects of chronic alcohol abuse on exocrine pancreatic secretion in man. *Dig. Dis. Sci.* 30:65–71.

58. Apte M.V., Norten I., Haber P., Applegate T., Korsten M., McCaughan G., Pirola R., and Wilson J.S. 1998. The effect of ethanol on pancreatic enzymes — a dietary artefact? *Biochim. Biophys. Acta.* 1379:314–324.

59. Haber P.S., Apte M.V., Applegate T.L., Norton I.D., Korsten M.A., Pirola R.C., and Wilson J.S. 1998. Metabolism of ethanol by rat pancreatic acinar cells. *J. Lab. Clin. Med.* 132:294–302.

60. Gukovskaya A.S., Mouria M., Gukovsky I., Reyes C.N., Kasho V.N., Faller L.D., and Pandol J. 2002. Ethanol metabolism and transcription factor activation in pancreatic acinar cells in rats. *Gastroenterology* 122:106–118.

61. Majumdar A.P., Vesenka G.D., Dubick M.A., Yu G.S., DeMorrow J.M., and Geokas M.C. 1986. Morphological and biochemical changes of the pancreas in rats treated with acetaldehyde. *Am. J. Physiol.* 250:G598–G606.

62. Nordback I.H., MacGowan S., Potter J.J., and Cameron J.L. 1991. The role of acetaldehyde in the pathogenesis of acute alcoholic pancreatitis. *Ann. Surg.* 214:671–678.

63. Foster J.R., Idle J.R., Hardwick J.P, Bars R., Scott P., and Braganza J.M. 1993. Induction of drug-metabolizing enzymes in human pancreatic cancer and chronic pancreatitis. *J. Pathol.* 169:457–463.

64. Norton I.D., Apte M.V., Haber P.S., McCaughan G.W., Pirola R.C., and Wilson J.S. 1998. Cytochrome P4502E1 is present in rat pancreas and is induced by chronic ethanol administration. *Gut* 42:426–430.

65. Laposata E.A. and Lange L.G. 1986. Presence of nonoxidative ethanol metabolism in human organs commonly damaged by ethanol abuse. *Science* 31(231):497–499.

66. Lange L.G. and Sobel B.E. 1983. Myocardial metabolites of ethanol. *Circ. Res.* 52:479–482.

67. Werner J., Laposata M., Fernandez-del Castillo C., Saghir M., Iozzo R.V., Lewandrowski K.B., and Warshaw A.L. 1997. Pancreatic injury in rats induced by fatty acid ethyl ester, a nonoxidative metabolite of alcohol. *Gastroenterology* 113:286–294.

68. Chance B., Sies H., and Boveris A. 1979. Hydroperoxide metabolism in mammalian organs. *Physiol. Rev.* 59:527–605.

69. Braganza J.M. 1998. A framework for the aetiogenesis of chronic pancreatitis. *Digestion* 59(Suppl. 4):1–12.

70. Schoenberg M.H., Buchler M., Pietrzyk C., Uhl W., Birk D., Eisele S., Marzinzig M., and Beger H.G. 1995. Lipid peroxidation and glutathione metabolism in chronic pancreatitis. *Pancreas* 10:36–43.

71. Mathew P., Wyllie R., Van Lente F., Steffen R.M., and Kay M.H. 1996. Antioxidants in hereditary pancreatitis. *Am. J. Gastroenterol.* 91:1558–1562.

72. Morris-Stiff G.J., Bowrey D.J., Oleesky D., Davies M., Clark G.W., and Puntis M.C. 1999. The antioxidant profiles of patients with recurrent acute and chronic pancreatitis. *Am. J. Gastroenterol.* 94:2135–2140.

73. Altomare E., Grattagliano I., Vendemiale G., Palmieri V., and Palasciano G. 1996. Acute ethanol administration induces oxidative changes in rat pancreatic tissue. *Gut* 38:742–746.

74. Grattagliano I., Palmieri V., Vendemiale G., Portincasa P., Altomare E., and Palasciano G. 1999. Chronic ethanol administration induces oxidative alterations and functional impairment of pancreatic mitochondria in the rat. *Digestion* 60:549–553.

75. Wilson J.S., Somer J.B., and Pirola R.C. 1984. Chronic ethanol feeding causes accumulation of serum cholesterol in rat pancreas. *Exp. Mol. Pathol.* 41:289–297.

76. Simsek H. and Singh M. 1990. Effect of prolonged ethanol intake on pancreatic lipids in the rat pancreas. *Pancreas* 5:401–407.

77. Zeki S., Miura S., Suzuki H., Watanabe N., Adachi M., Yokoyama H., Horie Y., Saito H., Kato S., and Ishii H. 2002. Xanthine oxidase-derived oxygen radicals play significant roles in the development of chronic pancreatitis in WBN/Kob rats. *J. Gastroenterol. Hepatol.* 17:606–616.

78. Gukovskaya A.S., Vaquero E., Zaninovic V., Gorelick F.S., Lusis A.J., Brennan M.L., Holland S., and Pandol S.J. 2002. Neutrophils and NADPH oxidase mediate intrapancreatic trypsin activation in murine experimental acute pancreatitis. *Gastroenterology* 122:974–984.

79. Seo J.Y., Kim H., Seo J.T., and Kim K.H. 2002. Oxidative stress induced cytokine production in isolated rat pancreatic acinar cells: effects of small-molecule antioxidants. *Pharmacology* 64:63–70.

80. Horwitz L.D. and Myers J.H. 1982. Ethanol-induced alterations in pancreatic blood flow in conscious dogs. *Circ. Res.* 50:250–256.

81. Foitzik T., Fernandez-del Castillo C., Rattner D.W., Klar E., and Warshaw A.L. 1995. Alcohol selectively impairs oxygenation of the pancreas. *Arch. Surg.* 130:357–360.

82. Foitzik T., Hotz H.G., Hot B., Kirchengast M., and Buhr H.J. 1998. Endothelin-1 mediates the alcohol-induced reduction of pancreatic capillary blood flow. *J. Gastrointestinal Surg.* 2:379–384.

83. Eibl G., Hotz H.G., Faulhaber J., Kirchengast M., Buhr H.J., and Foitzik T. 2000. Effect of endothelin and endothelin receptor blockade on capillary permeability in experimental pancreatitis. *Gut* 46:390–394.

84. McKim S.E., Uesugi T., Raleigh J.A., McClain C.J., and Arteel G.E. 2003. Chronic intragastric alcohol exposure causes hypoxia and oxidative stress in the rat pancreas. *Arch. Biochem. Biophys.* 417:34–43.

85. Schilling M.K., Redaelli C., Reber P.U., Friess H., Signer C., Stoupis C., and Buchler M.W. 1999. Microcirculation in chronic alcoholic pancreatitis: a laser Doppler flow study. *Pancreas* 19:21–25.

86. Toyama M.T., Patel A.G., Nguyen T., Ashley S.W., and Reber H.A. 1997. Effect of ethanol on pancreatic interstitial pH and blood flow in cats with chronic pancreatitis. *Ann. Surg.* 225:223–228.

87. Hartwig W., Werner J., Ryschich E., Mayer H., Schmidt J., Gebhard M.M., Herfarth C., and Klar E. 2000. Cigarette smoke enhances ethanol-induced pancreatic injury. *Pancreas* 21:272–278.

88. Ramo O.J., Jalovaara P., and Apaja-Sarkkinen M. 1987. Fat-rich diet induces inflammatory changes in the intact rat pancreas. *Histol. Histopathol.* 2:329–332.

89. Tsukamoto H., Towner S.J., Yu G.S., and French S.W. 1988. Potentiation of ethanol-induced pancreatic injury by dietary fat. *Am. J. Pathol.* 131:246–257.

90. Kono H., Nakagami M., Rusyn I., Connor H.D., Stefanovic B., Brenner D.A., Mason R.P., Arteel G.E., and Thurman R.G. 2001. Development of an animal model of chronic alcohol-induced pancreatitis in the rat. *Am. J. Physiol. Gastrointestinal Liver Physiol.* 280:G1178–G1186.

91. Sarles H., Sarles J.C., Camatte R., Muratore R., Gaini M., Guien C., Pastor J., and Le Roy F. 1965. Observations on 205 confirmed cases of acute pancreatitis, recurring pancreatitis, and chronic pancreatitis. *Gut* 6:545–559.

92. Pitchumoni C.S., Sonnenshein M., Candido F.M., Panchacharam P., and Cooperman J.M. 1980. Nutrition in the pathogenesis of alcoholic pancreatitis. *Am. J. Clin. Nutr.* 33:631–636.

93. Bachem M.G., Schneider E., Gross H., Weidenbach H., Schmid R.M., Menke A., Siech M., Beger H., Grunert A., and Adler G. 1998. Identification, culture, and characterization of pancreatic stellate cells in rats and humans. *Gastroenterology* 115:421–432.

94. Apte M.V., Haber P.S., Applegate T.L., Norton I.D., McCaughan G.W., Korsten M.A., Pirola R.C., and Wilson J.S. 1998. Periacinar stellate shaped cells in rat pancreas: identification, isolation, and culture. *Gut* 43:128–133.

95. Casini A., Galli A., Pignalosa P., Frulloni L., Grappone C., Milani S., Pederzoli P., Cavallini G., and Surrenti C. 2000. Collagen Type I synthesized by pancreatic periacinar stellate cells (PSC) co-localizes with lipid peroxidation-derived aldehydes in chronic alcoholic pancreatitis. *J. Pathol.* 192:81–89.

96. Apte M.V., Phillips P.A., Fahmy R.G., Darby S.J., Rodgers S.C., McCaughan G.W., Korsten M.A., Pirola R.C., Naidoo D., and Wilson J.S. 2000. Does alcohol directly stimulate pancreatic fibrogenesis? Studies with rat pancreatic stellate cells. *Gastroenterology* 118:780–794.

16

CHRONIC PANCREATITIS:
THE INITIAL TISSUE DAMAGE
AND THE RESULTING
FIBROSIS PATTERNS

Günter Klöppel

CONTENTS

16.1   INTRODUCTION

Fibrosis in the pancreas is caused by processes such as primary chronic pancreatitis or pancreatic duct obstruction by neoplastic (e.g., pancreatic ductal adenocarcinoma) or nonneoplastic lesions. Its net result is always the formation of extracellular matrix (ECM) in the interstitial spaces and the acinar tissue, which may lead to replacement of the acinar cells and changes in the structure of the ducts and the arrangement and compo-

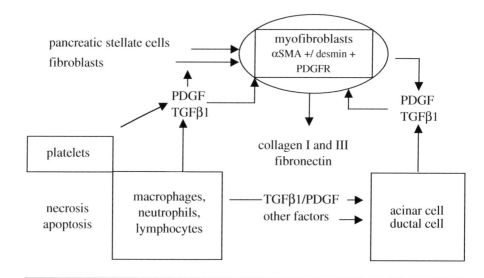

Figure 16.1   Mechanisms of pancreatic fibrosis.

sition of the islets. The initial event that induces fibrogenesis in the pancreas is an injury to certain cellular elements such as the interstitial mesenchymal cells, the duct cells, or the acinar cells. Damage to one of these tissue compartments of the pancreas is associated with transformation of fibroblasts and pancreatic stellate cells into myofibroblasts and the subsequent production and deposition of collagen Type I and Type III and fibronectin.[1–4]

The exact mechanisms initiating and maintaining the development of fibrosis in the pancreas are poorly understood, but may be viewed as a three-step process. The initial event is damage to one or all of the various tissue compartments or cell types of the pancreas leading to cell necrosis or apoptosis and subsequent release of cytokines (e.g., TGFβ1, PDGF) and C-C or C-X-C chemokines, either from immigrating inflammatory cells, especially macrophages, or nearby preexistent epithelial or mesenchymal cells.[5–8] In a second step, the damaged cells are phagocytosed by macrophages and the released cytokines cause proliferation of fibroblasts in the immediate vicinity of the original site of damage and induce them to transform into myofibroblasts or, as they are called today, activated pancreatic stellate cells (Figure 16.1).[9,10] In the last phase, fibroblasts and myofibroblasts produce and deposit ECM, which replaces the inflammatory infiltrate and affects the arrangement of the surviving pancreatic tissues.[11] At the end of this injurious process, the pancreas shows various patterns of fibrosis, which can be described as inter(peri)lobular, periductal, and intralobular types.[12–15] This article presents a working hypothesis that

Table 16.1 Etiologic Classification of Chronic Pancreatitis and Pancreatic Fibrosis

Chronic pancreatitis[a]
  Alcoholic
  Nonalcoholic
    Hereditary
    Metabolic (hypercalcemia, hyperlipidemia)
    Autoimmune
    Idiopathic
    Tropical
    Other forms
    Chronic pancreatitis associated with anatomic abnormalities[a]
    Obstructive chronic pancreatitis[b]
    Periampullary duodenal wall cysts
    Pancreas divisum
    Posttraumatic pancreatic scars
Pancreatic fibrosis not associated with chronic pancreatitis[a]
  Pancreatic fibrosis in the elderly
  Cystic fibrosis[c]
  Insulin-dependent diabetes mellitus
  Hemochromatosis

[a] Clinically defined by the occurrence of symptoms such as pain, pancreatic exocrine deficiency, and diabetes.

[b] Usually only associated with pancreatic insufficiency.

[c] Associated with pancreatic insufficiency.

Source: From Klöppel 1998.[17] Used with permission.

connects the etiologic type of chronic pancreatitis (i.e., alcoholic, hereditary, autoimmune) with the initial site of damage in the pancreas and the resulting pattern of fibrosis.

16.2 TYPE OF PANCREATITIS, INITIAL DAMAGE IN THE PANCREAS, AND FIBROSIS PATTERN

The etiological classification of chronic pancreatitis distinguishes between alcoholic, hereditary, autoimmune, idiopathic, tropical, and other rare types (Table 16.1).[16,17] Here the mechanisms that may lead to certain patterns of fibrosis in the pancreas will be discussed for alcoholic chronic pancreatitis, hereditary pancreatitis, and autoimmune pancreatitis.

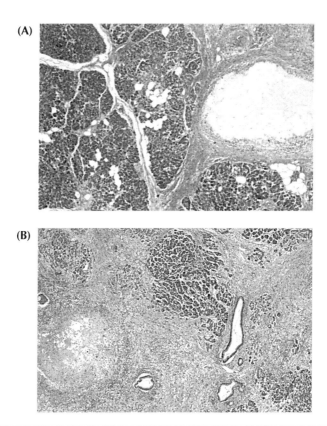

Figure 16.2 Development of fibrosis in relapsing severe acute pancreatitis caused by alcohol abuse. A: Interlobular fatty tissue necrosis demarcated from the adjacent preserved acinar tissue by inflammatory cells and a fibrotic band. B: Cellular interlobular fibrosis developing adjacent to an area of old autodigestive fatty tissue necrosis (bottom left).

16.2.1 Alcoholic Chronic Pancreatitis

It has been shown that the initial event in alcoholic pancreatitis is focal autodigestive necrosis of interstitial fat cells, which may involve adjacent vessels, especially veins, and other tissue compartments of the pancreas.[18] The inflammatory reaction and the hemorrhaging that follows the tissue necrosis initiate fibrosis (Figure 16.2).[19] As discussed above, the most important known mediators of fibrosis are cytokines. The cells from which fibrogenic cytokines such as transforming growth factor (TGF) β1 and platelet derived growth factor (PDGF)[6,7] may be released are macrophages and other inflammatory cells, which accumulate in large numbers around the necrotic areas, and platelets in areas of hemorrhagic necrosis. The cytokines generated by these cells mainly exert their paracrine effects on

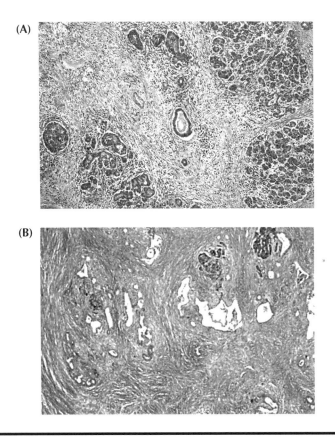

Figure 16.3   Fibrosis pattern in alcoholic chronic pancreatitis. A: Early stage of developing interlobular fibrosis, characterized by high cellularity. B: Late stage of fibrosis involving the ducts.

neighboring fibroblasts, but could also cause the surrounding ductal and acinar cells to produce and release further cytokines that affect fibroblasts. The latter assumption is supported by immunocytochemical studies demonstrating that TGFβ1, fibroblast growth factor (FGF) 1 and FGF2, and other factors are not only produced by fibroblasts, but also by epithelial cells of the pancreas.[6,9,20] During this process, many of the fibroblastic cells are transformed into myofibroblastic cells expressing smooth muscle actin.[15] This concerted action then leads to a local fibrogenic response at the site of the initial necrosis. Because the foci of autodigestive necrosis lie in the interstitial tissue in the vicinity of the acinar lobules and the interlobular ducts, the resulting fibrosis shows an inter(peri)lobular pattern (Figure 16.3).[13,18] If the interlobular fibrosis involves the interlobular duct, it gradually causes sacculations and strictures of the duct system. These ductal lesions may hamper the flow of pancreatic secretions and promote

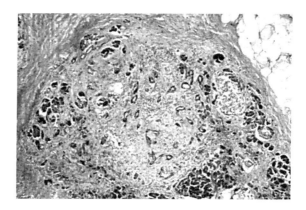

Figure 16.4   Advanced inter- and intralobular fibrosis in alcoholic chronic pancreatitis: remnants of acinar cells, ducts, and islets.

precipitation of proteins, with eventual calcification, as was experimentally demonstrated in dogs by partial pancreatic duct ligation for 4 months and longer.[21] Another effect of duct obstruction is the induction of acinar atrophy and intralobular fibrosis (Figure 16.3). When this occurs, the pancreas shows areas of diffuse fibrosis that only contain remnants of ducts and acini and irregularly distributed islets (Figure 16.4).

16.2.2   Hereditary Chronic Pancreatitis

It has recently been shown that the genetic alterations in hereditary chronic pancreatitis involve the cationic trypsinogen gene (*PRSS1*) or a serine protease inhibitor gene (*SPINK1*).[22–24] Mutations in these two genes seem to trigger the autoactivation of trypsinogen in the pancreas, which probably results in the early inappropriate conversion of pancreatic zymogens to active enzymes in pancreatic ducts with subsequent auto-digestive necrosis and inflammation. The site of the autodigestive necrosis may be either the acinar cell or, more likely, the duct wall. In our series of pancreatic resection specimens from six patients with hereditary chronic pancreatitis (G.K., unpublished observations), we found advanced chronic pancreatitis with severe periductal fibrosis and massively dilated ducts containing protein plugs and calculi staining for pancreatic stone protein (PSP). In one case, there was ductal necrosis in a number of medium-sized interlobular ducts that destroyed the duct epithelium and involved the periductal area (Figure 16.5A). This change, which we have also seen in a small number of rare cases of acute pancreatitis in patients after prolonged shock,[25] suggests that an autodigestive process may occur in the duct lumen, resulting in the necrosis of the lining duct cells and

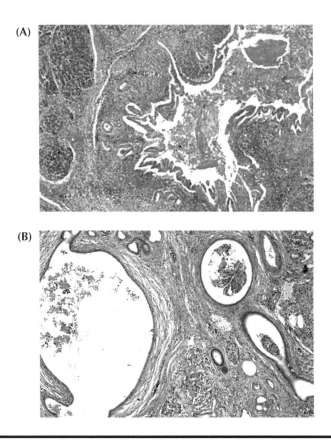

Figure 16.5 Hereditary pancreatitis. A: Interlobular duct showing granulocytes, epithelial destruction, and inflammatory infiltration of the duct wall. B: Dilated interlobular ducts with calculus fragments and periductal fibrosis.

the surrounding interstitial tissue. If we hypothesize that this autodigestive process remains more or less restricted to the duct system, it may induce dilatation of the involved duct segments, periductal fibrosis, and, to a lesser extent, intralobular fibrosis (Figure 16.5B).

16.2.3 Autoimmune (Duct Destructive) Chronic Pancreatitis

Recent studies have suggested that there is a special type of chronic pancreatitis characterized by lymphoplasmacellular infiltration of the pancreas with conspicuous involvement of the ducts and with autoimmune features.[26–31] Patients with this type of pancreatitis usually present with jaundice or abdominal pain, enlargement of the pancreas (usually the pancreas head), and an irregular narrowing of the main pancreatic duct and/or the bile duct. Most patients lack the symptoms of attacks of acute

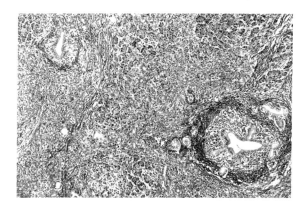

Figure 16.6 Autoimmune chronic pancreatitis: interlobular duct narrowed by a fibroinflammatory process that also involved the surrounding parenchyma.

pancreatitis and fail to show calcifications or pseudocysts. Histologically, the pancreatic tissue displays lymphoplasmacytic infiltration and fibrosis, which focus on medium-sized ducts, causing duct narrowing and occasionally destruction (Figure 16.6). In addition to pancreatitis, autoimmune disorders such as Sjögren's syndrome or ulcerative colitis may occur synchronously or metachronously. The pattern of inflammation in the pancreas in patients with autoimmune pancreatitis is so unique that it can be clearly distinguished from other types of chronic pancreatitis, such as alcoholic chronic pancreatitis[27] or hereditary chronic pancreatitis. Apart from the ducts, probably in a second step, the chronic inflammation also affects the acinar tissue leading to the disappearance of acinar cells. This course of the inflammation first results in periductal and interlobular fibrosis that is accompanied by intralobular fibrosis (Figure 16.6). At the end of this process, the affected region of the pancreas shows diffuse fibrosis (Figure 16.6). In some cases, the fibrosis may exhibit a myofibroblastic storiform pattern characterized by the presence of myofibroblasts, scattered lymphocytes and plasma cells, and dense collagen bundles. If such a fibrotic lesion is large, it may form an inflammatory pseudotumor.[29] Such lesions have been reported in the pancreatic head, but no connection with autoimmune pancreatitis was discussed. However, judging from the descriptions and illustrations of these cases, the lesions appear to be compatible with those seen in autoimmune pancreatitis. Therefore, the reported inflammatory pseudotumors of the pancreas might represent an advanced stage of autoimmune pancreatitis in which the fibrotic changes predominate.

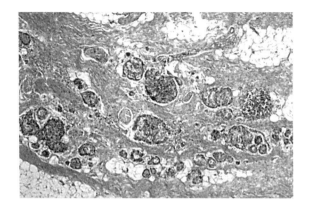

Figure 16.7   Obstructive chronic pancreatitis. End stage fibrosis with lobular aggregation of islets.

16.2.4   Other Types of Pancreatitis

As shown in Table 16.1, other types of pancreatitis include idiopathic chronic pancreatitis, tropical pancreatitis, chronic pancreatitis due to pancreas divisum or duodenal wall pancreatitis, and obstructive chronic pancreatitis.[17] With the exception of obstructive chronic pancreatitis, the morphologic substrate of the other types of pancreatitis has not yet been systematically studied.

16.2.5   Obstructive Chronic Pancreatitis

In obstructive chronic pancreatitis, there is a focal obstruction of the main pancreatic duct or one of the secondary ducts that lie in the interlobular spaces, which leads to ductal dilation proximal (upstream) to the stenosis, to atrophy of the acinar cells, and replacement by fibrous tissue (Figure 16.7). There are various possible causes for a duct obstruction, but the most important and common cause is ductal adenocarcinoma in the head of the pancreas occluding the main pancreatic duct. This process leads to a generalized involvement of the gland with interlobular fibrosis, which in long-standing cases is increasingly accompanied by intralobular fibrosis. Other causes include intraductal papillary-mucinous neoplasms, some cystic and endocrine neoplasms, acquired fibrous strictures of the pancreatic ducts, ductal papillary hyperplasia narrowing the duct lumen, and viscous mucin blocking the duct lumen.

The effects of all the listed duct obstructing mechanisms can be compared with those of duct ligation in the pancreas.[32,33] In the early

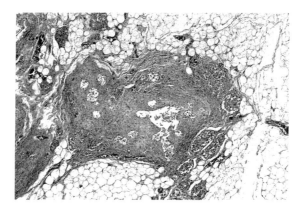

Figure 16.8   Advanced cystic fibrosis: diffuse fibrosis of the pancreas and lipomatosis.

phase after duct ligation, the acini are transformed into small ductal (tubular) complexes. In a next step, the acinar cells disappear, probably by apoptosis. These changes are associated with an inflammatory and fibrotic reaction involving numerous macrophages. The macrophages are the potential source of cytokines, which stimulate fibrogenesis by fibroblasts that acquire the properties of myofibroblasts. Because the inflammatory reaction takes place in all of the interlobular and intralobular areas of the pancreatic tissue that were once drained by the occluded duct, fibrosis develops in these regions at the same pace, producing interlobular and intralobular fibrosis in equal distribution.

A special situation of duct obstruction is encountered in cystic fibrosis of the pancreas and in pancreatic lobular fibrosis, which is frequently observed in elderly persons. Whereas the first condition causes duct obstruction due to clogging with viscous mucin, the second condition leads to narrowing of the duct lumen by papillary hyperplasia of the duct epithelium. In cystic fibrosis, complete or almost complete (inter- and intralobular) fibrosis develops slowly after birth,[34] after many years, fibrosis is replaced by fatty tissue (Figure 16.8),[35] a process that is not understood so far, but is of great interest for the resolution of fibrosis. In elderly persons the fibrosis that develops is mainly of the intralobular type and affects those lobes that are drained by ducts showing hyperplastic epithelial changes (Figure 16.9). The lobular fibrosis of the pancreas therefore shows a patchy pattern. Its degree varies from person to person; the most severe form and the highest incidence (up to 50%) are found in persons older than 60 years (Detlefsen, Feyerabend, and Klöppel, unpublished observations).

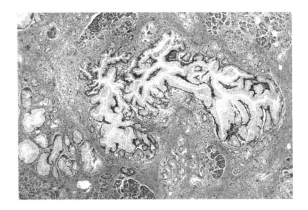

Figure 16.9 Lobular fibrosis of the pancreas. Interlobular duct showing mucinous hypertrophy of the epithelium and surrounded by fibrotic tissue.

Table 16.2 Pathogenesis of Pancreatic Fibrosis

Disease: chronic pancreatitis, duct obstruction
Causative factors: alcohol and other toxins, autoactivation of pancreatic
 enzymes, immune-mediated damage, intraductal pressure
Cell necrosis/apoptosis of mesenchymal or epithelial cells
Resorption of necrotic/apoptotic cells by macrophages
Induction of fibrosis by cytokines released by macrophages, platelets, and
 epithelial cells
Production of ECM by fibroblasts/stellate cells/myofibroblasts

16.3 CONCLUSION

Fibrosis of the pancreas is observed in primary chronic pancreatitis or duct obstructing processes. At its beginning, there is a tissue injury that induces a process of fibrogenesis that is similar to the development of fibrosis in other organs (Table 16.2). Most likely of importance for the pattern of fibrosis in the pancreas is the site where the process of fibrogenesis is initiated. It is our working hypothesis that this depends on the factors that cause necrosis or apoptosis of pancreatic cells. If it is primarily the interstitial mesenchymal cells that are injured, as in alcoholic pancreatitis, mainly perilobular fibrosis develops. If it is the duct epithelium and the duct wall that are injured, as in hereditary pancreatitis and autoimmune pancreatitis, the fibrosis mainly affects the periductal area. If it is the acinar cells that are injured, as in duct obstructing processes, mainly diffuse (inter- and intralobular) fibrosis results.

REFERENCES

1. Apte M.V., Haber P.S., Applegate T.L., Norton I.D., McCaughan G.W., Korsten M.A., Pirola R.C., and Wilson J.S. 1998. Periacinar stellate shaped cells in rat pancreas: identification, isolation, and culture. *Gut* 43:128–133.
2. Apte M.V., Haber P.S., Darby S.J., Rodgers S.C., McCaughan G.W., Korsten M.A., Pirola R.C., and Wilson J.S. 1999. Pancreatic stellate cells are activated by proinflammatory cytokines: implications for pancreatic fibrogenesis. *Gut* 44:534–541.
3. Gabbiani G. 2003. The myofibroblast in wound healing and fibrocontractive diseases. *J. Pathol.* 200:500–503.
4. Kuroda J., Suda K., and Hosokawa Y. 1998. Periacinar collagenization in patients with chronic alcoholism. *Pathol. Intl.* 48:857–868.
5. Andoh A., Takaya H., Saotome T., Shimada M., Hata K., Araki Y., Nakamura F., Shintani Y., Fujiyama Y., and Bamba T. 2000. Cytokine regulation of chemokine (IL-8, MCP-1, and RANTES) gene expression in human pancreatic periacinar myofibroblasts. *Gastroenterology* 119:211–219.
6. Ebert M., Kasper H.U., Hernberg S., Friess H., Büchler M.W., Roessner A., Korc M., and Malfertheiner P. 1998. Overexpression of platelet-derived growth factor (PDGF) B chain and type β PDGF receptor in human chronic pancreatitis. *Dig. Dis. Sci.* 43:567–574.
7. Luttenberger T., Schmid-Kotsas A., Menke A., Siech M., Beger H., Adler G., Grünert A., and Bachem M.G. 2000. Platelet-derived growth factors stimulate proliferation and extracellular matrixECM synthesis of pancreatic stellate cells: implication in pathogenesis of pancreas fibrosis. *Lab. Invest.* 80:47–55.
8. Saurer L., Reber P., Schaffner T., Büchler M.W., Buri C., Kappeler A., Walz A., Friess H., and Mueller C. 2000. Differential expression of chemokines in normal pancreas and in chronic pancreatitis. *Gastroenterology* 118:356–367.
9. Haber P.S., Keogh G.W., Apte M.V., Moran C.S., Stewart N.L., Crawford D.H.G., Pirola R.C., McCaughan G.W., Ramm G.A., and Wilson J.S. 1999. Activation of pancreatic stellate cells in human and experimental pancreatic fibrosis. *Am. J. Pathol.* 155:1087–1095.
10. Shek F.W.T., Benyon R.C., Walker F.M., McCrudden P.R., Pender S.L.F., Williams E.J., Johnson P.A., Johnson C.D., Bateman A.C., Fine D.R., and Iredale J.P. 2002. Expression of transforming growth factor-β1 by pancreatic stellate cells and its implications for matrix secretion and turnover in chronic pancreatitis. *Am. J. Pathol.* 160:1787–1798.
11. Izumi M., Suda K., Torii A., and Inadama E. 2001. Pancreatic ductal myofibroblasts. Proliferative patterns in various pathologic situations. *Virchows. Arch.* 438:442–450.
12. Ammann R.W., Heitz P.U., and Klöppel G. 1996. Course of alcoholic chronic pancreatitis: a prospective clinicomorphological long-term study. *Gastroenterology* 111:224–231.
13. Klöppel G. and Maillet B. 1992. The morphological basis for the evolution of acute pancreatitis into chronic pancreatitis. *Virchows Arch. A Pathol. Anat.* 420:1–4.
14. Martin E.D. 1984. Different pathomorphological aspects of pancreatic fibrosis, correlated with etiology: anatomical study of 300 cases. In *Pancreatitis: concepts and classification*. Gyr K.E., Singer M.V., and Sarles H. Eds. Amsterdam: Elsevier. pp. 77–82.

15. Suda K., Takase M., Takei K., Kumasaka T., and Suzuki F. 2000. Histopathologic and immunohistochemical studies on the mechanism of interlobular fibrosis of the pancreas. *Arch. Pathol. Lab. Med.* 124:1302–1305.
16. Etemad B. and Whitcomb D.C. 2001. Chronic pancreatitis: diagnosis, classification, and new genetic developments. *Gastroenterology* 120:682–707.
17. Klöppel G. 1998. Chronic pancreatitis. Etiology, pathophysiology, and pathology. In *Surgical diseases of the pancreas*. Howard J. et al. Eds. 3rd ed. Baltimore, MD: Williams & Wilkins. pp. 321–328.
18. Klöppel G. 1999. Progression from acute to chronic pancreatitis. A pathologist's view. *Surg. Clin. North Am.* 79:801–814.
19. Klöppel G. and Maillet B. 1991. Chronic pancreatitis: evolution of the disease. *Hepatogastroenterology* 38:408–412.
20. Friess H., Yamanaka Y., Büchler M., Beger H.G., Do D.A., Kobrin M.S., and Korc M. 1994. Increased expression of acidic and basic fibroblast growth factors in chronic pancreatitis. *Am. J. Pathol.* 144:117–128.
21. Konishi K., Izumi T., Kato O., Yamaguchi A., and Miyazaki I. 1981. Experimental pancreatolithiasis in the dog. *Surgery* 89:687–691.
22. Whitcomb D.C., Gorry M.C., Preston R.A., Furey W., Sossenheimer M.J., Ulrich C.D., Martin S.P., Gates Jr. L.K., Amann S.T., Toskes P.P., Liddle R., McGrath K., Uomo G., Post J.C., and Ehrlich G.D. 1996. Hereditary pancreatitis is caused by a mutation in the cationic trypsinogen gene. *Nat. Genet.* 14:141–145.
23. Witt H., Luck W., Hennies H.C., Classen M., Kage A., Lass U., Landt O., and Becker M. 2000. Mutations in the gene encoding the serine protease inhibitor, Kazal Type 1 are associated with chronic pancreatitis. *Nat. Genet.* 25:213–216.
24. Witt H. and Becker M. 2002. Genetics of chronic pancreatitis. *J. Pediatr. Gastroenterol. Nutr.* 34:125–136.
25. Klöppel G. 1997. Morphology of acute pancreatitis in relation to etiology and pathogenesis. In *Diagnostic procedures in pancreatic disease*. Malfertheiner P. et al. Eds. Berlin: Springer. pp. 13–20.
26. Abraham S.C., Leach S., Yeo C.J., Cameron J.L., Murakata L.A., Boitnott J.K., Albores-Saavedra J., and Hruban R.H. 2003. Eosinophilic pancreatitis and increased eosinophils in the pancreas. *Am. J. Surg. Pathol.* 27:334–342.
27. Ectors N., Maillet B., Aerts R., Geboes K., Donner A., Borchard F., Lankisch P., Stolte M., Lüttges J., Kremer B., and Klöppel G. 1997. Non-alcoholic duct destructive chronic pancreatitis. *Gut* 41:263–268.
28. Kawaguchi K., Koike M., Tsuruta K., Okamoto A., Tabata I., and Fujita N. 1991. Lymphoplasmacytic sclerosing pancreatitis with cholangitis: variant of primary sclerosing cholangitis extensively involving pancreas. *Hum. Pathol.* 22:387–395.
29. Klöppel G., Lüttges J., Löhr M., Zamboni G., and Longnecker D. 2003. Autoimmune pancreatitis: pathological, clinical, and imunological features. *Pancreas* 27:14–19.
30. Notohara K., Burgart L.J., Yadav D., Chari S., and Smyrk T.C. 2003. Idiopathic chronic pancreatitis with periductal lymphoplasmacytic infiltration: clinicopathologic features of 35 cases. *Am. J. Surg. Pathol.* 27:1119–1127.
31. Weber S.M., Cubukcu-Dimopulo O., Palesty J.A., Suriawinata A., Klimstra D., Brennan M.F., and Conlon K. 2003. Lymphoplasmacytic sclerosing pancreatitis: inflammatory mimic of pancreatic carcinoma. *J. Gastrointest. Surg.* 7:129–139.
32. Hultquist G.T. and Jönsson L.E. 1965. Ligation of the pancreatic duct in rats. *Acta. Soc. Med. Upsal.* 70:82–88.

33. Isaksson G., Ihse I., and Lundquist I. 1983. Influence of pancreatic duct ligation on endocrine and exocrine rat pancreas. *Acta. Physiol. Scand.* 117:281–286.

34. Seifert G. 1984. Cystic fibrosis and haemochromatosis. In *Pancreatic pathology*, chapter 4. Klöppel G. and Heitz P.U. Eds. Edinburgh: Churchill Livingstone. pp. 32–43.

35. Lack E.E. 2003. Cystic fibrosis and selected disorders with pancreatic insufficiency. In *Pathology of the pancreas, gallbladder, extrahepatic biliary tract, and ampullary region*. Lack E.E. Ed. Oxford: Oxford University Press. pp. 63–80.

17

CHEMICALLY INDUCED
EXOCRINE DEGENERATION,
ATROPHY, AND ARTERIAL
DISEASES IN RATS

*Abraham Nyska, Amy E. Brix, Michael P. Jokinen,
Donald M. Sells, Michael E. Wyde, Denise Orzech,
Joseph K. Haseman, Gordon Flake, and Nigel J. Walker*

CONTENTS

17.1   INTRODUCTION

In humans with portal hypertension, especially secondary to liver cirrhosis, pancreatic interstitial fibrosis may occur and the degree of fibrosis correlates with the severity of coronary artery disease.[1] There is a coincidence between the occurrence of thrombosis in coronary and pancreatic arteries in 16% of patients. Vascular pancreatitis, the so-called senile chronic pancreatitis, may be caused by severe arteriosclerosis with cholesterol microemboli.[2] Typical features of the vascular pancreatitis are occurrence beyond the fifth decade, predominance in males, frequency of pancreatic calcifications, and lack of pain.

Arteriolosclerosis of the pancreas is frequent in old patients and is associated with focal fibrosis or fatty replacement of atrophic acinar tissue. The so-called granular atrophy of the pancreas is considered the final stage of an arteriolosclerotic process resulting in focal fibrosis, acinar atrophy, and lipomatosis. Strangely enough, the islets of Langerhans are hardly affected by this vascular lesion.[3] In malignant hypertension (malignant nephrosclerosis) severe alterations in the pancreatic arteries in the form of fibrinoid necroses, thrombosis, inflammation, and angioblastic nodules may occur,[4–6] and can lead to ischemic pancreatic infarction.

The pancreas is affected in generalized panarteritis nodosa,[7–10] which can lead to partial ischemic infarction of the parenchyma and scarring of the pancreas mimicking chronic pancreatitis. Similar changes have also been described in cases of lupus erythematosis, scleroderma, rheumatic disease, and Wegener's granulomatosis.[11,12] Pancreatic cell necrosis has been described in cases of severe shock.[13] The presence of disseminated intravascular coagulation (DIC) in the pancreas in association with a variety of lesions in the exocrine and endocrine tissues characterizes the term *shock pancreatitis*.

In experimental animals, hypovolemic shock presents features that are similar to those seen in humans.[14,15] In this condition, an initial vacuolization of the cytoplasm is followed by degranulation of the apical cytoplasm of acinar cells and focal cytoplasmic dissolution. The cytoplasmic vacuoles could be detected before the occurrence of hyaline thrombi in the capillaries. Dilation of the endoplasmic reticulum, mitochondrial swelling, destruction of the cristae mitochondriales, spiraling myelin figures, and complete dissolution of enzyme granules are described. The short tolerance to ischemia of the exocrine pancreas could well be due to its high rate of protein metabolism.[16]

Spontaneous polyarteritis — synonymously termed periarteritis nodosa, polyarteritis nodosa, chronic arteritis, necrotizing arteritis, and necrotizing vasculitis — occurs occasionally in the various rat strains, including the SD rat, and is characterized in the acute stage by fibrinoid necrosis with intense inflammatory-cell infiltrate and then medial and adventitial fibrosis leading to a thickening of the arterial wall. The sites of predilection are the mesenteric, pancreatic, and spermatic arteries, but other muscular arteries sometimes are affected. The etiology of spontaneous polyarteritis is unknown, although it resembles immune-mediated arteritis in other species.[17]

Arteritis in the rat can be caused by treatment with various vasoactive compounds, in particular, vasodilator compounds.[18–20] Vasodilation has been suggested to damage the arterial wall by producing increased blood flow with increased intramural shear stress leading to an increase in endothelial permeability and medial necrosis.[21] The mesenteric and pancreatic arteries have been reported to be especially susceptible to the effects of vasodilators,[22,23] possibly due to a lack of anatomic supporting structures for these vessels or the presence of specific receptors for the vasodilator in the vessel wall.[20,24,25]

Polyhalogenated aromatic hydrocarbons (PHAHs) comprise a large class of compounds including polychlorinated dibenzodioxins (PCDDs), polychlorinated dibenzofurans (PCDFs), polychlorinated biphenyls (PCBs), polychlorinated naphthalenes (PCNs), and polybrominated diphenyl ethers (PBDEs). Certain PCDDs, PCDFs, and coplanar PCBs have the ability to bind to the aryl hydrocarbon (Ah) receptor and exhibit biological actions similar to 2,3,7,8-tetrachlorodibenzo-p-dioxin (TCDD); they are commonly referred to as dioxin-like compounds (DLCs). Exposure of female Sprague Dawley rats of the Spartan strain to 100 ng/kg of TCDD for 2 years was associated with atrophy, fibrosis, and periarteritis of the pancreas.[26] In a recent series of studies conducted with DCLs at the National Institute of Environmental Health Sciences, Research Triangle Park, North Carolina, these compounds have been found to induce cardiovascular alterations, in addition to toxic effects seen in the exocrine pancreas.[27,28]

17.2 PANCREATIC INJURY BY DLCS

Four DLCs were administered to female Sprague Dawley rats in food for a duration of 2 years. The compounds included TCDD; PCB126, which is a nonortho-substituted PCB; PeCDF, which is a dioxin-like polyhalogenated aromatic hydrocarbon; and mixture at equal ratios (1:1:1) of toxic equivalents (TEQs) for TCDD, PeCDF, and PCB126. Administration of 4 compounds was associated with increased incidences of nonneoplastic changes of the exocrine pancreas including cytoplasmic vacuolation, chronic active inflammation, atrophy, and arteritis, variably observed in the 14-, 31-, and 53-week interim sacrifices and seen in the 2-year studies.[28] In addition, low incidences of acinar adenoma and carcinoma were also seen in the TCDD, PeCDF, and toxic-equivalency-factor (TEF)-mixture studies. The general histological characteristics were comparable for all chemicals. The incidence of the pancreatic lesions observed in the rats treated for 2 years with TCDD are presented in Table 17.1.

Cytoplasmic vacuolation consisted of small, clear, discrete vacuoles within pancreatic acinar cells. Occasionally a single large vacuole was noted. The severity of the change was determined by the degree of vacuolization per cell and the amount of tissue involved.

Atrophy was a focal to multifocal to diffuse change consisting of a reduction in the amount of acinar tissue with an associated increase in

Table 17.1 Incidences and Average Severities of Selected Pancreatic Lesions in Female Sprague-Dawley Rats Treated for 2 Years with TCDD

Dose (ng/kg/d)	Vehicle Control	3	10	22	46	100	100 Stop
Probability of survival at end of the study (%; Kaplan-Meier determinations)	47	39	49	36	42	40	42
Number of organs examined	51	54	52	53	53	51	49
Acinus, vacuolation cytoplasmic	1[a] (2.0)[b]	0	0	1 (1.0)	15** (1.1)	42** (1.8)	0***
Acinus, atrophy	1 (1.0)	2 (1.5)	4 (1.5)	4 (1.5)	4 (1.5)	9* (2.2)	4 (1.8)
Inflammation, chronic active	0	0	2 (1.5)	2 (1.0)	3 (1.3)	6* (2.0)	4 (1.5)
Artery, inflammation, chronic active	0	1 (3.0)	1 (2.0)	2 (2.5)	2 (3.0)	7* (2.3)	2 (2.5)
Cases in which "inflammation, chronic active" or "Acinus, vacuolation cytoplasmic" or "Acinus, atrophy" were noted in the exocrine tissue and also "artery, inflammation, chronic active"	0	0	0	0	1	7*	1
Cases in which "inflammation, chronic active" or "Acinus, vacuolation cytoplasmic" or "Acinus, atrophy" were noted in the exocrine tissue, but no "artery, inflammation, chronic active" was noted in the same pancreas	2	1	4	6	21**	39**	7**
Acinus, adenoma[c]	0	0	0	0	0	1	0
Acinus, carcinoma[d]	0	0	0	0	0	2	1

* Significantly different (p < 0.05) from vehicle control group by the Poly-3 test. ** (p < 0.01). *** Significantly different (p < 0.01) from the 100 ng/kg core study group by the Poly-3 test.

[a] Number of animals with lesion.

[b] Average severity grade of lesions in affected animals: 1 = minimal, 2 = mild, 3 = moderate, 4 = marked.

[c] Historical incidence (pooled control incidence from the 4 studies) for 2-yr gavage studies with Sprague-Dawley vehicle control group; 1/207; range, 0 to 2%.

[d] Historical incidence: 0/207.

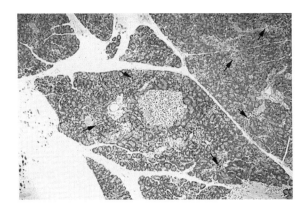

Figure 17.1 Degeneration of acini along the periductal and perivascular areas of the pancreas (arrows). Portion of the splenic artery with thickened wall is seen (lower right corner). Mild-to-moderate fibrosis is noticeable at this magnification. Hematoxylin and Eosin (H&E) × 20.[1] Note unaffected islet of Langerhans.

stromal fibrous connective tissue and dilatation of the ducts. Chronic active inflammation was generally seen in association with atrophy and consisted of an infiltrate of mononuclear cells with occasional neutrophils within the stroma. The islets of Langerhans were morphologically normal, dispersed throughout the affected acinar tissue and without reduction in their number.

Arterial chronic active inflammation was a focal to multifocal change characterized by a thick mantle of macrophages, lymphocytes, and plasma cells around the arteries, with infiltration into the muscular layers of the artery.[27] Fibrinoid necrosis of the vessel occurred often, and the tunica intima was frequently thickened. Endothelial cells were swollen or decreased in number. This inflammatory reaction often extended into the surrounding parenchyma.

In some of the high-dosed animals, the affected acini were primarily located along the blood vessels accompanying the ducts. At low magnification, the altered acini appeared to be arranged as radiating stripes confined to the periductal region, which showed mild to moderate fibrosis (Figure 17.1). Remarkably, the periinsular acini and islet cells were not affected (Figure 17.1 and Figure 17.2). Periductal fibrosis accompanied the alterations (Figure 17.2). The severity of acinar alterations increased with continued exposure and, due to periductal and perivascular fibrosis, ductal and ductular structures appeared prominent (Figure 17.3). Some peripancreatic, and especially intrapancreatic, arterioles exhibited severe thickening of the wall and fibrinoid degeneration of the luminal layer partially or totally occluding the lumen (Figure 17.4 and Figure 17.5). At

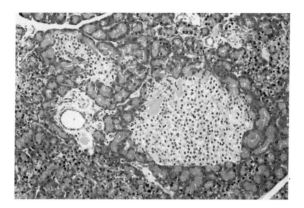

Figure 17.2 Higher power view of Figure 17.1 showing intact acini around the islets. Vacuolization of acinar cells can be seen in the periphery of the islets. Dilated blood vessels within the large islet. Note the lack of inflammation. H&E × 75.[1]

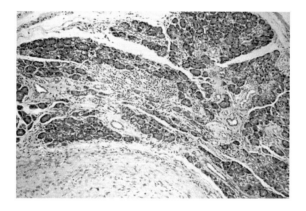

Figure 17.3 A more advanced stage of the alterations. Mild atrophy of the exocrine pancreas with prominent ductal and ductular structures due to the fibrosis. No signs of inflammation. The islets appear intact. Multifocal loss of acini and pseudoductular formations (lower right corner) are noticeable. A portion of severely fibrotic splenic artery is seen at the bottom. H&E × 20.[1]

this stage, the pancreas showed atrophic changes with a loss of acini and formation of pseudoductules. Contrary to chronic pancreatitis, there was no inflammatory reaction in the exocrine pancreas, but there was mild inflammation around the affected vessels. The restriction of fibrosis around ductules and blood vessels, which accompany the ductules, also distinguishes this alteration from chronic pancreatitis.

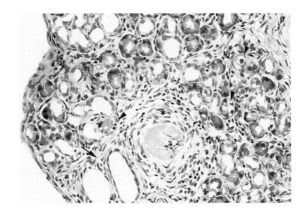

Figure 17.4 In advanced stage, most acinar cells are replaced by duct-like structures. The still remaining acinar cells show small vesicles, which are most likely fat droplets (arrowheads). The arteriole shows a thick layer of connective tissue and severe fibrinoid degeneration almost totally obliterating the lumen (*). Periductal fibrosis is also present (arrow). Note that the light inflammation is restricted to the wall of the vessel. H&E × 75.[1]

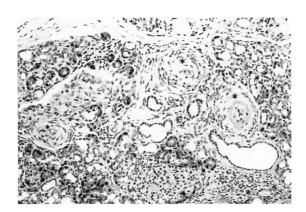

Figure 17.5 Another rat with atrophic pancreas showing several severely altered blood vessels (*) with partial or complete occlusion of the lumen. There are only a few acinar cells, especially around the islet (lower middle field). Little inflammation around the vessels but not around the ducts. H&E × 75.[1]

In the same rat DLC studies, where pancreatic pathology was reported, treatment-related increases in the incidence of degenerative cardiovascular lesions were noted.[26,27] In rabbits, TCDD induces preatherosclerotic lesions[29]; in AopE (-1) mice, it causes a trend toward an earlier onset and greater severity of atherosclerotic lesions.[30]

Dioxins have also been shown to induce cardiovascular alterations in fish and chick embryos. In zebra fish embryos, TCDD causes a reduced blood flow in the mesenteric vein, possibly due to the activation of arylhydrocarbon receptor and induction of P-450 (CYP 1A).[31]

It was striking that periinsular acini seemed to be resistant to degenerative changes. Based on the anatomy of the pancreatic blood vessels (see Chapter 4), where some part of the blood supply of the exocrine pancreas is provided by the efferent branches of the insular arteries, the lack of acinar cell alterations around the islets suggests that these particular arterial structures are not affected by the induced alterations. It is also possible that the higher concentration of insulin in the periinsular region protects the acinar cells. However, because islet cells even in the advanced stage, did not show any alteration, the first possibility is more likely.

In summary, the pancreas of humans and animals appears to be sensitive to vascular changes. Severe arteriosclerotic alterations can result in atrophy of the pancreas with its consequences. It appears that different branches of intrapancreatic arteries react differently to the causative agents and that the arteries nourishing the islets are protected from toxicity.

Our data indicate that the pancreatic exocrine acini are a target tissue of the DLCs, inducing mainly degenerative, inflammatory, and atrophic lesions and possibly also sporadic acinar adenomas and carcinomas. The generation of the DLC-related pathology in the pancreas may be related to a variety of factors, including the induction of drug-metabolizing enzymes such as the cytochromes P450, down-regulation of cholecystokinin (CCK), perturbations in the Vitamin A homeostasis, or other mechanisms, acting either independently or concomitantly to promote the development of pancreatic damage.[32–36]

A recent study in F344/N rats of theophylline, a nonspecific phosphodiesterase inhibitor that produces excessive vasodilation, also resulted in periarteritis of the mesenteric and pancreatic arteries that appeared morphologically similar to the arteritis in the TCDD studies.[20] The periadventitial acinar tissue close to the affected arteries was chronically inflamed. This finding suggests the possibility that the arteritis observed in our investigation was caused by the vasoactive effects of the compounds on the affected arteries, although the potential vasoactive effects of these compounds in rats are unknown. Dioxin has been reported to increase mean tail-cuff blood pressure in mice,[30] but this increase implies vasoconstrictor rather than vasodilator activity.

The mechanism by which the DLCs caused the occurrence of arteritis is undetermined and requires elucidation.

17.3 EPILOGUE

The potential contribution of the vascular pathology and ensuing ischemia to the development of the exocrine changes in the rats exposed to the TCDD was considered. However, in view of the fact that many animals exhibited inflammation and a trophy of acinar tissue without any identifiable vascular changes (Table 17.1), additional toxic effects of TCDD on the exocrine tissue can not be excluded. Nevertheless, it appears that in rats, as in humans, arterial diseases can result in damage to the exocrine pancreas.

NOTE

All figures are from female rats administered 1000 ng/kg PCB 126 per day by gavage for 2 years.

REFERENCES

1. Pollak O.J. 1968. Human pancreatic arteriosclerosis. *Ann. N.Y. Acad. Sci.* 149:928–931.
2. Ammann R. and Sulser H. 1976. Die 'senile' chronische Pankreatitis-eine neue nosologische Einheit? *Schweizerische medizinische Wochenschrift* 106:429–437.
3. Kloeppel G. and Heitz P.U. 1984. *Pancreatic pathology.* New York: Churchill Livingstone. pp. 75–77.
4. Hranilovich G.T. and Baggenstoss A.H. 1953. Lesions of the pancreas in malignant hypertension: Review of one hundred cases at necropsy. *Arch. Pathol.* 55:443–456.
5. Hughson M.D., Harley R.A., and Henninger G.R. 1982. Cellular arteriolar nodules. Their presence in heart, pancreas and kidneys of patients with malignant nephrosclerosis. *Arch. Pathol. Lab. Med.* 106:71–74.
6. Nager F. and Steiner H. 1965. Der Pankreasinfatkt bei maligner Hypertonie. *Schweizerische medizinische Wochenschrift* 95:119–124.
7. Aufdermaur M. 1947. Uber Pankreasnekrose als Folge generalisierter Arteriitis. *Gastroenterologia (Basel)* 72:81–95.
8. Becker V. 1973. Bauchspeicheldrüse (Inselsystem ausgenommen). In *Spezielle pathologische anatomie.* Doerr W. and Seifert G. Eds. Vol. 6. Berlin-Heidelberg: Springer.
9. Froboese C. 1949. Beitrag zur Stütze der rheumatischen Ätiologie der Periarteritis nodosa und zum subtotalen Pankreasinfarkt. *Virchows. Arch.* 317:430–448.
10. Seifert G. 1956. *Die pathologie des kindlichen pankreas.* Leipzig: Thieme.
11. Becker V. 1976. Allgemeine Pathologie der Bauchspeicheldruse. In *Handbuch der inneren medizin.* Forell M.M. Ed. Vol. 111/6 Pankreas. Berlin-Heidelberg: Springer. pp. 3–62.
12. Seifert G., Heinz N., and Ruffmann A. 1967. Pankreatitis bei viszeralem Lupus erythematodes. *Gastroenterologie* 107:317–327.

13. Jones R.T., Garcia J.H., Mergener W.J., Pendergrass R.E., Valigorsky J.M., Trump B.F. 1975. Effects of shock on the pancreatic acinar cell. Cellular and subcellular effects in humans. *Arch. Pathol.* 99:634–644.

14. Donath K., Mitschke H., and Seifert G. 1970. Ultrastrukturelle Veranderungen am Rattenpankreas beim hamorrhagischen schock. *Beiträge zur Pathologie* 141:33–51.

15. Jones R.T. and Trump B.F. 1975. Cellular and subcellular effects of ischemia on the pancreatic acinar cell. In vitro studies of rat tissue. *Virchows Arch. B Cell. Pathol.* 19:325–336.

16. Moser R., Meili H.U., and Largiader F. 1967. Die Ischämietoleranz des Pankreas. *Zeitschrift für die gesamte experimentelle Medizin.* 143:267–274.

17. Mitsumori K. 1990. Blood and lymphatic vessels. In *Pathology of the Fischer rat.* Boorman G.A., Eustis S.L., Elwell M.R., Montgomery Jr. C.A., and MacKenzie W.F. Eds. San Diego: Academic Press. pp. 473–484.

18. Joseph E.C. 2000. Arterial lesions induced by Phosphodiesterase III (PDE III) inhibitors and DA(1) agonists. *Toxicol. Lett.* 112–113:537–546.

19. Louden C. and Morgan D.G. 2001. Pathology and pathophysiology of drug-induced arterial injury in laboratory animals and its implications on the evaluation of novel chemical entities for human clinical trials. *Pharmacol. Toxicol.* 89:158–170.

20. Nyska A., Herbert R.A., Chan P.C., Haseman J.K., and Hailey J.R. 1998. Theophylline-induced mesenteric periarteritis in F344/N rats. *Arch. Toxicol.* 72:731–737.

21. Joseph E.C., Rees J.A., and Dayan A.D. 1996. Mesenteric arteriopathy in the rat induced by phosphodiesterase III inhibitors: An investigation of morphological, ultrastructural, and hemodynamic changes. *Toxicol. Pathol.* 24:436–450.

22. Johansson S. 1981. Cardiovascular lesions in Sprague-Dawley rats induced by long-term treatment with caffeine. *Acta Pathol. Microbiol. Scand.* [A] 89:185–191.

23. Kerns W.D., Joseph E.C., and Morgan E.D. 1991. Drug-induced lesions, arteries, rat. In *Cardiovascular and musculoskeletal systems.* Jones T.C., Mohr U., and Hunt R.C. Eds. Berlin: Springer Verlag. pp. 76–83.

24. Kerns W.D., Arena E., Macia R.A., Bugelski P.J., Matthews W.D., and Morgan D.G. 1989. Pathogenesis of arterial lesions induced by dopaminergic compounds in the rat. *Toxicol. Pathol.* 17:203–213.

25. Kerns W.A. 1996. Pathogenesis of drug-induced myocardial and arterial lesions: current concepts. *Vet. Pathol.* 33:574 (Abstract).

26. Kociba R.J., Keyes D.G., Beyer J.E., Carreon R.M., Wade C.E., Dittenber D.A., Kalnins R.P., Frauson L.E., Park C.N., Barnard S.D., Hummel R.A., and Humiston C.G. 1978. Results of a two-year chronic toxicity and oncogenicity study of 2,3,7,8-tetrachlorodibenzo-p-dioxin in rats. *Toxicol. Appl. Pharmacol.* 46:279–303.

27. Jokinen M.P., Walker N.J., Brix A.E., Sells D.M., Haseman J.K., and Nyska A. 2003. Cardiovascular pathology in female Sprague-Dawley rats following chronic treatment with dioxin-like compounds. *Cardiovasc. Toxicol.* (In press).

28. Nyska A., Jokinen M.P., Brix A.E., Sells D.M., Wyde M.E., Orzech D., Haseman J.K., Flake G., and Walker N.J. 2004. Exocrine pancreatic pathology in female Sprague Dawley rats following chronic treatment with 2,3,7,8-tetrachlorodibenzo-p-dioxin and dioxin-like compounds. *Environ. Health Perspect.* (In Press).

29. Brewster D.W., Bombick D.W., and Matsumura F. 1988. Rabbit serum hyper-triglyceridemia after administration of 2,3,7,8-tetrachlorodibenzo-p-dioxin (TCDD). *J. Toxicol. Environ. Health* 25:495–507.
30. Dalton T.P., Kerzee J.K., Wang B., Miller M., Dieter M.Z., Lorenz J.N., Shertzer H.G., Nerbert D.W., and Puga A. 2001. Dioxin exposure is an environmental risk factor for ischemic heart disease. *Cardiovasc. Toxicol.* 1:285–298.
31. Dong W., Teraoka Y., Yamazaki K., Tsukiyama S., Imani S., Imagawa T., Stegeman J.J., Peterson R.E., and Hiraga T. 2002. 2,3,7,8-tetrachlorodibenzo-p-dioxin toxicity in the zerbafish embryo: local circulation failure in the dorsal midbrain is associated with increased apoptosis. *Toxicol. Sci.* 69:191–201.
32. Baldwin G.S. 1995. The role of gastrin and cholecystokinin in normal and neoplastic gastrointestinal growth. *J. Gastroenterol. Hepatol.* 10:215–232.
33. Foster J.R., Idle J.R., Hardwick J.P., Bars R., Scott P., Braganza J.M. 1993. Induction of drug-metabolizing enzymes in human pancreatic cancer and chronic pancreatitis. *Pathol.* 169:457–463.
34. Schmidt C.K., Hoegberg P., Fletcher N., Nilsson C.B., Trossvik C., Hakansson H., and Nau H. 2003. 2,3,7,8-Tetrachlorodibenzo-p-dioxin (TCDD) alters the endogenous metabolism of all-trans-retinoic acid in the rat. *Arch. Toxicol.* 77:371–383.
35. Standop J., Ulrich A.B., Schneider M.B., Buchler M.W., and Pour P.M. 2002. Differences in the expression of xenobiotic-metabolizing enzymes between islets derived from the ventral and dorsal anlage of the pancreas. *Pancreatol.* 2:510–518.
36. Varga G., Kisfalvi K., Pelosini I., D'Amato M., and Scarpignato C. 1998. Different actions of CCK on pancreatic and gastric growth in the rat: effect of CCK(A) receptor blockade. *Br. J. Pharmacol.* 124:435–440.

18

CHRONIC PANCREATIC
TOXICITY — ROLE OF STEM
CELLS IN THE CONVERSION
OF PANCREAS TO LIVER

M. Sambasiva Rao and Janardan K. Reddy

CONTENTS

18.1   INTRODUCTION

Multicellular organisms are composed of a multitude of differentiated cells that exhibit unique structures and specialized functions. The development of differentiated cells from the fertilized egg through pluripotent embryonic stem cells is meticulously controlled by transcription factors, growth and differentiation factors, cell–cell interactions, and the extra-cellular matrix components.[1–3] Embryonic stem cells generate not only differentiated cells, but also reserves of tissue-specific stem cells (adult stem cells) that are known to persist in a variety of adult tissues.[3,4] It is generally considered that organ-specific adult stem cells have restricted differentiation potential and are capable of generating cells that are specific to the tissue in which they reside. Some studies during the past 5 years pointed to the developmental potential or plasticity of these adult organ-specific stem cells that might be greater than previously recognized in that they can differentiate into several other types of cells.[4,5] Adult stem cells and terminally differentiated cells have been shown to cross lineage boundaries and differentiate into a variety of other cell types.[3,6–8] The examples of adult stem cells differentiating into different cell types of unrelated lineages include differentiation of bone marrow stem cells into a variety of ectodermal, endodermal, and meso-dermal cells,[4,9] and generation of neurons, astrocytes, oligodendrocytes, and hematopoietic cells from neural stem cells.[10,11] Examples demonstrating plasticity of fully differentiated cells include conversion of pancreatic cells into hepatocytes and hepatocytes into exocrine and endocrine cells of pancreas.[7,8,12,13] The function and differentiation state of adult stem cells, like embryonic stem cells, are dictated by microenvironment and growth and differentiation factors.[3,14] Altered microenvironment resulting from either tissue injury or migration of cells to a different location may cause activation of different genes leading to the development of fully differentiated different cell type. These various studies on plasticity of adult stem cells have engendered tremendous enthusiasm and hope, because these cells will be a valuable source for cell-based therapy in a variety of genetic and acquired diseases.

18.1.1 Liver and Pancreatic Diseases — Use of Stem Cell Transplantation

Organ and cell transplantation has become a common clinical practice for treatment of acute and chronic diseases and saves the lives of thousands of patients every year. It is estimated that there are more than 150 million diabetics throughout the world that require continuous medical treatment and still develop a variety of chronic complications. Islet cell transplantation as a choice for the treatment of diabetes has been increasingly explored.[15] Although islet cell transplantation appears promising, it is still fraught with rejection, complications associated with long-term immunosuppression and short supply of donor islets. Similarly, liver transplantation has become an established treatment for fulminant liver failure and end-stage liver disease. Like islet cell transplantation, liver transplantation is also associated with complications and shortage of donor livers. To obviate these problems alternative treatment modalities are being constantly sought. Stem cells, because of their unique properties of proliferation and plasticity to differentiate into different cell types, are considered to be a practical and promising alternative to donor-dependent organ transplantation. The identification of stem cells in the pancreas and liver in adult rats provided an attractive model to study the mechanisms involved in cell differentiation in mammals.[16–18]

18.1.2 Lineage Switches of Cells — Definitions

When cells (tissues) are subjected to selection pressure, they can change their lineage and differentiate into different cell types. This lineage change can be through metaplasia or transdifferentiation. Although metaplasia and transdifferentiation are interchangeably used to signify the plasticity of cells, there is a fundamental difference between these two processes. *Metaplasia* represents a change in the commitment of a stem cell or a potential stem cell to differentiate along a new pathway generating cells not specific to the tissue.[19,20] *Transdifferentiation* is the conversion of one type of differentiated cell into a different type of fully differentiated cell with or without cell division, attesting to plasticity of terminally differentiated cells.[21,22] Metaplastic changes can occur in epithelium and connective tissue and are most common in tissues populated by steady-state renewing cell populations. Common examples of metaplasia are the replacement of ciliated columnar epithelium in the respiratory mucosa and mucous secreting cells in the endocervix by squamous epithelium and esophageal squamous mucosa by intestinal type of epithelium.[19,23,24] Examples of transdifferentiation include conversion of fibroblasts to myofibroblasts, adrenal chromaffin cells to sympathetic neurons, and retinal pigmented epithelial cells into different neural cell types.[21,25-27] The examples of metaplasia and transdifferentiation support the notion of selective gene activation in maintaining the differentiated state of cells rather than irreversible gene repression and plasticity of adult stem cells responding to cues from microenvironment.[3,28,29]

18.1.3   Pancreatic Hepatocytes

The initial description of the development of morphologically and functionally fully differentiated hepatocytes in the pancreas of adult hamsters treated with carcinogens and in rats rendered copper deficient has led to the new paradigm of plasticity of pancreatic cells and the concept of existence of dormant stem cells in the adult pancreas.[30-33] Following these studies, development of pancreatic hepatocytes was described in other species.[34,35] In addition, recently, direct conversion of pancreatic acinar cells (cell line derived from azaserine-induced acinar cell carcinoma) into hepatocytes *in vitro* after treatment with dexamethasone has been described.[36] Similarly, conversion of liver cells to pancreatic cells has also been demonstrated in *in vivo* and *in vitro* conditions. Pancreatic acinar tissue developed in livers of 41% of rats treated with polychlorinated biphenyls.[37,38] Human liver cell line (hepG2) transfected with activated form of PDX1 transcription factor differentiated into pancreatic exocrine and endocrine cells.[8] A rare example of exocrine pancreas in the human liver has also been described.[39] Observation of interconversion of pancreas to liver and liver to pancreas is not surprising considering the close embryonic relation of these two organs. Liver and pancreas develop from the foregut endoderm and epithelial–mesenchymal interactions, extracellular signaling molecules, and transcription factors influence specific differentiation into either liver or pancreas.[40-42] In this chapter, we will briefly summarize the experimental models of pancreatic hepatocytes and discuss the phenotypic features and pathogenesis of pancreatic hepatocytes induced in copper-deficiency model in rats (the model that has become a classic paradigm to study pancreatic hepatocyte metaplasia because of reproducibility) and induction of hepatocytes in large numbers in a high percentage of animals in a short time.

18.2   EXPERIMENTAL MODELS OF PANCREATIC HEPATOCYTES

18.2.1   Hamster Model

Male Syrian golden hamsters when given a single dose of N-nitrosobis(2-oxopropyl)-amine (30 mg/kg of body weight), a pancreatic carcinogen, during maximum pancreatic regeneration following ethionine-induced necrosis, developed multiple foci of hepatocytes.[30] The number of hepatocytes developed in the pancreas varied from animal to animal and appeared as early as 3 days after initiation of regeneration and increased in numbers thereafter.[30,31,43] Pancreatic hepatocytes exhibited identical light and ultrastructural features as parenchymal cells of the liver and contained glycogen and albumin. The pancreatic hepatocytes also showed peroxisome proliferation after administration of peroxisome proliferator, methyl clofenapate, and proliferation of smooth endoplasmic reticulum following phenobarbital administration.[44,45] These cells regenerated following partial hepatectomy. Based on initial studies, it was proposed that acinar cells are the precursors of hepatocytes. However, detailed electron microscopic studies showed that hepatocytes are derived from ductular cells.[31] Morphologically similar type of cells have been described in 4 to 8% of the pancreas of untreated (normal) male and female hamsters after the age of 30 weeks, under the designation of eosinophilic metaplasia.[46]

18.2.2 Mouse Model

Transgenic mice overexpressing keratinocyte growth factor (KGF) in the pancreatic β-cells under the transcriptional control of human insulin promoter developed hepatocytes associated with the islets between the ages of 1.5 to 12.5 months. Hepatocytes were observed in about one-fifth of the islets in 38% of the mice.[35] These hepatocytes expressed only liver-specific proteins such as α-fetoprotein and albumin, but not proteins or peptides produced by islet cells, acinar, or duct cells. KGF, a cytokine that belongs to the family of fibroblast growth factor, is known to play a role in cell proliferation and differentiation.[47,48] Development of hepatocytes in KGF transgenic mice indicates that KGF influences the differentiation of pancreatic stem cells to hepatocytes.

18.2.3 Rat Models

18.2.3.1 Ciprofibrate Model

Administration of ciprofibrate, a potent peroxisome proliferator, in diet (at a daily dose level of 10 mg/kg body weight) in male Fischer rats for 60 to 72 weeks resulted in the development of hepatocytes in 25% of the animals.[49,50] The number of hepatocytes differed from animal to animal and ranged from single focus of 10 to 15 cells to several foci. The hepatocytes were usually abutted against islets of Langerhans with extension into the surrounding acinar tissue.

18.2.3.2 Methionine-Deficient Diet Model

Male Fischer rats maintained on diets deficient in methionine alone or methionine and other dietary factors for extended periods (up to 52 weeks) developed pancreatic hepatocytes.[51] The incidence of hepatocyte development in different groups was 1.4, 11.5, and 25% in methionine-deficient group; methionine- and choline-deficient group; and methionine-, choline-, folic acid-, and cyanocobalamin-deficient group, respectively. Hepatocytes were usually localized adjacent to islets of Langerhans. In this model, however, the numbers of hepatocyte foci per pancreas were few.

18.2.3.3 Dichloro-P-Phenylenediamine Model

Male and female Fischer rats fed a diet containing 2,6-dichloro-p-phenylenediamine (dose range 1000 ppm to 6000 ppm) for 103 weeks developed pancreatic hepatocytes.[52] The incidence of hepatocytes ranged from 8 to 31% and depended on the dose level. At 2000 ppm dose level, the incidence of hepatocytes in males and females was 18 and 30%, respectively, indicating sensitivity of females. Hepatocyte clusters that ranged from 1 to 3 in number per pancreas were present most often around islet of Langerhans and rarely in the acinar parenchyma.

18.2.3.4 Cadmium Chloride Model

Male Wistar and Fischer rats were given 18 subcutaneous injections of cadmium chloride at weekly intervals (dose range 10 to 30 µmol/kg body weight) and sacrificed at 110 experimental weeks. Hepatocytes developed in 93% of Wistar rats at 20 µmol dose level and in 50% of Fischer rats at 30 µmol dose. The number of hepatocyte foci per pancreas was also higher in Wistar rats (4 foci compared to 2 foci in Fischer rats) indicating the susceptibility of this strain to cadmium. In both the strains, hepatocytes were present adjacent to islets.[53]

18.2.3.5 Spontaneous Hepatocyte Development

Spontaneous hepatocyte development was noted rarely. Chiu[54] reported 0.3% incidence of hepatocytes after analyzing pancreases of 4177 (age range 3 to 26 months) Sprague-Dawley rats. McDonald and Boorman[52] observed hepatocytes in one of 50 2-year-old female rats and none of 50 male rats.

18.2.3.6 Copper-Deficiency Model

Copper, an essential trace element, is vital to all cell functions. Both increased and decreased levels of copper cause disorders of multiple organ

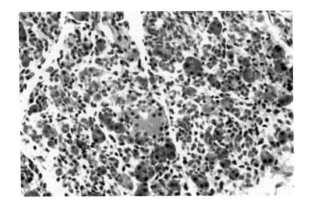

Figure 18.1   Pancreas from a rat maintained on copper-deficient diet for 6 weeks showing distortion of lobular architecture, loss of acinar tissue, and oval cell proliferation. H&E stain.

systems in animals and humans because more than 30 enzymes use copper as a cofactor.[55–57] Deficiency of copper causes systemic effects as well as specific morphological and biochemical changes in the pancreas of male rats.[58–60] Interestingly, copper deficiency has no deleterious effects on the pancreas and only minimal systemic effects in female rats (Rao M.S., personal observation).[61] Copper-deficiency-induced changes in the pancreas are sequential and progressive resulting in massive acinar cell loss associated with oval cell proliferation and development of hepatocytes.

18.2.3.6.1   Copper-Deficiency-Induced Pancreatic Changes

Male F-344 rats fed a copper-deficient diet supplemented with 0.6% triethylenetetramine tetrahydrochloride, a chelating agent, and allowed to drink deionized water for 8 weeks developed progressive morphological changes in the pancreas.[62–65] At 4 weeks, there was a slight increase in the apoptosis of acinar cells and a decrease in the number of zymogen granules. By 5 to 6 weeks, lobular architecture is disrupted associated with marked increase in acinar cell apoptosis (Figure 18.1). At 8 weeks, the pancreas weighed 20 to 30% of the control and histologically showed global acinar cell loss, resulting in complete collapse of lobular framework with approximation of islets of Langerhans and ducts (Figure 18.2 and Figure 18.3). In some animals, along with oval cell proliferation, there was marked increase in stromal fat at 8 weeks (Figure 18.4). Morphology of ductal cells and islets was unaffected by copper depletion. Acinar cell loss caused by copper depletion is due to programmed cell death as evidenced by characteristic morphological findings (light and electron

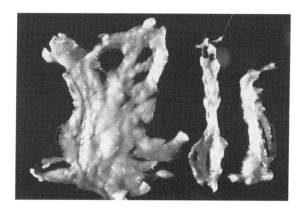

Figure 18.2 Gross picture of pancreas from a control rat (left) and a rat maintained on copper-deficient diet for 8 weeks (right). Pancreas from copper-deficient diet is much smaller and weighed 75% less than the control.

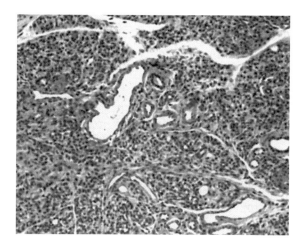

Figure 18.3 Pancreas from a rat maintained on copper-deficient diet for 8 weeks showing lobular acinar cell loss and marked oval cell proliferation. Scattered intact ducts are present. H&E stain.

microscopy) and corroborated by DNA fragmentation and sulfated glyco-protein-2 mRNA expression studies.[63,64,66] Concomitant with acinar cell loss, there is ductular and oval cell proliferation and stromal fatty infiltration.[16,67,68] Maximum oval cell proliferation occurs between 6 and 8 weeks of copper deficiency and during the first week of copper supplementation. Oval cells are arranged in small duct-like structures or present as isolated cells in the periductular and periinsular interstitium.[16]

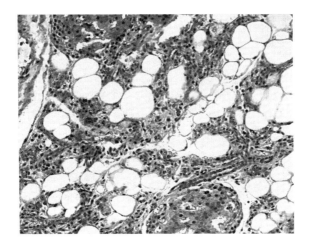

Figure 18.4 In some rats maintained on copper-deficient diet for 8 weeks oval cell proliferation is associated with marked stromal fatty infiltration. H&E stain.

Oval cells are morphologically similar to the oval cells described in the liver characterized by large nucleus and scant cytoplasm with poorly developed organelles.[16,69,70] Oval cells stain positively with OV-6 monoclonal antibodies raised against liver oval cells and stem cell factor (Figure 18.5).[72–74] Oval cells also express several liver-specific genes such as albumin, α-fetoprotein, α-1-antitrypsin, glucose 6-phosphatase, Met tyrosine kinase receptor, and liver enriched transcription factors (Figure 18.6).[16,74–77] Interestingly, oval cells strongly expressed pancreatic/duodenal homeobox-1 (PDX1) protein (Figure 18.7).[78] Oval cells cultured on col-

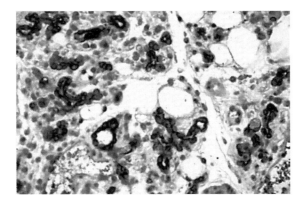

Figure 18.5 Ductules and oval cells in the pancreas of a rat fed copper-deficient diet for 8 weeks are strongly positive for OV-6 antigen. Immunoperoxidase stain.[71–74]

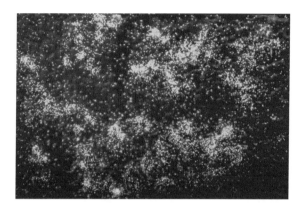

Figure 18.6 *In situ* hybridization of pancreas from a rat fed copper deficient diet for 8 weeks with ^{35}S-labeled albumin cDNA showing intense labeling of ductules and oval cells. Dark-field illumination.

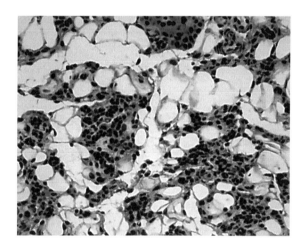

Figure 18.7 Oval cells and ductules in the pancreas of a rat fed copper deficient diet for 8 weeks showing strong nuclear positive stain for PDX1. Immunoperoxidase stain.

lagen gels retain their epithelial morphological features and grow as a monolayer. However, when cultured in collagen gels form duct-like structures and express albumin mRNA.[79,80] Oval cells cultured in the presence of hepatocyte growth factor or scatter factor showed increased DNA synthesis, migration, and tubule formation.[76] Oval cells maintained in culture spontaneously transformed during 13th passage exhibiting anchorage-independent growth in soft agar and formation of ductal type adenocarcinomas upon subcutaneous injection into nude mice.[76,81] In general, the morphological and phenotypic features of pancreatic oval cells are similar to that of liver oval cells.[82–85]

Oval cell proliferation in pancreas is a specific response to global acinar cell loss (Type 2 injury), such as caused by copper deficiency.[86] Oval cell proliferation does not occur after Type 1 injury, where there is focal or subtotal acinar cell loss. In Type 1 injury, enough numbers of acinar cells persist to repopulate the pancreas; whereas in Type 2 injury, only a few scattered acinar cells remain that are insufficient to proliferate and restore the acinar cell mass. It is postulated that creation of a cellular vacuum by global acinar cell loss exerts expansion pressure on dormant stem cells to proliferate, which results in perturbation in cell lineage and expansion of oval cells. It is interesting that no oval cell proliferation occurs in the pancreatic duct ligation model where there is complete loss of acinar tissue associated with stromal fatty infiltration (Rao M.S., personal observation).[87] Although there is rapid acinar cell loss secondary to apoptosis, the reasons for the lack of oval cell proliferation in pancreatic duct ligation model are not clear. Proliferation of pancreatic oval cells is analogous to oval cell proliferation in the liver where these cells proliferate only when there is extensive liver cell necrosis or impaired hepatocyte regeneration.[88] The possible origin of pancreatic oval cells includes ductular cells and periductular cells. Based on γ-glutamyl-transpeptidase negative property of oval cells, it is suggested that these cells are derived from periductular stem cells rather than duct or ductular cells that are strongly positive for this enzyme.

18.2.3.6.2 Pancreatic Changes during Copper Repletion Period

Feeding normal rat chow (containing copper) following 8 weeks of copper-deficient diet resulted in marked improvement in general condition of animals within a few days. Pancreas weight increased progressively over the next 2 to 3 weeks, reaching a maximum of 50% of normal weight. Histologically, beginning at 24 hours after copper substitution, the existing scattered acinar cells showed proliferative activity that increased gradually over the next 72 hours and decreased steadily thereafter. At 2 weeks, acinar cell regeneration was partial and acinar tissue occupied about 37% of the volume, compared to 84% in the normal pancreas.[62] Oval cell proliferation that began at 6 weeks of copper-deficiency phase continued during the first 2 weeks of copper repletion period. During the first week of recovery, in addition, single and small foci of hepatocytes developed in the pancreases of these animals. The number and size of the hepatic foci increased progressively; by 15 weeks, hepatocytes occupied 60 to 80% of the pancreas in some rats (Figure 18.8).[16,33,67] The remainder of

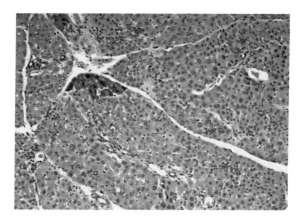

Figure 18.8 Extensive replacement of pancreas by sheets of hepatocytes in a rat fed copper-deficient diet for 8 weeks and fed control diet for several months. Only a few acinar cells are present. H&E stain.

the pancreas contained scattered foci of acinar tissue, fatty tissue, and normal complement of islets of Langerhans. Hepatocyte foci are randomly distributed throughout the pancreas and are often present in association with islets, ducts, and the stromal fat. Once pancreatic hepatocytes develop, they persist and proliferate throughout the life span of the animal.

18.2.3.6.3 Pancreatic Hepatocytes — Structural and Functional Features

Hepatocytes are arranged in nests of variable sizes and occasionally in sheets without acinar organization or sinusoidal vasculature. By light and electron microscopy, pancreatic hepatocytes exhibit morphological features identical to those of parenchymal cells of the liver, including well-developed bile canaliculi, endoplasmic reticulum, and peroxisomes containing urate oxidase rich nucleoids (Figure 18.9).[50,67,73,89] These cells express several integral membrane proteins specific to apical, lateral, and basolateral domains of the plasma membranes in spite of lack of organizational pattern as in liver.[90]

Functionally, pancreatic hepatocytes are fully differentiated and express several liver-specific proteins and their mRNAs. Albumin, α_2u-globulin, carbamoyl-phosphate synthase-1, glutamine synthetase, and urate oxidase are synthesized in these cells.[16,67,89,91,92] Male sex hormones regulate α_2u-globulin synthesis in these cells, similar to that in the normal liver. Functional heterogeneity exhibited by parenchymal cells of the liver is lost in pancreatic hepatocytes, particularly in relation to ammonia metabolizing enzymes. Glutamine synthetase and carbamoyl-phosphate syn-

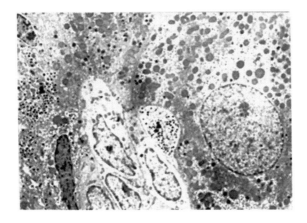

Figure 18.9   Electron micrograph of pancreatic hepatocytes adjacent to small duct and islet cells. Hepatocytes show well developed rough endoplasmic reticulum, round to oval mitochondria, and peroxisomes.

thase-1 genes are expressed in all pancreatic hepatocytes, unlike in the liver where the expression of these genes is mutually exclusive.[89,92] The loss of normal zonal pattern of gene expression in pancreatic hepatocytes may be due to the absence of structural organization and portal blood supply, which may contribute to the heterogeneity of parenchymal cells of the liver.[93] Pancreatic hepatocytes also proliferate in response to partial hepatectomy in rats and hamsters.[45]

In addition, pancreatic hepatocytes respond to xenobiotics and hepatocarcinogens analogous to hepatic parenchymal cells further attesting to the hepatocyte functional identity of these cells. Peroxisome proliferators, through steroid receptor mediated mechanisms, induce pleiotropic responses in the liver characterized by early adaptive changes (hepatomegaly, peroxisome proliferation, induction of peroxisomal and microsomal enzymes, etc.) and late carcinogenic effects.[94–96] Short-term dietary administration of peroxisome proliferators in rats and hamsters caused peroxisome proliferation and induced peroxisome-associated enzymes in pancreatic hepatocytes, but not in pancreocytes.[44,89,97] Chronic administration of peroxisome proliferators resulted in excessive accumulation of lipofuscin, a morphological feature of oxidative injury, and focal dysplastic changes in pancreatic hepatocytes.[17,49] Administration of single dose of hepatocarcinogens such as tannic acid, aflatoxin B_1, and lasiocarpine caused characteristic nucleolar segregation in pancreatic hepatocytes similar to that observed in parenchymal cells of the liver.[73,98] Chronic administration of hepatocarcinogen 2-acetylaminofluorene in diet resulted in proliferation of oval cells and emergence of glutathione S-transferase positive altered foci in the pancreases of rats with pancreatic hepatocytes.

18.3   ORIGIN OF PANCREATIC HEPATOCYTES

Hepatocyte development in the pancreas of adult rats although it appears peculiar is not illogical considering the embryological lineage of these two organs. Both pancreas and liver arise from the specified areas of foregut endoderm (pancreas from ventral and dorsal buds and liver from ventral bud adjacent to the ventral pancreatic bud).[99,100] Epithelial–mesenchymal interactions, extracellular signaling molecules, and transcription factors influence specific differentiation patterns of foregut endoderm into liver or pancreas. Sonic hedgehog signaling molecules, PDX1 transcription factor, and signals from notochord mesenchyme specify the commitment to pancreatic and liver fate.[41,101] Now there is clear evidence to suggest that all pancreatic cell lineages arise from PDX1 positive stem cells.[42,102] Once the endodermal stem cells are committed to pancreatic lineage, the subsequent events of pancreatic morphogenesis and differentiation depend on the mesenchyme that surrounds dorsal and ventral pancreatic buds. Both pancreatic buds give rise to exocrine and endocrine cells during embryogenesis without any constraints, and after birth, there is no anatomic segregation of pancreas to indicate dorsal and ventral bud derivatives. Mesenchymal factors and epithelial–mesenchymal interactions not only regulate the differentiation of pancreatic progenitor cells into exocrine and endocrine cells, but also ensure the appropriate balance of these different cell types.[103–105] TGF-β signaling components (activin, TGF-β1) and Notch signaling pathways promote endocrine differentiation; whereas, follistatin promotes exocrine development.[106–108] During embryogenesis, primitive duct cells that appear first function as stem cells and give rise to duct, acinar, and islet cells.[109] After fetal development, the duct cells retain the ability to differentiate into islet cells, although their capacity to differentiate into acinar cells is curtailed.[110–112] Contrary to the general belief, it appears that some of the embryonic stem cells persist postnatally and proliferate only under specific conditions, such as global acinar cell loss caused by copper deficiency.

18.3.1   Ductular and Oval Cells as the Progenitor Cells of Pancreatic Hepatocytes

During the late phase of copper deficiency (6 to 8 weeks) and early phase of recovery (1st week), there is marked oval and ductular cell proliferation as evidenced by increased mitosis and increased thymidine-labeling indices in these cells.[16,33] Both oval and ductular cells ultrastructurally showed features of undifferentiated cells similar to that observed in the protodifferentiated stage of embryonic pancreas. Sequential morphological, northern blot analysis, and *in situ* hybridization studies have shown that both oval and ductular cells are progenitor cells for hepatocytes (Figure 18.10).

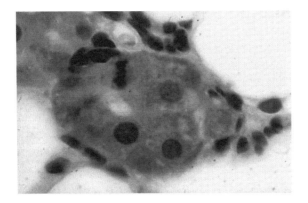

Figure 18.10   Fully differentiated hepatocytes in a small pancreatic duct. One of the hepatocytes is in mitosis. H&E stain.

Northern blot analysis of pancreatic RNA obtained during copper-deficiency phase, before the appearance of hepatocytes, showed expression of mRNAs for albumin and liver-enriched transcription factors CCAAT/enhancer binding protein α and β and hepatocyte nuclear factor -3β.[74,77] Albumin mRNA transcripts are localized in oval cells and ductular cells, but not in acinar, islet, or mesenchymal cells (Figure 18.6). Light and electron microscopic studies have shown progressive morphological changes in ductular and oval cells from undifferentiated transitional to fully differentiated hepatocytes.[16,33,75,97] Evidence for origin of hepatocytes from both ductular and oval cells is provided by studies showing hepatocytes with and without basement membrane, positive and negative staining for γ-glutamyl transpeptidase and presence or absence of India ink in the canaliculi of hepatocytes after injection into the pancreatic duct. Ductular cells are normally surrounded by basal lamina and positive for γ-glutamyl transpeptidase, but oval cells lack both these features.

18.3.2   Differentiation Potential of Oval and Ductular Cells — *In Vitro* and *In Vivo* Studies

Studies using isolated oval cells and ductal cells have provided further evidence for the pluripotent properties of these cells. Oval cells isolated from pancreas during the late phase of copper deficiency and grown in collagen gels formed ductular structures and expressed albumin mRNA. These cells also organized into duct-like structures after transplantation into subcutaneous tissue.[65,80] Similarly, transplantation of fragments of pancreas from rats maintained on copper-deficiency diet for 7 to 8 weeks into spleen or inguinal fat fads also showed formation of duct-like structures. In none of these *in vitro* and *in vivo* studies was hepatocyte differentiation observed indicating the absence of appropriate stimuli (intrinsic factors and microenvironment) for prompting these cells to differentiate into hepatocytes. Interestingly, epithelial progenitor cells isolated from the pancreases of rats maintained on copper-deficient diet for 8 to 10 weeks when injected into the liver differentiated into hepatocytes, which suggests the role of local factors for such differentiation.[113] Chen et al.[114] reported hepatocyte differentiation of ductal epithelial cells isolated from the pancreases of normal rats and embedded in basement membrane matrix after subcutaneous and intraperitoneal injection.

18.4 POSSIBLE MECHANISMS INVOLVED IN THE INDUCTION OF PANCREATIC HEPATOCYTES

The molecular and cellular mechanisms involved in reprogramming pancreatic stem cells to differentiate into hepatocytes are not clear. It is now well established that stem cells in adult tissues are highly plastic and can differentiate into unrelated cell lineages. This differentiation potential is influenced by a microenvironment that includes cell–cell interaction, the extracellular matrix, and cell growth and differentiation factors.[3] Microenvironment and cell–cell interactions in pancreas are completely distorted in copper deficiency because of global acinar cell loss. This creates not only expansion pressure on dormant stem cells to repopulate the pancreas, but also results in loss of cell differentiation stabilizing factors. In addition, copper deficiency causes marked decrease in free radical scavenging enzymes that may lead to oxidative stress and DNA changes. Liver enriched transcription factors, such as hepatocyte nuclear factors and CCAAT/enhancer binding protein, are expressed in the pancreas of copper-deficient animals. The role of transcription factors and growth factors in maintenance of differentiated state or metaplasia and transdifferentiation is well established. Transfection of pancreatic acinar cell line with C/EBPβ and HepG2 liver cell line with PDX1 transcription factors resulted in conversion to hepatocytes and pancreatic cells, respectively.[8,36] Overexpression of KGF, a member of the fibroblast growth factor family, in pancreatic β-cells causes hepatocyte differentiation in islets of Langerhans.[35] It is interesting to note that the PDX1 transcription factor that promotes pancreas development and suppresses liver development is strongly expressed in oval cells of the rat pancreas maintained on copper-deficient diet for 7 to 8 weeks.[78] Lack of down-regulation of PDX1 in the copper-deficient model indicates that for hepatocyte differentiation, expression of liver inducing transcription factors, such as fibroblast growth factor and C/EBP, etc., is probably more important in promoting differentiation of pancreatic stem cells into liver cells. It has been shown that there is a marked increase in the expression of C/EBPα, C/EBPβ, and

C/EBPδ and hepatocyte nuclear factor-3β mRNAs in the pancreas of rats maintained on copper-deficient diet for 4 to 6 weeks. Further studies are required to fully evaluate the alterations in other transcription factors and their role on induction of pancreatic hepatocytes in a copper-deficiency model in rats.

18.5 EPILOGUE

The copper-deficiency model clearly establishes the presence of pancreatic stem cells and their plasticity to differentiate into hepatocytes in adult rats. This experimental system provides a classical model to study the phenomenon of metaplasia and the molecular events responsible for such a process. Pancreatic stem cells can be easily isolated and maintained in culture systems. If the factors responsible for reprogramming pancreatic stem cells to liver cells or islet cells are identified, then pancreatic stem cells will be an attractive source of cell transplantation in patients with fulminant liver disease and diabetes.

ACKNOWLEDGMENTS

This research was supported by NIH Grant DK37958, NIH Grant NIH23750, and Adrian Meyer Pancreatic Cancer Research Fund.

REFERENCES

1. Nichols J., Zevnik B., Anastassiadis K., Niva H., Klewer-Nebenius D., Chambers I., Scholer H., and Smith A. 1998. Formation of pluripotent stem cells in the mammalian embryo depends on the POU transcription factor Oct4. *Cell* 95:379–391.
2. Hay E.D. 1991. Collagen and other matrix glycoproteins in embryogenesis. In *Cell biology of extracellular matrix*. Hay E.D. Ed. New York: Plenum Press. pp. 419–462.
3. Blau H.M., Brazelton T.R., and Weiman J.M. 2001. The evolving concept of a stem cell: entity or function. *Cell* 105:829–841.
4. Verfaillie C.M. 2002. Adult stem cells: assessing the case for pluripotency. *Trends Cell. Biol.* 12:502–508.
5. Triffitt J.T. 2002. Stem cells and the philosopher's stone. *J. Cell. Biochem.* 38:13–19 (Supplement).
6. Stocum D.L. 2002. A tail of transdifferentiation. *Science* 298:1901–1902.
7. Tosh D., Shen C., and Slack J.M.W. 2002. Differentiated properties of hepatocytes induced from pancreatic cells. *Hepatology* 36:534–543.
8. Horb M.E., Shen C., Tosh D., and Slack J.M.W. 2003. Experimental conversion of liver to pancreas. *Current Biol.* 13:103–115.
9. Krause D.S., Theise N.D., Collector M.I., Henegariu O., Hwang S., Gardner R., Neutzel S., and Sharkis S.J. 2001. Multi-organ, multi-lineage engraftment by a single bone marrow-derived stem cell. *Cell* 105:369–377.

10. Gage F.H. 2000. Mammalian neural stem cells. *Science* 287:1433–1438.

11. Bjornson C.R., Rietze R.L., Reynolds B.A., Magli M.C., and Vescovi A.L. 1999. Turning brain into blood: a hematopoietic fate adopted by adult neural stem cells in vivo. *Science* 283:534–537.

12. Scarpelli D.G., Reddy J.K., and Rao M.S. 1989. Metaplastic transformation of pancreatic cells to hepatocytes. In *The pathobiology of neoplasia*. Sirica A.E., ed. New York: Plenum Press, pp. 477–495.

13. Grompe M. 2003. Pancreatic-hepatic switches in vivo. *Mech. Develop.* 120:99–106.

14. Brivanlou A.H. and Darnell Jr. J.E. 2002. Signal transduction and control of gene expression. *Science* 295:813–818.

15. Ricordi C. 2003. Islet transplantation: a brave new world. *Diabetes* 52:1595–1603.

16. Rao M.S., Dwivedi R.S., Yeldandi A.V., Subbarao V., Tan X., Usman M.I., Shobha T., Nemali M.R., Kumar S., Scarpelli D.G., and Reddy J.K. 1989. Role of periductal and ductal epithelial cells of the adult rat pancreas in pancreatic hepatocyte lineage. *Am. J. Pathol.* 134:1069–1086.

17. Rao M.S. and Reddy J.K. 2003. Liver stem cells in pancreas. In *Disease progression and carcinogenesis in the gastrointestinal tract*. Galle P.R., Gerkin G., Schmidt W.E., and Wiedenmann B. Eds. Boston: Kluwer Academic Publishers. pp. 11–30.

18. Sell S. 1990. Is there a liver stem cell? *Cancer Res.* 50:3811–3815.

19. Rao M.S. and Reddy J.K. 1996. Cell and tissue adaptations to injury. In *Cellular and molecular pathogenesis*. Sirica A.E. Ed. Philadelphia: Lipponcott-Raven Publishers. pp. 57–77.

20. Slack J.M.W. and Tosh D. 2001. Transdifferentiation and metaplasia — switching cell types. *Curr. Opin. Gen. Develop.* 11: 581–586.

21. Okada T.S. Ed. 1991. *Transdifferentiation. Flexibility in cell differentiation*. Oxford: Clarendon Press.

22. Bresford W.A. 1990. Direct transdifferentiation: can cells change their phenotype without dividing? *Cell. Diff.* 29:81–93.

23. Lugo M. and Putong P.B. 1984. Metaplasia. *Arch. Pathol. Lab. Med.* 108:185–189.

24. Slack J.M.W. 1986. Epithelial metaplasia and the second anatomy. *Lancet* 2:268–271.

25. Schmitt-Graff A., Desmouliere A., and Gabbiani G. 1994. Heterogeneity of myofibroblast phenotypic features: an example of fibroblastic cell plasticity. *Virchows Archiv.* 425:3–24.

26. Anderson D.J. 1989. Cellular "neoteny": a possible developmental basis for chromaffin cell plasticity. *Trends Genet.* 5:174–178.

27. Eguchi G. and Kodama R. 1993. Transdifferentiation. *Curr. Opin. Cell Biol.* 5:1023–1028.

28. Blau H.M. and Baltimore D. 1991. Differentiation requires continuous regulation. *J. Cell Biol.* 112:781–783.

29. Caplan A.I. and Ordahl C.P. 1978. Irreversible gene repression model for control of development. *Science* 201:120–130.

30. Scarpelli D.G. and Rao M.S. 1981. Differentiation of regenerating pancreatic cells into hepatocyte-like cells. *Proc. Natl. Acad. Sci. U.S.A.* 78:2577–2581.

31. Makino T., Usuda N., Rao M.S., Reddy J.K., and Scarpelli D.G. 1990. Transdifferentiation of ductular cells into hepatocytes in regenerating hamster pancreas. *Lab. Invest.* 62:552–561.

32. Rao M.S. and Reddy J.K. 1986. Induction of hepatocytes in the pancreas of copper-depleted rats following copper repletion. *Cell. Diff.* 18:109–117.

33. Rao M.S., Yeldandi A.V., and Reddy J.K. 1990. Stem cell potential of ductular and periductular cells in the adult rat pancreas. *Cell. Diff. Develop.* 29:155–163.

34. Wolfe-Coote S., Louw J., Woodroof C., and Du Toit D.F. 1996. The non-human primate endocrine pancreas: development, regeneration potential and metaplasia. *Cell. Biol. Int.* 20:95–101.

35. Krakowski M.L., Kritzik M.R., Jones E.M., Krahl T., Lee J., Arnush M., Gu D., and Sarvetnick N. 1999. Pancreatic expression of keratinocyte growth factor leads to differentiation of islet hepatocytes and proliferation of duct cells. *Am. J. Pathol.* 154:683–691.

36. Shen C., Slack J.M.W., and Tosh D. 2000. Molecular basis of transdifferentiation of pancreas to liver. *Nat. Cell. Biol.* 2:879–887.

37. Kimbrough R.D. 1973. Pancreatic-type tissue in livers of rats fed polychlorinated biphenyls. *J. Natl. Cancer Inst.* 51:679–681.

38. Rao M.S., Bendayan M., Kimbrough R.D., and Reddy J.K. 1986. Characterization of pancreatic-type tissue in the liver of rat induced by polychlorinated biphenyls. *J. Histochem. Cytochem.* 34:197–201.

39. Wolf H.K., Burchette J.L., Garcia J.A., and Michalopoulos G. 1990. Exocrine pancreatic tissue in human liver: a metaplastic process? *Am. J. Surg. Path.* 14:590–595.

40. Pictet R. and Rutter W.J. 1972. Development of the embryonic endocrine pancreas. In *Handbook of physiology*. Steiner D.F. and Frenkel N. Eds. Washington D.C.: Williams and Wilkins. pp. 25–66.

41. Deutsch G., Jung J., Zheng M., Lora J., and Zaret K.S. 2001. A bipotential precursor population for pancreas and liver within the embryonic endoderm. *Development* 128:871–881.

42. Holland A.M., Hale M.A., Kagami H., Hammer R.E., and MacDonald R.J. 2002. Experimental control of pancreatic development and maintenance. *Proc. Natl. Acad. Sci. U.S.A.* 99:12236–12241.

43. Scarpelli D.G. and Rao M.S. 1981. Early changes in regenerating hamster pancreas following a single dose of N-nitrosobis (2-oxopropyl)amine (NBOP) administered at the peak of DNA synthesis. *Cancer* 47:1552–1561.

44. Rao M.S., Reddy M.K., Reddy J.K., and Scarpelli D.G. 1982. Response of chemically induced hepatocytelike cells in hamster pancreas to methyl clofenapate, a peroxisome proliferator. *J. Cell. Biol.* 95:50–56.

45. Rao M.S., Subbarao V., Leutteke N., and Sarpelli D.G. 1983. Further characterization of carcinogen-induced hepatocyte-like cells in hamster. *Am. J. Pathol.* 110:89–94.

46. Takahashi M. and Pour P. 1978. Spontaneous alterations in the pancreas of the aging Syrian hamster. *J. Natl. Cancer Inst.* 60:355–364.

47. Housley R.M., Morris C.F., Boyle W., Ring B., Blitz R., Tarpley J.E., Aukerman S.L., Devine P.L., Whitehead R.H., and Pierce G.F. 1994. Keratinocyte growth factor induces proliferation of hepatocytes and epithelial cells throughout the rat gastrointestinal tract. *J. Clin. Invest.* 94:1764–1777.

48. Guo L., Yu G-C., and Fuchs E. 1993. Targeting expression of keratinocyte growth factor to keratinocytes elicits striking changes in epithelial differentiation in transgenic mice. *EMBO J.* 12:973–986.

49. Reddy J.K., Rao M.S., Qureshi S.A., Reddy M.K., Scarpelly D.G., and Lalwani N.D. 1984. Induction and origin of hepatocytes in rat pancreas. *J. Cell. Biol.* 98:2082–2090.

50. Rao M.S., Scarpelli D.G., and Reddy J.K. 1986. Transdifferentiated hepatocytes in rat pancreas. *Curr. Top. Develop. Biol.* 20:63–78.

51. Hoover K.L. and Poirier L.A. 1986. Hepatocyte-like cells within the pancreas of rats fed methyl-deficient diets. *J. Nutr.* 116:1569–1575.

52. McDonald M.M. and Boorman G.A. 1989. Pancreatic hepatocytes associated with chronic 2,6-dichloro-p-phenylediamine administration in Fischer 344 rats. *Toxicol. Pathol.* 17:1–6.

53. Konishi N., Ward J.M., and Waalkes M.P. 1990. Pancreatic hepatocytes in Fischer and Wistar rats induced by repeated injections of cadmium chloride. *Toxicol. Appl. Pharmacol.* 104:149–156.

54. Chiu T. 1987. Focal eosinophilic hypertrophic cells of the rat pancreas. *Toxicol. Pathol.* 15:1–6.

55. Evans G.W. 1980. The role of copper in metabolic disorders. *Adv. Exp. Med. Biol.* 135:121–137.

56. Danks D.M. 1980. Copper deficiency in humans. In *Ciba Foundation Symposium 79, Biological roles of copper.* Amsterdam: Excerpta Medica. pp. 209–225.

57. Prohaska J.R. 1988. Biochemical functions of copper in animals. In *Essential and toxic trace elements in humans health and disease.* Prasad A.S. Ed. New York: Alan R. Liss, Inc. pp. 105–124.

58. Muller H.B. 1970. Der einfluss kupferarmer kost auf das pancreas. *Virchows. Arch. Pathol. Anat.* 350:353–367.

59. Smith P.A., Sunter J.P., and Case R.M. 1982. Progressive atrophy of pancreatic acinar tissue in rats fed a copper deficient diet supplemented with D-penicillamine or triethylene tetramine: morphological and physiological studies. *Digestion* 23:16–30.

60. Lewis C.G., Fields M., Craft N., Yang C., and Reiser S. 1988. Changes in pancreatic enzyme specific activities of rats fed a high fructose, low copper diet. *J. Am. Coll. Nutr.* 7:27–34.

61. Fields M., Lewis C.W., Beal T., Scholfield D., Patterson K., Smith J.C., and Reiser S. 1987. Sexual differences in the expression of copper deficiency in rats. *Proc. Soc. Exp. Biol. Med.* 186:183–187.

62. Rao M.S., Subbarao V., Yeldandi A.V., and Reddy J.K. 1987. Pancreatic acinar cell regeneration following copper deficiency-induced pancreatic necrosis. *Intl. J. Pancreatol.* 2:71–85.

63. Rao M.S., Yeldandi A.V., Subbarao V., and Reddy J.K. 1993. Role of apoptosis in copper deficiency-induced pancreatic involution in the rat. *Am. J. Pathol.* 142:1952–1957.

64. Kishimoto S., Iwamoto S., Masutani S., Yamamoto R., Jo T., Saji F., Terada N., Sasaki Y., Imaoka S., and Sugiyama T. 1993. Apoptosis of acinar cells in the pancreas of rats fed on a copper depleted diet. *Exp. Toxicol. Pathol.* 45:489–495.

65. Rao M.S. and Reddy J.K. 1997. Pancreatic hepatocytes: a unique adaptation of pancreatic ductular (oval) cells to severe acinar cell loss. In *Biliary and pancreatic ductal epithelia.* Sirica A.E. and Longnecker D.S. Eds. New York: Marcel Dekker, Inc. pp. 457–476.

66. Ide H., Yeldandi A.V., Reddy J.K., and Rao M.S. 1994. Increased expression of sulfated glycoprotein-2 and DNA fragmentation in the pancreas of copper-deficient rats. *Toxicol. Appl. Pharmacol.* 126:174–177.

67. Rao M.S., Dwivedi R.S., Subbarao V., Usman M.I., Scarpelli D.G., Nemali M.R., Yeldandi A., Thangada S., Kumar S., and Reddy J.K. 1988. Almost total conversion of pancreas to liver in the adult rat: a reliable model to study trans-differentiation. *Biochem. Biophys. Res. Commun.* 156:131–136.

68. Rao M.S. and Reddy J.K. 1995. Hepatic transdifferentiation in the pancreas. *Sem. Cell. Biol.* 6:151–156.

69. Grisham J.W. and Hartroft W.S. 1961. Morphological identification by electron microscopy of oval cells in experimental hepatic degeneration. *Lab. Invest.* 10:317–332.

70. Fausto N., Thompson N.L., and Braun L. 1987. Purification and culture of oval cells from rat liver. In *Cell separation methods and selected applications*. Pretlow I.I., T.G. and Pretlow T.P. Eds. Orlando: Academic Press. pp. 45–77.

71. Dunsford H.A. and Sell S. 1989. Production of monoclonal antibodies to preneoplastic liver cell populations induced by chemical carcinogens in rats and to transplantable Morris hepatoma. *Cancer Res.* 49:4887–4893.

72. Hixson D.C., Farris R.A., and Thompson N.L. 1990. An antigenic portrait of the liver during carcinogenesis. *Pathobiology* 58:65–77.

73. Rao M.S., Subbarao V., Sato K., and Reddy J.K. 1991. Alterations of hepatocytes in rats exposed to carcinogens. *Am. J. Pathol.* 139:1111–1117.

74. Rao M.S., Yukawa M., Omori M., Thorgeirsson S.S., and Reddy J.K. 1996. Expression of transcription factors and stem cell factor precedes hepatocyte differentiation in rat pancreas. *Gene Expression* 6:15–22.

75. Reddy J.K., Rao M.S., Yeldandi A.V., Tan X., and Dwivedi R.S. 1991. Pancreatic hepatocytes. An in vivo model for cell lineage in pancreas of adult rat. *Dig. Dis. Sci.* 36:502–509.

76. Jeffers M., Rao M.S., Rulong S., Reddy J.K., Subbarao V., Hudson E., Vande Woude G.F., and Resau J.H. 1996. Hepatocyte growth factor/scatter factor-Met signaling induces proliferation, migration, and morphogenesis of pancreatic oval cells. *Cell. Growth Diff.* 7:1805–1813.

77. Dabeva M.D., Hurston E., and Shafritz D.A. 1995. Transcription factor and liver-specific mRNA expression in facultative epithelial progenitor cells of liver and pancreas. *Am. J. Pathol.* 147:1633–1648.

78. Kashireddy P. and Rao M.S. 2004. Proliferation of PDX1 transcription factor positive oval cells in the pancreas of copper deficient rats but not after duct ligation. *Mod. Path.* In press.

79. Rao M.S. and Reddy J.K. 1991. Replicative culture in vitro of pancreatic epithelial oval cells derived from rats after copper deficiency-induced acinar cell depletion. *J. Tissue Cult. Methods* 15:121–124.

80. Ide H., Subbarao V., Reddy J.K., and Rao M.S. 1993. Formation of ductular strcture in vitro by rat pancreatic epithelial oval cells. *Exp. Cell Res.* 209:38–44.

81. Rao M.S., Pan J., Subbarao V., and Reddy J.K. 1996. Spontaneously transformed rat pancreatic epithelial oval cells give rise to ductal type adenocarcinomas. *Intl. J. Oncol.* 9:235–239.

82. Sell S. 2001. Heterogeniety and plasticity of hepatocyte lineage cells. *Hepatology* 33:738–750.

83. Mathis G.A., Walls S.A., and Sirica A.E. 1988. Biochemical characteristics of hyperplastic rat bile ductular epithelial cells cultured "on top" and "inside" different extracellular matrix substitutes. *Cancer Res.* 48:6145–6153.

84. Pack R., Heck R., Dienes H.P., Oesch F., and Steinberg P. 1993. Isolation, biochemical characterization, long-term culture, and phenotype modulation of oval cells from carcinogen-fed rats. *Exp. Cell Res.* 204:198–209.

85. Steinberg P., Steinbrecher R., Radaeva S., Schirmacher P., Dienes H.P., Oesch F., and Bannasch P. 1994. Oval cell lines OC/CDE 6 and OC/CDE 22 give rise to cholangio-cellular and undifferentiated carcinomas after transformation. *Lab. Invest.* 71:700–709.

86. Rao M.S., Yeldandi A.V., and Reddy J.K. 1990. Differentiation and cell proliferation patterns in rat exocrine pancreas: role of Type 1 and Type 2 injury. *Pathobiology* 58:37–43.

87. Walker N.I. 1987. Ultrastructure of the rat pancreas after experimental duct ligation. *Am. J. Pathol.* 126:439–451.

88. Thorgeirsson S. 1995. Stem cells and hepatocarcinogenesis. In *Liver regeneration and carcinogenesis.* Jirtle R.L. Ed. New York: Academic Press. pp. 99–112.

89. Usuda N., Reddy J.K., Hashimoto T., and Rao M.S. 1988. Immunocytochemical localization of liver-specific proteins in pancreatic hepatocytes of rat. *Eu. J. Cell. Biol.* 46:299–306.

90. Bartles J.R., Rao M.S., Zhang L., Fayos B.E., Nehme C.L., and Reddy J.K. 1991. Expression and compartmentalization of integral plasma membrane proteins by hepatocytes and their progenitors in the rat pancreas. *J. Cell. Sci.* 98:45–54.

91. Dwivedi R.S., Yeldandi A.V., Subbarao V., Feigelson P., Roy A.K., Reddy J.K., and Rao M.S. 1990. Androgen regulated expression of the α_2u-globulin gene in pancreatic hepatocytes of rat. *J. Cell. Biol.* 110:263–267.

92. Yeldandi A.V., Tan X., Dwivedi R.S., Subbarao V., Smith Jr. D.D., Scarpelli D.G., Rao M.S., and Reddy J.K. 1990. Coexpression of glutamine synthetase and carbamoylphosphate Synthase 1 genes in pancreatic hepatocytes of rat. *Proc. Natl. Acad. Sci. U.S.A.* 87:881–885.

93. Gumucio J.J. 1989. Hepatocyte heterogeneity: the coming of age from description of a biological curiosity to a partial understanding of its physiological meaning and regulation. *Hepatology* 9:154–160.

94. Reddy J.K. and Lalwani N.D. 1983. Carcinogenesis by hepatic peroxisome proliferators: evaluation of the risk of hypolipidemic drugs and industrial plasticizers to humans. *CRC Crit. Rev. Toxicol.* 12:1–58.

95. Reddy J.K. and Chu R. 1996. Peroxisome proliferator-induced pleiotropic responses: pursuit of a phenomenon. *Ann. N.Y. Acad. Sci.* 804:176–201.

96. Rao M.S. and Reddy J.K. 1987. Peroxisome proliferation and hepatocarcinogenesis. *Carcinogenesis* 8:631–636.

97. Rao M.S., Subbarao V., and Reddy J.K. 1986. Induction of hepatocytes in the pancreas of copper-depleted rats following copper repletion. *Cell Diff.* 18:109–117.

98. Svoboda D.J. and Reddy J.K. 1982. Some effects of carcinogens on cell organelles. In *Cancer, a comprehensive treatise.* Becker F. Ed. Vol 1. New York: Plenum Press. pp. 411–449.

99. Slack J.M.W. 1995. Developmental biology of the pancreas. *Development* 121:1569–1580.

100. Zaret K.S. 1996. Molecular genetics of early liver development. *Ann. Rev. Physiol.* 58:231–251.

101. Johansson K.A. and Grapin-Botton A. 2002. Development and disease of the pancreas. *Clin. Genet.* 62:14–23.

102. Percival A.C. and Slack J.M.W. 1999. Analysis of pancreatic development using a cell lineage label. *Exp. Cell. Res.* 247:123–132.

103. Golosow N. and Grobstein C. 1962. Epitheliomesenchymal interaction in pancreatic morphogenesis. *Dev. Biol.* 4:242–255.

104. Gittes G., Galante P., Hanahan D., Rutter W., and Debas H. 1996. Lineage specific morphogenesis in the developing pancreas: role of mesenchymal factors. *Development* 122:439–447.

105. Miralles F., Czernichow P., and Scharfman R. 1998. Follistatin regulates the relative proportions of endocrine versus exocrine tissue during pancreatic development. *Development* 125:1017–1024.

106. Kim S.K. and Hebrok M. 2001. Intercellular signals regulating pancreas development and function. *Genes Dev.* 15:111–127.

107. Massague J. and Chen Y.C. 2000. Controlling TGF-β signaling. *Genes Dev.* 14:627–644.

108. Gu G., Dubauskaite J., and Melton D.A. 2002. Direct evidence for the pancreatic lineage: NGN3+ are islet progenitors and are distinct from duct progenitors. *Development* 129:2447–2457.

109. Pictet L.R., Clark W.R., Williams R.H., and Rutter W.J. 1972. An ultrastructural analysis of the developing embryonic pancreas. *Dev. Biol.* 29:436–467.

110. Rosenberg L., Brown R.A., and Duguid W.P. 1983. A new approach to the induction of duct epithelial hyperplasia and nesidioblastosis by cellophane wrapping of the hamster pancreas. *J. Surg. Res.* 35:63–72.

111. Elsasser H., Lutcke H., and Kern H.F. 1986. Acinar and duct cell replication and regeneration. In *The exocrine pancreas*. Go V.W., Gardner J.D., Brooke F.P., Lebenthal E., DiMagno E.P., and Sheele A.E. Eds. New York: Raven Press. pp. 45–53.

112. Tsao M. and Duguid W.P. 1987. Establishment of propagable epithelial cell lines from normal adult rat pancreas. *Exp. Cell. Res.* 168:365–375.

113. Dabeva M.D., Seong-Gyu H., Vasa S.G., Hurston E., Novikoff P.M., Hixson D.C., Gupta S., and Shafritz D.A. 1997. Differentiation of pancreatic epithelial progenitor cells into hepatocytes following transplantation into rat liver. *Proc. Natl. Acad. Sci. U.S.A.* 94:7356–7361.

114. Chen J., Tsao M., and Duguid W.P. 1995. Hepatocyte differentiation of cultured rat pancreatic ductal epithelial cells after in vivo implantation. *Am. J. Pathol.* 147:707–717.

19

ENVIRONMENTAL FACTORS AND EXOCRINE PANCREATIC DISEASE

Albert B. Lowenfels and Patrick Maisonneuve

CONTENTS

19.1 INTRODUCTION

Until the last few decades, the inaccessibility of the pancreas, which is located in the depths of the abdominal cavity, has interfered with accurate studies of the exocrine portion of the gland. Radiographic studies have been limited, and until the past two decades, it has been difficult to perform endoscopic studies.

Functional studies, such as the secretin test or examination of the fat content of the stool, have also been time-consuming or required considerable patient compliance. With respect to malignant disease of the pancreas, in inoperable cases, gastroenterologists were reluctant to perform biopsies, because of the likelihood of inducing a pancreatic fistula.

With the widespread availability of computed tomography (CT) scans, and the frequent use of endoscopic retrograde cholangiopancreatography (ERCP), the diagnosis of pancreatic disease has become easier and much more accurate. In this chapter, we will review some of the important environmental aspects of acute pancreatitis, chronic pancreatitis, and adenocarcinoma of the pancreas, and update a previous review on this subject.[1]

19.2 ACUTE PANCREATITIS

There are multiple causes for acute pancreatitis, but two factors account for almost 90% of all attacks:

1. Gallstones, which cause acute biliary pancreatitis and are especially common in females
2. Alcohol, which is the most common cause of acute pancreatitis in males

Both of these forms of pancreatitis are related to environmental factors.

19.2.1 Biliary Pancreatitis

Gallstones are extremely common in developed countries, but the prevalence of gallstones varies widely throughout the world. They are more common in females than in males and become more common with increasing age. Gallstones form in the gallbladder, and when they are confined to this organ, they cause acute cholecystitis, rather than pancreatitis. But severe pancreatitis can occur if gallstones enter the common bile duct and obstruct the opening of the pancreatic duct. Fortunately, only a small percentage of patients with gallstone disease, perhaps 1 or 2%, ever develop pancreatitis. Small gallstones are often found in patients who develop biliary pancreatitis.

What are some of the environmental factors leading to the formation of gallstones? Numerous studies have determined that the major risk factor causing cholelithiasis is obesity. Both male and females who are overweight have an increased risk of cholelithiasis. In addition, weight reduction, perhaps by altering lipid excretion, can lead to the onset of symptomatic gallstone disease. In many countries, obesity is becoming a widespread problem, so it is reasonable to predict that the frequency of gallstones and biliary pancreatitis will increase.

If obesity is a major cause for gallstone disease, then exercise, an important weight control measure, should be beneficial. Leitzmann and coworkers investigated this hypothesis in a cohort study of 45,813 male health professionals.[2] Based on their findings, they estimated that approximately one-third of all cases of symptomatic gallstone disease could be prevented by about half an hour of exercise five times per week.

Many other risk factors have been investigated as a potential cause of cholelithiasis, but in most reports have not been found to be major contributory causes. In particular, smoking does not appear to be related to gallstone formation, whereas alcohol consumption may reduce the frequency of gallstones. Oral contraceptives do not seem to be related to gallstones.[3] Surprisingly, serum lipid levels may be negatively correlated with the presence of gallstones, although hyperlipidemia is sometimes associated with acute pancreatitis.

19.2.2   Alcoholic Pancreatitis

Excess alcohol consumption is the most common cause of acute pancreatitis in men (see Chapter 15). One or two episodes of acute alcoholic intoxication do not cause pancreatitis, but acute pancreatitis can be the first manifestation of pancreatic disease in heavy drinkers. Chronic pancreatitis is likely to develop, if these patients continue to drink. The exact mechanism by which alcohol damages the pancreas is unclear, although it is known that alcohol can stimulate pancreatic exocrine secretion.

19.2.3   Miscellaneous Causes of Acute Pancreatitis

In addition to gallstones and alcohol, there are several other environmental factors that occasionally cause acute pancreatitis. Parasitic disease such as ascariasis can cause acute pancreatitis, if, as sometimes happens, one of the worms migrates out of the digestive tract and enters the pancreaticobiliary system. Several viral infections such as mumps, infectious mononucleosis, Coxsackie B virus, and fulminant viral hepatitis sometimes cause acute pancreatitis. In addition, drug therapy for human immunodeficiency virus (HIV) infection can cause pancreatitis.

During the past few decades, ERCP has become a popular procedure for investigating the pancreatico-biliary system. Unfortunately, even when performed by a skilled endoscopist, this procedure can lead to acute pancreatitis in approximately 5 to 10% of patients.

Exposure to therapeutic drugs is another environmental cause for acute pancreatitis. Drugs known to cause pancreatitis can be classified in several categories, including antibiotics, diuretics, chemotherapeutic agents, and antiviral agents.[4] Sometimes a single exposure to an agent, such as tetracycline, causes pancreatitis; for other toxic drugs, the results are dose-dependent.

In addition to drug exposure, there have been a few cases of pancreatitis resulting from exposure to either industrial poisons, such as parathion, or natural toxins from spider bites or other insect stings.

19.3 CHRONIC PANCREATITIS

Chronic pancreatitis is a debilitating disorder characterized by pain, diarrhea, weight loss, jaundice, and in many patients, diabetes. Continued, prolonged symptoms often lead patients to seek surgical relief either by drainage procedures or, in severe cases, by partial or complete pancreatectomy. The diagnosis can be confusing, especially at the onset of the disease, and the frequency of the disease is likely to be underestimated. The frequency varies throughout the world, with an estimated range of 6/100,000/year in Germany,[5] to 45/100,000/year in Japanese males.[6]

19.3.1 Alcohol and Chronic Pancreatitis

There are several different causes for chronic pancreatitis, but in developed countries, by far the most common cause is heavy consumption of alcohol, accounting for about two-thirds of all cases of this disease (see Chapter 15). Moderate drinking or even occasional bouts of binge drinking do not seem to cause chronic pancreatitis, but prolonged regular intake in daily amounts of 80 grams of alcohol per day can produce chronic pancreatitis. One study performed in the U.K. revealed a nearly linear consumption between the frequency of chronic pancreatitis and the amount of alcohol consumed.[7] Heavy drinking is also the major risk factor for liver cirrhosis, but it is difficult to predict whether a particular alcohol consumer will develop cirrhosis, pancreatitis, or neither disease. It is unusual for a heavy drinker to develop both pancreatitis and alcoholic cirrhosis. Table 19.1, based upon data from a large multicenter study of chronic pancreatitis, lists some of the characteristics of patients with chronic pancreatitis.[5]

The exact mechanism by which alcohol damages the pancreas has not been firmly established and will be reviewed in Chapter 15 and Chapter

**Table 19.1 Characteristics of 2015
Subjects with Chronic Pancreatitis**

Characteristic	% Patients
Age at Diagnosis	
<40	35
40–59	50
≥60	15
Gender	
Male	79
Female	21
Type Pancreatitis	
Alcoholic	78
Nonalcoholic	22
Calcification	
Present	60
Absent	40
Estimated Alcohol Consumption	
<5 drinks/day	37
≥5 drinks/day	63
Smoking Status	
Ever	67
Never	33
Other Diseases	
Diabetes	45
Cirrhosis	7
Cancer	11
Pancreatic	3
Nonpancreatic	8

16. One recent theory, the necrosis-fibrosis theory, attempts to link acute and chronic pancreatitis with the gradual development of perilobular fibrosis, ductal stenosis, and ductal stenosis in response to many years of exposure to large quantities of alcohol.[8]

19.3.2 Smoking and Chronic Pancreatitis

Smoking has long been known to be a major risk factor for pancreatic cancer, and in recent years, several reports have appeared emphasizing the role of smoking as a risk factor for chronic pancreatitis. As shown in Table 19.1, a large proportion of patients with pancreatitis smoke and are usually alcohol consumers. Thus, there is the possibility that these two environmental exposures act synergistically to cause chronic pancreatitis.

Chowdhury and coworkers have studied the effects of nicotine on the pancreas.[9] In experimental animals, nicotine causes pancreatic damage, perhaps by leading to increased levels of calcium, causing cytotoxicity and eventual cell death. They also point out that polymorphic alleleles in the D2 dopamine receptor gene have been implicated in susceptibility to nicotine dependence and perhaps to alcohol dependence. This receptor gene has been studied in alcoholics and in patients with lung cancer; similar studies in patients with chronic pancreatitis would be worthwhile.

19.3.3   Nutrition and Chronic Pancreatitis

What role does nutrition play as a cause of chronic pancreatitis? A large proportion of the caloric intake consumed by patients with alcohol-related pancreatitis comes from the high alcohol content of their diet, resulting in a potential deficiency of other nutrient factors. Patients with chronic alcoholic pancreatitis have been reported to consume less protein, carbohydrate, and fats than control subjects.[10] Other reports suggest that an increased intake of items such as fat or protein can cause chronic pancreatitis.[11] In a Japanese case-control study, Lin reported that reduced intake of saturated fatty acids and Vitamin E both increased the risk of chronic pancreatitis.[12] Nutritional factors undoubtedly contribute to the development of chronic pancreatitis, but are not as well documented as either alcohol or smoking.

19.3.4   Tropical Pancreatitis

This form of pancreatitis is a common cause of chronic pancreatitis in southern India and in parts of Africa. Diabetes is often the initial manifestation of tropical pancreatitis, followed by additional signs and symptoms of exocrine pancreatic disease as the disease progresses. Pancreatic calcification is a common finding. The disease does not seem to have a strong familial component and the occasional occurrence of two family members with tropical pancreatitis is likely to be a chance finding. At one time, consumption of cassava root was thought to be a causative agent, but it is difficult to find compelling evidence for this hypothesis.[13] Because of the unusual geographic distribution, it is likely that this form of pancreatitis is related to some as yet unrecognized environmental exposure.

19.3.5   Hereditary Pancreatitis

Hereditary pancreatitis is a rare form of pancreatitis usually occurring in two or more generations, with onset at an early age, and a high risk of pancreatic cancer (see Chapter 19 and Chapter 25). The gene for this

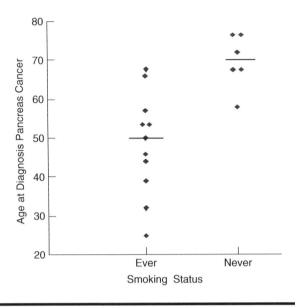

Figure 19.1 Age at diagnosis of pancreatic cancer in patients with hereditary pancreatitis, according to smoking status. Median age (horizontal lines) in ever smokers, 50 years; never smokers, 70 years (P = 0.02 Mann-Whitney test). (From Lowenfels et al. 2001.[14] Used with permission.)

disease has been located and is known to be inherited in an autosomal dominant fashion with a penetrance of about 80%. Although this type of pancreatitis is primarily caused by a genetic defect, smoking is known to interact with the defective gene, increasing the risk of pancreas cancer and causing the age of onset of pancreatic cancer to be significantly lower than in nonsmoking patients with hereditary pancreatitis who develop pancreas cancer (Figure 19.1).[14]

19.4 PANCREAS CANCER

Until now, pancreatic cancer has been unresponsive to both surgical and chemotherapeutic treatment, which explains why although the tumor is relatively rare, pancreas cancer ranks as the fourth most common cause of cancer mortality in the United States. Clinicians and patients know that afflicted patients rarely survive for 5 years.

Smoking is the major known cause of pancreatic cancer, accounting for about 25 to 30% of all cases. Numerous studies confirm this finding (Table 19.2) (see Chapter 22). The findings are highly consistent and reveal the following features:

Table 19.2 Relation between Smoking and Pancreatic Cancer in Various Studies

Author Year (Ref)	Country	Type Study	Rel Risk Smokers/Nonsmokers (95% CI*)	Remarks
Doll 1994 (33)	U.K.	Cohort	2.2 (1.5–3.0)	40-year follow-up of U.K. physicians.
Whittemore 1983 (34)	United States	Cohort	1.7 1.1–2.7	Male university graduates.
Coughlin 2000 (21)	United States	Cohort	Males 2.1 (1.9–2.4) Females 2.0 (1.8–2.3)	More than 1 million subjects with 14-years follow-up.
Wynder 1973 (35)	United States	Case-control	2.7**	Early study. Strong dose response.
Silverman 1994 (36)	Eastern United States	Case-control	1.4–5.2 1.7 (1.3–2.2)	Population-based. Direct interviews only.
Ghadirian 1991 (37)	Quebec Canada	Case-control	3.8 (1.8–7.8)	French speaking population.
Boyle 1996 (38)	Multicenter	Case-control	1.9 (1.5–2.4)	Includes Australia, Netherlands, Canada, Poland.
Mack 1986 (39)	United States California	Case-control	2.3** (1.3–4.2)	

CI = Confidence interval

**Unadjusted risks

Source: From Lowenfels and Maisonneuve 2003 (40). Used with permission.

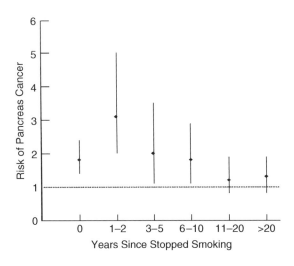

Figure 19.2   Risk of pancreas cancer in relation to time since cessation of smoking.[36] (From Lowenfels and Maisonneuve 2003.[40] Used with permission.)

- There is a time lag of at least two decades between the onset of smoking and the development of pancreas cancer.
- Smoking is a much weaker risk factor for pancreatic cancer than for lung cancer. For lung cancer, smoking results in at least a 10-fold increased risk of cancer, whereas smoking in most studies doubles the risk of pancreas cancer.
- There is a dose-response: the risk of pancreas cancer increases with the number of pack-years of exposure.
- Ex-smokers remain at increased risk of pancreas cancer for at least a decade (Figure 19.2).

The changing frequency of pancreatic cancer in many countries can be largely explained by changes in smoking prevalence. For example, in the United States, male smoking rates have decreased over the past two to three decades, as have incidence rates for pancreatic cancer. In contrast, female smoking rates have increased, causing an increasing incidence of female pancreas cancer, which now is nearly the same as in males.

19.4.1   Tobacco Carcinogens and Pancreatic Cancer

Tobacco smoke contains numerous carcinogens, and the ability or inability to detoxify such agents must be related to susceptibility to pancreatic cancer (see Chapter 7 and Chapter 22). Unfortunately, reports dealing with genetic susceptibility to smoking-induced pancreatic cancer are inconsistent. Some reports have not found a strong relationship between pancreatic cancer and polymorphisms of detoxifying genes such as P4501A1, P4502D6, P4502E1, GSTM1, GSTT1, and Nat1.[15–17] One study found that white female smokers carrying the GSTT1-null phenotype had a five-fold increased risk of pancreatic cancer, when compared to nonsmokers carrying the wild-type genotype.[18] In heavy smokers, the slow acetylator NAT2 genotype may be associated with an increased risk of pancreatic cancer.[16] Because of multiple endpoints, these observations need to be confirmed in other studies, but the role of polymorphic detoxifying genetic mechanisms is likely become increasingly important.

Uridine 5-diphosphate glucuronosyltransferases (UGTs) are a group of potentially important proteins for cellular defense. Their function is to detoxify many drugs, endogenous metabolites, and environmental carcinogens. In patients younger than 55 years, low levels of UGT1A7*3, which is found in the pancreas, have been associated with an increased risk of pancreas cancer: odds ratio = 4.7, 95% CI = 1.9–11.8.[19]

19.4.2 Diet and Pancreatic Cancer

Diet has been studied extensively with respect to causation or prevention of pancreatic cancer (see Chapter 20 and Chapter 22). Unlike tobacco, the results of dietary studies are less clear. Pancreatic cancer patients often die before being available for questioning in a case-control study, so that in many studies proxy respondents have provided information, which could be inaccurate. Caloric consumption is a major factor affecting the incidence of many benign and malignant diseases. Correlation studies reveal a strong association between calories, particularly those derived from animal sources, and pancreatic cancer.[20] Several studies have shown that persons with high caloric intakes are at higher risk of pancreas cancer than persons consuming fewer calories.[21–23] Exercise, which has the potential to prevent obesity by burning excess calories, appears to reduce pancreatic cancer risk.[24]

Many adults consume vitamins on a daily basis, but do supplemental vitamins reduce the risk of pancreatic cancer? In a large study of heavy-smoking males, long-term administration of alpha tocopherol or beta-carotene did not reduce the risk of pancreatic cancer.[25] In the same study, there was an inverse association between levels of serum folate and pyridoxine and pancreatic cancer, suggesting that maintaining adequate levels of these micronutrients could be beneficial.[26]

In 1981, a major case-control study appeared implicating coffee drinking as a risk factor for pancreatic cancer.[27] Nearly all subsequent studies have failed to find any link between coffee consumption and pancreatic cancer (see Chapter 20), but there may be a link between coffee and

K-ras positivity. Presumably, in the original study, there may have been confounding with smoking, or perhaps, the control subjects had a reduced consumption of coffee.

19.4.3 Occupational Factors

Occupational exposure has been found to have only a weak relationship to pancreatic cancer. The findings from several published studies are summarized in Table 19.3. Exposures to agents such as chlorinated hydrocarbons, pesticides, and organochlorine compounds have been investigated, but in most cases the evidence has been weak or conflicting (see Chapter 20). It is unlikely that occupational exposure accounts for more than a small proportion of all cases of pancreatic cancer.

19.4.4 Bacterial Infection and Pancreatic Disease

Infection with *helicobacter pylori* is a strong risk factor for gastric cancer. Could it also cause pancreatic cancer? (See Chapter 20.) One Austrian study found that *H. pylori* positivity doubled the risk of pancreatic cancer.[28] A second Finnish prospective study performed in male smokers again found a positive association between evidence of *H. pylori* infection and the eventual development of pancreas cancer: odds ratio = 1.9, 95% CI = 1.05–3.3.[29] However, international prevalence rates for *H. pylori* infection differ from the distribution of pancreatic cancer: high rates of *H. pylori* seropositivity do not correspond with areas with a high incidence of pancreatic cancer.

Typhoid infection increases the risk of biliary tract and has been reported as a causative factor for pancreatic cancer.[30] A possible mechanism might be that colonization of the biliary tract with typhoid organisms enhances the degradation of bile acids into carcinogens.

19.4.5 Ultraviolet Exposure and Pancreatic Cancer

Ultraviolet (UV) exposure is a powerful risk factor for skin cancer, but based upon ecologic studies, UV exposure may reduce the risk of several major human cancers.[31] A geographic gradient in the frequency of pancreatic cancer in Japan supports this hypothesis: pancreatic cancer is more frequent in the northern districts of the country and less frequent in southern regions.[32]

19.5 EPILOGUE

During the past three decades, because of newer diagnostic procedures, such as CT scans and ERCP, the diagnosis of pancreatic disorders has

Table 19.3 Occupational Exposure and Risk of Pancreas Cancer

Type Exposure	Author (Year)	Study Type (Country)	Findings
Chlorinated hydrocarbons	Ojajarvi et al. 2001 (41)	Meta-analysis (International)	Weak association
Formaldehyde	Collins et al. 2001. (42)	Meta-analysis (International)	Probably not a risk factor
Pesticides	Ji et al. 2001 (43)	Case-control (United States)	Pesticides may increase risk of pancreatic cancer
Pesticides	Kauppinen et al. 1995. (44)	Case-control (Finland)	Nonsignificant increased risk
Pesticides	Alguacil et al. 2000 (45)	Case-control (Spain)	Moderate risk, especially for arsenical pesticides
Organochlorines	Hoppin et al. 2000 (46)	Case-control (United States)	Moderate increased risk for polychlorinated biphenyls
Organochlorines	Porta et al. 1999 (47)	Case-control (Spain)	Increased risk possibly through modulation of K-ras activity.
Organochlorines	Fryzek et al. 1997 (48)	Case-control (United States)	Nonsignificant increased risk
Cadmium	Schwartz and Reis 2000 (49); Ojajärvi 2000 (50)	Reviews	Possible association, but results are conflicting

Source: Modified from Lowenfels and Maisonneuve 2003 (40). Used with permission.

Table 19.4 Summary of Major Environmental Risk Factors Suspected to Be Associated with Pancreatic Disorders

Risk Factor	Relationship to Pancreatic Disease
Smoking	Doubles risk of pancreatic cancer, accounting for 25 to 30% of all cases. A cofactor for chronic pancreatitis.
Alcohol	Most common cause of acute pancreatitis in males and for chronic pancreatitis in males and females.
Diet	High caloric intake produces gallstones, a frequent cause of acute pancreatitis. High caloric, high fat diets may be linked with pancreatic cancer.
Exercise	Counteracts excessive caloric intake and obesity, possibly reducing the frequency of pancreatic cancer.
Occupation	Not a major risk factor for any pancreatic disease; unlikely to cause more than 5 to 10% of pancreatic cancer.
Other factors	Infection with salmonella organs and possibly *H. pylori* increases the risk of pancreatic cancer. UV exposure possibly reduces risk of pancreatic cancer.

become much more accurate, enabling a more careful search for underlying causative factors. Table 19.4 lists some of the more common risk factors believed to be related to pancreatic disease. Although we can usually detect the cause of acute pancreatitis, the underlying cause of chronic pancreatitis is still obscure in approximately one-third of all patients. With respect to pancreatic cancer, smoking is the most common known risk factor, accounting for 25 to 30% of all cases and is the best public health target for reducing the burden of this lethal tumor.

ACKNOWLEDGMENTS

Supported in part by grants from the C.D. Smithers Foundation, Solvay Pharmaceuticals, Inc., and the Italian Association for Cancer Research.

REFERENCES

1. Longnecker D.S. 1977. Environmental factors and diseases of the pancreas. *Environ. Health Perspect.* 20:105–112.
2. Leitzmann M.F., Giovannucci E.L., Rimm E.B., Stampfer M.J., Spiegelman D., Wing A.L. et al. 1998. The relation of physical activity to risk for symptomatic gallstone disease in men. *Ann. Intern. Med.* 128(6):417–425.
3. Grodstein F., Colditz G.A., Hunter D.J., Manson J.E., Willett W.C., and Stampfer M.J. 1994. A prospective study of symptomatic gallstones in women: relation with oral contraceptives and other risk factors. *Obstet. Gynecol.* 84(2):207–214.

4. Steinberg W. and Tenner S. 1994. Acute pancreatitis. *N. Engl. J. Med.* 330:1198–1209.

5. Lowenfels A.B., Maisonneuve P., Cavallini G., Ammann R.W., Lankisch P.G., Andersen J.R. et al. 1994. Prognosis of chronic pancreatitis: an international multicenter study. International Pancreatitis Study Group. *Am. J. Gastroenterol.* 89(9):1467–1471.

6. Lin Y., Tamakoshi A., Matsuno S., Takeda K., Hayakawa T., Kitagawa M. et al. 2000. Nationwide epidemiological survey of chronic pancreatitis in Japan. *J. Gastroenterol.* 35(2):136–141.

7. Johnson C.D. and Hosking S. 1991. National statistics for diet, alcohol consumption, and chronic pancreatitis in England and Wales, 1960–88. *Gut* 32(11):1401–1405.

8. Ammann R.W., Heitz P.U., and Kloppel G. 1996. Course of alcoholic chronic pancreatitis: a prospective clinicomorphological long-term study. *Gastroenterology* 111(1):224–231.

9. Chowdhury P., MacLeod S., Udupa K.B., and Rayford P.L. 2002. Pathophysiological effects of nicotine on the pancreas: an update. *Exp. Biol. Med. (Maywood)* 227(7):445–454.

10. Mezey E., Kolman C.J., Diehl A.M., Mitchell M.C., and Herlong H.F. 1988. Alcohol and dietary intake in the development of chronic pancreatitis and liver disease in alcoholism. *Amer. J. Clin. Nutr.* 48:148–151.

11. Levy P., Matozaki T., Rosato F., Ruff S.J., and Bernades P. 1995. Multidimensional case-control study of dietary, alcohol, and tobacco habits in alcoholic men with chronic pancreatitis. *Pancreas* 10:231–238.

12. Lin Y., Tamakoshi A., Hayakawa T., Ogawa M., and Ohno Y. 2001. Associations of alcohol drinking and nutrient intake with chronic pancreatitis: findings from a case-control study in Japan. *Am. J. Gastroenterol.* 96(9):2622–2627.

13. Narendranathan M. and Cheriyan A. 1994. Lack of association between cassava consumption and tropical pancreatitis syndrome. *J. Gastroenterol. Hepatol.* 9(3):282–285.

14. Lowenfels A.B., Maisonneuve P., Whitcomb D.C., Lerch M.M., and DiMagno E.P. 2001. Cigarette smoking as a risk factor for pancreatic cancer in patients with hereditary pancreatitis. *JAMA* 286(2):169–170.

15. Liu G., Ghadirian P., Vesprini D., Hamel N., Paradis A.J., Lal G. et al. 2000. Polymorphisms in GSTM1, GSTT1 and CYP1A1 and risk of pancreatic adenocarcinoma. *Br. J. Cancer* 82(10):1646–1649.

16. Li D. and Jiao L. 2003. Molecular epidemiology of pancreatic cancer. *Intl. J. Gastrointest. Cancer* 33(1):3–14.

17. Bartsch H., Malaveille C., Lowenfels A.B., Maisonneuve P., Hautefeuille A., and Boyle P. 1998. Genetic polymorphism of N-acetyltransferases, glutathione S-transferase M1 and NAD(P)H:quinone oxidoreductase in relation to malignant and benign pancreatic disease risk. The International Pancreatic Disease Study Group. *Eur. J. Cancer Prev.* 7(3):215–223.

18. Duell E.J., Holly E.A., Bracci P.M., Liu M. et al. 2002. A population-based case-control study of polymorphisms in carcinogen-metabolizing genes, smoking and pancreatic adenocarcinoma risk. *J. Natl. Cancer Inst.* 94:297–306.

19. Ockenga J., Vogel A., Teich N., Keim V., Manns M.P., and Strassburg C.P. 2003. UDP glucuronosyltransferase (UGT1A7) gene polymorphisms increase the risk of chronic pancreatitis and pancreatic cancer. *Gastroenterology* 124(7):1802–1808.

20. Ghadirian P., Thouez J.P., and PetitClerc C. 1991. International comparisons of nutrition and mortality from pancreatic cancer. *Cancer Detect. Prev.* 15:357–362.

21. Coughlin S.S., Calle E.E., Patel A.V., and Thun M.J. 2000. Predictors of pancreatic cancer mortality among a large cohort of United States adults. *Cancer Causes Control* 11(10):915–923.

22. Howe G.R., Ghadirian P., Bueno de Mesquita H.B., Zatonski W., Baghurst P.A., Miller A.B. et al. 1992. A collaborative case-control study of nutrient intake and pancreatic cancer within the SEARCH programme. *Intl. J. Cancer* 51:365–372.

23. Silverman D.T., Swanson C.A., Gridley G., Wacholder S., Greenberg R.S., Brown L.M. et al. 1998. Dietary and nutritional factors and pancreatic cancer: a case-control study based on direct interviews. *J. Natl. Cancer Inst.* 90(22):1710–1719.

24. Michaud D.S., Giovannucci E., Willett W.C., Colditz G.A., Stampfer M.J., and Fuchs C.S. 2001. Physical activity, obesity, height, and the risk of pancreatic cancer. *JAMA* 286(8):921–929.

25. Rautalahti M.T., Virtamo J.R., Taylor P.R., Heinonen O.P., Albanes D., Haukka J.K. et al. 1999. The effects of supplementation with alpha-tocopherol and beta-carotene on the incidence and mortality of carcinoma of the pancreas in a randomized, controlled trial. *Cancer* 86(1):37–42.

26. Stolzenberg-Solomon R.Z., Albanes D., Nieto F.J., Hartman T.J., Tangrea J.A., Rautalahti M. et al. 1999. Pancreatic cancer risk and nutrition-related methyl-group availability indicators in male smokers. *J. Natl. Cancer Inst.* 91(6):535–541.

27. MacMahon B., Yen S., Trichopoulos D., Warren K., Nardi G., Nomura A. et al. 1981. Coffee drinkings and cancer of the pancreas. *Br. Med. J.* 283:628.

28. Raderer M., Wrba F., Kornek G., Maca T., Koller D.Y., Weinlaender G. et al. 1998. Association between Helicobacter pylori infection and pancreatic cancer. *Oncology* 55(1):16–19.

29. Stolzenberg-Solomon R.Z., Blaser M.J., Limburg P.J., Perez-Perez G., Taylor P.R., Virtamo J. et al. 2001. Helicobacter pylori seropositivity as a risk factor for pancreatic cancer. *J. Natl. Cancer Inst.* 93(12):937–941.

30. Caygill C.P., Hill M.J., Braddick M., and Sharp J.C. 1994. Cancer mortality in chronic typhoid and paratyphoid carriers. *Lancet* 343:83–84.

31. Grant W.B. 2003. Ecologic studies of solar UV-B radiation and cancer mortality rates. *Recent Results Cancer Res.* 164:371–377.

32. Kato I., Tajima K., Kuroishi T., and Tominaga S. 1985. Latitude and Pancreatic Cancer. *Jpn. J. Clin. Oncol.* 15:403–413.

33. Doll R., Peto R., Wheatley K., Gray R., and Sutherland I. 1994. Mortality in relation to smoking: 40 years' observations on male British doctors [see comments]. *Br. Med. J.* 309(6959):901–911.

34. Whittemore A.S., Paffenbarger R.S., Anderson K., and Halpern J. 1983. Early precursors of pancreatic cancer in college men. *J. Chron. Dis.* 36:251–256.

35. Wynder E.L., Mabuchi K., Maruchi N., and Fortner J.G. 1973. Epidemiology of cancer of the pancreas. *J. Natl. Cancer Inst.* 50:645–667.

36. Silverman D.T., Dunn J.A., Hoover R.N., Schiffman M., Lillemoe K.D., Schoenberg J.B. et al. 1994. Cigarette smoking and pancreas cancer: a case-control study based on direct interviews. *J. Natl. Cancer Inst.* 86(20):1510–1516.

37. Ghadirian P., Simard A., and Baillargeon J. 1991. Tobacco, alcohol, and coffee and cancer. A population-based case-control study in Quebec, Canada. *Cancer* 67:2664–2670.

38. Boyle P., Maisonneuve P., Bueno D.M., Ghadirian P., Howe G.R., Zatonski W. et al. 1996. Cigarette smoking and pancreas cancer: a case control study of the search programme of the IARC. *Intl. J. Cancer* 67(1):63–71.

39. Mack T.M., Yu M.C., Hanisch R., and Henderson B.E. 1986. Pancreas cancer and smoking, beverage consumption, and past medical history. *J. Natl. Cancer Inst.* 76:49–60.

40. Lowenfels A.B. and Maisonneuve P. 2003. Environmental factors and risk of pancreatic cancer. *Pancreatology* 3(1):1–7.

41. Ojajarvi A., Partanen T., Ahlbom A., Boffetta P., Hakulinen T., Jourenkova N. et al. 2001. Risk of pancreatic cancer in workers exposed to chlorinated hydrocarbon solvents and related compounds: a meta-analysis. *Am. J. Epidemiol.* 153(9):841–850.

42. Collins J.J., Esmen N.A., and Hall T.A. 2001. A review and meta-analysis of formaldehyde exposure and pancreatic cancer. *Am. J. Ind. Med.* 39(3):336–345.

43. Ji B.T., Silverman D.T., Stewart P.A., Blair A., Swanson G.M., Baris D. et al. 2001. Occupational exposure to pesticides and pancreatic cancer. *Am. J. Ind. Med.* 39(1):92–99.

44. Kauppinen T., Partanen T., Degerth R., and Ojajarvi A. 1995. Pancreatic cancer and occupational exposures. *Epidemiology* 6(5):498–502.

45. Alguacil J., Kauppinen T., Porta M., Partanen T., Malats N., Kogevinas M. et al. 2000. Risk of pancreatic cancer and occupational exposures in Spain. PANKRAS II Study Group. *Ann. Occup. Hyg.* 44(5):391–403.

46. Hoppin J.A., Tolbert P.E., Holly E.A., Brock J.W., Korrick S.A., Altshul L.M. et al. 2000. Pancreatic cancer and serum organochlorine levels. *Cancer Epidemiol. Biomarkers Prev.* 9(2):199–205.

47. Porta M., Malats N., Jariod M., Grimalt J.O., Rifa J., Carrato A. et al. 1999. Serum concentrations of organochlorine compounds and K-ras mutations in exocrine pancreatic cancer. PANKRAS II Study Group. *Lancet* 354(9196):2125–2129.

48. Fryzek J.P., Garabrant D.H., Harlow S.D., Severson R.K., Gillespie B.W., Schenk M. et al. 1997. A case-control study of self-reported exposures to pesticides and pancreas cancer in southeastern Michigan. *Intl. J. Cancer* 72(1):62–67.

49. Schwartz G.G. and Reis I.M. 2000. Is cadmium a cause of human pancreatic cancer? *Cancer Epidemiol. Biomarkers Prev.* 9(2):139–145.

50. Ojajarvi I.A., Partanen T.J., Ahlbom A., Boffetta P., Hakulinen T., Jourenkova N. et al. 2000. Occupational exposures and pancreatic cancer: a meta-analysis. *Occup. Environ. Med.* 57(5):316–324.

20

ENVIRONMENTAL DETERMINANTS OF EXOCRINE PANCREATIC CANCER

Eric J. Duell, Paige M. Bracci, and Elizabeth A. Holly

CONTENTS

20.1 INTRODUCTION

Adenocarcinoma of the exocrine pancreas has the lowest 5-year survival (<4%) of all cancers and is the fourth leading cause of cancer-related deaths in men and women in the United States.[1] A lack of reliable early markers for detection and an unclear understanding of etiology have made this one of the deadliest of human cancers. Sporadic pancreatic cancer is the most common form (>90%) and by definition, does not exhibit familial clustering. Thus, familial inheritance does not explain the majority of pancreatic cancer in the general population. Older age is one of the strongest predictors of risk with the majority of cases occurring after the age of 65.[2] Incidence rates are about 25 to 50% higher in men than women until the last decades of life when rates become nearly equivalent. African Americans have some of the highest incidence rates of pancreatic cancer in the world.[2] Generally, worldwide incidence rates of pancreatic cancer are the lowest in Asia, with the important exception of Japan where the rates are comparable to U.S. rates. The incidence of pancreatic cancer has increased in Japan over the past few decades and this observation has been used to suggest the importance of environmental and lifestyle risk factors in this disease (e.g., smoking, diet, and exercise). Some medical conditions have been associated with an increased risk of pancreatic cancer including Type II diabetes, pancreatitis, and peptic ulcer disease. For this review of environmental factors and pancreatic cancer, the conditions of diabetes, pancreatitis, and allergy as risk factors will not be discussed in detail (see Chapter 14 to Chapter 18, Chapter 19, and Chapter 29), except as they relate to dietary factors. Cigarette smoking is the only established environmental risk factor for pancreatic cancer and is estimated to account for 25 to 29% of cases.[3,4] Diets high in saturated fat and high glycemic load also have been implicated. Other potential and suspected environmental risk factors include other aspects of diet, obesity, alcohol, occupational exposures, and environmental pollutants. Studies of gene-environment interactions in sporadic pancreatic cancer will add to our understanding of the complex interplay between genetic risk factors and environmental and behavioral risk factors in this disease. We provide examples of potential gene-environment interactions and show how each may inform relevant pancreatic pathways.

20.2 SMOKING

Cigarette smoking is the only firmly established environmental risk factor for adenocarcinoma of the exocrine pancreas. Several, but not all, studies of smoking and pancreatic cancer have found higher relative risk (RR) estimates in women.[3,5,6] It is possible that there are both biological and nonbiological reasons for these observed differences.[7] Cigarette smoking lowers circulating estrogens, and they might reduce the risk for the development of pancreatic cancer.[8] Because circulating estrogens are higher in women than men, smoking could have more of a noticeable effect on lowering estrogens (and thus increasing risk) in women than in men. Men and women may differ in their ability to metabolize carcinogens and repair DNA.[9] Sex steroids such as estrogens may modify the DNA-damaging properties of cigarette smoke via the cytochrome P450 system.[10] (See Chapter 22 for additional information.) Sex-specific expression of gastrin-releasing peptide has been demonstrated in pancreatic cells and may be responsible for observed gender differences in smoking-related risk estimates for pancreatic cancer, as has been previously hypothesized for lung cancer.[8,11] Alternatively, the background rates for pancreatic cancer may differ between nonsmoking men and nonsmoking women (due to factors such as occupational exposures that may be more common in nonsmoking men) and thus contribute to the smaller observed RR estimates for smoking and pancreatic cancer among men than among women (see Section 20.9). This issue currently is unresolved.

The cellular mechanisms by which tobacco carcinogens affect the human pancreas and initiate or promote pancreatic cancer currently are unknown. A number of mechanisms have been proposed including direct and indirect effects of nitrosamines,[12] aromatic-amine intermediates,[13] other potential pancreatic carcinogens in cigarette smoke such as cadmium,[14] and metabolites related to oxidative stress and inflammation.[15]

20.3 DIETARY FACTORS

A number of case-control studies have evaluated the association between diet and risk of pancreatic cancer. Because of the retrospective assessment of diet in case-control studies, and the high fatality rate for this cancer (and thus, the use of proxy exposure data in many of these studies), case-control studies of diet and pancreatic cancer are considered to have a strong likelihood of bias. To date, there have been seven prospective studies of diet and pancreatic cancer. However, the methods of dietary assessment and the specific nutrients analyzed have varied and few of these studies had more than 100 cases.[16-19] We have separated several salient dietary factors for this review. However, it should be noted that dietary factors including specific foods consumed, macronutrient and energy intake, and level of physical activity are interrelated and may be difficult to separate as independent pancreatic cancer risk factors in many epidemiological studies.

20.3.1 Fat

High-fat diets promote duct-like tumors in animal models of pancreatic cancer[20] (Chapter 20 and Chapter 21). However, there is little prospective or experimental human data on the types of dietary fat and the risk of pancreatic cancer. Initial observations in several case-control studies suggested that diets high in animal fat were associated with increases in pancreatic cancer risk,[2] and two prospective studies have evaluated the association between fat intake and risk of pancreatic cancer.[18,19] A prospective analysis of diet and pancreatic cancer in Finnish male smokers (the Alpha-Tocopherol, Beta-Carotene or ATBC study cohort, based on 163 pancreatic cancer cases) found a positive association between saturated fat intake and the risk of developing pancreatic cancer.[18] Other types of fat were not associated with risk, and greater energy and carbohydrate intakes were associated with a reduced risk of pancreatic cancer. In a prospective analysis of U.S. women within the Nurse's Health Study (based on 178 cases), total fat and various types of fat including saturated fat all were unrelated to the risk of developing pancreatic cancer.[19] It was suggested by the authors that a lack of association between saturated fat and pancreatic cancer in the Nurse's Health Study may have been due to the substantially lower intake of fat and butter in that population compared with Finnish men in the ATBC cohort.[19] Moreover, in general, the Nurse's Health Study consists of former or never smokers, whereas the ATBC cohort consists of current smokers. The general health of the ATBC cohort may not be as favorable as that of women from the Nurse's Health Study.

The only two prospective epidemiologic studies to evaluate types of fat and pancreatic cancer risk had conflicting results.[18,19] Thus, this issue remains to be resolved. The data on dietary fat and human pancreatic cancer risk leave substantial room for interpretation and hypothesis generation regarding potential mechanisms of action. Some of the current hypotheses regarding dietary fat mechanisms and pancreatic cancer relate to:

■ Islet cell injury and proliferation[21] (Chapter 29).
■ Cholecystokinin (CCK)/gastrin stimulation.[22]
■ Long-term positive energy balance.
■ Insulin resistance.
■ Interaction of these potential mechanisms (see also Chapter 29).

The observed inverse effect of physical activity on risk may support the energy balance hypothesis (see Section 20.5), although experimental studies in hamsters did not support this hypothesis.[23] Underlying insulin resistance is likely to be an important determinant for pancreatic cancer risk, but in some human studies (as opposed to animal models) glycemic load may be a stronger determinant of insulin resistance and pancreatic cancer risk than a high-fat diet.[24,25]

20.3.2 Carbohydrates and Insulin Resistance

The presence of Type 2 diabetes (non-insulin-dependent diabetes mellitus) has been associated with pancreatic cancer risk (approximately two-fold increase in RR estimates) in many, but not all, epidemiologic studies.[26] Whether Type 2 diabetes is a risk factor for pancreatic cancer or is an early manifestation of the disease is discussed in Chapter 29. Insulin resistance is an important factor in the pathogenesis of Type 2 diabetes and may be an early event in pancreatic tumorigenesis.

Abnormal glucose tolerance (based on postload plasma glucose) in the United States is associated with increased overall cancer risk, including, although based on small numbers (n = 13), an excess of pancreatic cancers.[27] An association between high postload plasma glucose (measured at baseline) and increased risk of exocrine pancreatic cancer was confirmed among men and women with no history of diabetes at baseline.[24] In this study, 96 men and 43 women died of pancreatic cancer after an average of 25 years of follow-up among 20,475 men and 15,183 women. The authors concluded that factors associated with abnormal glucose metabolism are important in the development of pancreatic cancer.[24] A population-based case-control study from Canada[28] reported higher RRs with increasing body mass index (BMI) and lower RRs with increasing levels of physical activity, especially in men.[28] Because obesity promotes insulin resistance and physical activity is related to increased insulin sensitivity, these observations may be taken as indirect evidence supporting the hypothesis that insulin resistance is an etiologic factor in pancreatic cancer. In an animal model, improvement of peripheral insulin resistance with metformin prevented pancreatic cancer induction.[20]

In an analysis of two large prospective cohorts from the United States (the Health Professionals Follow-Up Study and the Nurses' Health Study contributing 140 men and 210 women diagnosed with pancreatic cancer), it was reported that obesity and physical inactivity both were associated with an increased risk of pancreatic cancer.[29] Only 24 women and 14 men were diabetic prior to their cancer diagnosis. In a recent analysis of the Nurses' Health Study by the same investigators, a diet high in glycemic load was associated with an increased risk of pancreatic cancer, especially among women who were more sedentary and had an elevated BMI.[25] These results were interpreted by the authors to suggest that diets high in glycemic load may increase risk, especially among women with evidence of underlying insulin resistance. Together, these epidemiologic studies suggest that factors related to prior insulin resistance, rather than the presence of diabetes itself, are important in the etiology of pancreatic cancer.

The etiology of insulin resistance is complex and multifactorial. In addition to genetic factors, environmental factors such as glycemic load, caloric intake, obesity, and physical activity are likely to contribute. Insulin resistance, obesity, and diabetes are related but distinct epidemiologic entities. For example, pancreatic cancer has been associated with a higher prevalence and history of insulin resistance and diabetes mellitus than comparable controls in many studies. However, most pancreatic cancer occurs in the absence of diabetes and is not clearly related to obesity, which is a risk factor for Type 2 diabetes and its complications.[26] Moreover, the location of excess adipose tissue (e.g., central or intraabdominal fat) appears to be important for the associations between obesity, insulin resistance and Type 2 diabetes, with nonabdominal fat deposits less strongly associated with insulin resistance and Type 2 diabetes.[30] The causal connection between insulin resistance and central adiposity and obesity is not well understood and could be a result of common environmental and genetic factors for both conditions.[30] Thus, the association between insulin resistance and pancreatic cancer is likely to be modified by obesity, physical activity, diet, history of diabetes, and genetic factors.

Animal models of pancreatic cancer suggest that a high-fat diet induces peripheral insulin resistance and has a promotional effect on pancreatic tumorigenesis.[21] In this model, a high-fat diet increases insulin levels without affecting glucose concentrations (i.e., promotes insulin resistance) and stimulates compensatory proliferation of islet cells that enhance the development of ductal adenocarcinoma of the exocrine pancreas.[20,21] As stated previously, normalization of insulin level by metformin prevented pancreatic cancer induction, indicating that either insulin *per se* or islet cell proliferation associated with peripheral insulin resistance is the underlying factor. Epidemiologic data from human studies are required to test whether these hypotheses generated from animal models explain the observed patterns of association.

20.3.3 Vegetables and Fruits

Although not always consistent, results from epidemiologic investigations of dietary factors and risk for pancreatic cancer generally show decreased risk estimates associated with increased consumption of vegetables and fruits.[31-38] Dietary effects may be related to variations in study design and small sample size, inherent variation in dietary data, inadequate adjustment for total energy intake, limited dietary assessment tools, analysis of nutrient values vs. consumption of specific foods or food groups, cooking methods, and estimation of other potential confounding factors such as smoking, BMI, and physical activity.[18,19,29,36,39-41] The reduced risk associated with fruit and vegetable consumption may be explained by several hypotheses, including exposure to plant hormones and to micronutrients including Vitamin C, Vitamin E, beta-carotene (antioxidant effects), and folic-acid levels in relation to DNA synthesis, repair and methylation.[42] Support for the role of folate is provided by evidence from a large cohort study of older male smokers in Finland.[43] In energy-adjusted models that included adjustment for potential confounding factors, a significant reduced trend for pancreatic cancer was associated with increasing levels of dietary folate, but not with other folate-related dietary nutrients or with folate supplements.[43] An earlier analysis within this cohort showed a decreased trend for pancreatic cancer risk with increasing levels of serum folate.[44] Detailed dietary investigations that include examination of genetic and molecular factors are needed to identify plausible biologic mechanisms that describe the role of these dietary factors in the development of pancreatic cancer.

20.3.4 Alcohol

Overall, there does not appear to be an association between regular alcohol consumption and increased risk of pancreatic cancer.[16,43,45-48] Results have been similar by sex despite data showing that women may be inherently more sensitive to the detrimental effects of alcohol due to sex differences in the pharmacokinetics of ethanol.[49,50] An exception to the consistent null associations are reports that suggest extremely heavy drinking may increase risk,[51,52] although these results have been based on a small number of cases, have had inconsistent definitions of heavy drinking, and risks have varied by the type of alcohol consumed. The relationship between pancreatic cancer and alcohol consumption also is unclear because of the association in some studies between alcohol and chronic pancreatitis, an inflammatory condition that may be an intermediary condition in the development of pancreatic cancer. One study that evaluated alcohol consumption, chronic pancreatitis, and pancreatic cancer among men reported that after adjustment for smoking, alcohol consumption was associated with chronic pancreatitis, but not with pancreatic cancer.[45] Although these estimates were based on few patients, results showed a nearly seven-fold increased RR estimate for alcohol consumption and any pancreatitis, a two-fold RR for chronic pancreatitis, and no association between alcohol consumption and risk of pancreatic cancer. Some studies of pancreatitis have shown little association with alcohol consumption, and several show no, or low, association with alcohol in women.[53] Additional large or pooled studies that include a sufficient number of participants with and without well-defined pancreatitis are needed to determine the role that alcohol consumption may play in the development of pancreatic cancer.

20.3.5   Coffee

Although there have been some conflicting reports in the literature on the possible association between coffee drinking and pancreatic cancer,[38,46–48,54–58] most have shown no association[38,46–48,55] and the current consensus is that coffee is not a risk factor for pancreatic cancer.[59] Results from the few earlier studies that suggested an elevated risk of pancreatic cancer with coffee drinking may have been due to residual (incomplete control of) confounding by smoking or selection bias.[59] Because some coffee drinkers also smoke cigarettes, earlier reports on coffee consumption that ignored smoking status were likely to have been confounded by smoking (see also Chapter 22).

20.3.6   Tea

Risk for pancreatic cancer has not been associated with tea consumption in several earlier epidemiologic investigations that evaluated this exposure,[47,54,56,57] although some recent studies have suggested a reduced risk with consumption of green tea.[60,61] Laboratory studies have shown that polyphenols in tea, especially (-)-epigallocatchin-3-gallate (EGCG), exhibit a variety of antitumor and cancer preventive activities.[62] The activity of these compounds in humans as they relate to dietary exposures requires further evaluation in epidemiologic studies.

20.4   OBESITY

Data from large cohorts of U.S. adults have shown that over the past decade, obesity and its related chronic conditions have been steadily increasing in the United States.[63,64] Incidence of pancreatic cancer may be affected by the increase in these conditions over time. Results from epidemiologic studies that have evaluated the association between BMI and pancreatic cancer have been mixed. The most consistent associations reported have shown elevated risk estimates among men and among obese men and women (BMI > 30).[28,29,32,65–67] A large cohort study of cancer mortality and BMI showed increased pancreatic cancer morality rates associated with increasing BMI both for women and for men, and for men and women who had not smoked during the 16 years of follow-up.[68] However, in a meta-analysis of data collected in epidemiologic studies of pancreatic cancer, BMI was only weakly associated with pancreatic cancer.[69] In two recent studies, increased risk estimates for pancreatic cancer were related to increasing BMI after adjustment for potential confounding factors including smoking, diabetes, physical activity, and total caloric intake.[28,29] Although the biologic mechanism to explain this potential relationship is unclear, it has been hypothesized that obesity's association with insulin resistance, glucose intolerance, and the resultant hyperinsulinemia may be an important factor in the development of pancreatic cancer. In animal studies, peripheral insulin resistance is associated with hyperinsulinemia and islet cell hyperplasia. However, it is unknown whether islet cell hyperplasia, which may make the pancreas more vulnerable to carcinogens, occurs in obese people.[20,21] Insulin resistance and hyperinsulinemia are associated with abdominal adiposity, a fat distribution pattern more common in men than in women that may partly explain the differences oberved by sex. Further research that can measure fat distribution in addition to other factors that influence obesity may provide etiologic clues to fat metabolism in relation to pancreatic carcinogenesis.

20.5 PHYSICAL ACTIVITY

Physical activity influences insulin levels and is a factor of interest for its independent association with pancreatic cancer and for its potential modifying effects with dietary factors and obesity. Comprehensive evaluations of physical activity and pancreatic cancer in humans have not been published and the experimental results are confusing. As stated earlier, in the hamster pancreatic cancer model, physical exercise had no effect on pancreatic carcinogenesis.[23] Most human studies have evaluated recent physical activity levels (1 to 2 years before interview or diagnosis), excluded occupational and household activities, and have used various instruments to assess physical activity levels. Some evaluations of exercise have shown an inverse association between exercise and pancreatic cancer,[28,29] especially among obese study participants,[29] although results have varied by sex and by intensity of activity. Recent results from analyses of physical activity and BMI data collected over time in a large cohort of women and men in the College Alumni Health Study showed no association between physical activity or BMI and pancreatic cancer mortality.[70] Physical activity improves glucose tolerance, and based on current evidence that implicates impaired glucose tolerance as an important factor in the development of pancreatic cancer,[24] the reported inverse associations between physical activity and pancreatic cancer may be attributable to improved glucose tolerance. Other proposed biological mechanisms by which physical activity may reduce risk for cancer include exercise's effect on antitumor immune function (increase in macrophage and killer cell activity), on antioxidant defense levels (increased activity of free-radical scavengers), and on reducing obesity.[71] Physical activity, diet, and obesity are interrelated, and investigations designed to collect extensive detailed history on these data are required to confirm the reported associations and to identify their independent and interactive role in pancreatic cancer etiology.

20.6 EXOGENOUS SEX HORMONES

The risk of pancreatic cancer associated with the use of exogenous hormones such as oral contraceptives (OCs) and hormone replacement therapy (HRT) has been evaluated in a limited number of epidemiologic studies. Of the studies that evaluated use of HRT and pancreatic cancer risk, two studies found inverse associations,[72,73] while two studies found positive associations with risk.[74–76] In a network of case-control studies,[76] HRT use for 2 or more years was associated with a decreased risk [odds ratio (OR = 0.7, 95% CI = 0.2–3.1)], while HRT use of less than 2 years was associated with an increased risk (OR = 2.0, 95% CI = 1.0–4.1), although CIs included unity. Of the studies that evaluated OC use and pancreatic cancer risk, two studies reported reduced risk estimates for pancreatic cancer with OCs,[73–74] while two studies reported a positive association.[66,75] In general, precision of the risk estimates was limited by small sample sizes.

Using data from our population-based case-control study of pancreatic cancer (n = 532 cases, 1701 controls) in the San Francisco Bay area, we studied the role of menstrual and reproductive factors including exogenous hormone use in the etiology of pancreatic cancer among women (241 cases, 818 controls). Our analysis of exogenous hormone use suggested that overall OC use (ever/never) is not associated with pancreatic cancer risk (OR = 0.95, 95% CI = 0.65–1.4), and that postmenopausal estrogen replacement therapy (ERT) (ever/never) probably is not associated with risk (OR = 0.84, 95% CI = 0.62–1.1).[77] Women who used ERT for 18 years or more had a somewhat reduced risk estimate for pancreatic cancer (OR = 0.72, 95% CI = 0.45–1.2), but estimates were imprecise. Age at menopause of 51 years and older was associated with a 2-fold increased risk of pancreatic cancer (OR = 1.9, 95% CI = 1.2–2.8), whereas age at menarche at ≤12 years was associated with a somewhat reduced risk estimate (OR = 0.82, 95% CI = 0.58–1.2), although with wide confidence limits. The estimated risk of pancreatic cancer for current smokers was higher among women who never had used OC or ERT (OR = 11.5, 95% CI = 3.5–3.8) compared with never smokers, and was lower among women who reported use of both OC and ERT (OR = 1.7, 95% CI = 0.56–5.0). Overall, our results do not show a consistent pattern with estrogen exposure and risk for the development of pancreatic cancer (see Section 20.11.4), although the potential modification of smoking related risk by OC and ERT may be noteworthy.[77]

20.7 *HELICOBACTER PYLORI* INFECTION

The presence of *Helicobacter pylori* bacteria or its antibodies is associated with peptic ulcer disease and in a small percentage of infected individuals (<3%), gastric adenocarcinoma.[78] The association with gastric adenocarcinoma is modified by other factors such as the strain of *H. pylori* and by host-related factors. Three pieces of evidence from studies in humans support a role for *H. pylori* in the development of pancreatic adenocarcinoma. The first study, from Austria, evaluated blood samples from 92 pancreatic cancer patients and 27 controls for the presence of IgG antibodies against *H. pylori*.[79] Antibody–positive blood samples were found in 65% of cases and 45% of controls. Pancreatic tumor specimens had no microscopic evidence of *H. pylori* infection. The second study evaluated antibody levels to *H. pylori* from sera obtained at baseline from a case-control analysis (121 cases, 226 controls) nested within the ATBC Cancer Prevention Study cohort of male Finnish smokers.[80] The prevalence of antibodies to *H. pylori* was 82% in cases and 73% in controls. Those with cytotoxin-associated gene A-positive (CagA+) strains prior to diagnosis had a 2-fold increased risk of developing pancreatic cancer (OR = 2.0, 95% CI = 1.1–3.7) compared with *H. pylori* seronegative participants.[80] The third study evaluated pancreatic biopsy specimens suspected of being neoplastic for the presence of *Helicobacter* species and *H. pylori* by bacterial culture, polymerase chain reaction (PCR), and DNA sequencing.[81] Five out of six ductal pancreatic cancer specimens were positive for *Helicobacter* genus by PCR although all ductal cancers were negative for *H. pylori* using species-specific primers. *Helicobacter* was not detected by PCR in two samples of benign pancreatic tissue and none of the tissue samples or biopsies could be cultured successfully for *Helicobacter* species.

The mechanism of action of *Helicobacter* in the pancreas currently is unknown although *H. pylori* can induce increased gastrin secretion within the gastric mucosa.[78] The hormone gastrin is known to have receptors in both normal and malignant pancreatic tissues and has been found to stimulate the growth of several human pancreatic cancer cell lines.[82,83] Taken together, these studies suggest that *Helicobacter* may play a minor role in pancreatic cancer, but this role probably depends on strain genetics, the presence of chronic inflammation, and other host-related factors.

20.8 NSAIDS

The use of nonsteroidal anti-inflammatory drugs (NSAIDS) has been associated with beneficial effects in the gastrointestinal tract including lowering the risk of colon cancer.[84,85] The effects on pancreatic cancer incidence have been less-well studied and the results have been mixed. An early study of aspirin use and incidence of pancreatic cancer in the National Health and Nutrition Examination Survey I (NHANES I) study found an inverse association (RR = 0.67, 95% CI = 0.33–1.4) based on 30 cases.[86] A clinic-based case-control study from the U.K. reported a dose-response association between the number of prescriptions for NSAIDS and pancreatic cancer risk (OR = 1.5, 95% CI = 1.02–2.2 for 7 or more prescriptions at 13 to 36 months before diagnosis).[87] A Danish study of low-dose prescription aspirin use and cancer did not find an association with pancreatic cancer incidence based on 62 cases.[88] A recent analysis of aspirin and other NSAID use and risk of pancreatic cancer from 1992 through 1999 was conducted within a prospective cohort of postmenopausal women in the Iowa Women's Health Study.[89] The RR for any use vs. no use of aspirin was 0.57 (95% CI = 0.36–0.90), based on 80 cases. Moreover, there was a trend of decreasing risk with increasing frequency of aspirin use (trend, p = 0.005).[89] In the Iowa study, the use of non-aspirin NSAIDS was not associated with reduced risk for pancreatic cancer. A similar analysis of aspirin use and pancreatic cancer in the Nurses' Health Study reported the opposite result from that found in the Iowa Women's Health Study.[90] Women who reported 20 or more years of regular aspirin use (2 tablets per week) in the Nurses' study experienced an increased RR of pancreatic cancer of 1.6 (95% CI 1.03–2.4). Further, among women who reported regular aspirin use on at least 2 of 3 biennial questionnaires, risk for pancreatic cancer increased monotonically with increasing number of aspirin tablets taken per week (p = 0.02).[90] The opposite results for these 2 prospective cohorts of women suggest that baseline risk of pancreatic cancer may differ between these 2 populations. For example, women who regularly used aspirin for heart disease prevention in the Nurses' Health Study at baseline may have been at higher risk of some cancers including pancreatic cancer, perhaps through an inflammatory mechanism. Any protective effect of aspirin would be difficult to detect in women already at increased risk. Other potential explanations for the mixed findings in these studies include differential follow-up, loss to follow-up, definitions of exposure (NSAIDS, aspirin, or other), differing dose and frequency criteria, and misclassification of exposure. Research on NSAIDS and risk of pancreatic cancer has begun recently and may identify biological and environmental factors to clarify the relationship.

20.9 OCCUPATION

A number of occupational cohort and case-control studies have evaluated occupation and pancreatic cancer mortality and incidence. Exposure is usually assessed by job title. One of the main weaknesses of these studies has been a lack of precision because of the small number of observed cases of pancreatic cancer. In general, the evidence linking occupations with pancreatic cancer has not been strong. However, a number of occupations and specific occupational exposures have emerged as potentially related to pancreatic cancer. Farming and pesticide exposures are included among the potentially related occupations and occupational exposures. These are covered in Section 20.10.

There is no convincing link between occupational radiation exposure, or asbestos exposure, among asbestos workers and pancreatic cancer.[2,91] Earlier studies of workers employed in rubber manufacturing have shown small increased risks, but a more recent cohort analysis of rubber workers exposed to nitrosamines did not find any excess risk of pancreatic cancer.[92–94] Some studies of occupations with exposure to petroleum products or products of incomplete combustion have found increased risks.[95–99] A number of studies have evaluated occupational solvents such as formaldehyde and chlorinated hydrocarbons in relation to pancreatic cancer mortality or incidence. A recent meta-analysis of formaldehyde exposure and pancreatic cancer found a small increased risk among embalmers [meta-relative risk (MRR) = 1.3, 95% CI = 1.0–1.6], pathologists, and anatomists (MRR = 1.3, 95% CI = 1.0–1.7), but no increased risk among industrial workers who typically have the highest exposures.[100] Another meta-analysis of workers exposed to chlorinated hydrocarbon solvents and pancreatic cancer found excess meta-RR estimates for metal degreasing (MRR = 2.0, 95% CI = 1.2–3.6) and dry cleaning (MRR = 1.4, 95% CI = 1.1–2.4).[101] An analysis from Spain evaluated organic solvent exposure data and activating K-*ras* mutations for the PANCRAS II Study Group using 83 mutated K-*ras* tumors and 24 wild-type K-*ras* tumors.[102] A positive association was found between K-*ras* mutations and solvent exposure in general including any hydrocarbon solvent, aromatic solvents, and aliphatic solvents, and more specifically, for benzene exposure.

20.10 PESTICIDES

Farmers and agricultural workers may be exposed to many different chemicals including insecticides, herbicides, fungicides, rodenticides, fertilizers, fuels, exhausts, dusts, and biological agents such as bacteria and animal viruses. Overall, occupational studies of farmers and agricultural workers have not indicated an excess of pancreatic cancer mortality or incidence associated with farming as an occupation.[103–107] However, a recent mortality analysis of crop and livestock farmers in the United States suggested that livestock farmers, but not crop farmers, had a significantly increased mortality from pancreatic cancer.[108] The authors concluded that livestock farmers may be exposed to more carcinogens than crop farmers. An earlier case-control study found an elevated RR estimate for ever having lived on a farm or for having had farm animals (OR = 1.6, p = 0.079).[109] Another study evaluated pancreatic cancer mortality among chemical manufacturing workers exposed to the organochlorine insecticide DDT (Dichloro-diphe-nyl-trichloroethane).[110] The analysis was based on 28 cases and 112 matched controls for whom pesticide exposure was determined from employment records, and lifestyle factors were determined by proxy interview with next of kin. Those with probable DDT exposure were about 5 times more likely to have died of pancreatic cancer than those with no DDT exposure (95% CI = 1.3–17.6).[110] Adjusting for smoking history based on next-of-kin data had little effect on RR estimates for DDT exposure.

Occupational exposure to pesticides based on job title and a job exposure matrix was evaluated within a population-based case-control study of pancreatic cancer.[111] This study found increased risks for pancreatic cancer associated with increasing pesticide exposure, in particular fungicides (OR = 1.5) and herbicides (OR = 1.6), but not insecticides.[111] Only one study has been published that has evaluated pancreatic cancer and serum organochlorine levels, such as insecticide metabolites like DDE (P,P'-Dichlorodiphenyldichloroethylene), and nonpesticide PCB (polychlorinated biphenyls) congeners.[112] This analysis was within the population-based San Francisco Bay Area study and was conducted on 108 pancreatic cancer cases and 82 control participants. Serum levels of DDE, PCBs, and trans-nonachlor were higher in cases than in controls, with PCBs showing the strongest associations after adjusting for the other organochlorines in multivariable regression models (OR = 4.0, 95% CI = 1.6–9.8). The potential impact of cancer cachexia on serum organochlorine levels in cases was evaluated and may have accounted for some, but not all, of the observed increased lipid burden of these chemicals among pancreatic cancer cases.[112]

In summary, despite some associations, there does not appear to be any one specific occupation or job title that consistently is associated with elevated mortality and incidence of pancreatic cancer. Petroleum products and pesticides, in particular some organochlorines in addition to PCBs, may be associated with pancreatic cancer risk, but larger prospective studies with more precise measures of exposure dose and frequency will be necessary before definitive conclusions or public health recommendations can be made.

20.11 PANCREATIC PATHWAYS — MOLECULAR EPIDEMIOLOGY AND THE STUDY OF GENE-ENVIRONMENT INTERACTIONS

Research on the role of common inherited mutations (polymorphisms) in genes for biochemical pathways that may be important in exposure-related pancreatic carcinogenesis has been expanding. (See also Chapter 24 and Chapter 25.) Certain polymorphisms in genes within biochemical pathways with functional relevance to the pancreas may have subtle effects on protein function or gene expression, and these effects in combination with exposures (gene-environment interactions) such as carcinogens in cigarette smoke, environmental pollution, or dietary factors may alter risk for the development of pancreatic cancer. Examples of some of the relevant pathways include carcinogen metabolism, oxidative stress, inflammation, DNA repair, and hormone metabolism. Other pathways and genes are likely to be relevant in pancreatic cancer, but due to space restrictions, will not be covered in this review.

20.11.1 Carcinogen Metabolism, Oxidative Stress

Two previous studies of genetic polymorphisms in metabolizing enzymes (*CYP1A1, CYP2D6, CYP2E1, NAT1, NAT2, GSTM1,* and *NQO1*) and pancreatic cancer analyzed main gene effects, did not examine gene-environment interactions, and found no associations.[113,114] These studies had small sample size that precluded statistical adjustment for potential confounders and evaluation of gene-environment interactions. Another case-control study of polymorphisms in *CYP1A1, GSTM1,* and *GSTT1* and pancreatic cancer did not find any statistically significant associations with pancreatic cancer risk, although small sample size may have hindered analyses of gene-environment interactions.[115] The glutathione S-transferases (GSTs) are a family of Phase II isoenzymes that are thought to protect cells from reactive intermediates and oxidative stress resulting from electrophilic xenobiotics and products of endogenous metabolism.[116] The spectrum of cytochrome P450 mono-oxygenase (CYP) and GST enzymes in the normal and diseased pancreas is given in Chapter 7. Expression of GST is variable in the human population and is sex- and tissue-specific.[117] Inheritance of null alleles with part or all of the gene deleted, for *GSTT1* and *GSTM1* is common in the population and has been correlated with loss of enzyme activity and cytogenetic damage.[118] We studied the risk of pancreatic cancer associated with deletion polymorphisms in *GSTT1* and *GSTM1*, and two single nucleotide polymorphisms (SNPs) in *GSTP1* that result in amino acid changes (Ile105Val and Ala114Val). We found no evidence for an association between *GSTM1*-

null (homozygous deletion) genotype and pancreatic cancer, nor did we find any evidence for interactions between cigarette smoking and the *GSTM1*-null genotype.[6] We did find evidence for interactions between cigarette smoking and *GSTT1*-null,[6] and between cigarette smoking and *GSTP1*-105Val (Duell and Holly, unpublished data) in relation to pancreatic cancer risk estimates. Interestingly, *GSTT1*/smoking interactions were stronger among women, whereas *GSTP1*/smoking interactions were stronger among men. These differences could reflect biological differences in expression and activity, or differences in smoking behavior or other exposures in nonsmokers (such as occupational exposures).

20.11.2 Inflammation

In human studies, obese and overweight individuals with insulin resistance exhibit low-grade, chronic systemic inflammation as measured by serum C-reactive protein.[119] Cytokines and other pro-inflammatory mediators have been implicated in inflammatory pancreatic diseases including pancreatitis and cancer. Insulin and other growth factors mediate the effects of pro-inflammatory cytokines such as TNF-α and IL-6.[120] TNF-α is a central mediator of inflammation and apoptosis and may possess both pro-tumor and anti-tumor activities. TNF-α may function to lower obesity by inducing insulin resistance.[121] The mechanism is not well established, but may involve the inhibition of tyrosine kinase activity at the insulin receptor.[122] Adipocytes, adipose tissue, and pancreatic tissue constituitively express TNF-α, and plasma levels are typically higher in obese patients compared with non-obese controls. Muscle TNF-α expression levels are elevated and similar in non-diabetic insulin resistant patients and in diabetic patients, respectively, but lower in insulin sensitive subjects.[121,123] A recent study of 50 nondiabetic lean and obese study participants found higher expression of TNF-α and interleukin 6 (IL-6) in adipose tissue from obese individuals than from lean individuals.[124]

A case-control study of pancreatic cancer (64 cases and 101 blood transfusion controls) evaluated the promoter polymorphism in *TNF*-α (G/A at position −308).[125] No association was found with case or control status, but those with the AA genotype had shorter survival. Small sample size and lack of control of confounding factors (e.g., age, sex, and smoking) may have contributed to the inconclusive findings. Recent research indicates the *TNF*-α promoter polymorphism −308G/A is not correlated with TNF-α transcription, but with lymphotoxin-α transcription, encoded by *LTA* and located just upstream of *TNF* locus.[126]

The chemokine RANTES and one of its receptors, CCR5, are believed to play a role in anti-tumor immunity through immune-cell recruitment. Polymorphisms have been identified in a number of cytokine genes, including the promoter for *RANTES* (–403 G/A) and a 32 bp deletion in its receptor *CCR5* (Δ32). Using questionnaire data obtained by in-person interviews and germ-line DNA collected in our population-based case-control study of pancreatic cancer, we conducted an analysis of cytokine/chemokine gene polymorphisms (*TNF-α, RANTES*, and *CCR5*) as risk factors for pancreatic cancer.[127] We used mass spectrometry and gel-based methods to genotype 308 cases and 964 population-based controls. We assessed potential interactions between these polymorphisms and pro-inflammatory conditions such as pancreatitis, ulcer, tobacco smoking, and other pro-inflammatory genes (gene–gene interactions) as risk factors for pancreatic cancer.[127] There was no overall association between pancreatic cancer risk and *TNF-α*, RANTES, and *CCR5* polymorphisms, nor was there evidence for two-locus interactions between these genes. However, among pancreatic cancer cases, we observed elevated OR estimates for: *TNF-α* –308 (GA + AA) and having a history of pancreatitis (OR = 3.2, 95% CI = 1.3–7.5), and *RANTES* 403 (GA + AA) and having a history of pancreatitis (OR = 2.3, 95% CI = 1.0–5.4). We did not observe interactions between cytokine/chemokine gene polymorphisms and having a history of duodenal or stomach ulcer. Results of analyses of tobacco smoking and cytokine/chemokine gene polymorphisms suggested the possibility of a positive interaction between *CCR5-del* genotypes and current active smoking. Our results lend some support for the hypothesis that proinflammatory gene polymorphisms, in combination with proinflammatory conditions, may influence the development of pancreatic cancer. Further study of proinflammatory cytokine gene polymorphisms, expression levels, and pancreatic cancer risk is warranted.

20.11.3 DNA Repair

X-ray repair cross-complementing Group 1 (XRCC1) is a base excision repair protein involved in the repair of damaged bases and single-strand breaks following exposure to endogenous and exogenous DNA damaging agents. A number of polymorphisms in *XRCC1* recently have been identified.[128] One of these polymorphisms (G/A) results in an arginine to glutamine change at Codon 399 within a breast cancer susceptibility Gene 1 product COOH terminus (BRCT) domain, a protein-binding motif present in several cell cycle and DNA repair proteins including BRCA1. The Arg399Gln polymorphism also occurs within a poly(ADP-ribose) polymerase (PARP) binding site. Thus, this polymorphism has the potential to affect XRCC1 protein function. Several studies have reported an association between the 399Gln allele and DNA damage phenotypes in human tissues.[129–132] We conducted an analysis of the Arg399Gln polymorphism in *XRCC1* using genomic DNA and questionnaire information from 308 cases of pancreatic cancer and 964 population-based controls from the San Francisco Bay Area.[7] We investigated interactions between the Arg399Gln polymorphism in *XRCC1* and cigarette smoking, age, obesity (elevated BMI), and polymorphisms in carcinogen-metabolizing genes (*GSTT1*, *GSTM1*, and *CYP1A1*) in relation to pancreatic cancer risk.[7] We used a mass spectrometry-based method to genotype individuals. Overall, we did not find evidence for a main effect of this polymorphism on risk of pancreatic cancer.[7] However, compared with nonsmokers with the Arg/Arg genotype, smokers with Gln/Gln or Arg/Gln genotypes were at a three- to four-fold increased risk of pancreatic cancer, almost double the risk of smokers with the Arg/Arg genotype, suggesting an interaction between smoking and the 399Gln allele associated with pancreatic cancer risk. We also found evidence for positive gene–gene interactions between *XRCC1* and *GSTT1* and *GSTM1* gene deletion polymorphisms.[7] We found no evidence for interactions between *XRCC1* and obesity as measured by BMI. In conclusion, our results suggested that the base excision repair protein XRCC1 plays a role in smoking- and exposure-related pancreatic cancer. Future analyses by our group will investigate other DNA repair gene polymorphisms and pathways in relation to pancreatic cancer risk.

20.11.4 Hormonal Pathways

CCK is a gastrointestinal hormone known to influence normal and malignant pancreatic growth and function (see also Chapter 6 and Chapter 14). A number of polymorphisms in genes for CCK and CCK receptors have been identified. Because of the importance of CCK in pancreatic growth and function, we hypothesized that a promoter polymorphism in CCK (–45 C/T) might influence risk. Using data from our San Francisco Bay Area study, we found evidence for a moderate interaction between the promoter polymorphism in *CCK* (CT+TT) and tobacco smoking with the highest risk among smokers with variant genotypes (Duell and Holly, unpublished data). We observed weak interactions between *CCK* –45 genotypes and calorie-adjusted saturated fat intake. There were no major modifications in estimates of RRs for *CCK* –45 and pancreatic cancer by sex. We also observed a possible weak interaction between *CCK* –45 genotype (CT+TT) and average daily alcohol consumption. Interestingly, ORs for pancreatic cancer were higher among those who consumed slightly less than one drink per day, an amount less than the median daily consumption of alcohol. Alcohol increases levels of sex steroid hormones such as estrogen. Thus, average estrogen levels may be higher in those who consumed alcohol regularly. Estrogen has been hypothesized to be protective for pancreatic cancer, thus moderate alcohol consumption may protect against pancreatic cancer. Interestingly, CCK has been shown to enhance the accumulation of nicotine and nicotine metabolites in rat pancreas cells,[133] thus it is plausible that a promoter polymorphism could influence smoking-associated risk of pancreatic cancer.

CYP17 encodes P450c17α, an enzyme with 17α-hydroxylase and 17,20-lyase activities at key branch points in estradiol biosynthesis. A polymorphism in the 5'UTR promoter region of *CYP17* (T27C) creates an Sp1 (CCACC box) promoter site and the A2 (27C) allele has been associated with higher levels of steroid hormones, including estradiol, and with breast cancer susceptibility in some studies.[134] Using data obtained in our population-based case-control study of pancreatic cancer carried out in the San Francisco Bay Area, we conducted an analysis of the *CYP17* polymorphism, as well as menstrual and reproductive risk factors for pancreatic cancer.[135] We used a mass spectrometry-based method to determine *CYP17* genotypes in 308 cases and 964 controls. Our results show a statistically significant inverse association between the A2 allele and pancreatic cancer risk: for A1/A2 heterozygotes, OR = 0.77, 95% CI = 0.57–1.0; for A2/A2 variant homozygotes, OR = 0.62, 95% CI = 0.42–0.94, relative to A1/A1 homozygotes with adjustment for age, race, sex, and pack-years of smoking. ORs for *CYP17* genotypes did not differ by sex. OC use and pancreatic cancer were inversely associated among those with the *CYP17* CC+TC genotypes compared with TT genotype. ERT use was inversely associated with pancreatic cancer risk and the associated ORs were reduced most among *CYP17* CC+TC genotypes. These data suggest that *CYP17* CC+TC genotypes may diminish the effect of exogenous hormone use on pancreatic cancer risk. In women, the *CYP17* genotype modified the effect of BMI on risk of pancreatic cancer such that ORs were increased with increased BMI among those with CC+TC genotypes and ORs were decreased with increased BMI among those with the TT genotype. In men, the *CYP17* genotype did not modify the association between pancreatic cancer and BMI. Among control women with genotype data (n = 432), the *CYP17* genotype was unrelated to age at menarche or age at menopause. Our results support the hypothesis that *CYP17-A2* may be associated with reduced risk estimates for the development of pancreatic cancer.

20.12 EPILOGUE

Further epidemiological research that includes a greater number of pancreatic cancer cases and continued follow-up of ongoing cohorts will undoubtedly increase our understanding of human pancreatic cancer. Specific research on dietary and environmental determinants and the molecular epidemiology of sporadic pancreatic cancer, including the comprehensive study of gene–environment interactions, will broaden our understanding of disease pathways and mechanisms and may have relevance for other exposure-related cancers. Elucidating pathways of pancreatic carcinogenesis will one day aid in prevention and therapy for this deadly disease.

REFERENCES

1. Jemal A., Tiwari R.C., Murray T., Ghafoor A., Samuels A., Ward E., Feuer E.J., and Thun M.J., Cancer statistics, 2004. *CA Cancer J Clin* 54(1):8–29.
2. Anderson K.E., Potter J.D., and Mack T.M., Pancreatic cancer, In *Cancer Epidemiology and Prevention*, S. D. and J.F. Fraumeni, Jr., Eds. 1996, Oxford University Press: New York. pp. 725–771.
3. Silverman D.T., Dunn J.A., Hoover R.N., Schiffman M., Lillemoe K.D., Schoenberg J.B., Brown L.M., Greenberg R.S., Hayes R.B., Swanson G.M., et al. 1994. Cigarette smoking and pancreas cancer: a case-control study based on direct interviews. *J Natl Cancer Inst* 86(20):1510–1516.
4. Fuchs C.S., Colditz G.A., Stampfer M.J., Giovannucci E.L., Hunter D.J., Rimm E.B., Willett W.C., and Speizer F.E. 1996. A prospective study of cigarette smoking and the risk of pancreatic cancer. *Arch. Intern. Med* 156:2255–2260.
5. Muscat J.E., Stellman S.D., Hoffman D., and Wynder E.L. 1997. Smoking and pancreatic cancer in men and women. *Cancer Epidemiology Biomarkers Prev* 6:15–19.
6. Duell E.J., Holly E.A., Bracci P.M., Liu M., Wiencke J.K., and Kelsey K.T. 2002. A population-based, case-control study of polymorphisms in carcinogen-metabolizing genes, smoking, and pancreatic adenocarcinoma risk. *J Natl Cancer Inst* 94(4):297–306.
7. Duell E.J., Holly E.A., Bracci P.M., Wiencke J.K., and Kelsey K.T. 2002. A population-based study of the Arg399Gln polymorphism in X-ray repair cross-complementing group 1 (XRCC1) and risk of pancreatic adenocarcinoma. *Cancer Res* 62(16):4630–4636.
8. Longnecker D.S. 1991. Hormones and pancreatic cancer. *Int J Pancreatol* 9:81–86.
9. Wei Q., Cheng L., Amos C.I., Wang L.E., Guo Z., Hong W.K., and Spitz M.R. 2000. Repair of tobacco carcinogen-induced DNA adducts and lung cancer risk: a molecular epidemiologic study. *J Natl Cancer Inst* 92(21):1764–1772.
10. Zhu B.T. and Conney A.H. 1998. Functional role of estrogen metabolism in target cells: review and perspectives. *Carcinogenesis* 19(1):1–27.
11. Shriver S.P., Bourdeau H..A., Gubish C.T., Tirpak D.L., Davis A..L., Luketich J.D., and Siegfried J.M. 2000. Sex-specific expression of gastrin-releasing peptide receptor: relationship to smoking history and risk of lung cancer. *J Natl Cancer Inst* 92(1):24–33.
12. Schuller H.M. 2002. Mechanisms of smoking-related lung and pancreatic adenocarcinoma development. *Nat Rev Cancer* 2(6):455–463.
13. Anderson K.E., Hammons G.J., Kadlubar F.F., Potter J.D., Kaderlik K.R., Ilett K.F., Minchin R.F., Teitel C.H., Chou H.C., Martin M.V., Guengerich F.P., Barone G.W., Lang N.P., and Peterson L.A. 1997. Metabolic activation of aromatic amines by human pancreas. *Carcinogenesis* 18(5):1085–1092.
14. Schwartz G.G. and Reis I.M. 2000. Is cadmium a cause of human pancreatic cancer? *Cancer Epidemiol Biomarkers Prev* 9(2):139–145.

15. Farrow B. and Evers B.M. 2002. Inflammation and the development of pancreatic cancer. *Surg Oncol* 10(4):153–169.

16. Isaksson B., Jonsson F., Pedersen N.L., Larsson J., Feychting M., and Permert J. 2002. Lifestyle factors and pancreatic cancer risk: a cohort study from the Swedish Twin Registry. *Int J Cancer* 98(3):480–482.

17. Hirayama T. 1989. Epidemiology of pancreatic cancer in Japan. *Jpn J Clin Oncol* 19(3):208–215.

18. Stolzenberg-Solomon R.Z., Pietinen P., Taylor P.R., Virtamo J., and Albanes D. 2002. Prospective study of diet and pancreatic cancer in male smokers. *Am J Epidemiol* 155(9):783–792.

19. Michaud D.S., Giovannucci E., Willett W.C., Colditz G.A., and Fuchs C.S. 2003. Dietary meat, dairy products, fat, and cholesterol and pancreatic cancer risk in a prospective study. *Am J Epidemiol* 157(12):1115–1125.

20. Schneider M.B., Matsuzaki H., Haorah J., Ulrich A., Standop J., Ding X.Z., Adrian T.E., and Pour P.M. 2001. Prevention of pancreatic cancer induction in hamsters by metformin. *Gastroenterology* 120(5):1263-1270.

21. Pour P.M. and K. Kazakoff 1996. Stimulation of islet cell proliferation enhances pancreatic ductal carcinogenesis in the hamster model. *Am J Pathol* 149(3):1017–1025.

22. Roebuck B.D., Kaplita P.V., Edwards B.R., and Praissman M. 1987. Effects of dietary fats and soybean protein on azaserine-induced pancreatic carcinogenesis and plasma cholecystokinin in the rat. *Cancer Res* 47(5):1333–1338.

23. Kazakoff K., Cardesa T., Liu J., Adrian T.E., Bagchi D., Bagchi M., Birt D.F., and Pour P.M. 1996. Effects of voluntary physical exercise on high-fat diet-promoted pancreatic carcinogenesis in the hamster model. *Nutr Cancer* 26(3):265–279.

24. Gapstur S.M., Gann P.H., Lowe W., Liu K., Colangelo L., and Dyer A. 2000. Abnormal glucose metabolism and pancreatic cancer mortality. *Jama* 283(19):2552–2558.

25. Michaud D.S., Liu S., Giovannucci E., Willett W.C., Colditz G.A., and Fuchs C.S. 2002. Dietary sugar, glycemic load, and pancreatic cancer risk in a prospective study. *J Natl Cancer Inst* 94(17):1293–1300.

26. Everhart J. and Wright D. 1995. Diabetes mellitus as a risk factor for pancreatic cancer. A meta-analysis. *Jama* 273(20):1605–1609.

27. Saydah S.H., Loria C.M., Eberhardt M.S., and Brancati F.L. 2003. Abnormal glucose tolerance and the risk of cancer death in the United States. *Am J Epidemiol* 157(12):1092–1100.

28. Hanley A.J., Johnson K.C., Villeneuve P.J., and Mao Y. 2001. Physical activity, anthropometric factors and risk of pancreatic cancer: results from the Canadian enhanced cancer surveillance system. *Int J Cancer* 94(1):140–147.

29. Michaud D.S., Giovannucci E., Willett W.C., Colditz G.A., Stampfer M.J., and Fuchs C.S. 2001. Physical activity, obesity, height, and the risk of pancreatic cancer. *Jama* 286(8):921–929.

30. Kahn B.B. and Flier J.S. 2000. Obesity and insulin resistance. *J Clin Invest* 106(4):473–481.

31. Bueno de Mesquita H.B., Maisonneuve P., Runia S., and Moerman C.J. 1991. Intake of foods and nutrients and cancer of the exocrine pancreas: a population-based case-control study in The Netherlands *Int. J. Cancer* 48(4):540–549.

32. Silverman D.T., Swanson C.A., Gridley G., Wacholder S., Greenberg R.S., Brown L.M., Hayes R.B., Swanson G.M., Schoenberg J.B., Pottern L.M., Schwartz A.G., Fraumeni J.F., Jr., and Hoover R.N. 1998. Dietary and nutritional factors and pancreatic cancer: a case-control study based on direct interviews. *J Natl Cancer Inst* 90(22):1710–1719.

33. Kalapothaki V., Tzonou A., Hsieh C.C., Karakatsani A., Trichopoulou A., Toupadaki N., and Trichopoulos D. 1993. Nutrient intake and cancer of the pancreas: a case-control study in Athens, Greece. *Cancer Causes Control* 4(4):383389.

34. Howe G.R., Jain M., and Miller A.B. 1990. Dietary factors and risk of pancreatic cancer: results of a Canadian population-based case-control study. *Int J Cancer* 45(4):604–608.

35. Fernandez E., La Vecchia C., and Decarli A. 1996. Attributable risks for pancreatic cancer in northern Italy. *Cancer Epidemiol Biomarkers Prev* 5(1):23–27.

36. Ji B.T., Chow W.H., Gridley G., McLaughlin J.K., Dai Q., Wacholder S., Hatch M.C., Gao Y.T., and Fraumeni, Jr., J.F. 1995. Dietary factors and the risk of pancreatic cancer: a case-control study in Shanghai China. *Cancer Epidemiol Biomarkers Prev* 4(8):885 J.F. 893.

37. La Vecchia C., Negri E., D'Avanzo B., Ferraroni M., Gramenzi A., Savoldelli R., Boyle P., and Franceschi S. 1990. Medical history, diet and pancreatic cancer. *Oncology* 47(6):463–466.

38. Shibata A., Mack T.M., Paganini-Hill A., Ross R.K., and Henderson B.E. 1994. A prospective study of pancreatic cancer in the elderly. *Int J Cancer* 58(1):46–49.

39. Potter J.D. 2002. Pancreas cancer — we know about smoking, but do we know anything else? *Am J Epidemiol* 155(9):793–795; discussion 796–797.

40. Anderson K.E., Sinha R., Kulldorff M., Gross M., Lang N.P., Barber C., Harnack L., DiMagno E., Bliss R., and Kadlubar F.F. 2002. Meat intake and cooking techniques: associations with pancreatic cancer. *Mutat Res* 506–507:225–231.

41. Ghadirian P., Baillargeon J., Simard A., and Perret C. 1995. Food habits and pancreatic cancer: a case-control study of the Francophone community in Montreal, Canada. *Cancer Epidemiol Biomarkers Prev* 4(8):895–899.

42. Steinmetz K.A. and Potter J.D. 1996. Vegetables, fruit, and cancer prevention: a review. *J Am Diet Assoc* 96(10):1027–1039.

43. Stolzenberg-Solomon R.Z., Pietinen P., Barrett M.J., Taylor P.R., Virtamo J., and Albanes D. 2001. Dietary and other methyl-group availability factors and pancreatic cancer risk in a cohort of male smokers. *Am J Epidemiol* 153(7):680–687.

44. Stolzenberg-Solomon R.Z., Albanes D., Nieto F.J., Hartman T.J., Tangrea J.A., Rautalahti M., Sehlub J., Virtamo J., and Taylor P.R. 1999. Pancreatic cancer risk and nutrition-related methyl-group availability indicators in male smokers. *J Natl Cancer Inst* 91(6):535–541.

45. Talamini G., Bassi C., Falconi M., Sartori N., Salvia R., Rigo L., Castagnini A., Di Francesco V., Frulloni L., Bovo P., Vaona B., Angelini G., Vantini I., Cavallini G., and Pederzoli P. 1999. Alcohol and smoking as risk factors in chronic pancreatitis and pancreatic cancer. *Dig Dis Sci* 44(7):1303–1311.

46. Villeneuve P.J., Johnson K.C., Hanley A.J., and Mao Y. 2000. Alcohol, tobacco and coffee consumption and the risk of pancreatic cancer: results from the Canadian Enhanced Surveillance System case-control project. Canadian Cancer Registries Epidemiology Research Group. *Eur J Cancer Prev* 9(1):49–58.

47. Michaud D.S., Giovannucci E., Willett W.C., Colditz G.A., and Fuchs C.S. 2001. Coffee and alcohol consumption and the risk of pancreatic cancer in two prospective United States cohorts. *Cancer Epidemiol Biomarkers Prev* 10(5):429–437.

48. Kalapothaki V., Tzonou A., Hsieh C.C., Toupadaki N., Karakatsani A., and Trichopoulos D. 1993. Tobacco, ethanol, coffee, pancreatitis, diabetes mellitus, and cholelithiasis as risk factors for pancreatic carcinoma. *Cancer Causes Control* 4(4):375–382.

49. Ramchandani V.A., Bosron W.F., and Li T.K. 2001. Research advances in ethanol metabolism. *Pathol Biol (Paris)* 49(9):676–682.

50. Sato N., Lindros K.O., Baraona E., Ikejima K., Mezey E., Jarvelainen H.A., and Ramchandani V.A. 2001. Sex difference in alcohol-related organ injury. *Alcohol Clin Exp Res* 25(5 Suppl ISBRA):40S–45S.

51. Silverman D.T., Brown L.M., Hoover R.N., Schiffman M., Lillemoe K.D., Schoenberg J.B., Swanson G.M., Hayes R.B., Greenberg R.S., Benichou J., et al. 1995. Alcohol and pancreatic cancer in blacks and whites in the United States. *Cancer Res* 55(21):4899–4905.

52. Partanen T.J., Vainio H.U., Ojajarvi I.A., and Kauppinen T.P. 1997. Pancreas cancer, tobacco smoking and consumption of alcoholic beverages: a case-control study. *Cancer Lett* 116(1):27–32.

53. Pezzilli R., Billi P., and Morselli-Labate A.M. 1998. Severity of acute pancreatitis: relationship with etiology, sex and age. *Hepatogastroenterology* 45(23):1859–1864.

54. Bueno de Mesquita H.B., Maisonneuve P., Moerman C.J., Runia S., and Boyle P. 1992. Lifetime consumption of alcoholic beverages, tea and coffee and exocrine carcinoma of the pancreas: a population-based case-control study in The Netherlands. *Int J Cancer* 50(4):514–522.

55. Farrow D.C. and Davis S. 1990. Diet and the risk of pancreatic cancer in men. *Am J Epidemiol* 132(3):423–431.

56. Gullo L., Pezzilli R., and Morselli-Labate A.M. 1995. Coffee and cancer of the pancreas: an Italian multicenter study. The Italian Pancreatic Cancer Study Group. *Pancreas* 11(3):223–229.

57. Harnack L.J., Anderson K.E., Zheng W., Folsom A.R., Sellers T.A., and Kushi L.H. 1997. Smoking, alcohol, coffee, and tea intake and incidence of cancer of the exocrine pancreas: the Iowa Women's Health Study. *Cancer Epidemiol Biomarkers Prev* 6(12):1081–1086.

58. MacMahon B., Yen S., Trichopoulos D., Warren K., and Nardi G. 1981. Coffee and cancer of the pancreas. *N Engl J Med* 304(11):630–633.

59. Tavani, A. and La Vecchia C. 2000 Coffee and cancer: a review of epidemiological studies, 1990-1999. *Eur J Cancer Prev* 9(4):241–256.

60. Ji B.T., Chow W.H., Hsing A.W., McLaughlin J.K., Dai Q., Gao Y.T., Blot W.J., and Fraumeni, Jr., J.F. 1997. Green tea consumption and the risk of pancreatic and colorectal cancers. *Int J Cancer* 70(3):255–258.

61. Bushman J.L. 1998. Green tea and cancer in humans: a review of the literature. *Nutr Cancer* 31(3):151–159.

62. Mukhtar H. and Ahmad N. 2000. Tea polyphenols: prevention of cancer and optimizing health. *Am J Clin Nutr* 71(6 Suppl):1698S–1702S; discussion 1703S–1704S.

63. Ford, E.S., W.H. Giles, and W.H. Dietz, Prevalence of the metabolic syndrome among US adults: findings from the third National Health and Nutrition Examination Survey. *Jama*, 2002. 287(3): p. 356-9.

64. Mokdad A.H., Bowman B.A., Ford E.S., Vinicor F., Marks J.S., and Koplan J.P. 2001. The continuing epidemics of obesity and diabetes in the United States. *Jama* 286(10):1195–1200.

65. Coughlin S.S., Calle E.E., Patel A.V., and Thun M.J. 2000. Predictors of pancreatic cancer mortality among a large cohort of United States adults. *Cancer Causes Control* 11(10):915–923.

66. Ji B.T., Hatch M.C., Chow W.H., McLaughlin J.K., Dai Q., Howe G.R., Gao Y.T., and Fraumeni, Jr., J.F. 1996. Anthropometric and reproductive factors and the risk of pancreatic cancer: a case-control study in Shanghai, China. *Int J Cancer* 66(4):432–43.

67. Eberle C.A., Bracci P.M., and Holly E.A. Anthropometric factors and pancreatic cancer in the San Francisco Bay Area. Submitted.

68. Calle E.E., Rodriguez C., Walker-Thurmond K., and Thun M.J. 2003. Overweight, obesity, and mortality from cancer in a prospectively studied cohort of U.S. adults. *N Engl J Med* 348(17):1625–1638.

69. Berrington de Gonzalez A., Sweetland S., and Spencer E. 2003. A meta-analysis of obesity and the risk of pancreatic cancer. *Br J Cancer* 89(3):519–523.

70. Lee I.M., Sesso H.D., Oguma Y., and Paffenbarger, Jr., R.S. 2003. Physical activity, body weight, and pancreatic cancer mortality. *Br J Cancer* 88(5):679–683.

71. Friedenreich C.M. and Orenstein M.R. 2002. Physical activity and cancer prevention: etiologic evidence and biological mechanisms. *J Nutr* 132(11 Suppl):3456S–3464S.

72. Adami H.O., Persson I., Hoover R., Schairer C., and Bergkvist L. 1989. Risk of cancer in women receiving hormone replacement therapy. *Int J Cancer* 44(5):833–839.

73. Bueno de Mesquita H.B., Maisonneuve P., Moerman C.J., and Walker A.M. 1992. Anthropometric and reproductive variables and exocrine carcinoma of the pancreas: a population-based case-control study in The Netherlands. *Int J Cancer* 52(1):24–29.

74. Fernandez E., La Vecchia C., D'Avanzo B., and Negri E. 1995. Menstrual and reproductive factors and pancreatic cancer risk in women. *Int J Cancer* 62(1):11–14.

75. Kreiger N., Lacroix J., and Sloan M. 2001. Hormonal factors and pancreatic cancer in women. *Ann Epidemiol* 11(8):563–567.

76. Fernandez E., Gallus S., Bosetti C., Franceschi S., Negri E., and La Vecchia C. 2003 Hormone replacement therapy and cancer risk: a systematic analysis from a network of case-control studies. *Int J Cancer* 105(3):408–412.

77. Duell E.J. and Holly E.A. 2005. Reproductive and menstrual risk factors for pancreatic cancer: a population-based study of San Francisco Bay Area women. *Am. J. Epidemiol* 161(8):741–747.

78. Peek, Jr., R.M., and Blaser M.J. 2002. Helicobacter pylori and gastrointestinal tract adenocarcinomas. *Nat Rev Cancer* 2(1):28–37.

79. Raderer M., Wrba F., Kornek G., Maca T., Koller D.Y., Weinlaender G., Hejna M., and Scheithauer W. 1998. Association between Helicobacter pylori infection and pancreatic cancer. *Oncology* 55(1):16–19.

80. Stolzenberg-Solomon R.Z., Blaser M.J., Limburg P.J., Perez-Perez G., Taylor P.R., Virtamo J., and Albanes D. 2001. Helicobacter pylori seropositivity as a risk factor for pancreatic cancer. *J Natl Cancer Inst* 93(12):937–941.

81. Nilsson H.O., Stenram U., Ihse I., and Wadstrom T. 2002. Re: Helicobacter pylori seropositivity as a risk factor for pancreatic cancer. *J Natl Cancer Inst* 94(8):632–633.

82. Smith J.P., Fantaskey A.P., Liu G., and Zagon I.S. 1995. Identification of gastrin as a growth peptide in human pancreatic cancer. *Am J Physiol* 268(1 Pt 2):R135–R141.

83. Weinberg D.S., Ruggeri B., Barber M.T., Biswas S., Miknyocki S., and Waldman S.A. 1997. Cholecystokinin A and B receptors are differentially expressed in normal pancreas and pancreatic adenocarcinoma. *J Clin Invest* 100(3):597–603.

84. La Vecchia C., Negri E., Franceschi S., Conti E., Montella M., Giacosa A., Falcini A., and Decarli A. 1997. Aspirin and colorectal cancer. *Br J Cancer* 76(5):675–677.

85. Morgan G. 1999. Beneficial effects of NSAIDs in the gastrointestinal tract. *Eur J Gastroenterol Hepatol* 11(4):393–400.

86. Schreinemachers D.M. and Everson R.B. 1994. Aspirin use and lung, colon, and breast cancer incidence in a prospective study. *Epidemiology* 5(2):138–146.

87. Langman M.J., Cheng K.K., Gilman E.A., and Lancashire R.J. 2000. Effect of anti-inflammatory drugs on overall risk of common cancer: case-control study in general practice research database. *Bmj* 320(7250):1642–1646.

88. Friis S., Sorensen H.T., McLaughlin J.K., Johnsen S.P., Blot W.J., and Olsen J.H. 2003. A population-based cohort study of the risk of colorectal and other cancers among users of low-dose aspirin. *Br J Cancer* 88(5):684–688.

89. Anderson K.E., Johnson T.W., Lazovich D., and Folsom A.R. 2002. Association between nonsteroidal anti-inflammatory drug use and the incidence of pancreatic cancer. *J Natl Cancer Inst* 94(15):1168–1171.

90. Schernhammer E.S., Kang J.H., Chan A.T., Michaud D.S., Skinner H.G., Giovannucci E., Colditz G.A., and Fuchs C.S. 2004. A prospective study of aspirin use and the risk of pancreatic cancer in women. *J Natl Cancer Inst* 96(1):22–28.

91. Selikoff I.J. and Seidman H. 1981. Cancer of the pancreas among asbestos insulation workers. *Cancer* 47(6 Suppl):1469–1473.

92. Delzell E., Louik C., Lewis J., and Monson R.R. 1981. Mortality and cancer morbidity among workers in the rubber tire industry. *Am J Ind Med* 2(3):209–216.

93. Delzell E. and Monson R.R. 1985. Mortality among rubber workers: IX. Curing workers. *Am J Ind Med* 8(6):537–544.

94. Straif K., Weiland S.K., Bungers M., Holthenrich D., Taeger D., Yi S., and Keil U. 2000. Exposure to high concentrations of nitrosamines and cancer mortality among a cohort of rubber workers. *Occup Environ Med* 57(3):180–187.

95. Norell S., Ahlbom A., Olin R., Erwald R., Jacobson G., Lindberg-Navier I., and Wiechel K.L. 1986. Occupational factors and pancreatic cancer. *Br J Ind Med* 43(11):775–778.

96. Pickle L.W. and Gottlieb M.S. 1980. Pancreatic cancer mortality in Louisiana. *Am J Public Health* 70(3):256–259.

97. Thomas T.L., Decoufle P., and Moure-Eraso R. 1980. Mortality among workers employed in petroleum refining and petrochemical plants. *J Occup Med* 22(2):97–103.

98. Lin R.S. and Kessler I.I. 1981. A multifactorial model for pancreatic cancer in man. Epidemiologic evidence. *Jama* 245(2):147–152.

99. Hanis N.M., Holmes T.M., Shallenberger G., and Jones K.E. 1982. Epidemiologic study of refinery and chemical plant workers. *J Occup Med* 24(3):203–212.

100. Collins J.J., Esmen N.A., and Hall T.A. 2001. A review and meta-analysis of formaldehyde exposure and pancreatic cancer. *Am J Ind Med* 39(3):336–345.

101. Ojajarvi A., Partanen T., Ahlbom A., Boffetta P., Hakulinen T., Jourenkova N., Kauppinen T., Kogevinas M., Vainio H., Weiderpass E., and Wesseling C. 2001. Risk of pancreatic cancer in workers exposed to chlorinated hydrocarbon solvents and related compounds: a meta-analysis. *Am J Epidemiol* 153(9):841–850.

102. Alguacil J., Porta M., Malats N., Kauppinen T., Kogevinas M., Benavides F.G., Partanen T., and Carrato A. 2002. Occupational exposure to organic solvents and K-ras mutations in exocrine pancreatic cancer. *Carcinogenesis* 23(1):101–106.

103. Falk R.T., Pickle L.W., Fontham E.T., Correa P., Morse A., Chen V., and Fraumeni, Jr., J.J. 1990. Occupation and pancreatic cancer risk in Louisiana. *Am J Ind Med* 18(5):565–576.

104. Blair A., Malker H., Cantor K.P., Burmeister L., and Wiklund K. 1985. Cancer among farmers. A review. *Scand J Work Environ Health* 11(6):397–407.

105. Franceschi S., Barbone F., Bidoli E., Guarneri S., Serraino D., Talamini R., and La Vecchia C. 1993. Cancer risk in farmers: results from a multi-site case-control study in north-eastern Italy. *Int J Cancer* 53(5):740–745.

106. Burmeister L.F. 1990. Cancer in Iowa farmers: recent results. *Am J Ind Med* 18(3):295–301.

107. Saftlas A.F., Blair A., Cantor K.P., Hanrahan L., and Anderson H.A. 1987. Cancer and other causes of death among Wisconsin farmers. *Am J Ind Med* 11(2):119–129.

108. Lee E., Burnett C.A., Lalich N., Cameron L.L., and Sestito J.P. 2002. Proportionate mortality of crop and livestock farmers in the United States, 1984-1993. *Am J Ind Med* 42(5):410–420.

109. Gold E.B., Gordis L., Diener M.D., Seltser R., Boitnott J.K., Bynum T.E., and Hutcheon D.F. 1985. Diet and other risk factors for cancer of the pancreas. *Cancer* 55(2):460–467.

110. Garabrant D.H., Held J., Langholz B., Peters J.M., and Mack T.M. 1992. DDT and related compounds and risk of pancreatic cancer. *J Natl Cancer Inst* 84(10):764–771.

111. Ji B.T., Silverman D.T., Stewart P.A., Blair A., Swanson G.M., Baris D., Greenberg R.S., Hayes R.B., Brown L.M., Lillemoe K.D., Schoenberg J.B., Pottern L.M., Schwartz A.G., and Hoover R.N. 2001. Occupational exposure to pesticides and pancreatic cancer. *Am J Ind Med* 39(1):92–99.

112. Hoppin J.A., Tolbert P.E., Holly E.A., Brock J.W., Korrick S.A., Altshul L.M., Zhang R.H., Bracci P.M., Burse V.W., and Needham L.L. 2000. Pancreatic cancer and serum organochlorine levels. *Cancer Epidemiol Biomarkers Prev* 9(2):199–205.

113. Bartsch H., Malaveille C., Lowenfels A.B., Maisonneuve P., Hautefeuille A., and Boyle P. 1998. Genetic polymorphism of N-acetyltransferases, glutathione S-transferase M1 and NAD(P)H:quinone oxidoreductase in relation to malignant and benign pancreatic disease risk. The International Pancreatic Disease Study Group. *Eur J Cancer Prev* 7(3):215–223.

114. Lee H.C., Yoon Y.B., and Kim C.Y. 1997. Association between genetic polymorphisms of the cytochromes P-450 (1A1, 2D6, and 2E1) and the susceptibility to pancreatic cancer. *Korean J Intern Med* 12(2):128–136.

115. Liu G., Ghadirian P., Vesprini D., Hamel N., Paradis A.J., Lal G., Gallinger S., Narod S.A., and Foulkes W.D. 2000. Polymorphisms in GSTM1, GSTT1 and CYP1A1 and risk of pancreatic adenocarcinoma. *Br J Cancer* 82(10):1646–1649.

116. Ketterer B. 1998. Glutathione S-transferases and prevention of cellular free radical damage. *Free Radic Res* 28(6):647–658.

117. Collier J.D., Bennett M.K., Hall A., Cattan A.R., Lendrum R., and Bassendine M.F. 1994. Expression of glutathione S-transferases in normal and malignant pancreas: an immunohistochemical study. *Gut* 35(2):266–269.

118. Wiencke J.K., Kelsey K.T., Lamela R.A., and Toscano, Jr., W.A. 1990. Human glutathione S-transferase deficiency as a marker of susceptibility to epoxide-induced cytogenetic damage. *Cancer Res* 50(5):1585–1590.

119. Visser M., Bouter L.M., McQuillan G.M., Wener M.H., and Harris T.B. 1999. Elevated C-reactive protein levels in overweight and obese adults. *Jama* 282(22):2131–2135.

120. Baumann H. and Gauldie J. 1994. The acute phase response. *Immunol Today* 15(2):74–80.

121. Kern P.A. 1997. Potential role of TNFalpha and lipoprotein lipase as candidate genes for obesity. *J Nutr* 127(9):1917S–1922S.

122. Dandona P., Weinstock R., Thusu K., Abdel-Rahman E., Aljada A., and Wadden T. 1998. Tumor necrosis factor-alpha in sera of obese patients: fall with weight loss. *J Clin Endocrinol Metab* 83(8):2907–2910.

123. Saghizadeh M., Ong J.M., Garvey W.T., Henry R.R., and Kern P.A. 1996. The expression of TNF alpha by human muscle. Relationship to insulin resistance. *J Clin Invest*, 97(4):1111–1116.

124. Kern P.A., Ranganathan S., Li C., Wood L., and Ranganathan G. 2001. Adipose tissue tumor necrosis factor and interleukin-6 expression in human obesity and insulin resistance. *Am J Physiol Endocrinol Metab* 280(5):E745–E751.

125. Barber M.D., Powell J.J., Lynch S.F., Gough N.J., Fearon K.C., and Ross J.A. 1999. Two polymorphisms of the tumour necrosis factor gene do not influence survival in pancreatic cancer. *Clin Exp Immunol* 117(3):425–429.

126. Knight J.C., Keating B.J., Rockett K.A., and Kwiatkowski D.P. 2003. In vivo characterization of regulatory polymorphisms by allele-specific quantification of RNA polymerase loading. *Nat Genet* 33(4):469–475.

127. Duell E.J., Casella D.P., Burk R.D., Kelsey K.T., and Holly E.A. Inflammation, genetic polymorphisms in pro-inflammatory genes RANTES, CCR5, and TNF-a, and risk of pancreatic adenocarcinoma. Submitted.

128. Mohrenweiser H.W., Xi T., Vazquez-Matias J., and Jones I.M. 2002. Identification of 127 amino acid substitution variants in screening 37 DNA repair genes in humans. *Cancer Epidemiol Biomarkers Prev*, 11(10 Pt 1):1054–1064.

129. Lunn R., Langlois R., Hsieh L., Thompson C., and Bell D. 1999. XRCC1 polymorphisms: effects on aflatoxin B1-DNA adducts and glycophorin A variant frequency. *Cancer Research* 59:2557–2561.

130. Duell E.J., Wiencke J.K., Cheng T.J., Varkonyi A., Zuo Z.F., Ashok T.D., Mark E.J., Wain J.C., Christiani D.C., and Kelsey K.T. 2000. Polymorphisms in the DNA repair genes XRCC1 and ERCC2 and biomarkers of DNA damage in human blood mononuclear cells. *Carcinogenesis* 21(5):965–971.

131. Hu J., Smith T., Miller M., Mohrenweiser H., Golden A., and Case L. 2001. Amino acid substitution variants of APE1 and XRCC1 genes associated with ionizing radiation sensitivity. *Carcinogenesis* 22:917–922.

132. Abdel-Rahman S. and El-Zein R. 2000. The 399Gln polymorphism in the DNA repair gene XRCC1 modulates the genotoxic response induced in human lymphocytes by the tobacco-specific nitrosamine NNK. *Cancer Lett* 159:63–71.

133. Doi R., Chowdhury P., Nishikawa M., Takaori K., Inoue K., Imamura M., and Rayford P.L. 1995. Carbachol and cholecystokinin enhance accumulation of nicotine in rat pancreatic acinar cells. *Pancreas* 10(2):154–160.

134. Haiman C.A., Hankinson S.E., Spiegelman D., Colditz G.A., Willett W.C., Speizer F.E., Kelsey K.T., and Hunter D.J. 1999. The relationship between a polymorphism in CYP17 with plasma hormone levels and breast cancer. *Cancer Res* 59(5):1015–1020.

135. Duell E.J. and Holly E.A.. 2003. A polymorphism in *CYP17*, reproductive factors, and pancreatic cancer risk. In *Proceedings from the American Association for Cancer Research*. Toronto, ON.

21

THE EFFECTS OF NUTRIENTS ON PANCREATIC MALIGNANCY

Parviz M. Pour

CONTENTS

21.1   INTRODUCTION

The complexity of epidemiological studies relative to the role of diet in pancreatic diseases has drawn focus on experimental studies. Because it is estimated that about 70% of human cancers are caused by chemical carcinogens, most experimental dietary studies have been performed on experimental cancer models. The models, in defined experimental conditions, can provide information on the dietary effect on the initiation or promotion of carcinogenesis processes that depend on a variety of factors.

Some factors may influence the carcinogenesis process at the initiation phase (i.e., DNA alkylation), whereas others modify the postinitiation phase as promoters or inhibitors. Based on epidemiological studies suggesting that a diet high in fat or carbohydrates increases the risk of many types of cancer, several studies focused on the effect of diet on pancreatic carcinogenesis. The involvement of diet in pancreatic carcinogenesis is obvious, because this tissue is the major nutritional machinery.

21.2 THE EFFECTS OF DIET ON SURVIVAL

In an initial study at the University of Nebraska Medical Center, it was found that the modification of the dietary ingredients influences the aging process both in hamsters and rats.[1] Syrian hamsters fed commercial chow diet had a longevity of about 40 weeks (short-lived) and suffered with a large spectrum of diseases, including cardiovascular, renal, and malignant diseases. Feeding a formulated semisynthetic diet with the same caloric concentration of fat, protein, and carbohydrate prolonged the survival to over 80 weeks (long-lived) and abolished or reduced many spontaneous diseases.[2,3] Both the short-lived and long-lived animals develop the same type and frequency of tumors, but at different times. In the short-lived hamsters, tumors appear much earlier than those in the long-lived hamsters. This was also the case when rats with usually long survival (about 2 years) were compared with the short-lived hamsters.[3] In both short-lived and long-lived hamsters, tumors appeared within the last quarter of their life. In the long-lived animals, the correlation between the longevity and tumorigenesis was shifted to a later time. This situation highlights a possible interpretation error in toxicological studies, where the experiment is terminated after the last short-lived animals, used as a control for the long-lived animals, dies. In this case, tumor prevention in long-lived animals may be assumed. The data also seems to be true for humans, as the increased life expectancy has shifted the appearance of certain diseases, such as cancers, to a later age.

Nevertheless, it is obvious that some dietary ingredients in the commercial diet can have an adverse effect on the health, possibly due to the contamination of the dietary source (grains) with pesticides, herbicides, or other chemicals. Note that the quality of the natural source for the preparation of the chow diet can change from year to year as does the quality of the chemicals used for crop treatment. Studies have also demonstrated that the level of amino acids in the protein, especially abnormal methyl metabolism, are important for the normal function of the pancreas and their alterations can exert significant toxicity on pancreatic cells. This issue is presented in Chapter 10.

21.3 THE EFFECTS OF DIET ON PANCREATIC CARCINOGENESIS

Confirming the epidemiological studies, experiments in the laboratories of Dr. Longnecker and at the University of Nebraska Medical Center have indeed confirmed the promoting effect of a high fat (HF) diet on pancreatic carcinogenesis in the two animal models.[3-6] In all these studies, corn oil was used as the source of fat. Feeding α-linolenic acid also promoted liver metastases in the hamster pancreatic cancer (PC) model.[7] The results of these studies, however, demonstrated the complexity of the effects of nutrients in pancreatic carcinogenesis. In addition to a specific nutrient many other factors, including the age, gender, the individual ingredients in the HF diet, the type of fat, the target organ, and even the type of cells within the same tissue presented confounding factors. Moreover, there seems to be a significant interaction between the individual nutrients within the same diet. This interaction was obvious between the HF diet and protein not only in pancreatic carcinogenesis, but in spontaneous diseases as well.[8,9] In the hamster PC model, which mimics the human disease in many aspects, an HF diet increased PC incidence only when it was fed with a medium or high level of protein,[8] whereas high fat, low protein (HF-LP) reduced the tumor yield and a protein-free diet inhibited tumor formation altogether.[10] Similar results were obtained in the other cancer models.[11-14] Speaking for the different susceptibly of different tissues, the combination of an HF diet and high protein (HP) worked differently in breast and urinary bladder cancer, where HF and medium level of protein had no effect on either side, whereas HF+LP was as effective as the combination of HF and HP in increasing the tumor incidence.[11-14]

Differences were also found in the initiation and promotional stages of carcinogenesis between the genders. When the HF diet was given prior to the carcinogen, the tumor incidence was reduced in females, but not in males; there were also remarkable differences in the incidence of benign and malignant lesions that were influenced by the diet. An HF+HP, a strong promoter of malignant lesions, did not similarly affect the incidence of the benign lesions in females.[4]

The type of fat and carbohydrate and the method of food processing and preparation also seemed to influence the toxicity.

21.4 MECHANISM OF THE EFFECT OF DIET ON CARCINOGENESIS

Discrepancies and many unknowns exist in the mechanism of these nutrients on carcinogenesis. It is believed that the promotional effect of an HF diet is related to either increased energy intake or the generation of free radicals. Studies performed to address these questions have provided conflicting results. In the rat PC model, like the mammary cancer model,[15–17] voluntary exercise reduced the growth rate of induced lesions,[16,18] whereas in the hamster-BOP (N-nitrosobis-2(oxopropyl)amine) model exercise had no effects on tumor yield.[19]

Several studies in the rat PC model examined the role of an HF diet on lipid peroxidation and the generation of free radicals. In the rat model, feeding high levels of Vitamin C, β-carotene, and selenium reduced the incidence of pancreatic tumors, whereas Vitamin E was ineffective.[20] In the hamster model, dietary supplementation of Vitamin C, β-carotene, Vitamin E, or selenium, either alone or in combination, had no effects in the hamster.[21] The results contrasted with another study showing a significant inhibitory effect of Vitamin A and Vitamin C[22] and a promoting effect of high doses of selenium on tumorigenesis in the hamster model.[23] Also, the effect of retinoids and feeding cabbage in the hamster model sharply contrasted with the protective view of these nutrients in carcinogenesis of other tissues.[24,25]

Studies at the University of Nebraska Medical Center observed consistent changes associated with feeding an HF diet and provided the concept that the enhancing effect of this diet on pancreatic carcinogenesis is due to the induction of peripheral insulin resistance. This condition, which is associated with hyperinsulinemia and islet cell proliferation, appeared to be the fundamental reason for the HF-promoted carcinogenesis.[19] This notion was in support of the primary development of tumors within pancreatic islets.[26–30] Supporting evidence was the preventive effect of Metformin, which inhibited peripheral insulin resistance and islet cell proliferation on tumorigenesis.[30] It was then hypothesized that any diet that can stimulate islet cell proliferation can promote any factor that can put pancreatic islets in rest, such as Metformin, protein-free diet, and exogenous insulin, can inhibit or prevent PC induction.[10,27,31,32] Clearly, studies are required to substantiate this possibility.

21.5 NUTRIENTS AS PREVENTIVE AGENTS

The natural history of PC with its uncontrollable progression and deadly outcome prompted preventive studies by using substances to overcome the toxicity of carcinogens. For example, butylated hydroxyanisole, α-tocopherol was found to inhibit tumorigenesis in the hamster model, whereas carbazole was carcinogenic.[33] The pretreatment of hamsters with freeze-dried aloe (*Aloe arborescence*) significantly reduced the incidence of the BOP-induced lesions associated with the reduced formation of DNA-adduct, O^6-methyldeoxyguanosine in pancreatic epithelial cells.[34] Oltipraz, which has been shown to inhibit tumorigenesis induced by a variety of carcinogens in several model systems, paradoxically increased the incidence of ductal cell dysplasia at a certain dosage and seemed to exert organ-dependant modifying effects on BOP-induced carcinogenesis in hamsters when given in the initiation stage.[35] In another study, nimesulide, a cyclooxygenase-2 inhibitor, protected against BOP-induced pancreatic tumors in hamsters.[36] The mechanism of the inhibition or promotion of pancreatic diseases is obscure. Among many other possibilities, activation or inhibition of drug-metabolizing enzymes has been suggested (see Chapter 7 and Chapter 8).[37]

21.6 NUTRIENTS AFFECTING THE NORMAL PANCREAS

There are a few studies on the effects of dietary ingredients on the nontumoral pancreas. One of the most striking observations is the effect of raw soy flour on the pancreas of several animal species. In the rat, it induces acinar cell hyperplasia, adenoma, or even carcinoma (see Chapter 14), whereas it is ineffective in this regard in any other mammalian species. In chicks, feeding raw soybeans has depressed growth and the enlarged the duodenum and pancreas.[38] Exogenous putrescine has overcome this toxicity.[38] Feeding raw lima beans (*Phaseolus lunatus L.*) and lima bean fractions to chicks produced serious histopathological changes within the pancreas and some other tissues.[39] The rat has also been shown to be exceptionally sensitive to gabapentin, which induces pancreatic acinar cell hyperplasia.[40] The herbal medicine keishi-to and its components were tested in the rat. Although keishi-to (TJ-45) had no detectable effect, one of its components, TJ-9, induced chronic pancreatitis.[41] Natural colorant extracted from onion has been found to induce hyperplasia of the pancreas and some other tissues of the mouse.[42] A decreased pancreatic weight was observed in chicks fed dried cassava root to broiler chicks,[43] but its effect on mammalian tissues is unknown. The relative weights of the liver and pancreas were found to be increased in the turkey fed the fumonisin B1 present in *Fusarium moniliforme*.[44] D-glucose fed to rats and hamsters in the form of 30 or 20% solution in tap water for 80 or 112 weeks showed a significantly higher body weight, as well as increased blood glucose and renal glucose excretion, indicating the development of non-insulin dependent diabetes mellitus.[45] The incidence of islet cell adenomas was increased significantly in male rats. This study suggested that the increase and decrease in benign neoplasms of hormone-sensitive tissues to be the nutritionally induced modulation of the these tissues, rather than the result of chronic glucose administration itself.[45]

21.7 EPILOGUE

How can experimental data be translated to humans when understanding the role of nutrients in pancreatic toxicity has been so variable? The essential problem with experimental data is that contrary to defined experimental conditions, humans do not follow standardized dietary practices and are willingly or unwillingly exposed to known and unknown environmental factors. Attempts to understand the role of variables in healthy and diseased tissues has been hampered by the enormous complexity of the effect of nutrients among the species. Even in the same species, the result of the same diet has shown variations. Reasons for these differences are still obscure. The species, strain, the breeding and housing environments, experimental designs, and the source of the given dietary ingredient are among the factors that can play a role.

REFERENCES

1. Pour P., It Y., and Althoff J. 1979. Comparative studies on spontaneous tumor incidence based on systemic histologic examination of rat and hamster strains of the same colony. *Prog. Exp. Tumor Res.* 24:199–206.
2. Pour P. and Birt D. 1979. Spontaneous diseases of Syrian hamsters — their implications in toxicological research: facts, thoughts and suggestions. *Prog. Exp. Tumor Res.* 24:145–156.
3. Birt D.F., Patil K., and Pour P.M. 1985. Comparative studies on the effects of semipurified and commercial diet on longevity and spontaneous and induced lesions in the Syrian golden hamster. *Nutr. Cancer* 7(3):167–177.
4. Birt D.F., Salmasi S., and Pour P.M. 1981. Enhancement of experimental pancreatic cancer in Syrian golden hamsters by dietary fat. *J. Natl. Cancer Inst.* 67(6):1327–1332.
5. Longnecker D.S., Roebuck B.D., and Kuhlmann E.T. 1985. Enhancement of pancreatic carcinogenesis by a dietary unsaturated fat in rats treated with saline or N-nitroso(2-hydroxypropyl)(2-oxopropyl)amine. *J. Natl. Cancer Inst.* 74(1):219–222.
6. Roebuck B.D., Yager Jr. J.D., Longnecker D.S., and Wilpone S.A. 1981. Promotion by unsaturated fat of azaserine-induced pancreatic carcinogenesis in the rat. *Cancer Res.* 41(10):3961–3966.
7. Wenger F.A., Jacobi C.A., Zieren J., Docke W., Volk H.D., and Muller J.M. 1999. Tumor size and lymph-node status in pancreatic carcinoma — is there a correlation to the preoperative immune function? *Langenbecks Arch. Surg.* 384(5):473–478.
8. Birt D.F., Stepan K.R., and Pour P.M. 1983. Interaction of dietary fat and protein on pancreatic carcinogenesis in Syrian golden hamsters. *J. Natl. Cancer Inst.* 71(2):355–360.
9. Birt D.F. and Pour P.M. 1985. Interaction of dietary fat and protein in spontaneous diseases of Syrian golden hamsters. *J. Natl. Cancer Inst.* 75(1):127–133.
10. Pour P.M. and Birt D.F. 1986. Effect of dietary protein on N-nitrosobis(2-oxopropyl)amine-induced carcinogenesis and on spontaneous diseases in Syrian golden hamsters. *J. Natl. Cancer Inst.* 76(1):67–72.
11. Frith C.H.N.M., Umholtz R., and Knapka J.J. 1980. Effect of dietary protein and fat levels on liver and urinary bladder neoplasia in mice fed 2-acetylaminoflourene. *J. Food Safety.* 30:835–840.

12. Clinton S.K., Mulloy A.L., Li S.P., Mangian H.J., and Visek W.J. 1997. Dietary fat and protein intake differ in modulation of prostate tumor growth, prolactin secretion and metabolism, and prostate gland prolactin binding capacity in rats. *J. Nutr.* 127(2):225–237.
13. Clinton S.K., Imrey P.B., Mangian H.J., Nandkumar S., and Visek W.J. 1992. The combined effects of dietary fat, protein, and energy intake on azoxymethane-induced intestinal and renal carcinogenesis. *Cancer Res.* 52(4):857–865.
14. Clinton S.K., Alster J.M., Imrey P.B., Simon J., and Visek W.J. 1988. The combined effects of dietary protein and fat intake during the promotion phase of 7,12-dimethylbenz(a)anthracene-induced breast cancer in rats. *J. Nutr.* 118(12):1577–1585.
15. Thompson H.J. 1992. Effect of amount and type of exercise on experimentally induced breast cancer. *Adv. Exp. Med. Biol.* 322:61–71.
16. Giles T.C. and Roebuck B.D. 1992. Effects of voluntary exercise and/or food restriction on pancreatic tumorigenesis in male rats. *Adv. Exp. Med. Biol.* 322:17–27.
17. Pariza M.W. 1987. Fat, calories, and mammary carcinogenesis: net energy effects. *Am. J. Clin. Nutr.* 45(1 Suppl.):261–263.
18. Roebuck B.D., McCaffrey J., and Baumgartner K.J. 1990. Protective effects of voluntary exercise during the postinitiation phase of pancreatic carcinogenesis in the rat. *Cancer Res.* 50(21):6811–6816.
19. Kazakoff K., Cardesa T., Liu J., Adrian T.E., Bagchi D., Bagchi M. et al. 1996. Effects of voluntary physical exercise on high-fat diet-promoted pancreatic carcinogenesis in the hamster model. *Nutr. Cancer* 26(3):265–279.
20. Woutersen R.A., Appel M.J., van Garderen-Hoetmer A., and Wijnands M.V. 1999. Dietary fat and carcinogenesis. *Mutat. Res.* 443(1–2):111–127.
21. Appel M.J. and Woutersen R.A. 1996. Effects of dietary beta-carotene and selenium on initiation and promotion of pancreatic carcinogenesis in azaserine-treated rats. *Carcinogenesis* 17(7):1411–1416.
22. Wenger F.A., Kilian M., Ridders J., Stahlknecht P., Schimke I., Guski H. et al. 2001. Influence of antioxidative vitamins A, C and E on lipid peroxidation in BOP-induced pancreatic cancer in Syrian hamsters. *Prostaglandins Leukot. Essential Fatty Acids* 65(3):165–171.
23. Birt D.F., Julius A.D., Runice C.E., White L.T., Lawson T., and Pour P.M. 1988. Enhancement of BOP-induced pancreatic carcinogenesis in selenium-fed Syrian golden hamsters under specific dietary conditions. *Nutr. Cancer* 11(1):21–33.
24. Birt D.F., Davies M.H., Pour P.M., and Salmasi S. 1983. Lack of inhibition by retinoids of bis(2-oxopropyl)nitrosamine-induced carcinogenesis in Syrian hamsters. *Carcinogenesis* 4(10):1215–1220.
25. Birt D.F., Pelling J.C., Pour P.M., Tibbels M.G., Schweickert L., and Bresnick E. 1987. Enhanced pancreatic and skin tumorigenesis in cabbage-fed hamsters and mice. *Carcinogenesis* 8(7):913–917.
26. Pour P.M., Weide L., Liu G., Kazakoff K., Scheetz M., Toshkov I. et al. 1997. Experimental evidence for the origin of ductal-type adenocarcinoma from the islets of Langerhans. *Am. J. Pathol.* 150(6):2167–2180.
27. Pour P.M. and Kazakoff K. 1996. Stimulation of islet cell proliferation enhances pancreatic ductal carcinogenesis in the hamster model. *Am. J. Pathol.* 149(3):1017–1025.

28. Ishikawa O., Ohigashi H., Imaoka S., Nakai I., Mitsuo M., Weide L. et al. 1995. The role of pancreatic islets in experimental pancreatic carcinogenicity. *Am. J. Pathol.* 147(5):1456–1464.

29. Pour P. 1978. Islet cells as a component of pancreatic ductal neoplasms. I. Experimental study: ductular cells, including islet cell precursors, as primary progenitor cells of tumors. *Am. J. Pathol.* 90(2):295–316.

30. Schneider M.B., Matsuzaki H., Haorah J., Ulrich A., Standop J., Ding X.Z. et al. 2001. Prevention of pancreatic cancer induction in hamsters by metformin. *Gastroenterology* 120(5):1263–1270.

31. Pour P.M., Duckworth W., Carlson K., and Kazakoff K. 1990. Insulin therapy prevents spontaneous recovery from streptozotocin-induced diabetes in Syrian hamsters. An autoradiographic and immunohistochemical study. *Virchows. Arch. A Pathol. Anat. Histopathol.* 417(4):333–341.

32. Pour P.M., Kazakoff K., and Carlson K. 1990. Inhibition of streptozotocin-induced islet cell tumors and N-nitrosobis(2-oxopropyl)amine-induced pancreatic exocrine tumors in Syrian hamsters by exogenous insulin. *Cancer Res.* 50(5):1634–1639.

33. Moore M.A., Tsuda H., Thamavit W., Masui T., and Ito N. 1987. Differential modification of development of preneoplastic lesions in the Syrian golden hamster initiated with a single dose of 2,2′-dioxo-N-nitrosodipropylamine: influence of subsequent butylated hydroxyanisole, alpha-tocopherol, or carbazole. *J. Natl. Cancer Inst.* 78(2):289–293.

34. Furukawa F., Nishikawa A., Chihara T., Shimpo K., Beppu H., Kuzuya H. et al. 2002. Chemopreventive effects of Aloe arborescens on N-nitrosobis(2-oxopropyl)amine-induced pancreatic carcinogenesis in hamsters. *Cancer Lett.* 178(2):117–122.

35. Son H.Y., Nishikawa A., Furukawa F., Kasahara K.I., Miyauchi M., Nakamura H. et al. 2000. Organ-dependent modifying effects of oltipraz on N-nitrosobis(2-oxopropyl)amine (BOP)-initiation of tumorigenesis in hamsters. *Cancer Lett.* 153(1–2):211–218.

36. Furukawa F., Nishikawa A., Lee I.S., Kanki K., Umemura T., Okazaki K. et al. 2003. A cyclooxygenase-2 inhibitor, nimesulide, inhibits postinitiation phase of N-nitrosobis(2-oxopropyl)amine-induced pancreatic carcinogenesis in hamsters. *Intl. J. Cancer* 104(3):269–273.

37. Rutishauser S.C., Ali A.E., Jeffrey I.J., Hunt L.P., Braganza J.M. 1995. Toward an animal model of chronic pancreatitis. Pancreatobiliary secretion in hamsters on long-term treatment with chemical inducers of cytochromes P450. *Intl. J. Pancreatol.* 18(2):117–126.

38. Mogridge J.L., Smith T.K., and Sousadias M.G. 1996. Effect of feeding raw soybeans on polyamine metabolism in chicks and the therapeutic effect of exogenous putrescine. *J. Anim. Sci.* 74(8):1897–1904.

39. Ologhobo A.D., Apata D.F., Oyejide A., and Akinpelu O. 1993. Toxicity of raw lima beans (*Phaseolus lunatus L.*) and lima bean fractions for growing chicks. *Br. Poult. Sci.* 34(3):505–522.

40. Dethloff L., Barr B., Bestervelt L., Bulera S., Sigler R., LaGattuta M. et al. 2000. Gabapentin-induced mitogenic activity in rat pancreatic acinar cells. *Toxicol. Sci.* 55(1):52–59.

41. Motoo Y., Su S.B., Xie M.J., Mouri H., Taga H., and Sawabu N. 2001. Effect of herbal medicine keishi-to (TJ-45) and its components on rat pancreatic acinar cell injuries in vivo and in vitro. *Pancreatology* 1(2):102–109.

42. Kojima T., Tanaka T., Mori H., Kato Y., and Nakamura M. 1993. Acute and subacute toxicity tests of onion coat, natural colorant extracted from onion (*Allium cepa L.*), in (C57BL/6 × C3H)F1 mice. *J. Toxicol. Environ. Health* 38(1):89–101.

43. Panigrahi S., Rickard J., O'Brien G.M., and Gay C. 1992. Effects of different rates of drying cassava root on its toxicity to broiler chicks. *Br. Poult. Sci.* 33(5):1025–1041.

44. Kubena L.F., Edrington T.S., Kamps-Holtzapple C., Harvey R.B., Elissalde M.H., and Rottinghaus G.E. 1995. Effects of feeding fumonisin B1 present in Fusarium moniliforme culture material and aflatoxin singly and in combination to turkey poults. *Poult. Sci.* 74(8):1295–1303.

45. Bomhard E., Bischoff H., Mager H., Krotlinger F., and Schilde B. 1998. D-glucose combined chronic toxicity and carcinogenicity studies in Sprague-Dawley rats and Syrian golden hamsters. *Drug Chem. Toxicol.* 21(3):329–353.

22

TOXICITY OF TOBACCO-SPECIFIC N-NITROSAMINES

Bogdan Prokopczyk, Dietrich Hoffmann, and Karam El-Bayoumy

CONTENTS

22.1 INTRODUCTION

Numerous epidemiological studies point to the serious toxicity of tobacco to the pancreas. The association between cigarette smoking and pancreatic cancer was first reported in the late 1960s.[1-3] Most of the studies reviewed by the International Agency for Research on Cancer in 1986[4] and those published subsequently reported a 1.5- to 2-fold increase in risk among smokers.[4-13] In 1973, Wynder et al. was the first to report a positive correlation between the number of cigarettes smoked per day and the increased risk for males; the female data showed an irregular dose-response pattern.[5] In this study, the authors have also demonstrated an association between cigar and pipe smoking with pancreatic cancer. This observation was later confirmed in a subsequent study.[13] Overall smokers have 1.5- to 5-fold increased risk for pancreatic cancer compared to nonsmokers depending on years of smoking and daily consumption of cigarettes. In 1991, in the United States, it was estimated that 28.6% of the mortality from pancreatic cancer in men and 33.3% of that in women was attributable to cigarette smoking.[9]

Several case-control studies have indicated that pancreatic cancer is more prevalent among certain families[11–16] due to their genetic makeup.[12] In these highly susceptible populations, smoking shortens the latency of this disease by as much as 20 years.[17] Our collaborative research has revealed that smoking also shortens the onset of pancreatic ductal adeno-carcinoma among patients with sporadic cancer by approximately 5 years.[18]

Although epidemiological studies suggest a possible link with various lifestyles, dietary habits, and other related factors, except for cigarette smoking, no other risk factor for pancreatic cancer has as yet been sufficiently established or unequivocally accepted.

22.2 ANIMAL BIOASSAYS — AGENTS THAT INDUCE PANCREATIC CANCER IN LABORATORY ANIMALS

Several animal model assays have been developed to study various aspects of pancreatic carcinogenesis. Historically, the first induction of pancreatic tumors was reported in 1968 when Druckrey et al. demonstrated that prolonged administration of N-methylnitrosourea and N-methyl-N-nitrosourethane in drinking water to outbred guinea pigs produced ade-nocarcinoma of the pancreas and stomach.[19] This report was followed by numerous investigations in which approximately 20 various chemicals were used to induce pancreatic cancer in rodents (Table 22.1). The most frequently reported models use either L-serine diazoacetate (azaserine) or synthetic N-nitrosamines such as N-nitrosobis(2-oxopropyl)amine (BOP), N-nitrosobis(2-hydroxopropyl)amine (BHP), N-nitroso(2-acetylpro-pyl)amine (BAP), and N-nitroso(2-hydroxypropyl)(2-oxopropyl)amine (HPOP).[20–23] Azaserine is highly carcinogenic to Wistar rats when admin-istered by multiple injections; it is not carcinogenic in the hamster. Other pancreatic carcinogens in rats include 7,12-dimethylbenzanthracene (DMBA),[24] 4-hydroxyaminoquinoline-1-oxide,[25] hypolipidemic drugs, namely nafenopin,[26] clofibrate,[27,28] and other peroxisome proliferators,[29] 2,6-dimethylnitrosomorpholine,[30] $N\delta$-(N-methyl-N-nitrosocarbamoyl)-L-ornithine[31]; the latter also induces pancreatic carcinomas in hamsters.[32]

Table 22.1 Animal Models of Chemically Induced Pancreatic Lesions

Carcinogen	Dose	Route of Administration	Frequency of Administration	Diet	Modifier	% Tumor Incidence (Multiplicity) Study Results	Ref. No.
Guinea Pig							
N-methylnitrosourea	2.5 mg/kg	Drinking water	Intermittent	Standard diet enhanced with hay and carrots	none	7.6	19
N-nitrosourethane	3.0 mg/kg	Drinking water	Intermittent	Standard diet enhanced with hey and carrots	none	44	19
N-methylnitrosourea	10 mg/kg	gavage	Daily, up to 27 weeks	Purina guinea pig chow	none	25 in males, 50 in females	147
Rat							
DMBA	2-3 mg/rat	Implantation into pancreas	once	Purina laboratory chow	none	80	24
Azaserine	30 mg/kg	i.p.	once	AIN-76A	20% unsaturated fat safflower oil	94 (4.7)	22
Azaserine	30 mg/kg	i.p.	once	AIN-76A	Essential fatty acid (EFS, linoleic acid)	94 (5.9) Number of pancreatic nodules increased from approx 4 to 18 nodules per pancreas with the increase of EFS content in the diet	148

Table 22.1 Animal Models of Chemically Induced Pancreatic Lesions (Continued)

Carcinogen	Dose	Route of Administration	Frequency of Administration	Diet	Modifier	% Tumor Incidence (Multiplicity) Study Results	Ref. No.
azaserine	30 mg/kg	i.p.	Once	AIN-76A	Various doses and types of fat	Approx 27 to 100% (0.4 to >10). Animals in lipotrope-deficient and unsaturated fat diet groups had significantly higher incidence of pancreatic neoplasms.	149
HPOP	160 mg/kg	i.p.	Once	RMH chow	none	86	152
IQ	0.4 mmole/kg	Gavage	3 times a week for weeks 1–8, once a week for next 25 weeks	NIH-07 supplemented with 15% corn oil	none	59 - Hyperplastic acinar lesions	36
NNK	0.5–5.0 ppm	Drinking water	Continuous	NIH-07	none	11.2 in 1.0 ppm group	33
NNAL	5.0 ppm					26.6	
NG	20.0 ppm					13.3	
NNK	2 ppm	Drinking water	Continuous	high fat (23.5 corn oil), low fat (5.0% corn oil)	Fat content	46.6 in high fat diet group, 31.6 in low fat diet group	37

Hamster

Compound	Route	Dose	Schedule	Diet	Modifier	Results	Ref.
BOP	s.c.	20 mg/kg	Once	Various experimental diets	Energy restriction	59 to 66 (0.9 to 1.7). Lack of inhibition of pancreatic carcinogenesis by energy restriction	23
BOP	s.c.	70 mg/kg	Once	Basal diet	Green tea extract	54 (1.0) in control; 33 (0.5) in green tea extract group	92
BOP	s.c.	10 mg/kg	Once	Basal diet	NNAL	18.5 (1.33) in BOP group and 14 (1.16) in high NNAL group	110
BOP	s.c.	20 mg/kg	3 times at weeks 6, 7, and 8	high fat (20%) cereal-based diet	β-carotene, Vitamin C, Vitamin E, and selenium	10 to 21. Vitamin C alone and in combination with β-carotene demonstrated limited inhibitory effects.	93
BOP	s.c.	20 mg/kg	Twice, one week apart	Basal diet	Phenethyl isocyanate (PEITC)	77(1.63) in BOP group; 0 (0) in high PEITC group	94
BOP	s.c.	15 mg/kg	Once weekly for 12 weeks	Basal diet	Oltipraz	89 (4.4) in BOP group; 60 (3.6) in high Oltipraz group	95

Table 22.1 Animal Models of Chemically Induced Pancreatic Lesions (Continued)

Carcinogen	Dose	Route of Administration	Frequency of Administration	Diet	Modifier	% Tumor Incidence (Multiplicity) Study Results	Ref. No.
BOP	70 mg/kg and 20 mg/kg	s.c.	Twice (70 mg/kg initially, 20 mg/kg on week 18)	Basal diet	Sulfonation inhibitors – dehydroepiandrosterone sulfate (DHAS) and 3'-phosphoadenosine 5'-phosphate (PAP)	93 (4.5) in BOP group, 80 (3.1) in high DHAS group, 86 (2.9) in high PAP group	150
BOP	10 mg/kg	s.c.	3 times, once a week for 3 weeks	Basal diet	Cigarette smoke	56.6 (0.8) in BOP group, 17.2 (0.24) in BOP + smoking group	112
BOP	10 mg/kg	s.c.	Once	Basal diet	NNK	3.4 (0.03) in BOP group, 4.7 (0.07) in BOP + NNK group	111
BOP	10 mg/kg	s.c.	3 times, once a week for 3 weeks	Basal diet	Caffeine, nicotine, ethanol, and sodium selenite	39 (0.4) in BOP group, 59 (1.0) in BOP/caffeine group, 43 (0.6) in BOP/nicotine group, 43 (0.7) in BOP/ethanol group, and 31 ().30 in BOP/selenite group	98

Compound	Dose	Route	Schedule	Diet	Modifier	Result	Ref
BOP	70 mg/kg	s.c.	Once	Basal diet	Cholesterol and cholestyramine	6.9 (0.69) in BOP group, 40 (1.6) in BOP/cholesterol group, and 30 (1.43) in BOP/cholestyramine group	151
BOP	10 mg/kg	s.c.	5 times, once a week for 5 weeks	Basal diet	Prostaglandin synthesis inhibitors (indomethacin, phenylbutazone, aspirin)	71.4 (1.29) in BOP group, 36.8 (0.58) in phenylbutazone group	96
BHP	500 mg/kg	s.c.	5 times, once a week for 5 weeks	Basal diet	Clofibrate	23.5 in BHP group, 5.6 in high clofibrate group	97
$N\delta$-(N-methyl-N-nitrosocarbamoyl)-L-ornithine	218 mg/kg	s.c	12 times, once a week for 12 weeks	RMH 3000 chow		71.4 males, 50 females	32
	654 mg/kg		6 times, once a week for 6 weeks				
2,6-dimethyl-nitrosomorpholine	9.1–73.4 mg/kg	gavage	Continuous, once weekly for life	Not reported	none	46 ductal carcinoma, 31 ductal adenoma	30

BOP, BHP, BAP, and HPOP are more specific carcinogens in the hamster inducing tumors with a ductal phenotype that are similar to those seen in men. However, all of the compounds listed above are synthetic chemicals. We have demonstrated that 4-(methylnitrosamino)-1-(3-pyridyl)-1-butanone (NNK) and 4-(methylnitrosamino)-1-(3-pyridyl)-1-butanol (NNAL) induce acinar and ductal-cell tumors in F344 rats.[33] Unlike BOP and azaserine, the most frequently used compounds to study pancreatic carcinogenesis, NNK and NNAL are environmental carcinogens to which humans are exposed. N-nitrosoguvacoline (NG) and a product of fried foods, 2-amino-3-methylimidazo[4,5-f]quinoline (IQ), are also environmental pancreatic carcinogens. NG is formed from *Arecoline* by nitrosation and has been detected in the saliva of betel quid chewers.[34,35] When given in drinking water, at a dose level of 20 ppm, NG induced 4 acinar cells adenomas in F344 rats.[33] IQ, given by gavage, induced atypical hyperplastic acinar cells in 19 of 32 Sprague-Dawley rats, one rat had an acinar cell adenoma, and another rat an islet cell adenoma.[36] Significantly, in all BOP, azaserine, and NNK models the higher incidence of ductal adenocarcinomas was observed among animals fed a high fat diet.[21,22,37]

Several other pancreatic cancer models utilized surgical implantation of pancreatic cancer cells or cancer tissue into the pancreas. These include the orthotropic xenograft of CaPan-1,[38] PAN-12,[39] and KCI-MOH1[40] cells in athymic mice, implantation of Panc02 cells into the pancreas of syngeneic C57BL/6 mice,[41] and transplantation of histologically intact pancreatic cancer specimens to the nude mouse pancreas.[42–44] These models provide tools for better understanding the biology of pancreatic cancer and in studying the therapeutic efficacy of various drugs and chemopreventive agents.

22.3 TOBACCO-DERIVED CARCINOGENS

22.3.1 Tobacco Components

There are approximately 69 carcinogens among the nearly 5000 chemical constituents identified in tobacco smoke.[45] The three major classes of carcinogens are the tobacco specific nitrosamines (TSNAs) (e.g., NNK and NNN), polycyclic aromatic hydrocarbons (PAH), and the aromatic and heterocyclic amines.[45]

NNK is one of the most common carcinogens in tobacco products with levels of 0.08–0.77 μg/cigarette in mainstream smoke.[45,46] NNK is among the most potent carcinogens tested in mice, rats, hamsters, and the mink.[47–49] Independent of the route of administration (i.e., oral swabbing, intraperitoneal, subcutaneous, or intravesicular injection), NNK induces primarily pulmonary adenomas and adenocarcinomas in rodents. Secondary sites are the nasal mucosa and the liver.[48,50–52] Importantly, when administered in drinking water, NNK also induces tumors of the exocrine pancreas in rats.[33] In rats fed a high fat diet, NNK elicits pancreatic tumors earlier and at a higher rate than in animals kept on low fat diet (see below for details).[37] The reductive metabolite of NNK, NNAL, is also a potent pulmonary carcinogen in the rat and mouse[33,47] and a pancreatic carcinogen in the rat.[33] Additionally, NNK is capable of transforming spontaneously immortal pancreatic cells into malignant phenotype[53] and, when administered with ethanol, it induces pancreatic tumors in the offspring of NNK-treated hamsters.[54]

A smoker who smokes 20 cigarettes per day inhales approximately 9.0 µg of NNK per day. Approximately 25% of this carcinogen is rapidly converted to NNAL by red blood cells. We have hypothesized that a portion of these carcinogens can be delivered into the pancreas either via the blood stream or bile. To test this hypothesis, we have analyzed human pancreatic juice for the presence of tobacco-derived compounds and have demonstrated that certain tobacco-derived products, namely cotinine and TSNA, are indeed present in pancreatic juice.[55] Cotinine was detected in all analyzed samples of pancreatic juice from smokers and was present in only two of nine samples of pancreatic juice from nonsmokers. NNK was detected in 15 of 18 samples from smokers at levels of 1.37–604 ng/ml pancreatic juice. In nine samples of pancreatic juice from nonsmokers, NNK ranged from undetectable level (in three samples) to 96.8 ng/ml juice. In pancreatic juice from smokers, the mean level of NNK was significantly higher than that from nonsmokers. N-nitrosonornicotine (NNN) was found in two samples of pancreatic juice; NNK was not detected in any other samples. NNAL was present in eight of 14 pancreatic juice samples from smokers and in three of nine samples from nonsmokers. The presence of other tobacco-derived carcinogens in human pancreatic juice that may be involved in the induction of pancreatic tumors (e.g., 4-aminobiphenyl) needs to be addressed. Nevertheless, these findings support the epidemiology-based association between cigarette smoking and cancer of the pancreas and provide a rationale for investigations aimed at the mechanism responsible for the induction of pancreatic neoplasms by TSNA.

22.3.2 Metabolic Activation and DNA Adducts

Metabolic activation of TSNA is required for the initiation of carcinogenesis. As a representative example, the metabolic activation of NNK is shown in Figure 22.1. Multiple forms of cytochrome P450, namely CYP 1A1, CYP 1A2, CYP 2A3, CYP 2A6, CYP 2B1, CYP 2B2, CYP 2B4, CYP 2D6, and CYP 2E1, are known to catalyze the metabolic activation of NNK *in vivo* (see Chapter 7 and Chapter 8).[47] The presence of CYP 1A1, CYP 1A2, CYP 2B6, CYP 2C8, CYP 2C9, CYP 2C19, CYP 2D1, CYP 2E1, CYP 3A1,

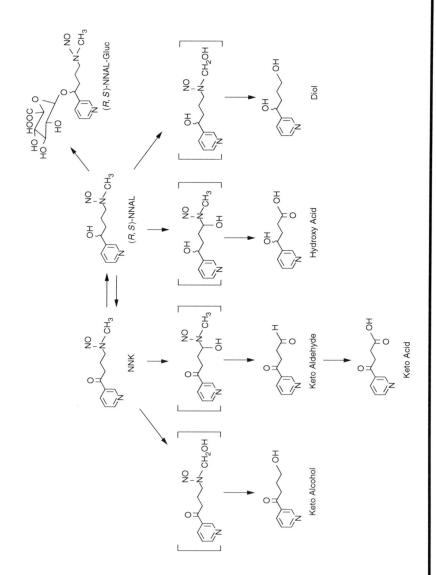

Figure 22.1   Metabolic pathways of NNK.

CYP 3A2, CYP 3A4, and CYP 11A in the pancreas of different species, including humans, has been reported.[56–64] Moreover, it appears that the expression of CYP 1A2, CYP 2C9, and CYP 2E1 is up-regulated in chronic pancreatitis patients.[64] Lipoxygenases[65] and cyclooxygenases[66] can also be involved in metabolic activation of NNK. In rodents, α-hydroxylation of NNK contributes to at least 50% of its total metabolites.[47,67,68] Most studies with human cells and human tissues indicate that the extent of oxidative NNK metabolism is significantly lower than that in rodents.[47,67–69] However, we showed that human cervical cells metabolize NNK to an extent comparable to that by the reductive pathway.[69]

Experiments with cultured human tissues and human liver and lung microsomes have pointed to carbonyl reduction as a major pathway of NNK metabolism.[67,68] This metabolic conversion results in the formation of NNAL. NNAL and NNAL-Gluc have been detected in the urine of active smokers[70] and passive smokers.[71] Although glucuronidation is generally considered a detoxification pathway, its role in carcinogen delivery to target organs cannot be excluded. In fact, α-hydroxylation products, indicative of metabolic activation as described below, have been detected in rat urine after treatment with NNAL-Gluc.[72] In the liver, for example, the glucuronidation process would enhance carcinogen elimination. However, in organs such as the colon, due to the presence of bacterial flora and the associated enzyme activity, NNAL-Gluc can be hydrolyzed to yield the parent nitrosamine by the action of β-glucuronidase. A similar process may occur in the pancreas because β-glucuronidase activity has, in fact, been demonstrated in human pancreatic juice.[73] Experiments aimed at assessing interindividual differences in β-glucuronidase activity are required. Studies addressing its expression in pancreatic juice of patients with pancreatic disorders are being conducted in our laboratory.

We have investigated the metabolism of NNK in human pancreatic ductal and islet cells as well as in microsomes and cytosols isolated from histologically normal pancreatic tissue (unpublished data). All incubations of NNK with ductal cells yielded α-hydroxylated products; these were formed at significantly higher rates than those observed in islet cells. Among metabolites derived from α-hydroxylation, 4-hydroxy-1-(3-pyridyl)-1-butanone (keto alcohol) was the major metabolite produced by islet cells. This compound was not detected in incubations with ductal cells. Both types of cells produced 4-hydroxy-4-(3-pyridyl)-1-butanoic acid (hydroxy acid) and 4-oxo-4-(3-pyridyl)-1-butanoic acid (keto acid). The chiral product of carbonyl reduction of NNK, NNAL was the major metabolite detected. Both cell types predominantly formed the (R)-enantiomer of NNAL. In both microsomal and cytosolic incubations with NNK, NNAL was the major metabolite formed. With regard to its formation, the microsomes fell into two groups. One group displayed a comparatively higher metabolizing capacity and formed primarily (R)-NNAL, the levels of which were only slightly reduced by glycyrrhizic acid (GA), an inhibitor of 11β-hydroxysteroid dehydrogenase, the enzyme that facilitates conversion of NNK to NNAL.[74,75] The second group displayed a lower reductive metabolic capacity, but formed primarily (S)-NNAL and its rates of formation were inhibited by 60 to 80% upon treatment with GA. By contrast to microsomes, no discernable pattern of NNAL formation in cytosolic preparations was found and (S)-NNAL was the predominant product. Menadione, an inhibitor of carbonyl reductase, blocked NNAL formation by approximately 95% in incubations of two cytosolic samples. These results demonstrate the capacity of the human pancreas to metabolically activate NNK and suggest the existence of individual differences in both the activity and stereochemical nature of carbonyl reduction.

Heterocyclic amines constitute another group of carcinogens present in tobacco smoke that may be relevant to pancreatic cancer in smokers. These carcinogens, represented by compounds such as 2-amino-3,4-dimethylimidazo[4,5-f]quinoline (MeIQ) or 2-amino-1-methyl-6-phenylimidazo[4,5-b]pyridine (PhIP), are formed during smoking of tobacco and also during heating or cooking of high protein diets. The human levels of exposure to heterocyclic amines and the nature of these compounds can vary significantly depending on the type of diet. PhIP is the most abundant heterocyclic aromatic amines (HAA) in tobacco smoke (11–23 ng/cigarette).[76] The levels of PhIP in cooked foods range from 15 to 65 ng/g.[77]

PAH mixtures are widely distributed in air, soil, and water. Major human exposure comes from PAH present in tobacco smoke and smoked foods. It is essentially impossible to avoid exposure to traces of these substances on a daily basis. Human daily intake of B[a]P by inhalation is estimated to be about 0.02 to 1.5 μg and by ingestion from food about 0.03 to 1.3 μg. The intake from cigarette smoke is calculated to be about 0.2 μg/day. PAHs have shown carcinogenic activity in laboratory animals when administered by various routes. The major sites are the lungs, skin, and mammary glands.[78–80]

The metabolic activation of carcinogens leads to formation of electrophiles that readily react with cellular macromolecules including DNA. TSNA-induced DNA adducts that have been identified to date include O^6-methylguanine, 7-methylguanine, and 4-methylthymidine. In the case of NNK and NNAL, there are also adducts resulting from pyridyloxobutylation and pyridylhydroxybutylation of DNA that remain largely uncharacterized.[47,67,68,81] NNK has also been shown to induce oxidative damage in lungs of A/J mice and F344 rats as illustrated by elevated levels of 8-hydroxy-2-deoxyguanosine.[68,82] It is possible that this lesion results from the formation of reactive oxygen species (ROS) and the latter may induce several transcriptional factors (e.g., NF-κB, AP-1), which, in turn, can activate several genes including COX-2. These genotoxic effects of NNK may also cause mutations in other critical genes known to be frequently mutated in pancreatic adenocarcinoma. Activation of various signal transduction pathways, induction of transcription and cell proliferation in the pancreas by NNK can also be caused by its epigenetic effects.[83] These effects are due to the function of NNK as an antagonist for β-adrenergic receptors. This process is known to activate various cellular events that result in the release of arachidonic acid, followed by the formation of its mitogenic metabolites. The arachidonic acid cascade is important in various diseases including pulmonary and pancreatic adenocarcinomas, both inducible by NNK. Activation of the arachidonic acid pathway results in the overexpression of the gene encoding COX-2. Up-regulation of COX-2 expression in human pancreatic cancer has been reported.[84,85] In fact, NNK is known to induce COX-2 in rats and mice.[86,87] Another frequent epigenetic event, which had been reported in pancreatic adenocarcinoma, is a hypomethylation of certain genes that are overexpressed in this type of cancer.[88]

Putative arylamine-DNA adducts have been detected in human pancreatic DNA samples using the ^{32}P-postlabeling technique.[89] A major adduct found in 8 of 29 samples was chromatographically identical to the 4-aminobiphenyl (ABP) DNA adduct, N-(deoxyguanosin-8-yl)-ABP. There was no apparent association between detection of this adduct and the current smoking status.

Alexandrov et al.[90] have analyzed 11 pancreatic and six colonic DNA samples for PAH-DNA and PhIP-DNA adducts using either gas chromatography/electron capture-mass spectrometry or the ^{32}P-postlabeling technique. Both methods provided evidence for PhIP-DNA adducts in two out of six colon samples. Neither PAH-DNA nor PhIP-DNA adducts have been detected in 11 human pancreatic samples, irrespective of whether the subjects were smokers or nonsmokers. Overall, the evidence that PAH are pancreatic carcinogens is weak.

22.4 MODIFIERS OF N-NITROSAMINE CARCINOGENESIS

Among compounds that have been tested as chemopreventive agents in azaserine- or synthetic N-nitrosamines-induced pancreatic cancers in Wistar rats and in hamsters are methionine[91]; green tea extracts[92]; beta-carotene, Vitamin C, Vitamin E, selenium[93]; phenyl isocyanate[94]; Oltipraz[95]; inhibitors of prostaglandin synthesis[96]; clofibrate[97]; and caffeine, nicotine, and ethanol.[98] The effects of dietary modifications, such as fat type and dose[20–22] and caloric restrictions[23,99,100] have also been studied (see Chapter 21). Preventive efficacy of several synthetic and naturally occurring compounds, given either alone or in combination, have been tested against

NNK-induced lung carcinogenesis. Compounds that inhibited NNK-induced lung tumorigenesis include 1,4-phenylenebis(methylene)seleno-cyanate (*p*-XSC), nonsteroidal anti-inflammatory drugs, tea and its active polyphenols, aromatic isothiocyanates and their *N*-acetylcysteine conjugates, myo-inositol, dexamethasone, indole-3-carbinol, aryl alkyls, 4-ipomeanol and its analogs, farnesyltransferase inhibitors, and d-limonene.[66–68,101–108] However, as reported by Witschi, most of these agents have no effect when tested against whole tobacco smoke; only myo-inositol and dexamethasone given in combination reduced lung tumor multiplicities and incidence.[109] None of these chemopreventive agents have been tested in NNK-induced pancreatic tumorigenesis. A literature search has revealed that there was not a single study addressing this issue, although, a few investigations assessed the promotional effects of NNK, NNAL, and tobacco smoke in BOP-induced pancreatic cancer.[110–112] Interestingly, our single study[37] has demonstrated that a dramatic, protective effect can be observed by the reduction of dietary fat content. In this study, groups of F344 rats were given NNK (2 ppm solution) in the drinking water. One group of rats was kept on a high fat diet (23.5% corn oil), while the fat content in the diet fed to the second group was reduced to 5.0% of corn oil. After 18 months, 11 of 60 animals on a high fat diet developed pancreatic tumors. By contrast, there was only a rat with pancreatic tumor in the low fat group (1 of 60). At termination of the experiment, there were 28 and 19 animals with pancreatic tumors in high and low fat diet groups, respectively.

22.5 GENETIC ALTERATION IN PANCREATIC CANCER AND POTENTIAL BIOMARKERS

The molecular basis for pancreatic carcinogenesis is still not well characterized. In recent years, various genetic alterations have been observed in human pancreatitis, pancreatic intraepithelial neoplasia, and in pancreatic cancer (see Chapter 21, Chapter 25, and Chapter 29). These alterations include activation of proto-oncogenes such as K-*ras* and HER2/nu,[113–115] inactivation of tumor suppressor genes such as p16, p53, DPC4/SMAD4, BRCA2, and SST2.[113–120] Frequencies and characteristics of their mutations have been established. The abnormal expressions of proapoptotic proteins such as Bcl-2, Bcl-x_L, Mcl-1, and of antiapoptotic proteins, Bax and Bcl-x_S, have been reported.[121–123] Much attention has recently been focused on SMAD family of genes.[124–127] SMAD4, a tumor suppressor gene in the transforming growth factor (TGF)-α family, is genetically inactivated in approximately 55% of pancreatic adenocarcinomas.[128] Loss of its expression in the late stage of human pancreatic intraepithelial neoplasia (PanIN-3) generally leads to rapid progression of this lesion to pancreatic carcinoma, an extremely invasive and lethal type of cancer.[129] Other genes have been identified by serial analysis of gene expression (SAGE)[130–132] and DNA microarray techniques[133–138] as potential candidates involved in the development of pancreatic neoplasia.

Whether tobacco-derived carcinogens can cause mutations in these genes in the pancreas of humans or animals is currently unknown. Indirect evidence, such as the relatively high proportion of G to A transitions in K-*ras* in pancreatic cancer and an increased risk for this malignancy in individuals with repair deficiency of the DNA alkylation, suggests the involvement of nitrosamines,[139] however, there is no direct proof that this type of damage results from the exposure to tobacco carcinogens. Interestingly, the most common mutations of the K-*ras* gene in mouse and hamster lung tumors induced by NNK are GGT to GAT transitions in Codon 12.[140–144] This type of mutation has been associated with O^6-methyl guanine,[141] whereas, 4-(acetoxymethylnitrosamino)-1-(3-pyridyl)-1-butanone (NNKOAc), a direct acting pyridyloxobutylating agent, causes GGT to TGT and GGT to GTT mutations in Codon 12 of this gene.[140] A higher frequency of K-*ras* Codon 12 mutations revealing G to T and G to C transversions, as well as G to A transitions in nonneoplastic pancreatata among individuals who smoked 1 to 2 packs of cigarettes per day for more than 20 years, has also been reported.[145]

22.6 EPILOGUE

Pancreatic cancer is one of the most deadly malignacies. As a lifestyle factor, cigarette smoking remains the most convincing etiological factor in the induction of this malignancy; the risk is enhanced in susceptible populations possibly due to their genetic makeup. However, factors that account for the differences in risk between races and genders remain ellusive. The evidence that TSNA may be responsible for mutation characteristics in pancreatic cancer genes (e.g., K-*ras*) is at best circumstantial and further elucidation of mutation frequency and mutation characteristics induced by select TSNA and other environmental carcinogens is highly recommended. To our knowledge, there are no biomarkers that are highly useful to assess early detection of pancreatic cancer. At the present time, development of such biomarkers still presents the largest and most important challenge for researchers. Synthetic carcinogens have been used extensively to induce pancreatic cancer in animal models including the hamster; the latter appears to be an appropriate model for inducing ductal carcinoma that is similar to that formed in humans. Limited studies are available on cancer induction by environmental carcinogens. The detection of cotinine and TSNA in pancreatic juice supports the association between cigarette smoking and pancreatic cancer. On the basis of data presented here, the search for other environmental carcinogens in pancreatic juice is highly recommended; 4-aminobiphenyl, IQ, as well as their metabolites, are potential candidates. Clearly, pancreatic tissue contains multiple cyto-chromes P450 that are capable of catalyzing activation of TSNA and other carcinogens to DNA damaging intermediates. Interindividual differences Phase I, Phase II, as well as hydrolases (e.g., β-glucuronidase) may account for cancer susceptibility among different populations. Studies in the area of cancer susceptibility are highly needed. Moreover, studies toward the unequivocal identification of DNA adducts in pancreatic tissue would provide requisite knowledge on the type of carcinogens that play an important role in the development of pancreatic cancer. Findings like those reported by us,[37] as well as several epidemiologic studies, underscore the importance of dietary factors in the induction of pancreatic tumors. More studies should be directed toward examinations of gene–environ-ment interactions. Finally, future studies should also be directed toward development of highly effective and less toxic chemopreventive agents against the development of pancreatic cancer induced by relevant envi-ronmental carcinogens, such as those found in tobacco smoke and diet.

REFERENCES

1. Hammond C.E. 1966. Smoking in relation to the death rates of one million men and women. *NCI Monogr.* 19:127–204.
2. Kahn H.A. 1966. The Dorn study of smoking and mortality among U.S. veterans. Report on eight and one-half years of observation. *NCI Monogr.* 19:1–125.
3. Hirayama T. 1967. Smoking in relation to the death rates of 265,118 men and women in Japan. Tokyo: National Cancer Center Institute.
4. International Agency for Research on Cancer. 1986. Tobacco Smoking. IARC Monographs on Evaluation of the Carcinogenic Risk of Chemicals to Humans. Lyon, France: IARC 38:421.
5. Wynder E., Mabuchi K., Maruchi N., and Fortner J.G. 1973. A case control study of cancer of the pancreas. *Cancer (Phila.)* 31:641–648.
6. Mack T.M., Yu M.C., Hanisch R., and Hendersen B.E. 1986. Pancreas cancer, smoking, beverage consumption, and past medical history. *J. Natl. Cancer Inst.* 76:49–60.
7. Falk R.T., Pickle L.W., Fontham E.T., Correa P., and Fraumeni Jr. J.F. 1988. Lifestyle risk factors for pancreatic cancer in Louisiana: a case-control study. *Am. J. Epidemiol.* 128:324–336.
8. Bueno de Mesquita H.B., Maisonneuve P., Moerman C.J., Runia S., and Boyle P. 1992. Lifetime history of smoking and exocrine carcinoma of the pancreas: a population-based case-control study in the Netherlands. *Intl. J. Cancer* 50:514–522.
9. Shopland D.R., Eyre H.J., and Pechacek T.F. 1991. Smoking-attributable cancer mortality in 1991: Is lung cancer now the leading cause of death among smokers in the United States? *J. Natl. Cancer Inst.* 83:1142–1148.

10. Zheng W., McLaughlin J.K., Gridley G., Bjelke E., Schuman L.M., Silverman D.T., Wacholder S., Co-Chien H.T., Blot W.J., and Fraumeni Jr. J.F. 1993. A cohort study of smoking, alcohol consumption, and dietary factors for pancreatic cancer (United States). *Cancer Causes Control* 4:477–482.

11. Ghadirian P., Lynch H.T., and Krewski D. 2003. Epidemiology of pancreatic cancer: an overview. *Cancer Det. Prev.* 27:87–93.

12. Michaud D.S. 2002. The epidemiology of pancreatic, gallbladder, and other biliary tract cancers. *Gastrointest. Endosc.* 56:S195–S200.

13. Muscat J.E., Stellman S.D., Hoffmann D., and Wynder E.L. 1997. Smoking and pancreatic cancer in men and women. *Cancer Epidemiol. Biomarkers Prev.* 6:15–19.

14. Jaffee E.M., Hruban R.H., Canto M., and Kern S.E. 2002. Focus on pancreas cancer. *Cancer Cell* 123:25–28.

15. Tersmette A.C., Petersen G.M., Offerhaus G.J., Falatko F.C., Brune K.A., Goggins M., Rozenblum E., Wilentz R.E., Yeo C.J., Cameron J.L., Kern S.E., and Hruben R.H. 2001. Increased risk of incident pancreatic cancer among first-degree relatives of patients with pancreatic cancer. *Clin. Cancer Res.* 7:738–744.

16. Coughlin S.S., Calle E.E., Patel A.V., and Thun M.J. 2000. Predictors of pancreatic cancer mortality among a large cohort of United States adults. *Cancer Causes Control* 11:915–923.

17. Maissonneuve P. and Lowenfels A.B. 2002. Chronic pancreatitis and pancreatic cancer. *Dig. Dis.* 20:32–37.

18. Leder G.H., Prokopczyk B., Ramadani M.A., Cunningham J., Siech M., Schwartz M., Poch B., Link K.H., Hoffmann D., El-Bayoumy K., and Beger H.G. 2001. Detection of tobacco derived carcinogen in pancreatic juice of smokers – a role for the onset of pancreatic cancer? *Pancreas* 23:448.

19. Druckrey H., Ivankovic S., Bcheler J., Preussmann R., and Thomas C. 1968. Erzeugung von Magen-und pancreas-krebs beim meerschweinchen durch methyl-nitroso-harnstoff und –urethan. *Z. Krebsforsch* 71:167–182.

20. Longnecker D. 1990. Experimental pancreatic cancer: role of species, sex and diet. *Bull. Cancer* 77:27–37.

21. Birt D.F., Salmasi S., and Pour P.M. 1981. Enhancement of experimental pancreatic cancer in Syrian golden hamsters by dietary fat. *J. Natl. Cancer Inst.* 67:1327–1332.

22. Roebuck B.D., Yager Jr. J.D., Longnecker D.S., and Wilpone S.A. 1981. Promotion by unsaturated fat of azaserine-induced pancreatic carcinogenesis in the rat. *Cancer Res.* 41:3961–3966.

23. Birt D.F., Pour P.M., Nagel D.L., Barnett T., Blackwood D., and Duysen E. 1997. Dietary energy restriction does not inhibit pancreatic carcinogenesis by N-nitrosobis-2-(oxopropyl)amine in the Syrian hamster. *Carcinogenesis* 18:2107–2111.

24. Dissin J., Mills L.R., Mains D.L., Black Jr. O., and Webster P.D. 1975. Experimental induction of pancreatic adenocarcinoma in rats. *J. Natl. Cancer Inst.* 55:857–864.

25. Hayashi Y. and Hasegawa T. 1971. Experimental pancreatic tumor in rat after intravenous injection of 4-hydroxyaminoquinoline 1-oxide. *Gann.* 62:329–330.

26. Reddy J.K. and Rao M.S. 1977. Malignant tumors in rats fed nafenopin, a hepatic peroxisome proliferator. *J. Natl. Cancer Inst.* 59:1645–1650.

27. Svoboda D.J. and Azarnoff D.L. 1979. Tumors in male rats fed ethylchlorophe-noxyisobutyrate, a hypolipidemic drug. *Cancer Res.* 39:3419–3428.

28. Reddy J.K. and Quereshi S.A. 1979. Tumorigenicity of the hypolipidemic peroxime prolifirator ethyl-p-chlorophenoxyisobutyrate (clofibrate) in rats. *Brit. J. Cancer* 40:476–482.

29. Obourn J.D., Frame S.R., Bell Jr. R.H., Longnecker D.S., Elliot G.S., and Cook J.C. 1997. Mechanisms for the pancreatic oncogenic effects of the peroxisome proliferator Wyeth-14,643. *Toxicol. Appl. Pharmacol.* 145:425–436.

30. Mohr U., Reznik G., Emminger E., and Lijinsky W. 1977. Brief communication: Induction of pancreatic duct carcinomas in the Syrian hamsters with 2,6-dimethylnitrosomorpholine. *J. Natl. Cancer Inst.* 58:429–432.

31. Longnecker D.S., Curphey T.J., Lilja H.S., French J.I., and Daniel D.S. 1980. Carcinogenicity in rats of the nitrosourea amino acid $N\delta$-(N-methyl-N-nitroso-carbamoyl)-L-orhithine. *J. Environ. Pathol. Toxicol.* 4:117–129.

32. Longnecker D.S., Curphey T.J., Kuhlmann E.T., and Schaeffer B.K. 1983. Exper-imental induction of pancreatic carcinoma in hamster with $N\delta$-(N-methyl-N-nitrosocarbamoyl)-L-orhithine. *J. Natl. Cancer Inst.* 71:1327–1336.

33. Rivenson A., Hoffmann D., Prokopczyk B., Amin S., and Hecht S.S. 1988. A stud of tobacco carcinogenesis XLII. Induction of lung and pancreas tumors in F344 rats by tobacco-specific and areca-derived N-nitrosamines. *Cancer Res.* 48:6912–6917.

34. Wenke G. and Hoffmann D. 1983. A study of betel quid carcinogenesis. I. On the *in vitro* N-nitrosation of arecoline. *Carcinogenesis* 4:169–172.

35. Nair J., Ohshima H., Friesen M., Croisy A., Bhide S.V., and Bartsch H. 1985. Tobacco-specific and betel nut-specific N-nitroso compounds: Occurrence in saliva and urine of betel quid chewers and formation by nitrosation of betel quid. *Carcinogenesis* 6:295–303.

36. Weisburger J.H., Barnes W.S., Lovelette C.A., Tong C., Tanaka T., and Williams G.M. 1986. Genotoxicity, carcinogenicity, and mode of action of the fried food mutagen 2-amino-3-methylimidazo[4,5-f]quinoline (IQ). *Environ. Health Per-spect.* 67:121–127.

37. Hoffman D., Rivenson A., Abbi R., and Wynder E.L. 1993. A study of tobacco carcinogenesis: effect of the fat content of the diet on the carcinogenic activity of 4-(methylnitrosamino)-1-(3-pyridyl)-1-butanone in F344 rats. *Cancer Res.* 53:2758–2761.

38. Alisauskus R., Wong G.Y., and Gold D.V. 1995. Initial studies of monoclonal antibody PAM4 targeting to xenografted orthotopic pancreatic cancer. *Cancer Res.* 5:5743s–5748s.

39. An Z., Wang X., Kubota T., Moossa A.R., and Hoffman R.M. 1996. A clinical nude mouse metastatic model for highly malignant human pancreatic cancer. *Anticancer Res.* 16:627–631.

40. Mohammad R.M., Dugan M.C., Mohamed A.N., Almatchy V.P., Flake T.M., Dergham S.T., Shields A.F., Al-Katib A.A., Vaitkevicius V.K., and Sarkar F.H. 1998. Establishment of human pancreatic tumor xenograft model: potential application for preclinical evaluation of novel therapeutic agents. *Pancreas* 16:19–25.

41. Wang B., Abbruzzese J.L., Xiong Q., Le X., and Xie K. 2001. A novel, clinically relevant animal model of metastatic pancreatic adenocarcinoma biology and therapy. *Intl. J. Pancreatol.* 29:37–46.

42. Fu X., Guadagni F., and Hoffman R.M. 1992. A metestatic nude-mouse model of human pancreatic cancer constructed orthotopically with histologically intact patient specimens. *Proc. Natl. Acad. Sci. USA* 89:5645–5649.

43. Cui J.H., Krueger U., Henne-Bruns D., Kremer B., and Kalthoff H. 2001. Orthotopic transplantation model of micrometastases. *World J. Gastroenterol.* 7:381–386.

44. Zervox E.E., Franz M.G., Salhab K.F., Shafii A.E., Menendez J., Gower W.R., and Rosemurgy A.S. 2000. Matrix metalloproteinase inhibition improves survival in an orthtopic model of human pancreatic cancer. *J. Gastrointest. Surg.* 4:614–619.

45. Hoffmann D. and Hoffmann I. 1997. The changing cigarette 1950–1995. *J. Toxicol. Environ. Health* 50:307–364.

46. Hoffmann D., Hoffmann I., and El-Bayoumy K. 2001. The less harmful cigarette: A controversial issue. A tribute to Ernst L. Wynder. *Chem. Res. Toxicol.* 14:765–790.

47. Hecht S.S. 1998. Biochemistry, biology, and carcinogenicity of tobacco-specific *N*-nitrosamines. Invited review. *Chem. Res. Toxicol.* 11:559–603.

48. Hoffmann D., Brunnemann K.D., Prokopczyk B., and Djordjevic M.D. 1994. Tobacco-specific *N*-nitrosamines and *Areca*-derived *N*-nitrosamines: chemistry, biochemistry, carcinogenicity, and relevance to humans. *J. Toxicol. Environ. Health* 41:1–52.

49. Koppang N., Rivenson A., Dahle H.K., and Hoffmann D. 1997. A study of tobacco carcinogenesis, LIII: carcinogenicity of *N*-nitrosonornicotine (NNN) and 4-(methylnitrosamino)-1-(3-pyridyl)-1-butanone (NNK) in mink (*Mustala vison*). *Cancer Lett.* 111:167–171.

50. Prokopczyk B., Rivenson A., and Hoffmann D. 1991. A study of betel quid carcinogenesis IX. Comparative carcinogenicity of 3-(methylnitrosamino)propionitrile (MNPN) and 4-(methylnitrosamino)-1-(3-pyridyl)-1-butanone (NNK) upon local application to mouse skin and rat oral mucosa. *Cancer Lett.* 60:153–157.

51. LaVoie E.J., Prokopczyk G., Rigotty J., Czech A., Rivenson A., and Adams J.D. 1987. Tumorigenic activity of tobacco-specific nitrosamines 4-(methylnitrosamino)-1-(3-pyridyl)-1-butanone (NNK), 4-(methylnitrosamino)-4-(3-pyridyl)-1-butanol (iso-NNAL) and *N*-nitrosonornicotine (NNN) on topical application to Sencar mice. *Cancer Lett.* 37:277–283.

52. Hecht S.S., Morse M.A., Amin S., Stoner G.D., Jordan K.G., Choi C.-I., and Chung F.-L. 1989. Rapid single-dose model for lung tumor induction in A/J mice by 4-(methylnitrosamino)-1-(3-pyridyl)-1-butanone and the effect of diet. *Carcinogenesis* 10:1901–1904.

53. Baskaran K., Laconi S., and Reddy M.K. 1994. Transformation of hamster pancreatic duct cells by 4-(methylnitrosamino)-1-(3-pyridyl)-1-butanone (NNK), *in vivo*. *Carcinogenesis* 15:2461–2466.

54. Schuller H.M., Jorquera R., Reichert A., and Castonguay A. 1993. Transplacental induction of pancreatic tumors in hamsters by ethanol and the tobacco-specific nitrosamine 4-(methylnitrosamino)-1-(3-pyridyl)-1-butanone. *Cancer Res.* 53:2498–2501.

55. Prokopczyk B., Hoffmann D., Bologna M., Cunnighamm A.J., Trushin N., Akerkar S., Boyiri T., Amin S., Desai D., Colosimo S., Pittman B., Leder G., Ramadani M., Henne-Bruns D., Beger H.G., and El-Bayoumy K. 2002. Identification of tobacco-derived compounds in human pancreatic juice. *Chem. Res. Toxicol.* 15:677–685.

56. Kessova I.G., DeCarli L.M., and Lieber C.S. 1998. Inducibility of cytochromes P-4502E1 and P-4501A1 in the rat pancreas. *Alcohol Clin. Exp. Res.* 22:501–504.

57. Zhang L., Weddle D.L., Thomas P.E., Zheng B., Castonguay A., Schuller H.M., Miller M.S. 2000. Low levels of expression of cytochrome P-450 in normal and cancerous fetal pancreatic tissues of hamsters treated with NNK and/or ethanol. *Toxicol. Sci.* 56:313–323.

58. Marek C.J., Cameron G.A., Elrick L.J., Hawksworth G.M., and Wright M.C. 2003. Generation of hepatocyte expressing functional cytochromes P450 from a pancreatic progenitor cell line in vitro. *Biochem. J.* 370:763–769.

59. Standop J., Ulrich A.B., Schneider M.B., Buchler M.W., and Pour P. 2002. Differences in the expression of xenobiotic-metabolizing enzymes between islets derived from the ventral and dorsal anlage of the pancreas. *Pancreatology* 2:510–518.

60. Duell E.J., Holly E.A., Bracci P.M., Liu M., Wiencke J.K., and Kelsey K.T. 2002. A population-based, case-control study of polymorphism in carcinogen-metabolizing genes, smoking, and pancreatic adenocarcinoma risk. *J. Natl. Cancer Inst.* 94:297–306.

61. Morales A., Cuellar A., Ramirez J., Vilchis F., and Diaz-Sanchez V. 1999. Synthesis of steroids in pancreas: evidence of cytochrome P-450scc activity. *Pancreas* 19:39–44.

62. Standop J., Schneider M.B., Ulrich A., Chauhan S., Moniaux N., Buchler M.W., Batra S.K., and Pour P.M. 2002. The pattern of xenobiotic-metabolizing enzymes in the human pancreas. *J. Toxicol. Environ. Health* 65:1379–1400.

63. Ulrich A.B., Standop J., Schmied B.M., Schneider M.B., Lawson T.A., and Pour P.M. 2002. Species differences in the distribution of drug-metabolizing enzymes in the pancreas. *Toxicol. Pathol.* 30:247–253.

64. Wacke R., Kirchner A., Prall F., Nizze H., Schmidt W., Fischer U., Nitschke F.P., Adam U., Fritz P., Belloc C., and Drewelow B. 1998. Up-regulation of cytochrome P450 1A2, 2C9, and 2E1 in chronic pancreatitis. *Pancreas* 16:521–528.

65. Smith T.J., Stoner G.D., and Yang C.S. 1995. Activation of 4-(methylnitrosamino)-1-(3-pyridyl)-1-butanone (NNK) in human lung microsomes by cytochrome P450, lipoxygenase, and hydroperoxidase. *Cancer Res.* 55:5566–5573.

66. Rioux N. and Castonguay A. 1998. Prevention of NNK-induced lung tumorigenesis in A/J mice by salicylic acid and NS-358. *Cancer Res.* 58:5354–5360.

67. Hecht S.S. 1996. Recent studies on mechanisms of bioactivation and detoxification of 4-(methylnitrosamino)-1-(3-pyridyl)-1-butanone (NNK), a tobacco-specific lung carcinogen. *Crit. Rev. Toxicol.* 26:163–181.

68. Hecht S.S. 1994. Metabolic activation and detoxification of tobacco-specific nitrosamines — a model for cancer prevention strategies. *Drug Met. Rev.* 26:373–390.

69. Prokopczyk B., Trushin N., Leszczynska J., Waggoner S.E., and El-Bayoumy K. 2001. Human cervical tissue metabolizes the tobacco-specific nitrosamine, 4-(methylnitrosamino)-1-(3-pyridyl)-1-butanone via α-hydroxylation and carbonyl reduction pathways. *Carcinogenesis* 22:107–114.

70. Carmella S.G., Akerkar S., and Hecht S.S. 1993. Metabolites of the tobacco-specific nitrosamine 4-(methylnitrosamino)-1-(3-pyridyl)-1-butanone in smokers' urine. *Cancer Res.* 53:721–724.

71. Hecht S.S., Carmella S.G., Murphy S.E., Akerkar S., Brunnemann K.D., and Hoffmann D. 1993. A tobacco-specific lung carcinogen in the urine of men exposed to cigarette smoke. *N. Engl. J. Med.* 329:1543–1546.

72. Atawodi S.E., Michelson K., and Richter E. 1994. Metabolism of a glucuronide of 4-(methylnitrosamino)-1-(3-pyridyl)-1-butanone in rats. *Arch. Toxicol.* 69:14–17.

73. Rinderknecht H., Renner I.G., and Koyama H.H. 1979. Lysosomal enzymes in pure pancreatic juice from normal healthy volunteers and chronic alcoholics. *Dig. Dis. Sci.* 24:180–186.

74. Maser E., Richter E., and Friebertshäuser J. 1996. The identification of 11β-hydroxysteroid dehydrogenase as carbonyl reductase of the tobacco-specific nitrosamine 4-(methylnitrosamino)-1-(3-pyridyl)-1-butanone. *Eur. J. Biochem.* 238:484–489.

75. Maser E. 1998. 11β-Hydroxysteroid dehydrogenase responsible for carbonyl reduction of the tobacco-specific nitrosamine 4-(methylnitrosamino)-1-(3-pyridyl)-1-butanone in mouse lung microsomes. *Cancer Res.* 58:2996–3003.

76. Manabe S., Tohyama K., Wada O., and Aramaki T. 1991. 2-Amino-1-methyl-6-phenylimidazo[4,5-b]pyridine (PhIP) in cigarette smoke condensate. *Carcinogenesis* 12:1945–1947.

77. El-Bayoumy K. 1992. Environmental carcinogens that may be involved in human breast cancer etiology. *Chem. Res. Toxicol.* 5:585–590.

78. International Agency for Research on Cancer. *Polycyclic aromatic compounds. Part 1. Chemical and environmental data.* Vol. 32. Lyon, France: IARC Monographs.

79. International Agency for Research on Cancer. *Polynuclear aromatic compounds. Part 3. Industrial exposure in aluminum production, coal gasification, coke production and iron and steel founding. Monographs on the evaluation of the carcinogenic risk of chemicals to humans.* Vol. 34. Lyon, France: IARC Monographs.

80. International Agency for Research on Cancer. *Overall evaluation of carcinogenicity: an updating of IARC monographs.* Vol. 1–42, suppl. 6 and 7. Lyon, France: IARC Monographs.

81. Upadhyaya P., Sturla S., Tretyakova N., Ziegel R., Vilalta P.W., Wang M., and Hecht S.S. 2003. Identification of adducts produced by the reaction of 4-(acetoxymethylnitrosamino)-1-(3-pyridyl)-1-butanol with deoxyguanosine and DNA. *Chem. Res. Toxicol.* 16:180–190.

82. Chung F.-L. and Xu Y. 1992. Increased 8-hydroxydeoxyguanosine in DNA of mice and rats treated with the tobacco-specific nitrosamine 4-(methylnitrosamino)-1-(3-pyridyl)-1-butanone. *Carcinogenesis* 13:1269–1272.

83. Schuller H.M. 2002. Mechanisms of smoking-related lung and pancreatic adenocarcinoma development. *Nat. Rev. Cancer* 2:455–463.

84. Tucker O.N., Dannenberg A.J., Yang E.K., Zhang F., Teng L., Daly J.M., Soslow R.A., Masferrer J.L., Woerner B.M., Koki A.T., Fahey 3rd. T.J. 1999. Cyclooxygenase-2 expression is up-regulated in human pancreatic cancer. *Cancer Res.* 59:987–990.

85. Yip-Schneider M.T., Barnard D.S., Billings S.D., Cheng L., Heilman D.K., Lin A., Marshall S.J., Crowell P.L., Marshall M.S., and Sweeney C.J. 2000. Cyclooxygenase-2 expression in human pancreatic adenocarcinomas. *Carcinogenesis* 21:139–146.

86. El-Bayoumy K., Iatropoulos M., Amin S., Hoffmann D., and Wynder E.L. 1999. Increased expression of cyclooxygenase-2 in rat lung tumors induced by the tobacco-specific nitrosamine 4-(methylnitrosamino)-1-(3-pyridyl)-1-butanone: the impact of a high fat diet. *Cancer Res.* 59:1400–1403.

87. Rioux N. and Castonguay A. 1999. Induction of COX expression by a tobacco carcinogen: implication in lung cancer chemoprevention. *Inflamm. Res.* 48:S136–S137.

88. Sato N., Maitra A., Fukushima N., van Heek N.T., Matsubayashi H., Iacobuzio-Danahue C.A., Rosty C., and Goggins M. 2003. Frequent hypomethylation of multiple genes overexpressed in pancreatic ductal adenocarcinoma. *Cancer Res.* 63:4158–4166.

89. Anderson K.E., Hammons G.J., Kadlubar F.F., Potter J.D., Kaderlick K.R., Ilett K.F., Minchin R.F., Teitel C.H., Chou H.-C., Martin M.V., Gungerich F.P., Barone G.W., Lang N.P., and Reterson L.A. 1997. Metabolic activation of aromatic amines by human pancreas. *Carcinogenesis* 18:1085–1092.

90. Alexandrov K., Rojas M., Kadlubar F.F., Lang N.P., and Bartsch H. 1996. Evidence of *anti*-benzo[a]pyrene diolepoxide-DNA adducts formation in human colon mucosa. *Carcinogenesis* 17:2081–2083.

91. Furukawa F., Nishikawa A., Lee I.S., Son H.Y., Nakamura H., Miyauchi M., Takahashi M., and Hirose M. 2000. Inhibition by methionine of pancreatic carcinogenesis in hamsters after initiation with *N*-nitrosobis(2-oxopropyl)amine. *Cancer Lett.* 152:163–167.

92. Hiura A., Tsutsumi M., and Satake K. 1997. Inhibitory effect of green tea extract on the process of pancreatic carcinigenesis induced by *N*-nitrosobis(2-oxopropyl)amine (BOP) and on tumor promotion after transplantation of *N*-nitrosobis(2-hydroxypropyl)amine (BHP)-induced pancreatic cancer in Syrian hamsters. *Pancreas* 15:272–277.

93. Appel M.J., van Garderen-Hoetmer A., and Woutersenm R.A. 1996. Lack of inhibitory effects of β-carotene, vitamin C, vitamin E and selenium on development of ductular adenocarcinomas in exocrine pancreas of hamsters. *Cancer Lett.* 103:157–162.

94. Nishikawa A., Furukawa F., Uneyama C., Ikezaki S., Tanakamaru Z., Chung F.L., Takahashi M., and Hayashi Y. 1996. Chemopreventive effects of phenyl isocyanate on lung and pancreatic tumorigenesis in *N*-nitrosobis(2-oxopropyl)amine-treated hamsters. *Carcinogenesis* 17:1381–1384.

95. Clapper M.L., Wood M., Leahy K., Lang D., Miknyoczki S., and Rugger B.A. 1995. Chemopreventive activity of Oltipraz against *N*-nitrosobis(2-oxopropyl)amine (BOP)-induced ductal pancreatic carcinoma development and effects on survival of Syrian golden hamsters. *Carcinogenesis* 16:2159–2165.

96. Takahashi M., Furukawa F., Toyoda H., Hasegawa R., Imaida K., and Hayashi Y. 1990. Effects of various prostaglandin inhibitors on pancreatic carcinogenesis in hamster after initiation with *N*-nitrosobis(2-oxopropyl)amine. *Carcinogenesis* 11:393–395.

97. Mizumoto K., Kitazawa S., Eguchi T., Nakajima A., Tsutsumi M., Ito S., Danda A., and Konishi Y. 1988. Modulation of *N*-nitrosobis(2-oxopropyl)amine-induced carcinogenesis by clofibrate in hamsters. *Carcinogenesis* 9:1421–1425.

98. Nishkawa A., Furukawa F., Imazawa T., Yoshimura H., Mitsomori K., and Takahashi M. 1992. Effects of caffeine, nicotine, ethanol and sodium selenite on pancreatic carcinogenesis in hamsters after initiation with *N*-nitrosobis(2-oxopropyl)amine. *Carcinogenesis* 13:1379–1382.

99. Hass B.S., Hart R.W., Gaylor D.W., Poirier L.A., and Lyn-Cook B.D. 1992. An *in vitro* pancreas acinar cell model for testing the modulating effects of caloric restriction and ageing on cellular proliferation and transformation. *Carcinogenesis* 13:2419–2425.

100. Craven-Giles T., Tagliaferro A.R., Ronan A.M., Baumgartner K.J., and Roebuck B.D. 1994. Dietary modulation of pancreatic carcinogenesis: calories and energy expenditure. *Cancer Res.* 54:1964s–1968s.

101. Hecht S.S. 1997. Approaches to chemoprevention of lung cancer based on carcinogens in tobacco smoke. *Environ. Health Prospect.* 105:955–963.

102. Hecht S.S. 1997. Approaches to cancer prevention based on understanding of nitrosamine carcinogenesis. *Proc. Soc. Exp. Biol. Med.* 216:181–191.

103. Wattenberg L.W. 1999. Chemoprevention of pulmonary carcinogenesis by myo-inositol. *Anticancer Res.* 19:3659–3661.

104. Lantry L.E., Zhang Z., Crist K.A., Wang Y., Hara M., Zeeck A., Lubet R.A., and You M. 2000. Chemopreventive efficacy of promising farnesyltransferase inhibitors. *Exp. Lung Res.* 26:773–790.

105. Chung F.L. 2001. Chemoprevention of lung cancer by isothiocyanates and their conjugates in A/J mouse. *Exp. Lung Res.* 27:319–330.

106. Hecht S.S., Upadhyaya P., Wang M., Bliss R.L., McIntee E.J., and Kenney P.M. 2002. Inhibition of lung tumorigenesis in A/J mice by *N*-acetyl-*S*-(*N*-2-phenylthiocarbamoyl)-L-cysteine and *myo*-inositol, individually and in combination. *Carcinogenesis* 23:1455–1461.

107. El-Bayoumy K. and Hoffmann D. 1999. Nutrition and tobacco-related cancer. Chapter 20. In *Nutritional oncology.* Heber D., Blackburn G. Eds. San Diego: Academic Press. pp. 299–324.

108. Prokopczyk B., Rosa J.G., Desai D., Amin S., Sohn O.S., Fiala E.S., and El-Bayoumy K. 2000. Chemoprevention of lung tumorigenesis induced by a mixture of benzo(*a*)pyrene and 4-(methylnitrosamino)-1-(3-pyridyl)-1-butanone by the organoselenium compound 1,4-phynylenebis(methylene)selenocyanate. *Cancer Lett.* 161:35–46.

109. Witschi H. 2000. Successful and not so successful chemoprevention of tobacco smoke-induced lung tumors. *Exp. Lung Res.* 26:743–755.

110. Furukawa F., Nishikawa A., Enami T., Mitsui M., Imazawa T., Tanakamaru Z., Kim H.-C., Lee I.-S., Kasahara T., and Takahashi M. 1997. Promotional effects of 4-(methylnitrosamino)-1-(3-pyridyl)-1-butanonol on *N*-nitrosobis(2-oxopropyl)amine (BOP)-induced carcinogenesis in hamsters. *Food Chem. Toxicol.* 35:387–392.

111. Furukawa F., Nishikawa A., Yoshimura H., Mitsui M., Imazawa T., Ikezaki S., and Takahashi M. 1994. Promotional effects of 4-(methylnitrosamino)-1-(3-pyridyl)-1-butanonone (NNK) on *N*-nitrosobis(2-oxopropyl)amine (BOP)-initiated carcinogenesis in hamsters. *Cancer Lett.* 86:75–82.

112. Nishikawa A., Furukawa F., Imazawa T., Yoshimura H., Ikezaki S., Hayashi Y., and Takahashi M. 1994. Effects of cigarette smoke on *N*-nitrosobis(2-oxopropyl)amine-induced pancreatic and respiratory tumorigenesis in hamsters. *Jpn. J. Cancer Res.* 85:1000–1004.

113. Rozenblum E., Schutte M., Goggins M., Hahn S.A., Panzer S., Zahurak M., Goodman S.N., Sohn T.A., Hruban R.H., Yeo C.J., and Kern S.E. 1997. Tumor-suppressive pathways in pancreatic carcinoma. *Cancer Res.* 57:1731–1734.

114. Moore P.S., Orlandini S., Zamboni G., Capelli P., Rigaud G., Falconi M., Bassi C., Lemoine N.R., and Scarpa A. 2001. Pancreatic tumours: molecular pathways implicated in ductal cancer are involved in ampullary but not in exocrine nonductal or endocrine tumorigenesis. *Br. J. Cancer* 84:253–262.

115. Moore P.S., Beghelli S., Zamboni G., and Scarpa A. 2003. Genetic abnormalities in pancreatic cancer. *Mol. Cancer* 2:7–14.

116. Lu X., Xu T., Qian J., Wen X., and Wu D. 2002. Detecting K-ras and p53 gene mutations from stool and pancreatic juice for diagnosis of early pancreatic cancer. *Chin. Med. J.* 115:1632–1636.

117. Luttges J., Galehdari H., Brocker V., Schwarte-Waldhoff I., Henne-Bruns D., Kloppel G., Schmiegel W., and Hahn S.A. 2001. Allelic loss is often the first hit in the biallelic inactivation of the p53 and DPC4 genes during pancreatic carcinogenesis. *Am. J. Pathol.* 158:1677–1683.

118. Biankin A.V., Morey A.L., Lee C.S., Kench J.G., Biankin S.A., Hook H.C., Head D.R., Hugh T.B., Sutherland R.L., and Henshall S.M. 2002. DPC4/Smad4 expression and outcome in pancreatic ductal adenocarcinoma. *J. Clin. Oncol.* 20:4531–4542.

119. Hahn S.A., Greenhalf B., Ellis I., Sina-Frey M., Rieder H., Korte B., Gerdes B., Kress R., Ziegler A., Raeburn J.A., Campra D., Grutzmann R., Rehder H., Rothmund M., Schmiegel W., Neoptolemos J.P., and Bartsch D.K. 2003. BRCA2 germline mutations in familial pancreatic cancer. *J. Natl. Cancer Inst.* 95:214–221.

120. Klein W.M., Hruben R.H., Klein-Szanto A.J., and Wilentz R.E. 2002. Direct correlation between proliferative activity and dysplasia in pancreatic intraepithelial neoplasia (PanIN); additional evidence for a recently proposed model of progression. *Mod. Pathol.* 15:441–447.

121. Miyamoto Y., Hosotani R., Wada M., Lee J.U., Koshiba T., Fujimoto K., Tsuji S., Nakajima S., Doi R., Kato M., Shimada Y., and Imamura M. 1999. Immunohistochemical analysis of Bcl-2, Bax, Bcl-X, and Mcl-1 expression in pancreatic cancers. *Oncology* 56:73–82.

122. Friess H., Lu Z., Graber H.U., Zimmermann A., Adler G., Korc M., Schmid R.M., and Buchler M.W. 1998. Bax, but not Bcl-3, influences the prognosis of human pancreatic cancer. *Gut* 43:414–421.

123. Campani D., Esposito I., Boggi U., Cecchetti D., Menicagli M., De Negri F., Colizzi L., Del Chiaro M., Mosca F., Fornaciari G., and Bevilacqua G. 2001. Bcl-2 expression in pancreas development and pancreatic cancer progression. *J. Pathol.* 194:444–450.

124. Sasaki S., Yamamoto H., Kaneto H., Ozeki I., Adachi Y., Takagi H., Matsumoto T., Itoh H., Nagakawa T., Miyakawa H., Muraoka S., Fujinaga A., Suga M., Itoh F., Endo T., and Imai K. 2003. Differential roles of alterations of p53, p16, and SMAD4 expression in the progression of intraductal papillary-mucinous tumors of the pancreas. *Oncol. Rep.* 10:21–25.

125. Biankin A.V., Morey A.L., Lee C.S., Kench J.G., Biankin S.A., Hook H.C., Head D.R., Hugh T.B., Sutherland R.L., and Henshall S.M. 2002. DPC4/SMAD4 expression and outcome in pancreatic ductal adenocarcinoma. *J. Clin. Oncol.* 20:4531–4542.

126. Tascilar M., Skinner H.G., Rosty C., Sohn T., Wilentz R.E., Offerhaus G.J.A., Adsay V., Abrams R.A., Cameron J.L., Kern S.E., Yeo C.J., Hruben R.H., and Goggins M. 2001. The SMAD4 protein and prognosis of pancreatic ductal adenocarcinoma. *Clin. Cancer Res.* 7:4115–4121.

127. Liu F. 2001. SMAD4/DPC4 and pancreatic cancer. *Clin. Cancer Res.* 7:3853–3856.

128. Hahn S.A., Schutte M., Hoque A.T., Moskaluk C.A., da Costa L.T., Rozenblum E., Weinstein C.L., Fischer A., Yeo C.J., Hruban R.H., and Kern S.E. 1996. DCP4, a candidate tumor suppressor gene at human chromosome 18q21.1. *Science (Wash, DC)* 271:350–353.

129. Su G.H., Bansal R., Murphy K.M., Montgomery E., Yeo C.J., Hruban R.H., and Kern S.E. 2001. ACVR1B (ALK4 activin receptor type 1B) genemutations in pancreatic carcinoma. *Proc. Natl. Acad. Sci. USA* 98:3254–3257.

130. Torrisani J. and Buscail L. 2002. Molecular pathways of pancreatic carcinogenesis. *Ann. Pathol.* 22:349–355.

131. Argani P., Rosty C., Reiter R.E., Wilentz R.E., Murugesan S.R., Leach S.D., Ryu B., Skinner H.G., Goggins M., Jaffee E.M., Yeo C.J., Cameron J.L., Kern S.E., and Hruben R.H. 2001. Discovery of new markers of cancer through serial analysis of gene expression: Prostate stem cell antigen is overexpressed in pancreatic adenocarcinoma. *Cancer Res.* 61:4320–4324.

132. Argani P., Iacobuzio-Donahue C., Ryu B., Rosty C., Goggins M., Wilentz R.E., Murugesan S.R., Leach S.D., Jaffee E., Yeo C.J., Cameron J.L., Kern S.E., and Hruben R.H. 2001. Mesothelin is overexpressed in the vast majority of ductal adenocarcinomas of the pancreas: identification of a new pancreatic cancer marker by serial analysis of gene expression (SAGE). *Clin. Cancer Res.* 7:3862–3868.

133. Iacobuzio-Donahue C.A., Maitra A., Olsen M., Lowe A.W., Van Heek N.T., Rosty C., Walter K., Sato N., Parker A., Ashfag R., Jaffee E., Ryu B., Jones J., Eshleman J.R., Yeo C.J., Cameron J.L., Kern S.E., Hruban R.H., Brown P.O., and Goggins M. 2003. Exploration of global expression patterns in pancreatic adenocarcinoma using cDNA microarrays. *Am. J. Pathol.* 162:1151–1162.

134. Tan Z.J., Hu X., Cao G., and Tang Y. 2003. Analysis of gene expression profile of pancreatic carcinoma using cDNA microarray. *World J. Gastroenterol.* 9:818–823.

135. Lagsdon C.D., Simeone D.M., Binkley C., Arumugam T., Greenson J.K., Giordano T.J., Misek D.E., and Hanash S. 2003. Molecular profiling of pancreatic adenocarcinoma and chronic pancreatitis identifies multiple genes differentially regulated in pancreatic cancer. *Cancer Res.* 63:2649–2657.

136. Nakajima F., Nishimori H., Hata F., Yasoshima T., Nomura H., Tanaka H., Ohno K., Yanai Y., Kamiguchi K., Sato N., Denno R., and Hirata K. 2003. Gene expression using cDNA macroarrays to clarify the mechanisms of peritoneal dissemination of pancreatic cancer. *Surg. Today* 33:190–195.

137. Sato N., Fukushima N., Maitra A., Matsubayashi H., Yeo C.J., Cameron J.L., Hruban R.H., and Goggins M. 2003. Discovery of novel targets for aberrant methylation in pancreatic carcinoma using high-throughput microarrays. *Cancer Res.* 63:3735–3742.

138. Maitra A., Iacobuzio-Donahue C., Rahman A., Sohn T.A., Argani P., Meyer R., Yeo C.J., Cameron J.L., Goggins M., Kern S.E., Ashfag R., Hruban R.H., and Wilentz R.E. 2002. Immunohistochemical validation of a novel epithelial and a novel stromal marker of pancreatic ductal adenocarcinoma identified by global expression microarrays: sea urchin fascin homolog and heat shock protein 47. *Am. J. Clin. Pathol.* 118:52–59.

139. Li D. 2001. Molecular epidemiology of pancreatic cancer. *Cancer J.* 7:259–265.
140. Ronai Z., Gradia S., Peterson L.A., and Hecht S.S. 1993. G to A transitions and G to T transversions in codon 12 of the K-*ras* oncogene isolated from mouse lung tumors induced by 4-(methylnitrosamino)-1-(3-pyridyl)-1-butanone (NNK) and related DNA methylating and pyridyloxobutylating agents. *Carcinogenesis* 14:2419–2422.
141. Belinsky S., Devereux T.R., Maronpot R.R., Stoner G.D., and Anderson M.W. 1989. The relationship between the formation of promutagenic adducts and the activation of the K-*ras* proto-oncogene in lung tumors from A/J mice treated with nitrosamines. *Cancer Res.* 49:5305–5311.
142. Chen B., Liu L., Castonguay A., Maronpot R.R., Anderson M.W., and You M. 1993. Dose-dependent *ras* mutation spectra in *N*-nitrosodimethylamine induced mouse liver tumors and 4-(methylnitrosoamine)-1-(3-pyridyl)-1-butanone induced mouse lung tumors. *Carcinogenesis* 14:1603–1608.
143. Oreffo V.I.C., Lin H.–W., Padmanabhan R., and Witschi H. 1993. K-*ras* and *p53* point mutations in 4-(methylnitrosamino)-1-(3-pyridyl)-1-butanone-induced hamster lung tumors. *Carcinogenesis* 14:451–455.
144. Steinberg W. 1990. The clinical utility of the CA 19-9 tumor-associated antigen. *Am. J. Gastroenterol.* 85:350–355.
145. Berger D.H., Chang H., Wood M., Huang L., Heath C.W., Lehman T., and Ruggeri B.A. 1999. Mutational activation of K-*ras* in nonneoplastic exocrine pancreatic lesions in relation to cigarette smoking status. *Cancer* 85:326–331.
146. Duffy M.J. 1998. CA19-9 as a marker for gastrointestinal cancers: a review. *Ann. Clin. Biochem.* 35:364–370.
147. Reddy J.K. and Rao M.S. 1975. Pancreatic adenocarcinoma in inbred guinea pig induced by *N*-methyl-*N*-nitrosourea. *Cancer Res.* 35:2269–2277.
148. Roebuck B.D., Longnecker D.S., Baumgartner K.J., and Thorn C.D. 1985. Carcinogen-induced lesions in the rat pancreas: effects of varying levels of essential fatty acid. *Cancer Res.* 45:5252–5256.
149. Roebuck B.D., Yager Jr. J.D., and Longnecker D.S. 1981. Dietary modulation of azaserine-induced pancreatic carcinogenesis in the rat. *Cancer Res.* 41:888–893.
150. Tsutsumi M., Noguchi O., Okita S., Horiguchi K., Kobayashi E., Tamura K., Tsujiuchi T., Denda A., Konishi Y., Iimura K., and Mori Y. 1995. Inhibitory effects of sulfonation inhibitors on initiation of pancreatic ductal carcinogenesis by *N*-nitrosobis-(2-oxopropyl)amine in hamsters. *Carcinogenesis* 16:457–459.
151. Ogawa T., Makino T., Kosahara K., Koga A., and Nakayama F. 1992. Promoting effects of both dietary cholesterol and cholestyramine on pancreatic carcinogenesis initiated by *N*-nitrosobis(2-oxopropyl)amine in Syrian golden hamsters. *Carcinogenesis* 13:2047–2052.
152. Longnecker D.S., Roebuck B.D., Kuhlmann E.T., and Curphey T.J. 1985. Induction of pancreatic carcinoma in rats with *N*-nitrosobis(2-hydroxypropyl) (2-oxopropyl)amine: histopathology. *J. Natl. Cancer Inst.* 74:209–217.

23

AUTOIMMUNE PANCREATITIS
— RECENT CONCEPT

Kazuichi Okazaki

CONTENTS

23.1 INTRODUCTION

Idiopathic pancreatitis, in which obvious causes are not detected, accounted for about 30 to 40% of chronic pancreatitis.[1] Since Sarles et al. observed a case of particular pancreatitis with hypergammaglobulinemia,[2] occasional coexistence of pancreatitis with other autoimmune diseases such as Sjögren's syndrome (SjS),[3] primary sclerosing cholangitis (PSC),[4,5] or primary biliary cirrhosis (PBC)[4] has been reported. These findings support the hypothesis that an autoimmune mechanism may be involved in the pathogenesis and pathophysiology in some patients with pancreatitis,[6–16] which leads us to the concept of an autoimmune-related pancreatitis,[15] so called "autoimmune pancreatitis" (AIP).[8] Although it has not been widely accepted as a new clinical entity,[17] "Diagnostic Criteria for Autoimmune Pancreatitis 2002" has been proposed by the Japan Pancreas Society.[18] The present chapter discusses the recent concept of AIP from the clinical and animal experimental aspects.

23.2   DEFINITION AND CONCEPT OF AIP

Although the pathogenesis and pathophysiology of AIP are still unclear, clinical aspects have been accumulated.[2–16] The characteristic findings in cases of AIP can be summarized as follows[2–16]:

■ Increased levels of serum gammaglobulin, IgG, or IgG4
■ Presence of autoantibodies
■ Diffuse enlargement of the pancreas
■ Diffusely irregular narrowing of the main pancreatic duct on endoscopic retrograde cholangiopancreatographic (ERCP) images
■ Fibrotic changes with lymphocyte infiltration
■ No symptoms or only mild symptoms, usually without acute attacks of pancreatitis
■ Rare pancreatic calcification
■ Rare pancreatic cysts
■ Occasional association with other autoimmune diseases
■ Effective steroid therapy

Other nomenclatures such as "chronic inflammatory sclerosis of the pancreas,"[2] "sclerosing pancreatitis,"[5] "pancreatitis showing the narrowing appearance of the pancreatic duct" (PNPD),[6] and tumefactive pancreatitis[17] have been proposed in the similar case to AIP. Occasional coexistence of pancreatitis with other systemic exocrinopathy has led to the concept of "a complex syndrome,"[3] "dry gland syndrome,"[4] or "autoimmune exocrinopathy." However, it is poorly understood whether the autoimmune mechanism of primary AIP without other autoimmune diseases is different from that of secondary AIP with other autoimmune diseases or not. The Diagnostic Criteria for Autoimmune Pancreatitis 2002 proposed by the Japan Pancreas Society contains three criteria — pancreatic imaging, laboratory data, and histopathological findings (Table 23.1).[3]

Table 23.1 Diagnostic Criteria for Autoimmune Pancreatitis 2002

1. Pancreatic imaging studies show diffuse narrowing of the main pancreatic duct with irregular wall (more than one-third length of the entire pancreas) and diffuse enlargement of the pancreas.
2. Laboratory data demonstrate abnormally elevated levels of serum gamma-globulin and/or IgG or the presence of autoantibodies.
3. Histopathological examination of the pancreas shows fibrotic changes with lymphocyte and plasma cell infiltration.

For diagnosis, Criterion 1 must be present, together with Criterion 2 and/or Criterion 3.

Proposed by the Japan Pancreas Society

23.3 EPIDEMIOLOGY OF AIP

The patients with AIP are rare, although the exact occurrence rate is still unknown. More than 300 cases have been reported as AIP or PNPD in the Japanese literature.[5,6,8–13,18–20] We experienced 21 cases of AIP in the totally 451 cases of chronic pancreatitis. High-aged males are usually predominant over females in Japan. Although the correct morbidity of primary or secondary AIP is unclear, more than half of the cases are primary.

23.4 ASSOCIATING DISEASES

It is still unclear whether the pathogenetic mechanism of secondary (or syndromic) AIP with other autoimmune diseases is different from primary AIP or not. Thirteen of our 21 cases were primary AIP without other diseases and 9 patients with secondary AIP have one or more associated diseases (Table 23.2). Nine patients (45%) were complicated with diabetes mellitus (DM), 6 sclerosing cholangitis (30%), 6 rheumatoid arthritis (30%),

Table 23.2 Autoimmune Pancreatitis and Associated Diseases

No complications	11 (55%)
Diabetes mellitus	9 (45%)
	(1 Type 1a, 8 Type 2)
Sclerosing cholangitis	6 (30%)
Rheumatoid arthritis	6 (30%)
Sjögren's syndrome	4 (20%)
Nephropathy	3 (15%)
Retroperitoneal fibrosis	2 (10%)

4 sialoadenitis (20%), 3 renal dysfunction (15%), and 2 retroperitoneal fibrosis (10%). Note that there is a possibility of development of systemic autoimmune diseases even in the patients diagnosed as primary AIP.[9] Patients with AIP often show narrowing of the common bile duct (CBD) mainly in the intrapancreatic area, which may result in dilatation of the upper biliary tract. The sclerosing changes of the extrapancreatic bile duct similar to PSC are often observed. Different from PSC, administration of steroid usually shows therapeutic effects on biliary lesions in AIP, which suggests that the mechanism of the development of biliary lesions in AIP may be different from typical PSC.

DM is often (43 to 68%) observed in patients with AIP and the majority shows Type 2 DM, some of which improve after steroid therapy.[4] Although the mechanism is obscure, cytokines from T cells and macrophages suppressing the function of islet β-cells may be down-regulated by steroids.

23.5 CLINICAL SYMPTOMS

The patients with AIP have usually slight or moderate discomfort in the epigastrium or back and symptoms related to other associated diseases.[2–16] Then, the clinical symptoms are different from those in the cases of acute or severe pancreatitis. In our 20 patients, 8 patients had jaundice (40%), 8 abdominal pain (40%), and 6 back pain (30%). Obstructive jaundice due to the stenosis of the CBD is characteristic for AIP, which is rare in other types of pancreatitis. Steroid therapy is usually effective for the stenosis of CBD associated with AIP as well as clinical and laboratory findings.[8–12]

23.6 LABORATORY DATA

Patients with AIP usually show increased levels of serum pancreatic enzymes, hypergammaglobulinemia, IgG, IgG4, and presence of several autoantibodies such as antinuclear, antilactoferrin (LF), anticarbonic anhydrase (CA-II) antibody, and rheumatoid factor.[19] However, antimitochondrial (M2) antibody specific for PBS is rarely observed (Figure 23.1). CA-II and LF are distributed in the ductal cells of several exocrine organs, including the pancreas, salivary gland, biliary duct, and distal renal tubules. Serum levels of IgG4 immune complexes and the IgG4 subclass of immune complexes are often increased in AIP.[20] Our patients with stenosis of CBD show increased serum IgG4 as well as abnormality of the serum bilirubin and hepato-biliary enzymes. In these cases, other liver diseases such as viral hepatitis, autoimmune hepatitis, or primary biliary cirrhosis (PBC) should be ruled out. The pancreatic exocrine function shows slightly or moderately abnormal. Eight of our 20 patients showed hypofunction (40%) by N-benzoyl-L-tyrosyl-para-aminobenzoic acid (BT-PABA) exocrine test. After steroid therapy, many abnormal laboratory findings were reversible.[9–13,19]

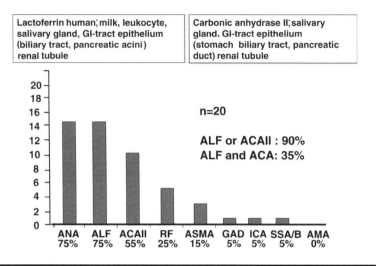

Figure 23.1   Prevalence of autoantibodies in AIP. Antinuclear (ANA), anti-lacto-ferrin (ALF), and anti-CA-II antibodies (ACAII) were identified in 75, 75, and 55% of 20 cases, respectively. Of the cases, 90% showed either antilactoferrin or anti-CA-II antibody, and 35% showed both antibodies, which suggests that immune responses against these proteins are heterogeneously activated. RF = rheumatoid factor, ASMA = anti-smooth muscle antibody, GAD = antiglutamic acid dehydrogenase, ICA = anti-islet cell antibody, SSA/B = anti-SSA-SSB antibody, AMS = antimitochondrial antibody.

23.7   PANCREATIC IMAGING

Computed tomography (CT), magnetic resonance imaging (MRI), or ultrasonography (US) shows the diffusely enlarged pancreas, so called sausage-like appearance, and a capsule-like rim that is low density on CT and hypo-intense on T2-weighted MR images and shows delayed enhancement on dynamic MR imaging (Figure 23.2).[21,22] Pancreatic calcification or pseudocyst is seldom observed. F-18 fluoro-2-deoxy-D-glucose (FDG)-positron emission tomography (PET) shows accumulative signals in the pancreatic lesions similar to the imaging of pancreatic cancer.[23] ERCP images in the AIP patients show segmental or diffuse narrowing of the main pancreatic duct (Figure 23.3 to Figure 23.5).[6,8] Although magnetic resonance cholangio-pancreatography (MRCP) poorly shows the stenosis of the pancreatic duct, it can well demonstrate the images of CBD. The patients with AIP often show stenosis of CBD mainly in the intrapancreatic but less extrapancreatic area, resulting in congestive dilatation of the upper-stream of the biliary tract. The sclerosing stenosis of the extrapancreatic CBD similar to primary sclerosing cholangitis (PSC) is sometimes observed (Figure 23.6).[13] Steroid therapy in AIP is usually effective for the

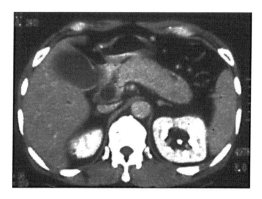

"sausage like"

Steroid
therapy →

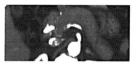

Figure 23.2   Computed tomogram of autoimmune pancreatitis diffusely enlarged pancreas similar to sausage is typical of autoimmune pancreatitis. Contrast-enhanced CT often demonstrates delayed enhancement. After steroid therapy, the size of the pancreas becomes normal, often atrophic.

stenosis of CBD as well as pancreatic duct,[12,19] but it is not effective for classic PSC.[24] These findings suggest that stenostic mechanism of CBD in AIP may be different from that in the classic PSC.

23.8   HISTOPATHOLOGY

It is usually difficult to obtain the pancreatic specimen. Microscopic findings, if obtained, show fibrotic changes with infiltration of lymphocytes and plasmacytes, sometimes eosinophils, mainly around the pancreatic duct (Figure 23.7).[8–11] By immuno-histochemical studies, major infiltrating cells around the pancreatic duct were T-cells although B-cells or plasma cells often infiltrate.[9] Human leukocyte antigen (HLA)–DR was expressed on pancreatic duct cells.[9,13] Histological features of sclerosing pancreatitis are similar to those of typical AIP, but extremely unique as follows[5]:

- Diffuse lymphoplasmacytic infiltration with pronounced acinar atrophy.
- Marked fibrosis of the contiguous soft tissue as well as the total pancreas.
- Obliterated phlebitis in and around the pancreas involving the portal vein.

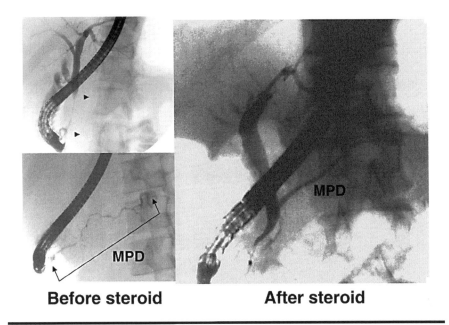

Before steroid   **After steroid**

Figure 23.3 ERCP images of autoimmune pancreatitis. In the lower-left panel, diffusely irregular narrowing of the main pancreatic duct is observed. In the upper-left panel, many patients show narrowing of intrapancreatic CBD as well as pancreatic duct. Narrowing of the terminal bile duct is supposed to be induced mainly by compression of the swollen pancreas. After steroid therapy, most patients remarkably improve as shown in the right picture.

- Inflammatory wall thickness of the CBD and gallbladder.
- The minor salivary gland in the lip biopsy bearing inflammation similar to the pancreatic lesion or that in Sjögren's syndrome.

It is skeptical whether such severe fibrosis or phlebitic thrombosis can be reversible by steroid therapy. The major infiltrating cells are lympho-plasmacytes, suggesting dominant B-cell lineage.

23.9 DIAGNOSIS AND DIFFERENTIAL DIAGNOSIS OF AIP

Although histological findings are most important for a final diagnosis, it is usually difficult to take a biopsy specimen from the pancreas. Therefore, it is important to make a diagnosis in combination with clinical, laboratory findings, and pancreatic imaging such as narrowing pancreatogram or diffusely enlarged pancreas. Recently, diagnostic criteria for autoimmune pancreatitis was proposed.[18] The differential diagnosis of diffuse pancreatic enlargement includes malignant lymphoma, plasmacytoma, metastases,

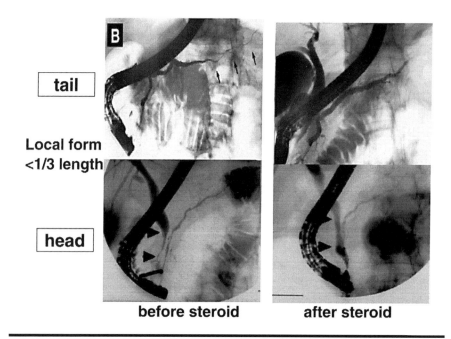

Figure 23.4 ERCP images of local type of autoimmune pancreatitis. ERCP images show the locally irregular narrowing main pancreatic duct in the pancreatic tail (the upper-left panel) and local narrowing in the pancreatic head (the lower-left panel). Irregular narrowing is observed in more than one-third length, but not the entire pancreas. After steroid therapy, most of these abnormal findings improve.

and diffuse infiltrative pancreatic carcinoma. Most of autoimmune pancreatitis can be distinguished from these other abnormalities with radiological imaging. However, it is often difficult to distinguish AIP from cancer of the pancreas head or diffuse pancreatic cancer.[13,17] Improvements on clinical findings by the steroid therapy may be useful in the differential diagnosis of AIP from pancreas cancer.[13,17]

23.10 TREATMENT AND PROGNOSIS

In the AIP patients with mild symptoms, usual treatments for acute pancreatitis such as fasting, protease inhibitors, and antibiotics are not necessarily required. In cases of jaundice, percutaneous transhepatic or endoscopic retrograde biliary drainage is often needed, especially in the complication of bacterial infection. Steroid therapy is usually effective for the stenosis of CBD as well as the pancreatic duct.[8–13] It is noted that some patients may spontaneously improve without any treatment.[13] Some patients associated with Type 2 diabetes mellitus may improve after steroid

Narrowing of MPD >1/3 (n=20)

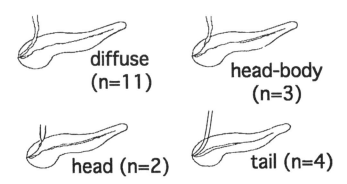

diffuse
(n=11)

head-body
(n=3)

head (n=2)

tail (n=4)

Figure 23.5 Pancreatogram of 20 patients with autoimmune pancreatitis. Out of 20, 11 cases showed diffuse narrowing pancreatic duct, 3 cases from head to body, 2 cases in the head, and 4 cases in the tail.

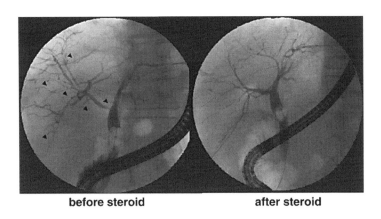

before steroid   after steroid

Figure 23.6 Sclerosing cholangitis associated with autoimmune pancreatitis. Cholangiogram shows a similar image to PSC. Different from PSC, steroid therapy is usually effective.

therapy.[12,25] In the unresponsive cases of CBD stenosis to steroid therapy, surgical operation is often necessary to be differentiated from malignancy as well as the relief of symptoms.[13] Eleven of our 20 patients were successfully treated with prednisolone, 2 with pancreatectomy, and 4 without medication. The long-term prognosis of AIP is unknown. As the clinical and laboratory findings of most cases are reversible after steroid therapy,[12,13,19] the prognosis of AIP may depend on the severity of complicated diseases, such as other autoimmune diseases or diabetes mellitus.

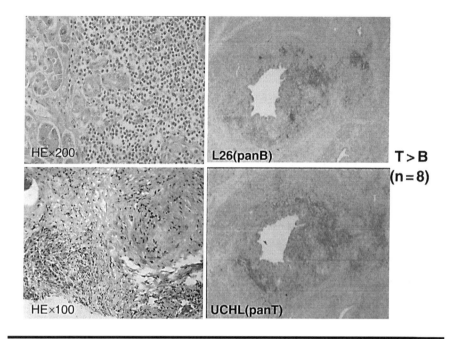

Figure 23.7 Microscopic findings of the pancreas in autoimmune pancreatitis. Massive infiltration of lymphocytes and plasmacytes (A) or fibrotic changes (B) were observed mainly around the pancreatic duct. In most cases, T lymphocytes (C) predominantly infiltrate over B lymphocytes (D) with occasional follicle formation.

23.11 PATHOPHYSIOLOGY OF AIP

23.11.1 Humoral Immunity and Target Antigens

Occasional coexistence of pancreatitis with other autoimmune diseases suggests that there may be common target antigens in the pancreas and other exocrine organs such as the salivary gland, biliary tract, and renal tubules. We observed that several autoantibodies such as antinuclear antibody (ANA), antilactoferrin antibody (ALF), anticarbonic anhydrase-II antibody (ACA-II), rheumatoid factor, and antismooth muscle antibody were frequently detected in patients with AIP. LF, a nonenzymatic protein, is also detected in the various human tissues, including the lactating breast; bronchial, salivary, and gastric glands; and the pancreatic acinus.[19] The high prevalence of these antibodies suggests that CA-II and LF may be the candidates for the target antigens in AIP. However, it is noted that these antibodies are not necessarily specific for AIP because ACA-II can be detected in some patients with SjS or systemic lupus erythematosus and ALF in ulcerative colitis or PSC.

23.11.2   Cellular Immunity and Effector Cells

Although the effector cells of AIP have been poorly understood, the activated CD4+ and CD8+ T-cells bearing HLA-DR and CD45RO were increased in the peripheral blood lymphocytes (PBLs) in the patients with AIP in comparison with those in other causes of pancreatitis such as alcoholic or gallstone-related pancreatitis.[26] CD4 or CD8 T-cells also predominantly infiltrate in the pancreas tissue over B-cells.[19,25] In some patients with AIP, HLA-DR antigens are expressed on the pancreatic duct cells as well as T-cells.[9]

23.11.3   Experimental Autoimmune Pancreatitis Using Animal Models

Although the pathogenesis of AIP is still unknown, several candidates for the target antigens have been proposed from the clinical aspects. To prove them, several animal models have been studied.[27-34] Our NTx-BALB/c mice, subcutaneously immunized with CA-II or LF, and synergetic nude mice, transferred with splenocytes from the disease-induced nTx-mice, developed pancreatitis as well as sialoadenitis and cholangitis,[34] but the normal BALA/c mice did not.

Several animal models using alymphoplasic (aly-/-) mice,[29] MRL/lpr,[30,31] spontaneously develop pancreatitis, in which effector cells were Th1 Type CD4+ T-cells. In our animal model using nTx-BALB/c mice and transferred nude mice, the CD4-positive cells were mainly involved in the development of pancreatitis, sialoadenitis, and cholangitis, but CD8+ T-cells were never pathogenic.[34] These findings suggested that MHC-Class II restricted-autoreactive CD4+ T-cells, which escape from the negative selection in the thymus, and depletion of regulatory T-cells such as CD25+ T-cells in the periphery take important roles in the development of autoimmune pancreatitis and exocrinopathy. Recent observations suggested a functional role of T-lymphocytes, such as cytotoxicity[35] or the neuroimmune interactions in acute pancreatitis as well.[36] In the mice with cerulein-induced experimental pancreatitis, CD4+ T-cells are required for the complete development of pancreatic lesions.[37] These CD4+ T-cells probably induce the activation of macrophage and further proinflammatory reactions during the early stage of acute pancreatitis as well as direct cytotoxicity effects through Fas ligand expression.[37] On the other hand, CD8+ T-cells may play roles as effector cells in pancreatitis of the major histocompatibility complex (MHC) Class II-deficient mice.[29] These animal models suggest that although CD8+ T-cells may be partially involved, CD4+ T-cells take major roles in the development of experimental pancreatitis, which is consistent with human AIP.

Table 23.3 Experimental Animal Models of Autoimmune Pancreatitis

Authors (year)	Animal	Antigens	Target Organs of Lesions	Effector Cells
Kanno et al. (1992)	MRL/lpr mice	?	pancreas	T cells
Tsubata et al. (1996)	aly-/aly- mice	?	pancreas, Pyer's patch deficiency	CD4+T cells
Uchida et al. (2002)	nTx-BALB/c mice and nude mice	CA-II	pancreas, salivary glands, bile duct, renal tubule	CD4+Th1 cells
Uchida et al. (2002)	nTx-BALB/c mice and nude mice	LF	pancreas, salivary glands, bile duct, renal tubule	CD4+Th1 cells
Vallance et al. (1998)	MHC-Class II deficient mice	?	pancreas, colon	CD8+T cells
Nishimori et al. (1995)	PL/J mice	CA-II	salivary glands	?
Ueno et al. (1998)	BALB/c mice	CA-II	bile duct	CD4+T cells
Haneji et al. (1996)	nTx-NFS/sld mice	α-fodrin	salivary glands	CD4+T cells
Mustafa et al. (1998)	MRL/lpr mice	?	salivary glands	Th1 cells

nTx: neonatal thymecomy
Question marks indicate "not identified."

23.12 EPILOGUE

In conclusion, recent studies support the concept of autoimmune-related pancreatitis, which appears to be a unique clinical entity. Although the pathogenesis of AIP is still unknown, several hypotheses have been proposed from the clinical and animal experimental aspects. The first step in the disease may be an antigenic alteration at the pancreatic duct or acinar cells such as an aberrant expression of HLA-DR. CD4+ T-cells might recognize the complex of HLA Class II and an autoantigenic peptide such as LF, CA-II or α-fodrin, and act as cytotoxic or helper cells. CD8+ T-cells might act as cytotoxic cells as well. Further studies are needed to clarify the pathogenesis.

ACKNOWLEDGMENTS

This study was supported by the following:

Grant-in-Aid for Scientific Research (C) of Ministry of Culture and Science of Japan (16590645)

Grant-in-Aid from the Research on Specific Disease (Intractable Disease of the Pancreas), Health and Labour Sciences Research Grants, Japanese Ministry of Health, Labour and Welfare.

REFERENCES

1. Steer M.L., Waxman I., and Freedman S. 1995. Chronic pancreatitis. *N. Engl. J. Med.* 332:1482–1490.
2. Sarles H., Sarles J.C., Muratore R. et al. 1961. Chronic inflammatory sclerosis of the pancreas – an autonomous pancreatic disease? *Am. J. Dig. Dis.* 6:688–698.
3. Montefusco P.P., Geiss A.C., Bronzo R.L., Randall S., Kahn E., and McKinley M.J. 1984. Sclerosing cholangitis, chronic pancreatitis, and SjS: a syndrome complex. *Am. J. Surg.* 147:822–826.
4. Epstein O., Chapman R.W.G., Lake-Bakaar G., Foo A.Y., Rosalki S.B., and Sherlock S. 1982. The pancreas in primary biliary cirrhosis and primary sclerosing cholangitis. *Gastroenterology* 83:117–182.
5. Kawaguchi K., Koike M., Tsuruta K., Okamoto A., Tabata I., and Fujita N. 1991. Lymphoplasmacytic sclerosing pancreatitis with cholangitis: a variant primary sclerosing cholangitis extensively involving pancreas. *Hum. Pathol.* 22:387–395.
6. Toki F., Kozu T., and Oi I. 1992. An unusual type of chronic pancreatitis showing diffuse irregular narrowing of the entire main pancreatic duct on ERC- A report of four cases. *Endoscopy* 24:640.
7. Lankisch P.G., Koop H., Seelig R., and Seeling H.P. 1981. Antinuclear and pancreatic acinar cell antibodies in pancreatic disease. *Digestion* 21:65–68.
8. Yoshida K., Toki F., Takeuchi T., Watanabe S., Shiratori K., and Hayashi N. 1995. Chronic pancreatitis caused by autoimmune abnormality. Proposal of concept of autoimmune pancreatitis. *Dig. Dis. Sci.* 40:1561–1568.
9. Ohana M., Okazaki K., Hajiro K., and Kobashi Y. 1996. Multiple pancreatic masses associated with autoimmunity. *Am. J. Gastroenterol.* 93:2607–2609.
10. Horiuchi A., Kaneki T., Yamamura M., Nagata A., Nakamura T., Akamatsu T., Mukawa K., Kawa S., and Kiyosawa K. 1996. Autoimmune chronic pancreatitis simulating pancreatic lymphoma. *Am. J. Gastroenterol.* 91:2607–2609.
11. Horiuchi A., Kawa S., Akamatsu T., Aoki Y., Mukawa K., Furuya N., Ochi Y., and Kiyosawa K. 1998. Characteristic pancreatic duct appearance in autoimmune chronic pancreatitis: a case report and review of the Japanese literature. *Am. J. Gastroenterol.* 93:99–102.
12. Ito T., Nakano I., Koyanagi S., Miyahara T., Migita Y., Ogoshi K., Sakai H., Matsunaga S., Yasuda O., Sumii T., and Nawata H. 1997. Autoimmune pancreatitis as a new clinical entity. Three cases of autoimmune pancreatitis with effective steroid therapy. *Dig. Dis. Sci.* 42:1458–1468.
13. Uchida K., Okazaki K., Konishi Y., Ohana M., Takakuwa H., Hajiro K., and Chiba T. 2000. Clinical analysis of autoimmune-related pancreatitis. *Am. J. Gastorenterol.* 95(10):2788–2794.
14. Coll J., Navarro S., Tomas R., Elena M., and Martinez E. 1989. Exocrine pancreatic function in Sjögren syndrome. *Arch. Intern. Med.* 149:848–852.

15. Sheikh S.H. and Shaw-Stiffel T.A. 1995. The gastrointestinal manifestations of Sjögren's syndrome. *Am. J. Gastroenterol.* 90:9–14.

16. Waldram R., Kopelman H., Tsantoulas D., and Williams R. 1975. Chronic pancreatitis, sclerosing cholangitis, and sicca complex in two siblings. *Lancet* 7906:550–552.

17. Pearson R.K., Longnecker D.S., Chari S.T., Smyrk T.C., Okazaki K., Frulloni L., and Cavallini G. 2003. Controversies in clinical pancreatology: autoimmune pancreatitis: does it exist? *Pancreas* 27(1):1–13.

18. The Japan Pancreas Society. 2002. Diagnostic criteria for autoimmune pancreatitis by the Japan Pancreas Society (2002). *J. Jpn. Pancreas Soc.* 17:587.

19. Okazaki K. and Chiba T. 2002. Autoimmune-related pancreatitis. *Gut* 51:1–4.

20. Hamano H., Kawa S., Horiuchi A. et al. 2001. High serum IgG4 concentrations in patients with sclerosing pancreatitis. *N. Engl. J. Med.* 344:732–738.

21. Irie H., Honda H., Baba S., Kuroiwa T., Yoshimitsu K., Tajima T., Jimi M., Sumii T., and Masuda K. 1998. Autoimmune pancreatitis: CT and MR characteristics. *Am. J. Roentgenol.* 170:1323–1327.

22. Furukawa N., Muranaka T., Yasumori K., Matsubayashi R., Hayashida K., and Arita Y. 1998. Autoimmune pancreatitis: radiologic findings in three histologically proven cases. *J. Comput. Assist. Tomogr.* 22:880–883.

23. Nakamoto Y., Sakahara H., Higashi T., Saga T., Sato N., Okazaki K., Imamura M., and Konishi J. 1999. Autoimmune pancreatitis with F-18 fluoro-2-deoxy-D-glucose PET findings. *J. Clin. Nucl. Med.* 24:778–780.

24. Angulo P. and Lindor K.D. 1999. Primary sclerosing cholangitis. *Hepatology* 30:325–332.

25. Tanaka S., Kobayashi T., Nakanishi K. et al. 2000. Corticosteroid-responsive diabetes mellitus associated with autoimmune pancreatitis. *Lancet.* 356:910–911.

26. Okazaki K., Uchida K., Ohana M., Nakase H., Uose S., Inai M., Matsushima Y., Katamura K., Ohmori K., and Chiba T. 2000. Autoimmune-related pancreatitis is associated with autoantibodies and Th1/Th2-type cellular immune response. *Gastroenterology* 118:1–10.

27. Nishimori I., Bratanova T., Toshkov I., Caffrey T., Mogaki M., Shibata Y., and Holingworth M.A. 1995. Induction of experimental autoimmune sialoadenitis by immunization of PL/J mice with carbonic anhydrase II. *J. Immunol.* 154:4865–4873.

28. Ueno Y., Ishi M., Takahashi S., Igarashi T., Toyoda T., and LaRusso F.N. 1998. Different susceptibility of mice to immune-mediated cholangitis induced by immunization with carbonic anhydrase II. *Lab. Invest.* 78:629–637.

29. Tsubata R., Tsubata T., Hiai H., Shinkura R., Matsumoto R., Sumida T., Miyawaki S., Ishida H., Kumagai S., Nakao K., and Honjo T. 1996. Autoimmune disease of exocrine organ immunodeficient model for Sjögren's syndrome. *Eur. J. Immunol.* 26:2742–2748.

30. Kanno H., Nose M., Itoh J., Taniguchi Y., and Kyogoku M. 1992. Spontaneous development of pancreatitis in the MRL/Mp strain of mice in autoimmune mechanism. *Clin. Exp. Immunol.* 89:68–73.

31. Mustafa W., Zhu J., Deng G., Diab A., Link H., Frithiof L., and Klinge B. 1998. Augmented levels of macrophage and Th1 cell-related cytokine mRNA in submandibular glands of MRL/lpr mice with autoimmune sialoadenitis. *Clin. Exp. Immunol.* 112:389–396.

32. Vallance B.A., Hewlett B.R., Snider D.P., and Collins S.M. 1998. T cell-mediated exocrine pancreatic damage in major histocompatibility complex class II-deficient mice. *Gastroenterology* 115:978–987.

33. Yamamichi M., Matsuoka N., Tomioka T., Eguchi K., Nagataki S., and Kanematsu T. 1997. Shared TCR Vbeta gene expression by the pancreas and salivary gland in immunodeficient alymphoplasic mice. *J. Immunol.* 159:427–432.

34. Uchida K., Okazaki K., and Chiba T. 1998. An animal model of autoimmune pancreatitis using neonatal thymectomized mice. *Gastroenterology (abstr.)* 114:A505.

35. Hunger R.E., Mueller C., Z'Graggen K., Friess H., and Buchler M.W. 1997. Cytotoxic cells are activated in cellular infiltrates of alcoholic chronic pancreatitis. *Gastroenterology* 112:1656–1663.

36. Di Sebastiano P., Fink T., Weihe E., Friess H., Innocenti P., Beger H.G., and Buchler M.W. 1997. Immune cell infiltration and growth-associated protein 43 expression correlate with pain in chronic pancreatitis. *Gastroenterology* 112:1648–1655.

37. Demols A., Moine O.L., Desalle F., Quertinmont E., Laethem J.L.V., and Deviere J. 2000. CD4+ T cells play an important role in acute experimental pancreatitis in mice. *Gastroenterology* 118:582–590.

24

ROLE OF GENETIC FACTORS IN INFLAMMATORY DISEASES OF THE PANCREAS

F. Ulrich Weiss and Markus M. Lerch

CONTENTS

24.1   INTRODUCTION

Many hypotheses have attempted to explain the origin and underlying pathophysiological mechanisms of acute and chronic pancreatitis (CP). To date, our understanding of the cellular mechanisms that trigger acute pancreatitis and induce a progression of the disease to a chronic condition is still limited. Recent progress in molecular techniques has intensified the search for genetic changes that play a role in the onset and progression of inflammatory pancreatic disorders. This chapter outlines the recent advances with respect to human inborn genetic varieties that have been found to be associated with pathophysiological disorders of the pancreas — namely acute and chronic pancreatitis.

Acute pancreatitis is a painful inflammatory disease that is initiated by events that cause acinar cell injury, the release of prematurely activated digestive enzymes, subsequent tissue and endothelial damage, and the development of a potentially lethal inflammatory response.[1] CP, on the other hand, is clinically characterized by acinar cell degeneration and fibrosis that may, over a variable period of time, lead to the destruction of exocrine and endocrine organ function and results in maldigestion and diabetes mellitus. Whether acute pancreatitis progresses to CP or whether the acute and chronic forms of the disease represent distinct entities is still a matter of controversy. Several underlying conditions appear to play a role in the pathogenesis of CP. Epidemiological studies have further established a link between inflammatory pancreatic disorders and pancreatic cancer and a variety of environmental and lifestyle-dependent risk factors such as alcohol and tobacco.

24.2 SUSCEPTIBILITY TO ENVIRONMENTAL TOXINS

A number of gene families determine the individual cellular settings of detoxifying enzymes and a person's capacity for enzymatic detoxification (see Chapter 7). This accounts for both an individual's susceptibility to environmental toxins and a predisposition for the development of environment-related diseases. The human super gene family of uridine 5'-diphosphate glucuronyl-transferases (UGT) is known to represent Phase II detoxifying enzymes that glucuronidate not only dietary by-products and endogenous metabolites, but also therapeutic drugs or polycyclic hydrocarbons and heterocyclic amines, which are known mutagens. The consequence of an impaired Phase II metabolism is an accumulation of cyto- and genotoxic xenobiotics that are generated by an adduct reaction with continuously produced reactive oxygen species from oxidative Phase I metabolic enzymes. The UGT1A7 gene encodes one specific UGT isoform that has been shown to glucuronidate several known carcinogens, including those from tobacco smoke. Transcripts of UGT1A are predominant in pancreatic tissue. Here the low detoxification activity of a specific UGT1A7*3 allele may represent a genetic risk factor and confer a predisposition for environmentally induced pancreatic injuries. In a genetic analysis of UGT1A polymorphisms in patients with pancreatic diseases by Ockenga et al.,[2] the UGT1A*3 allele was found to be associated with pancreatic carcinoma, but also with a subgroup of patients with chronic alcoholic pancreatitis, of which 89% were also smokers. Although the presence and relevance of certain UGT1A polymorphisms is presently being disputed (unpublished observations) the reported data would strongly support the concept that acute and chronic pancreatitis can result from a complex interaction of genetic and environmental factors.

The pathologic effects of alcohol on the pancreas are difficult to study in humans. Ethanol and its metabolites, however, appear to have complex short-term and long-term effects on acinar cell physiology, to cause damage to cell membranes, and to affect cellular signaling pathways (see Chapter 6 and Chapter 14). Animal models, which have been frequently used to investigate the effect of ethanol *in vivo,* have demonstrated that the pancreatic injury induced by ethanol exposure is likely to be multi-factorial. The mechanism seems to include some degree of ductal hypertension, decreased pancreatic blood flow, oxidative stress, direct acinar cell toxicity, changes in protein synthesis, an enhanced inflammatory response, or the stimulation of fibrosis. Acute administration of alcohol in the rat results in increased injury during pancreatitis induced by a combination of pancreatic duct obstruction and hormonal hyperstimulation. Rats under chronic ethanol feeding had also more severe pancreatitis. Although the generation of oxygen free radicals has been clearly demonstrated in the pancreas of rats under continuous ethanol feeding, ethanol alone (i.e., without an additional disease-inducing stimulus) does not cause pancreatitis. Generation of free radicals has been shown to cause depletion of intracellular antioxidants, such as glutathione, and accounts for subsequent oxidant damage of lipids, proteins, and nucleic acids. Therefore, some of the toxic effects of ethanol may be secondary to its effect on lipid metabolism and other metabolic pathways. The genetic predisposition toward environmental toxins including ethanol is therefore likely to involve one or several of the genes that regulate cellular detoxification but also those that control intracellular processes targeted by toxin-induced physiologic and metabolic changes.

24.3 METABOLIC DISORDERS

A variety of inborn errors of metabolism, such as disorders of specific amino acid transporters, branched chain amino acid degradation, or hemolysis, have been reported in some families to be associated with acute and chronic-recurrent episodes of pancreatitis. Although in most of these metabolic disorders, pancreatitis is not the most serious clinical manifestation of the underlying metabolic defect, some disorders are known to cause dramatically increased plasma concentrations of chylomicrons and triglycerides. Hyperlipidemia, as a consequence of familial disorders of lipid metabolism, is generally considered to put patients at a significant risk of developing pancreatitis.[3–5] A massive plasma accumulation of chylomicrons and triglycerides, for example, is found in patients with lipoprotein lipase deficiency or apolipoprotein C-II deficiency. Although lipoprotein lipase activity is located in the endothelial membrane, apolipoprotein C-II is contained within chylomicrons. Both proteins confer an important metabolic function in the hydrolysis of triglycerides and the production of free fatty acids. Therefore, genetic disorders in either of these genes can lead to a genetic predisposition for pancreatitis. Carriers of currently more than 30 disease-relevant lipoprotein lipase gene mutations can be identified by the reduced catalytic activity in postheparin plasma (heparin releases lipoprotein lipase into the blood stream) or by genetic testing. The mode of inheritance is autosomal recessive and the disease has a low incidence of 1 in 1 million. The first symptoms arise in early childhood and nearly 30% of patients with lipoprotein lipase deficiency develop recurrent episodes of pancreatitis. Apolipoprotein C-II deficiency is normally diagnosed at a later age during adolescence or in young adults, often with a similar clinical presentation of recurrent episodes of pancreatitis. More than 10 disease-relevant mutations have been identified in the apolipoprotein C-II gene so far, which apparently interfere with the biological function of apolipoprotein C-II as an activator of lipoprotein lipase. Up to 60% of affected patients develop pancreatitis as a frequent and severe complication that, in absence of lipid-lowering treatment, can result in chronic exocrine and endocrine pancreatic insufficiency.[6,7]

Several other disorders of lipid metabolism have been reported that can lead to chylomicronemia or hypertriglyceridemia and are independent of the lipoprotein lipase system. They represent a significant risk factor for the development of pancreatitis when plasma triglyceride levels rise above 2000 mg/dl. The incidence of such disorders is even higher than that of disorders in the lipoprotein lipase system and additional factors such as drug therapies including glucocorticoids, estrogens, diuretics or beta-adrenergic blockers, and also alcohol abuse or diabetes mellitus can contribute to a further increase of hypertriglyceridemia above the threshold level for developing pancreatitis. The most common familial disorders associated with chylomicronemia are Type I and Type V hyperlipoproteinemias (according to Levy and Fredrickson) that comprise a diverse family of disorders with moderate to severe hypertriglyceridemia.[8] A predisposition for increased plasma lipids is also found in patients with plasmocytoma, systemic lupus erythematous, and lymphomatous diseases.

24.4 HEREDITARY PANCREATITIS

As early as 1952, Comfort and Steinberg[9] reported a hereditary predisposition to chronic or recurrent pancreatitis that is independent of additional environmental factors in a number of families. Hereditary pancreatitis (HP) is an autosomal dominant disorder with a clinical manifestation that is indistinguishable from other etiological varieties of pancreatitis. In affected patients, HP begins with recurrent attacks of acute pancreatitis that usually start in childhood, but the age of onset of the disease can vary considerably and can sometimes be delayed until late in adulthood. The severity of acute attacks ranges from mild to complicated cases with progression to pancreatic necrosis and organ failure. Recurrent attacks of acute pancreatitis frequently progress to chronic disease at an early age and are associated with significant lifetime risk for the development of pancreatic cancer.

Although HP is not a common disease and to date only several hundred families have been identified worldwide, the studies addressing the onset of HP have permitted some recent breakthroughs in understanding the pathophysiology of acute and chronic pancreatitis in general.

The exocrine pancreas synthesizes and secretes large amounts of digestive proteases and a number of protective mechanisms are operative to prevent a self-destruction of the pancreas by premature protease activation. The induction of acute pancreatitis is believed to follow — independent of the different etiologic factors, such as gallstones, toxins, hyperlipidemia, hypercalcemia, inheritance, and others — a uniform mechanism that primarily affects the acinar cell and involves the activation of trypsinogen and other intracellular digestive enzymes. The initial triggering event in pancreatitis must therefore include a disruption of the protective mechanisms that prevent the primary activation of digestive enzymes (see Chapter 11). The intracellular zymogen activation represents an early event in the disease process and has to be looked at separately from other pathogenic events that include the generation of oxygen free radicals, the release of cytokines, the infiltration of pancreatic tissue by inflammatory cells, and in some cases the progression to necrotizing pancreatitis. Acute pancreatitis is associated with a proteolytic autodigestion of the pancreas and because trypsin is a known activator of other proteolytic enzymes in the gut, dysregulation of pancreatic trypsin activity has long been regarded as a key event in the onset of pancreatitis.

24.5 CATIONIC TRYPSINOGEN

The report in 1996 of a mutation in the cationic trypsinogen gene on Chromosome 7 in families with HP was a first success in the effort to understand the pathophysiology of acute and chronic pancreatitis on a molecular level.[10] This mutation leads to a substitution of the amino acid histidine (H) to arginine (R) at Position 122 and is now the most common (of several) mutation in the cationic trypsinogen in patients with HP. To determine whether a specific mutation is of clinical relevance, it is necessary to demonstrate not only the prevalence of that mutation in a defined patient group, but also to identify a change of function for the mutated gene in normal physiology. The pathophysiological mechanisms through which carriers of the R122H mutation develop pancreatitis are not fully explained, but it has been proposed that this amino acid substitution leads to a gain of trypsin function by either a more rapid or efficient intracellular trypsinogen activation or by an impaired inactivation and autolysis, once the trypsin is enzymatically activated. In either case, this would result in an extended trypsinogen activity that may initiate an autodigestive process within the acinar cell. The trypsin crystal structure[11] and recent biochemical data[12] are consistent with the hypothesis that R122 represents the trypsin autolysis site that may become disrupted in R122H mutation. This mutation affects an important fail-safe mechanism against premature or overactivation of trypsin within the pancreas and enhances or prolongs the trypsin protease function. Recently, three groups independently reported families with an arginine–cysteine exchange mutation within this same Codon 122 (R122C).[13–15] Interestingly, R122C trypsinogen appears to differ from the R122H mutant by a greatly reduced autoactivation and an increased resistance to autolysis. Activation of recombinantly expressed R122C mutant human trypsinogen by Cathepsin B is also greatly reduced, which is presumably due to a misfolding caused by disulfide mismatches. If this protein misfolding reflects the *in vivo* situation, a dramatic loss of cellular trypsin activity should be observed that raises the fundamental question whether a gain or a loss of trypsin function is crucial for the triggering mechanism in HP. Despite the remaining uncertainty about the molecular events that precede the induction of pancreatitis, it its remarkable that two different mutations as well as a silent polymorphism have been found at the same Codon 122 of cationic trypsinogen. This Arginine 122 therefore appears to represent a key site for the biochemical properties of trypsin and the events that determine the onset of pancreatitis. Support for a fundamental role of the Arginine 122 in the disease process also comes from the recent discovery of a spontaneous *de novo* R122H mutation at this site.[16] Without knowing the frequency of such spontaneous mutations, the finding suggests that the diagnosis of hereditary pancreatitis — as defined by a genetic predisposition — must be considered even in the complete absence of a familial history for pancreatitis.

Another disease-relevant trypsinogen mutation leading to a substitution of asparagine by isoleucine at Position 29 (N29I) has been suggested to result in a conformational change of the trypsin molecule that prevents proteolytic accession of its autolysis site and therefore renders trypsin more resistant to inactivation. Crystallographic data confirm that N29 is in short proximity to R122H on the protein surface, and the main support for this hypothesis comes from *in vitro* studies of recombinant rat cationic trypsinogen.[17] However, enhanced autoactivation at pH 5.0 has also been reported,[12] which suggests an alternative explanation for an increased trypsin function. A comparison of clinical data from R122H and N29I HP patients suggests that both have a similar clinical course.[18]

Furthermore, a mutation in the signal peptide cleavage site has been found associated with CP. A substitution of valine by alanine at Position 16 (A16V) is presumed to interfere with the intracellular processing of trypsinogen, but the pathophysiological mechanisms remain unclear. It has been assumed that A16V may disturb the segregation of trypsinogen from lysosomal Cathepsin B. As a result, the colocalized hydrolase Cathepsin B could activate trypsinogen more readily. A16V appears to be a relatively frequent mutation, whose penetrance seems lower than that of the mutations described above.

Other rare mutations have been identified within the trypsinogen activation peptide (K23R, D22G), which is cleaved from the N-terminus of trypsinogen during its proteolytic activation to trypsin. The pathophysiological mechanism suggested here would be an increased auto-activation of trypsinogen leading to elevated trypsin-levels that may initiate pancreatitis.

Further trypsinogen variants L104P, R116C, and C139F have been described in a mutational screening of patients with nonalcoholic CP. However, these mutations have been reported to date only in single individuals and therefore their relevance to the pathogenesis of pancreatitis remains open. A reported deletion of three nucleotides in the promotor region of cationic trypsinogen (delTCC), may either lead to an enhanced transcription and overproduction of cationic trypsinogen within the acinar cell or, as in the R122C mutation, to a loss of trypsin function.

The discovery of not only one single mutation, but of at least five independently acting mutations in the cationic trypsinogen gene is evidence for a genetic heterogeneity in HP (Table 24.1). Statistically, most hereditary disorders result from loss of function mutations that impair or decrease the normal function of a protein. In case of the human cationic trypsinogen, it would be unlikely that several independent mutations would all cause a gain of function and enhance or prolong the activity of trypsin intracellularly. The ultimate answer to the question how trypsinogen is involved in the onset of pancreatitis is still pending.

24.6   ANIONIC TRYPSINOGEN AND MESOTRYPSINOGEN

The human cationic trypsinogen gene is a member of the trypsinogen gene family that also includes two functional genes for anionic trypsinogen and mesotrypsinogen in addition to 5 nonfunctional relics or pseudogenes.

Table 24.1 Human Cationic Trypsinogen Mutations Associated with Hereditary Pancreatitis

Mutation	Suggested Pathogenic Mechanism
R122H	Prolonged activity, resistance to hydrolysis at R122 site
R122C	Prolonged activity, resistance to hydrolysis at R122 site or reduced activity by protein misfolding *in vivo*
N29I	Prolonged activity, hydrolysis prevented by conformational change
A16V	Disturbed intracellular processing, reduced activity
K23R	Enhanced autoactivation of trypsinogen
D22G	Enhanced autoactivation of trypsinogen
-28delTCC	Enhanced or decreased transcription of trypsinogen

Cationic and anionic trypsinogens are encoded by the protease serine (PRSS) Gene 1 and Gene 2, respectively, which are located on Chromosome 7q35 within the human T-cell receptor β locus (GenBank Accession: NG_001333), in a gene cluster with 6 other trypsinogen pseudogenes. Only one trypsinogen gene — the mesotrypsinogen gene PRSS3 — is not located within this gene cluster, but in a different location on Chromosome 9. Under physiological conditions, cationic trypsinogen is the major of the three secreted trypsinogen isoforms and constitutes approximately two-thirds of the trypsin content in pancreatic juice. One-third of this content consists of anionic trypsinogen, and mesotrypsinogen accounts for less than 5% of trypsinogens in the pancreatic juice. A characteristic feature of some pancreatic diseases such as CP is a selective up-regulation of anionic trypsinogen leading to a reversed proportion of cationic and anionic trypsinogen. To date, however, sequencing analysis of HP families gave no hint on an involvement of any mutations of anionic or mesotrypsinogen genes in the pathogenesis of HP.

From biochemical analysis, it is known that in the family of trypsinogens, only the cationic isoform has the ability to autoactivate and this may be the reason why only mutations in this specific isoform have been detected.

A sequence comparison between members of the trypsinogen genes/pseudogenes (T1–T8) within the T cell receptor β locus indicates that the missense mutations R122H (CGC>CAC), N29I (AAC>ATC), and A16V (GCC>GTC) in cationic trypsinogen are homologous to the corresponding sequence motifs in other trypsinogen pseudogenes. The most frequent R122H mutation in cationic trypsinogen (T4) resembles the nucleotide sequence of Codon 122 in Pseudogene T5. Similarly, the N29I mutation would be encoded by Codon 29 in the anionic trypsinogen gene T8 sequence, the A16V mutation by Codon 16 in Pseudogene T6. A gene conversion concept has therefore been proposed by Chen and Ferec[19] as an explanation for a nonreciprocal sequence exchange between allelic trypsinogen genes. Gene conversion is an important mechanism in the immune system that allows the generation of a great variety of different B cell immunoglobulins or T cell receptor genes and the location of the trypsinogen gene cluster within the T cell receptor β locus appears to explain a presumably higher susceptibility of trypsinogen genes for this kind of mutation mechanism. However, on closer analysis of the sequences, it becomes apparent that no other flanking sequence variations are evident in Pseudogenes T5 and the sequences neighboring the R122H mutation. Point mutations rather than gene conversion may therefore be a more likely explanation for the R122H mutation and the recent report of a spontaneous *de novo* R122H germ line mutation[16] may indicate a genetic hot spot in Codon 122 of the cationic trypsinogen gene.

24.7 THE PANCREATIC SECRETORY TRYPSIN INHIBITOR GENE (SPINK-1)

PSTI, a 59 amino acid long Kazal Type 1 serine protease inhibitor (SPINK-1), is synthesized in acinar cells as a 79 amino acid single chain polypeptide precursor, that is subsequently processed to the mature peptide and that is secreted into pancreatic ducts (see Chapter 11). It is regarded as a first line defense system that is capable of inhibiting up to 20% of total trypsin activity, which may result from accidental premature activation of trypsinogen to trypsin within acinar cells. First studies on the role of PSTI mutations in CP patients reported that some of these patients had a point mutation in Exon 3 of the PSTI gene that leads to the substitution of an asparagine by serine at Position 34 (N34S). Analysis of intronic sequences showed that the N34S mutation is in complete linkage disequilibrium with 4 additional sequence variants: IVS1-37TC, IVS2+268AG, IVS3-604GA, and IVS3-69insTTTT. Whether the N34S amino acid exchange or its association with these intronic mutations, which may confer splicing abnormalities, are causative in the context of PSTI pathophysiology is not clear at the moment. In a number of studies, further mutations and polymorphisms have been detected in PSTI, which include a methionine to threonine exchange that destroys the start codon of PSTI (1MT). These are a leucine to proline exchange in Codon 14 (L14P), an aspartate to glutamine exchange in Codon 50 (D50E), and a proline to serine exchange in Codon 55 (P55). Few studies have reported the frequencies of these mutations and they seem to be low in comparison to the N34S mutation. Most studies have therefore apparently restricted their analysis to this most frequent N34S mutation. N34S is present at a low percentage of 0.4 to 2.5% in the normal healthy population, but appears to be accumulated in selected groups of CP patients. Due to inconsistent selection criteria, different groups reported N34S mutations in 6%, 19%, 26%, or even 86% of alcoholic, hereditary, or familial idiopathic pancreatitis patient groups.[20–24] Considerable differences in these study results may be related not only to the absence of a generally accepted terminology for "familial" or "hereditary" and "idiopathic" pancreatitis, but could also be explained by the fact that determination of frequencies in some cases may involve several family members, whereas other studies counted unrelated patients only. Independent of different reports about the strength of this association with CP, the prevalence of N34S mutations appears to be increased in pancreatitis, but does not follow a clear-cut recessive or complex inheritance trait. In HP associated with mutations in the cationic trypsinogen gene, studies have demonstrated that an additional presence of SPINK-1 mutations does not affect the penetrance or the disease severity or the onset of a secondary diabetes mellitus.[25] Although this does not rule out that SPINK-1 is a weak risk factor for the onset of pancreatitis in general, it makes a role in onset of HP associated with strong PRSS1 mutations unlikely. Understanding of the clinical impact of PSTI mutations will require an extensive screening of a large number of patients according to well-defined selection criteria for idiopathic, familial, and hereditary pancreatitis.

In studies that analyzed the association of PSTI with tropical pancreatitis, an endemic variety of pancreatitis in Africa and Asia, several groups reported a strong association of N34S in populations in India and Bangladesh. Tropical pancreatitis is a type of idiopathic CP of so far unknown etiology that can be categorized by its clinical manifestations into either tropical calcific pancreatitis (TCP) or fibrocalculous pancreatic diabetes (FCPD). Although frequencies of the N34S mutation in the normal control population are comparable to previous reports from Europe and North America (1.3%), the mutation was found in 55% and 29% of FCDP-patients and in 20% and 36% of TCP-patients in Bangladesh and South India, respectively.[26,27]

Mutations in the PSTI gene may define a genetic predisposition for pancreatitis and apparently lowers the threshold for pancreatitis caused by other factors. A biochemical analysis of the protease inhibiting activity of PSTI by Kuwata et al., however, reported unchanged trypsin-inhibiting function of N34S-PSTI under both alkaline and acidic conditions.[28] At pH values between 5 and 9, recombinant N34S protein had the same inhibitory activity for trypsin as the wild type PSTI and also a variation of calcium concentrations revealed no differences of N34S function. The pathophysiology of N34S mutations, therefore, may follow mechanisms other than a decreased protease inhibitory activity due to a conformational change. Instead, the predisposition to pancreatitis in N34S patients may be caused by differences in PSTI expression levels possibly due to splicing defects. An analysis of PSTI protein expression levels in N34S patients will have to clarify this issue.

24.8 CYSTIC FIBROSIS TRANSMEMBRANE CONDUCTANCE REGULATOR

The role of cystic fibrosis transmembrane conductance regulator (CFTR) in pancreatitis is discussed in Chapter 26.

24.9 POLYMORPHISM IN GENES OF THE INFLAMMATORY RESPONSE SYSTEM

Proinflammatory and regulatory cytokines, such as tumor necrosis factor-α (TNF-α) and interleukin 1β (IL-1β), play an important role in the initial stages of the disease and in the development of severe acute pancreatitis. In addition to mutations that operate in concert with environmental risk factors in affecting disease susceptibility, some genetic polymorphisms have been shown to influence the severity of diseases by their control of the inflammatory response. Even though polymorphic alleles sometimes have a high frequency in the population and may be considered physiologically normal, they may have a moderate effect on the function of a gene product. Their disease-relevant mechanism may involve the transcription efficiency, the stability of mRNA or protein products or impaired protein–protein interactions that lead to a modulation in biological function. Powell et al.[29] performed a study in which they investigated cytokine genotypes in 190 pancreatitis patients and found no differences in polymorphism frequencies of TNF-α, IL-1α, and IL-1α receptor antagonist gene loci between pancreatitis patients and control individuals. In a similar report, Sargen et al.[30] conclude that TNF and IL-10 play no role in the determination of disease severity or susceptibility to acute pancreatitis. The same group, however, found that Allele 1 of the IL-1RN polymorphism was significantly increased in patients with acute pancreatitis and appears to determine the disease phenotype.[31] Also, polymorphisms in a-1-antitrypsin and a-2-macroglobulin have been suggested to play a moderating, but not dominant role in the course of alcoholic pancreatitis.[32] Also, sequence variations in the TNF-α promoter region have been analyzed and the variant TNF-238A has been reported to be a relevant risk factor for disease manifestation in families with HP.[33] To date, the hunt for new polymorphisms and genes is ongoing, but the task of sorting out the important candidates that either strongly influence susceptibility or open up new routes for interventional therapy is far from over.

24.10 GENETIC PREDISPOSITION IN ALCOHOLIC PANCREATITIS

There is good evidence that alcohol remains one of the most important toxic risk factors that are associated with CP (see Chapter 6 and Chapter 15). The risk for CP clearly has been shown to correlate with alcohol consumption in different parts of the world. In industrialized countries, long-term alcohol abuse accounts for approximately 70% of CP; whereas in 25%, the cause remains generally unknown. Although acute pancreatitis is in approximately 50% of cases caused by events unrelated to alcohol, alcohol-related pancreatitis presents in most cases as a chronic disease state, even though the clinical course is not uniform and may vary. Recurrent episodes of acute inflammatory conditions may lead over time to chronic inflammation and fibrosis. Other observations suggest that CP may also arise independent of acute disease recurrences. Interestingly, the correlation between alcohol abuse and CP is not strict, as it appears that less than 5% of alcoholics develop pancreatitis as a consequence of excessive ethanol consumption. Why the pancreas of one individual is more susceptible to alcohol than that of others and why the development of alcoholic pancreatitis appears to follow such different routes in individual alcoholics has prompted many investigators to study genetic predisposition. Therefore, the discovery of specific gene mutations in HP was an opportunity to search for a potential role of these mutations on the association between alcohol intake and CP.

A number of studies screened for HP-associated cationic trypsinogen mutations in alcoholic pancreatitis patients,[34–36] but the results clearly indicate that neither R122 nor N29I mutations — most frequently observed in HP — represent a major risk factor in alcoholic pancreatitis. The frequency of the SPINK N34S mutation in alcoholics was also analyzed by several groups and found to be slightly elevated compared to the control population.[22,23]

This low, but statistically significant frequency also rules out a major role of N34S mutations, but suggests some possible relevance for the intrapancreatic protease inhibitor function of SPINK-1 in certain individuals with alcoholic pancreatitis. So far, however, detailed reports about the clinical characteristics of patients with SPINK-1 N34S mutations and alcoholic pancreatitis are few and it remains unclear whether SPINK-1 mutations have an impact on the clinical course of alcoholic CP.

Further studies seeking to analyze mutations in some other genes such as the CFTR gene or the gene for alcohol dehydrogenase have been performed,[36–39] but revealed no conclusive evidence for the existence of an association with alcoholic CP. Even though the SPINK-1 N34S mutation may represent a minor genetic risk factor for the development of alcoholic pancreatitis, it does not appear to be the central or initiating cause.

24.11   EPILOGUE

A number of genetic mutations in a wide range of genes that relate to pancreatic function and to the regulation of inflammation appear to be important in the development of pancreatitis. These mutations are likely to determine each individual's susceptibility for the development of pancreatitis, the severity of the disease, and the disease progression or tissue regeneration processes. Although some mutations that are associated with a high disease penetrance lead to early disease onset and severe disease phenotypes, others may function in combination with additional gene defects or environmental cofactors such as toxins and lifestyle dependent risk factors like alcohol and tobacco. The identification of such additional environmental and genetic cofactors and the elucidation of their sometimes-complex interaction will be a target for future studies and help our understanding of the pathophysiology of pancreatitis — regardless of its clinical variety.

REFERENCES

1. Lerch M.M., Saluja A.K., Dawra R., Ramarao P., Saluja M., and Steer M.L. 1992. Acute necrotizing pancreatitis in the opossum: earliest morphological changes involve acinar cells. *Gastroenterology* 103:205–213.
2. Ockenga J., Vogel A., Teich N., Keim V., Manns M.P., and Strassburg C.P. 2003. UDP glucuronosyltransferase (UGT1A7) gene polymorphisms increase the risk of chronic pancreatitis and pancreatic cancer. *Gastroenterology* 124:1802–1808.
3. Hata A., Ridinger D.N., Sutherland S., Emi M., Shuhua Z., Myers R.L., Ren K., Cheng T., Inoue I., Wilson D.E. et al. 1993. Binding of lipoprotein lipase to heparin. Identification of five critical residues in two distinct segments of the amino-terminal domain. *J. Biol. Chem.* 268:8447–8457.
4. Brunzell J.D. and Schrott H.G. 1973. The interaction of familial and secondary causes of hypertriglyceridemia: role in pancreatitis. *Trans. Assoc. Am. Physicians* 86:245–254.
5. Siafakas C.G., Brown M.R., and Miller T.L. 1999. Neonatal pancreatitis associated with familial lipoprotein lipase deficiency. *J. Pediatr. Gastroenterol. Nutr.* 29:95–98.
6. Breckenridge W.C., Little J.A., Steiner G., Chow A., and Poapst M. 1978. Hypertriglyceridemia associated with deficiency of apolipoprotein C-II. *N. Engl. J. Med.* 298:1265–1273.
7. Cox D.W., Breckenridge W.C., and Little J.A. 1978. Inheritance of apolipoprotein C-II deficiency with hypertriglyceridemia and pancreatitis. *N. Engl. J. Med.* 299:1421–1424.

8. Levy R.I. and Fredrickson D.S. 1972. Familial hyperlipoproteinemia. In *The metabolic basis of inherited disease*. Stanbury J.B., Wyngaarden J.B., and Fredrickson D.S., Eds. 3rd ed. New York: McGraw-Hill. p. 545.

9. Comfort M.W. and Steinberg A.G. 1952. Pedigree of a family with hereditary chronic relapsing pancreatitis. *Gastroenterology* 21:54–63.

10. Whitcomb D.C., Gorry M.C., Preston R.A., Furey W., Sossenheimer M.J., Ulrich C.D., Martin S.P., Gates Jr. L.K., Amann S.T., Toskes P.P. et al. 1996. Hereditary pancreatitis is caused by a mutation in the cationic trypsinogen gene. *Nat. Genet.* 14:141–145.

11. Gaboriaud C., Serre L., Guy-Crotte O., Forest E., and Fontecilla-Camps J.C. 1996. Crystal structure of human trypsin 1: unexpected phosphorylation of Tyr151. *J. Mol. Biol.* 259:995–1010.

12. Sahin-Toth M. 2000. Human cationic trypsinogen. Role of Asn-21 in zymogen activation and implications in hereditary pancreatitis. *J. Biol. Chem.* 275:22750–22755.

13. Le Marechal C., Chen J.M., Quere I., Raguenes O., Ferec C., and Auroux J. 2001. Discrimination of three mutational events that result in a disruption of the R122 primary autolysis site of the human cationic trypsinogen (PRSS1) by denaturing high performance liquid chromatography. *BMC Genet.* 2:19.

14. Simon P., Weiss F.U., Sahin-Toth M., Parry M., Nayler O., Lenfers B., Schnekenburger J., Mayerle J., Domschke W., and Lerch M.M. 2002. Hereditary pancreatitis caused by a novel PRSS1 mutation (Arg-122 --> Cys) that alters autoactivation and autodegradation of cationic trypsinogen. *J. Biol. Chem.* 277:5404–5410.

15. Pfutzer R., Myers E., Applebaum-Shapiro S., Finch R., Ellis I., Neoptolemos J., Kant J.A., and Whitcomb D.C. 2002. Novel cationic trypsinogen (PRSS1) N29T and R122C mutations cause autosomal dominant hereditary pancreatitis. *Gut* 50:271–272.

16. Simon P., Weiss F.U., Zimmer K.P., Rand S., Brinkmann B., Domschke W., and Lerch M.M. 2002. Spontaneous and sporadic trypsinogen mutations in idiopathic pancreatitis. *JAMA* 288:2122.

17. Sahin-Toth M. 1999. Hereditary pancreatitis-associated mutation asn(21) --> ile stabilizes rat trypsinogen in vitro. *J. Biol. Chem.* 274:29699–29704.

18. Keim V., Bauer N., Teich N., Simon P., Lerch M.M., and Mossner J. 2201. Clinical characterization of patients with hereditary pancreatitis and mutations in the cationic trypsinogen. *Am. J. Med.* 111:622–626.

19. Chen J.M., and Ferec C. 2000. Gene conversion-like missense mutations in the human cationic trypsinogen gene and insights into the molecular evolution of the human trypsinogen family. *Mol. Genet. Metab.* 71:463–469.

20. Witt H., Luck W., Hennies H.C., Classen M., Kage A., Lass U., Landt O., and Becker M. 2000. Mutations in the gene encoding the serine protease inhibitor, Kazal Type 1 are associated with chronic pancreatitis. *Nat. Genet.* 25:213–216.

21. Pfutzer R.H., Barmada M.M., Brunskill A.P., Finch R., Hart P.S., Neoptolemos J., Furey W.F., and Whitcomb D.C. 2000. SPINK1/PSTI polymorphisms act as disease modifiers in familial and idiopathic chronic pancreatitis. *Gastroenterology* 119:615–623.

22. Threadgold J., Greenhalf W., Ellis I., Howes N., Lerch M.M., Simon P., Jansen J., Charnley R., Laugier R., Frulloni L. et al. 2002. The N34S mutation of SPINK1 (PSTI) is associated with a familial pattern of idiopathic chronic pancreatitis but does not cause the disease. *Gut* 50:675–681.

23. Drenth J.P., te Morsche R., and Jansen J.B. 2002. Mutations in serine protease inhibitor Kazal Type 1 are strongly associated with chronic pancreatitis. *Gut* 50:687–692.
24. Truninger K., Witt H., Kock J., Kage A., Seifert B., Ammann R.W., Blum H.E., and Becker M. 2002. Mutations of the serine protease inhibitor, Kazal Type 1 gene, in patients with idiopathic chronic pancreatitis. *Am. J. Gastroenterol.* 97:1133–1137.
25. Weiss F.U., Simon P., Witt H., Mayerle J., Hlouschek V., Zimmer K.P., Schnekenburger J., Domschke W., Neoptolemos J.P., and Lerch M.M. 2003. SPINK1 mutations and phenotypic expression in patients with pancreatitis associated with trypsinogen mutations. *J. Med. Genet.* 40:e40.
26. Schneider A., Suman A., Rossi L., Barmada M.M., Beglinger C., Parvin S., Sattar S., Ali L., Khan A.K., Gyr N. et al. 2002. SPINK1/PSTI mutations are associated with tropical pancreatitis and type II diabetes mellitus in Bangladesh. *Gastroenterology* 123:1026–1030.
27. Chandak G.R., Idris M.M., Reddy D.N., Bhaskar S., Sriram P.V., and Singh L. 2002. Mutations in the pancreatic secretory trypsin inhibitor gene (PSTI/SPINK1) rather than the cationic trypsinogen gene (PRSS1) are significantly associated with tropical calcific pancreatitis. *J. Med. Genet.* 39:347–351.
28. Kuwata K., Hirota M., Shimizu H., Nakae M., Nishihara S., Takimoto A., Mitsushima K., Kikuchi N., Endo K., Inoue M. et al. 2002. Functional analysis of recombinant pancreatic secretory trypsin inhibitor protein with amino-acid substitution. *J. Gastroenterol.* 37:928–934.
29. Powell J.J., Fearon K.C., Siriwardena A.K., and Ross J.A. 2001. Evidence against a role for polymorphisms at tumor necrosis factor, interleukin-1 and interleukin-1 receptor antagonist gene loci in the regulation of disease severity in acute pancreatitis. *Surgery* 129:633–640.
30. Sargen K., Demaine A.G., and Kingsnorth A.N. 2000. Cytokine gene polymorphisms in acute pancreatitis. *JOP* 1:24–35.
31. Smithies A.M., Sargen K., Demaine A.G., and Kingsnorth A.N. 2000. Investigation of the Interleukin 1 gene cluster and its association with acute pancreatitis. *Pancreas* 20:234–240.
32. Teich N., Walther K., Bodeker H., Mossner J., and Keim V. 2002. Relevance of variants in serum antiproteinases for the course of chronic pancreatitis. *Scand. J. Gastroenterol.* 37:360–365.
33. Beranek H., Teich N., Witt H., Schulz H.U., Mossner J., and Keim V. 2003. Analysis of tumour necrosis factor alpha and Interleukin 10 promotor variants in patients with chronic pancreatitis. *Eur. J. Gastroenterol. Hepatol.* 15:1223–1227.
34. Creighton J., Lyall R., Wilson D.I., Curtis A., and Charnley R. 1999. Mutations of the cationic trypsinogen gene in patients with chronic pancreatitis. *Lancet* 354:42–43.
35. Teich N., Mossner J., and Keim V. 1999. Screening for mutations of the cationic trypsinogen gene: are they of relevance in chronic alcoholic pancreatitis? *Gut* 44:413–416.
36. Monaghan K.G., Jackson C.E., KuKuruga D.L., and Feldman G.L. 2000. Mutation analysis of the cystic fibrosis and cationic trypsinogen genes in patients with alcohol-related pancreatitis. *Am. J. Med. Genet.* 94:120–124.

37. Norton I.D., Apte M.V., Dixson H., Trent R.J., Haber P.S., Pirola R.C., and Wilson J.S. 1998. Cystic fibrosis genotypes and alcoholic pancreatitis. *J. Gastroenterol. Hepatol.* 13:496–499.

38. Haber P.S., Norris M.D., Apte M.V., Rodgers S.C., Norton I.D., Pirola R.C., Roberts-Thomson I.C., and Wilson J.S. 1999. Alcoholic pancreatitis and polymorphisms of the variable length polythymidine tract in the cystic fibrosis gene. *Alcohol Clin. Exp. Res.* 23:509–512.

39. Haber P.S., Apte M.V., Applegate T.L., Norton I.D., Korsten M.A., Pirola R.C., and Wilson J.S. 1998. Metabolism of ethanol by rat pancreatic acinar cells. *J. Lab. Clin. Med.* 132:294–302.

25

GENETIC DISORDERS
ASSOCIATED WITH EXOCRINE
AND ENDOCRINE
PANCREATIC TUMORS

*Henry T. Lynch, Carolyn A. Deters,
Jane F. Lynch, and Randall Brand*

CONTENTS

25.1 INTRODUCTION

The projected incidence of pancreatic cancer (PC) in the United States is 32,180 for 2005, with an expected mortality of 31,800.[1] A recent review of the epidemiology of PC by Ghadirian et al.[2] indicates that approximately 170,000 new cases of PC (or around 2.1% of all cancers) occur worldwide every year.[3] The lifetime risk of PC in developed countries is approximately 1%.[4] PC is the fifth most common cause of cancer deaths among both males and females in Western countries, wherein 80 to 90% of cases are diagnosed in the nonresectable stage. As a result, the survival rate is extremely low, and the case–fatality ratio for the disease is approximately 0.9.[5,6]

It is estimated that between 5 to 10% of PC shows familial clustering. Therefore, using the above incidence figures, such familial clustering of PC in the United States would account for between 1609 and 3218 cases during 2005, which constitutes a formidable public health problem. Indeed, when considering PC's genotypic and phenotypic heterogeneity, these estimates of familial PC may be highly conservative.

25.2 EPIDEMIOLOGY

The incidence of PC worldwide is correlated with increasing age.[7] PC occurs rarely in patients younger than 25 years of age, and it is relatively uncommon for individuals younger than 45 years.[8] Although PC is more common in Western industrialized areas of the world, no specific industrial risk factor has been identified in its etiology (see Chapter 20).[9] PC shows wide variation in incidence throughout the world; its highest incidence occurs in men among New Zealand Maoris, native Hawaiians, and black Americans; conversely, the lowest reported incidence is in inhabitants of India and Nigeria.[10] Religious differences in PC incidence are also of interest. For example, individuals of Jewish extraction are at higher risk for PC than other religious groups and, in contrast, Seventh Day Adventists have a low risk.[11,12] PC occurs more often among single as opposed to married individuals, irrespective of sex or race.[13] There are urban–rural differences in PC's incidence, wherein its incidence is higher in urban populations.[11] Migration of individuals from one country to another may also affect incidence. For example, Japanese immigrants to the United

States have higher rates of PC than Japanese in Japan. Surprisingly, incidence rates are even higher in this group than among American whites.[14] There may also be socioeconomic factors affecting PC in that it appears to be more common among lower socioeconomic classes[15]; whereas in certain groups, higher socioeconomic status appears to play a positive role in PC's etiology.[16] These findings suggest a possible link between lifestyle, food habits, and other possible related factors in PC's etiology (see Chapter 20).

Bartsch et al.[17] reviewed familial PC and noted that the first systematic study of a cohort of familial PC families was published in 1989.[18] Lynch et al., in 1990,[19] and in 1996,[20] provided additional updated reviews of the subject in the search for host factors that may influence susceptibility to PC; attention was focused upon the manner in which genetic factors might interact with environmental exposures in PC's etiology.

25.3 GENETIC-ENVIRONMENTAL INTERACTION AND PC

The strongest epidemiologic risk factor for PC is cigarette smoking. It is suggested that a host-environmental phenomenon exists wherein cigarette smoking appears to interact with genetic susceptibility to PC. These findings were based upon a nested case-control study of 251 members of 28 PC-prone families.[21] Each family comprised 2 or more members with PC. The effects of smoking, young age of onset within the family, diabetes mellitus, sex, and number/standing of affected relatives with respect to the risk of PC were evaluated. Findings revealed that smoking posed an independent risk factor for familial PC, wherein the odds ratio (OR) was 3.7 and the 95% confidence interval (CI) was 1.8 to 7.6. The risk was greatest in males and in family members younger than age 50 (OR 5.2 and OR 7.6, respectively). Furthermore, "… Smokers developed cancer one decade earlier than nonsmokers (59.6 vs. 69.1 years; $P = 0.01$), and the number of affected first-degree relatives also increased risk (OR, 1.4; 95% CI 1.1–1.9 for each additional family member). Diabetes was not a risk factor for pancreatic cancer, although diabetes was associated with pancreatic dysplasia."[21] These authors concluded that cigarette smoking posed a significant risk factor in familial PC kindreds, which was particularly evident in males as well as those younger than age 50. Those family members who had multiple primary relatives with PC were also at increased risk.

25.4 CASE-CONTROL STUDIES OF FAMILIAL RISK FACTORS

Ghadirian et al.[22] in a population-based case-control study, found that 7.8% of the PC patients vs. only 0.6% of the controls reported a positive family history of PC, which represented a 13-fold difference. Interestingly, their study of environmental factors in a subset of the familial PC patients and controls revealed no difference between these groups.

Falk et al.[23] evaluated the family history of PC as part of an epidemiological study in Louisiana and found an elevated PC risk among persons reporting any cancer in a close relative (OR = 1.86; 95% CI = 1.42–2.44). The highest risk was seen in those subjects with a history of PC in a close relative (OR = 5.25; 95% CI = 2.08–13.21).

Lynch et al.[19] described 18 PC-prone nuclear families and observed that the sex ratio, age of onset, histologic type, and survival of PC affected were comparable to published data on patients from the general population. No pattern of extrapancreatic cancer association was found in this study. The data were primarily descriptive and they lacked a population-based case-control design. However, the findings did indicate the need to further investigate the role of familial factors in PC's etiology. It was reasoned that additional knowledge about familial PC, a disease whose prognosis is almost uniformly dismal, could, through identification of patients at high risk for PC, lead to improved cancer control measures. The phenotypic and genotypic heterogeneity of PC is extensive.

Our purpose in this review is to describe the cardinal features of familial and hereditary forms of exocrine and endocrine PC, including its genotypic and phenotypic variability; special attention will be given to pedigrees portraying germ line PC predisposing mutations (when known). The potential impact of host susceptibility factors will be discussed.

25.5 PANCREATIC CANCER — AN IMPORTANT MEDICAL AND GENETIC MODEL

Because PC has a late age of onset and an almost uniformly swift, fatal outcome, these factors may present obstacles to the acquisition of pertinent clinical, pathological, genetic, and DNA data. In addition, family members may be extremely apprehensive about their own PC risk, which collectively may contribute to their reticence to discuss the matter on a research basis. Consequently, they often develop a fatalistic outlook with remarks such as, "Members of our family who had pancreatic cancer all died in spite of surgery, chemotherapy, and radiation. So why should we cooperate with you, when trying to solve this problem is futile?"

Therefore, the families reported in this chapter are precious, and the results of family research studies are, potentially, highly informative. In certain circumstances, we fervently believe that the study of such families will motivate high-risk family members and their physicians to demand the development of better screening measures, given the fact that at this time the best possible assurance of potential cure is through surgery before metastasis takes place.

25.6 MOLECULAR GENETICS

The study of oncogenes such as K-*ras* and tumor suppressor genes such as *CDKN2A* (*p16*) may, once their chemistry and function are further unraveled, be used advantageously for diagnosis, chemoprevention strategies, as well as for the development of new designer drugs that may enable more targeted chemotherapy. Pertinent questions are "What occurs when K-*ras* is activated?" and "What is lost when *CDKN2A* is inactivated?"

To date, there is no drug to treat PC that in any way is comparable, for example, to Imatinib (Gleevec®, Novartis, Basel, Switzerland) in treating gastrointestinal stromal tumors (GIST).[24] GIST, like its PC counterpart, has a grave prognosis and occurs in certain families in accord with its autosomal dominant mode of inheritance,[25] in which the germ line mutation in the *KIT* gene predisposes to GIST.[26–28]

Table 25.1 describes the extant heterogeneity of PC-prone syndromes and, when known, those germ line mutations that contribute to PC. Selected syndromes that integrally include PC are given further attention to serve as medical genetic examples.

25.7 FAMILIAL AND HEREDITARY PANCREATIC CANCER — CLASSIFICATION

Classification of familial and hereditary varieties of PC embody certain limitations wherein the diagnostic *sine qua non* is the presence of a cancer-causing germ line mutation, such as the cationic trypsinogen gene[29,30] in hereditary pancreatitis.[31,32] When PC segregates in a pattern consistent with an autosomal dominant mode of inheritance transmission, in an extended pedigree, it may become extremely informative to the diagnostician. PC may be associated with a variety of hereditary cancer-prone syndromes, as in Peutz-Jeghers syndrome (PJS) wherein the germ line mutation is *STK11/LKB1* (Table 25.2) and in the familial atypical multiple mole melanoma (FAMMM) syndrome with the *CDKN2A* germ line mutation. Both FAMMM[33] and PJS[34] also harbor striking phenotypic cutaneous signs that, when coupled with known facts about their germ line mutations of note, will significantly aid in diagnosis and management.

25.8 PJS AND PC

Table 25.2 focuses on PJS and its association with PC. PJS is a rarely occurring disorder; yet the review of Yee et al.[34] indicates that there is an extensive knowledge base on PJS relevant to its multifaceted clinical and pathological features of PC and, when indicated, its association with other tumors. Review of Table 25.2 shows the extensive knowledge base accrued on PC's association with PJS.

Table 25.1 Genetic Syndromes with Inherited Predisposition to Pancreatic Cancer

Familial Syndrome[a]	Gene (Locus)	Relative Risk	Frequency[b]	References
Peutz-Jeghers syndrome	STK11 / LKB1 (19p13.3)	132	4%	97–102
Familial pancreatitis	Cationic trypsinogen (7q35)	50–60	Unknown	30,31,103
Familial pancreatic cancer syndrome	Unknown (4q32-34)	18–57	Unknown	104,105
FAMMM syndrome	p16^{INK4a}/MTS1 (9p21)	13–22	98%	46,106–108
Hereditary breast–ovarian cancer syndrome	BRCA2 (13q12)	10	7%	56,109,110
Familial adenomatous polyposis	APC (5q21)	5	40%	20,111–115
HNPCC[c]	hMSH2 (2p22-21) hMLH1 (3p21.3)	Unknown	4–11%	116–118
Ataxia telangiectasia	ATM (11q22-23)	Unknown	Unknown	41,119

[a] All of the above syndromes are inherited in an autosomal dominant fashion, except ataxia telangiectasia, which is an autosomal recessive disorder.

[b] Frequency refers to the frequency of genetic mutation found in sporadically occurring pancreatic carcinomas.

[c] HNPCC — hereditary nonpolyposis colorectal cancer, also known as cancer family syndrome or Lynch Syndrome II. HNPCC is typically caused by germ line mutations in DNA mismatch repair genes including hMSH2, hMLH1, hPMS2, and hMSH6. hMSH2 and hMLH1 mutations account for one-half to two-thirds of HNPCC. The frequency refers to a defect in any one of the five DNA mismatch repair genes causing HNPCC.

Source: Reproduced by permission from Yee et al. 2003. *Cancer Biol. Ther.* 2(1):38–47.

25.9 FAMMM SYNDROME AND PC

Historically, the first detailed report of a family with cutaneous features consistent with the FAMMM was by Norris in 1820.[35] Lynch et al.[36] and Clark et al.[37] provided the first detailed descriptions of the clinical and genetic (autosomal dominant) features of the FAMMM/BK mole syndrome.

Table 25.2   Clinical and Pathologic Features of PJS Patients with Pancreatic Tumors

Age[a]	Gender	Latency[b]	Clinical Presentation and Initial Detection of Pancreatic Tumors	Histopathology of Pancreatic Tumors	Associated Tumors	Refs.
71	F	64	Hematochezia, iron-deficiency anemia, weight loss; incidental CT[c] finding in pancreatic head	Adenocarcinoma, intraductal papillary neoplasm	Colonic adenocarcinoma	34
38	M	0	Nausea, vomiting, and abdominal distention due to enteric intussusception; incidental finding of retroperitoneal mass in laparotomy	Adenocarcinoma	None	120
15	M	1	Abdominal pain, palpable abdominal mass; US[d] finding of left retroperitoneal mass, autopsy finding of tumors in pancreatic body and tail with multiple metastases	Adenocarcinoma	None	121
47	M	25	Back pain; CT finding of mass in pancreatic head with dilatation of main pancreatic duct	Adenocarcinoma	None	122
60	M	0.2	Bowel obstruction, palpable abdominal mass; CT and US finding of mass in pancreatic body	Adenocarcinoma	None	123
60	M	5	Abdominal pain; US and CT finding of mass in pancreatic tail	Adenocarcinoma	None	124
40	M	1	NM[e]	Adenocarcinoma	None	125
50	F	44	NM	Adenocarcinoma	None	125
56	M	33	NM	Adenocarcinoma	None	125
60	M	3	NM	Adenocarcinoma	None	125

Table 25.2 Clinical and Pathologic Features of PJS Patients with Pancreatic Tumors (Continued)

Age[a]	Gender	Latency[b]	Clinical Presentation and Initial Detection of Pancreatic Tumors	Histopathology of Pancreatic Tumors	Associated Tumors	Refs.
31	M	0	Nausea, vomiting, and abdominal distention due to enteric intussusception; elevated serum amylase and lipase, US and CT finding of mass in pancreatic head	Mucinous cystadenocarcinoma	None	126
45	F	33	Abdominal pain due to enteric intussusception; incidental US finding of cyst in pancreatic body	Cystadenocarcinoma	None	127
50	F	NM	NM	IPMN[f], focal CIS[g] adenocarcinoma	None	102
35	M	NM	Asymptomatic; screening CT finding of cystic mass in pancreatic head	IPMN/dysplasia	None	123
48	M	32	Abdominal pain, weight loss, jaundice; postmortem finding of tumor in pancreatic duct proximal to ampulla of Vater	Papillary adenoma of pancreatic duct	Cholangiocarcinoma, tonsillar carcinoma	128
34	M	NM	NM	NM	None	129

a Age — age at diagnosis of pancreatic tumor.
b Latency — interval between diagnosis of PJS and diagnosis of pancreatic tumor in years.
c CT — computed tomography.
d US — ultrasonograph.
e NM — not mentioned in the original report.
f IPMN — intraductal papillary mucinous neoplasia of pancreas.
g CIS — carcinoma in situ.

Source: Table reproduced by permission from Yee et al. 2003. *Cancer Biol. Ther.* 2(1):38-47.

Members of FAMMM syndrome families are at risk of inheriting a predisposition to develop multiple atypical cutaneous nevi (>50), although not all patients with melanoma in these families display this phenotype.

In addition to an enormous lifetime proclivity to malignant melanoma, the FAMMM syndrome shows extant phenotypic and genotypic heterogeneity, as evidenced by its integral association with a variety of extramelanoma cancers, particularly PC.[33,38–43] A melanoma kindred apparently predisposed to PC was reported first in 1968 and a number of additional kindreds have been identified subsequently.[33,44]

Several studies of FAMMM kindreds have found an excess of nonmelanoma malignancies compared with the expected frequency of these malignancies in the general population. The risk of developing malignant disease in these kindreds appears to be increased 10-fold to 40-fold, and the cumulative risk of PC has been estimated at 17% by age 75 years.[45–50] Studies of families that show the FAMMM phenotype have demonstrated functional mutations in the *CDKN2A* gene on Chromosome 9p21, although other melanoma-prone families have mutations without clear functional significance. It is noteworthy that the increased risk of PC may be confined to kindreds with mutations that impair the function of the p16 protein.[46] In addition, FAMMM kindreds may be at increased risk of developing other carcinomas, including breast tumors, lung tumors, sarcoma, and digestive tract tumors. Thus, the pleiotropic genetic effects of *CDKN2A* mutations and the resulting neoplastic phenotypes have yet to be defined fully.

CDKN2A encodes p16, a low-molecular-weight protein that inhibits the cyclin D1-cyclin dependent kinase complex (CDK4). If it is not inhibited, the CDK4 complex, in turn, phosphorylates the retinoblastoma protein, allowing a cell to progress through the G1 phase of the cell cycle. Thus, p16 acts as a tumor suppressor protein, and mutations in *CDKN2A* can result in unregulated cell growth and neoplastic progression. Germ line *CDKN2A* mutations have been detected in up to 25% of melanoma-prone families worldwide.[49] Mutations in the *CDK4* gene on Chromosome 12q13 have been detected in only a handful of melanoma-prone families; the genetic defects in the other families are unknown.[51]

Rulyak et al.[52] performed segregation analysis on the offspring of parents affected with multiple cutaneous nevi, malignant melanoma, or pancreatic carcinoma (PC), from the Creighton study,[33] and found these to be significantly more likely to be affected themselves compared with the offspring of unaffected parents (48.9% vs. 16.7%; $p = 0.004$). This study provided additional evidence that multiple cutaneous nevi, melanoma, or PC known to be inherited as an autosomal-dominant trait in families with *CDKN2A* mutations and supports the thesis of an expanded tumor phenotype in such FAMMM-PC syndrome kindreds. This was evidenced in 8 families with the FAMMM-PC association in concert with a *CDKN2A* germ line mutation. Pedigrees of several of these families are shown in Figure 25.1. There was considerable diversity in cancer presentation within and among these families.[33] For example, malignant melanoma predominated in certain of the families, whereas PC predominated in others. PC occurred at earlier ages — 35, 45, 46, and 49 years — in some of the families and markedly later in others. There were 5 patients (III-2 in Family A; IV-5 in Family D; II-5 and III-8 in Family E; and III-6 in Family F) who manifested melanoma and PC as double primaries. One of them (III-6 in Family F) also had a third primary cancer, namely carcinoma of the breast. Another patient (III-2 in Family C) manifested sarcoma, esophageal carcinoma, and 2 primary melanomas, while his daughter (IV-3) had a sarcoma and was a harbinger of a *CDKN2A* mutation. It was concluded that these tumors, "... may collectively, in concert with *CDKN2A* mutations, constitute a 'new' putative hereditary carcinoma syndrome, referred to as FAMMM-PC. More clinical and molecular genetic research on additional families with pancreatic carcinoma in concert with the FAMMM will be required."[33]

To understand better the inheritance of neoplastic phenotypes in these families, we estimated segregation ratios for several neoplastic phenotypes using eight families that are at high risk for multiple nevi, melanomas, and PC and that had been determined in earlier studies to harbor *CDKN2A* mutations. We hypothesize that a predisposition to atypical nevi, melanoma, PC, and possibly other malignancies is inherited as an autosomal trait consistent with the presence of mutations in the tumor suppressor function of p16.

25.10 *BRCA2* MUTATIONS IN FAMILIAL PANCREATIC CANCER

Murphy et al.[53] investigated the role of germ line mutations in the etiology of PC in the search for mutations in four tumor suppressor candidate genes, namely *MAP2K4*, *MADH4*, *ACVR1B*, and *BRCA2*, through direct sequencing of constitutional DNA. The samples were selected from families containing three or more PC affecteds, wherein at least two were first-degree relatives. No mutations were identified in *MAP2K4*, *MADH4*, or *ACVR1B*, from which the authors conclude that, "... It is unlikely that germ-line mutations in these genes account for a significant number of inherited pancreatic tumors." Importantly, however, *BRCA2* gene sequencing disclosed five mutations (5 of 29, 17.2%) that were considered deleterious and one point mutation (M192T) that had not been previously reported. The authors concluded that, "... These findings confirm the increased risk of pancreatic cancer in individuals with *BRCA2* mutations and identified germ-line *BRCA2* mutations as the most common inherited genetic alteration yet identified in familial pancreatic cancer."

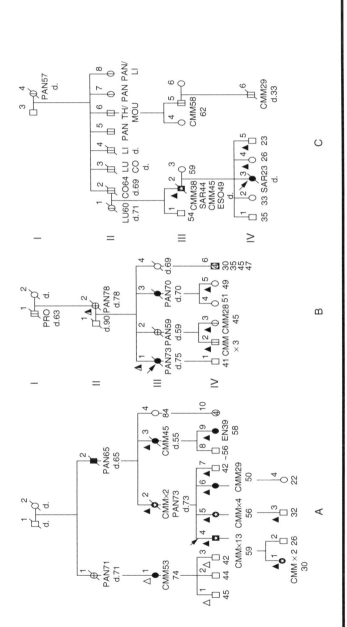

Figure 25.1 Pedigrees of several families with the FAMMM-PC association in concert with a *CDKN2A* germ line mutation, demonstrating the considerable diversity in cancer's phenotypic presentation.

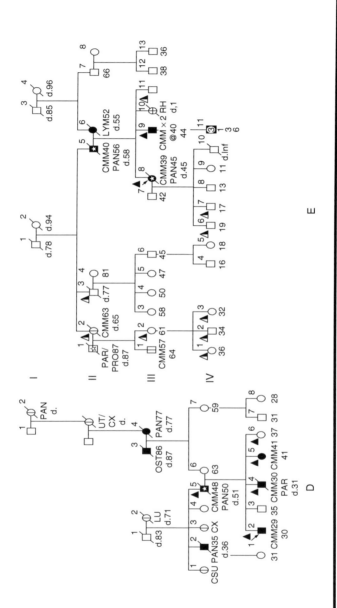

Figure 25.1 Continued.

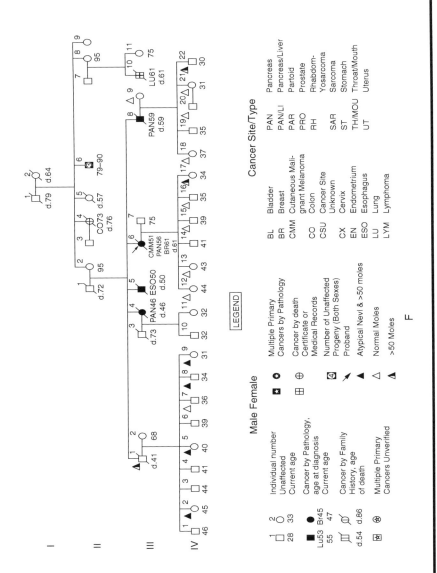

Figure 25.1 Continued.

White et al.[54] describe a remarkable German family showing a familial pattern of PC with a single case of breast cancer, wherein a protein-truncating *BRCA2* mutation was identified in two patients with PC, which strongly implicated *BRCA2* as playing a role in PC in this family. In this family, the proband manifested PC. Her sister developed ductal carcinoma of the breast at age 48 and four years later manifested a highly differentiated adenocarcinoma of the pancreas. A younger sister died at age 59 of PC, and a maternal uncle died at age 70 of PC. Furthermore, White et al. suggest that, in contrast to prior reports, the *BRCA2* mutation in this family appeared to be associated with a high penetrance for PC. The authors conclude that several lines of evidence implicate *BRCA2* mutations as predisposing to PC. Furthermore, they suggest that their findings complement prior observations that showed an increased relative risk for PC in *BRCA2*-positive families, as well as the previous association of germ line *BRCA2* mutations found in sporadic PC (cited by White et al.).[55–58] These authors also conclude that it is likely that additional modifying factors must be considered as predisposing to the specific risk for breast carcinoma vs. PC. Our own data have also shown PC to be associated with *BRCA2* mutation-prone families (H.T. Lynch, unpublished data, 2003).

25.11 PANCREATIC ACINAR CELL CARCINOMA

Rigaud et al.[59] studied pancreatic acinar cell (PAC) carcinoma, which is a rare pancreatic tumor that lacks any information pertinent to chromosomal and gene anomalies. As part of a genomewide allelotyping of nine PACs, these authors found that the allelotype of PAC is markedly different from that of either ductal or endocrine tumors of the pancreas. Furthermore, "… the involvement of chromosomes 4q and 16q appears to be characteristic of this tumor type. High-resolution mapping of the 12 frequently altered chromosomes in 5 cases with 222 markers permitted subchromosomal localization of regions of consensus loss on 5 chromosomes, including 1p36.31, 3p25.2, 4q26-31.1, 15q15-22.1, and 16q21-q22.1. … findings suggest that PAC tumorigenesis involves molecular pathways different from those occurring in more common pancreatic tumor types." This study emphasizes that progress in understanding the etiology and pathogenesis of PC may be achieved through meticulous attention to a selected subset of the PC story, in this case PAC.

25.12 PANCREATIC ENDOCRINE TUMORS

Pancreatic endocrine tumors make up about 2% of all pancreatic neoplasms. They occur sporadically or in context with the multiple endocrine neoplasia Type 1 (MEN 1) or in von Hippel-Lindau (VHL) disease. Many of these tumors are functional and secrete one or more biologically active peptides (hormones), which result in the development of clinical symptoms. At least five clinical types of endocrine PC tumors have been described — insulinoma, gastrinoma, vasoactive intestinal peptide-secreting tumors (VIPoma), glucagonoma, and somatostatinoma. There have been reports of other hormones produced by endocrine tumors including pancreatic polypeptide, ACTH, and growth hormone-releasing factor. Occasionally these tumors will secrete more than one peptide. A significant proportion of endocrine tumors is not associated with hormone overproduction (nonfunctional) and is diagnosed based on the symptoms.

25.13 CLINICAL FINDINGS

Many patients with pancreatic endocrine tumors present with signs and symptoms of excess hormone secretion. Clinical manifestations of the most common endocrine tumor, an insulinoma, are profound hypoglycemia with diaphoresis, confusion, and syncope. Classically, gastrinomas are the second most common tumor of islet-cell origin, and they present with peptic ulcer disease. Other common symptoms include diarrhea and esophagitis.

VIPomas are associated with a syndrome of watery diarrhea, hypokalemia, and achlorhydria (WDHA). Glucagonomas may present with diabetes, depression, deep-vein thrombosis, and dermatitis (necrolytic migratory erythema). Presenting symptoms of the rare somatostatinomas include cholelithiasis, diabetes, steatorrhea, diarrhea, weight loss, and abdominal pain.

Although the majority of these tumors present in a sporadic manner, it is important to identify those patients in whom the endocrine tumors occur in the setting of a hereditary syndrome, such as MEN 1 or VHL disease.

25.13.1 Multiple Endocrine Neoplasia Type 1 (MEN 1)

MEN 1 is characterized by neoplasia of the pancreatic islet cells, parathyroid glands, and anterior pituitary gland. In 1954, Wermer[60] first described this syndrome, with its autosomal dominant mode of inheritance, in a family. In 1988, Larsson et al.[61] were able to link the *MEN 1* gene to Chromosome 11q13. The *MEN 1* gene was subsequently identified by positional cloning about 10 years later.[62,63] This tumor suppressor gene consists of 10 exons, spans over 9 kb, and encodes for menin, a 610-amino acid protein, which appears to function as a nuclear factor interacting with the transcription factor JunD, an inhibitor of cell growth.[64]

The usual criteria for the clinical diagnosis of MEN 1 consists of an individual with involvement of at least two of the most commonly affected endocrine glands (parathyroid, pituitary, and pancreatic islet cells) along with a first-degree relative having at least one of the neoplastic lesions seen in MEN 1 or, within established kindreds, an individual with one of the major neoplastic lesions and concordance of an affected first-degree relative.[65,66] Multiple studies suggest a mean age at diagnosis ranging from 29 to 41,[66] and penetrance at age of 50 ranges from 82 to 100%.[66] These studies also demonstrate a range in the prevalence of islet cell tumors from 36 to 75%.

Gastrinomas are the most common functional tumor type followed by insulinomas.[66] Importantly, the malignant potential of MEN 1 gastrinomas ranges from 15 to 60%.[66] One report suggests that certain truncating mutations of the *MEN 1* gene have a greater potential for metastasis.[67] Insulinomas are commonly multiple and have a lower malignancy rate of 8% as compared to gastrinomas.[66,68] Although the other three endocrine tumor types have been seen with MEN 1, the third most common pancreatic lesion is a nonfunctional islet cell tumor.[66]

A comprehensive study by Carty et al.[69] of the penetrance and spectrum of manifestations of MEN 1 patients from western Pennsylvania and Ohio found that 13 of 34 (38%) patients presented initially with islet cell tumors. Noteworthy is that only 3 of these patients were diagnosed at the same time with primary hyperparathyroidism. The prevalence of MEN 1 islet cell tumors in these 34 patients was 71% and consisted of 16 gastrinomas, 3 insulinomas, 3 nonfunctional tumors, and 2 glucagonomas. Five of the 6 MEN 1 related deaths in this study occurred from metastatic islet cell tumors. The etiology and natural history of these deaths are consistent with results from a report from Doherty et al. who determined that pancreatic endocrine tumors were the most common cause of death in MEN 1 family members.[70] They found that 46% of deaths were at a median age of 47 years and were related to endocrine tumors.

These results suggest the importance of screening for pancreatic endocrine tumors. It is essential that a successful screening approach find small tumors. This is based on results of a study by Cadiot et al.[71] who found that 36% of patients with tumors greater than 3 cm had liver metastases as compared to 6% of patients with tumors that were less than 3 cm. One suggested screening method searches for hormonal evidence of endocrine tumors by measuring serum levels of gastrin, glucose, insulin, proinsulin, and pancreatic polypeptide.[72] Another screening approach could utilize imaging techniques such as computed tomography, magnetic resonance imaging, somatostatin receptor scintigraphy, and endoscopic ultrasound (EUS). This latter imaging modality has been shown to be the most sensitive technique available for the detection of small sporadic pancreatic endocrine tumors.[73] Indeed, a small retrospective study by Gauger et al.[72] found EUS to be useful for detecting endocrine tumors in asymptomatic patients with MEN 1.

25.13.2   Insulinomas

Bartsch et al.[74] note that the role molecular mechanisms play in contributing to the tumorigenesis of insulinomas remains elusive. They note, however, that inactivation of the $p16^{INK4a}$ tumor suppressor gene predisposes to gastrinomas and nonfunctioning endocrine pancreatic carcinomas. These authors studied the role of $p16^{INK4a}$ in the tumorigenesis of insulinomas. Findings showed that the $p16^{INK4a}$ tumor suppressor gene contributed to tumorigenesis in only a small set of insulinomas.

We believe that a reasonable pancreatic endocrine surveillance program should include both biochemical tests and imaging studies including computed tomography (CT) and EUS. CT scanning is recommended because of its ability to visualize other organs such as the liver and adrenal glands, whereas EUS is felt to be the best imaging modality for the pancreas.

25.13.3   VHL Disease

VHL disease is an autosomal dominant multisystem cancer syndrome characterized by the development of hemangioblastoma of the central nervous system, retinal angioma, renal cell carcinoma or cysts, pheochromocytoma, endolymphatic sac tumors, pancreatic cysts and neuroendocrine tumors. This highly penetrant syndrome results from a mutation in the VHL gene.[75,76] VHL is a tumor suppressor gene found on the short arm of Chromosome 3. The gene encodes for a 213-amino acid protein that is localized in the nucleus or cytoplasm. Loss of normal VHL protein function appears to effect ubiquitin-mediated degradation of large cellular proteins.

Pancreatic involvement in VHL consists of pancreatic cysts, serous cystadenomas, or neuroendocrine tumors.[75,77] Pancreatic neuroendocrine tumors arise in 12 to 17% of VHL patients.[77,78]

The association of endocrine tumors with VHL was first reported in 1979 in a family in which all four siblings developed lesions associated with VHL disease, and two of these siblings developed endocrine tumors.[79] Pancreatic neuroendocrine tumors appear to occur at a significantly younger age, namely a mean of 35 years[80] in VHL patients as compared to a mean age of 58 years for sporadic patients. Unlike MEN 1 patients, the pancreatic endocrine tumors are usually clinically nonfunctional and the majority did not appear to express pancreatic or gastrointestinal hormones by immunostaining.[80] A distinguishing histologic feature in the majority of these endocrine tumors in VHL patients was clear-cell morphology.[80]

Because renal cell carcinomas are the major malignancy in VHL, CT scan has been suggested as the imaging study of choice. One experienced group suggests starting screening for abdominal lesions at the age of 12 years using annual CT scanning.[81] Using this approach, they were able to identify 44 patients with a neuroendocrine tumor of the pancreas. Five of these patients were not candidates for resection based on the presence of metastatic disease. Twenty-five patients fulfilled their criteria for resection:

- No evidence of metastatic disease
- Lesion greater than 3 cm in the body or tail of the pancreas or greater than 2 cm in the head
- If the patient was undergoing a laparotomy for the resection of a kidney or adrenal lesion (another pathologic manifestation of VHL)

None of these 25 patients had developed metastatic disease with a median follow-up time of 32 months. Additionally, none of the patients with lesions or clinical scenarios that did not fulfill the above-mentioned criteria and were just followed with serial abdominal CT scans, developed metastatic disease.

25.14 GENETIC COUNSELING

There is a dire need for physicians to better comprehend the significance of hereditary PC and garner experience with genetic counseling,[82–84] particularly in these exceedingly complex PC-prone families. Unfortunately, there has been little experience with the implications of genetic counseling of hereditary PC patients. Therefore, we need to consider physician experience with genetic counseling in all hereditary cancer disorders. For example, Giardiello et al.[84] reported that physicians lacked an understanding of how mutations in the *APC* gene can confirm a diagnosis of familial adenomatous polyposis (FAP) or how a negative test result may be misinterpreted, thereby contributing to a patient's false sense of security.[85,86] Clearly, patients require highly accurate risk and management advice during genetic counseling. However, of at-risk patients who were tested for FAP predisposition, Giardiello et al.[84] found that only 18.6% were receiving such genetic counseling before being tested. The reluctance of physicians to order genetic testing for cancer susceptibility[87] can be overcome only by raising the awareness of the role that hereditary or genetic factors play in carcinogenesis.

25.15 EPILOGUE

Great strides have been accomplished in the comprehension of the genetic basis of human disease, particularly during the past decade. Therein, the most profound impact of this knowledge explosion has been in the area of cancer genetics.[88] Many high-penetrance cancer-causing germ line mutations have now been identified. However, in spite of these advances, we remain in the dark insofar as the contribution to hereditary cancer of low-penetrant genetic variants or polymorphisms are concerned; this knowledge deficit also applies to the risk of the more common sporadic cancer variants. It is, therefore, compelling that we continue to search for those putative germ line mutations that remain elusive, but which, nevertheless, confer increased susceptibility to cancer. Progress in this area will definitely be abetted through a better understanding of those complex somatic genetic events that take place in the emerging cancer cell.

Cowgill and Muscarella have recently reviewed the subject of hereditary exocrine and endocrine pancreatic tumors.[89] These authors appropriately stress the importance of investigating hereditary forms of pancreatic cancer to better characterize those molecular changes that play a role in tumor initiation and progression, so that cellular pathways in carcinogenesis, as well as for maintaining noncancerous states might be better elucidated. For example, in the case of endocrine pancreatic tumors, recent evidence has shown how MEN 1 and Rb-pathway alterations play a significant role as targets for intervention. These authors appropriately credit Zollinger, whose intense interest in the basic science of PCs contributed to the search for a better understanding of their pathophysiology through collection of gastrinoma specimens as part of the clinical database and tissue bank at Ohio State University.

25.15.1 Knowledge Gleaned from Study of Genetics of Breast and Colon Cancer and Their Influence on PC

These common malignancies are briefly discussed here because their study has many similarities to PC, with which clinical-genetic experience is more limited. The risk to first-degree relatives of patients affected with common malignances, such as carcinoma of the breast and colon, harbor an approximately two-fold increased risk for cancer of the same anatomic site.[90] Twin studies support much of this familial burden, particularly in the case of carcinoma of the breast.[91] However, when searching for an explanation for the familial risk association, mutations in breast cancer such as *BRCA1* and *BRCA2* cannot explain the bulk of this familial burden, because only about 20% of this two-fold excess in the relatives of affected patients will be attributable to *BRCA1* and *BRCA2*.[92]

In reviewing this entire matter, relevant to the familial burden of common cancers, Houlston and Peto[93] suggest that an important part of the remaining familial risk could be due to high penetrance mutations, "… in as yet unidentified genes, but a polygenic mechanism may provide a more plausible alternative explanation." For example, Houlston and Peto note that the investigation of striking multiple-case cancer-prone families has not been able to signify acceptable linkage to novel cancer susceptibility loci in recent investigations. However, employing the polygenic model, it is noteworthy that, "… a large number of alleles, each conferring a small genotypic risk (perhaps on the order of 1.5 to 2.0), combine additively or multiplicatively to confer a range of susceptibilities in the population. More than 100 such variants may contribute to susceptibility.[93–95] Individuals carrying few such alleles would be at reduced risk, and those with many might suffer a lifetime risk as high as 50%."[93–95]

25.15.2 Surveillance and Management of PC

It is important to develop methods that can detect pancreatic tumors at earlier stages, preferably at a severe dysplasia or carcinoma *in situ* stage (PanIN 3). These early stage lesions are currently undetectable with the most sensitive imaging modalities available. However, alterations in pancreatic tissue, such as focal lobular atrophy, inflammation, fibrosis, and ductal dilatations, which are large enough to be detected, are produced by precancerous pancreatic lesions and may allow for early detection of these lesions in some patients. This possibility was highlighted in a study of patients with a family history of pancreatic cancer.[96] In these patients, irregular ducts, poor filling of pancreatic ducts, narrowing or dilatation of ducts, and formation of cystic lesions were detected by endoscopic retrograde cholangiopancreatography (ERCP) and found to be indicative of precancerous dysplastic alterations. Furthermore, every patient in this study with an abnormal ERCP had an abnormal EUS that demonstrated echogenic foci, hypoechoic nodules, or an echogenic main duct. In general, both of these imaging findings are nonspecific and can be seen in chronic pancreatitis.

We counsel our PC-prone patients about the limitations for surveillance of PC. We suggest an EUS of the pancreas on an annual basis, starting at least 10 years before the earliest diagnosed PC case in the kindred. We prefer that the EUS be performed in a study situation because we perform a secretin-stimulated collection of duodenal contents (pancreatic juice) as part of a research protocol to utilize for the development of potentially new tumor markers. The pancreatic juice as well as a blood specimen is banked for future investigations of merit.

ACKNOWLEDGMENTS

We wish to acknowledge the kind assistance of Dr. Nelson S. Yee, University of Pennsylvania School of Medicine, in allowing us to use his tables in this chapter (Table 25. 1 and Table 25.2).

This chapter was supported by revenue from Nebraska cigarette taxes awarded to Creighton University by the Nebraska Department of Health and Human Services. Its contents are solely the responsibility of the authors and do not necessarily represent the official views of the State of Nebraska or the Nebraska Department of Health and Human Services.

Support was also received from NIH Grant #1U01 CA86389.

REFERENCES

1. Jemal A., Murray T., Ward E., et al., 2005. Cancer statistics, 2005. *CA Cancer J. Clin.* 55:10–30.
2. Ghadirian P., Liu G., Gallinger S., Schmocker B., Paradis A.-J., Lal G. et al. 2002. Risk of pancreatic cancer among individuals with a family history of cancer of the pancreas. *Intl. J. Cancer* 97:807–810.
3. Parkin D.M., Pisani P., and Ferlay J. 1999. Global cancer statistics. *CA Cancer J. Clin.* 49:33–64.
4. Chappuis P.O., Ghadirian P., and Foulkes W.D. 2001. The role of genetic factors in the etiology of pancreatic adenocarcinoma: an update. *Cancer Invest.* 19:65–75.
5. Rosewicz S. and Wiedenmann B. 1999. Pancreatic carcinoma. *Lancet* 15:786–792.
6. Yeo C.J. and Cameron J.L. 1994. Pancreatic cancer. *Curr. Probl. Surg.* 36:59–152.
7. Aoki K. and Ogawa H. 1978. Cancer of the pancreas, international mortality trends. *World Health Stat. Q.* 31:2–27.
8. Morgan R.G. and Wormsley K.G. 1977. Progress report, cancer of the pancreas. *Gut* 18:580–592.
9. Mack T.M., Peters J.M., and Yu M.C. et al. 1985. Pancreas cancer is unrelated to the workplace in Los Angeles. *Am. J. Indus. Med.* 7:253–266.
10. Boyle P., Hsieh C.-C., Maisonneuve P., La Vecchia C., Macfarlane G.J., Walker A.M. et al. 1989. Epidemiology of pancreas cancer (1988). *Intl. J. Pancreatol.* 5:327–346.
11. Mack T.M. 1982. Pancreas. In *Cancer epidemiology and prevention*. Schottenfeld D. and Fraumeni J.F. Eds. London: Saunders. pp. 638–677.
12. Phillips R.L., Garfinkel L., Kuzma J.W. et al. 1980. Mortality among California Seventh-Day Adventists and selected cancer sites. *J. Natl. Cancer Inst.* 65:1097–1108.
13. Ernster V.L., Sacks S.T., Selvin S. et al. 1979. Cancer incidence by marital status: U.S. Third National Cancer Survey. *J. Natl. Cancer Inst.* 63:567–585.
14. Haenszel W. and Kurihara M. 1968. Studies of Japanese migrants. Mortality from cancer and other diseases among Japanese in the United States. *J. Natl. Cancer Inst.* 40:43–68.
15. Dorn H.G. and Culter S.J. 1959. *Mortality from cancer in the United States, Part 1 and 2* (Public Health Monograph, No. 56). Washington, DC: Dept. HEW, U.S. Government Printing Office.
16. Williams R.R. and Horm J.W. 1977. Association of cancer sites with tobacco and alcohol consumption and socio-economic status of patients: Interview study from the Third National Cancer Survey. *J. Natl. Cancer Inst.* 58:525–547.
17. Bartsch D.K. 2003. Familial pancreatic cancer. *Br. J. Surg.* 90:386–387.
18. Lynch H.T., Lanspa S.J., Fitzgibbons Jr. R.J., Smyrk T., Fitzsimmons M.L., and McClellan J. 1989. Familial pancreatic cancer (Part 1): Genetic pathology review. *Nebr. Med. J.* 74:109–112.

19. Lynch H.T., Fitzsimmons M.L., Smyrk T.C., Lanspa S.J., Watson P., McClellan J. et al. 1990. Familial pancreatic cancer: clinicopathologic study of 18 nuclear families. *Am. J. Gastroenterol.* 85:54–60.

20. Lynch H.T., Smyrk T., Kern S.E., Hruban R.H., Lightdale C.J., Lemon S.J. et al. 1996. Familial pancreatic cancer: a review. *Semin. Oncol.* 23:251–275.

21. Rulyak S.J., Lowenfels A.B., Maisonneuve P., and Brentnall T.A. 2003. Risk factors for the development of pancreatic cancer in familial pancreatic cancer kindreds. *Gastroenterology* 124:1292–1299.

22. Ghadirian P., Boyle P., Simard A., Baillargeon J., Maisonneuve P., and Perret C. 1991. Reported family aggregation of pancreatic cancer within a population-based case-control study in the Francophone community in Montreal, Canada. *Intl. J. Pancreatol.* 10:183–196.

23. Falk R.T., Pickle L.W., Fontham E.T., Correa P., and Fraumeni Jr. J.F. 1988. Lifestyle risk factors for pancreatic cancer in Louisiana: a case-control study. *Am. J. Epidemiol.* 128:324–336.

24. Joensuu H., Roberts P.J., Sarlomo-Rikala M., Andersson L.C., Tervahartiala P., Tuveson D. et al. 2001. Effect of the tyrosine kinase inhibitor STI571 in a patient with a metastatic gastrointestinal stromal tumor. *N. Engl. J. Med.* 344:1052–1056.

25. Beghini A., Tibiletti M.G., Roversi G., Chiaravalli A.M., Serio G., Capella C. et al. 2001. Germline mutation in the juxtamembrane domain of the kit gene in a family with gastrointestinal stromal tumors and urticaria pigmentosa. *Cancer* 92:657–662.

26. Nishida T., Hirota S., Taniguchi M., Hashimoto K., Isozaki K., Nakamura H. et al. 1998. Familial gastrointestinal stromal tumours with germline mutation of the KIT gene. *Nat Genet.* 19:323–324.

27. Isozaki K., Terris B., Belghiti J., Schiffmann S., Hirota S., and Vanderwinden J.-M. 2000. Germline-activating mutation in the kinase domain of *KIT* gene in familial gastrointestinal stromal tumors. *Am. J. Pathol.* 157:1581–1585.

28. Hirota S., Nishida T., Isozaki K., Taniguchi M., Nishikawa K., Ohashi A. et al. 2002. Familial gastrointestinal stromal tumors associated with dysphagia and novel type germline mutation of KIT gene. *Gastroenterology* 122:1493–1499.

29. Gorry M.C., Gabbaizedeh D., Furey W., Gates Jr. L.K., Preston R.A., Aston C.E. et al. 1997. Mutations in the cationic trypsinogen gene are associated with recurrent acute and chronic pancreatitis. *Gastroenterology* 113:1063–1068.

30. Whitcomb D.C., Gorry M.C., Preston R.A., Furey W., Sossenheimer M.J., Ulrich C.D. et al. 1996. Hereditary pancreatitis is caused by a mutation in the cationic trypsinogen gene. *Nat Genet.* 14:141–145.

31. Lowenfels A.B., Maisonneuve P., Whitcomb D.C., and International Hereditary Pancreatitis Study Group. 2000. Risk factors for cancer in hereditary pancreatitis. *Med. Clin. North. Am.* 84:565–573.

32. Lowenfels A.B., Maisonneuve P., DiMagno E.P., Elitsur Y., Gates Jr. L.K., Perrault J. et al. 1997. Hereditary pancreatitis and the risk of pancreatic cancer. *J. Natl. Cancer Inst.* 89:442–446.

33. Lynch H.T., Brand R.E., Hogg D., Deters C.A., Fusaro R.M., Lynch J.F. et al. 2002. Phenotypic variation in eight extended *CDKN2A* germline mutation familial atypical multiple mole melanoma-pancreatic carcinoma-prone families: the familial atypical multiple mole melanoma-pancreatic carcinoma syndrome. *Cancer* 94:84–96.

34. Yee N.S., Furth E.E., and Pack M. 2003. Clinicopathologic and molecular features of pancreatic adenocarcinoma associated with Peutz-Jeghers syndrome. *Cancer Biol. Ther.* 2:38–47.

35. Norris W. 1820. Case of fungoid disease. *Edinburgh Med. Surg. J.* 16:562–565.

36. Lynch H.T., Frichot B.C., and Lynch J.F. 1978. Familial atypical multiple mole-melanoma syndrome. *J. Med. Genet.* 15:352–356.

37. Clark Jr. W.H., Reimer R.R., Greene M., Ainsworth A.M., and Mastrangelo M.J. 1978. Origin of familial malignant melanomas from heritable melanocytic lesions. "The B-K mole syndrome." *Arch. Dermatol.* 114:732–738.

38. Lynch H.T., Fusaro L., and Lynch J.F. 1992. Familial pancreatic cancer: a family study. *Pancreas* 7:511–515.

39. Lynch H.T., Fusaro L., Smyrk T.C., Watson P., Lanspa S.J., and Lynch J.F. 1995. Medical genetic study of eight pancreatic cancer-prone families. *Cancer Invest.* 13:141–149.

40. Lynch H.T. 1994. Genetics and pancreatic cancer. *Arch. Surg.* 129:266–268.

41. Lynch H.T., Brand R.E., Deters C.A., Shaw T.G., and Lynch J.F. 2001. Hereditary pancreatic cancer. *Pancreatology* 1:466–471.

42. Lynch H.T., Brand R.E., Deters C.A., and Fusaro R.M. 2001. Update on familial pancreatic cancer. *Curr. Gastroenterol. Rep.* 3:121–128.

43. Lynch H.T., Deters C.A., Lynch J.F., and Brand R.A. 2002. Challenging pancreatic cancer-prone pedigrees: a nosologic dilemma. *Am. J. Gastroenterol.* 97:3062–3070.

44. Lynch H.T. and Krush A.J. 1968. Heredity and malignant melanoma: implications for early cancer detection. *Can. Med. Assoc. J.* 99:17–21.

45. Lynch H.T., Fusaro R.M., Kimberling W.J., Lynch J.F., and Danes B.S. 1983. Familial atypical multiple mole-melanoma (FAMMM) syndrome: segregation analysis. *J. Med. Genet.* 20:342–344.

46. Goldstein A.M., Fraser M.C., Struewing J.P., Hussussian C.J., Ranade K., Zametkin D.P. et al. 1995. Increased risk of pancreatic cancer in melanoma-prone kindreds with p16[INK4] mutations. *N. Engl. J. Med.* 333:970–974.

47. Vasen H.F.A., Gruis N.A., Frants R.R., van der Velden P.A., Hille E.T.M., and Bergman W. 2000. Risk of developing pancreatic cancer in families with familial atypical multiple mole melanoma associated with a specific 19 deletion of *p16* (*p16-Leiden*). *Intl. J. Cancer* 87:809–811.

48. Borg Å., Sandberg T., Nilsson K., Johannsson O., Klinker M., Måsbäck A. et al. 2000. High frequency of multiple melanomas and breast and pancreas carcinomas in CDKN2A mutation-positive melanoma families. *J. Natl. Cancer Inst.* 92:1260–1266.

49. Goldstein A.M. and Tucker M.A. Screening for CDKN2A mutations in hereditary melanoma. 1997. *J. Natl. Cancer Inst.* 89:676–678.

50. Bartsch D.K., Sina-Frey M., Lang S., Wild A., Gerdes B., Barth P. et al. 2002. CDKN2A germline mutations in familial pancreatic cancer. *Ann. Surg.* 236:730–737.

51. Goldstein A.M., Struewing J.P., Chidambaram A., Fraser M.C., and Tucker M.A. 2000. Genotype-phenotype relationships in U.S. melanoma-prone families with CDKN2A and CDK4 mutations. *J. Natl. Cancer Inst.* 92:1006–1010.

52. Rulyak S.J., Brentnall T.A., Lynch H.T., and Austin M.A. 2003. Characterization of the neoplastic phenotype in the familial atypical multiple-mole melanoma-pancreatic carcinoma syndrome. *Cancer* 98:798–804.

53. Murphy K.M., Brune K.A., Griffin C., Sollenberger J.E., Petersen G.M., Bansal R. et al. 2002. Evaluation of candidate genes *MAP2K4, MADH4, ACVR1B,* and *BRCA2* in familial pancreatic cancer: deleterious *BRCA2* mutations in 17%. *Cancer Res.* 62:3789–3793.

54. White K., Held K.R., and Weber B.H.F. 2001. A BRCA2 germ-line mutation in familial pancreatic carcinoma. *Intl. J. Cancer* 91:742–744.

55. The Breast Cancer Linkage Consortium. 1999. Cancer risks in BRCA2 mutation carriers. *J. Natl. Cancer Inst.* 91:1310–1316.

56. Goggins M., Schutte M., Lu J., Moskaluk C.A., Weinstein C.L., Petersen G.M. et al. 1996. Germline *BRCA2* gene mutations in patients with apparently sporadic pancreatic carcinomas. *Cancer Res.* 56:5360–5364.

57. Smyth M.J., Takeda K., Hayakawa Y., Peschon J.J., van den Brink M.R.M., and Yagita H. 2003. Nature's TRAIL — on a path to cancer immunotherapy. *Immunity* 18:1–6.

58. Lal G., Liu G., Schmocker B., Kaurah P., Ozcelik H., Narod S.A. et al. 2000. Inherited predisposition to pancreatic adenocarcinoma: role of family history and germ-line *p16, BRCA1,* and *BRCA2* mutations. *Cancer Res.* 60:409–416.

59. Rigaud G., Moore P.S., Zamboni G., Orlandini S., Taruscio D., Paradisi S. et al. 2000. Alleotype of pancreatic acinar cell carcinoma. *Intl. J. Cancer* 88:772–777.

60. Wermer P. 1954. Genetic aspects of adenomatosis of endocrine glands. *Am. J. Med.* 16:363–371.

61. Larsson C., Skogseid B., Oberg K., Nakamura Y., and Nordenskjold M. 1988. Multiple endocrine neoplasia Type 1 gene maps to Chromosome 11 and is lost in insulinoma. *Nature* 332:85–87.

62. Chandrasekharappa S.C., Guru S.C., Manickam P., Olufemi S.E., Collins F.S., Emmert-Buck M.R. et al. 1997. Positional cloning of the gene for multiple endocrine neoplasia-Type 1. *Science* 276:404–407.

63. Lemmens I., Van de Ven W.J., Kas K., Zhang C.X., Giraud S., Wautot V. et al. 1997. Identification of the multiple endocrine neoplasia Type 1 (MEN1) gene. The European Consortium on MEN1. *Hum. Mol. Genet.* 6:1177–1183.

64. Agarwal S.K., Guru S.C., Heppner C., Erdos M.R., Collins R.M., Park S.Y. et al. 1999. Menin interacts with the AP1 transcription factor JunD and represses JunD-activated transcription. *Cell* 96:143–152.

65. Schussheim D.H., Skarulis M.C., Agarwal S.K., Simonds W.F., Burns A.L., Spiegel A.M. et al. 2001. Multiple endocrine neoplasia Type 1: new clinical and basic findings. *Trends in Endocrinol. Metab.* 12:173–177.

66. Glascock M.J. and Carty S.E. 2002. Multiple endocrine neoplasia Type 1: fresh perspective on clinical features and penetrance. *Surg. Oncol.* 11:143–150.

67. Bartsch D.K., Langer P., Wild A., Schilling T., Celik I., and Rothmund M. et al. 2000. Pancreaticoduodenal endocrine tumors in multiple endocrine neoplasia Type 1: surgery or surveillance? *Surgery* 128:958–966.

68. O'Riordain D.S., O'Brien T., van Heerden J.A., Service F.J., and Grant C.S. 1994. Surgical management of insulinoma associated with multiple endocrine neoplasia Type I. *World J. Surg.* 18:488–493.

69. Carty S.E., Helm A.K., Amico J.A., Clarke M.R., Foley T.P., Watson C.G. et al. 1998. The variable penetrance and spectrum of manifestations of multiple endocrine neoplasia Type 1. *Surgery* 124:1106–1113.

70. Doherty G.M., Olson J.A., Frisella M.M., Lairmore T.C., Well Jr. S.A., and Norton J.A. 1998. Lethality of multiple endocrine neoplasia Type 1. *World J. Surg.* 22:581–586.

71. Cadiot G., Lebtahi R., Sarda L., Bonnaud G., Marmuse J.P., Vissuzaine C. et al. 1996. Preoperative detection of duodenal gastrinomas and peripancreatic lymph nodes by somatostatin receptor scintigraphy. Groupe D-etude Du Syndrome De Zollinger-Ellison. *Gastroenterology* 111:845–854.

72. Gauger P.G., Scheiman J.M., Wamsteker E.J., Richards M.L., Doherty G.M., and Thompson N.W. 2003. Role of endoscopic ultrasonography in screening and treatment of pancreatic endocrine tumours in asymptomatic patients with multiple endocrine neoplasia Type 1. *Br. J. Surg.* 90:748–754.

73. Anderson M.A., Carpenter S., Thompson N.W., Nostrant T.T., Elta G.H., and Scheiman J.M. 2000. Endoscopic ultrasound is highly accurate and directs management in patients with neuroendocrine tumors of the pancreas. *Am. J. Gastroenterol.* 95:2271–2277.

74. Bartsch D.K., Kersting M., Wild A., Ramaswamy A., Gerdes B., Schuermann M. et al. 2000. Low frequency of p16$^{(INK4a)}$ alterations in insulinomas. *Digestion* 62:171–177.

75. Lonser R.R., Glenn G.M., Walther M., Chew E.Y., Libutti S.K., Linehan W.M. et al. 2003. von Hippel-Lindau disease. *Lancet* 361:2059–2067.

76. Latif F., Tory K., Gnarra J., Yao M., Duh F.-M., Orcutt M.L. et al. 1993. Identification of the von Hippel-Lindau disease tumor suppressor gene. *Science* 260:1317–1320.

77. Hammel P.R., Vilgrain V., Terris B., Penfornis A., Sauvanet A., Correas J.-M. et al. 2000. Pancreatic involvement in von Hippel-Lindaue disease. *Gastroenterology* 119:1087–1095.

78. Binkovitz L.A., Johnson D.C., and Stephens D.H. 1990. Islet cell tumors in von Hippel-Lindau disease: increased prevalence and relationship to the multiple endocrine neoplasias. *Am. J. Roentgenol.* 155:501–505.

79. Hull M.T., Warfel K.A., Muller J., and Higgins J.T. 1979. Familial islet cell tumors in von Hippel-Lindau's disease. *Cancer* 44:1523–1526.

80. Lubensky I.A., Pack S., Ault D., Vortmeyer A.O., Libutti S.K., Choyke P.L. et al. 1998. Multiple neuroendocrine tumors of the pancreas in von Hippel-Lindau disease patients: histopathological and molecular genetic analysis. *Am. J. Pathol.* 153:223–231.

81. Libutti S.K., Choyke P.L., Alexander H.R., Glenn G., Bartlett D.L., Zbar B. et al. 2000. Clinical and genetic analysis of patients with pancreatic neuroendocrine tumors associated with von Hippel-Lindau disease. *Surgery* 128:1022–1027.

82. Polednak A.P. 1998. Do physicians discuss genetic testing with family-history-positive breast cancer patients? *Conn. Med.* 62:3–7.

83. Cho M.K., Sankar P., Wolpe P.R., and Godmilow L. 1999. Commercialization of *BRCA1/2* testing: practitioner awareness and use of a new genetic test. *Am. J. Med. Genet.* 83:157–163.

84. Giardiello F.M., Brensinger J.D., Petersen G.M., Luce M.C., Hylind L.M., Bacon J.A. et al. 1997. The use and interpretation of commercial APC gene testing for familial adenomatous polyposis. *N. Engl. J. Med.* 336:823–827.

85. Wong N., Lasko D., Rabelo R., Pinsky L., Gordon P.H., and Foulkes W. 2001. Genetic counseling and interpretation of genetic tests in familial adenomatous polyposis and hereditary nonpolyposis colorectal cancer. *Dis. Colon Rectum* 44:271–279.

86. Rabelo R., Foulkes W., Gordon P.H., Wong N., Yuan Z.Q., MacNamara E. et al. 2001. Role of molecular diagnostic testing in familial adenomatous polyposis and hereditary nonpolyposis colorectal cancer families. *Dis. Colon Rectum* 44:437–446.

87. Wideroff L., Freedman A.N., Olson L., Klabunde C.N., Davis W., Srinath K.P. et al. 2003. Physician use of genetic testing for cancer susceptibility: results of a national survey. *Cancer Epidemiol. Biomarkers Prev.* 12:295–303.

88. Balmain A., Gray J., and Ponder B. 2003. The genetics and genomics of cancer. *Nat. Genet.* 33(Suppl.):238–244.

89. Cowgill S.M. and Muscarella P. 2003. The genetics of pancreatic cancer. *Am. J. Surg.* 186:279–286.

90. Peto J. and Houlston R.S. 2001. Genetics and the common cancers. *Eur. J. Cancer* 37(Suppl. 8):S88–S96.

91. Peto J. and Mack T.M. 2000. High constant incidence in twins and other relatives of women with breast cancer. *Nat. Genet.* 26:411–414.

92. The Anglian Breast Cancer Study Group. 2000. Prevalence of *BRCA1* and *BRCA2* mutations in a large population based series of breast cancer cases. *Br. J. Cancer* 83:1301–1308.

93. Houlston R.S. and Peto J. 2003. The future of association studies of common cancers. *Hum. Genet.* 112:434–435.

94. Ponder B.A. 2001. Cancer genetics. *Nature* 411:336–341.

95. Pharoah P.D., Antoniou A., Bobrow M., Zimmern R.L., Easton D.F., and Ponder B.A. 2002. Polygenic susceptibility to breast cancer and implications for prevention. *Nat. Genet.* 31:33–36.

96. Brentnall T.A., Bronner M.P., Byrd D.R., Haggitt R.C., and Kimmey M.B. 1999. Early diagnosis and treatment of pancreatic dysplasia in patients with a family history of pancreatic cancer. *Ann. Intern. Med.* 131:247–255.

97. Hemminki A., Tomlinson I., Markie D., Jarvinen H., Sistonen P., Bjorkqvist A.-M. et al. 1997. Localization of a susceptibility locus for Peutz-Jeghers syndrome to 19p using comparative genomic hybridization and targeted linkage analysis. *Nat. Genet.* 15:87–90.

98. Hemminki A., Markie D., Tomlinson I., Avizienyte E., Roth S., Loukola A. et al. 1998. A serine/threonine kinase gene defective in Peutz-Jeghers syndrome. *Nature* 391:184–187.

99. Jenne D.E., Reimann H., Nezu J., Friedel W., Loff S., Jeschke R. et al. 1998. Peutz-Jeghers syndrome is caused by mutations in a novel serine threonine kinase. *Nat. Genet.* 18:38–44.

100. Giardiello F.M., Brensinger J.D., Tersmette A.C., Goodman S.N., Petersen G.M., Booker S.V. et al. 2000. Very high risk of cancer in familial Peutz-Jeghers syndrome. *Gastroenterology* 119:1447–1453.

101. Su G.H., Hruban R.H., Bova G.S., Goggins M., Bansal R.K., Tang D.T. et al. 1999. Germline and somatic mutations of the *STK11/LKB1* Peutz-Jeghers gene in pancreatic and biliary cancers. *Am. J. Pathol.* 154:1835–1840.

102. Sato N., Rosty C., Jansen M., Fukushima N., Ueki T., Yeo C.J. et al. 2001. Peutz-Jeghers gene inactivation in intraductal papillary-mucinous neoplasms of the pancreas. *Am. J. Pathol.* 159:2017–2022.

103. Lowenfels A.B., Maisonneuve P., Cavallini G., Ammann R.W., Lankisch P.G., Andersen J.R. et al. 1993. Pancreatitis and the risk of pancreatic cancer. *N. Engl. J. Med.* 328:1433–1437.

104. Tersmette A.C., Petersen G.M., Offerhaus G.J.A., Falatko F.C., Brune K.A., Goggins M. et al. 2001. Increased risk of incident pancreatic cancer among first-degree relatives of patients with familial pancreatic cancer. *Clin. Cancer Res.* 7:738–744.

105. Eberle M.A., Pfützer R., Pogue-Geile K.L., Bronner M.P., Crispin D., Kimmey M.B. et al. 2002. A new susceptibility locus for autosomal dominant pancreatic cancer maps to Chromosome 4q32-34. *Am. J. Hum. Genet.* 70:1044–1048.

106. Lynch H.T. and Fusaro R.M. 1991. Pancreatic cancer and the familial atypical multiple mole melanoma (FAMMM) syndrome. *Pancreas* 6:127–131.

107. Caldas C., Hahn S.A., da Costa L.T., Redston M.S., Schutte M., Seymour A.B. et al. 1994. Frequent somatic mutations and homozygous deletions of the p16 (*MTS1*) gene in pancreatic adenocarcinoma. *Nat. Genet.* 8:27–32.

108. Schutte M., Hruban R.H., Geradts J., Maynard R., Hilgers W., Rabindran S.K. et al. 1997. Abrogation of the Rb/p16 tumor suppressive pathway in virtually all pancreatic carcinomas. *Cancer Res.* 57:3126–3130.

109. Wooster R., Neuhausen S.L., Mangion J., Quirk Y., Ford D., Collins N. et al. 1994. Localization of a breast cancer susceptibility gene, BRCA2, to Chromosome 13q12-13. *Science* 265:2088–2090.

110. Tulinius H., Olafsdottir G.H., Sigvaldason H., Tryggvadottir L., and Bjarnadottir K. 1994. Neoplastic diseases in families of breast cancer patients. *J. Med. Genet.* 31:618–621.

111. Kinzler K.W., Nilbert M.C., Su L.-K., Vogelstein B., Bryan T.M., Levy D.B. et al. 1991. Identification of FAP locus genes from Chromosome 5q21. *Science* 253:661–665.

112. Groden J., Thliveris A., Samowitz W., Carlson M., Gelbert L., Albertsen H. et al. 1991. Identification and characterization of the familial adenomatous polyposis coli gene. *Cell* 66:589–600.

113. Giardiello F.M., Offerhaus G.J., Lee D.H., Krush A.J., Tersmette A.C., Booker S.V. et al. 1993. Increased risk of thyroid and pancreatic carcinoma in familial adenomatous polyposis. *Gut* 34:1394–1396.

114. Horii A., Nakatsuru S., Miyoshi Y., Ichii S., Nagase H., Ando H. et al. 1992. Frequent somatic mutations of the APC gene in human pancreatic cancer. *Cancer Res.* 52:6696–6698.

115. Yashima K., Nakamori S., Murakami Y., Yamaguchi A., Hayashi K., Ishikawa O. et al. 1994. Mutations of the adenomatous polyposis coli gene in the mutation cluster region: comparison of human pancreatic and colorectal cancers. *Intl. J. Cancer* 59:43–47.

116. Lynch H.T., Voorhees G.J., Lanspa S.J., McGreevy P.S., and Lynch J.F. 1985. Pancreatic carcinoma and hereditary nonpolyposis colorectal cancer: a family study. *Br. J. Cancer* 52:271–273.

117. Jacob S. and Praz F. 2002. DNA mismatch repair defects: role in colorectal carcinogenesis. *Biochimie* 84:27–47.

118. Goggins M., Offerhaus G.J.A., Hilgers W., Griffin C.A., Shekher M., Tang D. et al. 1998. Pancreatic adenocarcinomas with DNA replication errors (RER⁺) are associated with wild-type *K-ras* and characteristic histopathology: poor differentiation, a syncytial growth pattern, and pushing borders suggest RER⁺. *Am. J. Pathol.* 152:1501–1507.

119. Swift M., Sholman L., Perry M., and Chase C. 1976. Malignant neoplasms in the families of patients with ataxia telangiectasia. *Cancer Res.* 36:209–215.

120. Thatcher B.S., May E.S., Taxier M.S., Bonta J.A., and Murthy L. 1986. Pancreatic adenocarcinoma in a patient with Peutz-Jeghers syndrome — a case report and literature review. *Am. J. Gastroenterol.* 81:594–597.

121. Bowlby L.S. 1986. Pancreatic adenocarcinoma in an adolescent male with Peutz-Jeghers syndrome. *Hum. Pathol.* 17:97.

122. Hirao S., Sho M., Kanehiro H., Hisanga M., Ikeda N., Tsurui H. et al. 2000. Pancreatic adenocarcinoma in a patient with Peutz-Jeghers syndrome: report of a case and literature review. *Hepatogastroenterology* 47:1159–1161.

123. Bardeesy N., Sharpless N.E., DePinho R.A., and Merlino G. 2001. The genetics of pancreatic adenocarcinoma. *Sem. Cancer Biol.* 11:201–218.

124. Hizawa K., Iida M., Matsumoto T., Kohrogi N., Kinoshita H., Yao T. et al. 1993. Cancer in Peutz-Jeghers syndrome. *Cancer* 72:2777–2781.

125. Giardiello F.M., Welsh S.B., Hamilton S.R., Offerhaus G.J.A., Gittelsohn A.M., Booker S.V. et al. 1987. Increased risk of cancer in the Peutz-Jeghers syndrome. *N. Engl. J. Med.* 316:1511–1514.

126. Pauwels M., Delcenserie R., Yzet T., Duchmann J.C., and Capron J.P. 1997. Pancreatic cystadenocarcinoma in Peutz-Jeghers syndrome. *J. Clin. Gastroenterol.* 25:485–492.

127. Yoshikawa A., Kuramoto S., Mimura T., Kobayashi K., Shimoyama S., Yasuda H. et al. 1998. Peutz-Jeghers syndrome manifesting complete intussusception of the appendix and associated with a focal cancer of the duodenum and a cystadenocarcinoma of the pancreas. Report of a case. *Dis. Colon Rectum* 41:517–521.

128. Bolwell J.S. and James P.D. 1979. Peutz-Jeghers syndrome with pseudoinvasion of hamartomatous polyps and multiple epithelial neoplasms. *Histopathology* 3:39.

129. Spigelman A.D., Murray V., and Phillips R.K.S. 1989. Cancer and the Peutz-Jeghers syndrome. *Gut* 30:1588–1590.

26

CYSTIC FIBROSIS OF THE PANCREAS — THE DISEASE AND ITS MANIFESTATIONS

Ajay P. Singh and Surinder K. Batra

CONTENTS

26.1 INTRODUCTION

Cystic fibrosis (CF) is a chronic, progressive, and one of the most common fatal genetic diseases in humans. The term *cystic fibrosis* was first used in 1938[1] for the fibrotic phenotype of the pancreas. However, the disease is now well recognized to present the pathophenotypes in many other organs of the body. It causes the body to produce thick, sticky mucus that clogs the lungs, leading to infection; reproductive ducts, causing infertility; and the pancreas, stopping digestive enzymes from reaching the intestine where they are required to digest food. The median age at diagnosis is 6 months and nearly two-thirds of the individuals are diagnosed before 1 year of age.[2] The poor prognosis is related to severe respiratory complication, malnutrition, and negligence of the disease condition; whereas, mild symptoms at diagnosis, pancreatic sufficiency, and atypical presentation indicate better prognosis. Life expectancy of the patients with CF has increased dramatically during the past years as we developed a better understanding about the disease presentation and advanced in the disease diagnosis and supportive therapy. Data from the CF patient registry collected by the Cystic Fibrosis Foundation show that the median age of survival has improved from 14 years in 1969 to a predicted 31.6 years in 2002.

In this chapter, we will discuss the CF pathology associated with the pancreas. CF is, by far, the most common inherited pancreatic disease of childhood and accounts for about 90% of childhood onset pancreatic disorders. Although some pancreatic disorders or injuries in CF appear as early as when the child is still in the mother's womb, others may appear later in life. The major complications observed in the pancreas of CF patients are pancreatic insufficiency (PI), CF related diabetes (CFRD), and pancreatitis. It has also been indicated that CF disease may pose higher risk toward pancreatic cancer development.

26.2 MOLECULAR BASIS OF CF

In 1989, the CF gene was identified on the long arm of Chromosome 7 (7q31.2) (Figure 26.1), covering a size of approximately 250 kb.[3,4] The gene consists of 27 exons that are joined together during RNA splicing to yield a 6.1 kb mRNA transcript. The mRNA is translated into a 1480 amino acid long integral membrane protein, cystic fibrosis transmembrane conductance regulator (CFTR), that has an approximate molecular weight of

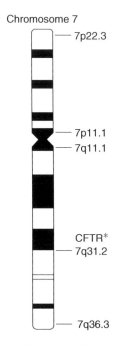

Chromosome 7

7p22.3

7p11.1
7q11.1

CFTR*
7q31.2

7q36.3

*Approximate gene location is based on chromosome 7 map from NCBI Entrez Map Viewer

Figure 26.1 Chromosomal location of the CFTR.

160 to 170 kD. Initially, CFTR was recognized for its function as a cyclic adenosine monophosphate (cAMP)-mediated chloride channel,[5] but later some additional functions (transport of other ions and regulating other ion channels, membrane trafficking, etc.) have also been attributed to this protein that may be relevant to the pathophysiology of CF and the multifunctionality of this protein.[6] The CFTR protein contains five functional domains: two membrane-spanning domains (MSD1 and MSD2), two nucleotide-binding domains (NBD1 and NBD2), and a regulatory domain (Figure 26.2). Each of the MSDs can be further subdivided into six transmembrane segments (TMs).[7] In addition to TMs, the MSDs also contain intra- and extracellular loops that connect the TMs on both surfaces of the cell membrane. The MSDs hold the ion channel in place within the membrane and form the walls of the ion channel. The chloride ion channel, CFTR usually opens, when it is activated by the binding of adenosine triphosphate (ATP) at the NBD regions. The NBDs hydrolyze the ATP to facilitate the movement of chloride ions along the concentration gradient. The function of the regulatory domain is not completely defined, but it is believed that it actually regulates the opening and closing of the ion channel by bringing about structural changes. Some also believe that

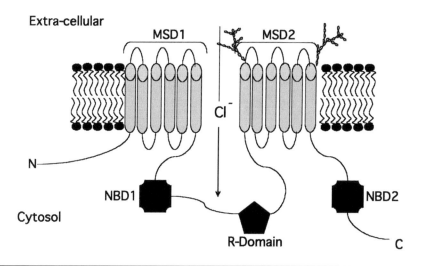

Figure 26.2 Diagrammatic representation of CFTR in the membrane. The CFTR protein contains five functional domains: two membrane-spanning domains (MSD1 and MSD2), two nucleotide-binding domains (NBD1 and NBD2), and a regulatory domain. Each of the MSDs can be further subdivided into six TMs. In addition to TMs, the MSDs also contain intra- and extracellular loops that connect the TMs on both surfaces of the cell membrane.

it enhances the interaction between ATP and NBDs.[8] In addition, there are certain regions in the CFTR protein, which bind chaperones and are distributed throughout the protein.[9] These sporadic regions ensure that the protein is correctly folded, processed, and trafficked.

The cAMP regulated chloride ion channel function of CFTR is essential for maintaining the luminal hydration in the ducts of many organs. Therefore, the gene mutations leading to defective CFTR production or reduced activity can lead to dysfunction in the lungs, exocrine pancreas, sweat glands, and reproductive ducts as manifestations of CF. Nearly 1000 mutations of the CFTR gene have been identified to date,[10] which lead to chloride channel dysfunction and impaired ion transport across epithelial cell membranes. The most common disease causing mutation, accounting for approximately 70% of CF chromosomes, is a 3-base pair deletion. This deletion corresponds to the loss of a single amino acid, phenyl-alanine, at Position 508 (F508) in the encoded CFTR protein.[3,4] It is believed that the F508 mutation leads to misfolding and decreased glycosylation of the protein and prevents the protein from being properly localized or transported to the plasma membrane. Instead, the protein becomes trapped in the endoplasmic reticulum (ER) and is degraded.

Based on the molecular fate of the CFTR protein, CFTR gene mutations have been categorized into five classes (Table 26.1). Class I, Class II, and

Table 26.1   Types of CFTR Mutations and Their Presentation

Class	Presentation
Class I	These mutations cause defects in CFTR synthesis resulting in complete absence of CFTR protein.
Class II	These mutations result in the production of abnormal CFTR (Δ F508) that fails to escape the endoplasmic reticulum.
Class III	These mutations lead to production of defective CFTR protein, which is trafficked to apical membrane, but is not functional due to disruption of activation and regulation at the cell membrane.
Class IV	These mutations also lead to defective CFTR protein expression, which translocates to the cell membrane, but the chloride conductance is reduced.
Class V	These mutations decrease membrane CFTR function by decreasing splicing of normal CFTR.

Class III mutations result in complete loss of CFTR function and, therefore, in severe disease symptoms. Class IV and Class V mutations retain some CFTR function and so, only lead to mild phenotypes.[11,12] It is not yet clear how mutations in a single gene coding for an ion channel can lead to so many diverse pathophenotypes. Studies are underway to understand the many of the yet unraveled functions of CFTR.

26.3   DISEASE INHERITANCE AND INCIDENCE

CF is an autosomal recessive life shortening disorder and, therefore, the heterozygous carriers are asymptomatic.[13a] The children with one of the parents carrying CFTR mutations are normal and only the offspring of the CF carrier parents are at risk of having this genetic defect by inheriting one defective allele from each parent. Accordingly, there is one in four chance of CF occurrence in a family of CF carrier parents based on simple Mendelian Inheritance.

The disease has the highest incidence (1:3300) in Caucasians with an average carrier frequency of 1:29. It also affects other races and ethnic groups with fairly high incidence in Hispanics (1:9500) and Native Americans (1:3970 in Pueblo people and 1:1500 among Zuni). It is rare in African and Asian Americans. The disease incidence data have been summarized in Table 26.2.

26.4   PATHOLOGICAL ALTERATIONS IN THE PANCREAS

The pancreatic pathology is one of the earliest features of CF. In fact, the alterations in the pancreas of the patients with CF led to the initial

Table 26.2 Incidence of CF[13b]

Group	Caucasians	Native Americans	Hispanics	African Americans	Asian Americans
Incidence	1/3300	1/3970 (Pueblo) 1/1500 (Zuni)	1/7–8000	1/15,300	1/32,100
Carrier	1/29	NA	1/49	1/60–65	1/90
%ΔF508	70	0	46	48	30

NA — not available.

identification and naming of the disease.[1] The pancreas is a large (25 cm long) gland lying inferior to the stomach and close to the duodenum. The anatomy of the pancreas reveals the presence of two basic types of tissues: the acini and the islets that carry out the exocrine and endocrine functions, respectively. The pancreas is abnormal in almost all the cases of CF. The most common abnormalities associated with the CF pancreas are pancreatic atrophy, cystic lesions, and fatty replacement of the tissue. Sometimes, stone formation or calcification within the ducts and body of the pancreas may also occur.[14] The occurrence of multiple microscopic or small macroscopic cysts lined by epithelium can also be observed.[15]

There is a spectrum of severity of the clinical findings in patients with CF at all ages.[16] The deposition of materials within the pancreatic ducts can be observed by the mid-trimester of human gestation.[17] Obstruction of the duct follows acinar autolysis as the acini differentiate and produce pancreatic enzymes by about 24 weeks of gestation. According to an earlier published report on autopsied cases of 146 individuals who died as a result of CF, there is a direct correlation between age and disease severity in the pancreas; however, pancreatic lesion may be absent in an individual irrespective of the age.[18] Daneman et al. (1983) reviewed autopsy findings in the pancreas of 27 CF patients and reported a diffuse fatty replacement in 15 (56%) patients, but the rest (44%) had minimal or no fat replacement.[19] The severity of pathological alterations in the CF pancreas also appears to correlate with pancreatic function. In the subset of patients with CF, who do not have impaired exocrine pancreatic function, only minimal pathologic lesions are found. Conversely, patients with severe, long-standing symptoms of malabsorption may show marked fibrosis, fatty replacement and cysts in the pancreas.[20] In a follow-up study over 9 years on 55 CF patients, a complete lipomatosis of the pancreas was reported in 23 patients, and 9 patients showed incomplete lipomatosis.[21] In addition to lipomatous involution, other pancreatic changes, such as pancreatic atrophy, cystic pancreas, and macrocyst formation were also present in 13, 5, and 1 patients, respectively.

A comparison of early pancreatic development in CF and normal subjects can be a better approach to understanding the progressive pathological changes occurring in the pancreas of a CF individual. Imrie and colleagues performed morphologic examination of the postnatal development in the pancreas of infants with CF using the comparative age-matched controls.[22] In their study, they included 60 autopsy specimens (29 normal and 31 CF subjects) from infants under the age of 4 months. Quantitative microscopy data revealed that there was an increase in fold-ratio of acinar vs. connective tissue from birth to 4 months of age in control cases. In contrast, the CF subjects demonstrated a decrease in normal acinar development and, therefore, acinar to connective tissue ratio showed a progressive decrease. King et al. (1986), in their study on premature infants that died because of CF, have described the focal dilation of the small pancreatic ductules as the earliest changes. These changes further lead to widespread involvement of the gland with epithelial flattening and plugging of the ductules and acini by inspissated secretion. In some infants, acinar atrophy and fibrosis were also observed.[23]

26.5 PATHOPHYSIOLOGY

Abnormal salt secretion in CF came into recognition long before its cause was determined.[24] Quite a few years later, a report came that a factor in the sweat of CF patients could stop the salt transport in normal saliva and sweat glands.[25] However, the first insight was brought on the pathophysiology of CF by the discovery of an increased bioelectrical potential difference across respiratory epithelium.[26] This pulled the idea that an abnormal ion-exchange mechanism may exist in CF patients.[27] Later, chloride ion (Cl⁻) impermeability was demonstrated to be the basic physiological CF defect.[28]

In the pancreas of CF patients, defective HCO_3^- transport is the hallmark of pathophysiological manifestations. Other salient features include the inhibition of endocytosis in acinar cells and an imbalance in membrane lipids.[29] CFTR-expressing ductal epithelia of the pancreas secrete an HCO_3^- rich liquid that flushes the enzymes into the duodenum. The secretion of bicarbonate-rich fluid takes place in response to the acidic chime that leaves the stomach and enters the duodenum, where it helps in neutralizing the acid and maintains an optimal pH for the digestive enzymes secreted by the pancreas.[30] According to the currently accepted model of pancreatic ductal HCO_3^- secretion, intracellular HCO_3^- is accumulated in response to the action of cytosolic carbonic anhydrase on the CO_2 that diffuses from the basolateral membrane. The G-protein coupled receptor (e.g., secretin and vasoactive intestinal peptide) activate cAMP-sensitive CFTR ion channel to secrete Cl⁻ in to the lumen. The resultant increase in luminal chloride ions drives Cl⁻/HCO$_3^-$ exchanger.[31] Experimental evidences suggest that reduced HCO$_3^-$ secretion from pancreatic ductal epithelium in CF patients alters intraductal pH, resulting in the precipitation of proteins secreted by the acinar cells. This further leads to the formation of protein plugs and disruption of vesicular trafficking in the apical domain in the acinar cells.[32]

26.6 DISEASE MANIFESTATIONS

Major complications of the CF disease occur in the respiratory tract, causing sinusitis and respiratory failure. Severe bacterial infestation of the respiratory passage is also a major problem in CF patients. The second most common complications arise in the gastrointestinal tract. Other organ systems that are involved and show complications are the sweat glands (salt loss syndrome) and reproductive system (male infertility and reduced female fertility) (Table 26.3). The disease presentation varies from patient to patient depending upon the functional status of the CFTR protein or by the type of mutation present in the CFTR gene. In addition, new evidences suggest roles for some modifier genes and environmental factors in the development of CF.[33,34] The pancreas is an important gland of the digestive and endocrine system of the body carrying out both the exocrine and endocrine function. The exocrine function is carried out by the acinar cells that produce the enzymes needed to digest food, and the islets perform the endocrine function and secrete three important hormones — insulin, glucagon, and somatostatin. These hormones are necessary for glucose homeostasis in the body. The damage caused to the pancreas by the diseased condition in a CF patient results in its impaired functioning and inflammation. Three main types of pathological manifestations associated with the pancreas — PI, diabetes, and pancreatitis — have been reported in CF patients.

26.6.1 Pancreatic Insufficiency

The pancreas makes more than 25 different digestive enzymes that are secreted into the intestine. It also secretes a large amount of sodium biocarbonate, which protects the duodenum by neutralizing the acid that comes from the stomach. The digestive enzymes are released from the cells of the acini and flow down through various channels into the pancreatic duct. The pancreatic duct joins the common bile duct at the sphincter of Oddi, where both flow into the duodenum. The enzymes secreted by the pancreas digest dietary proteins, fats, and carbohydrates into simpler molecules, so that they can be absorbed by the intestinal epithelial cell from the intestinal lumen. The pancreas has a tremendous

Table 26.3 Organ Systems Involved in CF and Their Complications

Organ System	Complications
Respiratory	
Upper:	
Nose	Nasal/sinus Polyposis
Sinus	Sinusitis
Lower:	
Bronchial tree	Cystic bronchiectasis
Lungs	Hemoptysis
	Pneumothorax
	Respiratory failure
Gastrointestinal	
Esophagus and stomach	Barrett's esophagus
	Esopageal varices
	Gastro-esophageal reflux disease (GERD)
Pancreas	Pancreatic insufficiency (PI)
	Cystic fibrosis related diabetes (CFRD)
	Pancreatitis
Intestinal tract	Meconium ileus
	Distal intestinal obstruction syndrome (DIOS)
	Rectal prolapse
Liver	Liver cirrhosis and liver failure
Others	
Sweat gland	Salt loss syndrome
Reproductive system	Male infertility
	Female reduced fertility
	Delayed sexual development
	Absence of/irregular menstruation
Blood and Circulatory system	Hypoelectrolemia
	Metabolic alkalosis
	Hypersplenism
	Portal Hypertension
	Macrolytic anemia

reserve capacity and, in fact, more than 95% of the function of the pancreas must be lost before the pancreas fails and symptoms of bloating and maldigestion develop.[35,36] The large pancreatic reserve, however, also means that children can have a severe pancreatic problem without experiencing any problem with digestion. At the time of childbirth, the exocrine pancreas is not fully developed and, therefore, does not have the same ability as the mature pancreas to produce enough enzymes, in particular, those needed to digest carbohydrates and fats.[37] Because of this, all healthy infants show some degree of maldigestion. The pancreas matures completely within 2 years after birth and starts functioning in the same way as an adult pancreas. The immature pancreas usually does not have any adverse effect on healthy children, but a major impact can be seen when children become malnourished or ill. Under the CF diseased condition, in some cases, the pancreas begins to be damaged even when the affected child is still in the mother's womb. The small ducts inside the pancreas, which allow digestive enzymes to reach the intestine, get blocked with mucus and proteins, and the pancreas becomes badly scarred and shrunken.[37]

Many children with CF have evidence of severe pancreatic failure immediately following birth, and by 2 years of age, 90% of CF cases are diagnosed, usually those with severe malnutrition. Approximately 85% of all people with CF have PI and need to take pancreatic enzymes with meals as well as additional fat-soluble vitamins. Incidences of PI in children with CF have been correlated with the type of CFTR gene mutations they carry. Of children having two classes of mutations (I and II/III), 85% suffer from PI. Later, these patients develop severe pulmonary complications, the males are infertile, and the average life expectancy is 30 years. On the other hand, the patients with Class IV and Class V mutations are pancreatic sufficient (PS), do not have steatorrhea, and, therefore, do not require pancreatic enzyme supplementation. Due to this fact, PS patients are diagnosed at a later stage, experience normal growth during childhood, and have a better prognosis than their PI counterparts. A severely affected CF phenotype is strongly associated with the presence of ΔF508, and 99% of those patients homozygous for this most common mutation have PI.[38,39] In a study by Kristidis et al., it was determined that most genotypes carrying CFTR mutations are associated either with PI or PS, but not both.[40] This might suggest the genetic association of the phenotype. Also, the mutations that confer the PS phenotype do so in a dominant fashion. In patients with the PS phenotype, progression of disease either does not occur or seems to be retarded. Based on this, CFTR mutations have been classified as severe or mild with respect to pancreatic function status, although a few indeterminate mutations seem to confer PS or PI phenotypes on a less consistent basis.[40] Although the course and prognosis of CF are mainly determined by the progression of pulmonary obstruction and infection, the assessment of pancreatic damage may be of clinical relevance in the screening of patients at high risk for early development of pulmonary infection. It has been reported that patients with PI may have earlier colonization of the airway by *Pseudomonas aeruginosa*, the major event in the clinical development of CF.[39]

26.6.2 CF-Related Diabetes

The pancreas produces three important hormones of the body's endocrine system, which are required for glucose homeostasis in the blood. These are insulin, which lowers the level of sugar (glucose) in the blood; glucagon, which raises the blood sugar level; and somatostatin, which prevents the other two hormones from being released. In general, diabetes mellitus is a metabolic condition that is either caused by the hyposecretion or hyposynthesis of insulin or by acquiring the insulin resistance (i.e., the insulin that is made does not work well enough). The insulin actually helps blood glucose get absorbed by the cells of the body for energy and convert the rest to glycogen reserves and triglycerides, particularly in muscles and adipose tissues, respectively. In the absence of insulin, the cells of the body starve due to the inability to absorb sugar, while the blood sugar reaches high levels. There are two common forms of diabetes reported in literature — Type 1 diabetes (previously referred to as insulin-dependent or juvenile onset) and Type 2 diabetes (previously referred to as non-insulin dependent or adult type).[41]

In CF patients, the pancreas may be blocked due to the excessive secretion and accumulation of mucus. Although this does not directly stop insulin from entering the bloodstream as it is internally secreted, it may have an indirect impact due to the pancreatic damage.[42] In CF patients, the damage is caused to the pancreas by its own enzymes, which are not getting out properly due to ductal blockade. These enzymes start digesting the pancreas itself, resulting in damage to the insulin-secreting cells and, thus, leads to decreased insulin production or secretion. Insulin resistance may also play a role in the development of CFRD, especially during acute infection.[43] Although, the majority of non-diabetic CF patients, in their usual state of health, appear to be insulin sensitive, they may also become insulin resistant during infection. CFRD presents common features of both Type 1 and Type 2 diabetes.[44] Most individuals with CFRD have the same primary problem of insulin deficiency as in Type 1 diabetes; however, the clinical presentation is more similar to Type 2 diabetes. The onset of CFRD is often asymptomatic, and in most cases it is undiagnosed for years.[45] The average age of CFRD onset is 18 to 21 years. About 20 to 30% of CF patients show impaired glucose tolerance (IGT) and only one-quarter of patients over 35 years has diabetes.[2] Furthermore, the percentage of those who develop diabetes mellitus increases with their extending life expectancy. According to the data registry of 22,732 CF patients maintained by the North American Cystic Fibrosis Foundation, the prevalence of diabetes was about 11.4% in 2001. The actual prevalence of diabetes, however, may be more in CF for the reason that CFRD is often asymptomatic.

The CFRD may cause a decline in the clinical status of CF patients and hyperglycemia often occurs in the patients being treated with immunosuppressive therapies after lung transplantation. Thus, diabetes is a growing concern in the follow-up of CF patients. The presentation of CFRD is different from diabetes in people not having CF and, therefore, it requires unique monitoring and management. There is a continuum of glucose tolerance in CF ranging from normal glucose tolerance to increasingly severe glucose intolerance to diabetes. In their baseline state of health, CF patients are generally insulin-sensitive, but during the stress of acute infection, pregnancy, or glucocorticoid therapy, patients develop insulin resistance. When the stress resolves, the insulin resistance usually goes away and the patient moves back to their baseline. Thus, it is common for CF patients to have a temporarily higher blood glucose level during physical stress. Malnutrition may be present in CF, which can impair the ability of the pancreas to make insulin. Malnutrition due to pancreatic exocrine insufficiency is common despite enzyme supplementation. Anorexia and nausea due to impaired digestion may also cause malnutrition. Abnormal intestinal movement may affect food absorption and thus influence blood sugar levels. The occurrence of liver disease, which is often present in CF patients, also influences how the body uses nutrients. The patients with CFRD are more underweight and have worse pulmonary function than those without diabetes. It is now recommended that anyone with CF, 14 years of age or older, be routinely screened for CFRD.

26.6.3 Pancreatitis

Pancreatitis is presented by the inflammation of the pancreas and may also lead to tissue damage, when active digestive enzymes begin attacking the pancreas. In some cases, there may be bleeding into the gland, infection, and cyst formation. Enzymes and toxins may enter the bloodstream and may cause injury to other organs. Pancreatitis occurs in two forms: the acute, which is sudden and may be severe or recurrent, and the chronic, which is long, severe, and often associated with weight loss and pain in the upper abdomen. In most cases, the cause of the disease is identifiable by the patient's history and other standard laboratory tests, but in certain cases where there is no apparent underlying cause for the disease, it is referred as idiopathic pancreatitis.

Usually, the clinical manifestations of CF in the pancreas do not resemble pancreatitis; however, many studies have indicated that mutations are present in the CFTR gene in patients with idiopathic pancreatitis.[46–48] In most cases, gene mutations on only one allele have been recorded with no significant pulmonary and other CF complications in patients. The molecular mechanisms that can associate the CF with pancreatitis are not yet defined. Nevertheless, it is believed that underlying mutations in CFTR predispose some CF patients to pancreatitis, when combined with other gene defects or when exposed to certain other environmental factors, such as the use of alcohol.[49–51] Recent studies have given the indications that the frequency of CFTR gene mutations is underestimated and that the analysis of such mutations may be of importance toward the etiology of idiopathic pancreatitis.[46,47] Investigators at veteran affairs and Duke University explored the relationship between CFTR mutations and idiopathic pancreatitis in 22 women and 5 men and referred to them for evaluation of idiopathic pancreatitis.[46] The patients were tested for 17 CFTR gene mutations and also for the 5T CFTR allele, which when present markedly reduces the level of functional gene product. Ten patients (37%) had at least one such abnormality, for a mutation frequency 11-fold greater than expected. In 3 patients, both CFTR alleles were affected, exceeding the expected frequency 80-fold. In a similar study, researchers at the Manchester Royal Infirmary in the U.K. studied 134 consecutive patients with chronic pancreatitis, including 71 cases (53%) with alcohol-related pancreatitis and 60 with idiopathic disease (45%).[47] They examined the 22 mutations (accounting for 95% of all mutations) and the 5T allele in Intron 8. They identified a mutation frequency nearly 2.5 times the expected figure. The frequency of the 5T allele was twice as high as expected. The CFTR mutations were found to be associated with idiopathic rather than alcohol-related disease.

Now, when the association of pancreatitis with CF has been well documented, the question arises what makes CFTR mutations susceptible to pancreatitis or is there any correlation among the various clinical manifestations? Analysis of 1075 patients with a diagnosis of CF and acute recurrent or chronic pancreatitis over a 30-year period (1966 to 1996) from the CF database have demonstrated that no patient with PI developed pancreatitis.[52] Of the 110 (≈10% of total) patients that were diagnosed as PS, 19 (17.3 or 1.7% of total) experienced one or more attacks of pancreatitis. The mean age at diagnosis of pancreatitis was 22.7 years with a wide range (10 to 35 years). The observation that PI patients do not have pancreatitis incidence can be attributed to the loss of functional acinar tissue at an early period of life or even *in utero*.[53] The presence of functional acinar cells is a prerequisite for pancreatitis and, therefore, the PS patients with impaired CFTR function are predisposed to pancreatitis. Although it remains a subject of concern and discussion whether or not the patients with pancreatitis should be routinely screened for CFTR mutations or if such pancreatitis incidences should be considered for the diagnosis of CF,[46–48] there is definitely a need to understand the relationship of CFTR mutations and pancreatitis.

26.7 DIAGNOSIS

In its classic form, the CF disease is easily diagnosed early in life by a combination of clinical evaluation and laboratory testing; however, in some cases, the disease presentation may appear a little later. Overall, 7% of CF patients are not diagnosed until 10 years of age, with a proportion of patients that does not show clear symptoms until age 15 years.[46] The diagnosis of CF is suspected by the presence of typical clinical features such as meconium ileus at birth, recurrent or persistent pneumonia, malabsorption causing weight loss and other growth abnormalities, salt-loss syndrome, etc.[54] The common diagnostic tests for CF have been described below.

26.7.1 Sweat Test

The sweat test is the most common test for CF. This test measures the amount of salt (sodium chloride) in sweat. Normal sweat chloride value in sweat is <40 mEq/L and a value of more than 60 mEq/L is considered abnormal. Conventionally, the diagnosis of CF is confirmed by a typical phenotype and increased sweat chloride measurements on repeat testing. Average sweat chloride values in CF patients are approximately 100 mEq/L.

26.7.2 Immunoreactive Trypsinogen Test

Immunoreative trypsinogen (IRT) test is used for those newborns who do not produce enough sweat. In this test, blood is drawn 2 to 3 days after birth and is analyzed for the presence of trypsinogen protein. It is, however, not a confirmatory test. CF is later confirmed by sweat and other tests.

26.7.3 Genetic Testing

This involves carrier screening and direct DNA analysis. It is hard to detect all of nearly 1000 CFTR gene mutations and, therefore, these tests are only 80 to 85% accurate.

26.7.4 Supportive Laboratory Tests

There are certain tests that are performed in the absence of the afore-mentioned evaluations or to further confirm the CF diagnosis. These are as follows.

26.7.4.1 Blood Biochemistry

Several abnormalities may exist in a CF patient's blood. There may be metabolic alkalosis and hypochloremia. A person may also be anemic with symptoms of hypoalbuminemia due to malabsorption.

26.7.4.2 Pancreatic Function Test

As the direct assessment of pancreatic function is difficult, the measurement of pancreatic elastase-1 in stool sample is performed. Semiquantitative estimates of intestinal fat malabsorption can be made using either fecal microscopy or fecal steatocrit.

26.7.4.3 Bacterial Colonization and Infection in Airways

Children with CF may have a severe bacterial infestation of *Pseudomonas aeruginosa* or *Staphylococcus aureus* in their airways. All CF-suspected children are tested to identify the respiratory pathogen by microbial culture of sputum specimens or cough swabs.

26.7.4.4 Obstructive Azoospermia

Sperms are usually absent in up to 98% of men with CF, and, therefore, a semen analysis for azoospermia can be carried out in postpubertal boys for diagnosing CF.

26.7.4.5 Radioimaging

Chest x-rays and a CT-scan may also be suggestive of CF. Hyperinflation, peribronchial thickening, and cystic changes in chest x-rays and cystic or varicose bronchiectasis, mucus impaction, and subpleural bullae formation in a CT-scan are common findings that may be indicative of CF, although these are not specific.

26.8 CF AND PANCREATIC CANCER

With increased life expectancy, the majority of the CF patients now survive at least to adolescence and almost one-third attain adulthood. This increased life span for CF patients, although good, has brought to our attention that people carrying the CFTR gene mutation may have an increased risk of developing cancer, particularly in the digestive tract.[55] The development of cancer is a multistep process that includes initiation due to gene mutation, promotion by accumulating the oncogenic mutation, and progression toward a malignant phenotype via further genetic and epigenetic changes. Although all people are equally prone to gene mutation, only a few get it in their lifetime. In fact, the development of cancer is a variable function of a person's genetic background, environmental exposure, diet, and lifestyle.

Pancreatic cancer, the fourth leading cause of cancer-related deaths in the United States,[56] has been discussed in Chapter 19, Chapter 21, and

Chapter 25. The association of pancreatic cancer with CF has long been suspected, however, there is no clear or direct evidence to prove such a linkage. In a cohort study conducted to evaluate the risk of cancer in CF, involving 412 patients, a significant excess of pancreatic and small-intestine cancers has been reported.[57] Later on, another investigation also reported similar findings (i.e., an increased risk of digestive tract cancers in CF patients); however, the overall risk of cancer among them was similar to that of the general population.[55]

Although we may consider that CF has a link with cancer development, the question arises whether it is directly related to some not yet recognized functions of CFTR or is due to chronic disease conditions. To date, we do not have direct evidence that can relate the CFTR defect with cancer development. Moreover, almost all the cases have been recognized later in life, denying a direct involvement of CFTR with carcinogenesis.[55,58,59] Therefore, it seems likely that cancer progression is secondary to the long-term disability of CFTR. The enhanced risks of organ-specific cancer (e.g., digestive tract cancer) in CF patients may be related to the differential expression patterns of CFTR gene in different body organs and to the susceptibility of that organ to damage by CFTR dysfunction. Pancreatic ducts and other digestive tissue (bile ducts, intestinal crypts, etc.) express high levels of the CFTR protein in comparison to several other body organ tissues. Furthermore, persistent pathologic alterations (e.g., tissue damage, cyst formation, etc.), leading to increased cell turnover, might also predispose these organ's cells toward oncogenic transformation. An observed deficiency of selenium, an antioxidant that offers protection against several diseases including cancer, may also increase the risk of carcinoma in CF patients.[60]

In a study on the expression of pancreatic tumor-associated mucin genes (MUC1 and MUC4) in the pancreatic tumor cell line derived from a CF patient (CFPAC1) and its CFTR corrected subline, CFPAC-PLJ-CFTR, we have reported that the overexpression of MUC4 is associated with CF phenotype.[61] Also, no CFTR expression was detected in 12 (75%) of the 16 pancreatic tumor cell lines tested, although pancreas is considered as a site of high CFTR expression. We have further confirmed the linkage between the CFTR defect and MUC4 overexpression, using short-interfering RNA (siRNA) mediated gene silencing of CFTR (Singh et al., unpublished data). MUC4 is a membrane bound mucin, which is aberrantly expressed in pancreatic tumor tissue and cancer cell lines, that has no detectable expression in the normal pancreas.[62] Recently, we have also shown that the knockdown of MUC4 in a metastatic pancreatic tumor cell line, CD18/HPAF, results in reduced tumor growth both *in vitro* and *in vivo* and decreases the tumor cell potential to metastasize.[63] Therefore, an association between the CFTR defect and MUC4 up-regulation, along with the observed CFTR down-regulation in pancreatic cancer cell lines may be of significance in understanding CF-linked risks of pancreatic tumor development and progression.

26.9 EPILOGUE

Since the identification of the disease in 1938 and then the discovery of the CF gene 50 years later in 1989, to our current understanding, we have come a long way. However, much more work still needs to be done to unravel the complexity of the disease presentation and to understand the molecular mechanisms that are responsible to the multiorgan, multiphenotype presentation of the disease. Such information may lie in the CFTR structure and its multiple functions (ion channel, ion channel modulator, interaction with other proteins like those having PDZ domain, etc.) or even beyond the involvement of a single molecule and may engross several other gene products. To maintain the proper cell functioning, a coordinated action of thousands of molecules and gene products are required; therefore, it should not come as a surprise if CFTR does not work alone to give rise to a diverse CF phenotype. In fact, it is now clear that defective CFTR affects many other regulatory molecules and enzymes at gene expression or function level. Hence, to identify these other gene products (mucins, ion channel proteins, etc.) that may potentially be involved in the diverse disease manifestations and to understand their association with the CFTR, would help in developing better ways to restore the ion transport defects, to prevent ductal blockage, and to find ways to resist and fight off infections.

The human genome sequence has been deduced. Having this information, microarray technology can be a useful tool for observing the expressional changes of thousands of genes simultaneously in response to CFTR dysfunction. Comparisons can be made on biopsied samples from CF and non-CF individuals to seek the gene expression profiles in a CF environment. Genes thus identified can be further studied to develop therapeutic strategies and even early diagnostic markers. High throughput screening technologies have been valuable in drug development, and efforts are in progress to test the drugs that can positively affect the ion channel functions. Future developments in this area may lead to a change in the treatment strategy from a symptomatic relief to direct targeting of the basic CFTR defect and other important etiological events of CF. While continuing efforts on these aspects, we should also focus on other exciting areas of research such as ion channel physiology, mucus secretion, gene modifiers, inflammation, and infection control. An enhanced understanding in these areas will not only be useful in developing supportive therapy of CF, but can also give insights to several other diseases.

In 1990, scientists succeeded to clone the CFTR gene and corrected the CFTR defect *in vitro* that restored the chloride channel transport mechanism.[64] Since then, studies have been focused on gene therapy with limited success *in vivo*. While this approach may be useful in the future for the treatment of respiratory manifestations of CF and may also provide an ultimate cure, it will remain a hard task to target the gene correction in the pancreas. Moreover, pancreatic injuries in CF may appear during prenatal development. Therefore, in this regard, efforts should be concentrated on supplemental enzyme therapy and developing better ways of disease management.

ACKNOWLEDGMENTS

The authors of this chapter were supported by the grants from the National Institutes of Health (RO1 CA78590) and the Nebraska Research Initiative. Ms. Kristi L.W. Berger, editor, Eppley Institute, is greatly acknowledged for editorial assistance.

REFERENCES

1. Anderson D.H. 1938. Cystic fibrosis of the pancreas and its relation to celiac disease. *Am. J. Dis. Child.* 56:344–399.
2. Cystic Fibrosis Foundation Patient Registry Annual Data Report, 2001.
3. Rommens J.M., Iannuzzi M.C., Kerem B., Drumm M.L., Melmer G., Dean M., Rozmahel R., Cole J.L., Kennedy D., and Hidaka N. 1989. Identification of the cystic fibrosis gene: chromosome walking and jumping. *Science* 245:1059–1065.
4. Riordan J.R., Rommens J.M., Kerem B., Alon N., Rozmahel R., Grzelczak Z., Zielenski J., Lok S., Plavsic N., and Chou J.L. 1989. Identification of the cystic fibrosis gene: cloning and characterization of complementary DNA. *Science* 245:1066–1073.
5. Welsh M.J. and Smith A.E. 1993. Molecular mechanisms of CFTR chloride channel dysfunction in cystic fibrosis. *Cell* 73:1251–1254.
6. Kunzelmann K. 2001. CFTR: interacting with everything? *News Physiol. Sci.* 16:167–170.
7. Sheppard D.N. and Welsh M.J. 1999. Structure and function of the CFTR chloride channel. *Physiol. Rev.* 79:S23–S45.
8. Wang W., He Z., O'Shaughnessy T.J., Rux J., and Reenstra W.W. 2002. Domain-domain associations in cystic fibrosis transmembrane conductance regulator. *Am. J. Physiol Cell. Physiol.* 282:C1170–C1180.
9. Meacham G.C., Lu Z., King S., Sorscher E., Tousson A., and Cyr D.M. 1999. The Hdj-2/Hsc70 chaperone pair facilitates early steps in CFTR biogenesis. *EMBO J.* 18:1492–1505.
10. Cystic Fibrosis Mutation Database; http://www.genet.sickkids.on.ca/cftr/.
11. Kulczycki L.L., Kostuch M., and Bellanti J.A. 2003. A clinical perspective of cystic fibrosis and new genetic findings: relationship of CFTR mutations to genotype-phenotype manifestations. *Am. J. Med. Genet.* 116A:262–267.

12. Ahmed N., Corey M., Forstner G., Zielenski J., Tsui L.C., Ellis L., Tullis E., and Durie P. 2003. Molecular consequences of cystic fibrosis transmembrane regulator (CFTR) gene mutations in the exocrine pancreas. *Gut* 52: 1159–1164.

13a. Kane K. 1988. Cystic fibrosis: recent advances in genetics and molecular biology. *Ann. Clin. Lab. Sci.* 18:289–296.

13b. Cutting G.R. Genetic epidemiology and genotype/phenotype correlations. In *Program and abstracts*. NIH Consensus Development Conference on Genetic Testing for Cystic Fibrosis. April 14–16, 1997.

14. Welsh M.J. and Smith A.E. 1995. Cystic fibrosis. *Sci. Am.* 273:52–59.

15. Grand R.J., Schwartz R.H., di Sant'Agnese P.A., and Gelderman A.H. 1966. Macroscopic cysts of the pancreas in a case of cystic fibrosis. *J. Pediatr.* 69:393–398.

16. Park R.W. and Grand R.J. 1981. Gastrointestinal manifestations of cystic fibrosis: a review. *Gastroenterology* 81:1143–1161.

17. Harris A. and Coleman L. 1987. Establishment of a tissue culture system for epithelial cells derived from human pancreas: a model for the study of cystic fibrosis. *J. Cell. Sci.* 87:695–703.

18. Oppenheimer E.H. and Esterly J.R. 1975. Pathology of cystic fibrosis review of the literature and comparison with 146 autopsied cases. *Perspect. Pediatr. Pathol.* 2:241–278.

19. Daneman A., Gaskin K., Martin D.J., and Cutz E. 1983. Pancreatic changes in cystic fibrosis: CT and sonographic appearances. *Am. J. Roentgenol.* 141:653–655.

20. Soyer P., Spelle L., Pelage J.P., Dufresne A.C., Rondeau Y., Gouhiri M., Scherrer A., and Rymer R. 1999. Cystic fibrosis in adolescents and adults: fatty replacement of the pancreas-CT evaluation and functional correlation. *Radiology* 210:611–615.

21. Feigelson J., Pecau Y., Poquet M., Terdjman P., Carrere J., Chazalette J.P., and Ferec C. 2000. Imaging changes in the pancreas in cystic fibrosis: a retrospective evaluation of 55 cases seen over a period of 9 years. *J. Pediatr. Gastroenterol. Nutr.* 30:145–151.

22. Imrie J.R., Fagan D.G., and Sturgess J.M. 1979. Quantitative evaluation of the development of the exocrine pancreas in cystic fibrosis and control infants. *Am. J. Pathol.* 95:697–707.

23. King A., Mueller R.F., Heeley A.F., and Robertson N.R. 1986. Diagnosis of cystic fibrosis in premature infants. *Pediatr. Res.* 20:536–541.

24. Gibson L.E. and Cooke R.E. 1959. Test for concentration of electrolytes in sweat in cystic fibrosis of the pancreas utilizing pilocarpine by iontophoresis. *Pediatr.* 23:545–549.

25. Mangos J.A. and McSherry N.R. 1967. Sodium transport: inhibitory factor in sweat of patients with cystic fibrosis. *Science* 158:135–136.

26. Knowles M., Gatzy J., and Boucher R. 1981. Increased bioelectric potential difference across respiratory epithelia in cystic fibrosis. *N. Engl. J. Med.* 305:1489–1495.

27. Quinton P.M. 1982. Suggestion of an abnormal anion exchange mechanism in sweat glands of cystic fibrosis patients. *Pediatr. Res.* 16:533–537.

28. Quinton P.M. 1983. Chloride impermeability in cystic fibrosis. *Nature* 301:421–422.

29. Freedman S.D., Blanco P., Shea J.C., and Alvarez J.G. 2000. Mechanisms to explain pancreatic dysfunction in cystic fibrosis. *Med. Clin. North Am.* 84:657–664.

30. Namkung W., Lee J.A., Ahn W., Han W., Kwon S.W., Ahn D.S., Kim K.H., and Lee M.G. 2003. Ca2+ activates cystic fibrosis transmembrane conductance regulator- and Cl−-dependent HCO3 transport in pancreatic duct cells. *J. Biol. Chem.* 278:200–207.

31. Shumaker H., Amlal H., Frizzell R., Ulrich C.D., and Soleimani M. 1999. CFTR drives Na+-nHCO-3 cotransport in pancreatic duct cells: a basis for defective HCO-3 secretion in CF. *Am. J. Physiol.* 276:C16–C25.

32. Scheele G.A., Fukuoka S.I., Kern H.F., and Freedman S.D. 1996. Pancreatic dysfunction in cystic fibrosis occurs as a result of impairments in luminal pH, apical trafficking of zymogen granule membranes, and solubilization of secretory enzymes. *Pancreas* 12:1–9.

33. Salvatore F., Scudiero O., and Castaldo G. 2002. Genotype-phenotype correlation in cystic fibrosis: the role of modifier genes. *Am. J. Med. Genet.* 111:88–95.

34. Merlo C.A. and Boyle M.P. 2003. Modifier genes in cystic fibrosis lung disease. *J. Lab. Clin. Med.* 141:237–241.

35. Durie P.R. 2000. Pancreatic aspects of cystic fibrosis and other inherited causes of pancreatic dysfunction. *Med. Clin. North Am.* 84:609–620.

36. Stormon M.O. and Durie P.R. 2002. Pathophysiologic basis of exocrine pancreatic dysfunction in childhood. *J. Pediatr. Gastroenterol. Nutr.* 35:8–21.

37. Durie P.R. 1997. Inherited causes of exocrine pancreatic dysfunction. *Can. J Gastroenterol.* 11:145–152.

38. Durie P.R. 1992. Pathophysiology of the pancreas in cystic fibrosis. *Neth. J. Med.* 41:97–100.

39. Kerem E., Corey M., Stein R., Gold R., and Levison H. 1990. Risk factors for Pseudomonas aeruginosa colonization in cystic fibrosis patients. *Pediatr. Infect. Dis. J.* 9:494–498.

40. Kristidis P., Bozon D., Corey M., Markiewicz D., Rommens J., Tsui L.C., and Durie P. 1992. Genetic determination of exocrine pancreatic function in cystic fibrosis. *Am. J. Hum. Genet.* 50:1178–1184.

41. Gannon M. 2001. Molecular genetic analysis of diabetes in mice. *Trends Genet.* 17:S23–S28.

42. Moran A., Doherty L., Wang X., and Thomas W. 1998. Abnormal glucose metabolism in cystic fibrosis. *J. Pediatr.* 133:10–17.

43. Mackie A.D., Thornton S.J., and Edenborough F.P. 2003. Cystic fibrosis-related diabetes. *Diabet. Med.* 20:425–436.

44. Hardin D.S., LeBlanc A., Para L., and Seilheimer D.K. 1999. Hepatic insulin resistance and defects in substrate utilization in cystic fibrosis. *Diabetes* 48:1082–1087.

45. Reisman J., Corey M., Canny G., and Levison H. 1990. Diabetes mellitus in patients with cystic fibrosis: effect on survival. *Pediatrics* 86:374–377.

46. Cohn J.A., Friedman K.J., Noone P.G., Knowles M.R., Silverman L.M., and Jowell P.S. 1998. Relation between mutations of the cystic fibrosis gene and idiopathic pancreatitis. *N. Engl. J. Med.* 339:653–658.

47. Sharer N., Schwarz M., Malone G., Howarth A., Painter J., Super M., and Braganza J. 1998. Mutations of the cystic fibrosis gene in patients with chronic pancreatitis. *N. Engl. J. Med.* 339:645–652.

48. Taylor C.J. 1999. Chronic pancreatitis and mutations of the cystic fibrosis gene. *Gut* 44:8–9.

49. Witt H. 2003. Chronic pancreatitis and cystic fibrosis. *Gut* 52 (Suppl. 2): 31–41.

50. Norton I.D., Apte M.V., Dixson H., Trent R.J., Haber P.S., Pirola R.C., and Wilson J.S. 1998. Cystic fibrosis genotypes and alcoholic pancreatitis. *J. Gastroenterol. Hepatol.* 13:496–499.

51. Gaia E., Salacone P., Gallo M., Promis G.G., Brusco A., Bancone C., and Carlo A. 2002. Germline mutations in CFTR and PSTI genes in chronic pancreatitis patients. *Dig. Dis. Sci.* 47:2416–2421.

52. Durno C., Corey M., Zielenski J., Tullis E., Tsui L.C., and Durie P. 2002. Genotype and phenotype correlations in patients with cystic fibrosis and pancreatitis. *Gastroenterology* 123:1857–1864.

53. Durie P.R. 1998. Pancreatitis and mutations of the cystic fibrosis gene. *N. Engl. J. Med.* 339:687–688.

54. Rosenstein B.J. and Cutting G.R. 1998. The diagnosis of cystic fibrosis: a consensus statement. Cystic Fibrosis Foundation Consensus Panel. *J. Pediatr.* 132:589–595.

55. Neglia J.P., FitzSimmons S.C., Maisonneuve P., Schoni M.H., Schoni-Affolter F., Corey M., and Lowenfels A.B. 1995. The risk of cancer among patients with cystic fibrosis. Cystic Fibrosis and Cancer Study Group. *N. Engl. J. Med.* 332:494–499.

56. Cullen J.J., Weydert C., Hinkhouse M.M., Ritchie J., Domann F.E., Spitz D., and Oberley L.W. 2003. The role of manganese superoxide dismutase in the growth of pancreatic adenocarcinoma. *Cancer Res.* 63:1297–1303.

57. Sheldon C.D., Hodson M.E., Carpenter L.M., and Swerdlow A.J. 1993. A cohort study of cystic fibrosis and malignancy. *Br. J. Cancer* 68:1025–1028.

58. Tsongalis G.J., Faber G., Dalldorf F.G., Friedman K.J., Silverman L.M., and Yankaskas J.R. 1994. Association of pancreatic adenocarcinoma, mild lung disease, and delta F508 mutation in a cystic fibrosis patient. *Clin. Chem.* 40:1972–1974.

59. McIntosh J.C., Schoumacher R.A., and Tiller R.E. 1988. Pancreatic adenocarcinoma in a patient with cystic fibrosis. *Am. J. Med.* 85:592.

60. Stead R.J., Redington A.N., Hinks L.J., Clayton B.E., Hodson M.E., and Batten J.C. 1985. Selenium deficiency and possible increased risk of carcinoma in adults with cystic fibrosis. *Lancet* 19(2):862–863.

61. Singh A.P., Andrianifahanana M., Chauhan S.C., Pandey K.K., Moniaux N., Hollingsworth M.A., and Batra S.K. 2002. MUC4 mucin overexpression is associated with the cystic fibrosis phenotype. *Pediatr. Pulmonol.* Suppl. 24:200 (Sixteenth Annual North American Cystic Fibrosis Conference, New Orleans (October 3–6, 2002)).

62. Andrianifahanana M., Moniaux N., Schmied B.M., Ringel J., Friess H., Hollingsworth M.A., Buchler M.W., Aubert J.P., and Batra S.K. 2001. Mucin (MUC) gene expression in human pancreatic adenocarcinoma and chronic pancreatitis: a potential role of MUC4 as a tumor marker of diagnostic significance. *Clin. Cancer Res.* 7:4033–4040.

63. Singh A.P., Moniaux N., Chauhan S.C., Meza J.L., and Batra S.K. 2004. Inhibition of MUC4 expression suppresses pancreatic tumor cell growth and metastasis. *Cancer Res.* (In Press).

64. Drumm M.L., Pope H.A., Cliff W.H., Rommens J.M., Marvin S.A., Tsui L.C., Collins F.S., Frizzell R.A., and Wilson J.M. 1990. Correction of the cystic fibrosis defect in vitro by retrovirus-mediated gene transfer. *Cell* 62:1227–1233.

27

TOXICOLOGY OF THE
ENDOCRINE PANCREAS

Mehmet Yalniz and Parviz M. Pour

CONTENTS

27.1 INTRODUCTION

Embedded within the exocrine pancreas, pancreatic islets distinguish themselves from the surrounding tissue in several aspects, including toxicology. As a gatekeeper tissue of the pancreas,[1] it is equipped with a greater concentration of drug-metabolizing enzymes than the exocrine tissue (see Chapter 7). Moreover, several P450 enzymes are selectively expressed in islet cells. Consequently, some toxins primarily or exclusively affect the islet cells. Although the numbers of endocrine cell toxins have not been sufficiently sought out, a few toxins have been the subject of intense investigation. Streptozotocin (STZ) and alloxan are the most studied compounds.

27.2 STREPTOZOTOCIN

STZ, derived from the soil microorganism *Streptomycetes achromogenes* in 1960, has been found to have a significant antimicrobial action for a wide spectrum of organisms.[2,3] However, during the preclinical toxicology studies, it was found that this compound causes hyperglycemia when given by intravenous administration in rats and dogs within a few hours.[4,5] Therefore, its use as an antimicrobial agent was abandoned. At the same time, its antitumor activity was demonstrated.[6]

27.2.1 Diabetogenic Action of STZ

Tumor studies in the murines[4,5] and toxicological studies in dogs and rhesus monkeys[4,7] demonstrated the temporary or permanent hyperglycemic and potent diabetogenic action of STZ.[4,8] Therefore, STZ is generally used to induce both insulin-dependent and non-insulin-dependent diabetes in animal studies.

Characteristically, STZ primarily damages β-cells, leading to alterations in blood glucose and insulin levels. Hyperglycemia is observed as early as 2 hours after STZ administration with a concomitant drop in blood insulin levels. However, about 6 hours later, the opposite situation occurs: blood glucose levels drop while insulin levels begin to increase. Eventually, blood insulin levels decrease and hyperglycemia occurs. The responsiveness of β-cells to glucose, which temporarily returns to normal after an initial abolishment, is lost permanently.[9] Immunohistochemical studies in STZ-treated animals revealed that the insulin-positive areas were decreased significantly and exhibited vacuoles of the remaining β-cells.[10–12] In multiple low-dose model, after administration of STZ, a rapid loss of islets occurs within 3 days.[13,14] There are conflicting reports about the islet size after STZ administration. Li et al.[14] have found the islet area increased to an average of about 50% of the original size 28 days after low-dose STZ administration. However, Bonnevie-Nielsen et al.[15] found a reduction of the islet area was down to 31% on Day 6 and a further decline to 1% of the original area at Day 14. Li et al.[14] reported that after multiple low-dose STZ treatments of C57BL/Ks male mice, they observed a fraction of the islets of Langerhans disappeared and in the remaining islet tissue an expansion of α-cells occurred.

The dose range of STZ is wider than alloxan, another diabetogenic agent (see below). The toxic action of STZ on islet β-cells generally induced in two different ways, with multiple low doses or with a single high dose. When given in multiple low doses (40 mg/kg body weight/day/5 days), it induces insulitis and progresses to nearly complete destruction of β-cells associated with atrophy of the islets and diabetes.[16,17] When given as a single high dose, generally between 40 and 60 mg/kg body weight in rats[18] and hamsters and between 75 and 150 mg/kg body weight in other species, it rapidly destroys islet β-cells.[19]

In some studies, inflammation in sites, other than the islets of Langerhans, has been noticed following a multiple-dose treatment. Papaccio et al.[20] observed that histologically the ductal cells in close proximity to islets were also affected by inflammatory cells that extend from the islets, whereas ducts far from islets were generally free from inflammation. These results suggest that the initial action of STZ is wider spread and is a less specific process that later undergoes restriction.

27.2.2 The Mechanisms of Streptozotocin Action

STZ is a monofunctional nitrosourea derivative and a member of alkylnitrosoureas, a group of alkylating antineoplastic drugs, which are clinically active against a broad range of tumors.[21]

β-cell toxicity of STZ requires its uptake into the cells.[22] STZ (2-deoxy-2-(3-(methyl 3-nitrosoureido)-D-glucopyranose) consists of a 2-deoxyglucose moiety substituted in position C-2 with nitrosourea[23] and is a D-glucopyranose derivative of N-methyl-N-nitrosourea (MNU). Although both STZ and MNU are potent alkylating agents,[24] highly toxic, and carcinogenic,[25,26] only STZ has selective β-cell toxicity.[27] It is generally believed that this selective β-cell toxicity is related to the glucose moiety in its chemical structure. This specific structure is believed to be responsible for its affinity to the β-cell via the low affinity glucose transporter GLUT2, which is not merely a structural protein specific for the β-cell membrane but is a crucial constituent for recognition and entry of glucose as well as glucose-like molecules, such as STZ, in the plasma membrane.[22,28,29] The observation that the RINm5f rat insulinoma cell line, which does not express this glucose transporter, resists STZ toxicity[30,31] and becomes sensitive to the toxic action of this compound only after the expression of the GLUT2 glucose transporter in this cell line[28] confirms this hypothesis. It is also demonstrated that in insulin-producing cells, which do not express GLUT2, the cellular uptake and the toxicity of STZ is low. Accordingly, it appears that the reduced expression of GLUT2 prevents the diabetogenic action of STZ.[28,32] However, it is also observed that STZ itself restricts GLUT2 expression both *in vivo* and *in vitro* when administrated in multiple doses.[33,34]

The exact mechanism underlying the diabetogenic action of STZ has not been fully understood. It is generally believed that its toxic action relates to the DNA alkylating activity of its MNU moiety[22,24,35]; particularly at the O^6 position of guanine (see also Chapter 9 and Chapter 27).[36,37] STZ damages DNA comprising its fragmentation that leads to apparent depletion of nicotinamide adenine dinucleotide (NAD)[+], which in turn, inhibits insulin biosynthesis and secretion, and finally causes β-cell death through adenosine triphosphate (ATP) depletion.[19] The demonstration of various methylated purines in the tissues of STZ-treated rats provides further evidence for the DNA damaging action of STZ.[24] Poly ADP-ribosylation, which is activated with the DNA damaging effect of STZ, has been proposed to lead to these unfavorable consequences in the β-cells.[38–41] This notion has been confirmed by studies showing that the inhibition of this process with 3-aminobenzamide and nicotinamide (poly ADP-ribose inhibitors) prevents the toxicity of STZ.[42,43]

Nitric oxide (NO), which carries biological information, is a free radical gas[44] and has been proposed as a possible mediator in the damage to the insulin-producing β-cells.[45,46] It has been found that STZ generates NO in aqueous solutions.[47] Kroncke et al.[48] also demonstrated NO generation during cellular metabolization of STZ. Hence, it was speculated that NO contributes to STZ-induced DNA damage.[47–49] It has also been shown that scavenging with NO protects against STZ-induced DNA strand breakage.[48] This issue, however, is controversial.[50,51] Moreover, in a recent study, it has been shown that multiple low doses and a high single dose of STZ does not stimulate NO production at islet cell levels.[52] It seems that NO at least partially takes part in STZ islet toxicity. The certain role of NO, however, still is not clear.

In addition to NO, STZ generates reactive oxygen species (ROS), which also contribute to DNA fragmentation in the β-cells.[53,54] STZ inhibits the Krebs cycle[47] and consequently decreases oxygen consumption by mito-chondria considerably.[41] The result is a strong limitation of mitochondrial ATP production and the depletion of this nucleotide in β-cells.[41,55] This process enhances O_2^- radical generation by the xanthine oxidase system of the pancreatic β-cell,[56,57] stimulates hydrogen peroxide and hydroxyl radicals generation, and causes DNA fragmentation in isolated rat pancre-atic islets.[53,56,58] It has been found that the inhibition of the formation of these radicals restricts the β-cells cytotoxicity of STZ *in vitro*.[58] Although there are a few reports that suggest that free radicals may not be involved in the DNA damage by STZ,[19,59] most of the studies support the idea that the free radicals may play a role in the DNA toxicity action of this agent.[60]

NO and ROS, however, can act separately or form the highly toxic peroxynitrite (ONOO⁻). Recent evidence claims that peroxynitrite (and not

NO) is the potent trigger of DNA strand breakage.[19] Peroxynitrite formation activates poly ADP-ribosylation and could play a role in the pathogenesis of islet cell damage in response to STZ or NO compounds.

STZ also induces cell death by apoptosis and necrosis in pancreatic islet cells.[49,61] Cell death by apoptosis was observed in cultured pancreatic β-cell HIT-T15 and RINm5F after treatment with STZ.[49] In addition, Saini et al.[62] showed that STZ at low doses induces apoptosis and, at high doses, causes necrosis in a murine pancreatic β-cell line, INS-1. STZ also causes β-cell apoptosis by the generation of toxic radicals.[59,63]

One of the primary actions of the diabetogenic chemicals (STZ, alloxan) is at the plasma membrane level of β-cells.[64,65] STZ modifies the molecular structure of phospholipids, particularly phosphatidylcholine, which is a major phospholipid of the outer leaflet of the plasma membrane, resulting in a decrease of membrane fluidity in β-cells.[66–68] Hence, it was suggested that STZ acts directly on the plasma membrane and induces alterations of islet cell membrane properties[64] and preservation of membrane integrity could protect β-cells from cytotoxic insults.

Recently, a new glycosylation pathway (other than the better-known N-linked pathway), O-linked protein glycosylation, has been described and implicated in diseases as diverse as cancer and Alzheimer's. The β-cells appear to be especially susceptible to disruption of the O-linked protein glycosylation pathway and an important link between β-cell O-linked protein glycosylation and β-cell apoptosis has been recently shown.[69] It is assumed that STZ irreversibly increases the O-glycosylation and subsequent β-cell apoptosis, as claimed in some recent reports that the diabetogenic agent STZ causes β-cell toxicity by this pathway.[70]

In brief, STZ induces DNA damage by alkylation of specific sites on DNA and that free radicals and NO generated during STZ metabolism seems to play a role in the mechanism by STZ. Severe DNA damage by STZ also results in cell death by apoptosis or necrosis. Furthermore, some other pathways like an increase in O-linked protein glycosylation and modification of the cellular membrane by STZ are also proposed as the mechanisms that underlie the toxic action of STZ.

27.2.3 Species Differences of STZ Action

STZ causes necrosis or marked degenerative changes in the β-cells with nuclear pyknosis and cytoplasmic vacuolization and produces permanent diabetes in different species, including the rat, mouse, guinea pig, Chinese hamster, Syrian golden hamster, and monkey.[8,71–73] Syrian hamsters have been shown to be more useful than rats as an animal model of human diabetes mellitus.[74] Most species indicate a high acute mortality to STZ injection.[75] Syrian hamsters also respond to a single dose of STZ with β-cell necrosis, but in contrast to other species, do not result in a higher percentage of permanent diabetes. β-cells regenerate and a high percentage of hamsters subsequently recover spontaneously from their diabetes.[76] Multiple daily doses of STZ, however, result in permanent diabetes in a large number of hamsters. Recovery from STZ-induced diabetes has also been reported in rats.[77] However, recovery required 8 to 18 months in rats and the induced diabetes was relatively mild.

We examined the regenerative properties and capacity of β-cells in STZ-treated hamsters, some of which received insulin.[10] A single dose (50 mg/kg body weight) of STZ induced diabetes in all hamsters, causing degeneration and depletion of β-cells, proliferation of glucagon and somatostatin cells, and their derangement within the islets as observed in other species.[78–80] However, 10 days after STZ treatment, degeneration stopped and pancreatic islet and ductal cells began to proliferate, which peaked 14 days post-STZ treatment (Figure 27.1A). These hamsters also recovered from their diabetes spontaneously. It was assumed that the regeneration of islet cells was triggered by hypoinsulinemia through a feedback mechanism. Confirming this mechanistic event was that insulin therapy prevented the spontaneous recovery and produced a persistent severe hyperglycemia.[10] Insulin also inhibited DNA synthesis in ductal, ductular, and acinar cells in STZ-pretreated hamsters, but not in normoglycemic control hamsters treated with insulin alone.[10] These results demonstrated a controversial deleterious effect of exogenous insulin in the course of STZ-induced diabetes in hamsters. In this study, the recovery and lack of recovery was associated with β-cell regeneration and lack of regeneration.

In a succeeding study, we found that β-cell regeneration in STZ-induced diabetic hamsters occurred primarily from undifferentiated cells within the islet (Figure 27.1A) and exogenous insulin inhibited this differentiation.[81] Nagasao et al.[82] have examined whether centroacinar and intercalated duct cells can serve as stem cells to induce recovery after the administration of STZ in rats. They found that rat pancreatic endocrine cells seemed to recover from newly generated cells derived from intercalated ductal and centroacinar cells. In a recent study, the recovery from Type 1 diabetes induced by multiple low doses of STZ in transgenic mice expressing the insulin-like growth Factor I (IGF-I) was studied (see also Chapter 13). It was found that the expression of IGF-I in β-cells restored β-cell mass and normoglycemia.[83] Thus, the authors speculated that IGF-I expression in β-cells of these transgenic mice might protect them against the oxidative and apoptotic effects of STZ.

The susceptibility of human islets to STZ damage, which is not well established, is a controversial issue. In a study, it was found that human islets are remarkably resistant to the diabetogenic effect of STZ *in vivo*.[84] Among the potential mediators of β-cell damage, cytokines and cytokine-

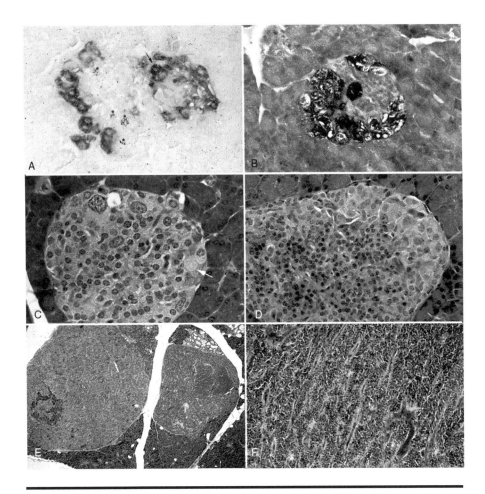

Figure 27.1   Streptozotocin-induced lesions. A: An islet of a hamster treated with a single dose of STZ 1 week earlier. Degeneration of the bulk of β-cells (center) surrounded by glucagon cells (black). Four cells in necrotic area show labeling with tritiated thymidine (*). Labeling is seen in one glucagons cell (arrow). Combined immunohistochemistry and autoradiography, ABC methods, ×75. B: The intact β-cells contain a large amount of glycogen (dark black). Islet of a STZ-treated hamster. ×50. C: The proliferative pattern of an islet 20 weeks after STZ-treatment. Note the presence of several cells with giant nuclei or foamy cytoplasm (arrow). One cell has a large opaque cytoplasm (upper middle field). Strikingly, regeneration seems to take place in the islet periphery (see Figure 27.1D). H&E ×75. D: In a STZ-treated hamster, the peripheral portion of an enlarged islet is replaced by atypical cells with foamy cytoplasm and pleomorphic nuclei. This hamster had an islet cell tumor in another part of the pancreas. H&E ×75. E: Two islet cell tumors in a hamster. Tumors show different size and morphology. The smaller tumor had hemorrhagic area. Ductular formation is present in the upper pole of the small tumor. This hamster was treated with a combination of STZ and BOP. H&E ×32. F: STZ-induced tumors generally show various histological architecture. They may show trabecular, medullary, or cystic patterns. The staining with antibodies against islet hormones is remarkably heterogeneous. H&E ×50.

induced NO production has received special attention. It has been suggested that these agents damage rat islets *in vitro*, but their effects on cultured human islets are less pronounced, despite the production of similar amounts of NO.[46] Eizirik and coworkers[85] investigated the differences between humans and rodents and have found that human β-cells are resistant to nitroprusside (a NO donor),[86] STZ, or alloxan (a generator of free radicals)[87] at concentrations that decrease survival and function of rat or mouse β-cells. It is reported that human fetal islets grafted into nude mice were not destroyed by injections of STZ, in spite of the adequate uptake of the drug by the human tissue.[88] It is well known that the time between the appearance of islet autoimmunity and the clinical onset of insulin-dependent diabetes mellitus (IDDM) is much longer in humans than in mice (non-obese diabetic mice) and, in particular, rats. Because the degree of resistance to various toxins suggested from the findings of Eizirik et al.[85] seems to follow a similar pattern (i.e., human→ mouse→rat), increased resistance to STZ injury may contribute to a longer period in humans. It appears that STZ does not produce any significant clinical diabetogenic effect in humans.[25] Rabbits have also shown to be highly resistant to STZ with little metabolic or histologic evidence of β-cell damage at doses up to 300 mg/kg and with higher doses having severe systemic toxicity.[89]

De Vos et al.[90] demonstrated interspecies differences in the expression of GLUT2 *in vitro*, which may explain the difference in sensitivity to toxic action of STZ. They found that human islet cells express little GLUT2 compared to GLUT1 or GLUT3 and that in the human β-cell the GLUT2 expression level is markedly lower than in rat β-cells. This low expression of GLUT2, which is a crucial factor for STZ uptake to β-cells, in the human β-cells could be responsible for the resistance to the toxic action of STZ in man. The exact mechanism of this process warrants further studies.

27.2.4 Tumorigenic Action of STZ

In addition to its antibiotic and diabetogenic properties, the genotoxic and tumorigenic actions of STZ have been demonstrated. This property, however, is not restricted to the pancreas as STZ induces tumors in other organs as well, including the liver and kidney.[91,92]

Pancreatic islets are also a target of the tumorigenic action of STZ. It has been found that STZ, in combination with nicotinamide, which prevents the acute toxicity of STZ, induces islet cell tumors with a high frequency after a long latency period.[93] The tumors have been described as a well-differentiated type and resemble the normal islet tissue, both morphologically and functionally, as they are rich in the typical β-cells and release insulin in response to glucose, both *in vivo* and in culture.[93]

In this study, it was also suggested that there was a different sensitivity of the different islet cell types to the chemically induced transformation. Doi[94] also observed functioning pancreatic islet cells tumors 407 days after STZ administration. However, Yoshino et al.[95] found that STZ induces two types of islet cell tumors: one is insulin-producing and insulin-secreting, whereas the other is insulin-producing but not insulin-secreting.

The activation of poly-ADP ribosylation due to fragmented DNA is described previously in the diabetogenic action of STZ. Although it has been shown that the prevention of poly-ribosylation with poly(ADP-ribose) synthetase inhibitors leads to the maintenance of β-cell functions normally, DNA strand breaks have not been prevented at all[96]; therefore, residual DNA damage may continue to affect the β-cells. In line with this concept, Okamoto and Yamamoto[96] demonstrated that about 1 year after the combined administration of alloxan or STZ with poly(ADP-ribose) synthetase inhibitors to rats, diabetes did not develop but islet β-cell tumors were found frequently. These results suggested that DNA breaks initiate two kinds of pathological states in β-cells — one is degenerative and the other is oncogenic. A human insulinoma case, after STZ therapy for metastatic gastrinoma, was observed by Bar et al.[97]

Yagihashi and Nagai[98] demonstrated that STZ-induced tumors mostly consisted of β-cells. Over half of the tumors examined showed mixed cellularity with considerable numbers of A cells and small numbers of D or PP cells. They claimed that the multiplicity of the endocrine cells of rat islet cell tumors might be an expression of cellular dedifferentiation of tumor cells, which could redifferentiate into the whole range of components of the endocrine pancreas.

The histologic patterns of islet cell tumors in hamsters are comparable to those induced in rats.[98] The induced islet cell tumors appear to be hormonally inactive, and no changes in blood glucose level were detectable in islet cell tumor-bearing hamsters. In a study in STZ-treated hamsters,[99] islet cell adenomas were primarily of a pleomorphic cell type and had pseudoinfiltrative tendencies. These neoplastic cells were found to develop from the most peripheral portion of islets, most probably from the periinsular ductules, which are believed to give rise to islet cells.[78,100] If this is the case, then, ductular cells also seem to be principal targets for STZ, which argues with the concept that STZ toxicity is β-cell specific. The induction of ductal and ductular lesions by STZ treatment alone and a significantly higher incidence of ductal-ductular carcinomas by STZ plus BOP (a pancreatic ductal cell cancer-producing nitrosamine) than by BOP alone[99] are further support for this possibility. In this context, Shepherd et al.[101] demonstrated the potent carcinogenic action of STZ on cultured rat pancreatic duct epithelial cells.

Because STZ and BOP have been shown to exert their oncogenic effect by the methylation of DNA[24,102] and large doses of BOP analogues cause islet cell necrosis,[103,104] it appears that STZ and BOP act by similar mechanisms. Yet, it is unclear why the affected ductular cells differentiate toward islet cells after STZ and to ductular cells after BOP. Theoretically, the quantity of DNA alkylation induced by STZ and BOP may be a determining factor in the differentiating process. Much of N-methylurea, the carcinogenic moiety of STZ, taken up by islet cells can decompose within the islet cells and generate cytotoxic substances, whereas only a smaller portion affects ductular cells. In fact, the uptake of STZ in the exocrine pancreas and its alteration at a subcellular level have been shown.[105] Not only the quantity, but also the quality of DNA alkylation may be important in the phenotypic expression of resulting tumors (STZ+BOP).

STZ has been found to be carcinogenic in rats, mice, and hamsters. In primary cultures of human and rat kidney cells, STZ induces neoplastic transformation.[106] Hence, STZ may be a potential carcinogen also in humans; although this possibility has been claimed,[97] it is not clear yet.

Differences seem to exist about the target cells of STZ on its acute and chronic effect. Although its acute toxicity is directed toward β-cells, its chronic action is on other cell types. Although it is believed that tumor cells derive from the regenerated or surviving β-cells, histological observations clearly pointed to the development of malignant cells in the islet periphery, where the cells initially show the presence of a large amount of glycogen (Figure 27.1B to Figure 27.1D). As in human tumors, the STZ-induced islet cell lesions show various morphological patterns even within the same animal with multiple tumors (Figure 27.1E and Figure 27.1F). Also, the immunoreactivity of the tumor cells to the antibodies against islet cell hormones shows great variations.

27.2.5 Antitumor Action of STZ

The potential antitumoral activity of STZ was demonstrated against mouse leukemia L1210.[6] This finding opened clinical trials for tumor therapy with STZ. It was used in several cancer types as an antitumor agent, particularly in metastatic insulinoma. In patients with insulinoma, tumor growth was significantly decreased.[107] Subsequently, the efficacy of STZ as a therapeutic agent on this rare tumor has been reported from numerous clinical studies.

Both β-cell and non-β-cell islet tumors that secrete various hormones (e.g., insulin, glucagons, gastrin) are suggested to be responsive to STZ, in both biochemical and tumor responses.[25] It has also been shown that STZ have similar effects on non-functioning islet cell tumors. A comparable response rate (36%) was found, regardless of the function of the tumor, by Moertel et al.[108] The antitumor activity of STZ against metastatic pancreatic islet tumors was also demonstrated in dogs.[109] In humans, interestingly, diabetes mellitus was not identified, though some patients had reversible glucose tolerance.[110] However, in dogs with insulinoma diabetes mellitus occurred following STZ treatment.[109]

In general, the overall success of STZ in the treatment of insulinoma is marginal because islet tumors are usually composed of mixed cell populations unresponsive to STZ.

Hyperinsulinemic hypoglycemic infancy, a rare genetic disorder, has two major different histologic appearances[111]: diffuse nesidioblastosis and a focal nodular form with a discrete regional adenomatous hyperplasia.[112] Due to selective β-cell toxicity, STZ was used in the treatment of this disease. Treatment of some patients with STZ has resulted in tumor cell damages ranging between 0 and 60% (Pour; unpublished observation).

Remarkably, STZ-treatment has been found to suppress or even abolish the development of pancreatic cancer in the BOP-hamster model.[113] Notably, the degree of this inhibition closely paralleled the severity of diabetes induced by STZ.[114] The inhibitory effect of STZ on pancreatic ductal cell adenocarcinoma occurred only when diabetes was present.[115] This observation supported the hypothesis that the origin of pancreatic ductal cell cancer resides in STZ-responsive islet cells.[116] The assumption that the protective effect of STZ was related to the absence of insulin, which has a cell growth-promoting effect, could not be confirmed[117] as the treatment with insulin did not overcome the inhibitory action of STZ on pancreatic carcinogenesis. Remarkably, although STZ-treated hamsters fully recovered from diabetes after 70 days, those treated with insulin remained hyperglycemic despite daily insulin treatment and showed a profound atrophy of islets.[10] The results indicated that the preventive effect of STZ on pancreatic cancer induction is unrelated to insulin or the action of insulin on tumor induction and growth is local or paracrine.[10,117] A single injection of STZ, at a dose that destroys only a portion of islet β-cells, inhibited pancreatic tumor induction; whereas multiple doses, leading to a complete destruction of islet cells, prevented pancreatic carcinogenesis, even when hamsters are treated with high doses of the carcinogen weekly.[118]

27.3 ALLOXAN

Alloxan (2,4,5,6-tetraoxyprimidine; 5,6-dioxyuracil), a uric acid derivative, was isolated in 1818 as the first pyrimidine derivative and was named by Wohler and Liebig who described its synthesis by uric acid oxidation.[119] The diabetogenic action of alloxan was shown by Dunn and coworkers in 1943,[120] who demonstrated its selective islet cell necrosis in rabbits.

Since then, alloxan has been widely used as an experimental diabetes model together with STZ.

In initial studies, it was observed that alloxan destroys the insulin-producing β-cells selectively,[121] while the α-cells have been shown to be resistant to this effect.[122] Diabetes induced with alloxan in animals displayed similar symptoms observed to those in human diabetics.[123]

The disadvantage of alloxan as a diabetogenic drug is that it loses its stability under physiological conditions.[124] The adjustment of its diabetogenic dose is difficult because several factors, such as fasting state and diet, age, the route of its administration, and species differences, interfere with its action.[125] It is also noteworthy that, even a small increase of its dose may be generally toxic, causing the loss of many experimental animals. This loss is mostly due to renal tubular cell toxicity, especially when high doses of alloxan are used.[126] Dunn et al.[120] found that alloxan in doses of 100 to 500 mg/kg is lethal to rabbits by specifically destroying the pancreatic islet cells and the epithelium of the renal convoluted tubules. Despite this, alloxan has been used intensively in studies of β-cell physiology and diseases.

Alloxan displays its diabetogenic action via parenteral way and its effective dose differs greatly upon the route of administration. In rats, its diabetogenic dose by intravenous injection is 65 mg/kg body weight,[127,128] whereas its effective intraperitoneal and subcutaneous dose is at least two or three times higher.[21,129] In dogs, doses between 50 to 75 mg/kg cause typical diabetic symptoms.[130] In the Syrian golden hamster, the effective dose is 60 mg/kg.[131]

Fasting state increases the susceptibility of pancreatic β-cells to alloxan toxicity.[129,132] Diets containing or lacking different nutrients also affect the alloxan toxicity.

Increased blood glucose prevents alloxan β-cell toxicity partially, which reverses with increasing the dose of alloxan.[132–134] A common recognition site on the β-cell for alloxan and glucose has been suggested to be responsible for this protection; however, the exact mechanism underlying the protective effect of glucose, is not clear yet and remains to be elucidated.

Young et al.[135] demonstrated that alloxan was almost ineffective in protein–calorie malnourished rats and in the control group it caused diabetes (100%). The finding was confirmed in a subsequent study[136] and led to the suggestion that a low-protein diet may lead to a defect in the alloxan–glucose recognition site.

Goldner et al.[130] examined the alloxan effect in rabbits, rats, guinea pigs, cats, dogs, and pigeons, all of which were found to be sensitive to alloxan in various degrees. They reported that permanent diabetes could be produced in rats, rabbits, and, with the most significant results, in dogs.

A single injection of alloxan caused a prompt onset of hyperglycemia and glycosuria within 24 to 48 hours in dogs.

Alloxan causes characteristic changes in blood sugar and insulin levels similar to the effect of STZ. After alloxan treatment, a typical three-phase blood sugar curve develops: initial hyperglycemia, followed by hypoglycemia, and permanent hyperglycemia.[137]

Weaver et al.[138] observed that alloxan evokes an abrupt rise in insulin secretion in isolated rat islets, regardless of the presence of glucose. Insulin release during the 5-min exposure to alloxan reached its maximum rate after 2 to 3 min of exposure and then rapidly declined. However, this process did not happen after repetitive exposures to alloxan. In this study, the preventive effects of 3-0-methyl-D-glucose, on the insulin release (stimulation and subsequent inhibition), during alloxan exposure was also demonstrated. Consequently, it has been suggested that alloxan and D-glucose are competing for a putative common site on the β-cells.

A high functional activity on pancreatic β-cells, along with inhibitory effects after alloxan administration, has also been observed in duct-ligated rats.[139] It was proposed that alloxan may have stimulatory effects on the pancreatic islets during the first hours of exposure. A short-term rapid increase in insulin release and subsequent complete suppression of the islet response to glucose, even though in high concentrations, was also demonstrated *in vitro*[140] and *in vivo*.[132] House et al.[141] found similar results in alloxan treated hamsters. An initial β-cell stimulatory effect of alloxan was found in the early phase reflected by increased granulation, vascular engorgement, and increased cytoplasmic nucleic acid. Additionally, they found that during initial hyperglycemia, liver glycogen was almost completely lost and nucleic acids were greatly reduced. The β-cells lost their ability to synthesize insulin between 2 and 5 hours, followed by a 95% degeneration of β-cells and the replacement by mononuclear cells between 5 and 24 hours.

Alloxan decreases the number of insulin-producing β-cells, but increases the number of glucagon-producing α-cells in the rat pancreas.[142] Rastogi and coworkers[143] found that 70% of islets were destroyed in dogs by alloxan. A marked increase in glucagon, but especially of somatostatin cells in residual islets, could explain the unchanged islet size despite the absence of β-cells. Also, in dogs,[13] a single dose of alloxan led to a complete degranulation of β-cells with shrinkage of the entire cell body. The number of the agranular cells gradually decreased, and by the end of the second week, few of the β-cells could be seen. The number of α-cells seemed to increase. The islets themselves became smaller and less numerous in the course of alloxan diabetes.

When the blood glucose levels were determined 2 weeks after alloxan treatment in hamsters, it was found that only 13% of the hamsters were diabetic.[144] This result contrasts with a similar study in the same species, where 58.1% of the hamsters were affected.[141] The rapid and complete metabolization of alloxan[145] may be the reason of the low percentage that was found in the former study.

Despite the intensive studies on the functional and morphological consequences of islet cell toxicity of alloxan, the results are still conflicting.

In addition to the islet cells, alloxan also affects the exocrine pancreatic tissue. Grossman and Ivy[146] reported that injury to the exocrine tissue was reflected by decreased sensitivity of the organ to secretin stimulation. Tiscornia et al.[147] studied sequential secretory processes in 11 dogs following the administration of alloxan. From 15 min to 24 hours after alloxan treatment, they found a significant increase in the rates of flow of bicarbonate secretion and amylase elaboration. Following the immediate increase in these parameters, the flow, the bicarbonate secretion, and the rate of enzyme secretion decreased for several days, and by the end of the first week, a recovery toward the control levels of secretion was observed. These secretory depressions were associated with mild ductular necrosis followed by pancreatic ductular recovery and subsequent glycogen deposition in the ductules. The changes in the flow rate of both bicarbonate and enzyme secretion supports the involvement of the exocrine tissue in alloxan toxicity. The most conspicuous feature of the pancreatic lesions observed in another study in alloxan-treated dogs was the extreme vacuolation of the epithelium of all intralobular ducts in the later stages of β-cell damage.[130]

27.3.1 The Mechanisms of Alloxan Action

Lenzen and Munday[148] found that alloxan is a hydrophilic and chemically unstable compound with a 1.5 min half-life at pH 7.4 and 37° in phosphate buffer. This short half-life contributes to difficulties in investigating and interpreting its mechanistic action. Because of the short half-life of alloxan, Szkudelski et al.[132] examined the insulin levels 2 min after alloxan administration in rats and found a sudden increase of insulin. These results indicated a direct effect of alloxan on pancreatic β-cells. The results also showed that in adequate doses, alloxan could reach the pancreas in deleterious amounts, albeit with short half-life.

Malaisse et al.,[149] using radioactive alloxan, also found the rapid uptake of alloxan by pancreatic β-cells. It has been shown that alloxan enters the intracellular space of islets and undergoes a subsequent internal decomposition.[150] Alloxan accumulates rapidly in islets and its uptake shows a time- and temperature-dependent manner. The rapid accumulation of alloxan in the liver[151] may explain the rapid metabolism of alloxan.

Boquist and colleagues[152] investigated the uptake of radiolabeled alloxan and found radioactivity in all organs with quantitative differences: endocrine pancreas greater than liver greater than exocrine pancreas and heart, regardless of the administration route. The same group of investigators also showed damage to mitochondria after alloxan treatment.[153] The earliest β-cell changes in the animals were localized within mitochondria, which showed swelling and disruption of inner and occasionally outer membranes. Later, many mitochondria were disintegrated, and the endoplasmic reticulum and golgi complexes disorganized. The secretory granules of β-cells were preserved, although sometimes with atypical configuration, in degenerating but nonnecrotic β-cells, suggesting that insulin stored in granules is not released until the cells are necrotic. Finally, frank necrosis was seen in some β-cells, whereas non-β-cells were unaffected. Based on these observations, the author proposed that the primary site of alloxan action in the mitochondria of β-cells, resulting in apoptosis, as was confirmed by others.[154]

In early studies, it has been shown that alloxan interacts with sulfhydryl enzymes, which are believed to be essential only for the function of the pancreatic β-cell.[155] The inhibition of the hexokinase enzyme by alloxan hypothesis has also been proposed.[125] Hexokinases are a group of isoenzymes that phosphorylate the glucose with a high affinity and are present in all cells of the organism.[119] Glucokinase, named also Hexokinase IV, is another glucose phosphorylating enzyme with low affinity and is present in a significant amount only in liver and pancreatic islets.[156] Hence, glucokinase is proposed as a possible target for alloxan in the pancreatic β-cells. In line with this possibility is another theory claiming that alloxan affects pancreatic β-cell function via interaction with sulfhydryl groups in the sugar binding site of the glucokinase.[119] Alloxan probably inhibits glucokinase reversibly through reaction with two adjacent sulfhydryl groups in the sugar binding site by the formation of a disulfide bond and concomitant inactivation of the enzyme.[157] This mechanism of interaction of alloxan with glucokinase, which couples changes in the blood glucose concentration to corresponding changes in the rate of insulin secretion, could provide an explanation for the ability of alloxan to inhibit glucose-induced insulin secretion, and may ultimately lead to the necrosis of the pancreatic β-cells.

There is also evidence that the generation of hydrogen peroxide, superoxide radicals, and hydroxyl radicals plays a critical role in the cytotoxicity of alloxan.[158–161] Alloxan is a mild oxidizing agent and it is chemically reduced to dialuric acid (5-OH-barbitric acid).[124] This reduced derivative dialuric acid generates superoxide anions and hydrogen peroxide.[162,163] Dialuric acid causes the formation of cytotoxic free radicals on reoxidation to alloxan.[158,159,164] Dialuric acid is rapidly reoxidized to alloxan by O_2. Hence, it has been strongly implicated in the mechanism of toxicity of alloxan through the formation of toxic-free radicals (O^{-2} and ^-OH). Presumably, the potential of alloxan to generate ROS is superior to the potential of STZ.[165] The obvious vulnerability of β-cells against oxidative stress, due to their low level of antioxidants compared to other tissues like lung, liver, and kidney,[166] facilitates alloxan toxicity.

Alloxan is reduced to dialuric acid by a thiol such as cysteine with the formation of cysteine.[124] In the presence of glutathione or certain protein thiols, it forms an additional product of unknown structure. It has been suggested that such modification of critical sulfhydryl groups in proteins is the cause of alloxan toxicity.[167] Thioredoxin, a hydrogen donor for the enzyme ribonucleotide reductase, an ubiquitous tissue and subcellular distribution in bovine, is essential for the regulation of the protein thiol-disulfide status of cells.[168] In keeping with this, Holmgren and Lyckeborg[162] showed *in vitro* that thioredoxin reductase catalyses rapid reduction of alloxan at diabetogenic concentrations of the drug and suggested that this enzyme is responsible for the rapid regeneration of dialuric acid from alloxan *in vivo*.

Alloxan decomposition is accompanied by the formation of superoxide radicals that undergo dismutation to form hydrogen peroxide.[169] Increased glutathione peroxide activity is one of the protective responses against oxidative stress. Thus, the augmentation of glutathione peroxide expression found in the liver of alloxan-treated rats provides further evidence to the presence and role of ROS during alloxan toxicity.[132]

In pancreatic β-cells, alloxan anion radicals are generated from alloxan probably mediated by the action of the cytochrome P-450 system (see also Chapter 7 and Chapter 8). These radicals have long half-lives and directly damage DNA *in vitro*. Some authors argued against the widely accepted hypothesis that the cause of alloxan-induced diabetes is attributable to O^{-2} radicals formed from alloxan.[57] They claimed that alloxan has potent scavenging effects against O^{-2} radicals and proposed that alloxan anion radicals seem to be directly related to the incidence of diabetes by alloxan. However, alloxan displays prooxidant or anti-oxidant features under the availability of suitable reducing agents. In the absence of reducing agents, alloxan is a scavenger of superoxide radicals formed by other reactions. Because of the high content of reducing compounds in the cell (e.g., glutathione), it is suggested that alloxan acts *in vivo* mainly as a generator of ROS.[170] Hence, despite the conflicting reports, it seems that ROS generated after the administration of alloxan seem to take part in the toxicity of this agent.

Okamoto and Yamamoto[96] showed *in vivo* and *in vitro* that the generation of hydroxyl radicals by alloxan attacks DNA to produce strand breaks. Subsequently, the fragmented DNA activates poly(ADP-ribose) synthetase, which depletes cellular NAD. Because NAD is the most abundant of cellular coenzymes and participates in many biological reactions in mammalian cells, the reduction in intracellular NAD to such a nonphysiological level may severely affect islet cell functions, including proinsulin synthesis. In keeping with this line of thought, Takasu et al.[53] demonstrated that alloxan stimulates hydrogen peroxide generation and causes DNA fragmentation with isolated rat islets *in vitro*. Presumably, the mechanism of DNA strand breaks due to alloxan differ from STZ, which causes DNA strand breaks via its alkylating property.[96,171]

The importance of O-linked protein glycosylation, a new protein glycosylation pathway, for pancreatic β-cells has been previously described. β-cells are vulnerable to the disruption of this pathway. It has been suggested that, alloxan, due to its uracil-like structure, inhibits the O-linked N-acetylglucosamine transferase enzyme, which is abundant in β-cells and causes β-cell toxicity.[69,172]

Certainly, the chemically related structure of alloxan with glucose is of eminent importance and may facilitate its specific recognition by β-cells.[173] Thus, in the early process of alloxan toxicity, the GLUT2 protein in the β-cell membrane, which is required for uptake of alloxan,[174] is a key target for early alloxan-induced damage that facilitates subsequent toxicity to GLUT2 and glucokinase mRNA. In this context, it has been shown that alloxan treatment reduces *ex vivo* mRNA expression of GLUT2 in murine pancreatic islets.[165] It has also been claimed that alloxan induces an increase in vascular permeability in mice. Jansson and Sandler[175] showed that alloxan causes an increase in islet vascular permeability, which appears to manifest at a later stage than the cytotoxic β-cell degeneration. This possibility was confirmed in a later study.[176]

Disturbances in intracellular calcium metabolism are also suggested to play a role in alloxan toxicity, *in vivo* and *in vitro* experiments have demonstrated that alloxan elevates cytosolic free Ca^2 concentration in pancreatic β-cells.[134,177] Alloxan-induced calcium influx from extracellular fluid exaggerated calcium mobilization from intracellular stores and its limited elimination from the cytoplasm. The calcium influx may result from the ability of alloxan to depolarize pancreatic β-cells.[178] It is proposed that the exaggerated concentration of this ion contributes to supraphysiological insulin release and, together with ROS, causes damage to pancreatic β-cells.[21] The protective effects of calcium canal blockers on the alloxan β-cell toxicity have confirmed the role of cytosolic calcium on the cytotoxic action of alloxan.[129,177]

Although the exact mechanisms of alloxan toxicity on islets is not clear yet, a combination of factors, including oxidation of sulfhydryl groups, inhibition of glucokinase, generation of free radicals, and disturbances in intracellular calcium homeostasis seem to take part in this process.

27.3.2 Species Differences with Alloxan Action

Like STZ, alloxan also exhibits species differences. Human β-cells were found to be less sensitive to alloxan than mouse and rat pancreatic islets.[85] The authors speculated that these marked interspecies differences could be due to different repair or defense mechanisms in β-cell destruction.

The lower content of GLUT2 in humans, which has also been claimed as one of the responsible factors for the resistance of human islets to STZ toxicity, is associated with a tenfold slower uptake of alloxan.[90] This leads to a lower intracellular exposure level to toxin and could well explain the resistance of human islets to alloxan toxicity. The importance of the GLUT2 transporter for pancreatic β-cell toxicity of alloxan and for resistance to this action has also been shown in other studies.[179,180] In these studies, it has been found that the extent of the toxicity of alloxan depended on the expression of the GLUT2 protein.

In addition to the cellular uptake of alloxan, the poor capacity of islet cells to protect themselves against the deleterious effects of peroxides is also proposed to be attributable to the species differences for alloxan toxicity.[181] This assumption, however, contradicts the evidence of the presence of a large number of drug-metabolizing enzymes in islet cells (see Chapter 7 and Chapter 8). Hence, it can be postulated that the presence, rather than the lack of enzymes is responsible for alloxan toxicity.

Both serum insulin and the volume fraction of β-cells in the pancreas of alloxan-treated guinea pigs were reduced by 70% 1 day after administration, but had returned to normal levels by Day 14.[182] These results indicate that guinea pigs recover from the diabetogenic effect of alloxan and the regeneration of the β-cells seems to be a factor. The same group of investigators also administrated a high dose (200 mg/kg) of alloxan into guinea pigs, and a significant decrease in insulin serum levels occurred within 24 hours following the treatment and returned to normal levels by 72 hours.[183]

Recovery of endocrine pancreatic function after alloxan treatment has also been shown in rats. De Haro-Hernandez et al.[184] showed that male Sprague-Dawley rats receiving a single intraperitoneal alloxan (120 mg/kg body weight) tended to normalize their endocrine function by Day 12. They also demonstrated that this process included both regeneration and neogenesis of pancreatic β-cells from either ductal or acinar cells.

In a comprehensive study, species differences for alloxan toxicity have been examined in mouse, rat, rabbit, dog, pig, human, and guinea pig by using the islets of these species in nude mice.[185] The results demonstrated that mouse and rat islet grafts were morphologically disturbed by alloxan and ROS. Rabbit and dog islet graft morphology was reasonably intact; and human, porcine, and guinea pig islet grafts were all well preserved. Guinea pig β-cells were affected by alloxan, but a regeneration process compensated for the observed apoptotic and necrotic cell death.

27.3.3 Tumorigenic Action of Alloxan

It has been shown that islet β-cell DNA is fragmented by STZ or alloxan,[171] which causes poly(ADP-ribose) synthetase activation and subsequent depletion in NAD. This depletion leads to an impairment of β-cell functions including the proinsulin synthesis.[19,59,186] These DNA strand breaks may also lead to an oncogenic process. As mentioned in the STZ section, about 1 year after the combined administration of alloxan or STZ with poly(ADP-ribose) inhibitors to rats, diabetes did not develop, but islet cell tumors were found frequently.[96]

Ten to 16 months after the combined administration of STZ or alloxan and poly(ADP) synthetase inhibitors induced islet β-cell tumors containing significant amounts of proinsulin mRNA sequences as well as B-granules with a high incidence in rats.[187] However, after the single injection of STZ or alloxan, islet cell tumors developed in 42 and 11% of surviving rats, respectively. The lower incidence of β-cell tumor development in alloxan-treated groups compared to STZ groups reflects some differences in the mode of action between these two agents. Alloxan has been shown to cause DNA strand breaks through generation of radical oxygen species, especially the hydroxyl radical.[59] However, STZ-induced DNA strand breaks seem to be mediated by alkylation of DNA bases may explain this remarkable difference.

Kazumi et al.[188] found pancreatic islet cell tumors in rats given alloxan and nicotinamide 10 to 14 months after treatment. These tumors were composed of at least three endocrine cell populations, although the majority of tumor cells were insulin-producing cells.

We studied the effect of alloxan on pancreatic carcinogenesis of BOP.[144] Treated hamsters developed islet cell adenomas as has also been the case in rats.[188] The incidence of islet cell adenomas in hamsters were within the spectrum of spontaneously occurring pancreatic lesions, although they usually occur in hamsters of a much older age.[189,190]

The available data is not sufficient to provide evidence for the neoplastic effect of alloxan.

27.3.4 Antitumor Action of Alloxan

Despite the toxicity and diabetes-inducing property of STZ and alloxan, they could be used with considerable benefit in the treatment of life-threatening insulinoma. In this context, the antitumor effect of STZ for various types of cancer, especially for islet cell tumors has been well established. However, alloxan does not exhibit an adequate antitumorigenic action. In contrast to normal pancreatic islet cells, rat tumoral islet cells have been reported to be less sensitive to the cytotoxic action of alloxan.[31,174] Rat pancreatic islets and insulin-producing cells of the RINm5F line were treated with [2-14C]alloxan and it has been found that tumoral insulin-producing cells are resistant to alloxan toxicity[174] either due to poor uptake or rapid detoxification of the diabetogenic agent.

In hamsters, alloxan has shown an effect on the induction of pancreatic ductal adenocarcinoma. In the hamster pancreatic cancer model, alloxan inhibited BOP-induced pancreatic cancer development in hamsters, when administrated at a diabetogenic dose (60 mg/kg body weight) 2 hours before BOP injection.[144] The development of a few neoplasms after alloxan was in accordance with the findings from alloxan diabetes that β-cells do not completely disappear,[139,191] and some even show a high functional activity.[139] However, BOP carcinogenesis was not affected when the BOP was given 2 weeks after alloxan administration.[144] Although a functional impairment of islet cells within 2 weeks has been reported,[139] it was assumed that the regenerated islet cells, which are believed to be the source of pancreatic ductal adenocarcinoma, regained metabolizing capacity within 2 weeks after alloxan damage. Additionally, this ineffectiveness could also be due to the rapid and complete metabolism of alloxan.[145] Although these results favor the role of islets in pancreatic ductal carcinogenesis, a possible tumor inhibitory effect of alloxan by other mechanisms cannot be ruled out entirely, because alloxan is also known to damage pancreatic ductular cells.[130,147,191]

Other studies, however, did not show any antitumorigenic effect of alloxan on islet cell tumors. Lenzen et al.[192] showed that in transplantable radiation-induced rat islet cell tumors, glucose metabolism is primarily based on a high affinity glucose phosphorylating enzyme, hexokinase. The activity of this enzyme is different from the normal pancreatic β-cells, the glucose metabolism of which is critically dependent on the availability of a low affinity glucose-phosphorylating enzyme, glucokinase. Hexokinase is less susceptible to inhibition by alloxan than glucokinase,[193] suggesting that the sulfhydryl groups in the sugar binding site of hexokinase are less accessible for alloxan than those of glucokinase. This plus the short half-life[145] may explain this ineffective antitumorigenic action of alloxan.

Flatt et al.[31] examined the effects of cytotoxic drugs and inhibitors *in vivo* in rats with a radiation-induced transplantable insulinoma and *in vitro* using cultured rat insulinoma cells and the RINm5F insulin-secreting cell line derived from it. They found that the administration of alloxan failed to affect circulating insulin or glucose concentrations. The ineffectiveness of alloxan in insulinoma-bearing rats was attributed to the high rate of decomposition of the drug *in vivo*.

Taken together, alloxan seems to have neither a sufficient antitumor-igenic nor tumorigenic action.

27.4 OTHER DRUGS

Despite the large number of anecdotal reports of drug-induced distur-bances of glucose metabolism, many of the so-called adverse drug reac-tions were either idiosyncratic or coincidental. However, few drugs were more frequently claimed to cause β-cell toxicity.

Cyclosporine and tacrolimus, potent immunosuppressive drugs used in transplantation, can cause abnormalities in glucose metabolism, in addition to other side effects like nephrotoxicity, hirsutism or alopecia, and tacrolimus. Histologically, cytoplasmic swelling, vacuolization, apo-ptosis, and abnormal immunostaining for insulin were reported in biopsies from patients receiving either tacrolimus or cyclosporine; islet cell damage was more severe in the group receiving tacrolimus.[194] The histological and clinical consequences of the treatment with these drugs were reversible with a reduction or discontinuation of the drug. Reports from clinical studies claimed that pentamidine, an antiprotozoal drug used to treat *Pneumocystis carinii* pneumonia in immunosuppressed patients, was fre-quently associated with dysglycemia due to its pancreatic β-cell cytotoxicity and has been considered as a new diabetogenic drug.[195,196] However, the mechanisms underlying the diabetogenic effect of these drugs remain to be elucidated.

27.5 ENDOGENOUS SUBSTANCES TOXIC TO PANCREATIC β-CELLS

Animal models studies[197,198] and human autopsy series[199–202] have provided evidence that, in addition to β-cell dysfunction, the reduction in the β-cell mass is involved in the pathogenesis of β-cell failure in Type 2 diabetes mellitus. Notably, in one of these reported autopsy series,[199] the islet β-cell mass was shown to be reduced in Type 2 diabetic subjects and was associated with increased rate of apoptosis. Accordingly, it appears that β-cell mass defects also exist in Type 2 diabetes they do in Type 1 diabetes. Studies from animals and man have suggested that long-term exposure to hyperglycemia and hyperlipidemia in diabetic conditions contribute to the deterioration of β-cell function, causes impaired insulin secretion, and even β-cell death.[203–206] (See also Chapter 9 and Chapter 28.)

During diabetic states, continual exposure to modest increases in blood glucose over a long time could have adverse irreversible effects on β-cells.[207] In the initial stages, this damage is characterized by defective insulin gene expression.[208] Robertson et al.[207] have suggested that the loss of insulin gene expression caused by chronic hyperglycemia and the resultant glucose toxicity in Zucker diabetic fatty rat was, in part, explained by chronic oxidative stress.

Studies have shown that pancreatic islet cells contain relatively small amounts of the antioxidant enzymes, such as CuZn-superoxide dismutase, Mn-superoxide dismutase, catalase, and glutathione peroxidase.[149,209] Due to the low level of antioxidant enzyme expression and activity, β-cells are at a greater risk of oxidative damage than other tissues with higher levels of antioxidant protection. Hence, the β-cell is an easy target for ROS. Thus, Robertson et al.[207] hypothesized that chronic oxidative stress is an important mechanism for glucose toxicity. The observations that hyperglycemia increases the intraislet peroxide levels provided further evidence for this idea.[210,211]

Increased free fatty acid (FFA) levels can also induce apoptosis in pancreatic β-cells,[212,213] as can elevated glucose levels.[214,215] Hence, a glucolipotoxicity hypothesis has also been proposed,[205,216] suggesting that the toxic actions of elevated FFAs on various tissues will become particularly apparent in the context of hyperglycemia. It is believed that high glucose decreases fat oxidation and consequently the cellular detoxification of fatty acids. Thus, glucose induces and activates enzymes and transcription factors involved in fat synthesis and storage and simultaneously switch off fat oxidation via the accumulation of Malonyl-Coenzyme A.[217,218] In a recent study, it has been shown that elevated glucose and FFAs together induce cell death by apoptosis in both rat and human islet β-cells with a marked synergistic effect.[219]

It is suggested that a sustained increase in circulating FFAs is responsible for a triglyceride accumulation in the islet cells and for elevated cellular free fatty acyl levels that are cytotoxic.[220] In isolated islets from prediabetic Zucker diabetic fatty rats, FFAs were shown to increase ceramide formation, leading to the expression of nitric oxide synthetase and NO-dependent β-cell apoptosis.[213] Cnop et al.[221] suggested that FFAs can be cytotoxic for normal islet β-cells as well as for normal noncells, leading to cell death by both necrosis and apoptosis in β-cells and mostly by apoptosis in noncells.

It has been shown that both rat and human β-cells express high affinity receptors for low density lipoprotein (LDL) and very low density lipoprotein (VLDL), which can internalize both lipoproteins.[222] The possible role of these lipoproteins was examined by the same group and was found that the uptake of LDL by islet β-cells and subsequent oxidative reactions can be damaging for the cells.[204] This LDL-induced necrosis of β-cells was not attributed to the production of toxic NO levels, as in the case of IL-1 induced necrosis by the authors, but was ascribed to an oxidative stress with production of free radicals.

Amylin, a 37 amino-acid polypeptide is the principal constituent of the amyloid deposits that form in the islets of Langerhans in patients with Type 2 diabetes mellitus. Lorenzo et al.[223] have shown that human amylin is toxic to insulin-producing β-cells of the adult pancreas of rats and humans. They proposed that this toxicity is mediated by the fibrillar form of the amylin peptide and requires direct contact of the fibrils with the cell surface. The mechanism of cell death involves RNA and protein synthesis and is characterized by plasma membrane blebbing, chromatin condensation, and DNA fragmentation. The authors concluded that amylin fibril formation in the pancreas might cause islet cell dysfunction and death in Type 2 diabetes mellitus.

During β-cell destruction some released cytokines, particularly IL-1, have also been implicated as effector molecules.[46] IL-1 and in combination with tumor necrosis factor-α and interferon-β, exerts a cytotoxic effect specifically on β-cells in the islets of Langerhans, in part via the induction of free radicals such as NO catalyzed by inducible nitric oxide synthase.[45,46,224,225] It has been postulated that cytokines may mediate the β-cell destructive process causing IDDM. However, this effect appears to not be sustained.[226] DNA is also proposed to be an early target of cytokine action in islet β-cells and is suggested as a mechanism of cytokine-induced β-cell destruction.[227]

REFERENCES

1. Pour P.M., Standop J., and Batra S.K. 2002. Are islet cells the gatekeepers of the pancreas? *Pancreatology* 2:440–448.
2. Lewis C. and Barbiers A.R. 1959. Streptozotocin, a new antibiotic. In vitro and in vivo evaluation. *Antibiot. Ann.* 7:247–254.
3. Varva J.J., DeBoer C., Dietz A., Hanka L.J., and Sokolski W.T. 1959–1960. Streptozotocin, a new antibacterial antibiotic. *Antibiot. Ann.* 7:230 –235.
4. Rakieten N., Rakieten M.L., and Nadkarni M.V. 1963. Studies on the diabetogenic action of streptozotocin (NSC-37917). *Cancer Chemother. Rep.* 29:91–98.
5. Evans J.S., Gerritsen G.C., Mann K.M., and Owen S.P. 1965. Antitumor and hyperglycemic activity of streptozotocin (NSC-37917) and its cofactor, U-15,774. *Cancer Chemother. Rep.* 48:1–6.
6. White F.R. 1963. Streptozotocin. *Cancer Chemother. Rep.* 30:49–53.
7. Carter S.K., Broder L., and Friedman M. 1971. Streptozotocin and metastatic insulinoma. *Ann. Intern. Med.* 74:445–446.
8. Arison R.N., Ciaccio E.I., Glitzer M.S., Cassaro J.A., and Pruss M.P. 1967. Light and electron microscopy of lesions in rats rendered diabetic with streptozotocin. *Diabetes* 16:51–56.
9. West E., Simon O.R., and Morrison E.Y. 1996. Streptozotocin alters pancreatic beta-cell responsiveness to glucose within six hours of injection into rats. *West Indian Med. J.* 45:60–62.

10. Pour P.M., Duckworth W., Carlson K., and Kazakoff K. 1990. Insulin therapy prevents spontaneous recovery from streptozotocin-induced diabetes in Syrian hamsters. An autoradiographic and immunohistochemical study. *Virchows. Arch. A Pathol. Anat. Histopathol.* 417:333–341.

11. Rosenberg L., Duguid W.P., Brown R.A., and Vinik A.I. 1988. Induction of nesidioblastosis will reverse diabetes in Syrian golden hamster. *Diabetes* 37:334–341.

12. Takatori A., Nishida E., Inenaga T., Horiuchi K., Kawamura S., Itagaki S., and Yoshikawa Y. 2002. Functional and histochemical analysis on pancreatic islets of APA hamsters with SZ-induced hyperglycemia and hyperlipidemia. *Exp. Anim.* 51:9–17.

13. Sandler S. 1984. Protection by dimethyl urea against hyperglycaemia, but not insulitis, in low-dose streptozotocin-induced diabetes in the mouse. *Diabetologia* 26:386–388.

14. Li Z., Karlsson F.A., and Sandler S. 2000. Islet loss and alpha cell expansion in Type 1 diabetes induced by multiple low-dose streptozotocin administration in mice. *J. Endocrinol.* 165:93–99.

15. Bonnevie-Nielsen V., Steffes M.W., and Lernmark A. 1981. A major loss in islet mass and B-cell function precedes hyperglycemia in mice given multiple low doses of streptozotocin. *Diabetes* 30:424–429.

16. Like A.A. and Rossini A.A. 1976. Streptozotocin-induced pancreatic insulitis: new model of diabetes mellitus. *Science* 193:415–417.

17. Kolb-Bachofen V., Epstein S., Kiesel U., and Kolb H. 1988. Low-dose streptozocin-induced diabetes in mice. Electron microscopy reveals single-cell insulitis before diabetes onset. *Diabetes* 37:21–27.

18. Ganda O.P., Rossini A.A., and Like A.A. 1976. Studies on streptozotocin diabetes. *Diabetes* 25:595–603.

19. Yamamoto H., Uchigata Y., and Okamoto H. 1981. Streptozotocin and alloxan induce DNA strand breaks and poly(ADP-ribose) synthetase in pancreatic islets. *Nature* 294:284–286.

20. Papaccio G., Strate C., and Linn T. 1994. Pancreatic duct infiltration in the low-dose streptozocin-treated mouse. *Histol. Histopathol.* 9:529–534.

21. Szkudelski T. 2001. The mechanism of alloxan and streptozotocin action in B cells of the rat pancreas. *Physiol. Res.* 50:537–546.

22. Elsner M., Guldbakke B., Tiedge M., Munday R., and Lenzen S. 2000. Relative importance of transport and alkylation for pancreatic beta-cell toxicity of streptozotocin. *Diabetologia* 43:1528–1533.

23. Wilson G.L. and Leiter E.H. 1990. Streptozotocin interactions with pancreatic beta cells and the induction of insulin-dependent diabetes. *Curr. Top. Microbiol. Immunol.* 156:27–54.

24. Bennett R.A. and Pegg A.E. 1981. Alkylation of DNA in rat tissues following administration of streptozotocin. *Cancer Res.* 41:2786–2790.

25. Weiss R.B. 1982. Streptozocin: a review of its pharmacology, efficacy, and toxicity. *Cancer Treat. Rep.* 66:427–438.

26. Schein P.S., O'Connell M.J., Blom J., Hubbard S., Magrath I.T., Bergevin P., Wiernik P.H. et al. 1974. Clinical antitumor activity and toxicity of streptozotocin (NSC-85998). *Cancer* 34:993–1000.

27. Gunnarsson R., Berne C., and Hellerstrom C. 1974. Cytotoxic effects of streptozotocin and N-nitrosomethylurea on the pancreatic B cells with special regard to the role of nicotinamide-adenine dinucleotide. *Biochem. J.* 140:487–494.

28. Schnedl W.J., Ferber S., Johnson J.H., and Newgard C.B. 1994. STZ transport and cytotoxicity. Specific enhancement in GLUT2-expressing cells. *Diabetes* 43:1326–1333.

29. Noel L.E. and Newgard C.B. 1997. Structural domains that contribute to substrate specificity in facilitated glucose transporters are distinct from those involved in kinetic function: studies with GLUT-1/GLUT-2 chimeras. *Biochemistry* 36:5465–5475.

30. Ledoux S.P. and Wilson G.L. 1984. Effects of streptozotocin on a clonal isolate of rat insulinoma cells. *Biochem. Biophys. Acta.* 804:387–392.

31. Flatt P.R., Swanston-Flatt S.K., Tan K.S., and Marks V. 1987. Effects of cytotoxic drugs and inhibitors of insulin secretion on a serially transplantable rat insulinoma and cultured rat insulinoma cells. *Gen. Pharmacol.* 18:293–297.

32. Thulesen J., Orskov C., Holst J.J., and Poulsen S.S. 1997. Short-term insulin treatment prevents the diabetogenic action of streptozotocin in rats. *Endocrinology* 138:62–68.

33. Wang Z. and Gleichmann H. 1995. Glucose transporter 2 expression: prevention of streptozotocin-induced reduction in beta-cells with 5-thio-D-glucose. *Exp. Clin. Endocrinol. Diabetes* 103 Suppl. 2:83–97.

34. Wang Z. and Gleichmann H. 1998. GLUT2 in pancreatic islets: crucial target molecule in diabetes induced with multiple low doses of streptozotocin in mice. *Diabetes* 47:50–56.

35. Delaney C.A., Dunger A., Di Matteo M., Cunningham J.M., Green M.H., and Green I.C. 1995. Comparison of inhibition of glucose-stimulated insulin secretion in rat islets of Langerhans by streptozotocin and methyl and ethyl nitrosoureas and methanesulphonates. Lack of correlation with nitric oxide-releasing or O6-alkylating ability. *Biochem. Pharmacol.* 50:2015–2020.

36. Goldmacher V.S., Cuzick Jr. R.A., and Thilly W.G. 1986. Isolation and partial characterization of human cell mutants differing in sensitivity to killing and mutation by methylnitrosourea and N-methyl-N′-nitro-N-nitrosoguanidine. *J. Biol. Chem.* 261:12462–12471.

37. Green M.H., Lowe J.E., Petit-Frere C., Karran P., Hall J., and Kataoka H. 1989. Properties of N-methyl-N-nitrosourea-resistant, Mex-derivatives of an SV40-immortalized human fibroblast cell line. *Carcinogenesis* 10:893–898.

38. Okamoto H. 1985. The role of poly(ADP-ribose) synthetase in the development of insulin-dependent diabetes and islet B-cell regeneration. *Biomed. Biochem. Acta.* 44:15–20.

39. Sandler S. and Swenne I. 1983. Streptozotocin, but not alloxan, induces DNA repair synthesis in mouse pancreatic islets in vitro. *Diabetologia* 25:444–447.

40. Heller B., Burkle A., Radons J., Fengler E., Jalowy A., Muller M., Burkart V. et al. 1994. Analysis of oxygen radical toxicity in pancreatic islets at the single cell level. *Biol. Chem. Hoppe Seyler* 375:597–602.

41. Nukatsuka M., Yoshimura Y., Nishida M., and Kawada J. 1990. Importance of the concentration of ATP in rat pancreatic beta cells in the mechanism of streptozotocin-induced cytotoxicity. *J. Endocrinol.* 127:161–165.

42. Masiello P., Novelli M., Fierabracci V., and Bergamini E. 1990. Protection by 3-aminobenzamide and nicotinamide against streptozotocin-induced beta-cell toxicity in vivo and in vitro. *Res. Commun. Chem. Pathol. Pharmacol.* 69:17–32.

43. Masiello P., Cubeddu T.L., Frosina G., and Bergamini E. 1985. Protective effect of 3-aminobenzamide, an inhibitor of poly (ADP-ribose) synthetase, against streptozotocin-induced diabetes. *Diabetologia* 28:683–686.

44. Knowles R.G. and Moncada S. 1994. Nitric oxide synthases in mammals. *Biochem. J.* 298 (Pt. 2):249–258.

45. Eizirik D.L., Flodstrom M., Karlsen A.E., and Welsh N. 1996. The harmony of the spheres: inducible nitric oxide synthase and related genes in pancreatic beta cells. *Diabetologia* 39:875–890.

46. Mandrup-Poulsen T. 1996. The role of interleukin-1 in the pathogenesis of IDDM. *Diabetologia* 39:1005–1029.

47. Turk J., Corbett J.A., Ramanadham S., Bohrer A., and McDaniel M.L. 1993. Biochemical evidence for nitric oxide formation from streptozotocin in isolated pancreatic islets. *Biochem. Biophys. Res. Commun.* 197:1458–1464.

48. Kroncke K.D., Fehsel K., Sommer A., Rodriguez M.L., and Kolb-Bachofen V. 1995. Nitric oxide generation during cellular metabolization of the diabetogenic N-methyl-N-nitroso-urea streptozotozin contributes to islet cell DNA damage. *Biol. Chem. Hoppe Seyler* 376:179–185.

49. Morgan N.G., Cable H.C., Newcombe N.R., and Williams G.T. 1994. Treatment of cultured pancreatic B-cells with streptozotocin induces cell death by apoptosis. *Biosci. Rep.* 14:243–250.

50. Eizirik D.L., Sandler S., Welsh N., Cetkovic-Cvrlje M., Nieman A., Geller D.A., Pipeleers D.G. et al. 1994. Cytokines suppress human islet function irrespective of their effects on nitric oxide generation. *J. Clin. Invest.* 93:1968–1974.

51. Papaccio G., Esposito V., Latronico M.V., and Pisanti F.A. 1995. Administration of a nitric oxide synthase inhibitor does not suppress low-dose streptozotocin-induced diabetes in mice. *Int. J. Pancreatol.* 17:63–68.

52. Papaccio G., Pisanti F.A., Latronico M.V., Ammendola E., and Galdieri M. 2000. Multiple low-dose and single high-dose treatments with streptozotocin do not generate nitric oxide. *J. Cell. Biochem.* 77:82–91.

53. Takasu N., Komiya I., Asawa T., Nagasawa Y., and Yamada T. 1991. Streptozocin- and alloxan-induced H_2O_2 generation and DNA fragmentation in pancreatic islets. H_2O_2 as mediator for DNA fragmentation. *Diabetes* 40:1141–1145.

54. Bedoya F.J., Solano F., and Lucas M. 1996. N-monomethyl-arginine and nicotinamide prevent streptozotocin-induced double strand DNA break formation in pancreatic rat islets. *Experientia* 52:344–347.

55. Sofue M., Yoshimura Y., Nishida M., and Kawada J. 1991. Uptake of nicotinamide by rat pancreatic beta cells with regard to streptozotocin action. *J. Endocrinol.* 131:135–138.

56. Nukatsuka M., Sakurai H., Yoshimura Y., Nishida M., and Kawada J. 1988. Enhancement by streptozotocin of O2- radical generation by the xanthine oxidase system of pancreatic beta-cells. *FEBS Lett.* 239:295–298.

57. Kawada J. 1992. New hypotheses for the mechanisms of streptozotocin and alloxan inducing diabetes mellitus. *Yakugaku Zasshi* 112:773–791.

58. Nukatsuka M., Yoshimura Y., Nishida M., and Kawada J. 1990. Allopurinol protects pancreatic beta cells from the cytotoxic effect of streptozotocin: in vitro study. *J. Pharmacobiodyn.* 13:259–262.

59. Uchigata Y., Yamamoto H., Kawamura A., and Okamoto H. 1982. Protection by superoxide dismutase, catalase, and poly(ADP-ribose) synthetase inhibitors against alloxan- and streptozotocin-induced islet DNA strand breaks and against the inhibition of proinsulin synthesis. *J. Biol. Chem.* 257:6084–6088.

60. Bolzan A.D. and Bianchi M.S. 2002. Genotoxicity of streptozotocin. *Mutat. Res.* 512:121–134.

61. Kaneto H., Fujii J., Seo H.G., Suzuki K., Matsuoka T., Nakamura M., Tatsumi H. et al. 1995. Apoptotic cell death triggered by nitric oxide in pancreatic beta-cells. *Diabetes* 44:733–738.

62. Saini K.S., Thompson C., Winterford C.M., Walker N.I., and Cameron D.P. 1996. Streptozotocin at low doses induces apoptosis and at high doses causes necrosis in a murine pancreatic beta cell line, INS-1. *Biochem. Mol. Biol. Int.* 39:1229–1236.

63. Uchigata Y., Yamamoto H., Nagai H., and Okamoto H. 1983. Effect of poly(ADP-ribose) synthetase inhibitor administration to rats before and after injection of alloxan and streptozotocin on islet proinsulin synthesis. *Diabetes* 32:316–318.

64. Orci L., Amherdt M., Malaisse-Lagae F., Ravazzola M., Malaisse W.J., Perrelet A., and Renold A.E. 1976. Islet cell membrane alteration by diabetogenic drugs. *Lab. Invest.* 34:451–454.

65. Norlund R., Grankvist K., Norlund L., and Taljedal I.B. 1984. Ultrastructure and membrane permeability of cultured pancreatic beta-cells exposed to alloxan or 6-hydroxydopamine. *Virchows. Arch. A Pathol. Anat. Histopathol.* 404:31–38.

66. Vecchini A., Del Rosso F., Binaglia L., Dhalla N.S., and Panagia V. 2000. Molecular defects in sarcolemmal glycerophospholipid subclasses in diabetic cardiomyopathy. *J. Mol. Cell. Cardiol.* 32:1061–1074.

67. Wieder T., Haase A., Geilen C.C., and Orfanos C.E. 1995. The effect of two synthetic phospholipids on cell proliferation and phosphatidylcholine biosynthesis in Madin-Darby canine kidney cells. *Lipids* 30:389–393.

68. Detmar M., Geilen C.C., Wieder T., Orfanos C.E., and Reutter W. 1994. Phospholipid analogue hexadecylphosphocholine inhibits proliferation and phosphatidylcholine biosynthesis of human epidermal keratinocytes in vitro. *J. Invest. Dermatol.* 102:490–494.

69. Konrad R.J. and Kudlow J.E. 2002. The role of O-linked protein glycosylation in beta-cell dysfunction. *Int. J. Mol. Med.* 10:535–539.

70. Konrad R.J., Mikolaenko I., Tolar J.F., Liu K., and Kudlow J.E. 2001. The potential mechanism of the diabetogenic action of streptozotocin: inhibition of pancreatic beta-cell O-GlcNAc-selective N-acetyl-beta-D-glucosaminidase. *Biochem. J.* 356:31–41.

71. Brosky G. and Logothetopoulos J. 1969. Streptozotocin diabetes in the mouse and guinea pig. *Diabetes* 18:606–611.

72. Pitkin R.M. and Reynolds W.A. 1970. Diabetogenic effects of streptozotocin in rhesus monkeys. *Diabetes* 19:85–90.

73. Wilander E. and Boquist L. 1972. Streptozotocin-diabetes in the Chinese hamster. Blood glucose and structural changes during the first 24 hours. *Horm. Metab. Res.* 4:426–433.

74. Ebara T., Hirano T., Mamo J.C., Sakamaki R., Furukawa S., Nagano S., and Takahashi T. 1994. Hyperlipidemia in streptozocin-diabetic hamsters as a model for human insulin-deficient diabetes: comparison to streptozocin-diabetic rats. *Metabolism* 43:299–305.

75. Ma P.T., Gil G., Sudhof T.C., Bilheimer D.W., Goldstein J.L., and Brown M.S. 1986. Mevinolin, an inhibitor of cholesterol synthesis, induces mRNA for low density lipoprotein receptor in livers of hamsters and rabbits. *Proc. Natl. Acad. Sci. U.S.A.* 83:8370–8374.

76. Phares C.K. 1980. Streptozotocin-induced diabetes in Syrian hamsters: new model of diabetes mellitus. *Experientia* 36:681–682.

77. Rakieten N., Gordon B.S., Beaty A., Cooney D.A., Schein P.S., and Dixon R.L. 1976. Modification of renal tumorigenic effect of streptozotocin by nicotinamide: spontaneous reversibility of streptozotocin diabetes. *Proc. Soc. Exp. Biol. Med.* 151:356–361.

78. Cantenys D., Portha B., Dutrillaux M.C., Hollande E., Roze C., and Picon L. 1981. Histogenesis of the endocrine pancreas in newborn rats after destruction by streptozotocin. An immunocytochemical study. *Virchows. Arch. B Cell. Pathol. Incl. Mol. Pathol.* 35:109–122.

79. Karunanayake E.H., Hearse D.J., and Mellows G. 1976. Streptozotocin: its excretion and metabolism in the rat. *Diabetologia* 12:483–488.

80. Katsilambros N., Rahman Y.A., Hinz M., Fussganger R., Schroder K.E., Straub K., and Pfeiffer E.F. 1970. Action of streptozotocin on insulin and glucagon responses of rat islets. *Horm. Metab. Res.* 2:268–270.

81. Tomioka T., Fujii H., Hirota M., Ueno K., and Pour P.M. 1991. The patterns of beta-cell regeneration in untreated diabetic and insulin-treated diabetic Syrian hamsters after streptozotocin treatment. *Int. J. Pancreatol.* 8:355–366.

82. Nagasao J., Yoshioka K., Amasaki H., and Mutoh K. 2004. Expression of nestin and IGF-1 in rat pancreas after streptozotocin administration. *Anat. Histol. Embryol.* 33:1–4.

83. George M., Ayuso E., Casellas A., Costa C., Devedjian J.C., and Bosch F. 2002. Beta cell expression of IGF-I leads to recovery from Type 1 diabetes. *J. Clin. Invest.* 109:1153–1163.

84. Yang H. and Wright Jr. J.R. 2002. Human beta cells are exceedingly resistant to streptozotocin in vivo. *Endocrinology* 143:2491–2495.

85. Eizirik D.L., Pipeleers D.G., Ling Z., Welsh N., Hellerstrom C., and Andersson A. 1994. Major species differences between humans and rodents in the susceptibility to pancreatic beta-cell injury. *Proc. Natl. Acad. Sci. U.S.A.* 91:9253–9256.

86. Feelisch M. and Noack E. 1987. Nitric oxide (NO) formation from nitrovasodilators occurs independently of hemoglobin or non-heme iron. *Eur. J. Pharmacol.* 142:465–469.

87. Malaisse W.J. 1982. Alloxan toxicity to the pancreatic B-cell. A new hypothesis. *Biochem. Pharmacol.* 31:3527–3534.

88. Tuch B.E., Turtle J.R., and Simeonovic C.J. 1989. Streptozotocin is not toxic to the human fetal B cell. *Diabetologia* 32:678–684.

89. Lazarus S.S. and Shapiro S.H. 1972. Streptozotocin-induced diabetes and islet cell alterations in rabbits. *Diabetes* 21:129–137.

90. De Vos A., Heimberg H., Quartier E., Huypens P., Bouwens L., Pipeleers D., and Schuit F. 1995. Human and rat beta cells differ in glucose transporter but not in glucokinase gene expression. *J. Clin. Invest.* 96:2489–2495.

91. Bell Jr. R.H., Hye R.J., and Miyai K. 1984. Streptozotocin-induced liver tumors in the Syrian hamster. *Carcinogenesis* 5:1235–1238.

92. Mauer S.M., Lee C.S., Najarian J.S., and Brown D.M. 1974. Induction of malignant kidney tumors in rats with streptozotocin. *Cancer Res.* 34:158–160.

93. Masiello P., Wollheim C.B., Gori Z., Blondel B., and Bergamini E. 1984. Streptozotocin-induced functioning islet cell tumor in the rat: high frequency of induction and biological properties of the tumor cells. *Toxicol. Pathol.* 12:274–280.

94. Doi K. 1975. Studies on the mechanism of the diabetogenic activity of streptozotocin and on the ability of compounds to block the diabetogenic activity of streptozotocin (author's transl.). *Nippon Naibunpi Gakkai Zasshi* 51:129–147.

95. Yoshino G., Kazumi T., Morita S., and Baba S. 1981. Insulin and glucagon in rats with islet cell tumors induced by small doses of streptozotocin. *Can. J. Physiol. Pharmacol.* 59:818–823.

96. Okamoto H. and Yamamoto H. 1983. DNA strand breaks and poly(ADP-ribose) synthetase activation in pancreatic islets — a new aspect to development of insulin-dependent diabetes and pancreatic B-cell tumors. *Princess Takamatsu Symp.* 13:297–308.

97. Bar M., Burke M., Isakov A., and Almog C. 1990. Insulinoma after streptozotocin therapy for metastatic gastrinoma: natural history or iatrogenic complication? *J. Clin. Gastroenterol.* 12:579–580.

98. Yagihashi S. and Nagai K. 1981. Immunohistochemical and ultrastructural studies on rat islet cell tumours induced by streptozotocin and nicotinamide. *Virchows. Arch. A Pathol. Anat. Histol.* 390:181–191.

99. Pour P.M. and Patil K. 1983. Modification of pancreatic carcinogenesis in the hamster model. X. Effect of streptozotocin. *J. Natl. Cancer Inst.* 71:1059–1065.

100. Dutrillaux M.C., Portha B., Roze C., and Hollande E. 1982. Ultrastructural study of pancreatic B cell regeneration in newborn rats after destruction by streptozotocin. *Virchows. Arch. B Cell. Pathol. Incl. Mol. Pathol.* 39:173–185.

101. Shepherd J.G., Chen J.R., Tsao M.S., and Duguid W.P. 1993. Neoplastic transformation of propagable cultured rat pancreatic duct epithelial cells by azaserine and streptozotocin. *Carcinogenesis* 14:1027–1033.

102. Lawson T.A., Gingell R., Nagel D., Hines L.A., and Ross A. 1981. Methylation of hamster DNA by the carcinogen N-nitroso-bis (2-oxopropyl)amine. *Cancer Lett.* 11:251–255.

103. Pour P.M., Wallcave L., and Nagel D. 1981. The effect of N-nitroso-2-methoxy-2,6-dimethylmorpholine on endocrine and exocrine pancreas of Syrian hamsters. *Cancer Lett.* 13:233–240.

104. Pour P.M. and Raha C.R. 1981. Pancreatic carcinogenic effect of N-nitrosobis (2-oxobutyl) amine and N-nitroso (2-oxobutyl) (2-oxopropyl) amine in Syrian hamster. *Cancer Lett.* 12:223–229.

105. Von Dorsche H.H., Krause R., Kohler E., Fiedler H., and Sulzmann R. 1975. Electronmicroscopy studies on the exocrine pancreas of Wistar-Rats following treatment with Streptozotocin. *Endokrinologie* 65:354–363.

106. Robbiano L., Mereto E., Corbu C., and Brambilla G. 1996. DNA damage induced by seven N-nitroso compounds in primary cultures of human and rat kidney cells. *Mutat. Res.* 368:41–47.

107. Murray-Lyon I.M., Eddleston A.L., Williams R., Brown M., Hogbin B.M., Bennett A., Edwards J.C. et al. 1968. Treatment of multiple-hormone-producing malignant islet-cell tumour with streptozotocin. *Lancet* 2:895–898.

108. Moertel C.G., Hanley J.A., and Johnson L.A. 1980. Streptozocin alone compared with streptozocin plus fluorouracil in the treatment of advanced islet-cell carcinoma. *N. Engl. J. Med.* 303:1189–1194.

109. Moore A.S., Nelson R.W., Henry C.J., Rassnick K.M., Kristal O., Ogilvie G.K., and Kintzer P. 2002. Streptozocin for treatment of pancreatic islet cell tumors in dogs: 17 cases (1989–1999). *J. Am. Vet. Med. Assoc.* 221:811–818.

110. Broder L.E. and Carter S.K. 1973. Pancreatic islet cell carcinoma. II. Results of therapy with streptozotocin in 52 patients. *Ann. Intern. Med.* 79:108–118.

111. Goossens A., Gepts W., Saudubray J.M., Bonnefont J.P., Nihoul F., Heitz P.U., and Kloppel G. 1989. Diffuse and focal nesidioblastosis. A clinicopathological study of 24 patients with persistent neonatal hyperinsulinemic hypoglycemia. *Am. J. Surg. Pathol.* 13:766–775.

112. Rahier J., Falt K., Muntefering H., Becker K., Gepts W., and Falkmer S. 1984. The basic structural lesion of persistent neonatal hypoglycaemia with hyper-insulinism: deficiency of pancreatic D cells or hyperactivity of B cells? *Diabetologia* 26:282–289.

113. Bell Jr. R.H. and Strayer D.S. 1983. Streptozotocin prevents development of nitrosamine-induced pancreatic cancer in the Syrian hamster. *J. Surg. Oncol.* 24:258–262.

114. Bell Jr. R.H., McCullough P.J., and Pour P.M. 1988. Influence of diabetes on susceptibility to experimental pancreatic cancer. *Am. J. Surg.* 155:159–164.

115. Bell Jr. R.H., Sayers H.J., Pour P.M., Ray M.B., and McCullough P.J. 1989. Importance of diabetes in inhibition of pancreatic cancer by streptozotocin. *J. Surg. Res.* 46:515–519.

116. Ishikawa O., Ohigashi H., Imaoka S., Nakai I., Mitsuo M., Weide L., and Pour P.M. 1995. The role of pancreatic islets in experimental pancreatic carcinogenicity. *Am. J. Pathol.* 147:1456–1464.

117. Pour P.M., Kazakoff K., and Carlson K. 1990. Inhibition of streptozotocin-induced islet cell tumors and N-nitrosobis(2-oxopropyl)amine-induced pancreatic exocrine tumors in Syrian hamsters by exogenous insulin. *Cancer Res.* 50:1634–1639.

118. Pour P.M. 1989. Experimental pancreatic cancer. *Am. J. Surg. Pathol.* 13 Suppl. 1:96–103.

119. Lenzen S. and Panten U. 1988. Alloxan: history and mechanism of action. *Diabetologia* 31:337–342.

120. Dunn J., Sheehan H., and McLethie N. 1943. Necrosis of islets of Langerhans produced experimentally. *Lancet* 1:484–487.

121. Dunn J., Kirkpatrick J., McLetchie N., and Telfer S. 1943. Necrosis of the islets of Langerhans produced experimentally. *J. Path. Bact.* 55:245–257.

122. Dunn J., Duffy E., Gilmour M., Kirkpatrick J., and McLetchie N. 1944. Further observations on the effects of alloxan on the pancreatic islets. *J. Physiol.* 103:233–243.

123. Goldner M. 1945. Alloxan diabetes. Its production and mechanism. *N.Y. Acad. Med. Bull.* 21:44–55.

124. Patterson J., Lazarow A., and Levey S. 1949. Alloxan and dialuric acid: their stabilities and ultraviolet absorption spectra. *J. Biol. Chem.* 177:187–196.

125. Cooperstein S. and Atkins D. 1981. Actions of toxic drugs on islet cells. In *The islet of Langerhans*. Watkins D. Ed. New York: Academic Press. pp. 387–425.

126. Lenzen S., Tiedge M., Jorns A., and Munday R. 1996. Alloxan derivatives as a tool for the elucidation of the mechanism of of the diabetogenic action of alloxan. In *Lessons from animal diabetes.* Shafrir E., Ed. Boston: Birkhauser. pp. 113–122.

127. Gruppuso P.A., Boylan J.M., Posner B.I., Faure R., and Brautigan D.L. 1990. Hepatic protein phosphotyrosine phosphatase. Dephosphorylation of insulin and epidermal growth factor receptors in normal and alloxan diabetic rats. *J. Clin. Invest.* 85:1754–1760.

128. Boylan J.M., Brautigan D.L., Madden J., Raven T., Ellis L., and Gruppuso P.A. 1992. Differential regulation of multiple hepatic protein tyrosine phosphatases in alloxan diabetic rats. *J. Clin. Invest.* 90:174–179.

129. Katsumata K., Katsumata Jr. K., and Katsumata Y. 1992. Protective effect of diltiazem hydrochloride on the occurrence of alloxan- or streptozotocin-induced diabetes in rats. *Horm. Metab. Res.* 24:508–510.

130. Goldner M. and Gomori G. 1943. Alloxan diabetes in the dog. *Endocrinology* 33:297–308.

131. Nace P.F., House E.L., and Tassoni J.P. 1956. Alloxan diabetes in the hamster: dosage and blood sugar curves. *Endocrinology* 58:305–308.

132. Szkudelski T., Kandulska K., and Okulicz M. 1998. Alloxan in vivo does not only exert deleterious effects on pancreatic B cells. *Physiol. Res.* 47:343–346.

133. Bansal R., Ahmad N., and Kidwai J.R. 1980. Alloxan-glucose interaction: effect on incorporation of 14C-leucine into pancreatic islets of rat. *Acta. Diabetol. Lat.* 17:135–143.

134. Park B.H., Rho H.W., Park J.W., Cho C.G., Kim J.S., Chung H.T., and Kim H.R. 1995. Protective mechanism of glucose against alloxan-induced pancreatic beta-cell damage. *Biochem. Biophys. Res. Commun.* 210:1–6.

135. Young J.K. and Dixit P.K. 1980. Lack of diabetogenic effect of alloxan in protein-calorie malnourished rats. *J. Nutr.* 110:703–709.

136. Dixit P.K. and Kaung H.L. 1985. Rat pancreatic beta-cells in protein deficiency: a study involving morphometric analysis and alloxan effect. *J. Nutr.* 115:375–381.

137. Schmidt R., Muller H., Glass P., Schneider S., and Unger E. 1990. Effect of alloxans on pancreatic B-cells with special regard to the alloxan-metal-complex theory. I. Effects of alloxan, alloxan-zinc chelates, dilauric acid and colchicine on blood sugar and rate of mitosis of B-cell Langerhans islets. *Acta. Histochem.* 88:29–46.

138. Weaver D.C., McDaniel M.L., Naber S.P., Barry C.D., and Lacy P.E. 1978. Alloxan stimulation and inhibition of insulin release from isolated rat islets of Langerhans. *Diabetes* 27:1205–1214.

139. Edstrom C. and Boquist L. 1973. Alloxan diabetes in duct-ligated rats. Light and electron microscopic findings. *Acta. Pathol. Microbiol. Scand. [A]* 81:47–56.

140. Kliber A., Szkudelski T., and Chichlowska J. 1996. Alloxan stimulation and subsequent inhibition of insulin release from in situ perfused rat pancreas. *J. Physiol. Pharmacol.* 47:321–328.

141. House E.L., Nace P.F., and Tassoni J.P. 1956. Alloxan diabetes in the hamster: organ changes during the first day. *Endocrinology* 59:433–443.

142. Aleeva G.N., Kiyasov A.P., Minnebaev M.M., Burykin I.M., and Khafiz'yanova R. 2002. Changes in the count of pancreatic beta- and alpha-cells and blood glucose level in rats with alloxan-induced diabetes. *Bull. Exp. Biol. Med.* 133:127–129.

143. Rastogi K.S., Lickley L., Jokay M., Efendic S., and Vranic M. 1990. Paradoxical reduction in pancreatic glucagon with normalization of somatostatin and decrease in insulin in normoglycemic alloxan-diabetic dogs: a putative mechanism of glucagon irresponsiveness to hypoglycemia. *Endocrinology* 126:1096–1104.

144. Pour P.M., Donnelly K., and Stepan K. 1983. Modification of pancreatic carcinogenesis in the hamster model. 3. Inhibitory effect of alloxan. *Am. J. Pathol.* 110:310–314.

145. Webb J. 1966. *Enzyme and metabolic inhibitors.* New York, London: Academic Press. 367–419.

146. Grossman M. and Ivy A. 1946. Effect of alloxan upon the external secretion of pancreas. *Proc. Soc. Exp. Biol. Med.* 63:62.

147. Tiscornia O.M., Janowitz H.D., and Dreiling D.A. 1968. The effect of alloxan upon canine exocrine pancreatic secretion. *Am. J. Gastroenterol.* 49:328–340.

148. Lenzen S. and Munday R. 1991. Thiol-group reactivity, hydrophilicity and stability of alloxan, its reduction products and its N-methyl derivatives and a comparison with ninhydrin. *Biochem. Pharmacol.* 42:1385–1391.

149. Malaisse W.J., Malaisse-Lagae F., Sener A., and Pipeleers D.G. 1982. Determinants of the selective toxicity of alloxan to the pancreatic B cell. *Proc. Natl. Acad. Sci. U.S.A.* 79:927–930.

150. Weaver D.C., McDaniel M.L., and Lacy P.E. 1978. Alloxan uptake by isolated rat islets of Langerhans. *Endocrinology* 102:1847–1855.

151. Bilic N. 1975. The mechanism of alloxan toxicity: an indication for alloxan complexes in tissues and alloxan inhibition of 4-acetamido-4'-isothiocyanato-stilbene-2,2'-disulphonic acid (SITS) binding for the liver cell membrane. *Diabetologia* 11:39–43.

152. Boquist L., Nelson L., and Lorentzon R. 1983. Uptake of labeled alloxan in mouse organs and mitochondria in vivo and in vitro. *Endocrinology* 113:943–948.

153. Boquist L. 1977. The endocrine pancreas in early alloxan diabetes. Including study of the alloxan inhibitory effect of feeding and some hexoses. *Acta. Pathol. Microbiol. Scand. [A]* 85A:219–229.

154. Sakurai K., Katoh M., Someno K., and Fujimoto Y. 2001. Apoptosis and mitochondrial damage in INS-1 cells treated with alloxan. *Biol. Pharm. Bull.* 24:876–882.

155. Lazarow A. 1947. Further studies of effect of sulphur compounds on production of diabetes with aloxan. *Proc. Soc. Exp. Biol. Med.* 66:4–7.

156. Iynedjian P.B., Mobius G., Seitz H.J., Wollheim C.B., and Renold A.E. 1986. Tissue-specific expression of glucokinase: identification of the gene product in liver and pancreatic islets. *Proc. Natl. Acad. Sci. U.S.A.* 83:1998–2001.

157. Lenzen S., Brand F.H., and Panten U. 1988. Structural requirements of alloxan and ninhydrin for glucokinase inhibition and of glucose for protection against inhibition. *Br. J. Pharmacol.* 95:851–859.

158. Heikkila R.E., Winston B., and Cohen G. 1976. Alloxan-induced diabetes-evidence for hydroxyl radical as a cytotoxic intermediate. *Biochem. Pharmacol.* 25:1085–1092.

159. Grankvist K., Marklund S., Sehlin J., and Taljedal I.B. 1979. Superoxide dismutase, catalase and scavengers of hydroxyl radical protect against the toxic action of alloxan on pancreatic islet cells in vitro. *Biochem. J.* 182:17–25.

160. Fischer L.J. and Hamburger S.A. 1980. Inhibition of alloxan action in isolated pancreatic islets by superoxide dismutase, catalase, and a metal chelator. *Diabetes* 29:213–216.

161. Heikkila R.E. and Cabbat F.S. 1980. The prevention of alloxan-induced diabetes by amygdalin. *Life Sci.* 27:659–662.

162. Holmgren A. and Lyckeborg C. 1980. Enzymatic reduction of alloxan by thioredoxin and NADPH-thioredoxin reductase. *Proc. Natl. Acad. Sci. U.S.A.* 77:5149–5152.

163. Winterbourn C.C. and Munday R. 1989. Glutathione-mediated redox cycling of alloxan. Mechanisms of superoxide dismutase inhibition and of metal-catalyzed OH. formation. *Biochem. Pharmacol.* 38:271–277.

164. Heikkila R.E. 1977. The prevention of alloxan-induced diabetes in mice by dimethyl sulfoxide. *Eur. J. Pharmacol.* 44:191–193.

165. im Walde S.S., Dohle C., Schott-Ohly P., and Gleichmann H. 2002. Molecular target structures in alloxan-induced diabetes in mice. *Life Sci.* 71:1681–1694.

166. Lenzen S., Drinkgern J., and Tiedge M. 1996. Low antioxidant enzyme gene expression in pancreatic islets compared with various other mouse tissues. *Free Radic. Biol. Med.* 20:463–466.

167. Watkins D., Cooperstein S.J., and Lazarow A. 1970. Effect of sulfhydryl reagents on permeability of toadfish islet tissue. *Am. J. Physiol.* 219:503–509.

168. Holmgren A. and Luthman M. 1978. Tissue distrubution and subcellular localization of bovine thioredoxin determined by radioimmunoassay. *Biochemistry* 17:4071–4077.

169. Takasu N., Asawa T., Komiya I., Nagasawa Y., and Yamada T. 1991. Alloxan-induced DNA strand breaks in pancreatic islets. Evidence for H2O2 as an intermediate. *J. Biol. Chem.* 266:2112–2114.

170. Bromme H.J., Ebelt H., Peschke D., and Peschke E. 1999. Alloxan acts as a prooxidant only under reducing conditions: influence of melatonin. *Cell. Mol. Life Sci.* 55:487–493.

171. Yamamoto H., Uchigata Y., and Okamoto H. 1981. DNA strand breaks in pancreatic islets by in vivo administration of alloxan or streptozotocin. *Biochem. Biophys. Res. Commun.* 103:1014–1020.

172. Konrad R.J., Zhang F., Hale J.E., Knierman M.D., Becker G.W., and Kudlow J.E. 2002. Alloxan is an inhibitor of the enzyme O-linked N-acetylglucosamine transferase. *Biochem. Biophys. Res. Commun.* 293:207–212.

173. Weaver D.C., McDaniel M.L., and Lacy P.E. 1978. Mechanism of barbituric-acid protection against inhibition by alloxan of glucose-induced insulin release. *Diabetes* 27:71–77.

174. Sener A. and Malaisse W.J. 1985. Resistance to alloxan of tumoral insulin-producing cells. *FEBS Lett.* 193:150–152.

175. Jansson L. and Sandler S. 1986. Alloxan-induced diabetes in the mouse: time course of pancreatic B-cell destruction as reflected in an increased islet vascular permeability. *Virchows. Arch. A Pathol. Anat. Histopathol.* 410:17–21.

176. Jansson L. and Sandler S. 1992. Alloxan, but not streptozotocin, increases blood perfusion of pancreatic islets in rats. *Am. J. Physiol.* 263:E57–E63.

177. Kim H.R., Rho H.W., Park B.H., Park J.W., Kim J.S., Kim U.H., and Chung M.Y. 1994. Role of Ca2+ in alloxan-induced pancreatic beta-cell damage. *Biochem. Biophys. Acta.* 1227:87–91.

178. Dean P.M. and Matthews E.K. 1972. The bioelectrical properties of pancreatic islet cells: effects of diabetogenic agents. *Diabetologia* 8:173–178.

179. Bloch K.O., Zemel R., Bloch O.V., Grief H., and Vardi P. 2000. Streptozotocin and alloxan-based selection improves toxin resistance of insulin-producing RINm cells. *Int. J. Exp. Diabetes Res.* 1:211–219.

180. Elsner M., Tiedge M., Guldbakke B., Munday R., and Lenzen S. 2002. Importance of the GLUT2 glucose transporter for pancreatic beta cell toxicity of alloxan. *Diabetologia* 45:1542–1549.

181. Malaisse-Lagae F., Sener A., and Malaisse W.J. 1981. Biochemical basis of a species difference in sensitivity to alloxan. *FEBS Lett.* 133:181–182.

182. Gorray K.C., Baskin D., Brodsky J., and Fujimoto W.Y. 1986. Responses of pancreatic b cells to alloxan and streptozotocin in the guinea pig. *Pancreas* 1:130–138.

183. Gorray K.C., Baskin D.G., and Fujimoto W.Y. 1986. Physiological and morphological changes in islet B cells following treatment of the guinea pig with alloxan. *Diabetes Res.* 3:187–191.

184. De Haro-Hernandez R., Cabrera-Munoz L., and Mendez J.D. 2004. Regeneration of beta-cells and neogenesis from small ducts or acinar cells promote recovery of endocrine pancreatic function in alloxan-treated rats. *Arch. Med. Res.* 35:114–120.

185. Tyrberg B., Andersson A., and Borg L.A. 2001. Species differences in susceptibility of transplanted and cultured pancreatic islets to the beta-cell toxin alloxan. *Gen. Comp. Endocrinol.* 122:238–251.

186. Okamoto H. 1981. Regulation of proinsulin synthesis in pancreatic islets and a new aspect to insulin-dependent diabetes. *Mol. Cell. Biochem.* 37:43–61.

187. Yamagami T., Miwa A., Takasawa S., Yamamoto H., and Okamoto H. 1985. Induction of rat pancreatic B-cell tumors by the combined administration of streptozotocin or alloxan and poly(adenosine diphosphate ribose) synthetase inhibitors. *Cancer Res.* 45:1845–1849.

188. Kazumi T., Yoshino G., and Baba S. 1980. Pancreatic islet cell tumors found in rats given alloxan and nicotinamide. *Endocrinol. Jpn.* 27:387–393.

189. Pour P., Mohr U., Althoff J., Cardesa A., and Kmoch N. 1976. Spontaneous tumors and common diseases in two colonies of Syrian hamsters. III. Urogenital system and endocrine glands. *J. Natl. Cancer Inst.* 56:949–961.

190. Takahashi M. and Pour P. 1978. Spontaneous alterations in the pancreas of the aging Syrian golden hamster. *J. Natl. Cancer Inst.* 60:355–364.

191. Trandaburu T. and Ionescu M. 1971. Ultrastructural alterations in the pancreatic B-cells of the newt (*Triturus vulgaris*) induced by alloxan. *Acta. Anat. (Basel)* 79:257–269.

192. Lenzen S., Freytag S., Panten U., Flatt P.R., and Bailey C.J. 1990. Alloxan and ninhydrin inhibition of hexokinase from pancreatic islets and tumoural insulin-secreting cells. *Pharmacol. Toxicol.* 66:157–162.

193. Lenzen S., Tiedge M., Flatt P.R., Bailey C.J., and Panten U. 1987. Defective regulation of glucokinase in rat pancreatic islet cell tumours. *Acta. Endocrinol. (Copenh.)* 115:514–520.

194. Drachenberg C.B., Klassen D.K., Weir M.R., Wiland A., Fink J.C., Bartlett S.T., Cangro C.B. et al. 1999. Islet cell damage associated with tacrolimus and cyclosporine: morphological features in pancreas allograft biopsies and clinical correlation. *Transplantation* 68:396–402.

195. Sai P., Boillot D., Boitard C., Debray-Sachs M., Reach G., and Assan R. 1983. Pentamidine, a new diabetogenic drug in laboratory rodents. *Diabetologia* 25:418–423.

196. Chan J.C., Cockram C.S., and Critchley J.A. 1996. Drug-induced disorders of glucose metabolism. Mechanisms and management. *Drug Saf.* 15:135–157.

197. Pick A., Clark J., Kubstrup C., Levisetti M., Pugh W., Bonner-Weir S., and Polonsky K.S. 1998. Role of apoptosis in failure of beta-cell mass compensation for insulin resistance and beta-cell defects in the male Zucker diabetic fatty rat. *Diabetes* 47:358–364.

198. Coleman D.L. and Hummel K.P. 1973. The influence of genetic background on the expression of the obese (Ob) gene in the mouse. *Diabetologia* 9:287–293.

199. Butler A.E., Janson J., Bonner-Weir S., Ritzel R., Rizza R.A., and Butler P.C. 2003. Beta-cell deficit and increased beta-cell apoptosis in humans with Type 2 diabetes. *Diabetes* 52:102–110.

200. Stefan Y., Orci L., Malaisse-Lagae F., Perrelet A., Patel Y., and Unger R.H. 1982. Quantitation of endocrine cell content in the pancreas of nondiabetic and diabetic humans. *Diabetes* 31:694–700.

201. Clark A., Wells C.A., Buley I.D., Cruickshank J.K., Vanhegan R.I., Matthews D.R., Cooper G.J. et al. 1988. Islet amyloid, increased A-cells, reduced B-cells and exocrine fibrosis: quantitative changes in the pancreas in Type 2 diabetes. *Diabetes Res.* 9:151–159.

202. Kloppel G., Lohr M., Habich K., Oberholzer M., and Heitz P.U. 1985. Islet pathology and the pathogenesis of Type 1 and Type 2 diabetes mellitus revisited. *Surv. Synth. Pathol. Res.* 4:110–125.

203. Roche E., Maestre I., Martin F., Fuentes E., Casero J., Reig J.A., and Soria B. 2000. Nutrient toxicity in pancreatic beta-cell dysfunction. *J. Physiol. Biochem.* 56:119–128.

204. Cnop M., Hannaert J.C., Grupping A.Y., and Pipeleers D.G. 2002. Low density lipoprotein can cause death of islet beta-cells by its cellular uptake and oxidative modification. *Endocrinology* 143:3449–3453.

205. Prentki M., Joly E., El-Assaad W., and Roduit R. 2002. Malonyl-CoA signaling, lipid partitioning, and glucolipotoxicity: role in beta-cell adaptation and failure in the etiology of diabetes. *Diabetes* 51 Suppl. 3:S405–S413.

206. Poitout V. and Robertson R.P. 2002. Minireview: Secondary beta-cell failure in Type 2 diabetes — a convergence of glucotoxicity and lipotoxicity. *Endocrinology* 143:339–342.

207. Robertson R.P., Harmon J., Tran P.O., Tanaka Y., and Takahashi H. 2003. Glucose toxicity in beta-cells: Type 2 diabetes, good radicals gone bad, and the glutathione connection. *Diabetes* 52:581–587.

208. Robertson R.P., Zhang H.J., Pyzdrowski K.L., and Walseth T.F. 1992. Preservation of insulin mRNA levels and insulin secretion in HIT cells by avoidance of chronic exposure to high glucose concentrations. *J. Clin. Invest.* 90:320–325.

209. Grankvist K., Marklund S.L., and Taljedal I.B. 1981. CuZn-superoxide dismutase, Mn-superoxide dismutase, catalase and glutathione peroxidase in pancreatic islets and other tissues in the mouse. *Biochem. J.* 199:393–398.

210. Wolff S.P. and Dean R.T. 1987. Glucose autoxidation and protein modification. The potential role of 'autoxidative glycosylation' in diabetes. *Biochem. J.* 245:243–250.

211. Kaneto H., Xu G., Song K.H., Suzuma K., Bonner-Weir S., Sharma A., and Weir G.C. 2001. Activation of the hexosamine pathway leads to deterioration of pancreatic beta-cell function through the induction of oxidative stress. *J. Biol. Chem.* 276:31099–31104.

212. Maedler K., Spinas G.A., Dyntar D., Moritz W., Kaiser N., and Donath M.Y. 2001. Distinct effects of saturated and monounsaturated fatty acids on beta-cell turnover and function. *Diabetes* 50:69–76.

213. Shimabukuro M., Zhou Y.T., Levi M., and Unger R.H. 1998. Fatty acid-induced beta cell apoptosis: a link between obesity and diabetes. *Proc. Natl. Acad. Sci. U.S.A.* 95:2498–2502.

214. Maedler K., Oberholzer J., Bucher P., Spinas G.A., and Donath M.Y. 2003. Monounsaturated fatty acids prevent the deleterious effects of palmitate and high glucose on human pancreatic beta-cell turnover and function. *Diabetes* 52:726–733.

215. Efanova I.B., Zaitsev S.V., Zhivotovsky B., Kohler M., Efendic S., Orrenius S., and Berggren P.O. 1998. Glucose and tolbutamide induce apoptosis in pancreatic beta-cells. A process dependent on intracellular Ca2+ concentration. *J. Biol. Chem.* 273:33501–33507.

216. Prentki M., Segall L., Roche E., Thumelin S., Brun T., McGarry J.D., Corkey B.E., Prentki M., Segall L., Roche E., Thumelin S., Brun T., McGarry J.D., Corkey B.E., and Assimacopoulos-Jeannet F. 1998. Gluco-lipotoxicity and gene expression in the pancreatic beta cell. *J. Ann. Diabetol. Hotel. Dieu.* 17–27.

217. Roche E., Farfari S., Witters L.A., Assimacopoulos-Jeannet F., Thumelin S., Brun T., Corkey B.E. et al. 1998. Long-term exposure of beta-INS cells to high glucose concentrations increases anaplerosis, lipogenesis, and lipogenic gene expression. *Diabetes* 47:1086–1094.

218. Jacqueminet S., Briaud I., Rouault C., Reach G., and Poitout V. 2000. Inhibition of insulin gene expression by long-term exposure of pancreatic beta cells to palmitate is dependent on the presence of a stimulatory glucose concentration. *Metabolism* 49:532–536.

219. El-Assaad W., Buteau J., Peyot M.L., Nolan C., Roduit R., Hardy S., Joly E. et al. 2003. Saturated fatty acids synergize with elevated glucose to cause pancreatic beta-cell death. *Endocrinology* 144:4154–4163.

220. Lee Y., Hirose H., Zhou Y.T., Esser V., McGarry J.D., and Unger R.H. 1997. Increased lipogenic capacity of the islets of obese rats: a role in the pathogenesis of NIDDM. *Diabetes* 46:408–413.

221. Cnop M., Hannaert J.C., Hoorens A., Eizirik D.L., and Pipeleers D.G. 2001. Inverse relationship between cytotoxicity of free fatty acids in pancreatic islet cells and cellular triglyceride accumulation. *Diabetes* 50:1771–1777.

222. Grupping A.Y., Cnop M., Van Schravendijk C.F., Hannaert J.C., Van Berkel T.J., and Pipeleers D.G. 1997. Low density lipoprotein binding and uptake by human and rat islet beta cells. *Endocrinology* 138:4064–4068.

223. Lorenzo A., Razzaboni B., Weir G.C., and Yankner B.A. 1994. Pancreatic islet cell toxicity of amylin associated with Type-2 diabetes mellitus. *Nature* 368:756–760.

224. Nerup J., Mandrup-Poulsen T., Helqvist S., Andersen H.U., Pociot F., Reimers J.I., Cuartero B.G. et al. 1994. On the pathogenesis of IDDM. *Diabetologia* 37 Suppl. 2:S82–S89.

225. Cetkovic-Cvrlje M. and Eizirik D.L. 1994. TNF-alpha and IFN-gamma potentiate the deleterious effects of IL-1 beta on mouse pancreatic islets mainly via generation of nitric oxide. *Cytokine* 6:399–406.

226. Sternesjo J., Bendtzen K., and Sandler S. 1995. Effects of prolonged exposure in vitro to interferon-gamma and tumour necrosis factor-alpha on nitric oxide and insulin production of rat pancreatic islets. *Autoimmunity* 20:185–190.

227. Rabinovitch A., Suarez-Pinzon W.L., Shi Y., Morgan A.R., and Bleackley R.C. 1994. DNA fragmentation is an early event in cytokine-induced islet beta-cell destruction. *Diabetologia* 37:733–738.

28

ALTERATIONS OF THE ENDOCRINE PANCREAS

Reid Aikin, Stephen Hanley, Mark Lipsett, and Lawrence Rosenberg

CONTENTS

28.1   INTRODUCTION

The adult human pancreas consists of three principal cell types — the exocrine, endocrine, and ductal cells (Figure 28.1). These cells encompass one organ with two distinct functions: the exocrine pancreas secretes digestive enzymes, and the endocrine pancreas produces the hormones responsible for glucose homeostasis. Briefly, the acinar cells secrete digestive enzymes into the ductal network, which produces bicarbonate and other components of pancreatic juice. The pancreatic juice is released, in a controlled manner, into the duodenum via the ampulla of Vater. The endocrine cells are typically grouped into organized clusters known as the islets of Langerhans, although individual cells or small clusters are also observed.

28.2   ISLET ANATOMY

28.2.1   Cell Types

The adult human pancreatic islet is a collection of a few cells to several thousand, with an islet diameter anywhere from 40 to 400 μm. Although

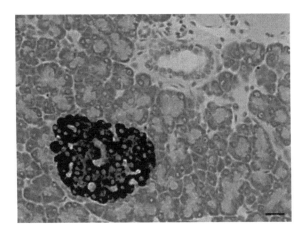

Figure 28.1 Photomicrograph of an insulin stained (dark) pancreatic tissue section in normal rat. Acinar cells surround the insulin-positive islet, with a pancreatic duct in the upper right. (Bar = 20 μm)

75% of islets are less than 160 μm in diameter, these only account for 15% of the total endocrine volume; whereas islets with diameters of greater than 250 μm account for only 15% of islet number, but 60% of islet volume.[1] There are four principal endocrine cell types found in the human pancreatic islet: the α-, β-, δ-, and PP-cells produce glucagon, insulin, somatostatin, and pancreatic polypeptide, respectively (see also Chapter 3, Chapter 9, and Chapter 28). In addition, there exist smaller numbers of other cell types, including the D_1-, EC, G_1-, and small cells.[1-3]

β-cells are by far the most common cells of the islet. These cells secrete insulin in response to the body's needs. Insulin is synthesized and packaged into immature secretory granules, which then mature as pro-insulin is cleaved by pro-protein convertases, producing insulin, consisting of an A-chain (21 amino acids) and a B-chain (30 amino acids) connected by two sets of disulfide bonds and C-peptide. Insulin is released in a glucose-responsive manner, which can also be potentiated by the administration of insulin secretagogues. Briefly, glucose enters the β-cell via the glucose transporter-2 (GLUT-2), whereby it is rapidly phosphorylated by glucokinase, resulting in glucose-6-phosphate. Glucose-6-phosphate is a substrate for the glycolytic reactions that produce two moles of pyruvate and a net conversion of eight moles of adenosine diphosphate (ADP) to adenosine triphosphate (ATP) for every mole of glucose catabolized. As a result, the ATP:ADP ratio within the β-cell shifts in favor of ATP. Under normal β-cell ATP:ADP ratios, ATP-sensitive potassium channels are constitutively open, allowing potassium outflow to set and maintain the β-cell's resting potential. The glycolytic shift in intracellular ATP levels induces a conformational change at the level of the potassium channel, inhibiting potassium conductance across the membrane, and inducing depolarization of the β-cell, yielding an action potential. This action potential travels along the cellular membrane via induced changes in voltage-gated ion channels. When the action potential reaches L-type voltage-sensitive calcium channels, it causes them to open, allowing an influx of calcium into the depolarized cell. This rapid influx of calcium initiates insulin exocytosis. Increased intracellular calcium causes degranulation by acting at the level of the vesicular fusion proteins as well as calmodium, which contain calcium-binding motifs. In this manner, glucose is directly responsible for the secretion of insulin; glucose acts as a signaling molecule via the glycolytic pathway.[4–6] Furthermore, β-cells within an islet are connected by gap junctions, which allow for electrochemical coupling, and as such, β-cells may act as a syncytium.[7–10] Of interest, however, is the presence of intra-islet variability with regards to glucose-stimulated insulin release: that is, not all β-cells respond equally to a given stimulus.[11–13] This fact reinforces the notion of individual β-cell recruitment in response to increased glucose loads. As well, a constitutive pathway in which insulin is released at a constant rate irrespective of the prevailing glucose levels also secretes insulin.[14]

The β-cell also produces amylin, or islet amyloid polypeptide (IAPP),[15–17] a 37 amino acid peptide whose function is poorly understood. Both increased age and type 2 diabetes mellitus result in a buildup of intra-islet amylin, which is related to defective β-cell secretory function. Amylin is found in immature insulin granules in a pro form, which is much more conducive to plaque formation than the mature molecule that is produced by pro-protein convertase-mediated cleavage. However, pro-amylin, and the subsequently formed plaques that affect islet function, are common in type 2 diabetics, who also release greater amounts of unprocessed pro-insulin than do normal patients.[18,19]

The α-cells are the second most common endocrine cell type in the adult islet. α-cells produce a variety of hormones via post-transcriptional modification of the glucagon gene product. Tissue-specific processing of prepro-glucagon yields glucagon, glucagon-like peptide-1 (GLP-1), and other C-terminally extended glucagon derivatives in the islet, and intestinal cells produce glicentin, GLP-1, and GLP-2.[20] All α-cell secretory products co-localize to the same secretory granules, but as the granules mature, the level of glicentin decreases while that of glucagon increases,[21] suggesting that glicentin is a precursor for other glucagon-related products. Glucagon, the 29 amino acid peptide hormone, is released by the α-cell in response to decreased blood glucose[22] and amino acid levels,[23] as well as nervous control,[24] and acts to increase blood glucose levels by inducing glycogenolysis as well as hepatic gluconeogenesis and other effects.[22,25,26]

The δ-cells produce somatostatin. This multifunctional peptide is known for its role as an inhibitor of insulin, glucagon, and growth hormone secretion, as well as that of other polypeptide hormones. Both the circular 14 and 28 amino acid sequences have the functional character of both hormone and neurotransmitter, acting in an endocrine manner when released into the circulation, and in a paracrine manner when released locally, both by δ-cells and neurons. The δ-cells are also recognizable by the somewhat dendritic appearance: they exhibit unipolar elongation of their cytoplasm, which extend to other cells as well as capillaries, and which are thought to be the site of somatostatin release.[1]

The fourth principal islet cell type, the PP-cell, secretes the 36 amino acid pancreatic polypeptide, whose function remains poorly understood.[27]

28.2.2 Organization

Islet cellular architecture is similar in mammals in that β-cells are found predominantly in the core of the islet, with a mantle composed of the three other endocrine cell types (see Chapter 3, Chapter 6, and Chapter 28). The relative proportion of endocrine cell types changes during maturation. At birth, the β:α:δ:PP cell ratio is 45:22:32:1%, although with maturation comes an increase in the relative proportion of β-cells. By adulthood, the final ratio sits at 68:20:10:2%.[28] The relative cell type frequencies within the islet also vary throughout the pancreas itself. The tail of the organ contains islets that are rich in α-cells and poor in PP-cells, while the head contains islets that have decreased α- and β-cell number, increased PP-cell number, and less well-defined borders.[28] Not surprisingly, data suggests islets in the head of the gland release less insulin in response to a glucose stimulus. A further level of heterogeneity is suggested by the notion that given an adult human pancreas contains roughly 1 million islets,[29] the majority are found within the tail of the organ. Finally, larger islets, which account for a minority of islet number but a majority of islet mass, are found predominantly in proximity with large ducts and blood vessels.[1,30]

28.2.3 Vasculature

Although consisting of only 2 or 3% of the total pancreatic mass, islets receive roughly 20% of the organ's arterial blood flow.[31] Systematic anatomical studies have shown that arterial flow is directed to the β-cell-containing islet core, where the arteriole opens into a system of highly fenestrated capillaries, as opposed to the few fenestrations found in acinar capillary beds. The capillary network passes through the islet mantle before coalescing into a series of postcapillary venules. In larger islets, these venules coalesce within the boundaries of the islet (see also Chapter 4);

in smaller islets, the capillaries persist into the acinar tissue prior to merging, although the proportion of acinar tissue that receives blood in this fashion is limited.[32] The endocrine cells are arranged in such a way as to maximize vascular exposure. Furthermore, β-cells appear to display a polarity with regards to the vasculature in that the secretory granules are directed toward the capillaries to maximize endocrine release.[1] Within the islet, the specific orientation of blood flow gives rise to a specific order of vascular exposure — β-to-α-to-δ-cell — that may relate to cell–cell interactions.[33,34] For details, see Chapter 4.

28.2.4   Innervation

Islets are richly innervated, receiving afferent innervation from cholinergic, adrenergic, and peptidergic neurons.[35] Pre-ganglionic parasympathetic nerves enter the pancreas, along with blood vessels, via the gland's intrinsic ganglia. These nerves, along with the post-ganglionic adrenergic fibers from the celiac plexus, then innervate the islets. While parasympathetic cholinergic nerves run to the β-cell core, and vagal stimulation promotes insulin release, adrenergic innervation is found primarily at the mantle, with sympathetic stimulation causing glucagon release and inhibiting insulin release.[27] The most common peptidergic nerves contain vasoactive intestinal peptide, distributed throughout the exocrine parenchyma, vessels, and islets,[36] and neuropeptide Y, found within adrenergic fibers.

28.2.5   Hormone Signaling and Regulation

The islet hormones induce a great number of systemic effects by ligation of cell surface receptors. Insulin signals via a receptor tyrosine kinase consisting of four polypeptide subunits arranged in an $\alpha_2\beta_2$ fashion. The extracellular α subunits bind insulin, resulting in transphosphorylation of the transmembrane β subunits at intracellular recruiting tyrosine residues. It is via these phosphotyrosine residues that the insulin receptor substrate (IRS) proteins (IRS-1, IRS-2) are recruited, which then act as docking sites for various signaling proteins — phosphatidylinositol 3'-kinase (PI3-K), SH2-containing phosphotyrosine phosphatase, and growth factor receptor-bound protein 2 (Grb-2). These enzymes then mediate insulin's downstream effects, which include regulation of glucose uptake as well as cellular and physiolgic growth.[37–39]

Conversely, glucagon acts via a G-protein-coupled receptor. This seven transmembrane cell surface receptor activates adenylyl cyclase following glucagon binding, resulting in increased intracellular cyclic adenosine monophosphate (cAMP) levels.[40] cAMP activates glycogen phosphorylase, resulting in glycogenolysis, and inhibits glycogen synthase, decreasing glycogen synthesis.[22] Similarly, somatostatin acts via its family of G-protein-coupled receptors.[41]

Not only do the islet cell hormones act in an endocrine manner, but these hormones also act in a juxtacrine fashion.[1,42] For example, it was previously stated that somatostatin down-regulates both insulin[43] and glucagon release.[44] Similarly, insulin appears to reduce somatostatin[24] and glucagon release.[45] Finally, glucagon is known to induce insulin and somatostatin secretion, although these interactions were shown using infused hormones, which do not take into account the naturally occurring unidirectional β-to-α-to-δ-cell blood flow.[34]

28.3 ISLET DEVELOPMENT

With regard to embryologic development, β-cells arise from precursor stem cells, now understood to reside in the pancreatic ducts.[46] The development of the exocrine and endocrine components of the pancreas are related and conserved (see Chapter 1, Chapter 9, and Chapter 28). Although β-cell development within the endocrine component parallels that of the exocrine portion, α-cells appear to develop earlier. Two developmental phases — biochemical and ultrastructural — have been identified. First, in the early stages of embryogenesis, the notochord and endoderm lie in close apposition. This interaction appears crucial to the development of the pancreas and its various cell types.[47] In humans, the pancreas arises from the endoderm of the fetal gastrointestinal tract over the course of the fifth gestational week. Subsequently, two evaginations form off the fetal gut: the duodenal, or dorsal, and hepatic, or ventral, diverticula. The ventral anlage develops from an evagination of the distal foregut at the site of insertion of the bile duct, with hepatic development having occurred prior to the growth of both the dorsal and ventral anlagen. During the seventh week of gestation, the ventral pancreatic bud rotates, resulting in fusion of the two anlagen, and formation of the fetal pancreas. Tissue derived from the dorsal pancreatic bud eventually forms the body and tail of the mature pancreas, while tissue from the ventral anlage forms the head. Both primordia retain their ductal systems, with the major ducts of each, that of Wirsung (ventral) or Santorini (dorsal or accessory), usually emptying separately into the gut lumen.[27]

Intercellular interactions appear to govern the development of the pancreas throughout this process. The pancreatic diverticulum associates closely with accumulating mesenchymal cells, which form a surrounding mesenchymal cap, and which must remain in proximity for further epithelial proliferation. A growth stage follows whereby lobulation of the ductal system occurs. Production of insulin and secretory enzymes remains low during the proto-differentiated state. Although β-cells and acini are not identifiable at this stage, a large number of cells are thought to express low levels of insulin or exocrine enzymes, and are already committed in this direction. The second phase of cytodifferentiation is marked by further cellular maturation: exocrine cells show a rapid increase in the number of rough endoplasmic reticulum and zymogen granules, while differentiated β-cells containing insulin secretory granules also appear. At this point, the acinar and β-cells have achieved their fully differentiated state. Differentiated α-cells are found by about 9 weeks of gestation. The mature δ-cell population appears next, at 10 weeks, followed finally by the β-cells at 11 weeks. Molecular phenotype is likely determined in advance of this timeline, however, as the appearance of detectable hormone levels precedes morphologic appearance by about 3 weeks.[27]

The developing human pancreas has also been found to express gastrin, although its expression is not found in the adult,[48] with the exception of the setting of certain disease states. Transient gastrin expression during development suggests a potential trophic role for this protein.[49] Endocrine tissue reaches its greatest density at the 30th week of gestation. At birth, islets make up 10% of the gland's mass, falling progressively to 7% in infancy and 2 to 3% in adulthood. This decrease in proportion is due to relative growth of the nonendocrine tissues.[50]

28.3.1 Genetic Regulation

The development of the pancreas, islets, and individual cell types appears to result from the differential expression and activity of a cascade of transcription factors (see Chapter 1, Chapter 9, and Chapter 28). Some of these appear to be involved earlier, that is, in the initial differentiation from a primitive precursor, and others have a role in the expression of the fully differentiated phenotype. The most important genes associated with β-cell growth are encoded by the homeobox family of transcription factors, namely *pdx-1, pax-4,* and *isl-1*.[51–55] Additionally, neurogenin-3, a member of the mammalian neurogenin gene family, has been proposed as a *bona fide* pro-endocrine gene.[56] The notch signaling pathway also controls the choice between differentiated endocrine and exocrine cell fates.[57–60] As such, lack of notch pathway signaling results in high neurogenin-3 levels, thereby promoting the endocrine fate.

Pancreatic and duodenal homeobox gene-1 (*pdx-1*) — also known as insulin promoter factor-1 (*ipf-1*),[51,52] islet/duodenum homeobox-1 (*idx-1*),[53] or somatostatin transactivating factor-1 (*stf-1*)[61] — was the first genetic element to be associated with pancreatic differentiation and is restricted to the gut endoderm from whence the dorsal and ventral pancreatic buds develop.[55] PDX-1 protein is found in the nuclei of 91% of β-cells and 15% of δ-cells.[62] *Pdx-1* knockout mice survive to birth, but die a few days thereafter, having exhibited growth retardation and dehydration.[63,64] These mice show a complete absence of normal pancreatic development, with but a primitive bud present, despite normal pancreatic mesenchymal growth.[55] It has been postulated, therefore, that PDX-1 controls development of the pancreatic epithelium and its normal response to mesenchymal signals.[47,55] Mice heterozygous for the *PDX-1* null mutation have a decreased β-cell mass, insulin, and GLUT-2 expression, but not diabetes.[65] To better characterize the effects of PDX-1 on β-cell maturation, β-cell-specific gene disruption was accomplished using the Cre-LoxP recombination system.[66,67] Islets in mutant mice showed a 40% decrease in β-cell number as well as decreased insulin and GLUT-2 expression, and the animals became diabetic.[65] PDX-1, therefore, is expressed in mature islets, duodenum, and distal stomach; is required for the development of pancreatic epithelium and subsequent exocrine and endocrine cells; and is, in the mature pancreas, a transactivator of the promoters for the *insulin, glucokinase, GLUT-2, amylin,* and *somatostatin* genes. Not only does PDX-1 activate *insulin* gene expression, but studies on transformed and normal islets now show that PDX-1 DNA-binding affinity is itself affected by glucose levels.[68] Furthermore, while high glucose acutely increases *pdx-1* expression and DNA-binding activity, chronic elevation of glucose causes down-regulation of both,[69–71] illustrating at least one mechanism by which PDX-1 may carry out the effects of glucotoxicity. More recent studies on islet cell lines demonstrate that *pdx-1* overexpression is associated with a decrease in *insulin* transcription and suggest that this gene may have a dual role as both an activator and an inhibitor of *insulin* gene expression.[72] As such, levels of *pdx-1* and related transcription factors may ultimately regulate their own action.

The homologous human *pdx-1* mutation is remarkably similar to the murine experimental condition in its phenotype, a case of pancreatic agenesis was reported and traced to a mutation in *pdx-1*.[73] This heterozygous condition has been associated with one variety of early onset type 1 diabetes, now referred to as maturity-onset diabetes of the young-4, or MODY-4. To rectify the discordant observations with regards to diabetes between the hetereozygous human condition and the murine models, it has been proposed that the mutated human PDX-1 protein retains its ability to inhibit *insulin* transcription.[74]

The homeodomain family transcription factor Isl-1 is distinguished from PDX-1 in its ability to activate the expression of *glucagon,*[75,76] *somatostatin,*[77] and *amylin,*[78] but not *insulin.*[79] In addition to all four islet cell types, *isl-1* is expressed in neurons, pituitary, lung, kidney, thymus, and ovary.[80] In islets, *isl-1* is expressed prior to the expression of specific hormones. Homozygous *isl-1* null mice have no dorsal pancreatic mesenchyme, nor are any endocrine cells observed, despite normal exocrine development.[81]

It is thought, therefore, that Isl-1 is involved primarily with islet cell differentiation from the pancreatic epithelium.

The *pax-4* gene belongs to a family of transcriptional regulators containing a paired helix-turn-helix DNA-binding domain as well as a homeodomain.[82,83] *Pax-4* is expressed early in the formation of the pancreatic bud, but is later restricted to the β-cells.[54] Mice homozygous for a *pax-4* null mutation show a lack of β- and δ-cells and an increased number of α-cells.[54] The *pax-6* gene, belonging to the same family, is expressed in islets as well as in the pituitary, central nervous system, and nasal epithelium.[84,85] In the pancreas of homozygous *pax-6* null mice, α-cells are absent, but β-, δ-, and PP-cells are scattered throughout the exocrine parenchyma without forming discrete islets.[86] Pax-6 may function, at least in part, through the up-regulation of the neural cell adhesion molecule (N-CAM) that, in islets, is found primarily on non-β-cells. Studies have suggested that the endocrine pancreas' N-CAM expression profile may explain the non-random distribution of cell types within the islet.[87,88] The effects of *pax-4* and *pax-6*, therefore, appear to direct endocrine cell differentiation to particular phenotypes. Mice lacking both *pax-4* and *pax-6* have no differentiated endocrine cells whatsoever, while *pdx-1*-positive cells are present, indicating that undifferentiated endocrine precursors remain. However, no cells are committed to an α-cell lineage as *pax-6* is missing, but β-, δ-, and PP-cells do not develop due to the lacking *pax-4*.[86] To further elucidate this process, the molecular targets of the Pax-4 protein must be identified.

Although less studied, Nkx-2.2 and Nkx-6.1 are nonetheless important to β-cell differentiation. By the end of endocrine differentiation, *nkx-2.2* is present in all β-cells, roughly 80% of α-cells, and most PP-cells, but no δ-cells.[89] Homozygous *nkx-2.2* null mice lack β-cells, have but 20% of the normal number of α-cells, and a slight decrease in the PP-cell population.[89] A significant number of cells, however, express no islet hormones, yet do express two β-cell markers — amylin and pro-protein convertase 1/3. These cells, therefore, appear to be incompletely differentiated β-cells. Meanwhile, *isl-1* and *pax-6* expression is unaffected, as is early *pdx-1* and *nkx-6.1* expression. However, the later expression of *pdx-1*, which is important to mature β-cell function, is inhibited. The role of Nkx-2.2, therefore, may be to activate both *pdx-1* and *nkx6.1*, which are essential for complete β-cell differentiation and which function as transactivators of insulin transcription.

NeuroD/β-2 which was initially identified as a mediator of neuronal differentiation,[90] is also involved in islet cell differentiation. It is expressed in the central nervous system, the endocrine cells of the intestine,[91] and the islets.[92] Expression in mice is detectable at the earliest stage of islet differentiation. In homozygous *neuroD/β-2* null mice, the number of β-cells decreases by 75%, with α- and δ-cells declining by 40 and 20%, respectively.[92] The micro-architecture of the islet also seems disturbed in these animals, as small clusters of endocrine cells form, but well-organized islets are absent.[92] Lack of *neuroD/β-2* appears to halt development at a stage normally characterized by β-cell proliferation. In this case, the islets show instead an increased number of apoptotic cells. In addition, enteroendocrine cells in the intestine are absent. As part of its actions as an *insulin* and *glucagon* gene transactivator,[93] NeuroD/β-2 functions synergistically with PDX-1, augmenting glucose-induced insulin secretion.[94] More recent evidence indicates that NeuroD/β-2 is capable of transactivating the *pdx-1* gene.[95]

28.4 ISLET MAINTENANCE

Even though fetal islet development in the context of pancreatic organogenesis has been studied extensively, maintenance of islet mass in the context of the adult organ is a relatively new area of interest. It has been observed that despite a relatively constant islet mass, the post-natal rodent pancreas undergoes significant islet neogenesis.[96] A mathematical model developed to understand the processes involved explained this apparent incongruity on the basis of an apoptotic wave during a period of endocrine remodeling.[50,96] Thus, it is likely that significant β-cell neogenesis, along with minimal levels of β-cell replication, effectively counter the apoptotic wave.[96] These data are consistent with the notion that maintenance of β-cell mass in the post-natal period is indeed an active process in which cell proliferation, differentiation, and death are implicated as part of a delicate balance providing tightly integrated homeostatic control of β-cell mass, at least in neonates. Whether there is a similar dynamic regulation of β-cell mass with aging remains to be clarified.

Dynamic control of β-cell mass in the adult pancreas has been suggested by studies of pregnancy during which the mother's β-cell mass expands greatly as metabolic demands increase in response to the altered hormonal milieu.[97,98] However, shortly after birth, when the hormonal environment reverts to the non-pregnant state, the expanded β-cell mass involutes, returning to normal levels through β-cell atrophy, increased β-cell apoptosis, and decreased β-cell replication.[97] Similar cell mass regulation has also been reported for lacrimal glands and breast tissue.[99–105] Thus, it appears that common control mechanisms in different tissues may serve to align cell mass with functional need, and it seems that such mechanisms are operative and particularly responsive not only in the neonatal, but in the adult pancreas as well.

Based on morphometric analyses of single β-cells and small β-cell clusters, Bouwens *et al.* suggested that islet neogenesis normally occurs in the adult human pancreas,[106] presumably contributing to the ongoing maintenance of β-cell mass. However, evidence suggests neogenesis may be under regulatory control that provides for a continual basal rate, but also allows for response to various stimuli. Butler *et al.* have provided evidence for an increase in β-cell mass in the presence of obesity, without increased β-cell replication.[18,107] This finding may suggest that neogenesis in the adult pancreas plays a physiologic role in response to certain critical stimuli or factors, including alterations in the local microenvironment. This notion seems highly likely in view of the recent observation that the development of type 2 diabetes may be correlated with a lack of a neogenic response in adult humans.[18] The underlying mechanisms for regulating homeostatic control and feedback inhibition of islet neogenesis remain to be elucidated.

28.4.1 Animal Models of Neogenesis

Adult islet neogenesis leading to β-cell mass expansion can be induced by partial pancreatic duct obstruction initiated through cellophane wrapping of the hamster[108,109] or monkey[110] pancreas. The new β-cell mass displays normal glucose-responsiveness with a normal counter-regulatory mechanism.[111] Moreover, the cellophane wrapping-induced neogenic islets are able to reverse hyperglycemia in streptozotocin (STZ)-treated hamsters.[112,113] β-cell mass expansion in this model is mediated by islet neogenesis associated protein (INGAP), an acinar cell protein.[113] It is noteworthy that some beneficial effects were observed using partial pancreatic ligation in children with type 1 diabetes over 70 years ago.[114–117] It is significant that *in vivo* studies have not produced any evidence of hyperfunctioning (e.g., hypoglycemia) or unchecked cell growth (e.g., tumor formation), suggesting that the induction of β-cell mass expansion in the normal adult pancreas may be regulated by inherent homeostatic control mechanisms.[118] Further studies have demonstrated that the administration of a biologically active 15 amino acid fragment of the native protein, termed INGAP peptide, is sufficient to induce the normalization of both β-cell mass and glycemia levels in STZ-treated mice.[118] As such, INGAP peptide appears to be of interest as a mediator of islet neogenesis, and INGAP binding has been suggested as a potential marker of progenitor cells within the ducts and islets.[119]

Another model that exists to examine the response in β-cell mass dynamics to increased metabolic demand is that of chronic glucose infusion.[120–125] In this model, animals receive a constant infusion of glucose via intravenous catheter. After 4 days of glucose infusion, the β-cell mass as much as doubles, accompanied by marked β-cell neogenesis. Furthermore, following glucose infusion, β-cell mass returns to basal values as a result of β-cell apoptosis and a decrease in β-cell replication, suggesting that homeostatic mechanisms exert dynamic control over the balance between neogenesis and apoptosis in the adult (rodent) pancreas.[120]

A third model of β-cell response to stimulation is the sucrose feeding model.[126–129] Five weeks of high-sucrose feeding of adult hamsters leads to a neogenesis-associated doubling of β-cell mass. In a variant model, sucrose feeding of pregnant mothers,[119] the offspring exhibit a significant decrease in β-cell apoptosis, together with a substantial increase in β-cell replication, islet neogenesis, and β-cell mass. INGAP is implicated in both cases,[119] thus supporting a role for INGAP in β-cell mass expansion through a modulation of islet neogenesis. Thus, a homeostatic balance between β-cell neogenesis and apoptosis is important in regulation of β-cell mass in the adult pancreas.[130]

The incretin GLP-1 (and its long-acting homologue exendin-4) increases β-cell mass in animal models of diabetes,[131,132] though the exact mechanism of β-cell mass expansion has not been fully clarified (i.e., anti-apoptotic effects of increased insulin secretion vs. effects of augmented insulin release on β-cell replication vs. effect on β-cell size). However, it has been suggested that GLP-1 expands β-cell mass directly by stimulating β-cell proliferation and inducing islet neogenesis in a PDX-1-dependent fashion in both young and old animals[133–139] and by inhibiting apoptosis.[140] These findings lend support to the view that the pancreas possesses control mechanisms to regulate β-cell mass throughout postnatal life. GLP-1 is an insulinotropic hormone secreted by enteroendocrine L-cells of the distal intestine in response to food ingestion,[136] and also regulates blood glucose by stimulating glucose-dependent insulin secretion, insulin biosynthesis, and β-cell proliferation.

Finally, partial pancreatectomy has been proposed as a means of inducing islet neogenesis.[141] In this model, in which 90% of the organ is excised, islet mass is restored to 45% of control at 8 weeks post-surgery.[142–144] The islet neogenesis observed is thought to occur via the proliferation of small ductules from putative precursors in the ductal epithelium, followed by the differentiation of cells within these ductules into the various endocrine phenotypes. In this model, these changes occur over the course of 7 to 10 days and are associated with the expression of various genes associated with islet formation. The first of these is transforming growth factor-β (TGF-β), which appears to be involved in the arrest of ductal proliferation and marks the beginning of the differentiation stage. During the latter stage, hepatocyte growth factor (HGF) and insulin-like growth factor-I (IGF-I) are expressed in association with ductal epithelium.[144] Furthermore, the Reg proteins (of which INGAP is a member) have been implicated in this instance of islet regeneration.[145]

There also exist other models of β-cell mass expansion, however the above were identified based on their ability to induce *neogenesis*, which is the novel formation of islets.

28.4.2 Stem/Progenitor Cells

Even though β-cell replication can account for a part of the novel islets required to maintain islet mass in the face of a certain amount of β-cell turnover, the aforementioned data strongly suggest that islet neogenesis plays a role in maintaining β-cell mass. As such, the persistence of a multipotent precursor cell throughout life, capable of replenishing the endocrine cell population, would be essential for the maintenance of islet homeostasis. However, although it seems certain that a stem cell or precursor for islet cells must exist at some point during the development of the organism[146] and likely still exists in adulthood, there has been little success in finding this cell, or identifying a distinct marker.

Ultimately, the therapeutic potential of β-cell neogenesis may rest in the ability to reliably generate functional β-cell mass, either *in vivo* or *in vitro* for transplantation. This necessarily relies on the ability to identify, and possibly to isolate, maintain, and manipulate a precursor cell type. Despite these observations, the true presence of a stem cell population within the duct or islet has not been conclusively proven. Although several candidates have been identified, a reliable marker for such a cell type has not been defined. Candidates have included neuronal markers such as tyrosine hydroxylase,[147] acid β-galactosidase,[148] glucokinase,[149] GLUT-2,[150] and the ductal markers cytokeratin-19 (CK-19) and cytokeratin-20 (CK-20).[151] Among more recently described markers, the intermediate filament protein nestin, thought to be expressed in neuronal stem cells,[152] has been identified in islet cells as well.[153,154] Other reports, however, question whether nestin expression may really represent intraislet fibroblasts.[155] Any persistent stem cell population within the adult islet may in fact be rather innocuous and unremarkable, making their molecular identification much more difficult. Small, primitive cells fitting this description have been described in isolated islet preparations.[3]

28.5 EFFECTS OF ISOLATION AND TRANSPLANTATION ON ISLET VIABILITY AND SURVIVAL

Understanding the cellular composition of the pancreas and the morphological relationships of the three principal cell compartments can be helpful when devising potential therapies for diabetes mellitus. Because diabetes results from a lack of adequate β-cell mass to meet the body's needs and current therapies of insulin injections, oral hypoglycemic agents, and diet and exercise are insufficient to prevent the devastating complications of the disease, replacement of the β-cell mass is a logical therapeutic option. Restoration of adequate β-cell mass can be achieved either through stimulation of endogenous-cell mass expansion or through transplantation of a glucose-responsive β-cell mass.

Islet cell transplantation has the potential to be an effective therapy for the treatment of type 1 diabetes.[156] However, one of the major hurdles that needs to be overcome is the need for multiple donor organ donors to meet the requirements of a single recipient.[156,157] One reason is due to the inefficient nature of the current islet isolation procedure, with average yields representing less than 50% of the original islet cell mass.[158,159] Nevertheless, patients who have undergone hemi-pancreatectomy generally do not display clinical diabetes,[160–162] suggesting that transplantation of much less than 100% of the total pancreatic islet cell mass could be sufficient to obtain normoglycemia. However, the precise number of islets required to obtain insulin independence is not known. Furthermore, the number of functional islets that survive and engraft following transplantation is also elusive. Even so, the current number of islets transplanted is greater than ever before (>11,000 IEQ/kg).[163] Previous attempts using significantly less tissue were plagued by early graft failure, with only 12% of patients from 1990 to 1998 remaining insulin-independent for more than 7 days.[164] It is conceivable that the current number of transplanted islets required to achieve normoglycemia includes a surplus of cells that are lost because of the isolation and transplantation process.

There are several possible causes of islet primary nonfunction following transplantation. Until recently, most islet graft failures were presumed to be due to rejection, in spite of adequate immunosuppression. However, the loss of human β-cells from islets transplanted into diabetic nude mice,[165] the failure of canine islet autografts,[166] and the eventual failure of encapsulated islets[167] indicated that perhaps non-immunological factors were responsible for graft failure. The harshness of the isolation procedure exposes islets to significant stress, which could contribute to decreased viability following isolation. Yet the degree to which primary non-function of islet grafts is due to islet cell death triggered by the isolation procedure remains unclear. Still, it is encouraging to note that a significant percentage of islets do survive the isolation and transplantation procedures to become functional grafts with long-term survival. The challenge is to identify the factors responsible for the early loss of islet cell mass. Our group has been investigating the mechanisms of islet cell death following isolation to design strategies to minimize islet cell loss and maximize the use of available donor tissue.

28.6 ISLET ISOLATION

Though many details have changed over the years, the basic steps of the present isolation procedure have changed little since being introduced by Ricordi.[168] The main pancreatic duct is cannulated and the pancreas is distended with cold Liberase HI (Roche) enzyme blend via the pancreatic duct. The distended pancreas is then cut into pieces and placed in the digestion chamber (Ricordi chamber) along with hollow steel balls that assist in mechanical dissociation. The chamber is attached to a closed circuit via a 450 μm mesh, allowing digested tissue to be released from the chamber. The circuit is warmed to activate the Liberase and the chamber is shaken to aid the mechanical disruption of the tissue. Once the tissue is sufficiently digested, the circuit is cooled and the digestate, which contains both exocrine and endocrine tissue, is collected and washed to remove the enzyme solution. Following a recovery period, islets are purified from exocrine cells using continuous density gradient centrifugation on a COBE 2991. The resulting islets are then placed in culture awaiting transplantation.

28.7 INSULTS OF ISOLATION ON ISLETS

Each step of the isolation incurs harm to the islets (Table 28.1). By extension, islet isolation success can be greatly affected by the methods used during pancreas procurement and the quality of the gland. The resection of the pancreas begins a period of ischemia. It has long been known that reducing warm ischemia time improves the success of solid organ transplants. Minimizing the degree of warm and cold ischemia is

Table 28.1 Isolation-Related Insults to Islets

Step	Related Insults
Organ procurement	Organ manipulation
	Ischemia
	Hypothermia
Digestion	Loss of extracellular matrix
	Cleavage of cell surface components
	Mechanical stress
	Growth factor withdrawal
	Ischemia
Purification	Sheer stress
	Osmotic stress
	Glucotoxicity

also imperative for the success of the isolation. By resecting the pancreas before other organs during multi-organ harvest, warm ischemia time can be minimized, however this is not always possible. The Edmonton group has shown that addition and replenishment of iced saline slush around the anterior and posterior aspects of the pancreas during the procurement of the liver and kidneys significantly lowered the core pancreas temperature and greatly improved islet yield and function.[169] In addition, care must be taken not to compromise the pancreatic capsule to prevent leakage of the digestion enzyme during distension. Additionally, loss of ductal integrity has been observed following cold storage of rat pancreas.[170] In young human donors, islet yields were significantly higher when the pancreas was distended immediately following procurement rather than after 3 hours of cold ischemia.[171] Conversely, the degree to which islets are affected by hypothermic injury sustained during organ procurement has not been examined.

Enzymatic digestion results in the loss of extracellular matrix (ECM), cell–cell contact, the loss of peri-insular basement membrane; cleavage of cell surface components, and mechanical stress — all of which are likely to affect islet viability. There is evidence that intrinsic pancreatic proteases may affect the success of the isolation procedure.[172,173] The Edmonton group has shown that inhibition of intrinsic serine protease activity within pancreata with cold ischemia times greater than 10 hours improves the isolation of viable islets.[174] A serine protease inhibitor, Pefabloc (Roche) was added to the Liberase solution prior to injection into the pancreatic duct and it was shown to be effective at blocking serine protease activity during the isolation procedure without affecting the digestion.[175]

Though the need for purification has been questioned,[176,177] the current procedure for clinical islet allotransplant involves purification of islets from the rest of the pancreas. Purification by density gradient places the islets under severe osmotic stress, sheer stress, and glucotoxicity.

28.8 ISLET RESPONSE TO ISOLATION

The isolation procedure exposes islets to a variety of stressors that are likely contributors to decreased viability following isolation. We will now consider the cell death mechanisms involved in islet cell death and the possible reasons for their activation (Figure 28.2).

28.9 CELL DEATH MECHANISMS

For a long time, eukaryotic cells were thought to die in two distinct ways — apoptosis or necrosis. *Apoptosis* is an energy-dependent process by

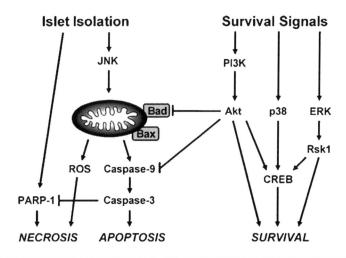

Figure 28.2 Schematic representation of intracellular signaling pathways implicated in islet cell death and survival following isolation.

which individual cells die by activating their own genetically programmed cell death mechanisms.[178] There are many external cues that can trigger a cell to undergo apoptosis. These external signals lead to activation of various cellular signaling pathways that are responsible for mobilizing the apoptotic machinery, which includes the caspase family of cysteine proteases.[179] During apoptosis, the effector caspases (i.e., caspase-3, caspase-6, caspase-7) cleave numerous cellular targets, which result in the systematic dismantling of the cell. Present in the cell as inactive zymogens, the effector caspases are activated by proteolytic cleavage by initiator caspases (i.e., caspase-8, caspase-9, caspase-10). The activation of the initiator caspases depends on interactions with adaptor molecules containing specific domains, such as death effector domains (DEDs) or caspase recruitment domains (CARDs).[180,181] Different adaptor molecules present in the cell can respond to a variety of upstream signals, allowing for caspase activation in response to a range of cellular stressors. Many apoptotic signals converge on the mitochondria, leading to mitochondrial membrane permeabilization and subsequent release of cytochrome c from the inter-membrane space.[182] Once in the cytosol, cytochrome c complexes with apoptosis-protease activating factor-1 (APAF-1) and, in the presence of dATP, leads to activation of caspase-9. The release of cytochrome c from the mitochondria is regulated by the Bcl-2 family of proteins.[183,184] In response to various stressors, pro-apoptotic members of the Bcl-2 family (i.e., Bax and Bad) translocate to the mitochondria and induce mitochondrial permeabilization.[184,185] The anti-apoptotic members of the Bcl-2 family (i.e., Bcl-2, Bcl-X$_L$) are often up-regulated by survival signals and act to inhibit proapoptotic Bcl-2 proteins by heterodimerization.[186] It is often suggested that the Bcl-2 family acts as a switch regulating the decision to undergo apoptosis dependent on the ratio of survival and death signals.[178,187]

Necrosis, on the other hand, does not require ATP and is much less regulated than apoptosis. Cells dying by necrosis typically swell and rupture, leading to a local inflammatory response. Necrosis is often associated with severe conditions such as ischemia, osmotic stress, or sheer stress. However, necrosis can also occur under normal physiological conditions, leading to the idea of programmed necrosis.[188] Necrosis can be induced by overactivation of poly(ADP-ribose) polymerase-1 (PARP-1), leading to depletion of nicotinamide adenine dinucleotide (NAD$^+$), impaired ATP production, and finally cell death.[189] In addition, permeabilization of the mitochondria can lead to necrotic-like cell death, which could be attributed to the release of reactive oxygen species (ROS) following disruption of the mitochondrial membrane.[188] Stimuli that cause an increase in nitric oxide have also been known to cause necrosis.[190]

Many insults can lead to both necrosis and apoptosis. Moreover, in many circumstances, not all dying cells display the typical hallmarks of apoptosis or necrosis.[191] This has led to the notion that perhaps apoptosis and necrosis represent two opposing ends of a complete range of intermediate forms of death.[192] In many cases, the decision by a cell to undergo apoptosis or necrosis is based on the availability of sufficient energy to carry out apoptosis. Therefore, a stimulus known to induce apoptosis may induce necrosis when insufficient energy (glucose) is present.

Our laboratory made the original observation that isolated human islets undergo apoptosis following routine isolation.[193,194] Similarly, high levels of apoptosis are observed in freshly isolated canine and rhesus monkeys islets.[195,196] We have also observed a significant increase in necrotic cell death with time in culture.[197,198] Isolated islet cell death results from a compound injury consisting of a barrage of several insults that, individually, are know to promote apoptosis, necrosis, or both. The procedure for isolation of islets from human and large animal pancreata is notably different from that used for small animals. For example, rodent organs are not distended with cold enzyme solution, but rather are minced and incubated with collagenase for a set period. Therefore, caution should be used when examining findings in rodent models that may not be relevant in a large-scale isolation paradigm. We will now consider some of the possible causes of isolated islet cell death.

28.9.1   Possible Causes of Isolated Islet Apoptosis

The survival of many cell types is dependent on specific interactions with the ECM. These interactions are primarily mediated by the integrin family of cell surface receptors, which are heterodimers of α and β subunits that can activate cellular signaling pathways in response to binding to the ECM.[199,200] The signaling pathways mediated by integrins regulate many cellular functions, including cell cycle progression, survival, growth, and shape. Loss of attachment to the ECM causes apoptosis in many cell types, a process known as *anoikis*.[201] Anoikis is important in development, tissue homeostasis, and the destruction of inappropriately located cells. Human islets are encapsulated by a peri-insular basement membrane that segregates endocrine tissue from exocrine.[202] Cells on the periphery of the islet form cell–matrix attachments with the basement membrane, but cells within the core of the islet are thought to depend on cell–cell interactions.[202] As mentioned earlier, enzymatic digestion destroys the basement membrane surrounding the islets, disrupting cell–matrix interactions, leading to apoptosis.[195,203] Furthermore, islet dispersion into single cells disrupts cell–cell interactions within the islet, inducing higher levels of cell death compared with intact islets.[204] Therefore, maintenance of cell–cell and cell–matrix interactions within the islet are important to prevent cell death.

Another consequence of islet isolation is the loss of trophic support by the surrounding acinar and ductal tissue. Growth factors act to modulate a broad range of cellular responses including growth, proliferation, metabolism, and survival.[205] Withdrawal of growth factors induces apoptosis in a variety of cell types.[206] As with anoikis, growth factor dependence is thought to regulate tissue homeostasis and prevent inappropriate cell placement because growth factor independence is a hallmark of a transformed cancerous cell.[207] In many cases, withdrawal of particular growth factors leads to activation of signaling pathways leading to apoptosis. It has been hypothesized that growth factors, like ECM, suppress a constitutive death signal that becomes uncovered upon their removal.[208] There are also significant metabolic changes that occur in cells following growth factor withdrawal that can trigger apoptosis.[209] Growth factor-starved cells can display reduced glucose transport,[210] lower glucose metabolism,[211] disruption of the electron transport chain,[212] and decreased ATP production.[213] Isolated islets are separated from the acinar and ductal tissue that may provide growth factor support. Indeed, cells of the duct epithelium secrete IGF-II, which functions in a paracrine manner to provide trophic support to the β-cell.[198] However, the extent to which acinar tissue provides trophic support for islets remains to be determined. On the other hand, there is also evidence of significant nitric oxide production by the ductal epithelium in response to cytokines, providing an argument for duct-free islet preparations.[214] It is important to note that many growth factors are secreted by islets themselves, and that this paracrine/autocrine support is likely to be maintained following isolation.

Oxygen deprivation is known to induce cell death. In situations of low oxygen (0 to 0.5%), cells will undergo apoptosis as long there is sufficient glucose to allow ATP production by glycolysis, otherwise necrosis will prevail.[215] This process depends on Bax/Bak-mediated release of cytochrome c from the mitochondria and the subsequent activation of caspase-9.[215,216] Hypoxia-induced cell death appears to solely depend on the mitochondrial pathway because overexpression of Bcl-2 or Bcl-X_L prevents apoptosis induced by oxygen deprivation.[217] There is also evidence that hypoxia-induced apoptosis is transcriptionally regulated. Hypoxia-inducible factor-1 (HIF-1) is a basic helix–loop–helix transcription factor involved in transcription of genes during oxygen deprivation. HIF-1 is a heterodimer consisting of the constitutively expressed HIF-1β and the hypoxia-induced HIF-1α subunits.[218] HIF-1 can regulate the transcription of genes involved in glycolysis, angiogenesis, and erythropoiesis.[219] In addition, HIF-1 can induce the transcription of apoptotic genes, such as BNIP3, a proapoptotic member of the Bcl-2 family.[220] Indeed, islets are highly vascularized structures whose function depends greatly on proper blood supply.[221] Following isolation islets are avascular, and revascularization is not complete until 10 to 14 days following transplantation.[222] During that time, oxygen delivery to the islets is limited to diffusion. Islets are exposed to varying degrees of ischemia throughout the isolation procedure, and culture techniques that are suitable for dispersed cell cultures may not be adequate for proper oxygen delivery throughout the entire islet. As mentioned previously, minimizing warm and cold ischemia times has a beneficial effect on islet yield and function, but it is difficult to attribute this effect solely to the prevention of apoptosis due to oxygen deficiency. Preservation of rat pancreata in University of Wisconsin (UW; Belzer, Viaspan,® DuPont) solution for 24 hours prior to isolation resulted in increased islet apoptosis and necrosis immediately following isolation.[223] Oxygen deprivation caused increased HIF-1α expression and apoptosis in cultured islets and in β-cell lines, and HIF-1α expression was thus proposed as a potential marker for islet ischemia.[224]

Endotoxin contamination has also been proposed as a cause of early islet failure.[225,226] Significant levels of endotoxin have been detected in collagenase preparations and in Ficoll (polysucrose),[227] and despite the introduction of low-endotoxin Liberase,[228] Ficoll is still a commonly used gradient for human islet isolation. Endotoxin (Lipid A) is the hydrophobic anchor of lipopolysaccharide (LPS), a glucosamine-based phospholipid found in the outer membranes of most Gram-negative bacteria.[229] After binding to CD14 on the surface of animal cells, endotoxin activates Toll-like receptor 4 (TLR-4), leading to activation of intracellular signaling pathways.[230] Endotoxin triggers cytokine production from macrophages as well as tissue factor production from endothelial cells, the later of which increases the risk of intraportal thrombosis following islet transplantation.[231] In addition, CD14 and TRL-4 expression has been demonstrated within isolated human islets and treatment of isolated rat islets with LPS resulted in increased production of TNF-α, IL-1β, and IL-6.[232] Human islet apoptosis is induced by pro-inflammatory cytokines (TNF-α, IL-1β, INF-γ) *in vitro*,[233] and it is therefore thought that endotoxin-induced cytokine production could contribute to apoptosis following isolation.[225,234] The use of endotoxin-free density gradients will hopefully eliminate the problem of endotoxin mediated islet failure.

Other stressors during islet isolation known to induce apoptosis include oxidative stress,[235] nitric oxide,[236] and hyperglycemia.[237,238] Additionally, islets are excitable cells, which could make them vulnerable to excitotoxicity, potentially mediated by calcium signaling.[239–241] Cells within the islet depend on neighboring cells for metabolic cues, which are imperative for proper control of hormone secretion. Disruption of the islet microenvironment may disturb this balance leading to deregulated hormone secretion. The extent to which cells within the islet depend on local hormone secretion for survival is not known.

28.9.2 Possible Causes of Isolated Islet Necrosis

Necrosis is often observed during situations of extreme environmental stress. Osmotic stress, mechanical stress, and sheer stress can all potentially lead to necrosis. As mentioned earlier, the decision by a cell to undergo apoptosis or necrosis is based on the availability of sufficient energy to carry out apoptosis. Thus, many of the aforementioned stimuli of apoptosis may induce necrosis if insufficient energy is present. Changes in ion concentrations, chronic decreased oxygen tension, and decreased ATP are all factors that could limit the availability of sufficient energy to ensure the execution of the apoptotic program.[242–244]

28.10 SIGNALING EVENTS FOLLOWING ISOLATION

The ability of cells to react to environmental changes depends on the cooperation of intracellular signal transduction pathways to coordinate the cellular response. The integration of various external cues leads to regulation of physiological processes such as proliferation, differentiation, and death. A better understanding of these signals can provide insight into the mechanisms involved in islet failure following isolation and may provide potential therapeutic targets to increase islet survival and function.

28.10.1 MAPK

The mitogen-activated protein kinases (MAPKs) have been shown to play an important regulatory role in a variety or cellular processes.[246] MAPKs are activated by a sequential cascade of protein phosphorylation in which MAPKs are phosphorylated by a MAPK kinase (MAPKK), which is itself activated by a MAPKK kinase (MAPKKK). This three-tiered kinase module is a common characteristic of all MAPK pathways. There are four main families of MAP — the c-jun NH_2-terminal kinases (JNK), the p38 kinases, the extracellular-signal regulated kinase (ERK) family, and the ERK5 family.[244] Little is known about the role of the ERK5 family in islets, therefore we will only consider the ERK, JNK, and p38 families in the current discussion.

28.10.2 JNK

The JNKs are a group of MAPKs that have been demonstrated to play a role in apoptosis, proliferation, survival, and embryonic morphogenesis.[245] JNK is activated when dually phosphorylated at Tyr185 and Thr183 by two dual-specificity MAPKKs — MKK4 (SEK1) and MKK7 (SEK2). Once activated, JNK can phosphorylate proteins in the nucleus and cytoplasm. Stress-induced JNK activation leads to phosphorylation of c-Jun and ATF-2, which heterodimerize to bind divergent AP-1 sites in the c-Jun promoter.[246] However, despite the fact that c-Jun expression has been shown to be required in some cases for apoptosis, little is know about which genes are involved in this process. JNK-mediated induction of the Fas-ligand (FasL) gene was proposed to be a possible mechanism of JNK-induced apoptosis[247,248]; however, blockage of Fas signaling does not prevent all forms of stress-induced apoptosis.[249] A more direct link between JNK and the apoptotic machinery was observed in JNK-deficient cells that failed to release cytochrome c from mitochondrial stores in response to ultraviolet (UV) radiation.[250] This effect was recently shown to be due to the ability of JNK to directly phosphorylate members of the BH3-only group of the Bcl-2 family (Bim and Bmf), leading to Bax-dependent apoptosis.[251,252] Therefore, JNK can regulate apoptosis by transcriptionally dependent and independent mechanisms that are only now beginning to be elucidated. In addition, the duration of JNK activation appears to be important in determining the physiological outcome. Sustained JNK activation, in contrast with transient activation, appears to regulate apoptosis in a variety of cell types.[253–257] Despite the association of JNK activation with stress-induced cell death, the precise role of JNK signaling in apoptosis has been unclear.[258] It is likely that the consequences of JNK activation depend largely on the type of stimulus and the cellular context.

JNK activation can be induced by growth factor withdrawal,[259] detachment from the ECM,[260] ischemia,[261] and by cytokines; all of which are possible consequences of islet isolation. We have demonstrated previously that freshly isolated islets display high JNK activity immediately following isolation, which diminishes over the first 48 hours following isolation.[264,265] This early activation of JNK coincides with the peak of islet apoptosis, which occurs 24 hours following isolation.[197,262,263] Addition of exogenous insulin to the culture media immediately following isolation reduced JNK activity and led to a decrease in DNA fragmentation after 24 hours in culture.[262] This finding suggested that JNK may mediate apoptosis following isolation and that perhaps insulin released by the islets themselves could act in an autocrine manner to suppress apoptosis. Preservation of rat pancreata for 24 hours by the two-layer method (TLM), which uses oxygenated perfluorocarbons to increase oxygen delivery to the organ, resulted in decreased apoptosis and lower JNK activity following isolation when compared to islets from pancreata stored in UW solution.[263] In addition, the same study showed that islets from pancreata that were processed immediately following resection had significantly lower JNK activation than islet from pancreata stored for 24 hours by TLM or in UW solution. The most convincing evidence implicating JNK in transplanted islet survival comes from transplantation of rat islets overexpressing dominant negative (DN) JNK into STZ-induced diabetic nude mice.[264] Mice receiving DN-JNK islets displayed lower blood glucose levels, which could be attributed to maintenance of *insulin* gene transcription despite the presence of oxidative stress at the graft site as well as increased survival of islets due to impaired apoptotic signaling by JNK. There is also considerable evidence supporting a role for JNK in cytokine-mediated β-cell apoptosis.[265–267] Taken together, these findings point toward a role for JNK in mediating islet apoptosis.

28.10.3 P38

The p38 family of MAPKs play a role in regulation of inflammation, apoptosis, differentiation, and cell growth.[268a] There are five isoforms of p38 that have been identified to date (α, β, β2, γ, and δ), which are activated to various degrees by MKK3, MKK4, and MKK6 in response to a range of stimuli including cytokines, growth factors, and environmental stress. Once active, p38 can target many substrates in both the nucleus and cytosol. As with JNK, the exact role of p38 in regulation of cell survival remains unclear as p38 appears to mediate both survival and apoptotic signals.[268a]

We observed that isolated islets display increasing p38 activity over the first 7 days in culture following isolation.[263] In porcine islets, we observed that high p38 activation following isolation correlated with decreased islet numbers over the first 36 hours following isolation, suggesting a role for p38 in mediating islet cell death.[263] However, pilot experiments with the p38 inhibitor SB203580 demonstrated a negative effect on islet survival.[268a] In addition, inhibition of p38 potentiated IL-1β-induced cell death in a β-cell line.[265] On the other hand, we observed an increase in p38 phosphorylation in islets treated with insulin, which can act as a survival stimulus in many cell systems.[262] This effect, however, is cell type dependent because insulin has been shown to inhibit p38 activation in primary neuronal cultures.[269] Nevertheless, IGF-I mediated survival has been shown to involve p38-dependent phosphorylation of cAMP-response element binding protein (CREB) and induction of Bcl-2 expression.[270,271] Taken together, these data suggest that p38 could play a role in regulating islet survival, possibly by mediating growth factor signaling (Figure 28.2). It is also plausible that activation of p38 in response to insulin represents an inhibitory feedback pathway to suppress insulin transcription.[272]

28.10.4 ERK

The ERK family of MAPKs regulate cellular processes such as mitosis, meiosis, survival, and cell growth.[244,273] The two ERK isoforms (ERK1 and ERK2) are ubiquitously expressed and become activated in response to signals emanating from receptor tyrosine kinases, G-protein coupled receptors, and integrins. ERK becomes activated upon dual phosphorylation by MEK, which is itself activated by Raf-1. Raf-1 can be activated by Ras, a small G-protein proto-oncogene. Ras mutations occur in several human cancers, leading to persistant activation of ERK and increased cellular proliferation. ERK can suppress apoptosis in many cell types by transcriptionally dependent and independent mechanisms.[274] ERK directly regulates the activity of many transcription factors such as Elk-1, Ets-1, c-Myc, and c-Jun.[275] Many of the downstream effects of ERK on survival are mediated by the Rsk family of protein kinases.[276] Activated Rsk is able to promote cell survival by phosphorylating and activating CREB, as well as by inhibitory phosphorylation of Bad.[274,277]

In isolated porcine islets, high ERK phosphorylation immediately following isolation correlated with increased islet recovery after 36 hours in culture.[263] In addition, inhibition of ERK activation by PD98059 or UO126 decreases isolated canine islet survival (unpublished observation). In contrast, ERK has been suggested to mediate IL-1β-induced NO production in β-cells and inhibition of ERK decreased cytokine-induced cell death in purified rat β-cells.[278,279] In addition, IGF-I mediated survival in INS-1 cells was not affected by PD98059 treatment despite the fact that inhibition of ERK activation decreased CREB phosphorylation.[280] Taken together, these results indicate that the role of ERK in islet survival is still unclear.

28.10.5   AKT

Akt (PKB) is a serine/threonine kinase that has been shown to block apoptosis and promote cell survival in response to growth factor or ECM stimulus.[281,282] Growth factor binding to cell surface receptors leads to the activation of several kinases, including PI3-K. PI3-K stimulation results in the recruitment of Akt to the plasma membrane where the constitutively active phosphoinositide-dependent kinase-1 (PDK-1) phosphorylates Akt on its Thr-308 residue.[283,284] Akt becomes fully activated upon autophosphorylation at its Ser-473 position.[285] Activated Akt can then protect against apoptosis by targeting proteins such as Bad,[286,287] caspase-9,[288] and FKHRL1 (Figure 28.2).[289] Akt is also thought to slow death by preventing a decline in cellular metabolism.[209]

Our laboratory has previously shown that Akt becomes highly activated within the first 24 hours following isolation of canine islets,[290] a finding that was recently demonstrated in human islets as well.[291] Though the underlying mechanism responsible for the activation of Akt in islets following isolation is not known, it could be responsible for the decrease in apoptosis, which is evident following 3 days in culture.[197] The significant rise in Akt activity has important implications for designing therapeutic interventions that are aimed at increasing islet survival. The ability of many growth factors and ECM proteins to prevent apoptosis has shown to be dependent on Akt activity. Akt can block apoptosis induced by many different stimuli (GF withdrawal, anoikis, ischemia)[284]; however, islet cell death appears to occur despite highly activated Akt. Akt has also been implicated in β-cell mitogenesis, and thus Akt activation could mediate both survival and proliferation in islets.[292,293]

Activation of Akt by simvastatin immediately following isolation, when Akt activity is normally low, was shown to reduce islet apoptosis by roughly 30% after 72 hours in culture.[291] Remarkably, STZ-induced diabetic mice receiving a marginal mass of transplanted islets cultured for 24 hours with simvastatin resulted in normoglycemia in 58% of mice 30 days after transplantation when compared to mice receiving untreated islets, which remained hyperglycemic. This finding would imply that treatments during islet culture have a profound effect on islet survival following transplantation, even weeks after the culture period. However, even simvastatin-treated islets showed increased levels of apoptosis in culture, suggesting that apoptosis can still occur in isolated islets despite high levels of Akt.

28.10.6 MKP-1

Inactivation of MAPKs occurs by dephosphorylation of the tyrosine and threonine residues by specific phosphatases. The mitogen activated protein kinase phosphatase (MKP) family are dual specificity phosphatases that are able to dephosphorylate both the tyrosine and threonine in the activation loop of MAPKs.[294] There are at least ten distinct MKP family members that can be divided roughly into two groups: the inducible nuclear MKPs and the more constitutively expressed MKPs found primarily in the cytosol. Each MKP displays selectivity toward certain MAPK family members. MKP-1 is an inducible phosphatase that is thought to preferentially dephosphorylate p38 and JNK,[295] yet in some cases it can also inactivate ERK.[296] The promoter of MKP-1 contains several regulatory sites (AP-1, AP-2, CRE, E-Box) that are modulated by a variety of signaling pathways, leading to MKP-1 gene induction. MKP-1 is upregulated by many stimuli such as growth factors, stress, phorbol esters, vasoactive peptides, increased intracellular cAMP, and increased Ca^{2+}.[294,297] Because MKP-1 can suppress p38 and JNK signaling, up-regulation of MKP-1 has been shown to have an anti-apoptotic effect.[298] Up-regulation of MKP-1 has also been observed in certain cancers, including pancreatic ductal cell adenocarcinoma.[299] We have shown that following islet isolation, MKP-1 expression increases over the first 72 hours following isolation.[263] This increase in MKP-1 coincides with the decrease in JNK phosphorylation and could therefore mediate negative feedback inhibition of the JNK pathway. MKP-1 up-regulation could also represent a survival signal by decreasing the sensitivity of the islet to apoptotic signals acting through p38 and JNK. As mentioned earlier, however, p38 activation also rises steadily with time in culture.[263] Thus, the exact role of MKP-1 in islet survival is still unclear.

28.10.7 Bcl-2 Proteins

Bax-related pro-apoptotic proteins are required for mitochondria-mediated apoptosis.[185] Cells deficient in Bax and Bak are resistant to apoptosis induced by staurosporine, ultraviolet radiation, growth factor deprivation, etoposide, thapsigargin, and tunicamycin.[185] In addition, Bax-related proteins are also essential for JNK-mediated apoptosis.[251] Human islets express high levels of Bax, indicating that mitochondria-mediated cell death could play an important role in isolated islets.[300] We have observed increasing levels of Bax in canine islets during the first 72 hours following isolation, suggesting the involvement of the mitochondrial pathway in islet cell death.[197] Overexpression of Bax alone has previously been shown to induce cell death, even in the presence of caspase inhibitors.[301] It is conceivable that high levels of Bax in isolated islets could be responsible for necrotic-like death, even in the presence of an Akt survival signal. The mechanisms regulating Bax expression, however, remain unclear. Bax is up-regulated via direct transcriptional control by p53.[302] In addition, c-Myc appears to lead to Bax up-regulation,[303] while NF-κB has an inhibitory effect on Bax expression.[304]

Anti-apoptotic Bcl-2 family members, on the other hand, are able to prevent mitochondria-mediated apoptosis. Real-time polymerase chain reaction analysis of isolated human islets demonstrated low expression of *bcl-2* relative to *bax*.[300] A decrease in Bcl-2 mRNA and immunostaining was observed in isolated human islets after 5 days in culture.[305] Likewise, a decrease in Bcl-2 immunoreactivity was observed in cultured rat islets after long-term (4 week) culture.[306]

28.11 IMPROVING ISLET SURVIVAL

Based on the cellular signaling events implicated in islets cell death, a therapeutic strategy can be designed to promote β-cell survival. We will now consider some of the approaches that could prove beneficial for improving islet survival following isolation.

28.11.1 Therapeutic Opportunities

In theory, as long as there is no chemical incompatibility between the desired agent and the particular solution in question, therapeutic interventions could be added to all steps of the isolation. However, some steps are more attractive for interventions than others (Table 28.2). Although the cold perfusion prior to the resection of the pancreas would be the

Table 28.2 Therapeutic Opportunities to Improve Islet Survival

Step	Solution	Conditions
Organ perfusion	UW	Minimum 1 hr
Organ preservation and transport	UW ± PFC	2–12 hr
Enzymatic digestion	Perfusion solution	10 min at 4°C 5–20 min at 37°C
Wash out	Dilution solution	45 min
Recovery	UW	1–2 hr
Density gradient centrifugation	Ficoll	10–60 min
Wash out	Dilution solution	15 min
Culture	Culture medium	1 hr–days
Engraftment	-	-

ideal step to begin treatment with a particular anti-apoptotic agent, the agent would be circulated throughout the body and would have to be compatible with all harvested organs. Therefore, the organ preservation solution provides the earliest opportunity for pretreating the organ prior to the enzymatic digestion. The enzymatic digestion solution is a more difficult environment to introduce certain compounds due to the presence of collagenases. The Ficoll gradient may also provide difficulty due to the high osmolarity and the importance of maintaining the gradient properties.

In many cases, the protective effect afforded by a particular agent is only observed when it is administered prior to the stressing agent.[307,308] However, due to the variable nature of the isolation procedure, it is difficult to ascertain the effects of treatments prior to the digestion because of the lack of an adequate control group. For example, if a pancreas were to be cut in two and separate digestions performed so as to eliminate donor variability between groups, they would still not be easily comparable because the digestion step could cause considerable variability between the two. The earliest point at which two groups can be separated and treated in an identical manner is following the digestion. Therefore, it is difficult to accurately assess the protective effect of a particular agent that may need to be added prior to the digestion to obtain a significant effect.

28.11.2   Pros and Cons of Culturing

Recently the recovery of islets in culture prior to transplantation has once again become common practice.[163] The logic holds that a majority of islet cell death due to the isolation occurs in the 48 hours following isolation, and thus it is advantageous to allow these cells to die off in culture so as not to provoke additional inflammation local to the graft site. As such, there is a much higher relative expression of inflammatory genes during the immediate post-isolation period when compared to islets cultured for 1 week.[309] However, current culture techniques for islets are likely insufficient for adequate delivery of oxygen to the center of the islet. Culturing islets also increases the chances of cellular contamination. Another drawback of islet culturing is the loss of islet mass by transdifferentiation in culture, as will be discussed later (see Section 28.12).[310,311]

28.11.3   Growth Factors

As mentioned earlier, one of the possible causes of islet cell death following isolation is the withdrawal of growth factors required for islet survival. The ability of many growth factors to suppress apoptosis depends on Akt activation, which is low following isolation. Despite the rise in

Akt activity as early as 16 hours following isolation, interventions that increase Akt immediately following isolation could have an impact on islet survival.[291] However, the therapeutic window can be significantly increased by exploiting of the potential to add growth factors prior to and during the isolation procedure. We will now consider some growth factors that could improve islet survival.

IGF-I and IGF-II suppress the intrinsic cell death machinery and thereby prevent apoptosis in different cell types.[281,312–314] IGF-I is a polypeptide trophic factor that plays an important role in promoting development, including that of the pancreas, during organogenesis.[315] Its role in the regulation of apoptosis in the postnatal pancreas has been highlighted,[316,317] and we have reported that islets co-cultured with IGF-II-secreting duct epithelial cells exhibit significantly less necrosis and apoptosis.[198] We have also demonstrated that IGF-I signaling through PI3-K can protect islets from cytokine-induced cell death.[290] IGF-I is currently being used in clinical islet transplantation (Hering B., personal communication).

Insulin itself is a potent survival factor for many cell types. Insulin shares many signaling pathways with IGF-I (Akt, ERK) and thus has similar anti-apoptotic effects.[318] We have demonstrated a decrease in nucleosomal fragmentation in canine islets treated with 100 ng/ml insulin for 24 hours following isolation.[262] The regulator effects of insulin on insulin secretion remains unclear, as there are studies demonstrating an inhibitory effect[319,320] and a potentiating effect on islet function.[321]

GLP-1 is another factor that regulates islet survival and function. Signaling from the GLP-1 receptor (GLP-1R) can protect β-cells from cytokine-mediated apoptosis in culture.[140] In Zucker diabetic rats, GLP-1 significantly reduced DNA fragmentation, caspase-3 expression, and increased Bcl-2 expression.[322] Importantly for islet transplant, culture of isolated human islets for 5 days in the presence of GLP-1 resulted in decreased apoptosis, caspase-3 immunostaining, and increased Bcl-2 levels.[305] Though significant levels of apoptosis were still detected in the presence of GLP-1, these findings support the notion that GLP-1 has a significant anti-apoptotic effect. GLP-1 also reduced H_2O_2-induced apoptosis in MIN6 insulin-secreting cells, an effect dependent on cAMP and PI3-K signaling.[323] Taken together, these recent findings indicate that GLP-1 is a potent β-cell survival factor with potential for use in preventing islet cell death caused by isolation. In addition, exendin-4, the GLP-1 analog isolated from the salivary glands of Gila monster lizards, has a longer half-life in culture and is therefore a better candidate for use during the isolation as well as in the culture media.[324]

Nerve growth factor (NGF) is a neurotropin that mediates the survival and differentiation of neuronal cells as well as nonneuronal cells. The two NGF receptors, TrkA and p75, have been characterized in the β-cells of human islets,[325] rat islets,[326] and β-cell lines.[327,328] Additionally, β-cells are able to synthesize and secrete NGF.[325,326] Treatment of βTC6-F7 cells, a mouse β-cell line, with an anti-NGF antibody led to increased nucleo-somal fragmentation, suggesting a dependence of β-cells on NGF.[325] However, the fact that β-cells can secrete NGF themselves would argue against NGF-withdrawal as a cause of islet cell death in culture.

Without a doubt, there are other growth factors that play a role in islet survival and could prove useful in islet transplant, such as erythro-poietin,[329] parathyroid-related protein (PTHrP),[330] HGF.[331] Identifying those factors that act by complementary mechanisms will be crucial in obtaining additive or synergistic improvements in islets survival.

28.11.4   Nicotinamide

Nicotinamide (Vitamin B$_3$) has a long history in the field of diabetes.[332] Nicotinamide is the amide derivative of nicotinic acid and acts to inhibit the DNA repair enzyme PARP-1. PARP-1 catalyzes the addition of long branched chains of poly(ADP)-ribose to a variety of nuclear proteins using NAD$^+$ as a substrate.[333] Under conditions that cause severe DNA damage, overactivation of PARP-1 leads to depletion of NAD$^+$, resulting in impaired ATP production and finally necrotic cell death.[189] PARP-1 is necessary for STZ-mediated toxicity because PARP-1$^{-/-}$ mice are resistant to STZ-induced diabetes, indicating β-cell protection from necrosis.[334] It has been suggested, however, that the protection against STZ afforded by PARP-1 inhibitors could be due to up-regulation of the multidrug resistance transporter (MDR-1) that occurs in the absence of PARP-1 activity.[335,336] Another PARP-1 inhibitor, PD128763, is effective at protecting islet cells from nitric oxide, oxygen radical generating compounds, and STZ at concentrations 100 times less than required for nicotinamide.[337] Addition of nicotinamide to isolated canine islet cultures caused a decrease in necrotic cell death after 3 days in culture.[197] However, an increase in caspase-3 processing was also observed, indicating a potential increase in apoptosis upon nicotinamide treatment. Nicotinamide also protected rat β-cells from necrosis induced by STZ or H$_2$O$_2$, but this effect was associated with delayed apoptosis.[338] The same study also demonstrated that nicotinamide can protect human β-cells against H$_2$O$_2$-induced necrosis without the occurrence of delayed apoptosis. Nicotinamide also induced apoptosis in rat insulinoma cells and fetal rat islets.[339] A similar increase in apoptosis upon nicotinamide treatment has been observed in a variety of cell systems.[340–344] Taken together, these data indicate that PARP-1 inhibition may target cells for an apoptotic form of cell death, depending on the insult and cellular context.

28.11.5 Caspase Inhibition

Caspase inhibitors have been successful in preventing the progression of apoptosis in a variety of circumstances.[179] However, despite the fact that caspases are responsible for many of the hallmarks of apoptosis, they are often not required for cell death.[191] We have shown that general caspase inhibition by benzyloxycarbonyl-Val-Ala-Asp-fluoromethylketone (Z-VAD) leads to increased caspase-independent cell death in isolated canine islets.[197] Nicotinamide co-treatment was able to abolish Z-VAD-induced necrosis, indicating that the increased necrosis was due to prevention of caspase-mediated PARP-1 cleavage. Cleavage of PARP-1 by caspase-3 to 89 kDa and 24 kDa fragments is a defining characteristic of classical apoptosis. One of the functions for cleavage of PARP-1 during apoptosis induced by insults such as alkylating agents is to prevent survival of the extensively damaged cells.[345] PARP-1 cleavage also ensures that energy stores are available for the execution of apoptosis.[189] Caspase inhibition has been shown to block oligonucleosomal fragmentation in dissociated islet cell cultures, however neither necrosis nor overall viability were assessed.[204] Caspase inhibition can also cause increased necrosis in a variety of other cell types.[346–349] The inability of caspase inhibition to prevent islet cell death may be due to the fact that the caspases lie downstream of any mitochondrial dysfunction, and thus their inhibition has no effect on mitochondria-mediated cell death.[182,350] Indeed, we and others have demonstrated increasing levels of Bax in islets following isolation, suggesting the involvement of the mitochondrial pathway.[197,300] Furthermore, overexpression of Bax alone has previously been shown to induce cell death, even in the presence of caspase inhibitors.[301] Taken together, these results indicate that caspase inhibition is likely too late of an intervention to improve islet viability.

28.11.6 JNK Inhibition

As described earlier, there is considerable evidence that JNK plays a significant role in islet cell death following isolation. JNK inhibitory peptides were recently described based on the 20-amino acid JNK binding domain of JIP-1 linked to the HIV-TAT translocation sequence.[266] These peptides are able to suppress IL-1β-induced apoptosis in the insulin-secreting βTC-3 cell line[266] as well as protect against cerebral ischemia.[351] In addition, a novel small molecule inhibitor of JNK has also been described.[352] SP600125, an anthrapyrazolone, is a reversible ATP-competitive inhibitor of JNK-1, -2, and -3.[352] An attractive complementary action of SP600125 is that it induced rapid CREB phosphorylation in the MIN6 mouse β-cell line.[353] Recent evidence indicates a pro-survival role for CREB in β-cells.[354]

28.11.7   Bax Inhibition

Unlike other molecules that can play a role in survival and apoptotic signaling, no prosurvival role for Bax has yet been described. Recently, a novel Bax-inhibitory peptide (BIP) has been synthesized based on the Bax-binding domain of Ku70.[355] Treatment with BIP blocked apoptosis induced by a variety of insults by preventing Bax translocation to the mitochondria. Based on the potential role of Bax in islet cell death, this novel inhibitor represents a viable approach to preventing the progression of apoptosis upstream of the mitochondria.

28.11.8   Three-Dimensional Support

We have previously shown that ECM proteins can improve the survival of isolated islets.[195] Both collagen and fibronectin were able to decrease apoptosis following isolation. A hydrogel containing collagen and laminin was shown to decrease isolated rat islet cell death.[356] One of the drawbacks of culturing cells in a matrix is that the gel needs to be digested to retrieve the cells, which creates additional stress on the cells. This problem could be overcome by the use of soluble matrix proteins or by the use of temperature-sensitive gels.

28.11.9   Gene Therapy

The process of introducing genetic materials into eukaryotic cells to treat a specific disorder is the basis of gene therapy. Genes can be introduced by three general approaches — viral, physical, and chemical. The viral method of transfection is still the most efficient, but immunogenicity, lower plasmid size, and the potential for carcinogenic and pathogenic complications are some of the major drawbacks of this system. Islets present a unique challenge for transfection because they are clusters of cells and they have low replication rates. Previous reports describe efficient transfection of isolated islet cells by adenoviral[357] or lentiviral vector,[358] yet these systems are not clinically viable approaches for gene delivery. Chemical delivery systems (i.e., polycations) are attractive because DNA delivery by this method is not incorporated into the host DNA, making this type of transfection reversible.[359] Therefore, islets could be treated with a particular anti-apoptotic gene following isolation, which would eventually be depleted and not pose any threat of carcinogenesis.

There are many potential candidate genes for transfer into isolated islets. Transfection of human pancreatic islets with an anti-apoptotic gene (Bcl-2) protects β-cells from cytokine-mediated cell death.[360,361] Adenoviral transfer of the anti-apoptotic protein A20 into isolated mouse islets improved their survival and function following transplantation.[362] Adenoviral transfer of IGF-I into human islets protected islets from IL-1β-mediated apoptosis and disfunction.[363] Nonviral transfection by Bcl-X$_L$ fused to a protein transduction domain (PTD) protected βTC-3 cells from TNF-α-induced apoptosis and did not affect insulin secretion in isolated rat islets.[364] Recently, a particular PTD (PTD-5) was shown to efficiently transduce human islets in culture as well as mouse islets when injected into the bile duct prior to collagenase digestion.[365] In this case, delivery of an IκB kinase (IKK) inhibitor prior to isolation improved mouse islet function following isolation. This study represents an important step in the treatment of islet cell death following isolation.

28.11.10 Future Directions

It appears that apoptosis and necrosis are complementary inter-dependent events that ensure the execution of a cell receiving a death signal. Islet cell death following isolation is certainly not unprovoked as the isolation procedure is exceptionally harsh. However, unless a new method of islet isolation is developed, new strategies must be adopted to decrease islet cell death caused by the enzymatic digestion and purification steps that are central to the current protocol. Indeed, islet isolation is a multifaceted insult that results in the activation of several cell death mechanisms. Therefore, therapies aimed at increasing islet survival following isolation should be designed to prevent multiple forms of death from occurring by targeting early events in cell death signaling.

Islet cell death is not the only consequence of islet isolation. Although islet cells were thought to be terminally differentiated cells, this does not appear to be the case. In the next section, we explore islet cell plasticity and the transformation of islets to duct epithelial structures. This phenomenon, best studied *in vitro*, takes on added significance when one recognizes that a similar *in vivo* interconversion has been implicated in the development of ductal adenocarcinoma of the pancreas.

28.12   ISLET PHENOTYPIC STABILITY FOLLOWING ISOLATION

The maintenance of cellular phenotype depends, at least in part, on the extracellular milieu, but also on the cytoplasmic signaling proteins and nuclear transcription factors that are activated by the ECM to maintain stable gene expression.[366,367] Islet isolation both destroys the islet–ECM relationship and disrupts many critical cellular inter-relationships.[368] Due to the loss of the islet–ECM interactions, transformation of islet phenotypic is a possible consequence of islet isolation.

28.12.1 Loss of ECM and Differentiation of Isolated Islets

The ECM is a dynamic complex of different molecules that serves as a cellular scaffold and regulates both cellular differentiation and survival.[195,356,366,369–378] Recently, Gittes *et al.* demonstrated that the default growth pathway for the embryonic pancreatic epithelium is to form islets.[379] However, in the presence of basement membrane constituents the pancreatic anlage epithelium appears to be programmed to form ductal epithelium.[379] This finding serves to re-emphasize the inter-relationship between ducts and islets, in the developing and adult pancreas, and suggests the important contribution of the ECM to the stabilization of cellular phenotype.

We have previously shown that when islets are embedded into a type-1 rat tail collagen or Matrigel (Becton-Dickinson) matrix and are exposed to substances that increase intracellular cAMP levels, they undergo a phenotypic switch into duct-like epithelial structures by a process known as *transdifferentiation* (Figure 28.3): a change from one differentiated cell phenotype to another.[311,378,380–382] Such changes reflect altered expression of morphological and functional phenotypic markers. Cellular transdifferentiation does not occur when islets are cultured in suspension with soluble laminin or fibronectin, in agarose (in the absence of matrix proteins), or in the absence of elevated cAMP levels.[311] These data suggest that the process of islet-to-duct differentiation depends not only on ECM proteins, but on an appropriate solid three-dimensional microenvironment in addition to growth and differentiation factors.

28.12.1.1 Growth Factors/Signaling

Phenotypic stability is not only altered by changes in the local microenvironment, but is also conferred by specific elements present in the microenvironment in the first instance.[369,375–377,383] Intracellular signal transduction pathways that mediate the effects of specific trophic/survival factors are associated with cell growth, death, and differentiation.[371,384]

Our ability to manipulate islet phenotype *in vitro* has offered an excellent opportunity to elucidate the signaling events that are implicated in the interconversion of islets to ducts. The process seems to occur in two steps: the first involves β-cell apoptosis and a loss of islet phenotypic markers. This is followed by a second stage of cellular transdifferentiation, characterized by the appearance of the duct–epithelial cell marker CK-19 and greatly increased proliferation.[380] During the first 3 days of culture when β-cell apoptosis is greatly elevated, there is an associated decrease in the phosphorylation of prosurvival signaling molecules ERK and Akt, along with an elevated activation of the pro-apoptotic molecules JNK and

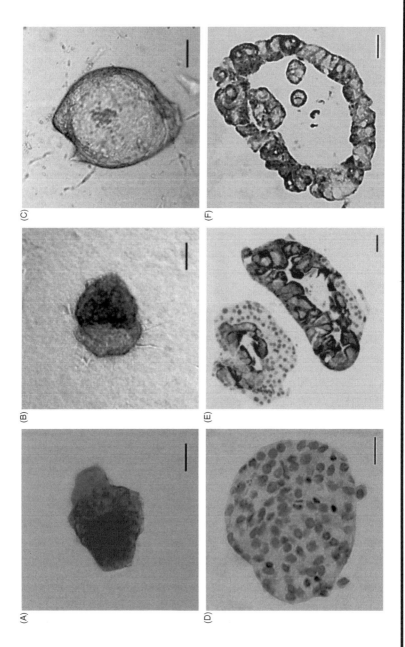

Figure 28.3 Inverted microscopic images (A, B, C) and CK-19 stained cell culture samples (D, E, F) demonstrating the *in vitro* progression of islets to duct epithelial cystic structures. (Bar = 20 μm)

caspase-3. In contrast, during the second stage of the transformation process, from day 6 to day 12, ERK and Akt phosphorylation are increased.

28.12.2 Phenotypic Alterations following Islet Isolation and as Part of the Progression of Pancreatic Disease

Phenotypic changes in adult pancreatic islets have been implicated in the development and progression of several disease states (e.g., diabetes mellitus, pancreatic cancer, pancreatitis[386–391]). For example, recognition of the critical roles played by cell death and differentiation following islet isolation are paramount when considering islet transplantation as a means to reverse the diabetic state. In fact, islet transplantation is associated with a high rate of graft failure, a problem that has yet to be fully understood and resolved.[223,392,393] It is likely that the destruction of the islet microenvironment and the loss of trophic support that occurs as a consequence of isolation are responsible to a significant extent for the poor outcome of islet transplantation to date.[394–398]

It is also noteworthy that the loss of the mature endocrine phenotype has also been linked to the initiation of pancreatic neoplasia in some models.[399–402] We will now briefly consider the potential role of islet-derived duct-like cells in the pathogenesis and progression of pancreatic adenocarcinoma.

28.12.3 Pancreatic Cancer

The vast majority of malignant pancreatic tumors are believed to arise from the ductal epithelium and have light microscopic features consistent with those of adenocarcinomas. Because the majority of pancreatic tumors are diagnosed when the cancer is well established (i.e., late in the natural course of the disease), it is difficult to precisely specify the phenotype of the cell giving rise to the tumor.[403,404] During the progression of pancreatic cancer, there is a localized loss of type IV collagen,[405,406] consistent with the notion that changes in the cell–ECM interactions could lead to an alteration of cell phenotype, as suggested by our *in vitro* model.[399,400,402,403,417–411]

28.12.3.1 Support for Islet-to-Duct-to-Cancer Paradigm

In 1978, Pour described the appearance of duct-like cells in islets during N-nitroso (2-oxopropyl) amine (BOP)-induced pancreatic carcinogenesis in hamsters[401] and suggested that islet-to-duct transformation may play a role in the development of pancreatic adenocarcinomas (Figure 28.4).[399–402] Studies subsequently have indicated that pancreatic cancer originates not only from pancreatic ductal cells, but also from within the islets of Langerhans.[407] Interestingly, following destruction of hamster pancreatic β-cells/islets with both alloxan and STZ, pancreatic carcinogenesis stimulated by BOP is significantly inhibited.[408–411] Islet-ductal cancers have been noted in humans, and because pancreatic cancer usually presents late in disease progression, it is indeed possible that the islet may be a key contributor to human pancreatic adenocarcinoma.[399,400,402–404,407]

Taken together, these data suggest that transplanted islets may have an increased potential to differentiate into duct-like structures because of changes in the islet–ECM relationship and that these transformed structures could possess an increased potential for neoplastic transformation, especially in the transplant setting of chronic immunosuppression.

There exists significant evidence that TGF-β plays a role as a tumor suppressor, and this correlates with observation that TGF-β_1 can inhibit islet-to-duct interconversion in a dose-responsive manner.[399,401,403,404,407–409,421] Furthermore, when duct-like cysts derived from isolated human islets were exposed to INGAP peptide, the cystic structures were stimulated to transform back into insulin producing islets, suggesting a potential role for INGAP peptide as a therapeutic tool in the prevention or treatment of pancreatic carcinoma.

Islet cell tumors are presented in Chapter 27.

28.13 EPILOGUE

The islets of Langerhans represent a complex community of cells within the pancreas. Although traditionally viewed as ostensibly an endocrine organ, it has become increasingly clear that islet biology is intimately and inextricably linked to that of the exocrine part of the gland, not only with respect to cellular ontogeny and regulation of pancreatic exocrine–endocrine function, but more importantly with respect to pathogenesis of pancreatic disease. There is much we still do not comprehend and many fertile areas of research remain to be pursued to develop a more complete understanding of pancreatic biology in health and disease.

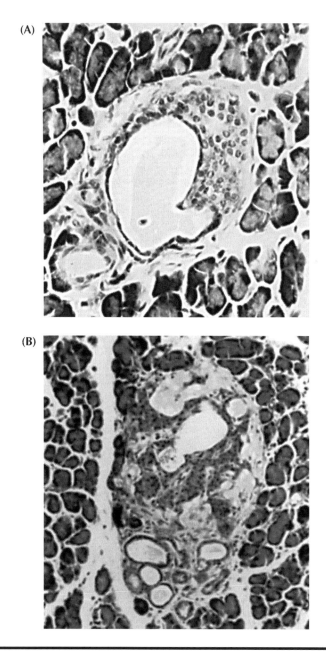

Figure 28.4 Photomicrographs demonstrating the early (A) and late (B) *in vivo* progression of islets to duct epithelial structures during BOP-induced pancreatic carcinogenesis in the hamster.

REFERENCES

1. Bonner-Weir S. 1991. Anatomy of the islet of Langerhans. In *The Endocrine Pancreas*. Samols E. Ed. New York: Raven Press. pp. 15–27.
2. Sundler F. and Hakanson R. 1976. D1 cell in search of a hormone. *Lancet* 2:1300.
3. Petropavlovskaia M. and Rosenberg L. 2002. Identification and characterization of small cells in the adult pancreas: potential progenitor cells? *Cell. Tissue Res.* 310:51–58.
4. Canaff L., Bennett H.P., and Hendy G.N. 1999. Peptide hormone precursor processing: getting sorted? *Mol. Cell. Endocrinol.* 156:1–6.
5. Halban P.A. and Irminger J.C. 1994. Sorting and processing of secretory proteins. *Biochem. J.* 299(Pt. 1):1–18.
6. Henquin J.C. 2000. Triggering and amplifying pathways of regulation of insulin secretion by glucose. *Diabetes* 49:1751–1760.
7. Meda P., Halban P., Perrelet A., Renold A.E., and Orci L. 1980. Gap junction development is correlated with insulin content in the pancreatic B cell. *Science* 209:1026–1028.
8. Meda P., Perrelet A., and Orci L. 1979. Increase of gap junctions between pancreatic B-cells during stimulation of insulin secretion. *J. Cell. Biol.* 82:441–448.
9. Meissner H.P. 1976. Electrophysiological evidence for coupling between beta cells of pancreatic islets. *Nature* 262:502–504.
10. Orci L., Unger R.H., and Renold A.E. 1973. Structural coupling between pancreatic islet cells. *Experientia* 29:1015–1018.
11. Pipeleers D. 1987. The biosociology of pancreatic B cells. *Diabetologia* 30:277–291.
12. Schuit F.C., In't Veld P.A., and Pipeleers D.G. 1988. Glucose stimulates proinsulin biosynthesis by a dose-dependent recruitment of pancreatic beta cells. *Proc. Natl. Acad. Sci. U.S.A.* 85:3865–3869.
13. Salomon D., and Meda P. 1986. Heterogeneity and contact-dependent regulation of hormone secretion by individual B cells. *Exp. Cell. Res.* 162:507–520.
14. Rhodes C.J. and Halban P.A. 1987. Newly synthesized proinsulin/insulin and stored insulin are released from pancreatic B cells predominantly via a regulated, rather than a constitutive, pathway. *J. Cell. Biol.* 105:145–153.
15. Westermark P., Wernstedt C., O'Brien T.D., Hayden D.W., and Johnson K.H. 1987. Islet amyloid in Type 2 human diabetes mellitus and adult diabetic cats contains a novel putative polypeptide hormone. *Am. J. Pathol.* 127:414–417.
16. Johnson K.H., O'Brien T.D., Hayden D.W., Jordan K., Ghobrial H.K., Mahoney W.C., and Westermark P. 1988. Immunolocalization of islet amyloid polypeptide (IAPP) in pancreatic beta cells by means of peroxidase-antiperoxidase (PAP) and protein A-gold techniques. *Am. J. Pathol.* 130:1–8.
17. Clark A., Cooper G.J., Lewis C.E., Morris J.F., Willis A.C., Reid K.B., and Turner R.C. 1987. Islet amyloid formed from diabetes-associated peptide may be pathogenic in Type-2 diabetes. *Lancet* 2:231–234.
18. Butler A.E., Janson J., Bonner-Weir S., Ritzel R., Rizza R.A., and Butler P.C. 2003. Beta-cell deficit and increased beta-cell apoptosis in humans with Type 2 diabetes. *Diabetes* 52:102–110.

19. Verchere C.B., D'Alessio D.A., Prigeon R.L., Hull R.L., and Kahn S.E. 2000. The constitutive secretory pathway is a major route for islet amyloid polypeptide secretion in neonatal but not adult rat islet cells. *Diabetes* 49:1477–1484.

20. Habener J.F., Drucker D.J., Mojsov S., Knepel W., and Philippe J. 1991. Biosynthesis of glucagon. In *The Endocrine Pancreas*. Samols E. Ed. New York: Raven Press. pp. 53–71.

21. Ravazzola M., Perrelet A., Unger R.H., and Orci L. 1984. Immunocytochemical characterization of secretory granule maturation in pancreatic A-cells. *Endocrinology* 114:481–485.

22. Gerich J.E., Schneider V., Dippe S.E., Langlois M., Noacco C., Karam J.H., and Forsham P.H. 1974. Characterization of the glucagon response to hypoglycemia in man. *J. Clin. Endocrinol. Metab.* 38:77–82.

23. Assan R., Attali J.R., Ballerio G., Boillot J., and Girard J.R. 1977. Glucagon secretion induced by natural and artificial amino acids in the perfused rat pancreas. *Diabetes* 26:300–307.

24. Samols E., Weir G.C., and Bonner-Weir S. 1983. Intra-islet insulin-glucagon-somatostatin relationships. In *Handbook of Experimental Pharmacology*. Lefebvre P.J. Ed. New York: Springer Verlag. pp. 133–162.

25. Aoki T.T., Muller W.A., Brennan M.F., and Cahill Jr. G.F. 1974. Effect of glucagon on amino acid and nitrogen metabolism in fasting man. *Metabolism* 23:805–814.

26. Cherrington A.D., Chiasson J.L., Liljenquist J.E., Jennings A.S., Keller U., and Lacy W.W. 1976. The role of insulin and glucagon in the regulation of basal glucose production in the postabsorptive dog. *J. Clin. Invest.* 58:1407–1418.

27. Bishop A., and Polak J. 1997. The anatomy, organization and ultrastructure of the islets of Langerhans. In *Textbook of diabetes*. Pickup J. and Williams G. Eds. Oxford: Blackwell Science. pp. 10.1.

28. Rahier J., Wallon J., and Henquin J.C. 1981. Cell populations in the endocrine pancreas of human neonates and infants. *Diabetologia* 20:540–546.

29. Hellerstrom C. 1984. The life story of the pancreatic B cell. *Diabetologia* 26:393–400.

30. Lammert E., Cleaver O., and Melton D. 2001. Induction of pancreatic differentiation by signals from blood vessels. *Science* 294:564–567.

31. Lifson N., Lassa C.V., and Dixit P.K. 1985. Relation between blood flow and morphology in islet organ of rat pancreas. *Am. J. Physiol.* 249:E43–E48.

32. Fujita T. and Murakami T. 1973. Microcirculation of monkey pancreas with special reference to the insulo-acinar portal system. A scanning electron microscope study of vascular casts. *Arch. Histol. Jpn.* 35:255.

33. Bonner-Weir S. and Orci L. 1982. New perspectives on the microvasculature of the islets of Langerhans in the rat. *Diabetes* 31:883–889.

34. Stagner J.I., Samols E., and Bonner-Weir S. 1988. Beta–alpha–delta pancreatic islet cellular perfusion in dogs. *Diabetes* 37:1715–1721.

35. Sundler F. and Bottcher G. 1991. Islet innervation, with special reference to neuropeptides. In *The Endocrine Pancreas*. Samols E. Ed. New York: Raven Press. pp. 29–52.

36. Sundler F., Alumets J., Hakanson R., Fahrenkrug J., and Schaffalitzky de Muckadell O. 1978. Peptidergic (VIP) nerves in the pancreas. *Histochemistry* 55:173–176.

37. Di Guglielmo G.M., Drake P.G., Baass P.C., Authier F., Posner B.I., and Bergeron J.J. 1998. Insulin receptor internalization and signalling. *Mol. Cell. Biochem.* 182:59–63.

38. Taha C., and Klip A. 1999. The insulin signaling pathway. *J. Membr. Biol.* 169:1–12.

39. White M.F. and Kahn C.R. 1994. The insulin signaling system. *J. Biol. Chem.* 269:1–4.

40. Brubaker P.L. and Drucker D.J. 2002. Structure-function of the glucagon receptor family of G protein-coupled receptors: the glucagon, GIP, GLP-1, and GLP-2 receptors. *Receptors Channels* 8:179–188.

41. Moller L.N., Stidsen C.E., Hartmann B., and Holst J.J. 2003. Somatostatin receptors. *Biochem. Biophys. Acta.* 1616:1–84.

42. Samols E. and Stagner J.I. 1991. Intraislet and islet-acinar portal systems and their significance. In *The Endocrine Pancreas*. Samols E. Ed. New York: Raven Press. pp. 93–124.

43. Alberti K.G., Christensen N.J., Christensen S.E., Hansen A.P., Iversen J., Lundbaek K., Seyer-Hansen K., and Orskov H. 1973. Inhibition of insulin secretion by somatostatin. *Lancet* 2:1299–1301.

44. Koerker D.J., Ruch W., Chideckel E., Palmer J., Goodner C.J., Ensinck J., and Gale C.C. 1974. Somatostatin: hypothalamic inhibitor of the endocrine pancreas. *Science* 184:482–484.

45. Maruyama H., Hisatomi A., Orci L., Grodsky G.M., and Unger R.H. 1984. Insulin within islets is a physiologic glucagon release inhibitor. *J. Clin. Invest.* 74:2296–2299.

46. Pictet R. and Rutter W. 1972. Development of the embryonic endocrine pancreas. In *Endocrinology: Endocrine Pancreas*. Freinkel N. and Steiner D. Eds. Washington: American Physiological Society. p. 25.

47. Yamaoka T. and Itakura M. 1999. Development of pancreatic islets (review). *Intl. J. Mol. Med.* 3:247–261.

48. Bardram L. 1990. Progastrin in serum from Zollinger-Ellison patients. An indicator of malignancy? *Gastroenterology* 98:1420–1426.

49. Brand S.J. and Wang T.C. 1988. Gastrin gene expression and regulation in rat islet cell lines. *J. Biol. Chem.* 263:16597–16603.

50. Finegood D.T., Scaglia L., and Bonner-Weir S. 1995. Dynamics of beta-cell mass in the growing rat pancreas. Estimation with a simple mathematical model. *Diabetes* 44:249–256.

51. Ohlsson H., Thor S., and Edlund T. 1991. Novel insulin promoter- and enhancer-binding proteins that discriminate between pancreatic alpha- and beta-cells. *Mol. Endocrinol.* 5:897–904.

52. Ohlsson H., Karlsson K., and Edlund T. 1993. IPF1, a homeodomain-containing transactivator of the insulin gene. *EMBO J.* 12:4251–4259.

53. Miller C.P., McGehee Jr. R.E., and Habener J.F. 1994. IDX-1: a new homeodomain transcription factor expressed in rat pancreatic islets and duodenum that transactivates the somatostatin gene. *EMBO J.* 13:1145–1156.

54. Sosa-Pineda B., Chowdhury K., Torres M., Oliver G., and Gruss P. 1997. The Pax4 gene is essential for differentiation of insulin-producing beta cells in the mammalian pancreas. *Nature* 386:399–402.

55. Ahlgren U., Jonsson J., and Edlund H. 1996. The morphogenesis of the pancreatic mesenchyme is uncoupled from that of the pancreatic epithelium in IPF1/PDX1-deficient mice. *Development* 122:1409–1416.

56. Sommer L., Ma Q., and Anderson D.J. 1996. Neurogenins, a novel family of atonal-related bHLH transcription factors, are putative mammalian neuronal determination genes that reveal progenitor cell heterogeneity in the developing CNS and PNS. *Mol. Cell. Neurosci.* 8:221–241.

57. Ma Q., Chen Z., del Barco Barrantes I., de la Pompa J.L., and Anderson D.J. 1998. Neurogenin1 is essential for the determination of neuronal precursors for proximal cranial sensory ganglia. *Neuron.* 20:469–482.

58. Fode C., Gradwohl G., Morin X., Dierich A., LeMeur M., Goridis C., and Guillemot F. 1998. The bHLH protein NEUROGENIN 2 is a determination factor for epibranchial placode-derived sensory neurons. *Neuron.* 20:483–494.

59. Apelqvist A., Li H., Sommer L., Beatus P., Anderson D.J., Honjo T., Hrabe de Angelis M., Lendahl U., and Edlund H. 1999. Notch signalling controls pancreatic cell differentiation. *Nature* 400:877–881.

60. Beatus P., Lundkvist J., Oberg C., and Lendahl U. 1999. The Notch 3 intracellular domain represses Notch 1-mediated activation through Hairy/Enhancer of split (HES) promoters. *Development* 126:3925–3935.

61. Leonard J., Peers B., Johnson T., Ferreri K., Lee S., and Montminy M.R. 1993. Characterization of somatostatin transactivating Factor-1, a novel homeobox factor that stimulates somatostatin expression in pancreatic islet cells. *Mol. Endocrinol.* 7:1275–1283.

62. Peers B., Leonard J., Sharma S., Teitelman G., and Montminy M.R. 1994. Insulin expression in pancreatic islet cells relies on cooperative interactions between the helix loop helix factor E47 and the homeobox factor STF-1. *Mol. Endocrinol.* 8:1798–1806.

63. Jonsson J., Carlsson L., Edlund T., and Edlund H. 1994. Insulin-promoter-Factor 1 is required for pancreas development in mice. *Nature* 371:606–609.

64. Offield M.F., Jetton T.L., Labosky P.A., Ray M., Stein R.W., Magnuson M.A., Hogan B.L., and Wright C.V. 1996. PDX-1 is required for pancreatic outgrowth and differentiation of the rostral duodenum. *Development* 122:983–995.

65. Ahlgren U., Jonsson J., Jonsson L., Simu K., and Edlund H. 1998. Beta-cell-specific inactivation of the mouse Ipf1/Pdx1 gene results in loss of the beta-cell phenotype and maturity onset diabetes. *Genes Dev.* 12:1763–1768.

66. Gu H., Zou Y.R., and Rajewsky K. 1993. Independent control of immunoglobulin switch recombination at individual switch regions evidenced through Cre-loxP-mediated gene targeting. *Cell* 73:1155–1164.

67. Gu H., Marth J.D., Orban P.C., Mossmann H., and Rajewsky K. 1994. Deletion of a DNA polymerase beta gene segment in T cells using cell type-specific gene targeting. *Science* 265:103–106.

68. Petersen H.V., Peshavaria M., Pedersen A.A., Philippe J., Stein R., Madsen O.D., and Serup P. 1998. Glucose stimulates the activation domain potential of the PDX-1 homeodomain transcription factor. *FEBS Lett.* 431:362–366.

69. Olson L.K., Sharma A., Peshavaria M., Wright C.V., Towle H.C., Rodertson R.P., and Stein R. 1995. Reduction of insulin gene transcription in HIT-T15 beta cells chronically exposed to a supraphysiologic glucose concentration is associated with loss of STF-1 transcription factor expression. *Proc. Natl. Acad. Sci. U.S.A.* 92:9127–9131.

70. Sharma A., Olson L.K., Robertson R.P., and Stein R. 1995. The reduction of insulin gene transcription in HIT-T15 beta cells chronically exposed to high glucose concentration is associated with the loss of RIPE3b1 and STF-1 transcription factor expression. *Mol. Endocrinol.* 9:1127–1134.

71. Moran A., Zhang H.J., Olson L.K., Harmon J.S., Poitout V., and Robertson R.P. 1997. Differentiation of glucose toxicity from beta cell exhaustion during the evolution of defective insulin gene expression in the pancreatic islet cell line, HIT-T15. *J. Clin. Invest.* 99:534–539.

72. Seijffers R., Ben-David O., Cohen Y., Karasik A., Berezin M., Newgard C.B., and Ferber S. 1999. Increase in PDX-1 levels suppresses insulin gene expression in RIN 1046-38 cells. *Endocrinology* 140:3311–3317.

73. Stoffers D.A., Zinkin N.T., Stanojevic V., Clarke W.L., and Habener J.F. 1997. Pancreatic agenesis attributable to a single nucleotide deletion in the human IPF1 gene coding sequence. *Nat. Genet.* 15:106–110.

74. Stoffers D.A., Stanojevic V., and Habener J.F. 1998. Insulin promoter Factor-1 gene mutation linked to early-onset Type 2 diabetes mellitus directs expression of a dominant negative isoprotein. *J. Clin. Invest.* 102:232–241.

75. Wang M. and Drucker D.J. 1994. The LIM domain homeobox gene Isl-1: conservation of human, hamster, and rat complementary deoxyribonucleic acid sequences and expression in cell types of nonneuroendocrine lineage. *Endocrinology* 134:1416–1422.

76. Wang M. and Drucker D.J. 1995. The LIM domain homeobox gene isl-1 is a positive regulator of islet cell-specific proglucagon gene transcription. *J. Biol. Chem.* 270:12646–12652.

77. Leonard J., Serup P., Gonzalez G., Edlund T., and Montminy M. 1992. The LIM family transcription factor Isl-1 requires cAMP response element binding protein to promote somatostatin expression in pancreatic islet cells. *Proc. Natl. Acad. Sci. U.S.A.* 89:6247–6251.

78. Wang M. and Drucker D.J. 1996. Activation of amylin gene transcription by LIM domain homeobox gene Isl-1. *Mol. Endocrinol.* 10:243–251.

79. Dandoy-Dron F., Deltour L., Monthioux E., Bucchini D., and Jami J. 1993. Insulin gene can be expressed in the absence of Isl-1. *Exp. Cell. Res.* 209:58–63.

80. Dong J., Asa S.L., and Drucker D.J. 1991. Islet cell and extrapancreatic expression of the LIM domain homeobox gene Isl-1. *Mol. Endocrinol.* 5:1633–1641.

81. Ahlgren U., Pfaff S.L., Jessell T.M., Edlund T., and Edlund H. 1997. Independent requirement for ISL1 in formation of pancreatic mesenchyme and islet cells. *Nature* 385:257–260.

82. Balczarek K.A., Lai Z.C., and Kumar S. 1997. Evolution of functional diversification of the paired box (Pax) DNA-binding domains. *Mol. Biol. Evol.* 14:829–842.

83. Jun S. and Desplan C. 1996. Cooperative interactions between paired domain and homeodomain. *Development* 122:2639–2650.

84. Walther C. and Gruss P. 1991. Pax-6, a murine paired box gene, is expressed in the developing CNS. *Development* 113:1435–1449.

85. Turque N., Plaza S., Radvanyi F., Carriere C., and Saule S. 1994. Pax-QNR/Pax-6, a paired box- and homeobox-containing gene expressed in neurons, is also expressed in pancreatic endocrine cells. *Mol. Endocrinol.* 8:929–938.

86. St-Onge L., Sosa-Pineda B., Chowdhury K., Mansouri A., and Gruss P. 1997. Pax6 is required for differentiation of glucagon-producing alpha-cells in mouse pancreas. *Nature* 387:406–409.

87. Rouiller D.G., Cirulli V., and Halban P.A. 1990. Differences in aggregation properties and levels of the neural cell adhesion molecule (NCAM) between islet cell types. *Exp. Cell. Res.* 191:305–312.

88. Cirulli V., Baetens D., Rutishauser U., Halban P.A., Orci L., and Rouiller D.G. 1994. Expression of neural cell adhesion molecule (N-CAM) in rat islets and its role in islet cell type segregation. *J. Cell. Sci.* 107(Pt. 6):1429–1436.

89. Sussel L., Kalamaras J., Hartigan-O'Connor D.J., Meneses J.J., Pedersen R.A., Rubenstein J.L., and German M.S. 1998. Mice lacking the homeodomain transcription factor Nkx2.2 have diabetes due to arrested differentiation of pancreatic beta cells. *Development* 125:2213–2221.

90. Lee J.E., Hollenberg S.M., Snider L., Turner D.L., Lipnick N., and Weintraub H. 1995. Conversion of Xenopus ectoderm into neurons by NeuroD, a basic helix-loop-helix protein. *Science* 268:836–844.

91. Mutoh H., Fung B.P., Naya F.J., Tsai M.J., Nishitani J., and Leiter A.B. 1997. The basic helix-loop-helix transcription factor BETA2/NeuroD is expressed in mammalian enteroendocrine cells and activates secretin gene expression. *Proc. Natl. Acad. Sci. U.S.A.* 94:3560–3564.

92. Naya F.J., Huang H.P., Qiu Y., Mutoh H., DeMayo F.J., Leiter A.B., and Tsai M.J. 1997. Diabetes, defective pancreatic morphogenesis, and abnormal enteroendocrine differentiation in BETA2/neuroD-deficient mice. *Genes Dev.* 11:2323–2334.

93. Dumonteil E., Laser B., Constant I., and Philippe J. 1998. Differential regulation of the glucagon and insulin I gene promoters by the basic helix-loop-helix transcription factors E47 and BETA2. *J. Biol. Chem.* 273:19945–19954.

94. Serup P., Petersen H.V., Pedersen E.E., Edlund H., Leonard J., Petersen J.S., Larsson L.I., and Madsen O.D. 1995. The homeodomain protein IPF-1/STF-1 is expressed in a subset of islet cells and promotes rat Insulin 1 gene expression dependent on an intact E1 helix-loop-helix factor binding site. *Biochem. J.* 310(Pt. 3):997–1003.

95. Sharma S., Jhala U.S., Johnson T., Ferreri K., Leonard J., and Montmin, M. 1997. Hormonal regulation of an islet-specific enhancer in the pancreatic homeobox gene STF-1. *Mol. Cell. Biol.* 17:2598–2604.

96. Scaglia L., Cahill C.J., Finegood D.T., and Bonner-Weir S. 1997. Apoptosis participates in the remodeling of the endocrine pancreas in the neonatal rat. *Endocrinology* 138:1736–1741.

97. Scaglia L., Smith F.E., and Bonner-Weir S. 1995. Apoptosis contributes to the involution of beta cell mass in the post partum rat pancreas. *Endocrinology* 136:5461–5468.

98. Parsons J.A., Bartke A., and Sorenson R.L. 1995. Number and size of islets of Langerhans in pregnant, human growth hormone-expressing transgenic, and pituitary dwarf mice: effect of lactogenic hormones. *Endocrinology* 136:2013–2021.

99. Djonov V., Andres A.C., and Ziemiecki A. 2001. Vascular remodelling during the normal and malignant life cycle of the mammary gland. *Microsc. Res. Tech.* 52:182–189.

100. Silberstein G.B. 2001. Postnatal mammary gland morphogenesis. *Microsc. Res. Tech.* 52:155–162.

101. Li M., Liu X., Robinson G., Bar-Peled U., Wagner K.U., Young W.S., Hennighausen L., and Furth P.A. 1997. Mammary-derived signals activate programmed cell death during the first stage of mammary gland involution. *Proc. Natl. Acad. Sci. U.S.A.* 94:3425–3430.

102. Furth P.A. 1999. Introduction: mammary gland involution and apoptosis of mammary epithelial cells. *J. Mammary Gland Biol. Neoplasia* 4:123–127.

103. Strange R., Metcalfe T., Thackray L., and Dang M. 2001. Apoptosis in normal and neoplastic mammary gland development. *Microsc. Res. Tech.* 52:171–181.

104. Li M., Hu J., Heermeier K., Hennighausen L., and Furth P.A. 1996. Apoptosis and remodeling of mammary gland tissue during involution proceeds through p53-independent pathways. *Cell Growth Differ.* 7:13–20.

105. Jerry D.J., Pinkas J., Kuperwasser C., Dickinson E.S., and Naber S.P. 1999. Regulation of p53 and its targets during involution of the mammary gland. *J. Mammary Gland Biol. Neoplasia* 4:177–181.

106. Bouwens L. and Pipeleers D.G. 1998. Extra-insular beta cells associated with ductules are frequent in adult human pancreas. *Diabetologia* 41:629–633.

107. Butler A.E. and Butler P.C. 2003. Regulation of beta cell mass in mice compared to humans, response to obesity. *Diabetes* 52:A334.

108. Rosenberg L., Brown R.A., and Duguid W.P. 1983. A new approach to the induction of duct epithelial hyperplasia and nesidioblastosis by cellophane wrapping of the hamster pancreas. *J. Surg. Res.* 35:63–72.

109. Rosenberg L. 1995. In vivo cell transformation: neogenesis of beta cells from pancreatic ductal cells. *Cell Transplant* 4:371–383.

110. Wolfe-Coote S.A., Louw J., Woodroof C.W., Heydenrych J.J., and du Toit D.F. 1998. Induction of cell proliferation and differentiation in the pancreas of the adult Vervet monkey (*Cercopithecus aethiops*): preliminary results. *Pancreas* 16:129–133.

111. Rosenberg L., Duguid W.P., and Vinik A.I. 1989. The effect of cellophane wrapping of the pancreas in the Syrian golden hamster: autoradiographic observations. *Pancreas* 4:31–37.

112. Rosenberg L., Schwartz R., Dafoe D.C., Clarke S., Turcotte J.G., and Vinik A.I. 1988. Preparation of islets of Langerhans from the hamster pancreas. *J. Surg. Res.* 44:229–234.

113. Rafaeloff R., Qin X.F., Barlow S.W., Rosenberg L., and Vinik A.I. 1996. Identification of differentially expressed genes induced in pancreatic islet neogenesis. *FEBS Lett.* 378:219–223.

114. De Takats G. 1930. Ligation of the tail of the pancreas in juvenile diabetes. *Endorinology* 14:255–264.

115. De Takats G. 1931. The effect of ligating the tail of the pancreas in juvenile diabetes. *Surg. Gynecol. Obstet.* 53:45–53.

116. De Takats G. and Cuthbert F. 1933. Surgical attempts at increasing sugar tolerance. *Arch. Surg.* 26:750–764.

117. De Takats G. and Wilder R. 1929. Isolation of tail of pancreas in a diabetic child. *JAMA* 93:606–610.

118. Rosenberg L., Lipsett M., Yoon J.W., Prentki M., Wang R., Jun H.S., Pittenger G.L., Taylor-Fishwick D., and Vinik A.I. 2004. A pentadecapeptide fragment of islet neogenesis-associated protein increases beta-cell mass and reverses diabetes in C57BL/6J mice. *Ann. Surg.* 240(5):875–84.

119. Gagliardino J.J., Del Zotto H., Massa L., Flores L.E., and Borelli M.I. 2003. Pancreatic duodenal homeobox-1 and islet neogenesis-associated protein: a possible combined marker of activateable pancreatic cell precursors. *J. Endocrinol.* 177:249–259.

120. Bernard C., Berthault M.F., Saulnier C., and Ktorza A. 1999. Neogenesis vs. apoptosis as main components of pancreatic beta cell mass changes in glucose-infused normal and mildly diabetic adult rats. *FASEB. J.* 13:1195–1205.

121. Lipsett M. and Finegood D.T. 2002. Beta-cell neogenesis during prolonged hyperglycemia in rats. *Diabetes* 51:1834–1841.

122. Bonner-Weir S., Deery D., Leahy J.L., and Weir G.C. 1989. Compensatory growth of pancreatic beta-cells in adult rats after short-term glucose infusion. *Diabetes* 38:49–53.

123. Steil G.M., Trivedi N., Jonas J.C., Hasenkamp W.M., Sharma A., Bonner-Weir S., and Weir G.C. 2001. Adaptation of beta-cell mass to substrate oversupply: enhanced function with normal gene expression. *Am. J. Physiol. Endocrinol. Metab.* 280:E788–E796.

124. Paris M., Bernard-Kargar C., Berthault M.F., Bouwens L., and Ktorza A. 2003. Specific and combined effects of insulin and glucose on functional pancreatic beta-cell mass in vivo in adult rats. *Endocrinology* 144:2717–2727.

125. Francini F., Del Zotto H., and Gagliardino J.J. 2001. Effect of an acute glucose overload on Islet cell morphology and secretory function in the toad. *Gen. Comp. Endocrinol.* 122:130–138.

126. Del Zotto H., Massa L., Gomez Dumm C.L., and Gagliardino J.J. 1999. Changes induced by sucrose administration upon the morphology and function of pancreatic islets in the normal hamster. *Diabetes Metab. Res. Rev.* 15:106–112.

127. Petrik J., Srinivasan M., Aalinkeel R., Coukell S., Arany E., Patel M.S., and Hill D.J. 2001. A long-term high-carbohydrate diet causes an altered ontogeny of pancreatic islets of Langerhans in the neonatal rat. *Pediatr. Res.* 49:84–92.

128. Del Zotto H., Gomez Dumm C.L., Drago S., Fortino A., Luna G.C., and Gagliardino J.J. 2002. Mechanisms involved in the beta-cell mass increase induced by chronic sucrose feeding to normal rats. *J. Endocrinol.* 174:225–231.

129. Del Zotto H., Massa L., Rafaeloff R., Pittenger G.L., Vinik A., Gold G., Reifel-Miller A., and Gagliardino J.J. 2000. Possible relationship between changes in islet neogenesis and islet neogenesis-associated protein-positive cell mass induced by sucrose administration to normal hamsters. *J. Endocrinol.* 165:725–733.

130. Bonner-Weir S. 2001. Beta-cell turnover: its assessment and implications. *Diabetes* 50(Suppl. 1):S20–S24.

131. De Leon D.D., Deng S., Madani R., Ahima R.S., Drucker D.J., and Stoffers D.A. 2003. Role of endogenous glucagon-like peptide-1 in islet regeneration after partial pancreatectomy. *Diabetes* 52:365–371.

132. Tourrel C., Bailbe D., Meile M.J., Kergoat M., and Portha B. 2001. Glucagon-like Peptide-1 and Exendin-4 stimulate beta-cell neogenesis in streptozotocin-treated newborn rats resulting in persistently improved glucose homeostasis at adult age. *Diabetes* 50:1562–1570.

133. Drucker D.J. 2003. Glucagon-like peptides: regulators of cell proliferation, differentiation, and apoptosis. *Mol. Endocrinol.* 17:161–171.

134. Pospisilik J.A., Martin J., Doty T., Ehses J.A., Pamir N., Lynn F.C., Piteau S., Demuth H.U., McIntosh C.H., and Pederson R.A. 2003. Dipeptidyl peptidase IV inhibitor treatment stimulates beta-cell survival and islet neogenesis in streptozotocin-induced diabetic rats. *Diabetes* 52:741–750.

135. Pospisilik J.A., Stafford S.G., Demuth H.U., Brownsey R., Parkhouse W., Finegood D.T., McIntosh C.H., and Pederson R.A. 2002. Long-term treatment with the dipeptidyl peptidase IV inhibitor P32/98 causes sustained improvements in glucose tolerance, insulin sensitivity, hyperinsulinemia, and beta-cell glucose responsiveness in VDF (fa/fa) Zucker rats. *Diabetes* 51:943–950.

136. Drucker D.J. 2001. Minireview: the glucagon-like peptides. *Endocrinology* 142:521–527.

137. Movassat J., Beattie G.M., Lopez A.D., and Hayek A. 2002. Exendin 4 up-regulates expression of PDX 1 and hastens differentiation and maturation of human fetal pancreatic cells. *J. Clin. Endocrinol. Metab.* 87:4775–4781.

138. Doyle M.E. and Egan J.M. 2001. Glucagon-like Peptide-1. *Recent Prog. Horm. Res.* 56:377–399.

139. Perfetti R., Zhou J., Doyle M.E., and Egan J.M. 2000. Glucagon-like Peptide-1 induces cell proliferation and pancreatic-duodenum homeobox-1 expression and increases endocrine cell mass in the pancreas of old, glucose-intolerant rats. *Endocrinology* 141:4600–4605.

140. Li Y., Hansotia T., Yusta B., Ris F., Halban P.A., and Drucker D.J. 2003. Glucagon-like Peptide-1 receptor signaling modulates beta cell apoptosis. *J. Biol. Chem.* 278:471–478.

141. Bonner-Weir S., Baxter L.A., Schuppin G.T., and Smith F.E. 1993. A second pathway for regeneration of adult exocrine and endocrine pancreas. A possible recapitulation of embryonic development. *Diabetes* 42:1715–1720.

142. Bonner-Weir S., Trent D.F., and Weir G.C. 1983. Partial pancreatectomy in the rat and subsequent defect in glucose-induced insulin release. *J. Clin. Invest.* 71:1544–1553.

143. Brockenbrough J.S., Weir G.C., and Bonner-Weir S. 1988. Discordance of exocrine and endocrine growth after 90% pancreatectomy in rats. *Diabetes* 37:232–236.

144. Bonner-Weir S., Stubbs M., Reitz P., Taneja M., and Smith F. 1997. Partial pancreatectomy as a model of pancreatic regeneration. In *Pancreatic Growth and Regeneration*. Sarvetnick N. Ed. Basel: Karger Landes. p. 138–153.

145. Terazono K., Uchiyama Y., Ide M., Watanabe T., Yonekura H., Yamamoto H., and Okamoto H. 1990. Expression of reg protein in rat regenerating islets and its co-localization with insulin in the beta cell secretory granules. *Diabetologia* 33:250–252.

146. Swenne I. and Eriksson U. 1982. Diabetes in pregnancy: islet cell proliferation in the fetal rat pancreas. *Diabetologia* 23:525–528.

147. Teitelman G., Joh T.H., and Reis D.J. 1981. Transformation of catecholaminergic precursors into glucagon (A) cells in mouse embryonic pancreas. *Proc. Natl. Acad. Sci. U.S.A.* 78:5225–5229.

148. Beattie G.M., Levine F., Mally M.I., Otonkoski T., O'Brien J.S., Salomon D.R., and Hayek A. 1994. Acid beta-galactosidase: a developmentally regulated marker of endocrine cell precursors in the human fetal pancreas. *J. Clin. Endocrinol. Metab.* 78:1232–1240.

149. Tu J., Tuch B.E., and Si Z. 1999. Expression and regulation of glucokinase in rat islet beta- and alpha-cells during development. *Endocrinology* 140:3762–3766.
150. Guz Y., Nasir I., and Teitelman G. 2001. Regeneration of pancreatic beta cells from intra-islet precursor cells in an experimental model of diabetes. *Endocrinology* 142:4956–4968.
151. Bouwens L., Wang R.N., De Blay E., Pipeleers D.G., and Kloppel G. 1994. Cytokeratins as markers of ductal cell differentiation and islet neogenesis in the neonatal rat pancreas. *Diabetes* 43:1279–1283.
152. Lendahl U., Zimmerman L.B., and McKay R.D. 1990. CNS stem cells express a new class of intermediate filament protein. *Cell* 60:585–595.
153. Hunziker E. and Stein M. 2000. Nestin-expressing cells in the pancreatic islets of Langerhans. *Biochem. Biophys. Res. Commun.* 271:116–119.
154. Zulewski H., Abraham E.J., Gerlach M.J., Daniel P.B., Moritz W., Muller B., Vallejo M., Thomas M.K., and Habener J.F. 2001. Multipotential nestin-positive stem cells isolated from adult pancreatic islets differentiate ex vivo into pancreatic endocrine, exocrine, and hepatic phenotypes. *Diabetes* 50:521–533.
155. Edlund H. 2001. Developmental biology of the pancreas. *Diabetes* 50(Suppl. 1):S5–S9.
156. Shapiro A.M., Lakey J.R., Ryan E.A., Korbutt G.S., Toth E., Warnock G.L., Kneteman N.M., and Rajotte R.V. 2000. Islet transplantation in seven patients with Type 1 diabetes mellitus using a glucocorticoid-free immunosuppressive regimen [see comments]. *N. Engl. J. Med.* 343:230–238.
157. Ryan E.A., Lakey J.R., Rajotte R.V., Korbutt G.S., Kin T., Imes S., Rabinovitch A., Elliott J.F., Bigam D., Kneteman N.M. et al. 2001. Clinical outcomes and insulin secretion after islet transplantation with the Edmonton protocol. *Diabetes* 50:710–719.
158. London N.J., Swift S.M., and Clayton H.A. 1998. Isolation, culture and functional evaluation of islets of Langerhans. *Diabetes Metab.* 24:200–207.
159. Morrison C.P., Wemyss-Holden S.A., Dennison A.R., and Maddern G.J. 2002. Islet yield remains a problem in islet autotransplantation. *Arch. Surg.* 137:80–83.
160. Robertson R.P., Lanz K.J., Sutherland D.E., and Seaquist E.R. 2002. Relationship between diabetes and obesity 9 to 18 years after hemipancreatectomy and transplantation in donors and recipients. *Transplantation* 73:736–741.
161. Seaquist E.R., Kahn S.E., Clark P.M., Hales C.N., Porte Jr. D., and Robertson R.P. 1996. Hyperproinsulinemia is associated with increased beta cell demand after hemipancreatectomy in humans. *J. Clin. Invest.* 97:455–460.
162. Kendall D.M., Sutherland D.E., Najarian J.S., Goetz F.C., and Robertson R.P. 1990. Effects of hemipancreatectomy on insulin secretion and glucose tolerance in healthy humans. *N. Engl. J. Med.* 322:898–903.
163. Shapiro A.M., Nanji S.A., and Lakey J.R. 2003. Clinical islet transplant: current and future directions towards tolerance. *Immunol. Rev.* 196:219–236.
164. International Islet Transplant Registry. 2000.
165. Davalli A.M., Ogawa Y., Ricordi C., Scharp D.W., Bonner-Weir S., and Weir G.C. 1995. A selective decrease in the beta cell mass of human islets transplanted into diabetic nude mice. *Transplantation* 59:817–820.
166. Alejandro R., Latif Z., Polonsky K.S., Shienvold F.L., Civantos F., and Mint D.H. 1988. Natural history of multiple intrahepatic canine islet allografts during and following administration of cyclosporine. *Transplantation* 45:1036–1044.

167. Lanza R.P., Sullivan S.J., and Chick W.L. 1992. Perspectives in diabetes. Islet transplantation with immunoisolation. *Diabetes* 41:1503–1510.

168. Ricordi C., Lacy P.E., and Scharp D.W. 1989. Automated islet isolation from human pancreas. *Diabetes* 38(Suppl. 1):140–142.

169. Lakey J.R., Kneteman N.M., Rajotte R.V., Wu D.C., Bigam D., and Shapiro A.M. 2002. Effect of core pancreas temperature during cadaveric procurement on human islet isolation and functional viability. *Transplantation* 73: 1106–1110.

170. Ohzato H., Gotoh M., Monden M., Dono K., Kanai T., and Mori T. 1991. Improvement in islet yield from a cold-preserved pancreas by pancreatic ductal collagenase distention at the time of harvesting. *Transplantation* 51:566–570.

171. Socci C., Davalli A.M., Vignali A., Pontiroli A.E., Maffi P., Magistretti P., Gavazzi F., De Nittis P., Di C.V., and Pozza G. 1993. A significant increase of islet yield by early injection of collagenase into the pancreatic duct of young donors. *Transplantation* 55:661–663.

172. Heiser A., Ulrichs K., and Muller-Ruchholtz W. 1994. Isolation of porcine pancreatic islets: low trypsin activity during the isolation procedure guarantees reproducible high islet yields. *J. Clin. Lab. Anal.* 8:407–411.

173. White S.A., Djaballah H., Hughes D.P., Roberts D.L., Contractor H.H., Pathak S., and London N.J. 1999. A preliminary study of the activation of endogenous pancreatic exocrine enzymes during automated porcine islet isolation. *Cell Transplant* 8:265–276.

174. Lakey J.R., Helms L.M., Kin T., Korbutt G.S., Rajotte R.V., Shapiro A.M., and Warnock G.L. 2001. Serine-protease inhibition during islet isolation increases islet yield from human pancreases with prolonged ischemia. *Transplantation* 72:565–570.

175. Rose N.L., Palcic M.M., Helms L.M., and Lakey J.R. 2003. Evaluation of Pefabloc as a serine protease inhibitor during human-islet isolation. *Transplantation* 75:462–466.

176. Gores P.F. and Sutherland D.E. 1993. Pancreatic islet transplantation: is purification necessary? *Am. J. Surg.* 166:538–542.

177. Downing R., Morrissey S., Kiske D., and Scharp D.W. 1986. Does the purity of intraportal islet isografts affect their endocrine function? *J. Surg. Res.* 41:41–46.

178. Hengartner M.O. 2000. The biochemistry of apoptosis. *Nature* 407:770–776.

179. Thornberry N.A. and Lazebnik Y. 1998. Caspases: enemies within. *Science* 281:1312–1316.

180. Tibbetts M.D., Zheng L., and Lenardo M.J. 2003. The death effector domain protein family: regulators of cellular homeostasis. *Nat. Immunol.* 4:404–409.

181. Weber C.H. and Vincenz C. 2001. The death domain superfamily: a tale of two interfaces? *Trends Biochem. Sci.* 26:475–481.

182. Green D.R. and Reed J.C. 1998. Mitochondria and apoptosis. *Science* 281:1309–1312.

183. Scorrano L. and Korsmeyer S.J. 2003. Mechanisms of cytochrome c release by proapoptotic BCL-2 family members. *Biochem. Biophys. Res. Commun.* 304:437–444.

184. Gross A., McDonnell J.M., and Korsmeyer S.J. 1999. BCL-2 family members and the mitochondria in apoptosis. *Genes Dev.* 13:1899–1911.

185. Wei M.C., Zong W.X., Cheng E.H., Lindsten T., Panoutsakopoulou V., Ross A.J., Roth K.A., MacGregor G.R., Thompson C.B., and Korsmeyer S.J. 2001. Proapoptotic BAX and BAK: a requisite gateway to mitochondrial dysfunction and death. *Science* 292:727–730.

186. Chao D.T. and Korsmeyer S.J. 1998. BCL-2 family: regulators of cell death. *Annu. Rev. Immunol.* 16:395–419.

187. Tsujimoto Y. and Shimizu S. 2000. Bcl-2 family: life-or-death switch. *FEBS Lett.* 466:6–10.

188. Proskuryakov S.Y., Konoplyannikov A.G., and Gabai V.L. 2003. Necrosis: a specific form of programmed cell death? *Exp. Cell Res.* 283:1–16.

189. Ha H.C. and Snyder S.H. 1999. Poly(ADP-ribose) polymerase is a mediator of necrotic cell death by ATP depletion. *Proc. Natl. Acad. Sci. U.S.A.* 96:13978–13982.

190. Leist M., Single B., Naumann H., Fava E., Simon B., Kuhnle S., and Nicotera P. 1999. Inhibition of mitochondrial ATP generation by nitric oxide switches apoptosis to necrosis. *Exp. Cell Res.* 249:396–403.

191. Leist M. and Jaattela M. 2001. Four deaths and a funeral: from caspases to alternative mechanisms. *Nat. Rev. Mol. Cell Biol.* 2:589–598.

192. Lee J.M., Zipfel G.J., and Choi D.W. 1999. The changing landscape of ischaemic brain injury mechanisms. *Nature* 399:A7–A14.

193. Paraskevas S., Maysinger D., Wang R., Duguid T.P., and Rosenberg L. 2000. Cell loss in isolated human islets occurs by apoptosis. *Pancreas* 20:270–276.

194. Paraskevas S., Duguid W.P., Maysinger D., Feldman L., Agapitos D., and Rosenberg L. 1997. Apoptosis occurs in freshly isolated human islets under standard culture conditions. *Transplant. Proc.* 29:750–752.

195. Wang R.N., and Rosenberg L. 1999. Maintenance of beta-cell function and survival following islet isolation requires re-establishment of the islet-matrix relationship. *J. Endocrinol.* 163:181–190.

196. Thomas F.T., Contreras J.L., Bilbao G., Ricordi C., Curiel D., and Thomas J.M. 1999. Anoikis, extracellular matrix, and apoptosis factors in isolated cell transplantation. *Surgery* 126:299–304.

197. Aikin R., Rosenberg L., Paraskevas S., and Maysinger D. 2004. Inhibition of caspase-mediated PARP-1 cleavage results in increased necrosis in isolated islets of Langerhans. *J. Mol. Med.* 82(6):389–97.

198. Ilieva A., Yuan S., Wang R.N., Agapitos D., Hill D.J., and Rosenberg L. 1999. Pancreatic islet cell survival following islet isolation: the role of cellular interactions in the pancreas. *J. Endocrinol.* 161:357–364.

199. Giancotti F.G. and Ruoslahti E. 1999. Integrin signaling. *Science* 285:1028–1032.

200. Hotz H.R., Biebinger S., Flaspohler J., and Clayton C. 1998. PARP gene expression: control at many levels. *Mol. Biochem. Parasitol.* 91:131–143.

201. Frisch S.M. and Screaton R.A. 2001. Anoikis mechanisms. *Curr. Opin. Cell Biol.* 13:555–562.

202. van Deijnen J.H., Hulstaert C.E., Wolters G.H., and van Schilfgaarde R. 1992. Significance of the peri-insular extracellular matrix for islet isolation from the pancreas of rat, dog, pig, and man. *Cell Tissue Res.* 267:139–146.

203. Rosenberg L., Wang R., Paraskevas S., and Maysinger D. 1999. Structural and functional changes resulting from islet isolation lead to islet cell death. *Surgery* 126:393–398.

204. Ris F., Hammar E., Bosco D., Pilloud C., Maedler K., Donath M.Y., Oberholzer J., Zeender E., Morel P., Rouiller D. et al. 2002. Impact of integrin-matrix matching and inhibition of apoptosis on the survival of purified human beta-cells in vitro. *Diabetologia* 45:841–850.

205. Sporn M.B. and Roberts A.B. 1988. Peptide growth factors are multifunctional. *Nature* 332:217–219.

206. Raff M.C. 1992. Social controls on cell survival and cell death. *Nature* 356:397–400.

207. Hanahan D. and Weinberg R.A. 2000. The hallmarks of cancer. *Cell* 100:57–70.

208. Collins M.K., Perkins G.R., Rodriguez-Tarduchy G., Nieto M.A., and Lopez-Rivas A. 1994. Growth factors as survival factors: regulation of apoptosis. *Bioessays* 16:133–138.

209. Plas D.R. and Thompson C.B. 2002. Cell metabolism in the regulation of programmed cell death. *Trends Endocrinol. Metab.* 13:75–78.

210. Vander Heiden M.G., Plas D.R., Rathmell J.C., Fox C.J., Harris M.H., and Thompson C.B. 2001. Growth factors can influence cell growth and survival through effects on glucose metabolism. *Mol. Cell Biol.* 21:5899–5912.

211. Garland J.M. and Halestrap A. 1997. Energy metabolism during apoptosis. Bcl-2 promotes survival in hematopoietic cells induced to apoptose by growth factor withdrawal by stabilizing a form of metabolic arrest. *J. Biol. Chem.* 272:4680–4688.

212. Vander Heiden M.G., Chandel N.S., Li X.X., Schumacker P.T., Colombini M., and Thompson C.B. 2000. Outer mitochondrial membrane permeability can regulate coupled respiration and cell survival. *Proc. Natl. Acad. Sci. U.S.A.* 97:4666–4671.

213. Whetton A.D. and Dexter T.M. 1983. Effect of haematopoietic cell growth factor on intracellular ATP levels. *Nature* 303:629–631.

214. Pavlovic D., Chen M.C., Bouwens L., Eizirik D.L., and Pipeleers D. 1999. Contribution of ductal cells to cytokine responses by human pancreatic islets. *Diabetes* 48:29–33.

215. McClintock D.S., Santore M.T., Lee V.Y., Brunelle J., Budinger G.R., Zong W.X., Thompson C.B., Hay N., and Chandel N.S. 2002. Bcl-2 family members and functional electron transport chain regulate oxygen deprivation-induced cell death. *Mol. Cell Biol.* 22:94–104.

216. Saikumar P., Dong Z., Patel Y., Hall K., Hopfer U., Weinberg J.M., and Venkatachalam M.A. 1998. Role of hypoxia-induced Bax translocation and cytochrome c release in reoxygenation injury. *Oncogene* 17:3401–3415.

217. Shimizu S., Eguchi Y., Kosaka H., Kamiike W., Matsuda H., and Tsujimoto Y. 1995. Prevention of hypoxia-induced cell death by Bcl-2 and Bcl-xL. *Nature* 374:811–813.

218. Wang G.L., Jiang B.H., Rue E.A., and Semenza G.L. 1995. Hypoxia-inducible Factor 1 is a basic-helix-loop-helix-PAS heterodimer regulated by cellular O2 tension. *Proc. Natl. Acad. Sci. U.S.A.* 92:5510–5514.

219. Semenza G.L. 1999. Regulation of mammalian O2 homeostasis by hypoxia-inducible Factor 1. *Annu. Rev. Cell Dev. Biol.* 15:551–578.

220. Sowter H.M., Ratcliffe P.J., Watson P., Greenberg A.H., and Harris A.L. 2001. HIF-1-dependent regulation of hypoxic induction of the cell death factors BNIP3 and NIX in human tumors. *Cancer Res.* 61:6669–6673.

221. Menger M.D., Yamauchi J., and Vollmar B. 2001. Revascularization and microcirculation of freely grafted islets of Langerhans. *World J. Surg.* 25:509–515.

222. Menger M.D., Jaeger S., Walter P., Feifel G., Hammersen F., and Messmer K. 1989. Angiogenesis and hemodynamics of microvasculature of transplanted islets of Langerhans. *Diabetes* 38(Suppl. 1):199–201.

223. Matsuda T., Suzuki Y., Tanioka Y., Toyama H., Kakinoki K., Hiraoka K., Fujino Y., and Kuroda Y. 2003. Pancreas preservation by the 2-layer cold storage method before islet isolation protects isolated islets against apoptosis through the mitochondrial pathway. *Surgery* 134:437–445.

224. Moritz W., Meier F., Stroka D.M., Giuliani M., Kugelmeier P., Nett P.C., Lehmann R., Candinas D., Gassmann M., and Weber M. 2002. Apoptosis in hypoxic human pancreatic islets correlates with HIF-1alpha expression. *FASEB J.* 16:745–747.

225. Vargas F., Vives-Pi M., Somoza N., Armengol P., Alcalde L., Marti M., Costa M., Serradel, L., Dominguez O., Fernandez-Llamazares J. et al. 1998. Endotoxin contamination may be responsible for the unexplained failure of human pancreatic islet transplantation. *Transplantation* 65:722–727.

226. Linetsky E., Inverardi L., Kenyon N.S., Alejandro R., and Ricordi C. 1998. Endotoxin contamination of reagents used during isolation and purification of human pancreatic islets. *Transplant. Proc.* 30:345–346.

227. Jahr H., Pfeiffer G., Hering B.J., Federlin K., and Bretzel R.G. 1999. Endotoxin-mediated activation of cytokine production in human PBMCs by collagenase and Ficoll. *J. Mol. Med.* 77:118–120.

228. Linetsky E., Bottino R., Lehmann R., Alejandro R., Inverardi L., and Ricordi C. 1997. Improved human islet isolation using a new enzyme blend, liberase. *Diabetes* 46:1120–1123.

229. Raetz C.R. and Whitfield C. 2002. Lipopolysaccharide endotoxins. *Annu. Rev. Biochem.* 71:635–700.

230. Beutler, B. 2000. Tlr4: central component of the sole mammalian LPS sensor. *Curr. Opin. Immunol.* 12:20–26.

231. Moberg L., Johansson H., Lukinius A., Berne C., Foss A., Kallen R., Ostraat O., Salmela K., Tibell A., Tufveson G. et al. 2002. Production of tissue factor by pancreatic islet cells as a trigger of detrimental thrombotic reactions in clinical islet transplantation. *Lancet* 360:2039–2045.

232. Vives-Pi M., Somoza N., Fernandez-Alvarez J., Vargas F., Caro P., Alba A., Gomis R., Labeta M.O., and Pujol-Borrell R. 2003. Evidence of expression of endotoxin receptors CD14, toll-like receptors TLR4 and TLR2 and associated molecule MD-2 and of sensitivity to endotoxin (LPS) in islet beta cells. *Clin. Exp. Immunol.* 133:208–218.

233. Rabinovitch A., Suarez-Pinzon W.L., Strynadka K., Lakey J.R., and Rajotte R.V. 1996. Human pancreatic islet beta-cell destruction by cytokines involves oxygen free radicals and aldehyde production. *J. Clin. Endocrinol. Metab.* 81:3197–3202.

234. Berney T., Molano R.D., Cattan P., Pileggi A., Vizzardelli C., Oliver R., Ricordi C., and Inverardi L. 2001. Endotoxin-mediated delayed islet graft function is associated with increased intra-islet cytokine production and islet cell apoptosis. *Transplantation* 71:125–132.

235. Bottino R., Balamurugan A.N., Bertera S., Pietropaolo M., Trucco M., and Piganelli J.D. 2002. Preservation of human islet cell functional mass by anti-oxidative action of a novel SOD mimic compound. *Diabetes* 51:2561–2567.

236. Ketchum R.J., Deng S., Weber M., Jahr H., and Brayman K.L. 2000. Reduced NO production improves early canine islet xenograft function: a role for nitric oxide in islet xenograft primary nonfunction. *Cell Transplant.* 9:453–462.

237. Donath M.Y., Gross D.J., Cerasi E., and Kaiser N. 1999. Hyperglycemia-induced beta-cell apoptosis in pancreatic islets of Psammomys obesus during development of diabetes. *Diabetes* 48:738–744.

238. Konrad R.J., Liu K., and Kudlow J.E. 2000. A modified method of islet isolation preserves the ability of pancreatic islets to increase protein O-glycosylation in response to glucose and streptozotocin. *Arch. Biochem. Biophys.* 381:92–98.

239. Zhou Y.P., Teng D., Dralyuk F., Ostrega D., Roe M.W., Philipson L., and Polonsky K.S. 1998. Apoptosis in insulin-secreting cells. Evidence for the role of intracellular Ca2+ stores and arachidonic acid metabolism. *J. Clin. Invest.* 101:1623–1632.

240. Bai J.Z., Saafi E.L., Zhang S., and Cooper G.J. 1999. Role of Ca2+ in apoptosis evoked by human amylin in pancreatic islet beta-cells. *Biochem. J.* 343(Pt. 1):53–61.

241. Nakata M., Uto N., Maruyama I., and Yada T. 1999. Nitric oxide induces apoptosis via Ca2+-dependent processes in the pancreatic beta-cell line MIN6. *Cell Struct. Funct.* 24:451–455.

242. Carlsson P.O., Kozlova I., Andersson A., and Roomans G.M. 2003. Changes in intracellular sodium, potassium, and calcium concentrations in transplanted mouse pancreatic islets. *Transplantation* 75:445–449.

243. Carlsson P.O., Palm F., Andersson A., and Liss P. 2001. Markedly decreased oxygen tension in transplanted rat pancreatic islets irrespective of the implantation site. *Diabetes* 50:489–495.

244. Widmann C., Gibson S., Jarpe M.B., and Johnson G.L. 1999. Mitogen-activated protein kinase: conservation of a three-kinase module from yeast to human. *Physiol. Rev.* 79:143–180.

245. Davis R.J. 2000. Signal transduction by the JNK group of MAP kinases. *Cell* 103:239–252.

246. Shaulian E. and Karin M. 2002. AP-1 as a regulator of cell life and death. *Nat. Cell Biol.* 4:E131–E136.

247. Faris M., Latinis K.M., Kempiak S.J., Koretzky G.A., and Nel A. 1998. Stress-induced Fas ligand expression in T cells is mediated through a MEK Kinase 1-regulated response element in the Fas ligand promoter. *Mol. Cell Biol.* 18:5414–5424.

248. Kasibhatla S., Brunner T., Genestier L., Echeverri F., Mahboubi A., and Green D.R. 1998. DNA damaging agents induce expression of Fas ligand and subsequent apoptosis in T lymphocytes via the activation of NF-kappa B and AP-1. *Mol. Cell* 1:543–551.

249. Yeh W.C., Pompa J.L., McCurrach M.E., Shu H.B., Elia A.J., Shahinian A., Ng M., Wakeham A., Khoo W., Mitchell K. et al. 1998. FADD: essential for embryo development and signaling from some, but not all, inducers of apoptosis. *Science* 279:1954–1958.

250. Tournier C., Hess P., Yang D.D., Xu J., Turner T.K., Nimnual A., Bar-Sagi D., Jones S.N., Flavell R.A., and Davis R.J. 2000. Requirement of JNK for stress-induced activation of the cytochrome c-mediated death pathway. *Science* 288:870–874.

251. Lei K., Nimnual A., Zong W.X., Kennedy N.J., Flavell R.A., Thompson C.B., Bar-Sagi D., and Davis R.J. 2002. The Bax subfamily of Bcl2-related proteins is essential for apoptotic signal transduction by c-Jun NH(2)-terminal kinase. *Mol. Cell Biol.* 22:4929–4942.

252. Lei K. and Davis R.J. 2003. JNK phosphorylation of Bim-related members of the Bcl2 family induces Bax-dependent apoptosis. *Proc. Natl. Acad. Sci. U.S.A.* 100:2432–2437.

253. Roulston A., Reinhard C., Amiri P., and Williams L.T. 1998. Early activation of c-Jun N-terminal kinase and p38 kinase regulate cell survival in response to tumor necrosis factor alpha. *J. Biol. Chem.* 273:10232–10239.

254. Chen Y.R., Wang X., Templeton D., Davis R.J., and Tan T.H. 1996. The role of c-Jun N-terminal kinase (JNK) in apoptosis induced by ultraviolet C and gamma radiation. Duration of JNK activation may determine cell death and proliferation. *J. Biol. Chem.* 271:31929–31936.

255. Guo Y.L., Baysal K., Kang B., Yang L.J., and Williamson J.R. 1998. Correlation between sustained c-Jun N-terminal protein kinase activation and apoptosis induced by tumor necrosis factor-alpha in rat mesangial cells. *J. Biol. Chem.* 273:4027–4034.

256. Mansouri A., Ridgway L.D., Korapati A.L., Zhang Q., Tian L., Wang Y., Siddik Z.H., Mills G.B., and Claret F.X. 2003. Sustained activation of JNK/p38 MAPK pathways in response to cisplatin leads to Fas ligand induction and cell death in ovarian carcinoma cells. *J. Biol. Chem.* 278:19245–19256.

257. Tobiume K., Matsuzawa A., Takahashi T., Nishitoh H., Morita K., Takeda K., Minowa O., Miyazono K., Noda T., and Ichijo H. 2001. ASK1 is required for sustained activations of JNK/p38 MAP kinases and apoptosis. *EMBO Rep.* 2:222–228.

258. Lin A. 2003. Activation of the JNK signaling pathway: breaking the brake on apoptosis. *Bioessays* 25:17–24.

259. Xia Z., Dickens M., Raingeaud J., Davis R.J., and Greenberg M.E. 1995. Opposing effects of ERK and JNK-p38 MAP kinases on apoptosis. *Science* 270:1326–1331.

260. Frisch S.M., Vuori K., Kelaita D., and Sicks S. 1996. A role for Jun-N-terminal kinase in anoikis; suppression by bcl-2 and crmA. *J. Cell Biol.* 135:1377–1382.

261. He H., Li H.L., Lin A., and Gottlieb R.A. 1999. Activation of the JNK pathway is important for cardiomyocyte death in response to simulated ischemia. *Cell Death. Differ.* 6:987–991.

262. Paraskevas S., Aikin R., Maysinger D., Lakey J.R., Cavanagh T.J., Agapitos D., Wang R., and Rosenberg A. 2001. Modulation of JNK and p38 stress activated protein kinases in isolated islets of Langerhans: insulin as an autocrine survival signal. *Ann. Surg.* 233:124–133.

263. Paraskevas S., Aikin R., Maysinger D., Lakey J.R., Cavanagh T.J., Hering B., Wang R., and Rosenberg L. 1999. Activation and expression of ERK, JNK, and p38 MAP-kinases in isolated islets of Langerhans: implications for cultured islet survival. *FEBS Lett.* 455:203–208.

264. Kaneto H., Xu G., Fujii N., Kim S., Bonner-Weir S., and Weir G.C. 2002. Involvement of c-Jun N-terminal kinase in oxidative stress-mediated suppression of insulin gene expression. *J. Biol. Chem.* 277:30010–30018.

265. Ammendrup A., Maillard A., Nielsen K., Aabenhus A.N., Serup P., Dragsbaek M.O., Mandrup-Poulsen T., and Bonny C. 2000. The c-Jun amino-terminal kinase pathway is preferentially activated by Interleukin-1 and controls apoptosis in differentiating pancreatic beta-cells. *Diabetes* 49:1468–1476.

266. Bonny C., Oberson A., Negri S., Sauser C., and Schorderet D.F. 2001. Cell-permeable peptide inhibitors of JNK: novel blockers of beta-cell death. *Diabetes* 50:77–82.

267. Mandrup-Poulsen T. 2001. Beta-cell apoptosis: stimuli and signaling. *Diabetes* 50(Suppl. 1):S58–S63.

268a. Kyriakis J.M. and Avruch J. 2001. Mammalian mitogen-activated protein kinase signal transduction pathways activated by stress and inflammation. *Physiol. Rev.* 81:807–869.

268b. Aikin R., Maysinger D., and Rosenberg, L. 2004. Cross-talk between phosphatidylinositol 3-kinase/AKT and c-jun NH2-terminal kinase mediates survival of isolated human islets. *Endocrinology* 145(10):4522–31.

269. Heidenreich K.A. and Kummer J.L. 1996. Inhibition of p38 mitogen-activated protein kinase by insulin in cultured fetal neurons. *J. Biol. Chem.* 271:9891–9894.

270. Pugazhenthi S., Miller E., Sable C., Young P., Heidenreich K.A., Boxer L.M., and Reusch J.E. 1999. Insulin-like growth Factor-I induces bcl-2 promoter through the transcription factor cAMP-response element-binding protein. *J. Biol. Chem.* 274:27529–27535.

271. Pugazhenthi S., Boras T., O'Connor D., Meintzer M.K., Heidenreich K.A., and Reusch J.E. 1999. Insulin-like growth Factor I-mediated activation of the transcription factor cAMP response element-binding protein in PC12 cells. Involvement of p38 mitogen-activated protein kinase-mediated pathway. *J. Biol. Chem.* 274:2829–2837.

272. Kemp D.M. and Habener J.F. 2001. Insulinotropic hormone glucagon-like Peptide 1 (GLP-1) activation of insulin gene promoter inhibited by p38 mitogen-activated protein kinase. *Endocrinology* 142:1179–1187.

273. Chang F., Steelman L.S., Lee J.T., Shelton J.G., Navolanic P.M., Blalock W.L., Franklin R.A., and McCubrey J.A. 2003. Signal transduction mediated by the Ras/Raf/MEK/ERK pathway from cytokine receptors to transcription factors: potential targeting for therapeutic intervention. *Leukemia* 17:1263–1293.

274. Bonni A., Brunet A., West A.E., Datta S.R., Takasu M.A., and Greenberg M.E. 1999. Cell survival promoted by the Ras-MAPK signaling pathway by transcription-dependent and -independent mechanisms [see comments]. *Science* 286:1358–1362.

275. Chang F., Steelman L.S., Shelton J.G., Lee J.T., Navolanic P.M., Blalock W.L., Franklin R., and McCubrey J.A. 2003. Regulation of cell cycle progression and apoptosis by the Ras/Raf/MEK/ERK pathway (Review). *Intl. J. Oncol.* 22:469–480.

276. Frodin M. and Gammeltoft S. 1999. Role and regulation of 90 kDa ribosomal S6 kinase (RSK) in signal transduction. *Mol. Cell Endocrinol.* 151:65–77.

277. Fang X., Yu S., Eder A., Mao M., Bast Jr. R.C., Boyd D., and Mills G.B. 1999. Regulation of BAD phosphorylation at Serine 112 by the Ras-mitogen-activated protein kinase pathway. *Oncogene* 18:6635–6640.

278. Larsen C.M., Wadt K.A., Juhl L.F., Andersen H.U., Karlsen A.E., Su M.S., Seedorf K., Shapiro L., Dinarello C.A., and Mandrup-Poulsen T. 1998. Interleukin-1 beta-induced rat pancreatic islet nitric oxide synthesis requires both the p38 and extracellular signal-regulated kinase 1/2 mitogen-activated protein kinases. *J. Biol. Chem.* 273:15294–15300.

279. Pavlovic D., Andersen N.A., Mandrup-Poulsen T., and Eizirik D.L. 2000. Activation of extracellular signal-regulated kinase (ERK)1/2 contributes to cytokine-induced apoptosis in purified rat pancreatic beta-cells. *Eur. Cytokine Netw.* 11:267–274.

280. Liu W., Chin-Chance C., Lee E.J., and Lowe Jr. W.L. 2002. Activation of phosphatidylinositol 3-kinase contributes to insulin-like growth Factor I-mediated inhibition of pancreatic beta-cell death. *Endocrinology* 143:3802–3812.

281. Dudek H., Datta S.R., Franke T.F., Birnbaum M.J., Yao R., Cooper G.M., Segal R.A., Kaplan D.R., and Greenberg M.E. 1997. Regulation of neuronal survival by the serine-threonine protein kinase Akt [see comments]. *Science* 275:661–665.

282. Datta S.R., Brunet A., and Greenberg M.E. 1999. Cellular survival: a play in three Akts. *Genes Dev.* 13:2905–2927.

283. Alessi D.R., James S.R., Downes C.P., Holmes A.B., Gaffney P.R., Reese C.B., and Cohen P. 1997. Characterization of a 3-phosphoinositide-dependent protein kinase which phosphorylates and activates protein kinase Balpha. *Curr. Biol.* 7:261–269.

284. Alessi D.R., Deak M., Casamayor A., Caudwell F.B., Morrice N., Norman D.G., Gaffney P., Reese C.B., MacDougall C.N., Harbison D. et al. 1997. 3-Phosphoinositide-dependent protein kinase-1 (PDK1): structural and functional homology with the Drosophila DSTPK61 kinase. *Curr. Biol.* 7:776–789.

285. Toker A. and Newton A.C. 2000. Akt/protein kinase B is regulated by autophosphorylation at the hypothetical PDK-2 site. *J. Biol. Chem.* 275:8271–8274.

286. Datta S.R., Dudek H., Tao X., Masters S., Fu H., Gotoh Y., and Greenberg M.E. 1997. Akt phosphorylation of BAD couples survival signals to the cell-intrinsic death machinery. *Cell* 91:231–241.

287. del Peso L., Gonzalez-Garcia M., Page C., Herrera R., and Nunez G. 1997. Interleukin-3-induced phosphorylation of BAD through the protein kinase Akt. *Science* 278:687–689.

288. Cardone M.H., Roy N., Stennicke H.R., Salvesen G.S., Franke T.F., Stanbridge E., Frisch S., and Reed J.C. 1998. Regulation of cell death protease Caspase-9 by phosphorylation [see comments]. *Science* 282:1318–1321.

289. Brunet A., Bonni A., Zigmond M.J., Lin M.Z., Juo P., Hu L.S., Anderson M.J., Arden K.C., Blenis J., and Greenberg M.E. 1999. Akt promotes cell survival by phosphorylating and inhibiting a Forkhead transcription factor. *Cell* 96:857–868.

290. Aikin R., Rosenberg L., and Maysinger D. 2000. Phosphatidylinositol 3-kinase signaling to Akt mediates survival in isolated canine islets of Langerhans. *Biochem. Biophys. Res. Commun.* 277:455–461.

291. Contreras J.L., Smyth C.A., Bilbao G., Young C.J., Thompson J.A., and Eckhoff D.E. 2002. Simvastatin induces activation of the serine-threonine protein kinase AKT and increases survival of isolated human pancreatic islets. *Transplantation* 74:1063–1069.

292. Bernal-Mizrachi E., Wen W., Stahlhut S., Welling C.M., and Permutt M.A. 2001. Islet beta cell expression of constitutively active Akt1/PKB alpha induces striking hypertrophy, hyperplasia, and hyperinsulinemia. *J. Clin. Invest.* 108:1631–1638.

293. Garofalo R.S., Orena S.J., Rafidi K., Torchia A.J., Stock J.L., Hildebrandt A.L., Coskran T., Black S.C., Brees D.J., Wicks J.R. et al. 2003. Severe diabetes, age-dependent loss of adipose tissue, and mild growth deficiency in mice lacking Akt2/PKB beta. *J. Clin. Invest.* 112:197–208.

294. Haneda M., Sugimoto T., and Kikkawa R. 1999. Mitogen-activated protein kinase phosphatase: a negative regulator of the mitogen-activated protein kinase cascade. *Eur. J. Pharmacol.* 365:1–7.

295. Franklin C.C. and Kraft A.S. 1997. Conditional expression of the mitogen-activated protein kinase (MAPK) phosphatase MKP-1 preferentially inhibits p38 MAPK and stress-activated protein kinase in U937 cells. *J. Biol. Chem.* 272:16917–16923.

296. Camps M., Nichols A., and Arkinstall S. 2000. Dual specificity phosphatases: a gene family for control of MAP kinase function. *FASEB J.* 14:6–16.

297. Keyse S.M. 2000. Protein phosphatases and the regulation of mitogen-activated protein kinase signalling. *Curr. Opin. Cell Biol.* 12:186–192.

298. Franklin C.C., Srikanth S., and Kraft A.S. 1998. Conditional expression of mitogen-activated protein kinase phosphatase-1, MKP-1, is cytoprotective against UV-induced apoptosis. *Proc. Natl. Acad. Sci. U.S.A.* 95:3014–3019.

299. Liao Q., Guo J., Kleeff J., Zimmermann A., Buchler M.W., Korc M., and Friess H. 2003. Down-regulation of the dual-specificity phosphatase MKP-1 suppresses tumorigenicity of pancreatic cancer cells. *Gastroenterology* 124:1830–1845.

300. Thomas D., Yang H., Boffa D.J., Ding R., Sharma V.K., Lagman M., Li B., Hering B., Mohanakumar T., Lakey J. et al. 2002. Proapoptotic Bax is hyperexpressed in isolated human islets compared with anti-apoptotic Bcl-2. *Transplantation* 74:1489–1496.

301. Xiang J., Chao D.T., and Korsmeyer S.J. 1996. BAX-induced cell death may not require Interleukin 1 beta-converting enzyme-like proteases. *Proc. Natl. Acad. Sci. U.S.A.* 93:14559–14563.

302. Miyashita T. and Reed J.C. 1995. Tumor suppressor p53 is a direct transcriptional activator of the human bax gene. *Cell* 80:293–299.

303. Mitchell K.O., Ricci M.S., Miyashita T., Dicker D.T., Jin Z., Reed J.C., and El Deiry W.S. 2000. Bax is a transcriptional target and mediator of c-myc-induced apoptosis. *Cancer Res.* 60:6318–6325.

304. Bentires-Alj M., Dejardin E., Viatour P., Van Lint C., Froesch B., Reed J.C., Merville M.P., and Bours V. 2001. Inhibition of the NF-kappa B transcription factor increases Bax expression in cancer cell lines. *Oncogene* 20:2805–2813.

305. Farilla L., Bulotta A., Hirshberg B., Li C.S., Khoury N., Noushmehr H., Bertolotto C., Di Mario U., Harlan D.M., and Perfetti R. 2003. Glucagon-like Peptide 1 inhibits cell apoptosis and improves glucose responsiveness of freshly isolated human islets. *Endocrinology* 144:5149–5158.

306. Hanke J. 2001. Apoptosis in cultured rat islets of Langerhans and occurrence of Bcl-2, Bak, Bax, Fas and Fas ligand. *Cells Tissues Organs* 169:113–124.

307. Okubo Y., Blakesley V.A., Stannard B., Gutkind S., and Le Roith D. 1998. Insulin-like growth Factor-I inhibits the stress-activated protein kinase/c-Jun N-terminal kinase. *J. Biol. Chem.* 273:25961–25966.

308. Mabley J.G., Belin V., John N., and Green I.C. 1997. Insulin-like growth Factor I reverses interleukin-1beta inhibition of insulin secretion, induction of nitric oxide synthase and cytokine-mediated apoptosis in rat islets of Langerhans. *FEBS Lett.* 417:235–238.

309. Johansson U., Olsson A., Gabrielsson S., Nilsson B., and Korsgren O. 2003. Inflammatory mediators expressed in human islets of Langerhans: implications for islet transplantation. *Biochem. Biophys. Res. Commun.* 308: 474–479.

310. Schmied B.M., Ulrich A., Matsuzaki H., Ding X., Ricordi C., Weide L., Moyer M.P., Batra S.K., Adrian T.E., and Pour P.M. 2001. Transdifferentiation of human islet cells in a long-term culture. *Pancreas* 23:157–171.

311. Wang R., Li J., and Rosenberg L. 2001. Factors mediating the transdifferentiation of islets of Langerhans to duct epithelial-like structures. *J. Endocrinol.* 171:309–318.

312. Valentinis B., Morrione A., Peruzzi F., Prisco M., Reiss K., and Baserga R. 1999. Anti-apoptotic signaling of the IGF-I receptor in fibroblasts following loss of matrix adhesion. *Oncogene* 18:1827–1836.

313. Kulik G., Klippel A., and Weber M.J. 1997. anti-apoptotic signalling by the insulin-like growth Factor I receptor, phosphatidylinositol 3-kinase, and Akt. *Mol. Cell Biol.* 17:1595–1606.

314. Kulik G. and Weber M.J. 1998. Akt-dependent and -independent survival signaling pathways utilized by insulin-like growth Factor I. *Mol. Cell Biol.* 18:6711–6718.

315. Hill D.J., Petrik J., and Arany E. 1998. Growth factors and the regulation of fetal growth. *Diabetes Care* 21(Suppl. 2):B60–B69.

316. Hill D.J., Strutt B., Arany E., Zaina S., Coukell S., and Graham C.F. 2000. Increased and persistent circulating insulin-like growth Factor II in neonatal transgenic mice suppresses developmental apoptosis in the pancreatic islets. *Endocrinology* 141:1151–1157.

317. Hill D.J., Petrik J., Arany E., McDonald T.J., and Delovitch T.L. 1999. Insulin-like growth factors prevent cytokine-mediated cell death in isolated islets of Langerhans from pre-diabetic non-obese diabetic mice. *J. Endocrinol.* 161:153–165.

318. Alessi D.R. and Downes C.P. 1998. The role of PI 3-kinase in insulin action. *Biochem. Biophys. Acta* 1436:151–164.

319. Draznin B., Goodman M., Leitner J.W., and Sussman K.E. 1986. Feedback inhibition of insulin on insulin secretion in isolated pancreatic islets. *Endocrinology* 118:1054–1058.

320. Elahi D., Nagulesparan M., Hershcopf R.J., Muller D.C., Tobin J.D., Blix P.M., Rubenstein A.H., Unger R.H., and Andres R. 1982. Feedback inhibition of insulin secretion by insulin: relation to the hyperinsulinemia of obesity. *N. Engl. J. Med.* 306:1196–1202.

321. Clayton H., Turner J., Swift S., James R., and Bell P. 2001. Supplementation of islet culture medium with insulin may have a beneficial effect on islet secretory function. *Pancreas* 22:72–74.

322. Farilla L., Hui H., Bertolotto C., Kang E., Bulotta A., Di Mario U., and Perfetti R. 2002. Glucagon-like Peptide-1 promotes islet cell growth and inhibits apoptosis in Zucker diabetic rats. *Endocrinology* 143:4397–4408.

323. Hui H., Nourparvar A., Zhao X., and Perfetti R. 2003. Glucagon-like Peptide-1 inhibits apoptosis of insulin-secreting cells via a cyclic 5'-adenosine monophosphate-dependent protein kinase A- and a phosphatidylinositol 3-kinase-dependent pathway. *Endocrinology* 144:1444–1455.

324. Goke R., Fehmann H.C., Linn T., Schmidt H., Krause M., Eng J., and Goke B. 1993. Exendin-4 is a high potency agonist and truncated exendin-(9-39)-amide an antagonist at the glucagon-like Peptide 1-(7-36)-amide receptor of insulin-secreting beta-cells. *J. Biol. Chem.* 268:19650–19655.

325. Pierucci D., Cicconi S., Bonini P., Ferrelli F., Pastore D., Matteucci C., Marselli L., Marchetti P., Ris F., Halban P. et al. 2001. NGF-withdrawal induces apoptosis in pancreatic beta cells in vitro. *Diabetologia* 44:1281–1295.

326. Rosenbaum T., Vidaltamayo R., Sanchez-Soto M.C., Zentella A., and Hiriart M. 1998. Pancreatic beta cells synthesize and secrete nerve growth factor. *Proc. Natl. Acad. Sci. U.S.A.* 95:7784–7788.

327. Scharfmann R., Tazi A., Polak M., Kanaka C., and Czernichow P. 1993. Expression of functional nerve growth factor receptors in pancreatic beta-cell lines and fetal rat islets in primary culture. *Diabetes* 42:1829–1836.

328. Kanaka-Gantenbein C., Dicou E., Czernichow P., and Scharfmann R. 1995. Presence of nerve growth factor and its receptors in an in vitro model of islet cell development: implication in normal islet morphogenesis. *Endocrinology* 136:3154–3162.

329. Fenjves E.S., Ochoa M.S., Cabrera O., Mendez A.J., Kenyon N.S., Inverardi L., and Ricordi C. 2003. Human, nonhuman primate, and rat pancreatic islets express erythropoietin receptors. *Transplantation* 75:1356–1360.

330. Porter S.E., Sorenson R.L., Dann P., Garcia-Ocana A., Stewart A.F., and Vasavada R.C. 1998. Progressive pancreatic islet hyperplasia in the islet-targeted, parathyroid hormone-related protein-overexpressing mouse. *Endocrinology* 139:3743–3751.

331. Garcia-Ocana A., Takane K.K., Reddy V.T., Lopez-Talavera J.C., Vasavada R.C., and Stewart A.F. 2003. Adenovirus-mediated hepatocyte growth factor expression in mouse islets improves pancreatic islet transplant performance and reduces beta cell death. *J. Biol. Chem.* 278:343–351.

332. Kolb H. and Burkart V. 1999. Nicotinamide in Type 1 diabetes. Mechanism of action revisited. *Diabetes Care* 22(Suppl. 2):B16–B20.

333. Smith S. 2001. The world according to PARP. *Trends Biochem. Sci.* 26:174–179.

334. Masutani M., Suzuki H., Kamada N., Watanabe M., Ueda O., Nozaki T., Jishage K., Watanabe T., Sugimoto T., Nakagama H. et al. 1999. Poly(ADP-ribose) polymerase gene disruption conferred mice resistant to streptozotocin-induced diabetes. *Proc. Natl. Acad. Sci. U.S.A.* 96:2301–2304.

335. Wurzer G., Herceg Z., and Wesierska-Gadek J. 2000. Increased resistance to anticancer therapy of mouse cells lacking the poly(ADP-ribose) polymerase attributable to up-regulation of the multidrug resistance gene product P-glycoprotein. *Cancer Res.* 60:4238–4244.

336. Chiarugi A. 2002. Poly(ADP-ribose) polymerase: killer or conspirator? The 'suicide hypothesis' revisited. *Trends Pharmacol. Sci.* 23:122–129.

337. Burkart V., Blaeser K., and Kolb H. 1999. Potent beta-cell protection in vitro by an isoquinolinone-derived PARP inhibitor. *Horm. Metab. Res.* 31:641–644.

338. Hoorens A. and Pipeleers D. 1999. Nicotinamide protects human beta cells against chemically-induced necrosis, but not against cytokine-induced apoptosis. *Diabetologia* 42:55–59.

339. Saldeen J. and Welsh N. 1998. Nicotinamide-induced apoptosis in insulin producing cells is associated with cleavage of poly(ADP-ribose) polymerase. *Mol. Cell Endocrinol.* 139:99–107.

340. Virag L. and Szabo C. 2001. Purines inhibit poly(ADP-ribose) polymerase activation and modulate oxidant-induced cell death. *FASEB J.* 15:99–107.

341. Tentori L., Balduzzi A., Portarena I., Levati L., Vernole P., Gold B., Bonmassar E., and Graziani G. 2001. Poly (ADP-ribose) polymerase inhibitor increases apoptosis and reduces necrosis induced by a DNA minor groove binding methyl sulfonate ester. *Cell Death. Differ.* 8:817–828.

342. Rosenthal D.S., Simbulan-Rosenthal C.M., Liu W.F., Velena A., Anderson D., Benton B., Wang Z.Q., Smith W., Ray R., and Smulson M.E. 2001. PARP determines the mode of cell death in skin fibroblasts, but not keratinocytes, exposed to sulfur mustard. *J. Invest. Dermatol.* 117:1566–1573.

343. Filipovic D.M., Meng X., and Reeves W.B. 1999. Inhibition of PARP prevents oxidant-induced necrosis but not apoptosis in LLC-PK1 cells. *Am. J. Physiol.* 277:F428–F436.

344. Moroni F., Meli E., Peruginelli F., Chiarugi A., Cozzi A., Picca R., Romagnoli P., Pellicciari R., and Pellegrini-Giampietro D.E. 2001. Poly(ADP-ribose) polymerase inhibitors attenuate necrotic but not apoptotic neuronal death in experimental models of cerebral ischemia. *Cell Death. Differ.* 8:921–932.

345. Halappanavar S.S., Rhun Y.L., Mounir S., Martins L.M., Huot J., Earnshaw W.C., and Shah G.M. 1999. Survival and proliferation of cells expressing caspase-uncleavable Poly(ADP-ribose) polymerase in response to death-inducing DNA damage by an alkylating agent. *J. Biol. Chem.* 274:37097–37104.

346. Lemaire C., Andreau K., Souvannavong V., and Adam A. 1998. Inhibition of caspase activity induces a switch from apoptosis to necrosis. *FEBS Lett.* 425:266–270.

347. Los M., Mozoluk M., Ferrari D., Stepczynska A., Stroh C., Renz A., Herceg Z., Wang Z.Q., and Schulze-Osthoff K. 2002. Activation and caspase-mediated Inhibition of PARP: a molecular switch between fibroblast necrosis and apoptosis in death receptor signaling. *Mol. Biol. Cell* 13:978–988.

348. Tafani M., Schneider T.G., Pastorino J.G., and Farber J.L. 2000. Cytochrome c-dependent activation of Caspase-3 by tumor necrosis factor requires induction of the mitochondrial permeability transition. *Am. J. Pathol.* 156:2111–2121.

349. Chang L.K. and Johnson Jr. E.M. 2002. Cyclosporin A inhibits caspase-independent death of NGF-deprived sympathetic neurons: a potential role for mitochondrial permeability transition. *J. Cell Biol.* 157:771–781.

350. Hirsch T., Marchetti P., Susin S.A., Dallaporta B., Zamzami N., Marzo I., Geuskens M., and Kroemer G. 1997. The apoptosis-necrosis paradox. Apoptogenic proteases activated after mitochondrial permeability transition determine the mode of cell death. *Oncogene* 15:1573–1581.

351. Borsello T., Clarke P.G., Hirt L., Vercelli A., Repici M., Schorderet D.F., Bogousslavsky J., and Bonny C. 2003. A peptide inhibitor of c-Jun N-terminal kinase protects against excitotoxicity and cerebral ischemia. *Nat. Med.* 9:1180–1186.

352. Bennett B.L., Sasaki D.T., Murray B.W., O'Leary E.C., Sakata S.T., Xu W., Leisten J.C., Motiwala A., Pierce S., Satoh Y. et al. 2001. SP600125, an anthrapyrazolone inhibitor of Jun N-terminal kinase. *Proc. Natl. Acad. Sci. U.S.A.* 98:13681–13686.

353. Vaishnav D., Jambal P., Reusch J.E., and Pugazhenthi S. 2003. SP600125, an inhibitor of c-jun N-terminal kinase, activates CREB by a p38 MAPK-mediated pathway. *Biochem. Biophys. Res. Commun.* 307:855–860.

354. Jhala U.S., Canettieri G., Screaton R.A., Kulkarni R.N., Krajewski S., Reed J., Walker J., Lin X., White M., and Montminy M. 2003. cAMP promotes pancreatic beta-cell survival via CREB-mediated induction of IRS2. *Genes Dev.* 17:1575–1580.

355. Sawada M., Hayes P., and Matsuyama S. 2003. Cytoprotective membrane-permeable peptides designed from the Bax-binding domain of Ku70. *Nat. Cell Biol.* 5:352–357.

356. Nagata N.A., Inoue K., and Tabata Y. 2002. Co-culture of extracellular matrix suppresses the cell death of rat pancreatic islets. *J. Biomater. Sci. Polym. Ed.* 13:579–590.

357. Weber M., Deng S., Kucher T., Shaked A., Ketchum R.J., and Brayman K.L. 1997. Adenoviral transfection of isolated pancreatic islets: a study of programmed cell death (apoptosis) and islet function. *J. Surg. Res.* 69:23–32.

358. Giannoukakis N., Mi Z., Gambotto A., Eramo A., Ricordi C., Trucco M., and Robbins P. 1999. Infection of intact human islets by a lentiviral vector. *Gene Ther.* 6:1545–1551.

359. Liu F. and Huang L. 2002. Development of non-viral vectors for systemic gene delivery. *J. Control Release* 78:259–266.

360. Rabinovitch A., Suarez-Pinzon W., Strynadka K., Ju Q., Edelstein D., Brownlee M., Korbutt G.S., and Rajotte R.V. 1999. Transfection of human pancreatic islets with an anti-apoptotic gene (bcl-2) protects beta-cells from cytokine-induced destruction. *Diabetes* 48:1223–1229.

361. Liu Y., Rabinovitch A., Suarez-Pinzon W., Muhkerjee B., Brownlee M., Edelstein D., and Federoff H.J. 1996. Expression of the bcl-2 gene from a defective HSV-1 amplicon vector protects pancreatic beta-cells from apoptosis. *Hum. Gene Ther.* 7:1719–1726.

362. Grey S.T., Longo C., Shukri T., Patel V.I., Csizmadia E., Daniel S., Arvelo M.B., Tchipashvili V., and Ferran C. 2003. Genetic engineering of a suboptimal islet graft with a20 preserves Beta cell mass and function. *J. Immunol.* 170:6250–6256.

363. Giannoukakis N., Mi Z., Rudert W.A., Gambotto A., Trucco M., and Robbins P. 2000. Prevention of beta cell dysfunction and apoptosis activation in human islets by adenoviral gene transfer of the insulin-like growth Factor I. *Gene Ther.* 7:2015–2022.

364. Embury J., Klein D., Pileggi A., Ribeiro M., Jayaraman S., Molano R.D., Fraker C., Kenyon N., Ricordi C., Inverardi L. et al. 2001. Proteins linked to a protein transduction domain efficiently transduce pancreatic islets. *Diabetes* 50:1706–1713.

365. Rehman K.K., Bertera S., Bottino R., Balamurugan A.N., Mai J.C., Mi Z., Trucco M., and Robbins P.D. 2003. Protection of islets by in situ peptide-mediated transduction of the Ikappa B kinase inhibitor Nemo-binding domain peptide. *J. Biol. Chem.* 278:9862–9868.

366. Ekblom P. and Timpl R. 1996. Cell-to-cell contact and extracellular matrix. A multifaceted approach emerging. *Curr. Opin. Cell. Biol.* 8:599–601.

367. Okada T.S. 1983. Recent progress in studies of the transdifferentiation of eye tissue in vitro. *Cell. Differ.* 13:177–183.

368. Rosenberg L. 1998. Clinical islet cell transplantation. Are we there yet? *Intl. J. Pancreatol.* 24:145–168.

369. Arias A.E. and Bendayan M. 1993. Differentiation of pancreatic acinar cells into duct-like cells in vitro. *Lab. Invest.* 69:518–530.

370. Beattie G.M., Rubin J.S., Mally M.I., Otonkoski T., and Hayek A. 1996. Regulation of proliferation and differentiation of human fetal pancreatic islet cells by extracellular matrix, hepatocyte growth factor, and cell-cell contact. *Diabetes* 45:1223–1228.

371. Bergmann A., Tugentman M., Shilo B.Z., and Steller H. 2002. Regulation of cell number by MAPK-dependent control of apoptosis: a mechanism for trophic survival signaling. *Dev. Cell.* 2:159–170.

372. Jiang F.X., Cram D.S., DeAizpurua H.J., and Harrison L.C. 1999. Laminin-1 promotes differentiation of fetal mouse pancreatic beta-cells. *Diabetes* 48:722–730.

373. Jiang F.X., Naselli G., and Harrison L.C. 2002. Distinct distribution of laminin and its integrin receptors in the pancreas. *J. Histochem. Cytochem.* 50:1625–1632.

374. Lucas-Clerc C., Massart C., Campion J.P., Launois B., and Nicol M. 1993. Long-term culture of human pancreatic islets in an extracellular matrix: morphological and metabolic effects. *Mol. Cell. Endocrinol.* 94:9–20.

375. Meredith Jr. J.E., Fazeli B., and Schwartz M.A. 1993. The extracellular matrix as a cell survival factor. *Mol. Biol. Cell.* 4:953–961.

376. Miralles F., Battelino T., Czernichow P., and Scharfmann R. 1998. TGF-beta plays a key role in morphogenesis of the pancreatic islets of Langerhans by controlling the activity of the matrix metalloproteinase MMP-2. *J. Cell. Biol.* 143:827–836.

377. Streuli C. 1999. Extracellular matrix remodelling and cellular differentiation. *Curr. Opin. Cell. Biol.* 11:634–640.

378. Yuan S., Rosenberg L., Paraskevas S., Agapitos D., and Duguid W.P. 1996. Transdifferentiation of human islets to pancreatic ductal cells in collagen matrix culture. *Differentiation* 61:67–75.

379. Gittes G.K., Galante P.E., Hanahan D., Rutter W.J., and Debase H.T. 1996. Lineage-specific morphogenesis in the developing pancreas: role of mesenchymal factors. *Development* 122:439–447.

380. Jamal A.M., Lipsett M., Hazrati A., Paraskevas S., Agapitos D., Maysinger D., and Rosenberg L. 2003. Signals for death and differentiation: a two-step mechanism for in vitro transformation of adult islets of Langerhans to duct epithelial structures. *Cell. Death Differ.* 10:987–996.

381. Yuan S., Paraskevas S., Duguid W.P., and Rosenberg L. 1995. Phenotypic transformation of isolated human islets in collagen matrix culture. *Transplant. Proc.* 27:3364.

382. Jamal A.M., Lipsett M., Sladek R., Hanley S., and Rosenberg L. 2005. Morphogenetic plasticity of adult human islets of Langerhans. *Cell Death Differ.* In press.

383. Hart A.W., Baeza N., Apelqvist A., and Edlund H. 2000. Attenuation of FGF signalling in mouse beta-cells leads to diabetes. *Nature* 408:864–868.

384. Lee S.E., Woo K.M., Kim S.Y., Kim H.M., Kwack K., Lee Z.H., and Kim H.H. 2002. The phosphatidylinositol 3-kinase, p38, and extracellular signal-regulated kinase pathways are involved in osteoclast differentiation. *Bone* 30:71–77.

385. Hazrati A., Jamal A.M., and Rosenberg L. 2002. The role of transforming growth factor-beta in the transdifferentiation of islets of Langerhans to duct-like epithelial structures. *Diabetes* 51:A376–A377.

386. Jaikaran E.T. and Clark A. 2001. Islet amyloid and Type 2 diabetes: from molecular misfolding to islet pathophysiology. *Biochem. Biophys. Acta.* 1537:179–203.

387. Janssen S.W., Hermus A.R., Lange W.P., Knijnenburg Q., van der Laak J.A., Sweep C.G., Martens G.J., and Verhofstad A.A. 2001. Progressive histopathological changes in pancreatic islets of Zucker diabetic fatty rats. *Exp. Clin. Endocrinol. Diabetes* 109:273–282.

388. Szynaka B., Zimnoch L., and Puchalski Z. 2002. Ultrastructural observations of intermediate cells in chronic pancreatitis. *Hepatogastroenterology* 49: 1120–1123.

389. Iovanna J.L., Lechene de la Porte P., and Dagorn J.C. 1992. Expression of genes associated with dedifferentiation and cell proliferation during pancreatic regeneration following acute pancreatitis. *Pancreas* 7:712–718.

390. Suen K.C. 1991. Cytology of head and neck tumors, liver, and pancreas. *Clin. Lab. Med.* 11:317–356.

391. Lechene de la Porte P., Iovanna J., Odaira C., Choux R., Sarles H., and Berger Z. 1991. Involvement of tubular complexes in pancreatic regeneration after acute necrohemorrhagic pancreatitis. *Pancreas* 6:298–306.

392. Brandhorst D., Brandhorst H., Zwolinski A., Nahidi F., and Bretzel R.G. 2001. Prevention of early islet graft failure by selective inducible nitric oxide synthase inhibitors after pig to nude rat intraportal islet transplantation. *Transplantation* 71:179–184.

393. Shapiro A.M., Hao E.G., Lakey J.R., Yakimets W.J., Churchill T.A., Mitlianga P.G., Papadopoulos G.K., Elliott J.F., Rajotte R.V., and Kneteman N.M. 2001. Novel approaches toward early diagnosis of islet allograft rejection. *Transplantation* 71:1709–1718.

394. Murakami M., Satou H., Kimura T., Kobayashi T., Yamaguchi A., Nakagawara G., and Iwata H. 2000. Effects of micro-encapsulation on morphology and endocrine function of cryopreserved neonatal porcine islet-like cell clusters. *Transplantation* 70:1143–1148.

395. Lundgren T., Bennet W., Tibell A., Soderlund J., Sundberg B., Song Z., Elgue G., Harrison R., Richards A., White D.J. et al. 2001. Soluble complement Receptor 1 (TP10) preserves adult porcine islet morphology after intraportal transplantation into cynomolgus monkeys. *Transplant Proc.* 33:725.

396. Jaeger C., Wohrle M., Federlin K., and Bretzel R.G. 1995. Pancreatic islet xenografts at two different transplantation sites (renal subcapsular versus intraportal): comparison of graft survival and morphology. *Exp. Clin. Endocrinol. Diabetes* 103(Suppl 2.):123–128.

397. Rutzky L., Kloc M., Bilinski S., Phan T., Zhang H., Stepkowski S.M., and Katz S. 2001. Microgravity culture conditions decrease immunogenicity but maintain excellent morphology of pancreatic islets. *Transplant Proc.* 33:388.

398. Ling Z., Van de Casteele M., Eizirik D.L., and Pipeleers D.G. 2000. Interleukin-1 beta-induced alteration in a beta-cell phenotype can reduce cellular sensitivity to conditions that cause necrosis but not to cytokine-induced apoptosis. *Diabetes* 49:340–345.

399. Pour P.M. 1997. The role of Langerhans islets in pancreatic ductal adenocarcinoma. *Front. Biosci.* 2:D271–D282.

400. Pour P.M., Weide L., Liu G., Kazakoff K., Scheetz M., Toshkov I., Ikematsu Y., Fienhold M.A., and Sanger W. 1997. Experimental evidence for the origin of ductal-type adenocarcinoma from the islets of Langerhans. *Am. J. Pathol.* 150:2167–2180.

401. Pour P. 1978. Islet cells as a component of pancreatic ductal neoplasms. I. Experimental study: ductular cells, including islet cell precursors, as primary progenitor cells of tumors. *Am. J. Pathol.* 90:295–316.

402. Pour P.M. and Schmied B.M. 1999. One thousand faces of Langerhans islets. *Intl. J. Pancreatol.* 25:181–193.

403. Permert J., Mogaki M., Andren-Sandberg A., Kazakoff K., and Pour P.M. 1992. Pancreatic mixed ductal-islet tumors. Is this an entity? *Intl. J. Pancreatol.* 11:23–29.

404. Pour P.M. and Schmied B. 1999. The link between exocrine pancreatic cancer and the endocrine pancreas. *Intl. J. Pancreatol.* 25:77–87.

405. Lee C.S., Montebello J., Georgiou T., and Rode J. 1994. Distribution of Type IV collagen in pancreatic adenocarcinoma and chronic pancreatitis. *Intl. J. Exp. Pathol.* 75:79–83.

406. Wang Z.H., Manabe T., Ohshio G., Imamura T., Yoshimura T., Suwa H., Ishigami S., and Kyogoku T. 1994. Immunohistochemical study of heparan sulfate proteoglycan in adenocarcinomas of the pancreas. *Pancreas* 9:758–763.

407. Schmied B.M., Ulrich A., Matsuzaki H., Li C.H., and Pour P.M. 1999. In vitro pancreatic carcinogenesis. *Ann. Oncol.* 10(Suppl. 4):41–45.

408. Pour P.M., Donnelly K., and Stepan K. 1983. Modification of pancreatic carcinogenesis in the hamster model. 3. Inhibitory effect of alloxan. *Am. J. Pathol.* 110:310–314.

409. Pour P.M. and Patil K. 1983. Modification of pancreatic carcinogenesis in the hamster model. X. Effect of streptozotocin. *J. Natl. Cancer Inst.* 71:1059–1065.

410. Bell Jr. R.H., Sayers H.J., Pour P.M., Ray M.B., and McCullough P.J. 1989. Importance of diabetes in inhibition of pancreatic cancer by streptozotocin. *J. Surg. Res.* 46:515–519.

411. Bell Jr. R.H., McCullough P.J., and Pour P.M. 1988. Influence of diabetes on susceptibility to experimental pancreatic cancer. *Am. J. Surg.* 155:159–164.

412. Ishizaka S., Takeuchi H., Kimoto M., Kanda S., and Saito S. 1998. Fosfomycin, an antibiotic, possessed TGF-beta-like immunoregulatory activities. *Intl. J. Immunopharmacol.* 20:765–779.

413. Kuklin N.A., Daheshia M., Chun S., and Rouse B.T. 1998. Immunomodulation by mucosal gene transfer using TGF-beta DNA. *J. Clin. Invest.* 102:438–444.

414. McCarthy P.L. 1994. Down-regulation of cytokine action. *Baillieres Clin. Haematol.* 7:153–177.

415. Lenner R., Padilla M.L., Teirstein A.S., Gass A., and Schilero G.J. 2001. Pulmonary complications in cardiac transplant recipients. *Chest* 120:508–513.

416. Fung J.J., Jain A., Kwak E.J., Kusne S., Dvorchik I., and Eghtesad B. 2001. De novo malignancies after liver transplantation: a major cause of late death. *Liver Transpl.* 7:S109–S118.

417. Hiddemann W. 1989. What's new in malignant tumors in acquired immunodeficiency disorders? *Pathol. Res. Pract.* 185:930–934.

418. Biemer J.J. 1990. Malignant lymphomas associated with immunodeficiency states. *Ann. Clin. Lab. Sci.* 20:175–191.

419. Vallejo R., Hord E.D., Barna S.A., Santiago-Palma J., and Ahmed S. 2003. Perioperative immunosuppression in cancer patients. *J. Environ. Pathol. Toxicol. Oncol.* 22:139–146.

420. Touloumi G., Hatzakis A., Potouridou I., Milona I., Strarigos J., Katsambas A., Giraldo G., Beth-Giraldo E., Biggar R.J., Mueller N. et al. 1999. The role of immunosuppression and immune-activation in classic Kaposi's sarcoma. *Intl. J. Cancer* 82:817–821.

421. Birt D.F., Stepan K.R., and Pour P.M. 1983. Interaction of dietary fat and protein on pancreatic carcinogenesis in Syrian golden hamsters. *J. Natl. Cancer Inst.* 71:355–360.

29

PANCREATIC CANCER AND DIABETES — CELLULAR ORIGIN OF ADEONOCARCINOMA

Nicolas Moniaux, Krishan K. Pandey, Parviz M. Pour, and Surinder K. Batra

CONTENTS

29.1 INTRODUCTION

In 2003, pancreatic cancer became the fourth leading cause of cancer-related deaths in the United States.[1,2] The survival time for patients diagnosed with pancreatic cancer ranges from 3 to 6 months on an average, with a 5% chance of 5-year survival.[3] The highest cure rate occurs if the tumor is detected early and localized to the pancreas only; however, this stage of disease accounts for fewer than 20% of cases.[4] Unfortunately, the signs of early stages of pancreatic cancer are vague and often mistakenly attributed to other problems by both patients and physicians. More specific symptoms tend to develop after the tumor has grown to invade other organs or blocked the bile ducts. Patients are usually diagnosed at an advanced stage, with a high incidence of associated metastases spread all over the body.

The association between pancreatic cancer and diabetes mellitus is reported extensively.[5] However, a controversy remains as to whether pancreatic cancer is the cause of abnormal glucose metabolism or if hyperinsulinemia is a risk factor for pancreatic cancer. This important question finds its relevance in the unknown cellular origin of pancreatic cancer. The answer of this question lies in complexity of the nature of the pancreas, which is a multifunctional gland with endocrine and exocrine functions. Its exocrine function is the release of enzymes and salts that take part in digestion. Its endocrine function is the hormonal regulation of blood glucose levels. The multiplicity of the functions is associated with the multiplicity of the cells that compose the pancreas.

29.2 ANATOMY OF THE PANCREAS

The pancreas is formed from two pancreatic anlagen that bud from the gut endoderm to the surrounding splanchnic mesoderm. The extension of these two buds leads to branched structures that develop independently, forming both endocrine and exocrine tissues that end up merging to form the pancreas (Figure 29.1).

The pancreas is an elongated gland (12 to 15 cm in length), lying laterally to the rear of the upper right hand side of the abdominal cavity and is divided into a head, body, and tail. The main pancreatic duct runs through the length of the gland up to the upper duodenum and converges with the duct coming from the liver and the gallbladder. The exocrine and endocrine functions of the gland are carried out by two histologically distinct subunits. The exocrine portion (95% of the pancreas), which is organized as a tubuloalveolar gland, is composed of both the acini (85%) and the ductal epithelium cells (10%). The acini cells secrete trypsin, chymotrypsin, aminopeptidases, elastase, amylases, lipase, phospholi-

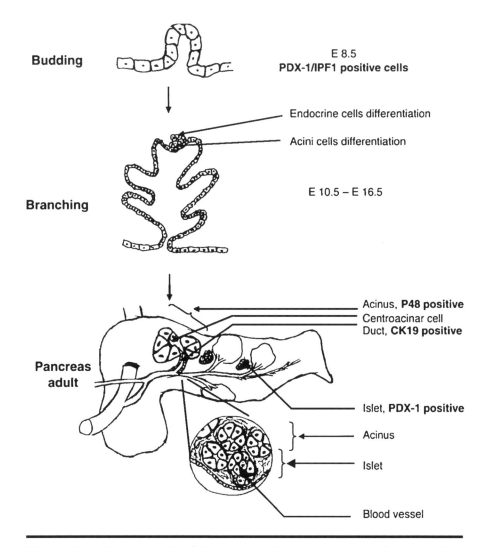

Budding   E 8.5
PDX-1/IPF1 positive cells

Endocrine cells differentiation

Acini cells differentiation

E 10.5 – E 16.5

Branching

Acinus, **P48 positive**
Centroacinar cell
Duct, **CK19 positive**

Pancreas adult

Islet, **PDX-1 positive**

Acinus

Islet

Blood vessel

Figure 29.1 Organogenesis of the pancreas. The pancreas arises from an initial bud that emerges from dorsal and ventral primordial. The bud is characterized by an early expression of the *PDX1* gene. The PDX1 expression is required for the expansion and the branching steps. Differentiation of the bud formation begins in parallel. Endocrine cells are initially found in the epithelium and migrate in the mesenchyme to aggregate forming the islets of Langerhans. The adult pancreas is composed of both endocrine and exocrine cells. The PDX1, Cytokeratine 19 (CK19), and p48 markers are expressed specifically in endocrine, ductal, and acinar cells, respectively.

pases, and nucleases. The endocrine function is performed by clusters of cells, dispersed all over the pancreas, called islets of Langerhans. Islets are separated from the tubuloalveolar components by a loose reticular capsule. The islets have a rich blood supply through an intraislet portal system. The islet cells are composed of four types of endocrine cells, with each type of cell secreting a specific hormone. The type of α-cells secrete glucagon, which is associated with insulin in glucose homeostasis in blood. It is secreted in response to the low blood glucose level and causes other body cells to release glucose in blood. The β-cells secrete insulin that is responsible for maintaining the blood glucose level. It is secreted in response to a high blood glucose level. The δ-cells and PP-cells secrete somatostatin and pancreatic polypeptide (PP), respectively. The PP cells constitute about 10% of total islet cells. The function and origin of pancreatic polypeptides are still uncertain, although the hormones may influence gastrointestinal function and promote intraislet homeostasis. Pancreatic polypeptide is released into the plasma when stimulated by the ingestion of food. PP inhibits the stimulation of gastric and pancreatic exocrine secretions.

29.3 PANCREATIC CANCER

Cancers of the pancreas are commonly identified by the site of involvement within the pancreas: duct cell carcinoma, acinar cell carcinoma, papillary mucinous carcinoma, signet ring carcinoma, adenosquamous carcinoma, undifferentiated carcinoma, mucinous carcinoma, giant cell carcinoma, mixed type (ductal-endocrine or acinar-endocrine), small cell carcinoma, cystadenocarcinoma (serous and mucinous types), unclassified, pancreatoblastoma, papillary-cystic neoplasm, mucinous cystic tumor with dysplasia, intraductal papillary mucinous tumor with dysplasia, pseudopapillary solid tumor, and endocrine carcinoma (insulinoma, glucagonoma, etc.).

29.3.1 Adenocarcinomas of the Pancreas

Up to 95% of pancreatic cancer cases are ductal cell carcinoma or adenocarcinoma. It is not clear at this time if these different pancreatic cancers are originates from unique or distinct cell subtypes, but each cancer presents a specific phenotype, genotype, and level of malignancy.

The pancreatic adenocarcinoma cells present strong resemblance to pancreatic duct cells, with cuboid architecture, expression of ductal specific markers, and differentiation into tubular structures. Therefore, it is thought that duct cells give rise to pancreatic adenocarcinoma. However, the expression of endocrine factors as well as pancreatic enzymes indicates certain plasticity in the development of this type of cancer that can originate from precursor cells within the pancreas or transdifferentiation of pancreatic cells. The most aggressive stage of pancreatic adenocarcinoma is its infiltrative form. The infiltrative adenocarcinoma is thought to arise from several pancreatic lesions such as the pancreatic intraepithelial neoplasia (PanIN),[6] mucinous cystic neoplasm,[7] and intraductal papillary mucinous neoplasm (IPMN and IPMT).

The PanIN lesions progress from flat to papillary without atypia, to papillary with atypia, and to lesions with severe atypia and are classified as PanIN-1A to PanIN-1B, PanIN-2, and PanIN-3.[6] Atypical intraductal lesions can progress to invasive adenocarcinoma.[8] PanINs are frequently found associated with pancreatic cancer.[9] More importantly, patients diagnosed with PanIN lesions usually develop pancreatic cancer,[8] and genetic abnormalities found in adenocarcinoma are also present in PanINs.

Mucinous cystadenocarcinomas (MCNs) also progress through different stages, going from mucinous cystadenoma to borderline mucinous cystic neoplasm to mucinous cystic neoplasm with *in situ* carcinoma.[10] Mucinous cystadenomas contain a single layer of mucin-producing, columnar epithelium lacking significant atypia. Borderline mucinous cystic neoplasms contain cells with moderate atypia. Mucinous cystic neoplasms with *in situ* carcinoma show significant architectural and cytological atypia.[10] If fully resected before its final stage, MCN do not recur. Fifty percent of patients with adenocarcinoma arising from MCN present a long-term survival.[11]

In a similar way, the IPMN or IPMT lesions can give rise to pancreatic adenocarcinoma; if detected before the infiltrating stage, they do not recur and do not metastasize. As in the two other classes of lesions, IPMNs progress from intraductal papillary mucinous adenoma to borderline IPMN to IPMN with *in situ* carcinoma.

The presence of numerous genetic deletions, mutations, and amplifications acquired during the transformation and progression processes have been reported. These types of alterations provide a signature to pancreatic cancer. The majority of mutations are found in *K-ras*, *p16*, *p53*, and *DPC4* genes (Table 29.1).

The K-ras mutation is found in 90 to 95% of pancreatic cancer patients. Mutation appears in a single codon, Codon 12, which changes a glycine to aspartic acid or valine.[12] K-ras is a 21 kDa monomeric membrane-bound guanosine triphosphate/guanosine diphosphate (GTP/GDP), binding the G protein and transferring signals from the membrane to the nucleus. Ras is activated by a variety of extracellular stimuli and, in turn, regulates various cellular processes like cell proliferation, differentiation, and apoptosis.[13] Mutations in *K-ras* are considered to be early events in pancreatic tumorigenesis. Mutated Ras genes encode constitutively activated proteins

Table 29.1   Mutation Causing Activation of Oncogenes and Inactivation of Tumor Suppressor Genes in Established Human Pancreatic Tumors

Gene	Tumors Examined	Mutation	Homozygous Deletion	Total % Activation/ Inactivation	References
K-ras	22	21		95	92
	82	68		83	93
P53	27	19		70	94
P16	37	14	15	78	17
DPC4	84	—	30	52	23
	27	6	—		
BRCA2	41	*4		*10	95
MMK4	92	—	2		96
	45	1	—	4	
RB	30	2		7	97

*Three out of four showed germline mutation.

that have been implicated in tumorigenesis.[14] Studies have shown that the mutated allele of K-ras is responsible for uncontrolled cell growth. The Codon 12 mutation abolishes the GTP hydrolysis property of K-ras and maintains its activation. The p16 gene code for an inhibitor of the cyclin D-cyclin dependant kinase 4 (cdk4) complexes.[15] It is a tumor-suppressor gene and loss of its function comes from homozygous deletion, mutation, or aberrant methylation of its promoter.[16] The inactivation of p16 is found in 95% of pancreatic cancer cases.[17] The p53 gene also codes for a tumor-suppressor protein that is inactivated in 60% of pancreatic adenocarcinoma.[18] Like p16, p53 regulates cell-cycling but also controls apoptosis after DNA damage. p53 is inactivated mainly by point mutation, biallelic mutation being not necessary for complete p53 lost of function. Mutation in K-ras has been shown to be associated with p53 mutation.[19] DPC4 or Smad4 is a key player of the TGFβ signaling pathway.[20,21] Smad4 is a member of the SMAD family and acts both as a transcriptional activator and repressor.[22] DPC4 is mutated or deleted in 50 to 55% of pancreatic cancers.[23,24]

In addition to the above four common genetic alterations detected in pancreatic cancer, other markers also are known to be differentially expressed, such as the MUC4 mucin. MUC4 is a O-glycoprotein belonging to the membrane-bound mucin family (Figure 29.2).[25] An aberrant expression of the MUC4 gene is associated with pancreatic cancer development and progression.[26–28] MUC4 is highly expressed in human pancreatic tumors and pancreatic tumor cell lines (70 to 80% of the pancreatic

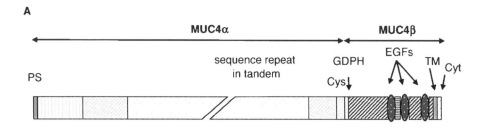

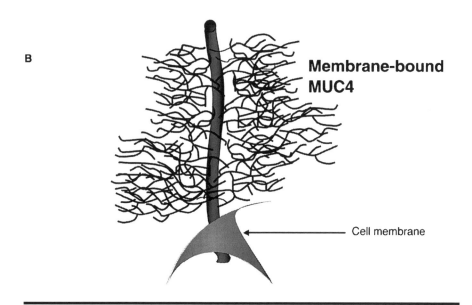

Figure 29.2   MUC4 mucin. A: A schematic representation of the MUC4 protein.
The *MUC4* RNA, 27 kb-long, codes for a heterodimeric O-glycoprotein, composed
of a mucin-type MUC4α and a membrane-bound growth factor-like MUC4β
subunits. MUC4β possesses 3 EGF-like domains that are suspected to interact
with the members of the receptor tyrosine kinase (RTK). B: Schematic representation of MUC4 as it is presented on the cell surface with an extended conformation maintained by the numerous O-glycan chains.

adenocarcinoma), but is minimally or not expressed in the normal pancreas
or chronic pancreatitis.[26,29–31] The trans-elements responsible for MUC4
overexpression have remained unclear until now. However, endogenic
MUC4 aberrant expression was shown to be dependent on cellular
environment[32] and serum factors, such as retionoids.[32,33] MUC4 is associated
with an increase of tumorigenicity and metastasis of pancreatic cancer
cells.[34] As for the regulation of its expression, the mechanistic aspects

associated with MUC4 functions are still unknown, but might be related to its heterodimerization with the proto-oncogene HER2.[35,36]

The aforementioned genetic alterations are already detected in the early precursor of pancreatic adenocarcinoma; however, each epigenetic mutation presents a specific temporal pattern. In the PanIN lesions, K-ras mutation is characterized as early events already detected in PanIN-1A lesion.[37] The loss of p16 is an intermediate event (detected for PanIN-2),[37,38] and the loss of p53[39] and DPC4[39,40] are late occurring events in PanIN-3 lesions. Abnormality within K-ras,[41] p16,[42] and p53[42,43] are also detected in IPMN/IPMT; however, their frequency and temporal distribution vary from those detected in PanIN. Inactivation of p53 and DPC4/Smad4 occur less frequently in IPMNs than in infiltrating pancreatic ductal adenocarcinoma. The pattern of mucin expression is also specific to the lesion and different from PanIN to IPMN/IPMT. MUC4 is expressed by metaplastic ducts and its expression increases with a higher grade in PanINs.[27,28] Similarly, MUC1 expression also increases with the progression of PanIN lesions, whereas MUC2 expression remains unchanged.[44] MUC2 and MUC5AC levels were shown to be high in IPMT compared to adenocarcinomas.[45,46] Terris et al.[47] showed altered expression of MUC genes (overexpression of MUC2 and MUC5AC) in IPMTs and correlated the expression profile with types and lesions of IPMTs. The main difference between PanIN and IPMT is an increase of MUC2 expression associated with a decrease of MUC1 expression in IPMT vs. PanIN. In both cases, MUC4 is highly aberrantly expressed. In MCNs, p53 and K-ras mutations are associated with severe epithelial dysplasia, but not in early MCNs.[48] The loss of DPC4 is detected at a late stage (invasive cystadenocarcinoma).[49]

If PanIN, IPMN (IPMT), and MCN lesions can give rise to invasive adenocarcinoma, they are histologically and genetically distinct during their progression. At this point, it is unknown if this difference is related to a distinct cellular type implicated in the initiation process of the transformation.

29.3.2 Familial Pancreatic Cancer

Numerous reports suggest that there is an inherited component to pancreatic cancer in 3 to 16% of the cases. There are three main classes for familial pancreatic cancer:

1. Patients with germline mutation in genes known to give rise to cancer
2. Patients with familial pancreatitis
3. Patients with two or more first-degree relatives with pancreatic cancer, but without any association to any cancer syndrome[50]

For the germline mutations, several known syndromes have been associated with an increased risk for pancreatic cancer:

- BRCA2 germline mutation (3.5- to 10-fold increased risk)[51]
- FAMMM (familial atypical multiple mole melanoma) syndrome with a p16 germline mutation (12- to 20-fold increased risk)[52]
- Peutz-Jeghers syndrome with mutation in the Serine/Threonine Kinase 11 (STK 11) (130-fold increased risk)[53,54]
- HNPCC (hereditary nonpolyposis colorectal cancer) with a mutation in one of the DNA repair genes *bMSH1* or *bMSH2*[55]

The hereditary pancreatitis is also caused by a germline autosomal dominant mutation of the cationic trypsinogen gene, Protease Serine 1 (PRSS).[56] Mutation in this gene represents 70% of the cases. Another mutation in the gene Kazal Type 1 (SPINK 1) was also identified as a cause for familial pancreatitis. Patients with familial pancreatitis have a 40% increased chance of developing pancreatic cancer before the age of 70 years old.[57] Many authors define *familial pancreatic cancer* as occurring in a patient with two or more family members diagnosed with pancreatic cancer, but without another hereditary cancer syndrome.[58] In a recent study, it was estimated that the risk of developing pancreatic cancer for an individual with two first-degree relatives diagnosed with pancreatic cancer was increased to 18-fold and could go up to an increase of 57-fold when three first-degree relatives of the family were diagnosed.[58]

29.3.3   Endocrine Cancer of the Pancreas

Endocrine cancers of the pancreas are extremely rare, account for less than 1% of all pancreatic neoplasms, and are often curable tumors. Endocrine cancers of the pancreas are also called islet tumors. Islet tumors may either be functional (produce one or more hormones) or nonfunctional. The majority of functioning tumors that produce insulin are benign; however, 90% of nonfunctioning tumors are malignant. Nonfunctional islet cell cancers produce symptoms from tumor bulk or metastatic dissemination. Islet cell carcinoma can be regrouped into different neoplasms depending on the type of the islet cells that initiated the tumor formation. When these tumors are functional, they produce unique metabolic and clinical characteristics. Therefore, these types of tumors — insulinoma, gastrinoma, glucagonoma, somatostatinoma, etc. — are usually classified by the type of hormone that they synthesize. Of all these tumors, insulinoma and gastrinoma are the most common. Studies involving the genes most frequently altered in exocrine cancer (i.e., *p53*, *K-ras*, *p16*, and *DPC4*) show that the endocrine cancers arise from specific pathways unrelated to ductal cancers.[59] Mutations in K-ras and p53 are rare and mutations or alterations in p16 and DPC4 are not detected. Four genetic syndromes are reported as associated with endocrine pancreatic tumors, but with a frequency that varies with the type of neoplasias. These syndromes are the multiple endocrine neoplasia Type I syndrome (MEN-1), von Hippel-Lindau (VHL), von Recklinghausen's disease (NF-1), and tuberous sclerosis.[60–62] The MEN-1 syndrome harbors the strongest association with the development of pancreatic cancer. An average of 30 to 50% of patients with MEN-1 syndrome develop nonfunctioning pancreatic endocrine tumors.[63] Twenty-five percent develop gastrinoma, but only 7% develop insulinoma. The MEN-1 is an autosomal dominant inherited genetic disease, which is characterized by parathyroid hyperplasia. MEN-1 syndrome develops from both germline[64] or sporadic[65] mutations and germline or sporadic loss of heterozygosity (LOH)[66,67] of the MEN-1 gene coding for the menin. No real function is known for this gene that acts as a tumor suppressor. Insulinomas are the most common type of endocrine cancer. The vast majority of insulinomas remain benign, however, around 6% become malignant and are associated with local invasion of the surrounding tissues and lymph nodes and liver metastases.[68] In contrast, most gastrinomas are malignant, however they progress slowly. Survival for patients with gastrinoma is long compared to pancreatic adenocarcinoma.[69]

29.3.4 Acinar Cell Carcinoma

Acinar cell carcinomas are malignant tumors with acinar differentiation (production of digestive enzymes trypsin, lipase, chymotrypsin, and, in some cases, amylase).[70] Acinar cell pancreatic cancer is not a common type of cancer. Although acinar cells comprise a significant majority of the exocrine portion of the pancreas, they only account for 1% of cancers that arise from the exocrine pancreas. They can be large when found and can have already metastasized. The mean survival for patients with resected acinar cell carcinoma is 18 months.[71] Genetically, *K-ras* mutation is rare; *p53*, *p16*, and *DPC4* alterations have not been detected so far.[59] Up to 70% of the cases possess a loss of heterozygosity on the following chromosomes — 1p, 4q, 17p, 11q, 13q, 15q, and 16q.[59,72] Twenty-five percent of the cases show alteration in the APC/β-catenin pathway.[73]

29.4 DIABETES MELLITUS AND PANCREATIC CANCER

Pancreatic cancer is not a unique disease, but regroup a large family of neoplasm. These neoplasms have distinct phenotypes as well as genotypes. This multiplicity is one of the main reasons for the actual lack of knowledge regarding the cellular origin of pancreatic cancer. Moreover, as mentioned in the introduction, even if endocrine cancers of the pancreas represent only 1 to 3% of the cases, pancreatic cancer is closely related with diabetes mellitus. The apparent link between pancreatic cancer and diabetes mellitus maintains the hypothesis for the islet origin of pancreatic cancer. Even though the islet cells in the pancreas constitute just 2% of the volume of the total pancreas, they are involved in crucial functions in the body. Any deviation from normal course during the differentiation process and events resulting in decreased number of islet cells results in diseases. The major disease associated with islet cells, specifically β-cells, is diabetes. Although diabetes was known to humans for centuries, its association with the pancreas became clear much later. Bouchardat (1845) first suggested that diabetes could be a disease of the pancreas. The experimental proof came in 1889 from Von Mering and Minkowski, when they removed the pancreas from dogs and these dogs developed severe diabetes. In 1893, Laguesse confirmed that the defects in the islet of Langerhans, affecting the endocrine function of the pancreas, cause diabetes.

Diabetes is a fairly common disease in advanced countries. In the United States about 6% of its population (approximately 17 million people) suffers from diabetes. In reality, this number is much higher as a number of people are unaware of their diabetes. In the diabetic condition, the patient does not produce sufficient insulin or does not properly use insulin. Insulin is a hormone produced by β-cells in the pancreas, which is needed to convert sugar, starches, and other food into the energy needed for daily life. Insulin is the major hormone capable of lowering the blood glucose level and maintaining it at a specific level by counteracting the effects of hyperglycemic hormones (e.g., glucagon or epinephrin). In a healthy adult, the blood glucose level is maintained between 80 to 130 mg/dl, despite considerable variations in individual diets. The failure of islet cells to sense the variation in the blood glucose level and to modulate the insulin biosynthesis and release depending upon the requirement leads to chronic hyperglycemia.

There are three types of diabetes. In Type I diabetes, the number of islet β-cells falls in the body, causing an insufficient production of insulin. The patient with Type 2 diabetes cannot use the insulin made by their β-cells efficiently, resulting in insulin resistance. The compensatory mechanisms activated in the β-cells to secrete more insulin are not enough to meet the requirement and hence the blood glucose level goes up. The third type of diabetic condition, gestational diabetes, is observed in women when they are pregnant. This condition affects about 135,000 women in the United States.

Type 1 diabetes mellitus (T1DM) is an old disease and diseases having similar symptoms have been described in ancient Greek and Egyptian writings. Type 1 diabetes, previously known as insulin-dependent diabetes mellitus (IDDM) or juvenile diabetes, is an autoimmune disease involving the destruction of β-cells of the islets of Langerhans in the pancreas. The exact mechanisms by which this disease progresses, however, are not well understood. Recent studies using non-obese diabetes (NOD) mice have established the role of T-cell mediated autoimmunity in the development of T1DM. Defects in islet differentiation, increased β-cells apoptosis, and a higher number of immature islets were present in NOD mice at the time of birth, indicating the possibility that abnormal islet development might somehow be involved with autoimmunity. T1DM is primarily a T-lympho-cyte-mediated autoimmune disease. NOD mice having no T-cells (genet-ically athymic), or even the mice from which the thymus was removed, at birth do not develop T1DM.

The presence of susceptibility genes, diet, and environmental factors also play an important role in the development of T1DM. Immunological studies have confirmed the presence of a high percentage of T-cells and macrophages in the leukocytes infiltrating in islet cells. Furthermore, the injection of T-cells from a diabetic donor in NOD mice causes the recipient to become diabetic. The NOD mice show insulitis at 3 to 5 weeks of age, which by 12 to 15 weeks results in diabetes. At this stage, the number of β-cells is low in these mice. It is not clear what causes the islet cells of NOD mice to become a target of its own immune system. We do not know how the T-cells get access to the β-cell antigens. Generally, the T-lymphocytes do not come in direct contact with the tissue, but they circulate freely in the blood and lymphoid organs. It is suggested that, somehow, β-cells are taken by antigen presenting cells (APCs), such as DC, in islets, stimulating the maturation of APCs. When these activated APCs reach draining pancreatic lymph nodes, they come in contact with β-cell-reactive T-cells in circulation and activate them. Activated T-cells can pass through the tissue and encounter the cognate antigen in islet cells, causing their reactivation and ultimately insulitis, the first stage of T1DM. The most characteristic feature of the pancreas at this stage is a total lack of β-cells, even though the other islet cells α, δ, and PP remain unaffected. The activated T-cells could be of CD4+ as well as CD8+ type as both are successful in inducing T1DM in normal recipient mice.[74] It is still not clear how the β-cell antigens become available to APCs to start the T-cell activation. It might be due to the death of β-cells through normal physiological processes (e.g., apoptosis). β-cells in the NOD mice islet and in the patients of T1DM show extensive similar morphological features of apoptotic cell death including fragmented DNA and condensed nuclei.

Type 2 diabetes mellitus (T2DM), previously known as non-insulin dependent diabetes mellitus (NIDDM), is a heterogeneous, multifactorial, polygenic disease characterized by a high level of glucose concentration in the blood resulting from insulin resistance and β-cell dysfunction. Type 2 diabetes is the most common form of diabetes, constituting about 90% of total diabetic patients. Currently there are approximately 150 million patients with Type 2 diabetes worldwide, with approximately 16 million in the United States alone, making it a major public health issue in this country. The estimated disease-related cost in the United States is approximately $100 billion annually and the number of Type 2 diabetic patients is growing each year as a result of an increase in obesity, a sedentary lifestyle, and an increased awareness with a concomitant diagnosis of the disease. In Type 2 diabetes, either the body does not produce enough insulin or the cells become resistant to insulin. Obesity is the most common cause of insulin resistance in humans.

Type 2 diabetes in children, teenagers, and adolescents is a serious new aspect to the epidemic and an emerging public health problem of significant proportions. Although Type 1 diabetes remains the main form of the disease in children worldwide, it seems possible that Type 2 diabetes will be the predominant form within 10 years in many ethnic groups. A number of molecular defects have been associated with insulin resistance. These include reduced expression of insulin receptors on the surface of insulin-responsive cells; alterations in the signal-transduction pathways that become activated after insulin binds to its receptor; and the abnormalities in biological pathways normally stimulated by insulin, including glucose transport and glycogen synthesis. Mutations in the insulin-receptor gene are responsible for insulin resistance in a limited number of people, but the molecular basis of insulin resistance in the vast majority of people with NIDDM is still unclear.

In addition to the insulin resistance in muscles, liver, and adipose tissues, another characteristic feature of T2DM is increased visceral fat, which might be due to the exposure of tissues to excessive levels of free fatty acids and excessive hepatic gluconeogenesis. The resultant hyperglycemia exacerbates the insulin resistance and β-cell dysfunction. The failure of glucose-stimulated insulin secretion (GSIS) in β-islet cells helps to sustain the elevations of serum glucose and free fatty acids, which reinforce the failure of GSIS, possibly by inhibiting the expression of the transcription factor PDX-1/IDX-1/IPF-1. Diabetes occurs when the β-cells fail to cope with the insulin requirements of the body due to increasing insulin resistance.

29.5 DIABETES: CAUSE OR EFFECT OF PANCREATIC CANCER

Despite the great progress in understanding the progression of diabetes, it is still not clear whether it is the precursor to pancreatic adenocarcinoma or the other way around. There have been reports blaming each of these for the other disease. An Italian study involving 720 pancreatic cancer patients and an equal number of control patients concluded that there was no definite proof suggesting diabetes to be a cause of pancreatic cancer.[75] This study also suggested diabetes to be an aftereffect of pancreatic cancer in these patients. The initiation of pancreatic cancer causes β-cells to become more prone to diabetes. About 80% of pancreatic cancer patients have glucose intolerance.[76] However, a meta-analysis involving 30 studies published over a period of 20 years suggested diabetes to be a risk factor for pancreatic cancer.[77] In May 2004, a new publication with questions and responses still argues the implication of diabetes and pancreatic cancer[78]; therefore, extreme caution needs to be taken regarding this question, and more analysis is needed.

Reports supporting that pancreatic cancer might be the causative agent for diabetes gain confidence from *in vitro* experiments using tumor cell lines. Coculturing the islet cells with tumor cells caused decreased insulin production and a higher IAPP/insulin ratio. Also, the cancer patients who underwent surgery to get their tumor removed showed improvement in glucose tolerance and normalization of IAPP levels in the circulation. Therefore, the increased IAPP release seen in pancreatic cancer patients may be responsible, at least, in part, for the islet dysfunction.[79] On the other hand, treatment of pancreatic cancer cell lines with insulin did not induce their growth, though the hamster pancreatic cancer cell line and rat acinar cells did show growth stimulation on insulin treatment. The number of epidemiological studies found a modest, but a definite link between diabetes and pancreatic cancer. The situation is more complex, and we are still not sure if pancreatic cancer is the cause or the effect.

29.6 ORIGIN OF PANCREATIC ADENOCARCINOMA

As summarized in the previous sections, the pancreas is a multifunctional gland composed of exocrine and endocrine cells. The nature of the cells that compose the pancreas is distinct, even if they are believed to rise from common precursor cells. There are close functional and physical interactions between the endocrine and exocrine cells. The necessary cooperation between the endocrine and exocrine cells explains the plasticity of the pancreas to adjust to the internal demands and external stimuli. Hence, each type of pancreatic cell has the ability to provide a function that is dictated by the physiological demand in an organized fashion. Because of its complex structure, it is not surprising to have distinct forms of pancreatic cancer. If, in a simple point of view, it may seem logical to think that exocrine cells develop pancreatic adenocarcinoma and endocrine cells develop endocrine cancer, several lines of evidence suggest that the problem is much more complex. The first evidence is the link between pancreatic cancer and diabetes mellitus. Up to 95% of the cancers from the pancreas are adenocarcinomas; however, about 80% of the pancreatic cancer patients have glucose intolerance.[76] Diabetes is associated with endocrine cells.

For over a decade, Pour and collaborators[80,81] have reported in favor of the endocrine origin of pancreatic cancer. The islets are the major cells of the pancreas that express drug-metabolizing enzymes.[82] Islets have direct contact with the circulation via the insuloportal vessels. Any xenobiotic that reaches the pancreas would be first in contact with the islets cells, where it would be metabolized to either a non-toxic or a proximate carcinogen, implying a greater involvement of islet cells in the metabolism of xenobiotics within the pancreas. One of the main risk factors recognized for pancreatic cancer is cigarette smoke. Numerous authors showed that the nitrosamines present in tobacco smoke were powerful carcinogens for the pancreas. Hamsters treated with such kinds of compounds develop with time pancreatic cancer.[80] CYP2E1 is the main enzyme responsible for the process of nitrosamine. CYP2E1 is mainly expressed within the islets (PP cells). PP cells are present in large numbers in the ductal epithelium. Therefore, PP cells within the ductal epithelium might be the reason for the localization of the pancreatic adenocarcinoma in the duct.

In addition, islets appear to be altered in all pancreatic cancers. This alteration of the islets is validated first by the glucose intolerance; the reduction in the number of β-cells; the expression of ductal markers such as CA-19-9, DUPAN-2, and TAG-72 by islet cells; and finally by the development of intrainsular ductular structures that express the aforementioned ductal markers. The experimental destruction of β-cells in hamsters by the streptozotocin pretreatment inhibited cancer development following action of N-nitrosobis(2-oxopropyl)amin (BOP) as compared to hamsters not treated with streptozotocin.[83,84] Finally, transplantation of islets into the submandibular gland of hamster followed by BOP treatment resulted in the development of ductal-like adenocarcinoma within the submandibular gland of the hamster.[85]

Several lines of evidence suggest that ductal cells are the origin of pancreatic adenocarcinomas. The fact that the majority of pancreatic cancers are composed of cells presenting a ductal-type structure is an overwhelming reason for this hypothesis. Most importantly, these cells express the ductal markers, such as the aforementioned CA-19-9, DUPAN-2, TAG72, and the new markers of pancreatic cancer like MUC4. Pancreatic cancers are usually localized in the pancreatic duct. In addition, it was shown that patients diagnosed previously with atypical papillary duct lesions later develop invasive adenocarcinoma. The apparent ability of ductal cells to differentiate into other cell types, including squamous, mucinous, and pyloric cells further supports the ability of ductal cells to undergo malignant transformation. The most recent information that confirmed the ductal origin of pancreatic cancer is an experimental model developed by Tuveson's group.[86] The conditional expression of the activated form of K-ras, K-ras^{v12}, in the pancreas of mice generated progressive PanIN lesions that gave rise at low frequency to the development of invasive and metastatic adenocarcinoma. PanIN lesions have been known for a long time and usually present in the pancreas of pancreatic adenocarcinoma patients. No direct link between PanIN lesions and invasive adenocarcinoma is known so far. The report of Tuveson and collaborators was the first direct evidence that PanIN lesions, going from PanIN-1A to PanIN-3 progress until the development of pancreatic adenocarcinoma. This experimental report and the fact that there is progression within the genetic alteration detected in the PanIN model of progression bring strength to the ductal origin hypothesis for pancreatic cancer.

The idea of acinar cells, as the tumor progenitor of pancreatic cancer, emerges from the rat experimental model, as well as in humans where transdifferentiation of acinar cells to duct-like structures (pseudoductules) have been reported. In transgenic mouse models, PanIN-like lesions can develop in the context of TGFα-driven acinar-to-ductal metaplasia.[87,88] In transgenic mice overexpressing TGFα under the control of acinar-specific elastase-1 promoter, groups of acinar cells convert into metaplastic epithelia, from which PanIN-like lesions and pancreatic adenocarcinoma develop.

29.7 TRANSDIFFERENTIATION OF PANCREATIC CELLS

All the pieces of information that tend to prove the islet or ductal origin of pancreatic cancer are minimized by one important property of the pancreatic cells, which is the transdifferentiation. Indeed, pancreatic cells either ductal, islets, or acinar have the property to transdifferentiate to any other pancreatic cellular types. *Transdifferentiation* is defined as an irreversible switch of one differentiated cellular type to another. The transdifferentiation does not require dedifferentiation, which means that the markers of both cells type (original cell and final cell) can coexist during intermediate phase. The best documented case of transdifferentiation for pancreatic cells is the passage from pancreas to liver or liver to pancreas. The transdifferentiation of pancreas to liver (and vice versa) has been observed in animal experiments and in certain human pathologies. As this is not in the focus of this article, we will not develop more on this point. It is important to note that understanding the mechanisms that trigger the transdifferentiation from liver to pancreas is important for development of new therapeutic for diabetes, ultimately using hepatic cell to develop new endocrine insulin positive pancreatic cells.

Transdifferentiation is also reported in between the exocrine and endocrine cells from the pancreas. For instance, using cytokeratin as a marker, Bouwens established the differentiation of acinar and islet from duct cells during the fetal pancreas development and acini to duct (dedifferentiation) or ducts to islet cells (neogenesis) in adult pancreas.[89] More recently, Song and collaborators show that pancreatic acini isolated from Sprague-Dawley rats when maintained in suspension culture converted into duct-like cells with a loss of amylase expression. More importantly, few insulin-positive cells were identified in between the duct-type cells, suggesting that the spontaneous transdifferentiation of acinar to duct cells continued further toward a final differentiation of new duct cells to islets β-cells.[90] Similarly, Wang and collaborators show that freshly isolated islet cells can transdifferentiate into duct-like cells, and the process required an appropriate integrin-matrix support.[91]

This property of pancreatic cells to transdifferentiate from one cell type to another makes the research for the origin of pancreatic cancer a challenge. One can argue that cells in the process of differentiation, when they express markers of duct, islet, or acinar cells, might be more sensitive to alteration and could give rise to pancreatic cancer. More importantly, some authors hypothesize now that the pancreas might contain some non-differentiated stem cells that could differentiate in any pancreatic cellular type.

29.8 CONCLUSIONS AND PERSPECTIVES

Pancreatic cancer is a common and usually lethal disease. Understanding the biology of pancreatic cancer is crucial for making inroads in this devastating disease. The etiology of pancreatic cancer is heterogeneous and remains elusive. Nevertheless, recent discoveries in the molecular genetics of certain hereditary pancreatic cancer-prone syndromes have contributed to a growing understanding of the genetic defects that predispose these individuals to cancer. Genetic lesions detected in pancreatic carcinomas, include alterations in genes encoding K-ras, the *INK4a* (p16 and p14), p53, and members of the TGFβ pathway (Smad4 or p15). The activation of K-ras and the loss of p16, found in approximately 90% of pancreatic adenocarcinomas, are believed to represent early lesions. The MUC4 mucin is associated with progression and metastasis of pancreatic cancer. A limited number of transgenic and carcinogen-induced pancreatic cancer models are known.

The nature of the cells from which pancreatic adenocarcinomas develop is still a matter of controversy. Pancreatic adenocarcinomas are classified as ductal on the basis of their histological appearance, as the well-differentiated tumors display foci of duct-like structures and are positive for markers of the pancreatic ductal epithelial cells, Cytokeratin 19. There are observations that pancreatic adenocarcinomas show evidence of multipotency, suggested by their expression of markers of multiple cell lineages — including acinar cells, islet cells, gastric epithelial cells, gastroduodenal mucopeptic cells, colorectal epithelial cells, and small intestinal goblet cells — marks the stem cells as the origin of pancreatic adenocarcinomas. Detailed morphological and molecular biological studies are needed to settle the controversy. The conditional activation of oncogenes (K-ras) or inactivation of tumor suppressor genes (p16, p53, DPC4) that are frequently associated with the development of pancreatic adenocarcinomas, in specific cell-lineages using specific cell or specific promoters (PDX1, CK-19, elastase, or P48) (Figure 29.1), may provide some clear evidence. Nevertheless, considering the plasticity of pancreatic cells, tumor cells can derive from any pancreatic cells.

ACKNOWLEDGMENTS

The authors of this chapter were supported by grants from the National Institutes of Health (RO1 CA78590) and the Nebraska Research Initiative. Ms. Kristi L. W. Berger, editor, Eppley Institute, is greatly acknowledged for editorial assistance.

REFERENCES

1. Jemal A., Murray T., Samuels A., Ghafoor A., Ward E., and Thun M.J. 2003. Cancer statistics. *CA Cancer J. Clin.* 53:5–26.
2. Jemal A., Tiwari R.C., Murray T., Ghafoor A., Samuels A., Ward E., Feuer E.J., and Thun M.J. 2004. Cancer statistics. *CA Cancer J. Clin.* 54:8–29.
3. Cooperman A.M. 2001. Pancreatic cancer: the bigger picture. *Surg. Clin. North Am.* 81:557–574.
4. Cleary S.P., Gryfe R., Guindi M., Greig P., Smith L., Mackenzie R., Strasberg S., Hanna S., Taylor B., Langer B., and Gallinger S. 2004. Prognostic factors in resected pancreatic adenocarcinoma: analysis of actual 5-year survivors. *J. Am. Coll. Surg.* 198:722–731.
5. Yalniz M. and Pour P.M. 2005. Diabetes mellitus: a risk factor for pancreatic cancer? *Langenbecks Arch. Surg.* 390:66–72.
6. Hruban R.H., Adsay N.V., Albores-Saavedra J., Compton C., Garrett E.S., Goodman S.N., Kern S.E., Klimstra D.S., Kloppel G., Longnecker D.S., Luttges J., and Offerhaus G.J. 2001. Pancreatic intraepithelial neoplasia: a new nomenclature and classification system for pancreatic duct lesions. *Am. J. Surg. Pathol.* 25:579–586.
7. Compagno J. and Oertel J.E. 1978. Microcystic adenomas of the pancreas (glycogen-rich cystadenomas): a clinicopathologic study of 34 cases. *Am. J. Clin. Pathol.* 69:289–298.
8. Brat D.J., Lillemoe K.D., Yeo C.J., Warfield P.B., and Hruban R.H. 1998. Progression of pancreatic intraductal neoplasias to infiltrating adenocarcinoma of the pancreas. *Am. J. Surg. Pathol.* 22:163–169.

9. Cubilla A.L. and Fitzgerald P.J. 1976. Morphological lesions associated with human primary invasive nonendocrine pancreas cancer. *Cancer Res.* 36:2690–2698.

10. Wilentz R.E., Albores-Saavedra J., and Hruban R.H. 2000. Mucinous cystic neoplasms of the pancreas. *Semin. Diagn. Pathol.* 17:31–42.

11. Wilentz R.E., Albores-Saavedra J., Zahurak M., Talamini M.A., Yeo C.J., Cameron J.L., and Hruban RH. 1999. Pathologic examination accurately predicts prognosis in mucinous cystic neoplasms of the pancreas. *Am. J. Surg. Pathol.* 23:1320–1327.

12. Hruban R.H., Van Mansfeld A.D., Offerhaus G.J., Van Weering D.H., Allison D.C., Goodman S.N., Kensler T.W., Bose K.K., Cameron J.L., and Bos J.L. 1993A. K-ras oncogene activation in adenocarcinoma of the human pancreas. A study of 82 carcinomas using a combination of mutant-enriched polymerase chain reaction analysis and allele-specific oligonucleotide hybridization. *Am. J. Pathol.* 143:545–554.

13. Shields J.M., Pruitt K., McFall A., Shaub A., and Der C.J. 2000. Understanding Ras: 'it ain't over 'til it's over.' *Trends Cell. Biol.* 10:147–154.

14. Shih C., Padhy L.C., Murray M., and Weinberg R.A. 1981. Transforming genes of carcinomas and neuroblastomas introduced into mouse fibroblasts. *Nature* 19(290):261–264.

15. Serrano M., Hannon G.J., and Beach D. 1993. A new regulatory motif in cell-cycle control causing specific inhibition of cyclin D/CDK4. *Nature* 366:704–707.

16. Huang L., Goodrow T.L., Zhang S.Y., Klein-Szanto A.J., Chang H., and Ruggeri B.A. 1996. Deletion and mutation analyses of the P16/MTS-1 tumor suppressor gene in human ductal pancreatic cancer reveals a higher frequency of abnormalities in tumor-derived cell lines than in primary ductal adenocarcinomas. *Cancer Res.* 56:1137–1141.

17. Caldas C., Hahn S.A., Da Costa L.T., Redston M.S., Schutte M., Seymour A.B., Weinstein C.L., Hruban R.H., Yeo C.J., and Kern S.E. 1994. Frequent somatic mutations and homozygous deletions of the p16 (MTS1) gene in pancreatic adenocarcinoma. *Nat. Genet.* 8:27–32.

18. Scarpa A., Capelli P., Mukai K., Zamboni G., Oda T., Iacono C., and Hirohashi S. 1993. Pancreatic adenocarcinomas frequently show p53 gene mutations. *Am. J. Pathol.* 142:1534–1543.

19. Kalthoff H., Schmiegel W., Roeder C., Kasche D., Schmidt A., Lauer G., Thiele H.G., Honold G., Pantel K., Riethmuller G. et al. 1993. p53 and K-RAS alterations in pancreatic epithelial cell lesions. *Oncogene* 8:289–298.

20. Lagna G., Hata A., Hemmati-Brivanlou A., and Massague J. 1996. Partnership between DPC4 and SMAD proteins in TGF-beta signalling pathways. *Nature* 383:832–836.

21. Zhang Y., Feng X., We R., and Derynck R. 1996. Receptor-associated Mad homologues synergize as effectors of the TGF-beta response. *Nature* 383:168–172.

22. Zawel L., Dai J.L., Buckhaults P., Zhou S., Kinzler K.W., Vogelstein B., and Kern S.E. 1998. Human Smad3 and Smad4 are sequence-specific transcription activators. *Mol. Cell.* 1:611–617.

23. Hahn S.A., Schutte M., Hoque A.T., Moskaluk C.A., Da Costa L.T., Rozenblum E., Weinstein C.L., Fischer A., Yeo C.J., Hruban R.H., and Kern S.E. 1996. DPC4, a candidate tumor suppressor gene at human chromosome 18q21.1. *Science* 271:350–353.

24. Hahn S.A., Hoque A.T., Moskaluk C.A., Da Costa L.T., Schutte M., Rozenblum E., Seymour A.B., Weinstein C.L., Yeo C.J., Hruban R.H., and Kern S.E. 1996. Homozygous deletion map at 18q21.1 in pancreatic cancer. *Cancer Res.* 56:490–494.

25. Moniaux N., Escande F., Porchet N., Aubert J.P., and Batra S.K. 2001. Structural organization and classification of the human mucin genes. *Front. Biosci.* 6:D1192–D1206.

26. Andrianifahanana M., Moniaux N., Schmied B.M., Ringel J., Friess H., Hollingsworth M.A., Buchler MW, Aubert J.P., and Batra S.K. 2001. Mucin (MUC) gene expression in human pancreatic adenocarcinoma and chronic pancreatitis: a potential role of MUC4 as a tumor marker of diagnostic significance. *Clin. Cancer Res.* 7:4033–4040.

27. Park H.U., Kim J.W., Kim G.E., Bae H.I., Crawley S.C., Yang S.C., Gum Jr. J., Batra S.K., Rousseau K., Swallow D.M., Sleisenger M.H., and Kim Y.S. 2003. Aberrant expression of MUC3 and MUC4 membrane-associated mucins and sialyl lex antigen in pancreatic intraepithelial neoplasia. *Pancreas* 26:48–54.

28. Swart M.J., Batra S.K., Varshney G.C., Hollingsworth M.A., Yeo C.J., Cameron J.L., Willentz R.E., Hruban R.H., and Argani P. 2002. MUC4 Expression increases progressively in pancreatic intraepithelial neoplasia (PanIN). *Am. J. Clin. Pathol.* 117:791–796.

29. Balague C., Gambus G., Carrato C., Porchet N., Aubert J.P., Kim Y.S., and Real F.X. 1994. Altered expression of MUC2, MUC4, and MUC5 mucin genes in pancreas tissues and cancer cell lines. *Gastroenterology* 106:1054–1061.

30. Choudhury A., Moniaux N., Winpenny J.P., Hollingsworth M.A., Aubert J.P., and Batra S.K. 2000. Human MUC4 mucin cDNA and its variants in pancreatic carcinoma. *J. Biochem.* 128:233–243.

31. Hollingsworth M.A., Strawhecker J.M., Caffrey T.C., and Mack D.R. 1994. Expression of MUC1, MUC2, MUC3 and MUC4 mucin mRNAs in human pancreatic and intestinal tumor cell lines. *Intl. J. Cancer* 57:198–203.

32. Choudhury A., Moniaux N., Ulrich A.B., Schmied B.M., Standop J., Pour P.M., Gendler S.J., Hollingsworth M.A., Aubert J.P., and Batra S.K. 2004. MUC4 mucin expression in human pancreatic tumours is affected by organ environment: the possible role of TGFbeta2. *Br. J. Cancer* 90:657–664.

33. Choudhury A., Singh R.K., Moniaux N., El-Metwally T.H., Aubert J.P., and Batra S.K. 2000. Retinoic acid dependent transforming growth factor-beta2-mediated induction of muc4 mucin expression in human pancreatic tumor cells follows retinoic acid receptor-alpha signaling pathway. *J. Biol. Chem.* 275:33929–33936.

34. Singh A.P., Moniaux N., Chauhan S.C., Meza J.L., and Batra S.K. 2004. Inhibition of MUC4 expression suppresses pancreatic tumor cell growth and metastasis. *Cancer Res.* 64:622–630.

35. Carraway K.L., Ramsauer V.P., Haq B., and Carothers Carraway C.A. 2003. Cell signaling through membrane mucins. *Bioessays* 25:66–71.

36. Ramsauer V.P., Carraway C.A., Salas P.J., and Carraway K.L. 2003. Muc4/sialomucin complex, the intramembrane ErbB2 ligand, translocates ErbB2 to the apical surface in polarized epithelial cells. *J. Biol. Chem.* 278:30142–30147.

37. Moskaluk C.A., Hruban R.H., and Kern S.E. 1997. p16 and K-ras gene mutations in the intraductal precursors of human pancreatic adenocarcinoma. *Cancer Res.* 57:2140–2143.

38. Fukushima N., Sato N., Ueki T., Rosty C., Walter K.M., Wilentz R.E., Yeo C.J., Hruban R.H., and Goggins M. 2002. Aberrant methylation of preproenkephalin and p16 genes in pancreatic intraepithelial neoplasia and pancreatic ductal adenocarcinoma. *Am. J. Pathol.* 160:1573–1581.

39. Luttges J., Galehdari H., Brocker V., Schwarte-Waldhoff I., Henne-Bruns D., Kloppel G., Schmiegel W., and Hahn S.A. 2001. Allelic loss is often the first hit in the biallelic inactivation of the p53 and DPC4 genes during pancreatic carcinogenesis. *Am. J. Pathol.* 158:1677–1683.

40. Wilentz R.E., Iacobuzio-Donahue C.A., Argani P., McCarthy D.M., Parsons J.L., Yeo C.J., Kern S.E., and Hruban R.H. 2000. Loss of expression of Dpc4 in pancreatic intraepithelial neoplasia: evidence that DPC4 inactivation occurs late in neoplastic progression. *Cancer Res.* 60:2002–2006.

41. Sessa F., Solcia E., Capella C., Bonato M., Scarpa A., Zamboni G., Pellegata N.S., Ranzani G.N., Rickaert F., and Kloppel G. 1994. Intraductal papillary-mucinous tumours represent a distinct group of pancreatic neoplasms: an investigation of tumour cell differentiation and K-ras, p53 and c-erbB-2 abnormalities in 26 patients. *Virchows. Arch.* 425:357–367.

42. Sasaki S., Yamamoto H., Kaneto H., Ozeki I., Adachi Y., Takagi H., Matsumoto T., Itoh H., Nagakawa T., Miyakawa H., Muraoka S., Fujinaga A., Suga T., Satoh M., Itoh F., Endo T., and Imai K. 2003. Differential roles of alterations of p53, p16, and SMAD4 expression in the progression of intraductal papillary-mucinous tumors of the pancreas. *Oncol. Rep.* 10:21–25.

43. Kawahira H., Kobayashi S., Kaneko K., Asano T., and Ochiai T. 2000. p53 protein expression in intraductal papillary mucinous tumors (IPMT) of the pancreas as an indicator of tumor malignancy. *Hepatogastroenterology* 47:973–977.

44. Adsay N.V., Merati K., Andea A., Sarkar F., Hruban R.H., Wilentz R.E., Goggins M., Iocobuzio-Donahue C., Longnecker D.S., and Klimsta D.S. 2002. The dichotomy in the preinvasive neoplasia to invasive carcinoma sequence in the pancreas: Differential expressions of MUC1 and MUC2 supports the existence of two separate pathways of carcinogenesis. Mod. Pathol. 15:1087–1095.

45. Terada T. and Nakanuma Y. 1996. Expression of mucin carbohydrate antigens (T, Tn and sialyl Tn) and MUC-1 gene product in intraductal papillary-mucinous neoplasm of the pancreas. *Am. J. Clin. Pathol.* 105:613–620.

46. Yonezawa S., Sueyoshi K., Nomoto M., Kitamura H., Nagata K., Arimura Y., Tanaka S., Hollingsworth M.A., Siddiki B., Kim Y.S., and Sato E. 1997. MUC2 gene expression is found in noninvasive tumors but not in invasive tumors of the pancreas and liver: its close relationship with prognosis of the patients. *Hum. Pathol.* 28:344–352.

47. Terris B., DuBois S., Buisine M.P., Sauvanet A., Ruszniewski P., Aubert J.P., Porchet N., Couvelard A., Degott C., and Flejou J.F. 2002. Mucin gene expression in intraductal papillary-mucinous pancreatic tumours and related lesions. *J. Pathol.* 197:632–637.

48. Jimenez R.E., Warshaw A.L., Z'Graggen K., Hartwig W., Taylor D.Z., Compton C.C., and Fernandez-Del Castillo C. 1999. Sequential accumulation of K-ras mutations and p53 overexpression in the progression of pancreatic mucinous cystic neoplasms to malignancy. *Ann. Surg.* 230:501–509.

49. Iacobuzio-Donahue C.A., Wilentz R.E., Argani P., Yeo C.J., Cameron J.L., Kern S.E., and Hruban R.H. 2000. Dpc4 protein in mucinous cystic neoplasms of the pancreas: frequent loss of expression in invasive carcinomas suggests a role in genetic progression. *Am. J. Surg. Pathol.* 24:1544–1548.

50. Rieder H. and Bartsch D.K. 2004. Familial pancreatic cancer. *Fam. Cancer* 3:69–74.

51. Murphy K.M., Brune K.A., Griffin C., Sollenberger J.E., Petersen G.M., Bansal R., Hruban R.H., and Kern S.E. 2002. Evaluation of candidate genes MAP2K4, MADH4, ACVR1B, and BRCA2 in familial pancreatic cancer: deleterious BRCA2 mutations in 17%. *Cancer Res.* 62:3789–3793.

52. Borg A., Sandberg T., Nilsson K., Johannsson O., Klinker M., Masback A., Westerdahl J., Olsson H., and Ingvar C. 2000. High frequency of multiple melanomas and breast and pancreas carcinomas in CDKN2A mutation-positive melanoma families. *J. Natl. Cancer Inst.* 92:1260–1266.

53. Boardman L.A., Thibodeau S.N., Schaid D.J., Lindor N.M., McDonnell S.K., Burgart L.J., Ahlquist D.A., Podratz K.C., Pittelkow M., and Hartmann L.C. 1998. Increased risk for cancer in patients with the Peutz-Jeghers syndrome. *Ann. Intern. Med.* 128:896–899.

54. Giardiello F.M., Welsh S.B., Hamilton S.R., Offerhaus G.J., Gittelsohn A.M., Booker S.V., Krush A.J., Yardley J.H., and Luk G.D. 1987. Increased risk of cancer in the Peutz-Jeghers syndrome. *N. Engl. J. Med.* 316:1511–1514.

55. Goggins M., Offerhaus G.J., Hilgers W., Griffin C.A., Shekher M., Tang D., Sohn T.A., Yeo C.J., Kern S.E., and Hruban R.H. 1998. Pancreatic adenocarcinomas with DNA replication errors (RER+) are associated with wild-type K-ras and characteristic histopathology. Poor differentiation, a syncytial growth pattern, and pushing borders suggest RER+. *Am. J. Pathol.* 152:1501–1507.

56. Whitcomb D.C., Gorry M.C., Preston R.A., Furey W., Sossenheimer M.J., Ulrich C.D., Martin S.P., Gates Jr. L.K., Amann S.T., Toskes P.P., Liddle R., McGrath K., Uomo G., Post J.C., and Ehrlich G.D. 1996. Hereditary pancreatitis is caused by a mutation in the cationic trypsinogen gene. *Nat. Genet.* 14:141–145.

57. Etemad B. and Whitcomb D.C. 2001. Chronic pancreatitis: diagnosis, classification, and new genetic developments. *Gastroenterology* 120:682–707.

58. Tersmette A.C., Petersen G.M., Offerhaus G.J., Falatko F.C., Brune K.A., Goggins M., Rozenblum E., Wilentz R.E., Yeo C.J., Cameron J.L., Kern S.E., and Hruban R.H. 2001. Increased risk of incident pancreatic cancer among first-degree relatives of patients with familial pancreatic cancer. *Clin. Cancer Res.* 7:738–744.

59. Moore P.S., Orlandini S., Zamboni G., Capelli P., Rigaud G., Falconi M., Bassi C., Lemoine N.R., and Scarpa A. 2001. Pancreatic tumours: molecular pathways implicated in ductal cancer are involved in ampullary but not in exocrine nonductal or endocrine tumorigenesis. *Br. J. Cancer* 84:253–262.

60. Jensen R.T. 1999. Pancreatic endocrine tumors: recent advances *Ann. Oncol.* 10(Suppl. 4):170–176.

61. Kondo K. and Kaelin Jr. W.G. 2001. The von Hippel-Lindau tumor suppressor gene. *Exp. Cell. Res.* 264:117–125.

62. Verhoef S., Diemen-Steenvoorde R., Akkersdijk W.L., Bax N.M., Ariyurek Y., Hermans C.J., Van Nieuwenhuizen O., Nikkels P.G., Lindhout D., Halley D.J., Lips K., and Van Den Ouweland A.M. 1999. Malignant pancreatic tumour within the spectrum of tuberous sclerosis complex in childhood. *Eur. J. Pediatr.* 158:284–287.

63. Radford D.M., Ashley S.W., Wells Jr. S.A., and Gerhard D.S. 1990. Loss of heterozygosity of markers on Chromosome 11 in tumors from patients with multiple endocrine neoplasia syndrome Type 1. *Cancer Res.* 50:6529–6533.

64. Giraud S., Zhang C.X., Serova-Sinilnikova O., Wautot V., Salandre J., Buisson N., Waterlot C., Bauters C., Porchet N., Aubert J.P., Emy P., Cadiot G., Delemer B., Chabre O., Niccoli P., Leprat F., Duron F., Emperauger B., Cougard P., Goudet P., Sarfati E., Riou J.P., Guichard S., Rodier M., Calender A. et al. 1998. Germ-line mutation analysis in patients with multiple endocrine neoplasia Type 1 and related disorders. *Am. J. Hum. Genet.* 63:455–467.

65. Zhuang Z., Vortmeyer A.O., Pack S., Huang S., Pham T.A., Wang C., Park W.S., Agarwal S.K., Debelenko L.V., Kester M., Guru S.C., Manickam P., Olufemi S.E., Yu F., Heppner C., Crabtree J.S., Skarulis M.C., Venzon D.J., Emmert-Buck M.R., Spiegel A.M., Chandrasekharappa S.C., Collins F.S., Burns A.L., Marx S.J., Lubensky I.A. et al. 1997. Somatic mutations of the MEN1 tumor suppressor gene in sporadic gastrinomas and insulinomas. *Cancer Res.* 57:4682–4686.

66. Debelenko L.V., Zhuang Z., Emmert-Buck M.R., Chandrasekharappa S.C., Manickam P., Guru S.C., Marx S.J., Skarulis M.C., Spiegel A.M., Collins F.S., Jensen R.T., Liotta L.A., and Lubensky I.A. 1997. Allelic deletions on Chromosome 11q13 in multiple endocrine neoplasia Type 1-associated and sporadic gastrinomas and pancreatic endocrine tumors. *Cancer Res.* 57:2238–2243.

67. Larsson C., Skogseid B., Oberg K., Nakamura Y., and Nordenskjold M. 1988. Multiple endocrine neoplasia Type 1 gene maps to Chromosome 11 and is lost in insulinoma. *Nature* 332:85–87.

68. Broder L.E. and Carter S.K. 1973. Pancreatic islet cell carcinoma. I. Clinical features of 52 patients. *Ann. Intern. Med.* 79:101–107.

69. Yu F., Venzon D.J., Serrano J., Goebel S.U., Doppman J.L., Gibril F., and Jensen R.T. 1999. Prospective study of the clinical course, prognostic factors, causes of death, and survival in patients with long-standing Zollinger-Ellison syndrome. *J. Clin. Oncol.* 17:615–630.

70. Caruso R.A., Inferrera A., Tuccari G., and Barresi G. 1994. Acinar cell carcinoma of the pancreas. A histologic, immunocytochemical and ultrastructural study. *Histol. Histopathol.* 9:53–58.

71. Klimstra D.S., Heffess C.S., Oertel J.E., and Rosai J. 1992. Acinar cell carcinoma of the pancreas. A clinicopathologic study of 28 cases. *Am. J. Surg. Pathol.* 16:815–837.

72. Rigaud G., Moore P.S., Zamboni G., Orlandini S., Taruscio D., Paradisi S., Lemoine N.R., Kloppel G., and Scarpa A. 2000. Allelotype of pancreatic acinar cell carcinoma. *Intl. J. Cancer* 88:772–777.

73. Abraham S.C., Wu T.T., Hruban R.H., Lee J.H., Yeo C.J., Conlon K., Brennan M., Cameron J.L., and Klimstra D.S. 2002. Genetic and immunohistochemical analysis of pancreatic acinar cell carcinoma: frequent allelic loss on Chromosome 11p and alterations in the APC/beta-catenin pathway. *Am. J. Pathol.* 160:953–962.

74. Hoglund P., Mintern J., Waltzinger C., Heath W., Benoist C., and Mathis D. 1999. Initiation of autoimmune diabetes by developmentally regulated presentation of islet cell antigens in the pancreatic lymph nodes. *J. Exp. Med.* 189:331–339.

75. Gullo L., Pezzilli R., and Morselli-Labate A.M. 1994. Diabetes and the risk of pancreatic cancer. Italian Pancreatic Cancer Study Group. *N. Engl. J. Med.* 331:81–84.

76. Permert J., Ihse I., Jorfeldt L., Von Schenck H., Arnquist H.J., and Larsson J. 1993. Improved glucose metabolism after subtotal pancreatectomy for pancreatic cancer. *Br. J. Surg.* 80:1047–1050.

77. Everhart J. and Wright D. 1995. Diabetes mellitus as a risk factor for pancreatic cancer. A meta-analysis. *JAMA* 273:1605–1609.

78. Bonelli L., De Micheli A., and Pugliese V. 2004. Diabetes and pancreatic cancer: reply. *Pancreas* 28:451–452.

79. Wang F., Adrian T.E., Westermark G., Gasslander T., and Permert J. 1999. Dissociated insulin and islet amyloid polypeptide secretion from isolated rat pancreatic islets cocultured with human pancreatic adenocarcinoma cells. *Pancreas* 18:403–409.

80. Pour P.M., Weide L., Liu G., Kazakoff K., Scheetz M., Toshkov I., Ikematsu Y., Fienhold M.A., and Sanger W. 1997. Experimental evidence for the origin of ductal-type adenocarcinoma from the islets of Langerhans. *Am. J. Pathol.* 150:2167–2180.

81. Pour P.M., Pandey K.K., and Batra S.K. 2003. What is the origin of pancreatic adenocarcinoma? *Mol. Cancer* 2:13–23.

82. Standop J., Ulrich A.B., Schneider M.B., Buchler M.W., and Pour P.M. 2002. Differences in the expression of xenobiotic-metabolizing enzymes between islets derived from the ventral and dorsal anlage of the pancreas. *Pancreatology* 2:510–518.

83. Pour P.M., Kazakoff K., and Carlson K. 1990. Inhibition of streptozotocin-induced islet cell tumors and N-nitrosobis(2-oxopropyl)amine-induced pancreatic exocrine tumors in Syrian hamsters by exogenous insulin. *Cancer Res.* 50:1634–1639.

84. Tomioka T., Fujii H., Hirota M., Ueno K., and Pour P.M. 1991. The patterns of beta-cell regeneration in untreated diabetic and insulin-treated diabetic Syrian hamsters after streptozotocin treatment. *Intl. J. Pancreatol.* 8:355–366.

85. Fienhold M.A., Kazakoff K., and Pour P.M. 1997. The effect of streptozotocin and a high-fat diet on BOP-induced tumors in the pancreas and in the submandibular gland of hamsters bearing transplants of homologous islets. *Cancer Lett.* 117:155–160.

86. Hingorani S.R., Petricoin E.F., Maitra A., Rajapakse V., King C., Jacobetz M.A., Ross S., Conrads T.P., Veenstra T.D., Hitt B.A., Kawaguchi Y., Johann D., Liotta L.A., Crawford H.C., Putt M.E., Jacks T., Wright C.V., Hruban R.H., Lowy A.M., and Tuveson D.A. 2003. Preinvasive and invasive ductal pancreatic cancer and its early detection in the mouse. *Cancer Cell* 4:437–450.

87. Greten F.R., Wagner M., Weber C.K., Zechner U., Adler G., and Schmid R.M. 2001. TGF alpha transgenic mice. A model of pancreatic cancer development. *Pancreatology* 1:363–368.

88. Wagner M., Luhrs H., Kloppel G., Adler G., and Schmid R.M. 1998. Malignant transformation of duct-like cells originating from acini in transforming growth factor transgenic mice. *Gastroenterology* 115:1254–1262.

89. Bouwens L. 1998. Cytokeratins and cell differentiation in the pancreas. *J. Pathol.* 184:234–239.

90. Song K.H., Ko S.H., Ahn Y.B., Yoo S.J., Chin H.M., Kaneto H., Yoon K.H., Cha B.Y., Lee K.W., and Son H.Y. 2004. In vitro transdifferentiation of adult pancreatic acinar cells into insulin-expressing cells. *Biochem. Biophys. Res. Commun.* 316:1094–1100.

91. Wang R., Li J., and Rosenberg L. 2001. Factors mediating the transdifferentiation of islets of Langerhans to duct epithelial-like structures. *J. Endocrinol.* 171:309–318.

92. Almoguera C., Shibata D., Forrester K., Martin J., Arnheim N., and Perucho M. 1988. Most human carcinomas of the exocrine pancreas contain mutant c-K-ras genes. *Cell* 53:549–554.

93. Hruban R.H., Van Mansfeld A.D., Offerhaus G.J., Van Weering D.H., Allison D.C., Goodman S.N., Kensler T.W., Bose K.K., Cameron J.L., and Bos J.L. 1993B. K-ras oncogene activation in adenocarcinoma of the human pancreas. A study of 82 carcinomas using a combination of mutant-enriched polymerase chain reaction analysis and allele-specific oligonucleotide hybridization. *Am. J. Pathol.* 143:545–554.

94. Redston M.S., Caldas C., Seymour A.B., Hruban R.H., Da Costa L., Yeo C.J., and Kern S.E. 1994. p53 mutations in pancreatic carcinoma and evidence of common involvement of homocopolymer tracts in DNA microdeletions. *Cancer Res.* 54:3025–3033.

95. Goggins M., Schutte M., Lu J., Moskaluk C.A., Weinstein C.L., Petersen G.M., Yeo C.J., Jackson C.E., Lynch H.T., Hruban R.H., and Kern S.E. 1996. Germline BRCA2 gene mutations in patients with apparently sporadic pancreatic carcinomas. *Cancer Res.* 56:5360–5364.

96. Su G.H., Hilgers W., Shekher M.C., Tang D.J., Yeo C.J., Hruban R.H., and Kern S.E. 1998. Alterations in pancreatic, biliary, and breast carcinomas support MKK4 as a genetically targeted tumor suppressor gene. *Cancer Res.* 58:2339–2342.

97. Huang L., Lang D., Geradts J., Obara T., Klein-Szanto A.J., Lynch H.T., and Ruggeri B.A. 1996. Molecular and immunochemical analyses of RB1 and cyclin D1 in human ductal pancreatic carcinomas and cell lines. *Mol. Carcinog.* 15:85–95.

INDEX

A

AACN, 168–169

Accessory pancreas, 27, 38–39

Acetaldehyde, 133, 315–316, 320

N-acetyl-5-hydroxytryptophyl-5-hydroxytryptophan amide, 126

2-acetylaminofluorene, 367

Acetylcholine, 116, 118–119, 122, 128, 132, 286–287, 300–301

N-acetylcysteine, 446

N-acetylgalactosamine, 226–228

N-acetylglucosamine, 557

N-acetyltransferases, 148

Achlorhydria, 505

Acinar cells, 57–59

  AACN and, 168–169

  acetylcholine and, 286–287, 300–301

  acute pancreatitis, 59

  age and, 103, 344

  alcohol and, 133, 312–317, 320

  alloxan and, 558

  amino acids and, 58, 192–194

  azaserine and, 168–169

  bicarbonate and, 178

  blood type and, 93–94, 95, 103, 106

  blood vessels and, 181, 583–584

  BOP and, 167

  cancer and, 28, 98–103, 167, 436, 440, 504, 654, 660

  CCK and, 285–302

  centroacinar cells and, 60, 92–105

  choline and, 196

  in chromatin, 92

  copper and, 197–198, 361–362, 365–366, 368–370

  CYP and, 143, 152, 165–167

  cystic fibrosis and, 524–526

  diet and, 192–194

  differentiation of, 28, 65, 103, 660–661

  DLCs and, 345–350

  duct formation and, 59–60, 64

  EGF and, 257–260

  embryonic development of, 6, 11–15, 64, 177, 179, 368, 586

  epoxide hydrolase and, 148

  ethionine and, 196

  extracellular matrix and, 58

  FGF and, 263

  fibrosis and, 329–330, 333–334, 336–338

  genes and, 654

  GSTs and, 146–147

  hepatocytes and, 358, 359–360, 369

  IGFs and, 182, 250

  insulin and, 246–247

  islet cells and, 128, 246–247, 598

  KGF and, 264

  Lewis gene and, 93–94, 95

  liver and, 358

  magnesium and, 198

  markers for, 647

  methionine and, 195

  methyl deficiency and, 195

  in mucin, 224

  NA-OR and, 152

  nerves and, 41, 69

  neurotrophins and, 267–268

  pancreatitis and, 59, 103, 244, 291–300, 333–334, 336–338, 464

  PanIN lesions and, 660

  PAR and, 207

  protein and, 58–59, 192–194, 285–290, 293

  PSTI and, 208–212

  schema of, 92, 289

  secretin and, 287, 301

Printed and bound by CPI Group (UK) Ltd, Croydon, CR0 4YY

01/11/2024

01782640-0001